Volume

	in³	ft³	yd³	m³	qt	liter	barrel	gal. (U.S.)
1 in³ =	1	—	—	—	—	0.02	—	—
1 ft³ =	1,728	1	—	.0283	—	28.3	—	7.480
1 yd³ =	—	27	1	0.76	—	—	—	—
1 m³ =	61,020	35.315	1.307	1	—	1,000	—	—
1 quart (qr) =	—	—	—	—	1	0.95	—	0.25
1 liter (l) =	61.02	—	—	—	1.06	1	—	0.2642
1 barrel (oil) =	—	—	—	—	168	159.6	1	42
1 gallon (U.S.) =	231	0.13	—	—	4	3.785	0.02	1

Energy and Power

1 kilowatt-hour = 3,413 Btus = 860,421 calories

1 Btu = 0.000293 kilowatt-hour = 252 calories = 1,055 joule

1 watt = 3.413 Btu/hr = 14.34 calories/min

1 calorie = the amount of heat necessary to raise the temperature of 1 gram (1 cm³) of water 1 degree Celsius

1 quadrillion Btu = (approximately) 1 exajoule

 D0216806

Mass and Weight

1 pound = 453.6 grams = 0.4536 kilogram = 16 ounces

1 gram = 0.0353 ounce = 0.0022 pound

1 short ton = 2,000 pounds = 907.2 kilograms

1 long ton = 2,240 pounds = 1,008 kilograms

1 metric ton = 2,205 pounds = 1,000 kilograms

1 kilogram = 2.205 pounds

Temperature

F is degrees Fahrenheit

C is degrees Celsius (centigrade)

$$F = \frac{9}{5}C + 32$$

Fahrenheit		Celsius
	Freezing of H₂O	
32	(Atmospheric Pressure)	0
50	————	10
68	————	20
86	————	30
104	————	40
122	————	50
140	————	60
158	————	70
176	————	80
194	————	90
212	Boiling of H₂O (Atmospheric Pressure)	100

ENVIRONMENTAL GEOLOGY

CRC LIBRARY

Edward A. Keller

University of California, Santa Barbara

▼

▼

▼

ENVIRONMENTAL GEOLOGY
Sixth Edition

Macmillan Publishing Company
New York
Maxwell Macmillan Canada
Toronto
Maxwell Macmillan International
New York Oxford Singapore Sydney

CFCC LIBRARY

Editor: Robert McConnin
Production Editor: Mary M. Irvin
Art Coordinator: Vincent A. Smith
Photo Research: John Schultz / Par NYC
Text Designer: Anne Flanagan
Cover Designer: Robert Vega
Production Buyer: Patricia A. Tonneman

This book was set in Garamond by The Clarinda Company
and was printed and bound by Von Hoffman. The cover was
printed by Lehigh Press.

Copyright © 1992 by Macmillan Publishing Company, a
division of Macmillan, Inc.

Printed in the United States of America

All rights reserved. No part of this book may be reproduced
or transmitted in any form or by any means, electronic or
mechanical, including photocopy, recording, or any
information storage and retrieval system, without permission
in writing from the Publisher.

Earlier edition(s), entitled Environmental Geology copyright
© 1988, 1985, 1982, 1979, and 1976 by Merrill Publishing
Company.

Macmillan Publishing Company
866 Third Avenue
New York, New York 10022

Macmillan Publishing Company is part of the
Maxwell Communication Group of Companies.

Maxwell Macmillan Canada, Inc.
1200 Eglinton Avenue East, Suite 200
Don Mills, Ontario M3C 3N1

Library of Congress Cataloging-in-Publication Data
Keller, Edward A., 1942—
  Environmental geology / Edward A. Keller.—6th ed.
    p.   cm.
  Includes bibliographical references and index.
  ISBN 0-02-363270-4
  1. Environmental geology.   I. Title.
QE38.K45   1992
304.2—dc20   91-24084
                                                          CIP

Printing: 2   3   4   5   6   7   8   9   Year: 2   3   4   5

Cover photos: Tim Alipalo / Reuters / Bettman Archive
              Bruno Barbey / Magnum
              Heinz Plenge / Peter Arnold
              John Nakata / U.S. Geological Survey

For Sarah and Jamila

PREFACE

▼

▼

▼

Environmental Geology is the branch of applied geology that focuses on the entire spectrum of possible interactions between people and the physical environment. The sixth edition of *Environmental Geology,* as with previous editions, is intended as an introduction to the study of applied geology. Students who become keenly interested in the subject may then take advanced courses in areas such as engineering geology and hydrogeology.

Under ideal circumstances, the study of environmental geology is facilitated by previous exposure to either physical geology or geography. However, unless they have majored in geology, students may not have the latitude in their undergraduate curriculum to take more than a single geology course. Therefore, *Environmental Geology* is designed to allow students to study the subject without having had previous exposure to the geological sciences. This book attempts to present case histories and subject matter that are relevant to a wide spectrum of students including those in the traditional scientific disciplines such as chemistry, biology, geology, physical geography, and physics; liberal arts students majoring in disciplines such as art, economics, environmental studies, human geography, literature, and sociology; and students who may be preparing for professional schools such as engineering, architecture, and planning.

The organization of the sixth edition is very similar to the fifth with the exception that one new chapter is presented. Chapter 16 introduces earth system science, global change, and the air environment. The new chapter reflects our concern for global environmental problems including ozone depletion and potential global warming. In particular, Chapter 16 focuses on the role of geology in better understanding global processes and how we use geology to measure change.

The book is arranged into five parts. Part One introduces physical and fundamental principles important in the study of environmental geology. The purpose is to unite the cultural and physical environments and introduce important aspects of geology necessary to understand the remainder of the book. Part One also sets a philosophical framework for the remainder of the text.

Part Two provides an overview of the major natural processes (hazards) that continue to make life on earth occasionally difficult for people. These include flooding, landslides, earthquakes, volcanoes, and coastal erosion. Material has been added in appropriate places in these chapters to better present some of the fundamental principles concerning these processes. New case histories have also been added including: the 1989 Loma Prieta (San Francisco) earthquake; hurricane Hugo, which struck the eastern coast of the United States in 1989; the Allentown Pennsylvania sinkhole; and 1991 volcanic eruptions in Japan and the Philippines. In addition, many case histories have been updated.

Part Three is concerned with human interactions with the environment, and includes detailed discussions of water resources, waste management, and geologic aspects of environmental health. In particular, the material in Chapter 12 on waste management has been completely revised to reflect new thinking in areas such as hazardous waste management and integrated waste management. In Chapter 13 an entire new section is included on the radon gas hazard.

Part Four presents discussions of minerals, energy, and environmental issues associated with resource utilization and management. A new case history of the Star Fire Mine in eastern Kentucky is presented, as well as new material on solar energy. Also included is a more detailed discussion on the debate between energy policy that centers on whether we follow the hard path that relies on more traditional energy sources, or seek to make an energy transformation to the soft paths that are more decentralized and often renewable sources.

Part Five introduces subjects such as global change and earth system science, land-use and decision making, and environmental law. An entire new chapter (Chapter 16) is presented to cover the area of earth system science and global change. The landscape evaluation chapter, which includes land-use planning, has been extensively revised to introduce important topics such as comprehensive planning. The chapter on environmental law has also been revised to include important new areas of environmental law such as mediation.

Special Note To Students

Environmental Geology was written at a level that will challenge you to think carefully about environmental problems and their solutions. I have heard comments that today's college students cannot read as well as students of a few years ago, and therefore we need to write books at an easier reading level. I do not believe this statement, and I am convinced that students are interested in obtaining the best possible education and in fact expect to be challenged. I have made no attempt to dodge difficult subjects such as hydraulic conductivity, moment magnitude of earthquakes, and fluid pressure in rocks and soils, among others. I have tried to write these difficult subjects in a way that students will be able to better understand what is going on. Often, however, these are difficult subjects and study is necessary to thoroughly understand them. In addition, more advanced courses in areas such as hydrogeology and seismology are often offered for students who wish to pursue these areas further.

I hope you enjoy reading the textbook. I certainly enjoyed preparing it and learned a great deal myself.

ACKNOWLEDGEMENTS

▼
▼
▼

Successful completion of a textbook that includes hundreds of photographs and illustrations as well as case histories would not be possible without the cooperation of many individuals, companies and agencies. In particular, I greatly appreciate the work of agencies such as the United States Geological Survey, which has an extensive environmental program as well as individual state geological surveys which have provided a great deal of information and concepts important in the development of the subject of environmental geology. To all those individuals who are so helpful in this endeavor I offer my sincere appreciation. Reviews of chapters by Roger J. Bain, Douglas G. Brookings, William Chadwick, P. Thompson Davis, Thomas L. Davis, John G. Drost, Richard V. Fisher, Stanley T. Fisher, Cal Janes, Donald L. Johnson, Ernest K. Johnson, Ernest Kastning, Harold L. Krivoy, Robert Mathews, Robert M. Norris, James R. Lauffer, Gill LaFreniere, John S. Pomeroy, James Dennis Rash, Derrick J. Rust, Samual E. Swanson, and William S. Wise are acknowledged and appreciated. Edward M. Bert, Daniel Botkin, Laurence R. Davis, James Kennett, Hugo Loaiciga, Mel Manalis, Marc McGinnes, June A. Oberdorfer, Charles J. Ritter, Taz Talley, and reviewers chosen by Macmillan: P. Thompson Davis, Bentley College; Anne Erdmann, University of Minnesota; Edward B. Evenson, Lehigh University; Robert B. Furlong, Wayne State University; H. G. Goodell, University of Virginia; Gilbert LeFreniere, Williamette University; Gene W. Lene, St. Mary's University; Gary L. Millhollen, Fort Hays Kansas State University; Patricia Miller, West Virginia University; Roderic A. Parnell, North Arizona State University; Paul T. Ryberg, Union College; William C. Sherwood, James Madison University; Robert Shuster, University of Nebraska-Omaha; Mark T. Steward, University of South Florida; and Russel O. Utgard, Ohio State University.

I am particularly indebted to my editors at Macmillian including Robert McConnin, Mary Irvin, Vince Smith, and Rebecca Bobb.

The Environmental Studies Program and the Department of Geological Sciences at the University of California, Santa Barbara, continue to provide a stimulating environment in which to do research and write. I would like to thank the many staff members who have readily given their time in thoughtful discussion and help in preparation of many aspects of *Environmental Geology*. In particular, I would like to acknowledge the help of Barbara Hollins and Carla Lease.

Edward A. Keller
Santa Barbara, California

CONTENTS

▼ PART FIVE
Global Change, Land Use, and
Decision Making   417

CHAPTER SIXTEEN
Global Change, Earth Systems Science,
and Urban Air   418

CHAPTER SEVENTEEN
Landscape Evaluation   456

E verything has a beginning and an end. Our earth began approximately 5 billion years ago when a cloud of interstellar gas known as a *solar nebula* collapsed, forming protostars and planetary systems; life on earth began about 2 billion years later, or 3 billion years ago. Since then, a multitude of different types of organisms have emerged, prospered, and died out, leaving only their fossils to mark their place in earth's history. Several million years ago, on one of the more recent pages in earth history, our ancestors set the stage for the eventual dominance of the human race. As certainly as our sun will eventually die, we, too, will disappear. The impact of humanity on earth history may not be significant, but to us living now, our children and theirs, our environment is significant indeed.

Environmental geology is applied geology. Specifically, it is the application of geologic information to solving conflicts, minimizing possible adverse environmental degradation, or maximizing possible advantageous conditions resulting from our use of the natural and modified environment. This includes evaluation of *natural hazards* such as floods, landslides, earthquakes, and volcanic activity to minimize loss of human life and property damage; evaluation of the *landscape* for site selection, land-use planning, and *environmental impact analysis;* and evaluation of *earth materials* (such as elements, minerals, rocks, soils, and water) to determine their potential use as resources or waste disposal sites and their effects on human health, and to assess the need for conservation practices. In a broader sense, environmental geology is that branch of earth science that emphasizes the entire spectrum of human interactions with the physical environment.

Environment may be considered as the total set of circumstances that surround an individual or a community. It may be defined to include two parts: first, physical conditions, such as air, water, gases, and landforms that affect the growth and development of an individual or a community; and second, social and cultural aspects, such as ethics, economics, and aesthetics, that affect the behavior of an individual or a community. Therefore, a complete introduction to environmental geology involves consideration of those philosophical and cultural aspects that influence how we perceive and react to our landscape, as well as the physical earth processes, resources, and landforms that may be more readily recognized by the observant earth scientist.

Chapters 1 through 4 provide the philosophical framework for the remainder of the book. Chapters 1 and 2 integrate the influence of cultural and physical activities into our total environment, and Chapter 3 introduces the physical environment through the geological cycle. The term *cycle* emphasizes that most earth materials, such as air, water, minerals, and rock, although changed physically and chemically and transported from place to place, are constantly being reworked, conserved, and renewed by natural earth processes. Chapter 3 also introduces basic earth science terminology and engineering properties of earth materials excluding soil, while Chapter 4 introduces soil in terms of its development, classification, engineering properties, and other factors important to land-use planning.

PART ONE

▼
▼
▼

Philosophy
and
Fundamental
Principles

Photo by Ken Graham/AllStock

1

CHAPTER ONE

▼
▼
▼

Cultural Basis for Environmental Awareness

The cultural aspect of environmental awareness involves the entire way of life that we have transmitted from one generation to another. Therefore, to uncover the roots of our present condition, we must look to the past and consider various functional categories and social institutions that have developed. The functional categories of society that are especially significant in environmental studies are ethical, economic, political, aesthetic, and, perhaps, religious. The interactions between individuals and the institutions responsible for maintaining these functions are intimately associated with the way we perceive and respond to our physical environment.

To solve environmental problems such as overpopulation, disposal of hazardous waste, global warming, and resource depletion, we must look to the future and determine how ethical, economic, and political institutions can contribute to solutions. Fundamental changes in how society works at both the personal and institutional level will be necessary. The magnitude of adjustment may be similar to past major changes, such as the industrial revolution, which changed the relationship between people and the environment by producing ever-increasing demands on resources and releasing toxic waste into the environment. Global environmental issues are now serious political issues, perhaps for the first time. Industrial countries and developing countries must work together. For example, we cannot expect South American countries to better manage the rain forest if the United States and other industrial countries continue to place heavy economic pressure on South America to export tremendous quantities of resources. How can we expect poor, developing countries to respect the environment when the industrial countries with apparent wealth remain unwilling to seriously address their own environmental problems? Fortunately, public concern for the environment appears to be increasing; if so, the study of political, environmental

issues will evolve into real progress in solving environmental problems.

▼ ENVIRONMENTAL ETHICS

What started as the "quiet crisis" of the 1960s has evolved into what Stewart Udall, statesman and conservationist, refers to as the "crisis of survival" (1). More important than the certainty of a crisis is whether society believes there is a crisis. In other words, is there a new awareness that is destined to change our life-style, morals, ethics, and institutions, or is environmental concern just another prestigious fad that interests the intellectual community?

The evolution of ethics (Figure 1.1) is an important environmental trend. Aldo Leopold emphasizes the lack of ethics regarding property through the story of Odysseus, who, upon returning from Troy, hanged a dozen slave women for suspected misbehavior during his absence. His right to do this was unquestioned; the women were property, and the disposal of property was a matter of expediency, much as it is today. Although concepts of right and wrong were present in Greece three thousand years ago, these ethical values did not extend to slaves (2). Since that time, ethical values have been extended to many other areas of human behavior; but, apparently, only within this century has the relationship between civilization and its physical environment begun to emerge as a relationship with moral considerations.

Ecological ethics involve limitations on social as well as individual freedom of action in the struggle for existence in our stressed environment (2). A land ethic assumes that we are ethically responsible not only to other individuals and society but also to the total environment, that larger community consisting of plants, animals, soil, atmosphere, and so forth. The environmental ethic proposed by Leopold affirms the right of all resources, including plants, animals, and earth materials, to continued existence and, at least in certain locations, continued existence in a natural state. This ethic effectively changes our role from that of conqueror of the land to that of citizen and protector of the environment. This role change obviously requires us to revere and love our land and not, for instance, to allow economics to determine all land use.

A possible dichotomy or source of confusion exists between an ideal and a realistic land ethic. To give rights to the plants, animals, and landscape might be interpreted as granting to individual plants and animals the fundamental right to live. If we are to be part of the environment, however, we must extract the energy necessary to survive. Therefore, although the land ethic assigns rights for animals such as deer, cattle, or chickens to survive as a *species,* it does not necessarily assign rights to an *individual* deer, cow, or chicken. The same argument may be given to justify the use of stream gravel

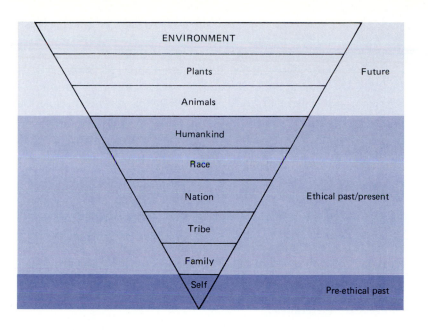

Figure 1.1
The evolution of ethics. (After Roderick Nash, "Do Rocks Have Rights?" *The Center Magazine*, November–December, 1977).

for construction material, or to mine and use the other resources necessary for our well-being. However, unique landscapes with high aesthetic value, like endangered species, are in need of complete protection within our ethical framework.

The implications of environmental ethics and moral responsibility are restated by Stewart Udall. Each generation has its own rendezvous with the land, for despite our fee titles and claims of ownership, we are all brief tenants on this planet. By choice or by default, we will carve out a land for our heirs. We can misuse the land and diminish the usefulness of resources, or we can create a world in which physical affluence and spiritual affluence go hand in hand (1).

The resounding message is that humanity is an integral part of the environment. A person is no more than any other being, and has a moral obligation to those beings who will follow. This obligation is to ensure that they will also have the opportunity to experience the pleasure of belonging to and cooperating with the entire land community.

▼ ECONOMIC AND POLITICAL SYSTEMS

Arriving in late fall of 1620, after two months on the stormy North Atlantic, 73 men and 29 women from the *Mayflower* confronted what they considered a wild and savage land. The colonists were not equipped with the skills and knowledge necessary to adapt quickly to their new environment. Regardless of these shortcomings and despite their fear of the wilderness, they brought three things that assured their success in the New World. First, they brought a new technology. Reportedly, when the Pilgrims landed, they did not even have a saw, but they did have Iron Age skills necessary to ensure relentless subjugation of the land and its earliest inhabitants, Native

Americans. In the long run, the ax, gun, and wheel asserted their supremacy. Second, the colonists brought with them the blueprints to remake the New World. They knew how to organize work, use work animals, and sell their surplus to overseas markets. Third, they brought with them a concept of land ownership completely different from that of the Native Americans, whose bonds to the land were religious and held by kinship with nature rather than exclusive possession. The colonists' idea of ownership involved an absolute title to land regardless of who worked the land or how far away the owner was. After the Native Americans were displaced, land use or abuse depended entirely on the attitude of the owner.*

America now, as in its early years, suffers greatly from the *myth of superabundance*. This myth assumes that the land and resources in America are inexhaustible and that management of resources is therefore unnecessary. Management of the people and their society, however, continues to reflect the deep roots transplanted from the Old World.

Stewart Udall writes that the land myth was instrumental in environmental degradation from the "birth of land policy" in the eighteenth century throughout the "raid on resources" that lasted into the twentieth century. Even young Thomas Jefferson, who in later life was to become aware of the value of conservation, said there was such a great deal of farmland that it could be wasted as he pleased. However, the real raid on resources probably began with the mountain men and their beaver trapping in the 1820s. This was only the beginning; there followed the invention of machines that were capable of large-scale removal of resources and landscape alter-

---

*Stewart Udall, *The Quiet Crisis*, pp. 25–27.

ation. The inventions included the sawmills that precipitated the destruction of the American forests, and the "Little Giant" hose nozzle that could tear up an entire hillside in the search for California gold.

The "Great Giveaway" of land which resulted in the destruction of forests and consequent soil erosion eventually ended, and in 1884, hydraulic mining was outlawed. However, the effects of these repugnant land-use practices are still visible today. Similar examples of the effects of the raid on resources are the plights of the fur seal, the buffalo, and the passenger pigeon, and the Dust Bowl of the 1930s (1).

The seeds of conservation were planted in the latter part of the nineteenth century by men such as Carl Schurz, Secretary of the Interior, and John Wesley Powell, geologist and explorer. Their messages concerning conservation of resources and land-use planning, although largely ignored when first introduced, today stand as landmarks in perceptive and innovative conservation (1).

The historical roots of our landscape heritage, while not a pretty picture, are full of lessons. For example, we learned the hard way that our resources are not infinite and that land and water management is necessary for meaningful existence. This conclusion has become even more significant over the years, as American society continues to urbanize and consume resources at an ever-increasing rate.

The convergence of available resources with the needs of society, along with an ever-growing production of waste, has produced what is popularly referred to as the *environmental crisis*. This impending crisis in America, according to Lewis W. Moncrief, is a result of individual and institutional inability in our democratic system to organize technology, conservation, urbanization, and the capitalistic mission to the betterment of our landscape (3). Moncrief further contends that the present condition is characterized by three features that tend to prevent a quick solution to environmental problems: first, the absence of individual and personal moral direction concerning the way we treat our natural resources; second, the inability of our social institutions to make adjustments to reduce environmental stress; and third, an abiding faith in technology.

Overpopulation, urbanization, and industrialization, combined with little ethical regard for our land and inadequate institutions (or perhaps too many institutions stumbling over one another) to cope with environmental stress, may well be the immediate source of the crisis. The overpopulation, urbanization, and industrialization factors, although part of the political and economic systems, currently tend to transcend those systems.

Political and economic theorists are often surprised to learn that disruption of the environment is as serious a problem in the USSR as in the United States. Goldman reports that the Soviets have greatly misused their natural resources. For example, Soviet cities have air pollution problems, and water pollution there has resulted in massive fish kills, in turn resulting in an increase in mosquitoes and malaria peril. Land-use problems in the USSR include a system of dams, reservoirs, and canals that have diverted so much water that there is serious concern for the future of the Caspian Sea and its famous caviar fisheries. The reservoirs have also increased evaporation, which is disrupting natural moisture patterns and changing rainfall cycles. Furthermore, seepage from unlined canals has caused the water table to rise in normally dry areas, facilitating the deposition of harmful salts in the soil. Mining of beach deposits for construction material in conjunction with a decreased supply of sediment to the beaches (the reservoirs are holding back the natural flow of sediment) has resulted in serious coastal erosion; without the sand and gravel to impede the impact of waves, the coastline is subject to rapid erosion (4).

Water diversion for agriculture in southern USSR has in a period of only 30 years nearly eliminated the Aral Sea. The area of the sea in 1960 was about 67,000 square kilometers. Diversion of the two main rivers that fed the sea has resulted in a drop in surface elevation of over 20 meters and loss of about 28,000 square kilometers of surface area. Towns that were once fishing centers on the shore of the lake are today about 30 kilometers inland. The Aral Sea was a potential tourist vacation spot in 1960. It is now a dying sea surrounded by thousands of square kilometers of salt flats, and the change is permanently damaging the economic base of the region. Loss of the sea is also changing the regional climate: the winters are colder and the summers warmer. This results because the sea had a moderating effect on the climate. Wind storms pick up salty dust and spread it over a vast area, damaging the land and polluting the air. The real lesson from the Aral Sea is how quick regional change from environmental damage can be. What worries many people is that what people have done to the Aral region is symptomatic of what we may be doing on many fronts on a global scale (5).

The factors responsible for environmental problems in the USSR, as in America, are population explosion and rapid industrialization. Furthermore, the Soviet government, as sole owner of the productive resources, has been no more successful than other countries in regulating or controlling environmental degradation. In fact, the national commitment to centralized control of industry has given rise to unique environmental problems resulting from uniform regulations for industry regardless of local conditions (4).

The 1970s was a period of environmental awareness in both the Soviet Union and the United States. The Soviets passed pollution abatement and other environmental laws, as did the United States. However, the Soviet emphasis on high production rates, coupled with institutional inflexibility, has reduced the laws' effectiveness, as is the case, to a lesser extent, in America (6). The environmental situation in the Soviet Union in the late 1980s was similar to that in the United States during the late 1960s, when the Environmental Protection Agency was established. In 1988 the State Committee for Environmental Protection (SCEP) was established in the USSR. There are pollution laws in the Soviet Union but, as is the case in the United States, they are often difficult to enforce. Grass-root environmental programs are being established in the USSR today and the political situation is changing rapidly. The large government ministries administered from Moscow are perceived to have formed the policies responsible for much of the air and water pollution episodes in the USSR. There is increased public resentment concerning environmental degradation, and regional governments are attempting to deal with environmental problems. It will be interesting to see whether new policies will reflect the rise in consciousness concerning environmental problems and find solutions to those problems. Some people have argued that the serious environmental degradation in parts of Eastern Europe and the USSR was partly responsible for the major political and economic changes that began in 1989.

The conclusion from consideration of political systems is that not private enterprise, but rather industrialization, urbanization, economic consideration, and lack of a land ethic are primarily responsible for environmental degradation. Therefore, the salvation of the landscape community involves social, economic, and ethical behavior on the part of individuals rather than by political systems as they exist today.

Optimistically, it appears that the emerging environmental ethics inherent in the spirit of individual actions and new legislation will facilitate the types of changes needed. These changes are possible because our democratic system, with private ownership of resources and free enterprise, has the flexibility necessary to allow meaningful change. We emphasize, however, that the system cannot be expected to react until individual citizens are willing to practice environmental ethics, and the political, economic, and legal institutions are willing to seriously consider solving environmental problems. Fortunately, this seems to be happening in the 1990s.

▼ AESTHETIC PREFERENCE AND JUDGMENT

Environmental intangibles, such as the pleasures of private outdoor experiences in nature, are extremely difficult to evaluate. The hunter in a blind on a crisp autumn morning; fishermen in icy mountain streams as the day gives way to darkness; the nature photographer about to culminate the search for an elusive subject on a lonely mountaintop; hikers showing their three-year-old child a snail shell; picnickers relaxing; or a motorist out for a Sunday drive in the country—all perceive to a lesser or greater extent various aspects of the landscape and react to it with various types of behavior. Their experiences and memories cannot readily be equated to economic value, but to the individual they may be priceless.

An understanding of beauty was originally studied in the branch of philosophy known as *aesthetics*. Today, little attention is given to philosophers, and beauty is defined by artists and art critics (7). There appears to be an important distinction between the philosopher who studies aesthetics to establish evaluative judgments, similar to verdicts and findings, and the artist and art critic who may be more concerned with appreciative aesthetic judgments that express preferences such as affection or antipathy (8). Given a distinction between evaluation and preference, the former constitutes a more objective approach.

One perplexing problem of aesthetic evaluation is the impact of personal preference. For example, one person may appreciate a meandering river in an isolated swamp; another may prefer a bubbling mountain stream; a third would rather visit a public park in an urban area. Regardless of such preferences, if we are going to consider aesthetic factors in local, regional, and national land-use planning, we must develop a method of aesthetic evaluation for landscapes that is easy to understand and is quantitative, credible, and predictive.

Three basic criteria necessary to judge aesthetic quality have been recognized: unity, vividness, and variety (9). *Unity* refers to the quality of wholeness of the perceived landscape, not as an assemblage but as a single harmonious unit. *Vividness* refers to that quality of the landscape that reflects a visually striking scene. This is nearly synonymous with *intensity*, *novelty*, or *clarity*. *Variety* refers to how different one landscape is from another. *Diversity* and *uniqueness* also have similar meanings. Greater diversity, however, is not necessarily an indication of higher aesthetic value.

▼ IMPACT OF RELIGION

The role of religion in causing, perpetuating, or condoning environmental disruption and degradation is vigorously debated. One school of thought holds that the Judeo-Christian heritage of Western civilization is responsible for the way we treat the environment. A second school refutes this and argues that human treatment of the land is a characteristic that transcends religious and cultural teaching.

Arguing that the Judeo-Christian heritage is responsible for Western attitudes and behavior with respect to the environment, Lynn White, Jr., cites the following evidence. First, Christianity is the most anthropocentric of the major world religions. It establishes a dualism of humanity and nature, insisting that it is God's will that the human race exploit nature. Second, by destroying pagan animism, which tended to unite humanity with nature, Christianity made it possible for us to degrade the environment in a way completely indifferent to the rights or feelings of natural objects. Third, Western science, technology, and industrialization are a natural result of Judeo-Christian dogma of creation, which teaches that humankind was created after plants, animals, and fishes as their rightful monarch. Fourth, great Western scientists from the thirteenth century to the eighteenth explained their motivation in religious terms, indicating that science was seeking to understand God's mind by discovering how creation operates. Before the thirteenth century, on the other hand, our predecessors studied nature because, since God created nature, nature consequently revealed the divine mentality. Nature was valued not for itself but because every aspect of nature was believed to contain a "hidden message"—symbols that must be decoded and interpreted in order to understand God's communication. In the West, this changed in the thirteenth century when scientists, in the name of religious progress, began investigating physical processes of light and matter (10).

The second school argues that environmental degradation is not a religious problem. They criticize White's thesis on several grounds. First, prehistoric people, through the use of fire and water, also caused considerable environmental disruption. White also recognizes this. Second, before the birth of Christ, the early Greeks and Romans both imposed their will on the environment. Third, the triumph of Christianity over paganism brought no revolutionary change to the relation between society and nature. Fourth, although the ideals of some cultures may suggest that land is sacred, there is a considerable gap between ethical ideals and actual land-use practices (3, 11).

One can conclude that religious attitudes and beliefs are not a primary cause of the environmental crisis. This does not suggest that religious activities are not responsible for considerable environmental disruption. On the contrary, numerous examples, such as timber shortages in China caused by Buddhists' cremation of the dead and deforestation in Japan associated with construction of huge wooden Buddhist halls and temples, suggest that religious activity can promote environmental problems.

The implication that one religion, one culture, or one political system is responsible for the way people treat the land cannot be rigorously defended. Therefore, White's model (Figure 1.2a) is not entirely correct; likewise, Moncrief's modification of White's model (Figure 1.2c) to include capitalism and democracy, with or without the Judeo-Christian start, is not a complete and accurate progression. Since environmental degradation apparently transcends both religious belief and political system, we must look further for the primary cause of our present condition. A simpler explanation (Figure 1.2b) assumes that our environmental problems result from a pattern of human development that began when the earliest people attempted to use tools to better their chances for survival. A product of harsh times, early *Homo sapiens,* like other animals, extended its niche as far as restraints allowed. Therefore, each innovation not only asserted the individual but also assured that everyone who followed had an easier time. As a result, increased populations created greater demand on resources as well as demanded more innovations. This spiral has continued to the present, when there are signs that we may be on a collision course with our environment. In a small way, the condition may be analogous to what sometimes happens to deer when, through our artificial management of animals, their numbers exceed

FIGURE 1.2.
Models showing possible paths leading to environmental degradation.

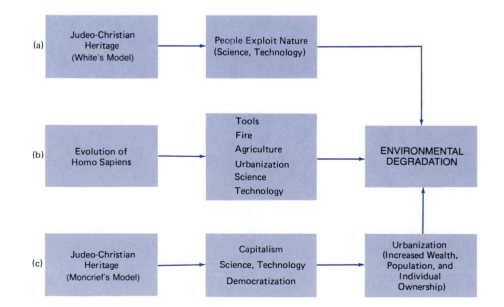

the carrying capacity of the land. Deprived of their natural enemies, the deer herds increase until these "artificial deer" eat all available food, causing a serious shortage of winter feed and little reproduction of food plants for the following spring. Everything from wild-flowers to trees is gradually impoverished, and the deer either become dwarfed from malnutrition or starve (2). Are we, with our increasingly artificial environment, doing to ourselves what we have done to the deer?

▼ ▼ ▼ SUMMARY AND CONCLUSIONS

Functional categories of society that are significant in environmental studies and are the cultural bases for environmental degradation are ethical, economic, political, aesthetic, and, perhaps, religious.

Our ethical framework appears to be slowly expanding and will eventually include the total environment in a land ethic. This ethic affirms the right of all resources, including plants, animals, and earth materials, to continued existence, and, at least in certain locations, continued existence in a natural state (2).

The immediate cause of environmental degradation is overpopulation, urbanization, and industrialization combined with, as yet, little ethical regard for our land and inadequate institutions to cope with environmental stress. These problems are not unique to a particular political system, and, therefore, we conclude that the salvation of the landscape community necessitates changing social, economic, and ethical behavior that transcends political systems.

Aesthetic factors are now being considered in local, regional, and national land-use planning, and scenery is considered a natural resource. A problem remains in finding a method of aesthetic evaluation that is easy to understand and is quantitative, credible, and predictive. Until we do have a satisfactory methodology, it will remain difficult to balance aesthetic and economic costs and benefits.

The role of religion in causing, perpetuating, or condoning environmental degradation remains a much-debated issue. Some authors argue that the Judeo-Christian heritage is responsible for Western man's attitudes and behavior toward the environment. The argument is that the Judeo-Christian teachings and practices destroyed pagan animism, which had previously tended to unite nature and humanity, and thereby made it possible for humans to degrade the environment with complete indifference. This argument cannot be rigorously defended. Prehistoric humans and modern peoples of both Eastern and Western religions have exploited and disrupted their land. One can conclude that religious institutions have indeed been responsible for some environmental problems, but that the general tendency for degradation of the environment is a more universal problem that transcends religious teachings.

▼ ▼ ▼ REFERENCES

1. UDALL, S. L. 1963. *The quiet crisis.* New York: Avon Books.
2. LEOPOLD, A. 1949. *A Sand County almanac.* New York: Oxford University Press.
3. MONCRIEF, L. W. 1970. The cultural basis for our environmental crisis. *Science* 170: 508–12.
4. GOLDMAN, M. I. 1971. Environmental disruption in the Soviet Union. In *Man's impact on environment,* ed. T. R. Detwyler, pp. 61–75. New York: McGraw-Hill.
5. ELLIS, W. S. 1990. A Soviet sea lies dying. *National Geographic* 177, no. 2: 73–92.
6. PRYDE, P. R. 1983. The "decade of the environment" in the U.S.S.R. *Science* 220: 274–79.
7. FLORMAN, S. C. 1968. *Engineering and the liberal arts.* New York: McGraw-Hill.
8. ZUBE, E. H. 1973. Scenery as a natural resource. *Landscape Architecture* 63: 126–32.
9. LITTON, R. B. 1973. Aesthetic dimensions of the landscape. In *Natural environments,* ed. J. V. Kantilla, pp. 262–91. Baltimore: Johns Hopkins University Press.
10. WHITE, L., JR. 1967. The historical roots of our ecological crisis. *Science* 155: 1203–7.
11. YI-FU, T. 1970. Our treatment of the environment in ideal and actuality. *American Scientist* 58: 244–49.

CHAPTER TWO

▼

▼

▼

Fundamental Concepts

In this chapter we will discuss concepts basic to the understanding and study of environmental geology. Although these concepts do not constitute a complete list, they provide the philosophical framework of this book. They are not to be memorized. An understanding of the general thesis of each concept will be a significant help in comprehending and evaluating philosophical and technical material throughout the remainder of the text.

▼ CONCEPT ONE

The number-one environmental problem is increase in human population.

The number-one environmental problem is the ever-growing human population. Garrett Hardin, well known human ecologist, has stated that the total environmental impact is equal to the product of the impact per person times population. Therefore, as population increases the total impact must also increase. As population increases, more resources are needed, and given our present technology, greater environmental disruption results. From 1830 to 1930, the world population doubled from 1 to 2 billion people, an annual growth rate of less than 1 percent. By 1970 it had nearly doubled again (annual growth rate of about 1.8 percent for the 40-year period), and by the year 2000, it is expected that there will be more than 6 billion people on earth.

Overpopulation has been a problem in some areas for at least several hundred years. It is now apparent that it is becoming a global problem. Sometimes the problem is called the *population bomb* because the growth, as shown on Figure 2.1, is exponential; that is, the rate of increase is a constant percentage of the current size. As an example of exponential growth, consider the student who, upon taking a job for one month, requests from the employer a payment of 1 cent for the first day of work, 2 cents for the second day, 4 for the third day and so on. In other words, the payment would double each day. What would be the total? It would take the student 8 days to earn a wage of more than $1 per day, and by the eleventh day, earnings would be more than $10 per day. Payment for the sixteenth day of the month would be more than $300, and on the last day of the thirty-one-day month (Figure 2.2a, p. 10) the student's earnings for that one day would be more than $10 million! Exponential growth is clearly a very dynamic process; it will be discussed further under Concept Two, when we consider systems and change.

Because the earth's population is increasing exponentially (Figure 2.2b) many scientists are concerned that it will be impossible to supply resources and quality environment for the expected billions of people who may be added to the present world population in the twenty-first century. Increasing population at local, regional, and global levels compounds nearly all environmental geology problems, including pollution of ground and surface waters; production and management of hazardous waste; and exposure of people and human structures to natural processes (hazards), such as floods, landslides, volcanic eruptions, and earthquakes.

When resource and other environmental data are combined with population growth data, the conclusion is clear: it is impossible, in the long run, to match exponential population growth with a finite resource base. Therefore, one of the primary goals of environmental work is to ensure that we are able to defuse any potential environmental population bomb. Pessimistic scientists believe that population growth will take care of itself through disease and other catastrophes such as famine. Optimistic people and scientists hope that we will be able to control the population of the world within the limits of our available resources, space, and other environmental needs.

▼ CONCEPT TWO

The earth is essentially a closed system, and an understanding of rates of change and feedback in systems is critical to solving environmental problems.

A **system** may be considered any part of the universe that is isolated in thought or in fact for the purpose of studying or observing changes that take place under various imposed conditions (1). Examples of systems might include a planet, a volcano, or an ocean basin. Most systems contain various component parts which mutually adjust, and each part exerts partial control on the others. For example, the earth may be considered a system with four parts: the **atmosphere**; the **hydrosphere**; the **biosphere**; and the **lithosphere**. The mutual interaction of these parts is responsible for the surface features of the

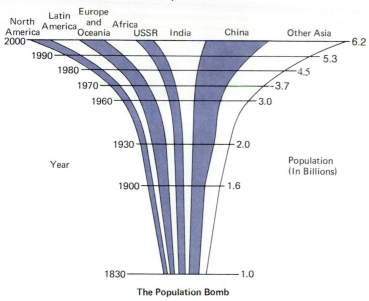

Figure 2.1
The population bomb. (Modified after U.S. Department of State.)

We know that the earth is not static; rather, it is a dynamic, evolving system in which material and energy are constantly changing. Such dynamics might be considered evidence that the earth is an open system with no boundaries of energy or material. This interpretation is true as long as the sun continues to impart energy to the earth, the earth radiates energy to space, meteors fall to earth, and earth material escapes to space. However, considering natural earth cycles such as the water and rock cycles in which there is a continual recycling of earth materials, we can best think of the earth as a closed system or, in reality, a coalition of a large number of closed systems (2). For example, the rain that falls today will eventually return to the atmosphere, and the sediment deposited yesterday will be transformed into solid rock. Therefore, although the earth is currently, and seemingly forever will be, an open system in terms of energy and material, it is essentially a closed system in terms of natural earth cycles.

Feedback

There are two types of feedback cycles in systems: positive feedback and negative feedback. *Positive feedback* is often known as the vicious cycle, whereas *negative feedback* is self-regulating, inducing the system to approach an equilibrium or steady state. Many processes in nature exhibit feedback cycles. For example, off-road vehicle use may be a positive cycle because as vehicle use increases, the number of plants that are uprooted increases, which increases erosion. As this occurs, still more plants are damaged, which further increases erosion until eventually an area intensively used by off-road vehicles may be completely denuded of all vegetation and have a very high erosion rate. On the other hand, systems such as rivers often display a negative feedback such that a rough steady state is formed. That is, as rivers change in response to an increase in regional rainfall or urbanization, the channel and floodplain system will change to accommodate the new, increased amount of water or sediment, and within a relatively short time, a new steady state may be established.

Growth Rates

Growth rates are important in changes that take place in systems. As described under Concept One, *exponential growth* is particularly significant. Notice that both curves in Figure 2.2 are shaped like the letter *J;* in the early stages the growth may be quite slow, but then it increases rapidly and then very rapidly. Growth of many systems, both human-induced and natural, may approach the form of the J-curve. For example, the increase in world population of people and increased use of resources follow the J-curve. Involved in exponential growth are

earth today. Furthermore, any change in the magnitude or frequency of processes in one part will affect the other parts. The propensity for change in the various parts of the environment is known as the *principle of environmental unity*—that is, everything affects everything else. For example, a change (increase) in the magnitude of the processes that uplift mountains may affect the atmosphere by causing regional changes in precipitation patterns as a new rain shadow is produced. This in turn affects the local hydrosphere as more or less runoff reaches the ocean basins. Biospheric changes as a result of changes in the environment also can be expected, and eventually, the steeper slopes will also affect the lithosphere by facilitating increased erosion, which in turn will change the rate and types of sediments produced and, thus, the types of rocks produced from the sediments. These interactions among the variables in systems are not random and may be understood by examining each variable to determine how it interacts with other variables and how it varies spatially over a site, area, or region. In the hydrosphere, for example. the spatial distribution of the oceans with respect to the sunlight received affects the evaporation process of ocean waters, which in turn affects atmospheric conditions by increasing or decreasing the amount of water in the atmosphere.

FIGURE 2.2
Exponential growth. (a) Example
of a student's pay, beginning at
1 cent for the first day of work
and doubling daily for 31 days.
(b) World population. Notice
both curves have the characteristic *J* shape, with a slow initial increase followed by a rapid increase. (Population data from U.S.
Department of State.)

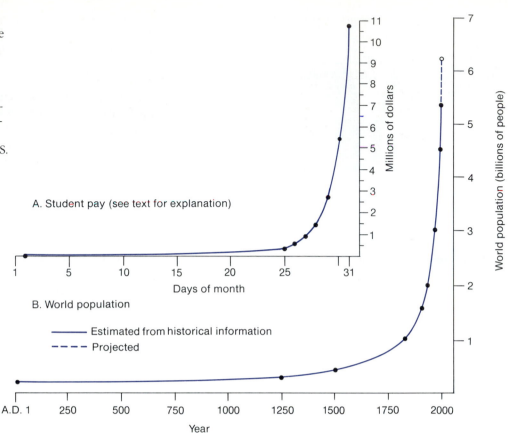

two important factors: the rate of growth measured as a percentage, and the doubling time in years, which is the time necessary for the quantity of whatever is being measured to double. A general rule of thumb is that the doubling time is approximately equal to 70 divided by the growth rate. This rule applies to growth rates up to approximately 10 percent. Beyond that, the errors may become quite large.

Many systems in nature display exponential growth for some periods of time, so it is important that we be able to recognize it. In particular, it is important to recognize exponential growth with positive feedback, as it may be very difficult to stop positive feedback cycles. Negative feedback, on the other hand, tends toward a steady state and thus is easier to control.

Changes in Systems

Changes in natural systems may be predictable and should be recognized by anyone looking for solutions to environmental problems (Figure 2.3). Where the input into the system is equal to the output (Figure 2.3a), a rough steady state is established and no change occurs. Examples of an approximate steady state may be on a global scale—the balance between incoming solar radiation and outgoing radiation from the earth, or the system of plate tectonics in which new lithosphere is being created and destroyed at about the same rate—or

on the smaller scale of a duck farm in which ducks are brought in and harvested at a constant rate. Another example of change is where the input into the system is less than the output (Figure 2.3b). Examples of this would be the use of resources such as groundwater or the harvest of certain plants or animals. If the input is much less than the output, then the groundwater may be completely used or the plants and animals may die out. In a system where input exceeds output (Figure 2.3c), positive feedback may occur, and the stock of whatever is being measured will increase. Examples are the buildup of heavy metals in lakes or the pollution of soil and water. By evaluating rates of change or input/output analysis of systems, we can derive an average *residence time*. The average residence time is a measure of the time it takes for the total stock or supply of a particular material, such as a resource, to be cycled through the pool. To compute the average residence time, we simply take the total size of the stock or pool and divide it by the average rate of transfer through that pool or stock.

An understanding of changes in systems is primary in many problems in environmental studies because very small growth rates may yield incredibly large numbers in modest periods of time. On the other hand, with other systems it may be possible to compute a residence time for a particular resource and, knowing this, apply the information to develop sound management principles. Recognizing positive and negative feedback systems and

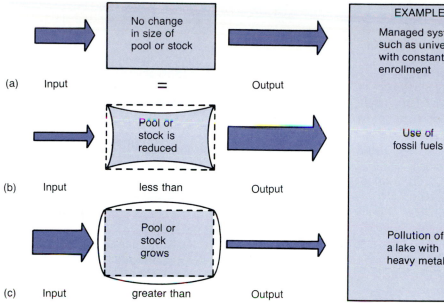

EXAMPLE

Managed system
such as university
with constant
enrollment

Use of
fossil fuels

Pollution of
a lake with
heavy metals

Figure 2.3
Major ways in which a pool or stock of some material may change. (Modified after P. R. Ehrlich, A. H. Ehrlich, and J. P. Holdren, *Ecoscience: Population, resources, environment,* 3rd ed., W. H. Freeman, 1977.)

calculating growth rates and residence times, then, enable us to make predictions concerning resource management.

As more and more demands are made on the earth and its limited resources, it becomes increasingly important for us to understand the magnitude and frequency of the processes that maintain earth cycles. For example, if we hope to manage the water resources of a region, we must know the nature and extent to which natural processes will supply groundwater and surface water. Or, if we are concerned with disposal of dangerous chemicals in a disposal well, we must know how the disposal procedure will interact with natural cycles to ensure that we or our heirs will not be exposed to hazardous chemicals. This becomes especially critical in dealing with radioactive wastes, which must be contained for periods ranging from several centuries to as long as a quarter of a million years. Therefore, it is exceedingly important to recognize earth cycles and determine the length of time involved in various parts of specific cycles. Tables 2.1 and 2.2 list the residence times of selected earth materials and rates of some natural processes.

Our discussion concerning input-output analysis of systems provides a basis for interpretation of some of the changes that affect systems. The model that has been presented and defended for many parts of our natural environment is that natural systems untampered with by human activity tend toward some sort of equilibrium. Thus we hear such ideas as "the balance of nature." Although it is true that many aspects of systems do tend toward an equilibrium due to negative feedback, it is worthwhile to raise the question as to how accurate the equilibrium model really is. As we look at natural systems in greater detail over a spectrum of time scales, it is evident that equilibrium is seldom obtained or maintained for a very long period of time. Instead, changes in

systems are related to such ideas as complex response, thresholds, and disturbance. As an example, consider a river system that periodically experiences floods. A large flood can be viewed as a disturbance within the river system that can cause considerable change to both the physical and biological environment. Usually, however, a flood must be of a certain magnitude before change occurs. For example, the banks of a river are often composed of sand, gravel, silt, and clay, bound together by frictional and cohesive forces related to these materials as well as by roots of the vegetation growing on the banks. Erosion of the banks will not occur unless the power of the moving floodwaters exceeds the resistance of the banks. The resistance of the banks may be looked upon as a threshold that, if crossed, will result in change. In this case the change will be bank erosion. Suppose that a high-magnitude storm occurs in the headwater portions of a large river system. The threshold of bank erosion is crossed and erosion occurs, carrying eroded materials from the slopes and river banks down to the lower part of the drainage system. If in the lower part of the drainage system the river is unable to carry all of the sediment delivered, then deposition will occur. Thus, in the upper part of the drainage system erosion dominates the activity, while deposition may dominate what happens in the lower portion downstream. During subsequent storms the eroded channel upstream behaves differently from the portion downstream where deposition occurred. This is the essence of complex response, which is characterized by changes that occur within systems due to internal processes within the system itself. The underlying principle is that systems change and that the changes often are not related to maintaining a balance or equilibrium but in fact are complex, depending upon disturbance and crossing of thresholds. Thus the lesson to be learned is that disturbance in natural

Table 2.1.
Residence times of some selected materials.

Earth Materials	Some Typical Residence Times
Atmosphere circulation	
Water vapor	10 days (lower atmosphere)
Carbon dioxide	5 to 10 days (with sea)
Aerosol particles	
Stratosphere (upper atmosphere)	Several months to several years
Troposphere (lower atmosphere)	One week to several weeks
Hydrosphere circulation	
Atlantic surface water	10 years
Atlantic deep water	600 years
Pacific surface water	25 years
Pacific deep water	1,300 years
Terrestrial groundwater	150 years [above 760 m depth]
Biosphere circulation[a]	
Water	2,000,000 years
Oxygen	2,000 years
Carbon dioxide	300 years
Seawater constituents[a]	
Water	44,000 years
All salts	22,000,000 years
Calcium ion	1,200,000 years
Sulfate ion	11,000,000 years
Sodium ion	260,000,000 years
Chloride ion	Infinite

[a]Average time it takes these materials to recycle within the atmosphere and hydrosphere.

Source: *The Earth and Human Affairs* by the National Academy of Sciences. Copyright © 1972 by the National Academy of Sciences (Canfield Press). By permission of Harper & Row, Publishers.

TABLE 2.2.
Rates of some natural processes.

Earth Processes	Some Typical Rates
Erosion	
Average U. S. erosion rate[a]	6.1 cm per 1,000 years
Colorado River drainage area	16.5 cm per 1,000 years
Mississippi River drainage area	5.1 cm per 1,000 years
N. Atlantic drainage area	4.8 cm per 1,000 years
Pacific slope (Calif.)	9.1 cm per 1,000 years
Sedimentation[b]	
Colorado River	281 million metric tons per year
Mississippi River	431 million metric tons per year
N. Atlantic coast of U.S.	48 million metric tons per year
Pacific slope (Calif.)	76 million metric tons per year
Tectonism	
Sea-floor spreading	
N. Atlantic	2.5 cm per year
E. Pacific	7 to 10 cm per year
Faulting	
San Andreas (Calif.)	1–5 cm per year
Mountain uplift	
Cajon Pass, San Bernardino Mts. (Calif.)	1 cm per year

[a]Thickness of the layer of surface of the continental United States eroded per 1,000 years.

[b]Includes solid particles and dissolved salts.

Source: *The Earth and Human Affairs* by the National Academy of Sciences. Copyright © 1972 by the National Academy of Sciences (Canfield Press). By permission of Harper & Row, Publishers.

systems is common and in fact probably necessary for the systems to function and provide diversity in the physical and biological environments. Furthermore, human activity is only one type of disturbance. Events such as hurricanes, floods, volcanic eruptions, and earthquakes also cause natural systems to change in complex ways as thresholds of change are exceeded.

Earth Systems Science

Earth systems science is an emerging field of study that focuses on understanding the entire planet earth in terms of how components such as the atmosphere, hydrosphere, biosphere, and lithosphere have formed, evolved, been maintained, and functioned, and how they will continue to evolve in the next decade to century (3). The challenge is to learn to predict changes likely to be important to society, and then to develop management strategies to minimize adverse environmental impacts. For example, study of atmospheric chemistry suggests that our atmosphere has changed. Trace gases such as carbon dioxide and methane have increased by about 25 percent and 100 percent respectively since 1850. Chlorofluorocarbons (CFCs) released at the surface have migrated to the stratosphere where they react with energy from the sun, causing ozone depletion. The important topics of global change and earth systems science will be discussed in Chapter 16 of this book, following topics such as natural hazards, energy resources, and waste management.

▼ CONCEPT THREE

The earth is the only suitable habitat we have, and its resources are limited.

The place of humanity in the universe is well stated in the *Desiderata*: "You are a child of the universe, no less than the trees and the stars; you have a right to be here. And whether or not it is clear to you, no doubt the universe is unfolding as it should" (4).

Leo F. Laporte, senior author of *The Earth and Human Affairs*, believes the context of Concept Three includes two fundamental truths: first, that this earth is indeed the only place to live that is now accessible to us; and second, that our resources are limited, and while some resources are renewable, many are not. Therefore, we eventually will need large-scale recycling of many materials, and a large part of our solid and liquid waste-disposal problems could be alleviated if these wastes were recycled. In other words, many things that are now considered pollutants could be considered resources out of place.

There are at least two dichotomous views on natural resources. One school holds that finding resources is not so much a problem as is finding ways to use them. In other words, the entire earth, including the ocean and atmosphere, has raw materials that can be made useful if we can develop the necessary ingenuity and skill (5). The basic assumption is that as long as there is freedom to think and innovate, we will be able to produce sufficient energy and locate sufficient resources to meet our needs. There is evidence to support this line of reasoning: first, efficient and intelligent use of materials has historically been a successful venture; second, we know more about extracting minerals and fuel than we did in the past and so can find new resources faster and mine lower-grade mineral deposits; and third, new work with atomic power and recycling of resources can help us meet the needs of the future.

The second school holds that "cornucopian premises" such as that outlined above are fallacious on grounds that an exponential increase of people and mineral products on a finite resource base is impossible. Furthermore, Preston Cloud claims that we are in a resource crisis for a number of reasons: first, improvements in medical technology contributing to overpopulation of the earth; second, an unrealistic view of the necessity of an ever-increasing gross national product based on obsolescence and waste; third, the finite nature of the earth's accessible minerals; and fourth, increased risk of irreversible damage to the environment as a result of overpopulation, waste, and the necessity of larger and larger mining operations to obtain ever smaller proportions of useful minerals (6).

The history of *Homo sapiens* can be traced back only several million years. Geologically, this is a very short time. Dinosaurs, for example, ruled the land for more than 100 million years. Evidence from earth history suggests that more species have become extinct than have survived! What then will be our history, and who will write it? We can hope that we will be something more in the geologic record than a good index fossil indicating a brief time in earth history when the human race flourished.

▼ CONCEPT FOUR

Today's physical processes are modifying our landscape and have operated throughout much of geologic time. However, the magnitude and frequency of these processes are subject to natural and artificially induced change.

The concept that understanding the present processes that form and modify our landscapes will facilitate the development of inferences concerning the geologic history of a landscape is known as the doctrine of **uniformitarianism**. Stated simply as "the present is the key to the past," uniformitarianism was first suggested by James Hutton in 1785, elegantly restated by John Playfair in 1802, and popularized by Charles Lyell in the early part of the nineteenth century. Today it is heralded as a fundamental concept of the earth sciences.

Uniformitarianism does not demand or even suggest that the magnitude and frequency of natural processes remain constant with time. Furthermore, it is obvious that the principle cannot be extended back throughout all of geologic time because the processes operating in the oxygen-free environment of the first 2 billion years of earth history were quite different from today's processes. However, as long as past continents, oceans, and atmosphere were similar to those of today, we can infer that the present processes also operated in the past. For example, if we have studied present alpine glaciers and the characteristic erosional and depositional landforms associated with alpine glaciation, we can then infer that valleys with similar landforms were at one time glaciated even if no glacial ice is present today. Similarly, if one finds ancient gravel deposits with all the characteristics of stream gravel on the top of a mountain, then uniformitarianism can be used to suggest that a stream must have flowed there at one time. In other words, what was originally a stream valley has been changed by differential erosion and/or uplift to a mountain top, known as *inversion of topography.* Phenomena such as this would be difficult to interpret correctly if it were not for the principle of uniformitarianism.

We must understand the effects of human activity on increasing or decreasing the magnitude and frequency of natural earth processes. For example, rivers will flood regardless of human activities, but the magnitude and frequency of flooding may be greatly increased or decreased because of human activities. Therefore, to predict the long-range effects of a certain process such as flooding, we must be able to determine how our future activities will change the rate of the process. In this case, the present may have to be the key to the future. We can assume that the same processes will operate but that rates will vary as the environment adjusts to human activity. Furthermore, we must conclude that ephemeral landforms such as beaches and lakes will appear and disappear in response to natural processes, and human influence may be small in comparison.

Although the effects of human activity may be small on a global scale, they are very pronounced in a local area. One year of erosion at a construction site may exceed many decades of erosion from an equivalent tract of woodland or even agricultural land (7). This erosion results from exposure of the soil following the removal of vegetation. Therefore, to maximize the value of geologic knowledge in land-use planning, we must be able to use our understanding of natural earth processes in both a historical and a predictive mode. For example, when environmental geologists examine recent mudflow (flowage of saturated heterogeneous debris) deposits in an area designated to become a housing development, they must use uniformitarianism to infer where there will be future mudflows, as well as to predict what effects urbanization will have on the magnitude and frequency of future flows.

▼ CONCEPT FIVE

There have always been earth processes that are hazardous to people. These natural hazards must be recognized and avoided where possible, and their threat to human life and property must be minimized.

Our discussion of uniformitarianism established that present processes have been operating a good deal longer than humankind has been on the earth. Therefore, we have always been obligated to contend with processes that tend to make our lives difficult. Surprisingly, however, *Homo sapiens* appears to be a product of the Ice Age, one of the harshest of all environments.

Early in the history of the human race, its struggle with natural earth processes was probably a day-to-day experience. However, its numbers were neither great nor concentrated, and, therefore, losses from hazardous earth processes were not very significant. As people developed and learned to produce and maintain a constant food supply, both population and (we can guess) the effects of hazardous earth processes increased, because population centers probably became local centers of pollution and disease. The concentration of population and resources also increased the impact of periodic earthquakes, floods, and other natural disasters. This trend has continued until many people today live in areas that are likely to be damaged by hazardous earth processes or that are susceptible to the adverse impact of such processes in adjacent areas.

Natural earth processes are *exogenetic* if they operate at or near the surface of the earth or *endogenetic* if they operate within or below the earth's crust. Exogenetic processes include weathering (physical or chemical changes in rocks at or near the surface of the earth that cause the rock to break into fragments or chemically decompose), mass wasting (movement of soil or rock down slopes, as in landslides), and either erosion or deposition by such agents as running water, wind, or ice. Volcanic activity and diastrophism (processes that produce mountains, continents, ocean basins, and so on) are common endogenetic processes. The work of organisms, including people, are primarily exogenetic processes; however, we are now able to cause some endogenetic processes, such as earthquakes.

Many processes continue to cause loss of life and property damage, including flooding, earthquakes, volcanic activity, mass-wasting phenomena such as landslides and mudflows, and weathering. The magnitude and frequency of these processes depend on such factors as a region's climate, geology, and vegetation. For example, the effects of running water as an erosional or depositional process depend on the intensity of rainfall; the frequency of storms; how much and how fast the rainwater is able to infiltrate rock or soil; the rate of evaporation and transpiration of water back into the atmosphere; the nature and extent of the vegetation; and topography. The endogenetic processes associated with volcanoes, earthquakes, and tsunamis (very large sea waves, usually incorrectly referred to as tidal waves) are located primarily in response to geologic conditions. For example, the "ring of fire" consisting of the circum-Pacific area is well known for its high incidence of earthquakes and volcanic activity resulting from dynamic, global geologic processes originating deep within the earth.

From our discussion of natural earth processes, we can conclude that many processes can be recognized and predicted by considering climatic, biologic, and geologic conditions. After earth scientists have identified potentially hazardous processes, they should make the information available to planners and decision makers who can then formulate various alternatives to avoid or minimize any threat to human life or property.

▼ CONCEPT SIX

Land- and water-use planning must strive to obtain a balance between economic considerations and the less tangible variables such as aesthetics.

Scenery is now considered a natural resource, and the aesthetic evaluation of a site or landscape before modification may be an important part of the "environmental impact" statement. We find it refreshing that consideration is now given to the less tangible variables such as aesthetics, as well as the more traditional benefits-cost analysis. Until this revolution, justification for a project was weighed by comparing the financial benefits over a period of time with the cost. The assumption now is that there are varying *scenic* values, just as there are varying economic values, associated with a landscape and proposed modification (8).

Most of the research on landscape aesthetics has been concerned with unique landscape nearly untouched by human activity. Since most of us now live in an urban environment, however, we should evaluate the aesthetic resources of these areas. Evaluation would facilitate land use by identifying various alternatives. Evaluation might temporarily interfere with the supply and demand aspects of urbanization, but the result might be worth the extra trouble.

Balancing economic criteria with aesthetic criteria is ambitious and optimistic as well as difficult. The problem is, on what basis can the two be compared? The one logical solution is a hierarchical ranking of the economic alternatives compared to a similar ranking of the aesthetic evaluation. Trade-offs and compromise can then be considered. Pragmatically, the problem of comparing economic considerations with aesthetic evaluation is not particularly difficult, provided we can develop a generally agreed-upon rating scale for aesthetic evaluation; develop a reliable quantitative method to analyze the data and hierarchically rank the alternatives; and develop techniques to map our scenic resources.

▼ CONCEPT SEVEN

The effects of land use tend to be cumulative, and, therefore, we have an obligation to those who follow.

Several million years ago, when early hominids roamed the grasslands, marshy deltas, and adjacent forests of ancient Lake Rudoff along the Great Rift Valley system of East Africa, these prehistoric people were completely dependent upon their immediate environment. Their effect on that environment was probably insignificant as they hunted game and were in turn hunted by predators. This relationship between people and the environment probably existed until about 800,000 years ago, when they developed skill in the use of fire.

The use of fire brought new effects that differed from earlier impacts on the environment. First, fire was capable of affecting large areas of forest or grasslands. Second, it was a repetitive process capable of damaging the same area at rather frequent intervals. Third, it was a rather selective process, in that certain species were locally exterminated while other species that exhibited a resistance to or rapid recovery from fire were favored (9). Early use of fire for protection and hunting probably had a significant effect on the environment, and as people became more and more dependent on an increasing variety of resources for clothing, lodging, and hunting, they also increased their capacity to observe and test the environment. This early experimentation probably led to the use of plants and primitive agriculture about 7,000 B.C.

The emergence of agriculture was the first instance of an artificial land use capable of modifying the natural environment. It also set the stage for the development of a more or less continuously occupied site or cluster of sites that introduced further modification of the environment, such as shelter for living space, primitive latrines, and protective barriers against predators and other people. Furthermore, these early sites probably became the first areas to experience pollution problems resulting from disposal of waste, and soil erosion problems resulting from removal of indigenous vegetation (9). The innovation of agriculture also supplied the necessary nutrients for an increasing population that necessitated the clearing of additional land. This activity certainly influenced an area's ecological balance as some species were domesticated or cultivated and others were removed as pests, so it is not surprising that the increase in human population is paralleled by an increase in the number of extinctions among birds and mammals (Figure 2.4).

The significant point of the entire developmental process of the human race through time is that as cities and farms increase, demand for diversification of land use increases, and the effects tend to be cumulative with time (10). If this is the case, then from an ethical and moral standpoint, we need to examine the effects of land use in a historical framework, if only to ensure that our children and their children can survive in the environment they inherit. This is especially critical because it has been determined that at least since the beginning of civilization, 6,000 years ago, and perhaps as far back as 15,000 years ago, the entire surface of the earth has been altered by human activity. In other words, little, if any, land can be considered original or untouched (2). Furthermore, our ability to cause further changes is increasing at a rapid rate. In defense of human activity, it can be noted that, compared to the energy of mountain building, volcanic activity, and erosion power of streams, human impact is small. It is not insignificant, however, especially in the large metropolitan areas where there are many negative aspects of urbanization.

The lessons in land use from the Old World are explicit. Where sound conservation practices were used, there were successful adjustments of population, but where wasteful exploitation of resources was practiced, the results varied from gullied fields and alluvial plains on rocky hills and steep slopes, to silted-up irrigation reservoirs and canals, to ruins of prosperous cities. Three examples from Lowdermilk's study will emphasize this point (10).

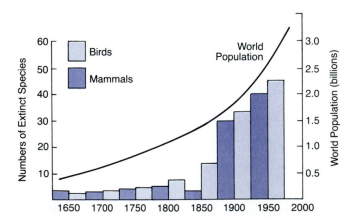

Figure 2.4
Increase in the human population paralleled by increase in the extinction of birds and animals. (Reproduced, by permission, from V. Ziswiler, *Extinct and Vanishing Species* [New York: Springer-Verlag, 1967].)

Ancient Phoenicia and Slope Farming

About 5,300 years ago, the Phoenicians migrated from the desert to settle along the eastern coast of the Mediterranean Sea and establish the coastal towns of Tyre and Sidon, Beyrouth and Byblos. The land is mountainous, with a relief of about 3,000 meters, and at that time was heavily forested with the famous cedars of Lebanon. These trees became the timber supply for the alluvial plains of the Nile and Mesopotamia. As the limited flat land along the coast was populated to its carrying capacity, people moved to the slopes. As the slopes were cleared and cultivated, they were subject to soil erosion. Today a large number of terrace walls, in various states of repair, indicate that the ancient Phoenician farmers attempted to control erosion with rock walls across the slope as many as 40 to 50 centuries ago.

The cedars of Lebanon retreated under the ax, until today very little of the original forest of approximately 2,600 square kilometers remains. Evidence suggests that, given present climatic conditions, the forests would grow where soil has escaped the process of erosion. Today the bare limestone slopes strewn with remnants of former terrace walls are testimony to the results of erosion and the decline and loss of a country's resources.

Dead Cities of Syria

In northern Syria are a number of formerly prosperous cities that are now practically dead. Before the invasion of the Persians and Arabs, cities such as El Bare prospered from the conversion of forest to farmland, which resulted in exports of olive oil and wine to Rome. After the invasion, which destroyed the agriculture, as much as two meters of soil eroded from the slopes, and today the 13 centuries of neglect can be seen in the nearly complete destruction of the land. What is left of the formerly productive land is an artificial desert generally lacking vegetation, water, and soil.

Palestine Story

The Promised Land described by Moses on Mount Neba approximately 3,000 years ago was a land of streams and springs, a land of wheat, barley, and vines, a land of abundant resources. The Promised Land at the time of Lowdermilk's investigation was a sad commentary on human use of the land. Soil erosion and accelerated runoff from barren slopes had caused many of the hills to become greatly depopulated.

Although the land in Palestine cannot be restored to its original productivity, efforts to redeem the land in recent years show what can be done at great cost. Lowdermilk reported that as early as 1960, Israel had more than doubled its cultivated land to about 4,000 square kilometers, and established range cover on vast amounts of uncultivated land for its livestock industry. The results of Israel's efforts are significant in that the country had at that time nearly attained agricultural self-sufficiency with an export-import balance in food (11).

Histories of long-populated areas suggest that soil erosion is a serious problem that has destroyed land and retarded the progress of civilization. Therefore, conservation of our soils must remain a national interest, for, as stated by Lowdermilk, "One generation of people replaces another, but productive soils destroyed by erosion are seldom restorable and never replaceable."

Potential for Catastrophe and Future Generations

A *catastrophe* is an event that causes great loss of life, suffering, and/or destruction. Increased population with stress on global resources in developing countries, along with industrialization and the arms race in others, is increasing the potential for and chances of a human-induced regional or even global catastrophe. On one front are problems of overgrazing, deforestation, and desertification; on another are hazardous chemical wastes and acid precipitation. However, the expected magnitude of disruption from these, while large, seems almost insignificant compared to the projected environmental consequences of large-scale nuclear war. The potential global impact of multiple nuclear explosions might be to produce a new "dark age" in human history. That event might destroy a significant part, if not all, of the biological and physical support systems necessary for civilization in the Northern Hemisphere. Those people, animals, and plants who were to survive the initial blast, fire, and radiation might be immediately subjected to a period of cold and darkness lasting several months or more that might severely alter global ecosystems for years (12, 13). Although results of studies that have modeled effects of nuclear war disagree on the magnitude and extent of climatic disruption, the potential effects are so terrible that its prevention must be given top priority over all other potential environmental problems.

▼ CONCEPT EIGHT

The fundamental component of every person's environment is the geologic factor, and understanding this environment requires a broad-based comprehension and appreciation of the earth sciences and other related disciplines.

Concept Eight arises from the fact that all geology is environmental, and since we all live on the surface of the earth, we are both directly and indirectly affected by geologic processes (14). Therefore, an understanding of our complex environment requires considerable knowledge of such disciplines as **geomorphology**, the study of

landforms and surface processes; **petrology**, the study of rocks and minerals; **sedimentology**, the study of environments of depositions of sediments; **tectonics**, the study of processes that produce continents, ocean basins, mountains, and other large structural features; **hydrogeology**, the study of surface and subsurface water; **pedology**, the study of soils; **economic geology**, the application of geology to locating and evaluating mineral materials; and **engineering geology**, the application of geologic information to engineering problems. Beyond this, the serious earth scientist should also be aware of the contributions to environmental research from such areas as biology, conservation, atmospheric science, chemistry, environmental law, architecture, and engineering, as well as physical, cultural, economic, and urban geography. Environmental geology is the domain of the generalist with strong interdisciplinary interest. This in no way refutes the significant contributions of specialists in various aspects of environmental studies, or the importance of the generalist's consideration of specific problems or specialty areas for research. It merely suggests that, although our research interest may be specialized, we should be generalists in terms of awareness of other disciplines and their contribution to environmental geology. Also, many projects may be studied best by an interdisciplinary team of scientists—but teams must be well disciplined.

The importance of the interdisciplinary nature of environmental geology becomes apparent when we explore the nature of environmental problems. Most projects are complex and involve many different facets that may be generalized into three categories: physical, biological, and of human use and interest. These are essentially the same categories used by Leopold to evaluate river valleys, and their extension to other areas of environmental research seems appropriate (15). The category of physical factors includes such considerations as physical geography, geologic processes, hydrologic processes, rock and soil types, and climatology. Biologic factors include consideration of the nature of plant and animal activity, changes in biologic conditions or processes, and spatial analysis of biologic information. Human use and interest factors include such attributes as land use, economics, aesthetics, interaction between human activity and the physical and biological realms, and environmental law.

Obviously, no one project or research interest will involve all possible factors in each of the three categories, and there is considerable interaction among the categories. Projects such as waste-disposal operations, highway construction, mass transit systems, urban land-use planning, and mining of resources may be concerned with all of the categories. For example, the planning, construction, and operation of a sanitary landfill site is concerned with physical factors, such as physical location, topography, soil type, and hydrologic conditions; biologic processes, which determine the rate of decay of the organic refuse, as well as any contamination of the biologic realm in the vicinity of the site; and human interests, such as compliance with laws, regulations, and good engineering practice.

▼ ▼ ▼ SUMMARY AND CONCLUSIONS

There are eight fundamental concepts basic to the understanding of environmental geology. The number-one environmental problem is the increasing world population. The earth is essentially a closed system, and understanding of feedback and rates of change is critical to solving environmental problems. The earth is the only suitable habitat we have, and its resources are limited. Today's physical processes are modifying our landscape and have operated throughout much of geologic time, but the magnitude and frequency of these processes are subject to natural and artificially induced change. Earth processes that are hazardous to people have always existed. These natural hazards must be recognized and avoided where possible, and their threat to human life and property must be minimized. Land- and water-use planning must strive to obtain a balance between economic considerations and the less tangible variables such as aesthetics. The effects of land use tend to be cumulative, and, therefore, we have an obligation to those who follow. The fundamental component of every person's environment is the geologic factor, and understanding this environment requires a broad-based comprehension and appreciation of the earth sciences and other related disciplines.

Although they do not constitute a complete list, these concepts establish a philosophical framework from which to investigate and discuss environmental geology.

▼ ▼ ▼ REFERENCES

1. EHLERS, E. G. 1968. *Phase equilibria: Laboratory studies in mineralogy.* San Francisco: W. H. Freeman.
2. NATIONAL RESEARCH COUNCIL. 1971. *The earth and human affairs.* San Francisco: Canfield Press.
3. EARTH SYSTEMS SCIENCE COMMITTEE. 1988. *Earth Systems Science.* Washington, D.C.: National Aeronautics and Space Administration.
4. EHRMANN, M. 1927. *Desiderata.* Terre Haute, Indiana.
5. HOLMAN, E. 1952. Our inexhaustible resources. *Bulletin of the American Association of Petroleum Geologists* 6: 1323–29.
6. CLOUD, P. E., JR. 1968. Realities of mineral distribution. In *Man and his physical environment,* ed. G. D. McKenzie, and R. O. Utgard, pp. 194–207. Minneapolis, Minnesota: Burgess.
7. WOLMAN, M. G., and SCHICK, A. P. 1967. Effects of construction on fluvial sediment, urban and suburban areas of Maryland. *Water Resources Research* 3: 451–64.
8. ZUBE, E. H. 1973. Scenery as a natural resource. *Landscape Architecture* 63: 126–32.
9. NICHOLSON, M. 1970. Man's use of the earth: historical background. In *Man's impact on environment,* ed. T. R. Detwyler, pp. 10–21. New York: McGraw-Hill.
10. LOWDERMILK, W. C. 1943. *Lessons from the Old World of the Americans in land use.* Smithsonian Report for 1943, pp. 413–28.
11. ———. 1960. The reclamation of a man-made desert. *Scientific American* 202: 54–63.
12. TURCO, R. P.; TOON, O. B.; ACKERMAN, T. P.; POLLACK, J. B.; and SAGAN, C. 1983. Nuclear winter: Global consequences of multiple nuclear explosions. *Science* 222: 1203–92.
13. EHRLICH, P. R.; HARTE, J.; HARWELL, M. A.; RAVEN, P. H.; SAGAN, C.; WOODWELL, G. M.; AYENSU, E. S.; EHRLICH, A. H.; EISNER, T.; GOULD, S. J.; GROVER, H. D.; HERRERA, R.; MAY, R. M.; MAYR, E.; MCKAY, C. P.; MOONEY, H. A.; MYERS, N.; PIMENTAL, D.; and TEAL, J. M. 1983. Long-term biological consequences of nuclear war. *Science* 222: 1293–1300.
14. OAKESHOTT, G. B. 1970. Controlling the geologic environment for human welfare. *Journal of Geological Education* 18: 193.
15. LEOPOLD, L. B. 1969. *Quantitative comparison of some aesthetic factors among rivers.* U.S. Geological Survey Circular 620.

▼ GEOLOGIC CYCLE

Throughout the 4.5 billion years of earth history, the materials on or near the earth's surface have been created, maintained, and destroyed by numerous physical, chemical, and biochemical processes. Except during the early history of our planet, the processes that produce the earth materials necessary for our survival have periodically reproduced new materials. Collectively, the processes are referred to as the **geologic cycle** (Figure 3.1), which is really a group of subcycles (1). Two of the more important subcycles are the **tectonic** and **hydrologic cycles**.

The hydrologic cycle is the movement of water from the oceans, to the atmosphere, and back to the oceans, by way of precipitation, evaporation, stream runoff, and groundwater flow. Only a very small amount of the total water in the cycle is active near the earth's surface at any one time, and yet this small amount of water is tremendously important in facilitating the movement and sorting of chemical elements in solution (geochemical cycle), sculpturing the landscape, weathering rocks, transporting and depositing sediments, and providing our water resources. Therefore, we should learn more about the hydrologic cycle to better understand and use this important resource.

Tectonic processes are driven by forces deep within the earth. They deform the earth's crust, producing external forms such as ocean basins, continents, and mountains. These processes are collectively known as the tectonic cycle. We now know that the outer layer of the earth, containing the continents and oceans, is about 100 kilometers thick. Called the **lithosphere**, it is not a continuous, uniform layer. Rather, the lithosphere is broken into at least twelve large parts called *plates* that move relative to one another (Figure 3.2) (2). As the lithospheric plates move over the asthenosphere, which is thought to be a more or less continuous layer of little strength below the lithosphere, the continents also move (Figure 3.3) (3). This moving of continents is called **continental drift**. It is believed that the most recent episode of drift started about 200 million years ago, when a supercontinent called *Pangaea* broke up.

The boundaries between plates are geologically active areas where most earthquakes and volcanic activities occur. The three types of boundaries are divergent, convergent, and transform fault (4). **Divergent boundaries** occur at spreading ridges where plates are moving away from each other and producing new lithosphere. **Convergent boundaries (subduction zones)** occur when one plate dives beneath the leading edge of another plate. However, if both leading edges are composed of relatively light continental material, it is more difficult for subduction to start, and a special type of convergent plate boundary called a *continental collision boundary* may develop. This produces linear mountain systems such as

CHAPTER THREE
▼
▼
▼

Earth Materials and Processes

the Alps and the Himalayas. **Transform fault boundaries** occur where one plate slides past another, as, for example, the San Andreas fault in California (Figures 3.2, p. 21, and 3.4, p. 22).

Rates of plate motion relative to each other are shown on Figure 3.2. In general the rates are about as fast as your fingernails grow, but vary from about 2 to 15 centimeters per year. The San Andreas fault moves on average about 5 centimeters per year, which means that Los Angeles is slowly moving toward San Francisco, and in 10 to 12 million years the cities will be side by side. The movement is not steady from year to year at all locations: several meters of movement may occur violently during great earthquakes. Fortunately such events only occur at any one location every few hundred years.

The importance of the tectonic cycle (Figure 3.1) or plate tectonics to environmental geology cannot be overstated. For the first time, earth scientists have a global, unifying theory to better explain how the earth works. The theory of plate tectonics is to geology what Darwin's theory of the origin of species was to biology. We now have an understanding, in the discovery of DNA, of the mechanism of biology. In geology we are still seeking the exact mechanism that drives plate tectonics. Nevertheless, everything living on the earth is affected by plate tectonics. As the plates slowly move a few centimeters per year, so do the continents and ocean basins, producing zones of resources (oil, gas, and minerals), earthquakes, and volcanoes. Furthermore, plate motion over long periods of time (millions of years) changes or modifies flow patterns in the oceans, influencing global climate and regional-local variation in precipitation, and thus the productivity of the land and where people wish to live.

Two other subcycles of the geologic cycle are the **geochemical** and **rock cycles**. Geochemistry is the study

Figure 3.1
The geologic cycle is composed of subcycles, including the hydrologic and tectonic cycles.

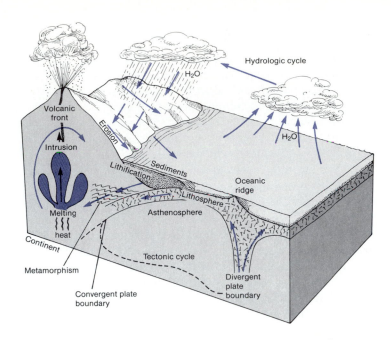

of the distribution and migration of elements in earth processes, and the geochemical cycle is the migratory path of elements during geologic changes. This cycle involves the chemistry of the lithosphere, asthenosphere, hydrosphere, and biosphere. The rock cycle (Figure 3.5, p. 23) is a sequence of processes that produces the three rock families: igneous, sedimentary, and metamorphic. The geochemical and rock cycles are closely related to each other and intimately related to the hydrologic cycle, which provides the water necessary for many chemical and physical processes. The tectonic cycle is also intimately related to the other cycles as it provides water from volcanic processes as well as heat and energy to form and change many earth materials.

Our discussion of the geologic cycle has established that earth materials such as minerals, rocks, soil, and water are constantly being created, changed, and destroyed by internal and external earth processes in numerous subcycles. In the remainder of this chapter we will discuss the different earth materials and various aspects of their occurrence and geology that have environmental significance.

▼ MINERALS

Minerals are naturally occurring, solid, crystalline substances with physical and chemical properties that vary within known limits. (For selected minerals, see color plate.) Although there are over 2,000 minerals, only a few are necessary to identify most rocks. Nearly 75 percent by weight of the earth's crust is oxygen and silicon. These two elements in combination with a few other elements (aluminum, iron, calcium, sodium, potassium, and magnesium) account for the chemical composition of minerals that make up about 95 percent of the earth's crust. Minerals that include the elements silicon and oxygen in

their chemical composition are called *silicates*; these are the most abundant of the rock-forming minerals. The three most important rock-forming silicate minerals or mineral groups are quartz, feldspar, and ferromagnesian.

Quartz, one of the most abundant minerals in the crust of the earth, is a hard, resistant mineral composed of silicon and oxygen. It is often white, but, because of impurities, may also be rose, purple, black, or another color. It is usually recognized by its hardness, which is greater than that of glass, and the characteristic way it fractures—conchoidally (like a clam shell). Because it is very resistant to natural processes that lead to the breakdown of most minerals, quartz is the common mineral in river and most beach sands.

Feldspars, the most abundant and perhaps the most important group of rock-forming minerals in the crust of the earth, are aluminosilicates of potash, soda, and lime. They are generally white, gray, or pink and are fairly hard. Feldspars are very abundant and are important commercial minerals in the ceramics and glass industries. Feldspars weather chemically to form clays, which are hydrated aluminosilicates. This process has important environmental implications. All rocks are fractured, and water, which facilitates chemical weathering, enters the fractures. The feldspars and other minerals then weather. If clay forms along the fracture, it can greatly reduce the strength of the rock and increase the chance of a landslide.

Ferromagnesian minerals are a group of silicates in which the silicon and oxygen combine with iron and magnesium. These are the dark minerals in most rocks. They are not very resistant to weathering and erosional processes. Thus, they tend to be altered or removed relatively quickly. Because they weather quickly to oxides such as limonite (rust), clays, and soluble salts, these minerals, when abundant, may produce weak rocks.

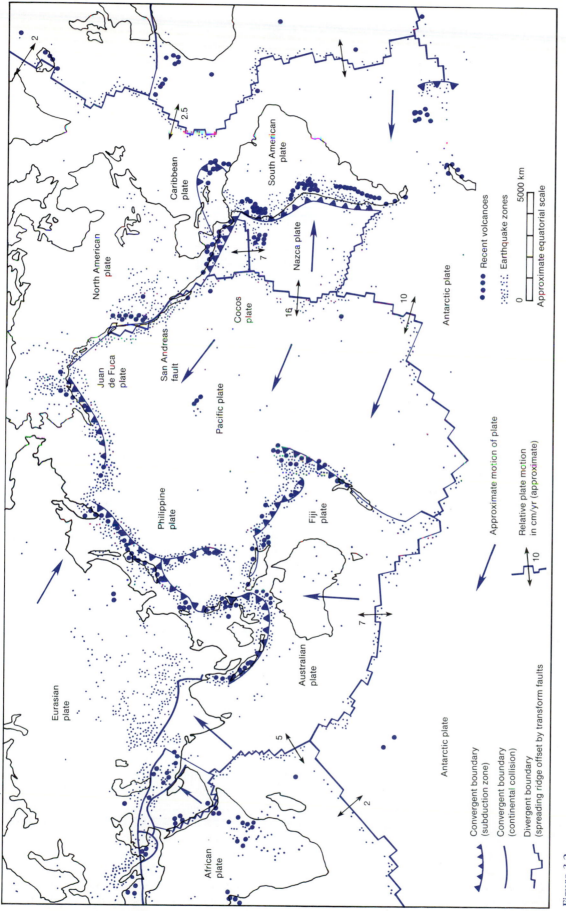

Figure 3.2

Lithospheric plates that form the earth's outer layer. Three types of plate junctions are shown: *spreading ridges*, forming divergent boundaries; *subduction zones*, forming convergent plate boundaries; and more rarely, *transform fault plate boundaries*, such as the San Andreas fault in California, where one plate is sliding by another. (Modified after B. A. Bolt, *Earthquakes* [New York: W. H. Freeman, 1988], p. 11. Rates after J. B. Minster and T. H. Jordan, *J. Geophysical Res.* 83 [1978]: 5331–54.)

Eurasian plate

African plate

Antarctic plate

Australian plate

Fiji plate

Philippine plate

Pacific plate

San Andreas fault

Juan de Fuca plate

North American plate

Caribbean plate

South American plate

Cocos plate

Nazca plate

Antarctic plate

Convergent boundary (subduction zone)

Convergent boundary (continental collision)

Divergent boundary (spreading ridge offset by transform faults)

Approximate motion of plate

Relative plate motion in cm/yr (approximate)

Recent volcanoes

Earthquake zones

0   5000 km

Approximate equatorial scale

2

2.5

7

16

10

10

7

5

2

Figure 3.3
Diagram of the model of sea floor spreading which is thought to drive the movement of the lithospheric plates. New lithosphere is being produced at the spreading ridge (divergent plate boundary). The lithosphere then moves laterally and eventually returns down to the interior of the earth at a convergent plate boundary (subduction zone). This process produces ocean basins and provides a mechanism that moves continents.

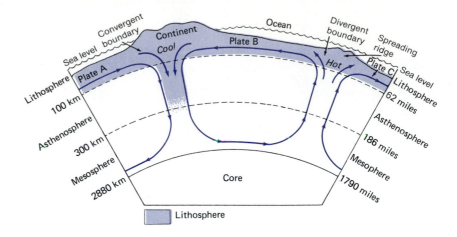

Caution must be used when evaluating construction sites, such as for highways, tunnels, and reservoirs, that contain rocks that are high in ferromagnesians.

Other groups of minerals important in environmental studies include the oxides, carbonates, sulfides, and native elements.

Iron and aluminum are probably the most important metals in our industrial society. The most important iron ore (hematite) and the most important aluminum ore (bauxite) are both oxides. Magnetite, also an iron oxide but economically less important than hematite, is common in many rocks. It is a natural magnet (lodestone) that will attract and hold iron particles. Where particles of magnetite are abundant, they may produce a black sand in streams or beach deposits.

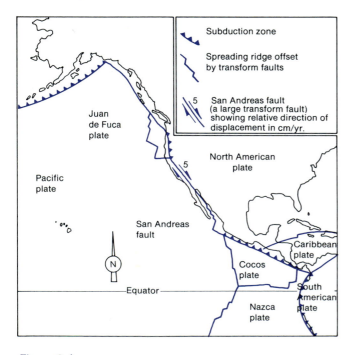

Figure 3.4
Detail of boundary between the North American and Pacific plates. (Modified from P. R. Vogt, *Geological Society of America Bulletin* 101 [1989], 1226.)

Environmentally, the most important carbonate mineral is **calcite**, which is calcium carbonate. This mineral is the major constituent of limestone and marble, two very important rock types. Weathering by solution of this mineral from such rocks often produces caverns, sinkholes (surface pits), and other unique features. Cavern systems carry groundwater, and water pollution problems in urban areas over limestone bedrock are well known. In addition, construction of highways, reservoirs, and other engineering structures is a problem where caverns or sinkholes are likely to be encountered.

The sulfide minerals, such as pyrite, or iron sulfide (fool's gold), are sometimes associated with environmental degradation. This occurs most often when roads, tunnels, or mines cut through coal-bearing rocks that contain sulfide minerals. The minerals oxidize to form compounds such as ferric hydroxide and sulfuric acid. This is a major problem in the coal regions of Appalachia.

Native elements such as gold, silver, copper, and diamonds have long been sought as valuable minerals. They usually occur in rather small accumulations but occasionally are found in sufficient quantities to justify mining. As we continue to mine these minerals in ever lower-grade deposits, the environmental impact will continue to increase.

▼ ROCKS

Rocks are aggregates of one or more minerals (see color plate), and the rock cycle is the largest of the earth cycles. For this cycle to operate, the tectonic cycle is required for energy, the geochemical cycle for materials, and the hydrologic cycle for water. Water is used in the processes of weathering, erosion, transportation, deposition, and lithification of sediments.

The general model of the rock cycle is that the three rock families—igneous, metamorphic, and sedimentary—are involved in a worldwide recycling process. The elements of the cycle are shown in Figure 3.5. Internal heat from the tectonic cycle drives the rock cycle and with crystallization produces igneous rocks from

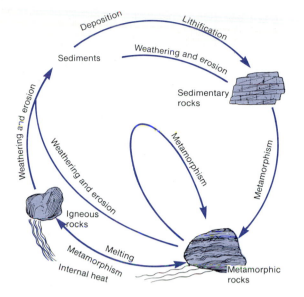

Figure 3.5
The rock cycle.

molten materials. These may crystallize beneath or on the earth's surface. Rocks at or near the surface break down chemically and physically by weathering processes to form sediments that are transported by wind, water, and ice. The sediments accumulate in depositional basins, such as the ocean, where they are eventually transformed into sedimentary rocks. After the sedimentary rocks are buried to sufficient depth, they may be altered by heat, pressure, or chemically active fluids to produce metamorphic rocks, which may then melt to begin the cycle again.

Possible variations of the idealized sequence are indicated by the arrows in Figure 3.5. For example, either igneous or metamorphic rocks may be altered by metamorphism into a new metamorphic rock without ever being exposed to weathering or erosion processes.

The tectonic cycle provides several environments for the rock cycle (Figure 3.1). Generally, new lithosphere is produced at the spreading ridges. This material moves laterally away from the ridges, which are divergent plate boundaries, until it dives beneath another plate at a convergent plate boundary. These environments and associated rock-forming processes are shown in Figure 3.1. Transform fault and collision boundaries also produce rock-forming processes that fit in the generalized rock cycle. Recycling of rock and mineral material is the most important aspect of the rock cycle, while the processes that drive and maintain the cycle are important in determining the properties of resulting rocks. Therefore, interest is more than academic because it is upon these earth materials that we build our homes, industries, roads, and other structures. Understanding of the various aspects of this cycle, such as the igneous-rock-forming

processes or the soil-forming processes, facilitates the best use of these resources.

Other aspects of the rock cycle are responsible for concentrating as well as dispersing materials. These aspects are extremely important in mining minerals. If it were not for igneous, sedimentary, and metamorphic processes that concentrate minerals, it would be difficult indeed to extract resources. Thus, we take resources that are concentrated by one aspect of the rock cycle, transform these resources through industrial activities, and then return them to the cycle in a diluted form where they are further dispersed by continuing earth processes (5). Once this process has taken place, the resource cannot be concentrated again within a useful frame of time. Take lead, for example, once widely used in automobile fuel. Lead is mined in a concentrated form; transformed and diluted in fuel; and further dispersed by traffic patterns, air currents, and other processes. Eventually, lead may become abundant enough to contaminate soil and water but seldom sufficiently concentrated to be recycled. Similar examples may be cited for many other resources used in paints, solvents, and other industrial products.

Important aspects of human use of earth materials produced by the rock cycle include properties that affect engineering design. Therefore, the terminology of applied geology is somewhat different from that used in traditional geologic investigations. For example, the term *rock* can be reserved for earth materials that cannot be removed without blasting, and *soil* can be defined as those earth materials that can be excavated with normal earth-moving equipment. Therefore, a very friable (loosely compacted or poorly cemented) sandstone may be considered a soil, but a well-compacted clay may be called a rock. Although confusing at first, this pragmatic approach has advantages over conventional terminology in that more useful information is conveyed to planners and designers. Clay, for example, is generally unconsolidated and easily removed. Thus, a contractor might assume, without further information, that clay is a soil and bid low for an excavation job in the belief that it can be removed without blasting. If the clay turns out to be well compacted, however, he may have to blast it, which is much more expensive. This kind of error is avoided if a proper preliminary investigation is made and the contractor forewarned that such a clay should be considered rock. (Soils and their properties will be considered in detail in Chapter 4.)

Strength of Rocks

The strength of earth materials varies with composition, texture, and location. Thus, weak rocks, such as those containing many altered ferromagnesian minerals, may creep or nearly flow under certain conditions and be very difficult to tunnel through. On the other hand,

TYPE
OF
STRESS

RESULTING
STRAIN

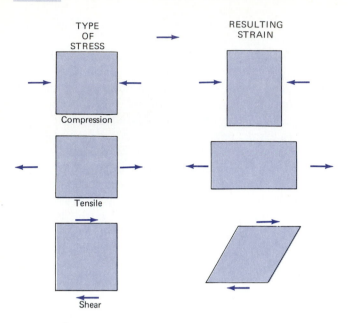

Compression

Tensile

Shear

Figure 3.6
Common types of stress and resulting strain.

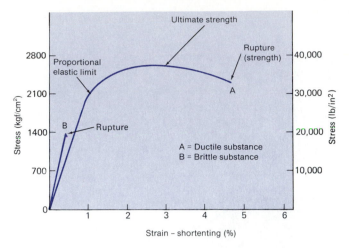

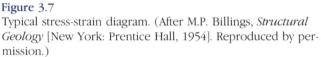

Figure 3.7
Typical stress-strain diagram. (After M.P. Billings, *Structural Geology* [New York: Prentice Hall, 1954]. Reproduced by permission.)

granite, a very common igneous rock, is generally a strong rock that needs little or no support. If granite were placed deep within the earth, however, it might also flow and be deformed. Hence, the strength of a rock may be quite different under different types and different amounts of stress.

Stress is measured as force per unit area that develops within rocks to resist external forces, and it is generally measured in kilograms (force) per square centimeter, or kgf/cm^2 (previously, pounds per square inch or lb/in^2). Three types of stress, *compressive, tensile,* and *shear,* are shown in Figure 3.6. **Strain** is defined as deformation induced by a stress. A material under stress may deform in an **elastic** or a **plastic** manner. Elastic deformation is like a rubber band; that is, the deformed material returns to its original shape after the stress is removed. Plastic deformation is characterized by permanent strain; that is, the material does not return to its original shape after the stress is removed. Materials that rupture before any plastic deformation are **brittle**, and those that rupture after elastic and plastic deformation are **ductile** (Figure 3.7) (6).

The strength of an earth material is usually described as the compressive, shear, or tensile stress necessary to break a sample of it. We must remember, however, that rocks are neither homogeneous nor isotropic, and both conditions are necessary before completely reliable strength values can be assigned to any material. Thus, we assign a safety factor (SF) to ensure that a rock will perform as expected; that is, the rock is loaded only to a fraction of its assumed strength. For example, if a rock has a compressive strength of 700 kgf/cm^2, then with a safety factor of 10, the rock should be loaded only to 70 kgf/cm^2. The testing necessary to establish strengths of

rocks can be extremely expensive, so this information is usually obtained only for large engineering structures. For small structures, the experienced earth scientist can render a judgment consistent with good engineering practice.

Avoiding extensive testing to determine the strength of rocks at a site does not extend to a similar treatment of soils. The cost of a good soil survey to determine the strength characteristic of soils is well worth the initial cost compared to the possible damages and costs that can result from not considering these properties. Even small structures, such as an apartment building with a swimming pool and retaining wall, are subject to considerable structural damage (cracking of walls or steps pulling away from the buildings) by differential settling on unstable soils. With proper use of a soil survey before buildings are designed, however, damage caused by adverse soil conditions can be minimized (see Chapter 4).

The strength of rocks is greatly affected by the frequency and orientation of fractures, joints, and shear zones, which can range in size from small hairline fractures to huge shear zones such as the San Andreas fault zone (Figure 3.8) which scars the California landscape for hundreds of kilometers. (A complete discussion of faulting, which involves displacement of rocks across fractures, is given in Chapter 8 on earthquakes.) When large structures are planned, shear zones and fracture systems in rocks must be considered. One active shear zone in the proposed immediate vicinity of the foundation of a large masonry dam may well be sufficient reason to look for another site. It has been concluded that the Baldwin Hills Reservoir failure in 1963 in southern California, which claimed five lives and caused $11 million in damages, resulted from gradual movement addition, a series of periodic inspections was initiated to ensure the safety of the operation. Even with these precautions, however, the failure came quite suddenly. Were it not for fortunate circumstances that gave several hours warning and enabled officials to act quickly, the loss of life might have been much greater (7).

The problems and dangers associated with fractures, shear zones, and faults are not confined to "active" systems. Once a fracture has developed, it is subject to weathering, which may produce clay minerals. These minerals may be unstable and may become a lubricated surface facilitating landslides, or the clays may wash out and create open conduits for water to move through the rocks. Therefore, although active faulting is obviously a hazard to the stability of engineered structures, even small, inactive fractures, faults, and other shear zones must be inspected and carefully evaluated to maximize structural safety.

Types of Rocks

Our discussion of the generalized rock cycle established that there are three rock families: igneous, sedimentary, and metamorphic. Table 3.1 lists the common rock types. Although rocks are primarily classified according to both mineralogy and texture, texture is most significant in environmental geology. **Rock texture**—the size, shape, and arrangement of grains—along with fractures, shear zones, and faults, determines the strength and utility of rock. This should not be interpreted to mean that the mineralogy is not important. On the contrary, in specific cases it can be extremely significant. In considering groups of rocks such as the intrusive igneous rocks, however, mineralogy is of secondary importance to the texture and structure. Granitic rocks, for example, can be given a number of different names based upon their mineralogy, but the engineering properties of all of them will be nearly the same. More important in practical applications are the texture and nature of fractures and related alteration zones.

Igneous rocks. **Igneous rocks** are rocks that have crystallized from a naturally occurring, mobile mass of quasi-liquid earth material known as **magma**. If magma crystallizes below the surface of the earth, the resulting igneous rock is called **intrusive**. Therefore, whenever intrusive igneous rocks such as granite are exposed at the surface of the earth, we may conclude that erosional processes have removed the original cover material. Magma is probably generated in the upper asthenosphere or the lithosphere. As it moves up, it displaces the rock it intrudes, often breaking off portions of the rocks into which the magma moves and incorporating them into the mass of moving magma. These foreign blocks, known as *inclusions,* are good evidence of forcible intrusion.

Figure 3.8
Aerial view of the San Andreas fault, looking in northerly direction along Elkhorn scarp, San Luis Obispo County. The irregular dark line at left is tumbleweeds piled against a fence. Arrows show direction of displacement along the fault. (Photo courtesy R. E. Wallace, U.S. Geological Survey.)

along shear zones (faults). The movement fractured the tile drain, foundation, and asphaltic membrane of the dam and reservoir floor. After the reservoir had drained, long cracks in the asphaltic concrete pavement were observed. The cracks extended continuously the length of the reservoir and in line with the deep gash in the dam (Figure 3.9). The dam was 71 meters high and 198 meters long. When construction began in 1947, it incorporated the most advanced knowledge available. The fracture, along with several others, was discovered during construction and was examined and judged not dangerous to the stability of the dam. Because of a previous experience with shear zones (the 1928 failure of the St. Francis dam), the design of the Baldwin Hills dam included some provisions for the fractures in the foundation material. In

(a)

(b)

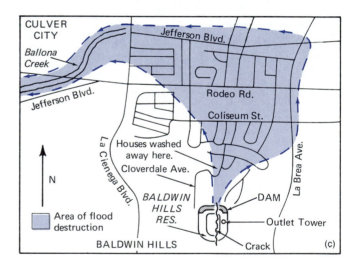

Figure 3.9
Failure of the Baldwin Hills Reservoir, aerial and ground views (a, b) and map showing area flooded (c). (Photograph [a] courtesy of the California Department of Water Resources. Drawing [c] from Walter E. Jessup, "Baldwin Hills Dam Failure." Reprinted with permission from the February 1964 issue of *Civil Engineering—ASCE*, official monthly publication of the American Society of Civil Engineers, vol. 34, 1964. Photo [b] courtesy of Los Angeles Department of Water and Power.)

Intrusive igneous rocks are generally strong and have a relatively high unconfined compressive strength varying from about 700 to 2,800 kgf/cm² (8). The strength of an individual granite depends upon the grain size and the degree of fracturing. In general, fine-grained granites are stronger than coarse-grained. Fresh, unweathered intrusive rocks, unless extensively fractured, are usually satisfactory for all types of engineering construction and operations (9). If they can be economically quarried, they are often good sources of construction material—crushed stone for concrete aggregate, dimension stone

for facing or veneer, broken stone for fill material, or riprap to protect slopes against erosion. Even with the general favorable physical properties of intrusive rocks, however, it is not safe to assume that all these rocks are satisfactory for their intended purpose. This is especially true for large, heavy structures whose stability depends upon a solid, safe foundation. In this case, an entire project may be jeopardized when the site is studied and evaluated and a shear zone or an alteration zone is identified. Therefore, although most intrusive rock *is* satisfactory for nearly all purposes, one should not

Type	Texture	Materials
Igneous		
Intrusive		
Granitic[a]	Coarse[b]	Feldspar, quartz
Ultrabasic	Coarse	Ferromagnesians, ±quartz
Extrusive		
Basaltic[c]	Fine[d]	Feldspar, ±ferromagnesians, ±quartz
Volcanic breccia	Mixed—coarse and fine	Feldspar, ±ferromagnesians, ±quartz
Welded tuff	Fine volcanic ash	Glass, feldspar, ±quartz
Metamorphic		*Parent Material*
Foliated		
Slate	Fine	Shale or basalt
Schist	Coarse	Shale or basalt
Gneiss	Coarse	Shale, basalt, or granite
Nonfoliated		
Quartzite	Coarse	Sandstone
Marble	Coarse	Limestone
Sedimentary		*Materials*
Detrital		
Shale	Fine	Clay
Sandstone	Coarse	Quartz, feldspar, rock fragments
Conglomerate	Mixed—very coarse and fine	Quartz, feldspar, rock fragments
Chemical		
Limestone	Coarse to fine	Calcite, shells, calcareous algae
Rocksalt	Coarse to fine	Halite

Table 3.1.
Common rocks (engineering geology terminology)

[a]Textural name used by engineers for a group of coarse-grained, intrusive igneous rocks, including granite, diorite, and gabbro.

[b]Individual mineral grains can be seen with naked eye.

[c]Textural name used by engineers for a group of fine-grained, extrusive igneous rocks, including rhyolite, andesite, and basalt.

[d]Individual mineral grains cannot be seen with naked eye.

assume that it is. Field evaluation, including mapping, drilling, and laboratory tests, is necessary before large structures are designed.

Extrusive igneous rocks form when magma reaches the surface and is blown out of a volcano as pyroclastic debris or flows out as lava. In either case, the resulting igneous rock is extrusive. The variety and composition of extrusive rocks are considerable, and, therefore, generalizations concerning their suitability for a specific purpose are difficult. Again, when one is working with these rocks, careful field examination, including detailed mapping and drilling, along with laboratory testing, is always necessary before large engineering structures are designed (9).

Extrusive rocks crystallized from lava flows may be stronger than granite (with an unconfined compressive strength exceeding 2,800 kgf/cm^2) (8). However, if the lava flows are mixed with volcanic breccia (angular fragments of broken lava and other material) or with thick vesicular or scoriaceous zones (produced as gas escapes from cooling lava), then the strength of the resulting rocks may be greatly reduced. Furthermore, after cooling and solidifying, lava flows often exhibit extensive columnar jointing (Figure 3.10) which may lower the strength of the rock. Solidified lava flows also may have subterranean voids known as *lava tubes,* which may either collapse from the weight of the overlying material or carry large amounts of groundwater, either of which may cause problems during the planning, design, or construction phases of a project.

Pyroclastic debris ejected from a volcano produces a variety of extrusive rocks. The most common material is **tephra**, a comprehensive term for any clastic material ejected from a volcano. Volcanic ash consists of rock fragments and glass shards less than 4 mm in diameter. When it is compacted, cemented, or welded together, it is called **tuff**. Although the strength of a tuff depends upon how well cemented or welded it is, generally it is a soft,

Figure 3.10
Devil's Postpile. Characteristic columnar joints that have formed because of contraction during cooling of lava. (Photo by Cecil W. Stoughton, courtesy of the U.S. Department of the Interior, National Park Service.)

weak rock that may have an unconfined compressive strength of less than 350 kgf/cm^2 (8). Some tuff may be altered to a clay known as **bentonite**, an extremely unstable material. When it is wet, bentonite expands to many times its original volume. Pyroclastic activity also produces larger fragments which, when mixed with ash and cemented together, form **volcanic breccia** or **agglomerate**. The strength of this material may be comparable to that of intrusive rocks or may be quite weak, depending upon how the individual particles are held together. In general, however, volcanic breccia and agglomerates make poor concrete aggregates because of the large and variable amount of fine materials produced when the rock is crushed.

From the discussion of extrusive igneous rocks, and especially pyroclastic debris, we can see that planning, design, and construction of engineering projects in or on these rocks can be complicated and risky (9). This was tragically emphasized on June 5, 1976, when the Teton Dam in Idaho failed, killing 14 people and inflicting approximately $1 billion in property damage. The causes of the failure had strong geologic aspects, namely, highly fractured volcanic rocks over which the dam was constructed and highly erodible wind-deposited clay-silts used in construction of the dam interior or core. Open

fractures in the volcanic rocks were probably not completely filled with a cement slurry (grout) during construction, and while the reservoir was filling, water began moving under the foundation area of the dam. When the moving water came into contact with the highly erodible material of the core, it quickly eroded a tunnel through the base of the dam, which would explain the observed whirlpool several meters across that formed in the reservoir just prior to failure. In other words, development of a vortex of water draining out of the reservoir near the dam strongly suggested the presence of a subsurface tunnel of free-flowing water below the dam. The final failure of the dam came minutes later, and a wall of water up to 20 meters high rushed downstream, destroying homes, farms, equipment, animals, and crops along a 160-kilometer reach of the Teton and Snake Rivers.

Sedimentary rocks. **Sedimentary rocks** form when sediments are transported, deposited, and then lithified by natural cement, compression, or other mechanism. There are two types of sedimentary rocks: *detrital* sedimentary rocks, which form from broken parts of previously existing rocks; and *chemical* sedimentary rocks, which form from chemical or biochemical processes that remove material carried in chemical solution.

Detrital sedimentary rocks include shale, sandstone, and conglomerate. Of the three, **shale** includes about 50 percent of all sedimentary rocks and is by far the most abundant. Shale also causes considerable environmental problems, and the existence of shale is a red flag to the applied earth scientist.

There are two types of shale: *compaction* shale and *cementation* shale. Compaction shale is held together primarily by molecular attraction of the fine clay particles. It is a very weak rock with the following environmental problems. First, depending upon the type of clay (mineralogy), it may have a high potential to absorb water and swell. Second, depending upon the bonding between the depositional layers (bedding planes), it may have a high potential to slide even on gentle slopes. Third, it has a high potential to slake; that is, contact with water will cause the surface to break away and curl up. This problem seriously retards the making of a firm bond between the rock and engineering structures such as foundations of buildings and dams. Fourth, because of the elastic nature of clays, these rocks tend to rebound if stress conditions change, making the rock a very poor foundation material. Fifth, these rocks have an unconfined compressive strength as low as 1.8 kgf/cm^2.

Depending upon the degree and type of cementing material, cemented shales can be very stable, strong rock suitable for all engineering purposes; but the presence of shale is a danger sign, calling for a close look to determine the type and extent of the rock and its physical properties.

Sandstones and **conglomerates** are coarse-grained and make up about 25 percent of all sedimentary rock. Depending on the type of cementing material, these rocks may be very strong and stable for engineering purposes. Common cementing materials are silica, calcium carbonate, and clay. Of these, silica is the strongest; calcium carbonate tends to dissolve in weak acid; and clay may be unstable and tend to wash away. It is always advisable to evaluate carefully the strength and stability of cementing materials in the detrital sedimentary rocks.

Limestones make up about 25 percent of all sedimentary rocks and are by far the most abundant of the chemical sedimentary rocks. Limestone is commonly composed of the mineral calcite. Human use and activity generally do not mix well with limestone. Although it may have sufficient strength, it weathers easily to form subsurface cavern systems and solution pits. In such limestone areas, most of the streams may be diverted to subterranean routes. These routes can easily become polluted by contaminated runoff that enters the groundwater. The mineralogy of limestone is such that little or no natural purification of contaminated waters in contact with them occurs. In addition, construction may be hazardous in areas with abundant caverns and sinkholes.

Another important chemical sedimentary rock is **rocksalt,** which is composed primarily of halite (sodium chloride). Rocksalt forms when shallow seas or lakes dry up. As water evaporates, a series of salts, one of which is halite, is precipitated. The salts may later be covered up with other types of sedimentary rocks as the area again becomes a center of deposition of sediments. It is from these sedimentary basins that salt is mined, often by solution mining, a process in which water is pumped down into the salt and then flows or is pumped out again supersaturated with salt. The process leaves huge holes in the salt deposits. Because rocksalt is abundant, usually completely isolated from circulating groundwater, and located in areas with low earthquake risk, it has been suggested as an acceptable place to dispose of radioactive wastes (10).

Geologic structures, such as folds and unconformities, have important relationships to human interest and activity. **Folds** or bends (Figure 3.11) are produced when rocks are deformed under stress. Geologic events that fold rocks are usually associated with the tectonic cycle. Intense folding occurs at depths of several kilometers or more but can also occur near the surface in response to crustal shortening accompanied by earthquakes. **Unconformities** are buried erosion surfaces (Figure 3.12). Our present landscape is an erosion surface produced by the action of running water, wind, and other processes. If the sea were to rise and cover part of the landscape, sediments would be deposited over the surface and bury it, and an unconformity would be formed.

Folds and unconformities are important in environmental work for several reasons. First, they tend to influence the flow of groundwater in the rocks; that is, water usually migrates down the limbs of folds and along unconformities. Second, petroleum reservoirs are often found in folded sedimentary rocks and in association with unconformities. Third, folding produces typical fracture systems characterized by open tension fractures on the crest of anticlines and closed compression fractures in the troughs of synclines (Figure 3.13). This may be significant in the siting of reservoirs in areas of folded sedimentary rocks such as limestone, where solutional weathering enlarges the fractures. Synclines are likely to hold the water better than anticlines and,

(a)

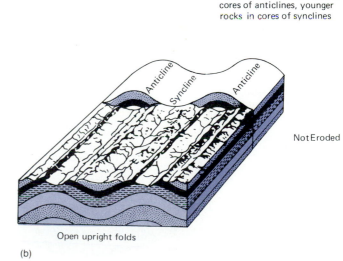

Erosion exposes older rocks in cores of anticlines, younger rocks in cores of synclines

Anticline
Syncline
Anticline

Not Eroded

Open upright folds

(b)

Figure 3.11
(a) Series of folds in the Rocky Mountains of southern British Columbia, and (b) the effects of erosion of folded rocks. (Photo courtesy of the Geological Survey of Canada.)

Figure 3.12
Unconformity (buried surface of erosion) near Morro Bay, California. The light rock is a steeply inclined shale that has been eroded and covered with more recent sediments that are nearly horizontal.

therefore, may be favored sites (9). Fourth, erosion of folded rocks may expose a variety of rock types at different inclinations and hence affect the design of an engineering structure for that area.

Metamorphic rocks. **Metamorphic rocks** are changed rocks. Heat, pressure, and chemically active fluids produced in the tectonic cycle may change the mineralogy and texture of rocks, in effect producing new rocks. The two types of metamorphic rocks are *foliated,* in which the elongated or flat mineral grains have a preferential parallel alignment or banding of light and dark minerals; and *nonfoliated,* without preferential alignment or segregation of minerals.

Foliated metamorphic rocks, such as slate, schist, and gneiss, have a variety of physical and chemical properties,

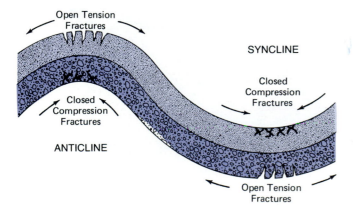

Figure 3.13
Diagram of an anticline and a syncline and the accompanying tension and compression fractures formed during folding.

so it is difficult to generalize about the usefulness of such rocks for engineering projects. **Slate** is generally an excellent foundation material. **Schist,** when composed of soft minerals, is a poor foundation material for large structures. **Gneiss** is usually a hard, tough rock suitable for most engineering purposes.

Foliation planes of metamorphic rocks are potential planes of weakness. The strength of the rock, its potential to slide, and the movement of water through the rock all vary with the orientation of the foliation. Consider, for example, the construction of road cuts and dams in terrain where foliated metamorphic rocks are common. In the case of road cuts, foliation planes inclined toward the road result in unstable blocks that might fall or slide downslope toward the road. Also, groundwater will tend to flow down the foliation, causing a drainage problem at the road. The preferred orientation is with the foliation planes dipping away from the road cut (Figure 3.14). For construction of dams, the preferred orientation is with nearly vertical foliation planes parallel to the axis of the structure (Figure 3.15) (8). This position minimizes the chance of leaks and unstable blocks.

Important nonfoliated metamorphic rocks include quartzite and marble. **Quartzite** is a metamorphosed sandstone. It is a hard, strong rock suitable for many engineering purposes. The engineering properties of marble are similar to those of its parent rock (limestone), and cavern systems and surface pits should be expected.

On the night of March 12, 1928, more than 500 lives were lost and $10 million in property damage was done as ravaging flood waters raced down the San Francisquito Canyon near Saugus, California. This disaster did more than any previous event to focus public attention on the need for geologic investigation as part of siting reser-

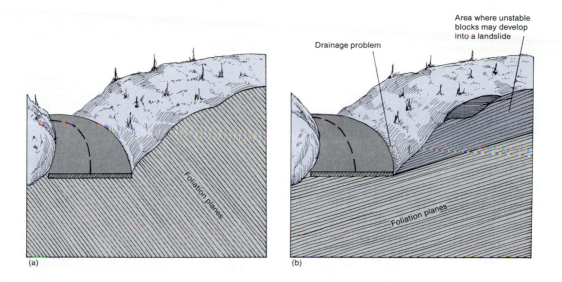

Figure 3.14
Two possible orientations of foliation in metamorphic rock and the effect on highway stability. Where the foliation is inclined away from the road (a), there is less likelihood that unstable blocks will fall on the roadway. Where the foliation is inclined toward the road (b), unstable blocks of rock above the road may produce a landslide hazard.

voirs. The St. Francis Dam, 63 meters high with a main section 214 meters long and holding 47 million cubic meters of water, had failed. The cause was clearly geologic. Adverse geologic conditions (Figure 3.16) included these: first, the east canyon wall is metamorphic rock (schist) with foliation planes parallel to the wall. Before the failure, both recent and ancient landslides indicated the instability of the rock. Second, the rocks of the west

side of the dam are sedimentary and form prominent ridges, suggesting that they were strong and resistant. Under semiarid conditions, this was true; when the rocks became wet, however, they disintegrated. This characteristic was not discovered and tested until after the dam had failed. Third, the contact between the two rock types is a fault with approximately a 1.5-meter-thick zone of crushed and altered rock. The fault was shown on

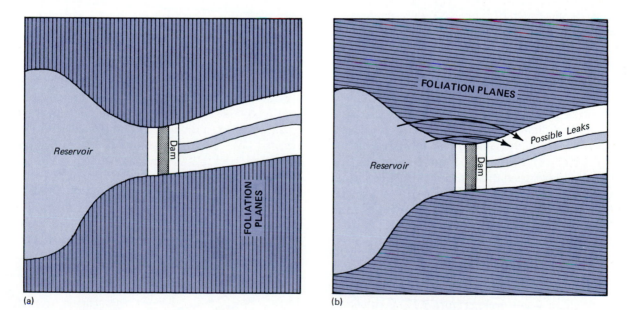

Figure 3.15
Two possible orientations of foliation in metamorphic rocks at a dam and reservoir site. The most favorable orientation of the foliation is shown in part (a), where the foliation is parallel to the axis of the dam. The least favorable orientation is where the foliation planes are perpendicular to the axis of the dam, as shown in (b).

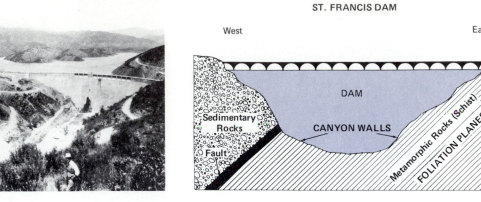

(a)

Figure 3.16
St. Francis Dam (a) prior to failure; (b) geology along the axis of the dam; and (c) after failure. (Photos courtesy of Los Angeles Department of Water and Power.)

(c)

California's 1922 fault map but was either not recognized or ignored. The causes of the dam failure were a combination of slipping of the metamorphic rock, disintegration and sliding of the sedimentary rock, and leakage of water along the fault zone that washed out the crushed rock (7, 8). These three causes together destroyed the bond between the concrete and the rock and precipitated failure. This tragic event clearly illustrated the need to investigate carefully the properties of earth materials before construction of large engineering structures. Such investigations are now standard procedure.

▼ WATER

The hydrologic or water cycle (Figure 3.17) is driven by solar energy and supplies nearly all our water resources. Of the water on earth, approximately 97 percent is in the oceans. A small percentage is locked up in glacier icecaps, and only a small fraction of 1 percent is in the

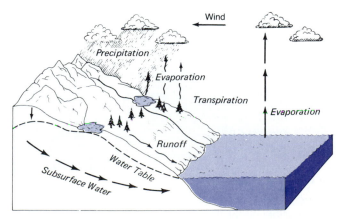

Figure 3.17
The water cycle.

entire atmosphere. Nevertheless, the freshwater phase of the hydrologic cycle is dependent upon this small portion of total water resources.

Types of Water

Of the many types of water that can be described, we will consider six. The first is **meteoric water**, the water in or derived from the atmosphere. The second is **connate water**, water that is no longer in circulation or contact with the present water cycle. Connate water is usually saline water trapped during the deposition of sediments and may be considered as "fossil water." The third is **juvenile water**, water derived from the interior of the earth that has not previously existed as atmospheric or surface water. The slow release of juvenile water, usually by volcanic activity today, is believed to be the source of most of the water in the hydrologic cycle. The fourth is **surface water**, the waters above the surface of the lithosphere (river, lake, ocean waters). The fifth is **subsurface water**, all the waters within the lithosphere. These include soil, capillary, and connate waters, and groundwater. **Groundwater**, the sixth, is that part of the subsurface waters within the zone of saturation. These waters may be fresh or saline. Fresh groundwater is an important part of our potable water supply.

On land, meteoric and surface water processes, and, to a lesser extent, groundwater processes, are responsible for erosion and deposition of earth materials that form most of our landscape. Even in arid areas, running water produces most of the landforms. In addition, the same water processes interact and provide the water resources necessary for our existence.

The hydrologic cycle shown in Figure 3.17 is that of a humid region. In humid climates, processes of precipitation balance with those of evaporation, transpiration, runoff, and groundwater flow, so the entire system forms a water budget. The situation is simplified in arid areas, where evaporation may exceed precipitation, and the flow of water is essentially one-way from the land up to the atmosphere (1).

Processes involved with the movement of surface waters and groundwaters are intimately related. In the contiguous United States, approximately 30 percent of the precipitation enters into the surface-subsurface flow system. Of this, approximately 1 percent reaches the ocean by way of groundwater flow (1). This is grossly oversimplified, however, because there are so many possible interactions where surface water enters into the groundwater flow, and conversely. The rate of surface runoff or infiltration into the groundwater system depends upon the distribution and amount of precipitation; the types of soils and rocks; the slope of the land; the amount and type of vegetation; and the amount of rejected recharge, or water that cannot enter the ground because the soil is already saturated. Principles of river flow are discussed with flooding in Chapter 6, and groundwater is discussed in Chapter 11 with water resources.

▼ WIND AND ICE

Wind- and ice-related processes are responsible for the erosion, transport, and deposition of tremendous quantities of surficial earth materials. Furthermore, these processes both modify and create a substantial number of landforms in environmentally sensitive areas—coastal, desert, arctic, and subarctic.

Wind

Windblown deposits are generally subdivided into two groups: *sand deposits,* mainly dunes; and *loess* (windblown silt), further divided into *primary loess,* which is essentially unaltered since deposition, and *secondary loess,* which has been transported and reworked over a short distance by water or has been intensely weathered in place (8).

Extensive deposits of windblown sand and silt cover thousands of square kilometers in the United States (Figure 3.18). Sand dunes and related deposits are found along the coasts of the Atlantic and Pacific oceans and the Great Lakes. Inland sand is found in areas of Nebraska, southern Oregon, southern California, Nevada, and northern Indiana, and along large rivers flowing through semiarid regions, as, for example, the Columbia and Snake rivers in Oregon and Washington. The majority of loess is located adjacent to the Mississippi Valley, but some is also found in the Pacific Northwest and Idaho.

Sand dunes are constructed from sand moving close to the ground. They have a variety of sizes and shapes and develop under a variety of conditions (Figures 3.19, 3.20, and 3.21). Regardless of where they are located, how they form, or whether they are active or relic or otherwise stabilized, they tend to cause environmental problems. Migrating sand is particularly troublesome, and stabilization of sand dunes is a major problem in construction and maintenance of highways and railroads that cross sandy areas of deserts. The complex group of sand dunes shown in Figure 3.22a (p. 36) is encroaching on Highway 95 near Winnemucca, Nevada. The rate of movement is about 12 meters per year. The sand is removed about three times a year (Figure 3.22b), and approximately 1,500 to 4,000 cubic meters are removed each time. Attempts to stabilize the dunes by planting several varieties of grass in test plots have not been successful because there is simply not enough precipitation to support the grass.

Building and maintaining reservoirs in sand dune terrain are even more troublesome and tend to be extremely expensive. These reservoirs should be constructed only if very high water loss can be tolerated.

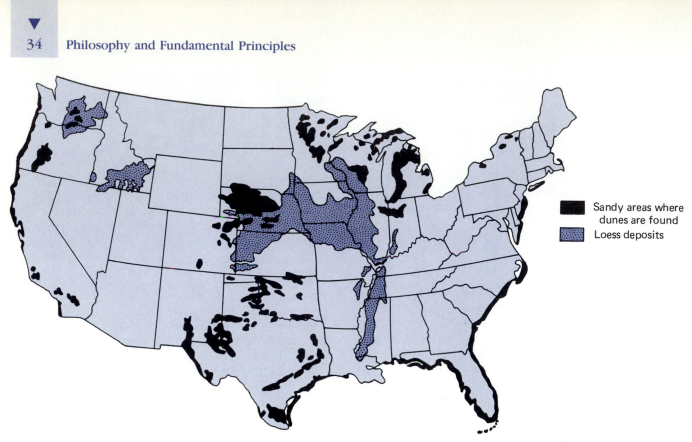

Figure 3.18
Distribution of windblown deposits within the conterminous United States. (Reproduced, by permission, from Douglas S. Way, *Terrain Analysis* [Stroudsburg, Pennsylvania: Dowden, Hutchinson & Ross, 1973].)

Canals in sandy areas should be lined to hold water and control erosion (8).

In contrast to sand, which seldom moves more than a meter off the ground, windblown silt and dust can be carried in huge dust clouds thousands of meters in altitude (Figure 3.23). A typical dust storm 500 to 600 kilometers in diameter may carry over 100 million tons of silt and dust, sufficient to form a pile 30 meters high and 3 kilometers in diameter (11). Terrible dust storms in the 1930s probably exceeded even this, perhaps carrying over 58 thousand tons of dust per square kilometer.

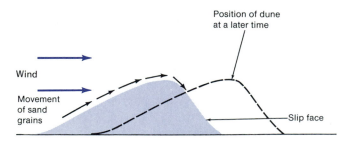

Figure 3.19
Movement of a sand dune. The wind moves the individual sand grains along the surface of the gentle windward side of the dune until they fall down the steeper leeward face. In this manner, the dune slowly migrates in the direction of the wind. (After Robert J. Foster, *General Geology,* 4th ed. [Columbus: Charles E. Merrill, 1983].)

Loess, or windblown silt, in the United States derived primarily during the Pleistocene Ice Ages from glacial outwash in the vicinity of major streams that carried the glacial meltwater from the ice front. Retreat of the ice left large, unvegetated areas adjacent to rivers; these areas were highly susceptible to wind erosion. We know this because loess generally decreases rapidly in thickness with distance from the major rivers.

Loess is a mixture of fine sand, silt, and clay in which the grains are arranged in an open framework. Loess is porous, with a vertical permeability greater than its horizontal. This difference results in part from the presence of long, vertical tubes in the loess that are probably casts of plant roots that developed as the loess accumulated. Although loess may form nearly vertical slopes (Figure 3.24), it rapidly consolidates when subjected to a load (such as a building) and wetted, a process called **hydroconsolidation,** which results as clay films or calcium carbonate cement around the silt grains washes away. Loess may therefore be a dangerous foundation material. Settling and cracking of a house reportedly took place overnight when water from a hose was accidentally left on. On the other hand, when loess is properly compacted and remolded, it acquires considerable strength and resistance to erosion, and, therefore, a more reliable platform for building footings may be constructed (8).

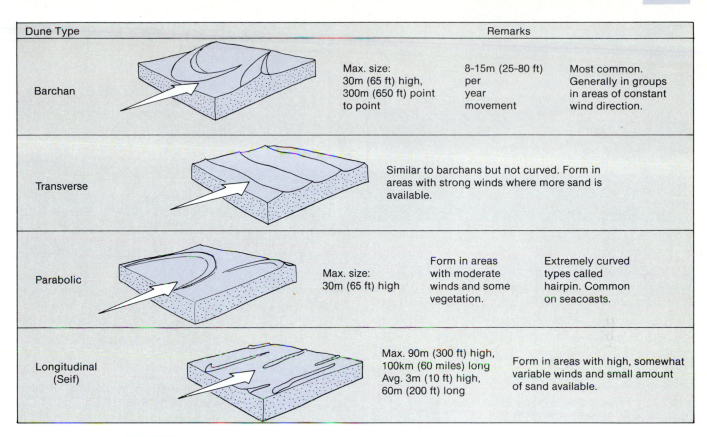

Dune Type		Remarks		
Barchan		Max. size: 30m (65 ft) high, 300m (650 ft) point to point	8-15m (25-80 ft) per year movement	Most common. Generally in groups in areas of constant wind direction.
Transverse		Similar to barchans but not curved. Form in areas with strong winds where more sand is available.		
Parabolic		Max. size: 30m (65 ft) high	Form in areas with moderate winds and some vegetation.	Extremely curved types called hairpin. Common on seacoasts.
Longitudinal (Seif)		Max. 90m (300 ft) high, 100km (60 miles) long Avg. 3m (10 ft) high, 60m (200 ft) long	Form in areas with high, somewhat variable winds and small amount of sand available.	

Figure 3.20

Types of sand dunes. (After Foster, *General Geology,* 4th ed. [Columbus: Charles E. Merrill, 1983].)

Ice

The cold-climate phenomenon of ice has become an important environmental topic. As more people live and work in the higher latitudes, we will have to learn more about how to ensure the best use of these sometimes fragile environments.

Only a few thousand years ago, the most recent continental glaciers retreated from the Great Lakes region of the U.S. Figure 3.25 (p. 37) shows the maximum extent of the Pleistocene ice sheets. Several times in the last 2 million years, the ice advanced southward, and scientists are still speculating as to whether it will advance once again. We may, indeed, still be in the Ice Age.

Glacial ice today covers about 10 percent of the land area on earth. During the Pleistocene, glacial ice expanded and covered as much as 30 percent of the land area of earth including the present sites of cities such as New York and Chicago. Most of the glacial ice today is located in the Antarctic ice sheet, with lesser amounts in the Greenland ice sheet and glaciers in such locations as Alaska, southern Norway, the Alps in Europe, and the Southern Alps in New Zealand. Glaciers in Alaska are known to show considerable irregularity in terms of rates of advance and retreat. For example, the Black Rapids Glacier in Alaska advanced several kilometers down its valley in a period of only five months in 1936 and 1937 and then started to retreat. Such a rapid advance is

Figure 3.21
Barchan dunes in the Columbia River Valley, Oregon. (Photo by G. K. Gilbert, courtesy of the U.S. Geological Survey.)

(a)   (b)

Figure 3.22
(a) Complex group of sand dunes encroaching on Highway 95 near Winnemucca, Nevada.
(b) Removal of the sand from the highway is a continuous problem. (Photos courtesy of
J. O. Davis [a] and D. T. Trexler [b].)

Figure 3.23
Dust storm caused by cold front
at Manteer, Kansas, 1935. (Photo
courtesy of Environmental Science Services Administration.)

Figure 3.24
Nearly vertical exposure of loess near Chester, Illinois. (Photo
courtesy of James Brice.)

termed a *glacial surge,* and such events can radically
change local environments. For example, in 1971 the
Hubbard Glacier in southern Alaska started to advance
more rapidly. By the summer of 1986 the glacier was
surging at a rate of 30 or more meters per day, a rate that
slowed to approximately 6 meters per day by September
1986. In contrast, glaciers generally move less than 1
meter per day, and many move only a few centimeters
per day.

The surge of the Hubbard Glacier in the summer of
1986 blocked Russell Fjord behind an ice dam, forming
Russell Lake. The advancement was so rapid that it
trapped seals, porpoises, and other marine animals in the
new lake (Figure 3.26). People concerned for the
porpoises tried to capture and transport them, and some
of the seals attempted to walk around the ice dam and
gain access back to the sea. Water rising in Russell Lake
precipitated concern that it might spill over into the Situk
River, greatly increasing the discharge and possibly

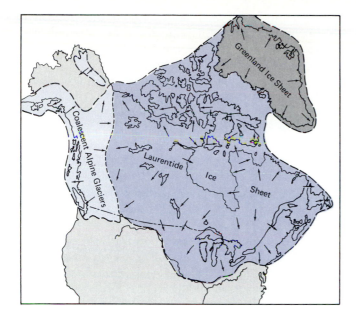

Figure 3.25
Maximum extent of ice sheets during the Pleistocene glaciation. (From Foster, *General Geology,* 4th ed. [Columbus: Charles E. Merrill, 1983].)

damaging the ecosystem there. Located nearby on glacial deposits that are about 1000 years old is the village of Yakutat. People in that village make a living through fishing in the ocean and in the Situk River, which is well known for its salmon fisheries. The fear was that if the lake overflowed into the river, it would bring with it large quantities of silt that would damage the spawning beds in the stream. People in the village, concerned that their livelihood would be lost, suggested to various local, state, and federal authorities that a canal should be constructed to partially drain the lake and thus maintain the water quality in the river.

On October 8, 1986, the ice dam ruptured, allowing the trapped marine animals to escape and temporarily eliminating the threat of overflow into the Situk River. However, if the glacier continues to advance, it may, as it has done in the past, completely fill Yakutat Bay and make Lake Russell a more permanent feature. In which case, as one of the scientists studying the glacial advance has noted, although the people in the village today are concerned with glacial water entering the Situk River, people in generations to come, might be equally alarmed when the glacier again retreats, Russell Lake is turned into a fjord, and the discharge to the river is drastically reduced. On the other hand, the glacier may continue to

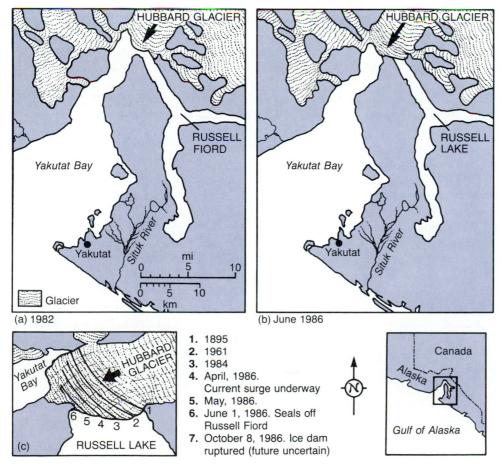

(a) 1982

(b) June 1986

(c)

1. 1895
2. 1961
3. 1984
4. April, 1986. Current surge underway
5. May, 1986.
6. June 1, 1986. Seals off Russell Fiord
7. October 8, 1986. Ice dam ruptured (future uncertain)

Figure 3.26
1986 surge of the Hubbard Glacier. Position of the glacier is shown for (a) 1982, (b) June 1986, and (c) various stages over the past century. (Data from U.S. Forest Service.)

retreat. Since the final outcome is unknown, a strategy to mitigate potential impacts is yet to be worked out.

Scientists are gathering valuable information from the recent rapid advance of the Hubbard Glacier, which should be applicable to future changes in glaciers in other parts of the world. As more people move into areas that are still partly glaciated, we will have more situations in which the human use of the land comes into conflict with natural processes.

The effects of recent glacial events are easily seen in the landscape. The flat, nearly featureless ground moraine or till plains of central Indiana are composed of **till**, material carried and deposited by continental glaciers, which buried preglacial river valleys. It is hard to believe that beneath the glacial deposits is a topography formed by running water much like the hills and valleys of southern Indiana where glaciers never reached.

Continental and mountain glaciers produced a variety of erosional and depositional landforms. Because of the variable types of forms and deposits, the environmental geology in a recently glaciated area may be complex. Glacial-related deposits include sands and gravels from streams in, on, under, and in front of the ice (outwash), as well as heterogeneous material (till) deposited directly by the ice. In addition, numerous lakes were formed when ice blocks buried by glacial deposits melted. These lakes often contained peat and other organic material forming high-organic soils. The wide variety of possibilities of earth materials in glaciated areas requires that detailed evaluation of the physical properties of surficial and subsurface materials be conducted before planning, designing, and building structures such as dams, highways, and large buildings.

In the higher latitudes, **permafrost**, permanently frozen ground, is a widespread natural phenomenon that still underlies about 20 percent of the land area in the world (Figure 3.27). Two main types of permafrost are defined by the areal extent of the phenomenon: *discontinuous* permafrost and *continuous* permafrost. At higher latitudes, discontinuous permafrost is characterized by scattered islands of thawed ground that exist in a predominantly frozen area. Toward the southern border of the permafrost, however, the percentage of unfrozen ground increases until all the ground is unfrozen. In continuous permafrost areas, the only ice-free areas are beneath deep lakes or rivers. The distribution of these types of permafrost for Alaska, where about 85 percent of the state is underlain by frozen ground, is shown in Figure 3.28. Known thickness of the permafrost in Alaska varies from about 400 meters in northern Alaska to less than 0.3 meters at the southern margin of the frozen ground (12).

A cross section through permafrost (Figure 3.29) shows an upper active layer that thaws during the summer, and unfrozen layers within the permafrost below the active layer. The thickness of the active layer depends upon such factors as exposure, slope, amount of water, and particularly the presence or absence of a vegetation cover which greatly affects the thermal conductivity of the soil.

Special engineering problems are associated with the design, construction, and maintenance of structures such as roads, railroads, airfields, pipelines, and buildings in permafrost areas. Lack of knowledge about permafrost has led to very high maintenance costs and relocation or abandonment of highways, railroads, and other structures. Figure 3.30 shows a gravel road with severe differential subsidence caused by thawing of permafrost. Figure 3.31 shows a tractor trail constructed by bulldozing off the vegetation cover over permafrost on the North Slope of Alaska. The small ponds on the abandoned trail formed during the first summer following the trail construction. They will continue to grow deeper and wider as the permafrost continues to thaw (12).

The entire range of engineering problems in permafrost areas is extensive and beyond the scope of our discussion. Specific types of problems occur with different types of earth materials. In general, the major problems are associated with permafrost that occurs in fine-grained, poorly drained, frost-susceptible materials. In these materials, a lot of ice melts if the thermal regime changes. Melting produces unstable materials, resulting in settling, subsidence, landslides, and lateral or downslope flowage of saturated sediment. This thawing of permafrost and subsequent frost heaving and subsidence caused by freezing and thawing of the active layer are responsible for many of the engineering problems in the arctic and subarctic regions (12).

Experience has shown that two basic methods of construction can be used on permafrost: the *active* method and the *passive* method. The active method is used where the permafrost is thin or discontinuous or contains a relatively small amount of ice. The method consists of thawing the permafrost, and if the thawed material has sufficient strength, conventional construction is used. This works best in coarse-grained, well-drained soils that are not particularly frost-susceptible. The passive method is used where it is impractical to thaw the permafrost. The basic principle is to keep the permafrost frozen and not upset the natural quasi-equilibrium of environmental factors. This equilibrium is so sensitive that even the passage of a tracked vehicle that destroys the vegetation cover will upset it and initiate melting which, once started, is nearly impossible to reverse. Special design of structures and foundations to minimize melting of the permafrost is the key to the passive method (12).

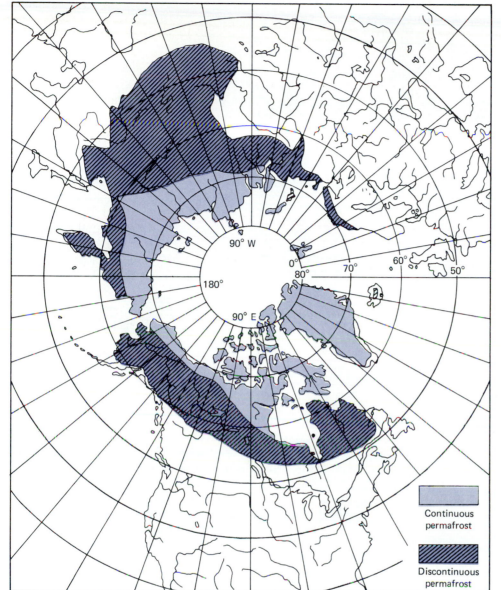

Continuous
permafrost

Discontinuous
permafrost

Figure 3.27
Extent of permafrost zones in the
Northern Hemisphere. (From Ferrians, Kachadoorian, and Greene,
*Permafrost and Related Engineer-
ing Problems in Alaska,* U.S. Geological Survey Professional Paper
678, 1969.)

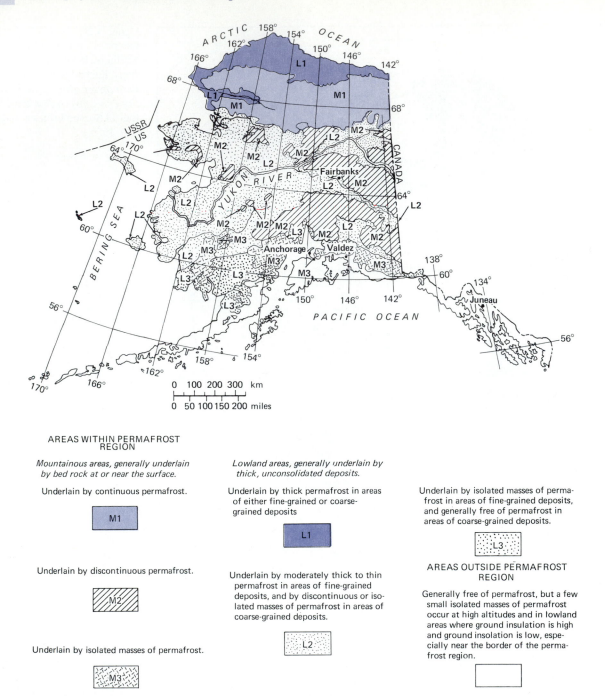

AREAS WITHIN PERMAFROST REGION

Mountainous areas, generally underlain by bed rock at or near the surface.

Underlain by continuous permafrost.

| M1 |

Underlain by discontinuous permafrost.

| M2 |

Underlain by isolated masses of permafrost.

| M3 |

Lowland areas, generally underlain by thick, unconsolidated deposits.

Underlain by thick permafrost in areas of either fine-grained or coarse-grained deposits

| L1 |

Underlain by moderately thick to thin permafrost in areas of fine-grained deposits, and by discontinuous or isolated masses of permafrost in areas of coarse-grained deposits.

| L2 |

Underlain by isolated masses of permafrost in areas of fine-grained deposits, and generally free of permafrost in areas of coarse-grained deposits.

| L3 |

AREAS OUTSIDE PERMAFROST REGION

Generally free of permafrost, but a few small isolated masses of permafrost occur at high altitudes and in lowland areas where ground insulation is high and ground insolation is low, especially near the border of the permafrost region.

| |

Figure 3.28

Extent and distribution of permafrost in Alaska. (From Ferrians, Kachadoorian, and Greene, *Permafrost and Related Engineering Problems in Alaska,* U.S. Geological Survey Professional Paper 678, 1969.)

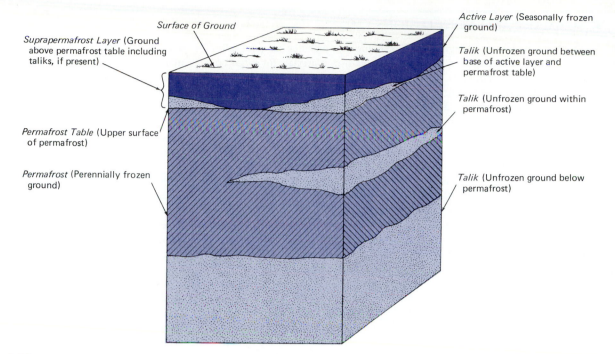

Surface of Ground

Active Layer (Seasonally frozen ground)

Suprapermafrost Layer (Ground above permafrost table including taliks, if present)

Talik (Unfrozen ground between base of active layer and permafrost table)

Talik (Unfrozen ground within permafrost)

Permafrost Table (Upper surface of permafrost)

Permafrost (Perennially frozen ground)

Talik (Unfrozen ground below permafrost)

Figure 3.29
Block diagram of permafrost morphology. (From Ferrians, Kachadoorian, and Greene, *Permafrost and Related Engineering Problems in Alaska,* U.S. Geological Survey Professional Paper 678, 1969.)

Figure 3.30
Gravel road in Alaska showing severe differential subsidence caused by thawing of permafrost. (Photo by O. J. Ferrians, Jr., courtesy of the U.S. Geological Survey.)

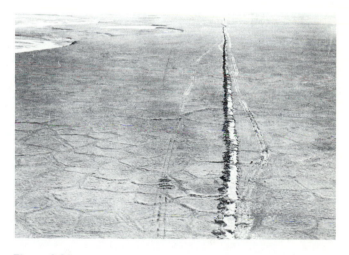

Figure 3.31
Tractor trail on the North Slope of Alaska. The small ponds (light patches above the straight road) are created by thawing of the permafrost in the roadway. The polygons are a type of "pattern ground" formed by vertical ice wedges in frost-susceptible materials. (Photo by O. J. Ferrians, Jr., courtesy of the U.S. Geological Survey.)

▼ ▼ ▼ SUMMARY AND CONCLUSIONS

Processes that create, maintain, change, and destroy earth materials (water, minerals, rocks, and soil) are collectively referred to as the *geologic cycle.* More correctly, the geologic cycle is a group of subcycles, including the *tectonic, hydrologic, rock,* and *geochemical* cycles (1). In various activities, people use earth materials found in different parts of these cycles. The materials may be found in a concentrated or "pure" state; but after various artificial processes, they are returned to the earth and are dispersed, contaminated, or polluted in another part of the geologic cycle (5). Once dispersed or otherwise altered, these materials cannot be concentrated or made available for human use in any reliable timetable.

Application of geological information to environmental problems requires an understanding of both geological and engineering properties of earth materials. Different minerals and rocks behave predictably, but they behave differently for various land uses.

Thus, compaction shales are generally poor foundation materials for large engineering structures, whereas granite with few fractures is satisfactory for most purposes. In addition, the strength of rocks is greatly influenced by active fracture systems, or old fractures that have been weathered and altered to unstable minerals.

Most continental landforms are produced by running water, and an understanding of both surface and subsurface hydrogeologic processes is necessary to work on many environmental problems.

Windblown sand, silt, and dust cover many thousands of square kilometers. Major deposits of sand are concentrated along coastal and interior areas, and deposits of windblown silt, or loess, are concentrated along major rivers that carried the meltwater from continental glaciers during the Pleistocene glaciation.

Engineering problems caused by migrating and stabilized sand dunes involve expensive construction and maintenance costs for highways, buildings, and hydraulic structures.

Loess deposits, unless properly compacted, make hazardous foundations for engineering purposes because loess may hydroconsolidate and settle when it becomes wet.

Geologically, recent continental glaciation has produced a variety of different types of earth materials such as glacial till, highly organic soils, and water-transported sediment, all of which may be located in a given area. As a result, careful evaluation of surface and subsurface deposits is necessary before planning, designing, and building.

Today permafrost underlies about 20 percent of the world's land area, producing a fragile and sensitive environment. Special engineering procedures are necessary to minimize adverse effects from artificial thawing of fine-grained, poorly drained, frozen ground.

▼ ▼ ▼ REFERENCES

1. LONGWELL, C. L.; FLINT, R. F.; and SANDERS, J. E. 1969. *Physical geology.* New York: John Wiley & Sons.
2. LE PICHON, X. 1968. Sea-floor spreading and continental drift. *Journal of Geophysical Research* 73:3661–97.
3. ISACKS, B.; OLIVER, J.; and SYKES, L. 1968. Seismology and the new global tectonics. *Journal of Geophysical Research* 73:5855–99.
4. DEWEY, J. F. 1972. Plate tectonics. *Scientific American* 22:56–68.

5. Committee on Geological Sciences. 1972. *The earth and human affairs.* San Francisco: Canfield Press.
6. BILLINGS, M. P. 1954. *Structural geology.* 2nd ed. Englewood Cliffs, New Jersey: Prentice-Hall.
7. JESSUP, W. E. 1964. Baldwin Hills dam failure. *Civil Engineering* 34:62–64.
8. KRYNINE, D. P., and JUDD, W. R. 1957. *Principles of engineering geology and geotechnics.* New York: McGraw-Hill.
9. SCHULTZ, J. R., and CLEAVES, A. B.

1955. *Geology in engineering.* New York: John Wiley & Sons.
10. HOLDEN, C. 1971. Nuclear waste: Kansas riled by AEC plans for atom dump. *Science* 172:249–50.
11. WAY, D. S. 1973. *Terrain analysis.* Stroudsburg, Pennsylvania: Dowden Hutchinson & Ross.
12. FERRIANS, O. J., JR.; KACHADOORIAN, R.; and GREENE, G. W. 1969. *Permafrost and related engineering problems in Alaska.* U. S. Geological Survey Professional Paper 678.

Soils may be defined in several ways. Soil scientists, for example, define soil as solid earth material that has been altered by physical, chemical, and organic processes, such that it can support rooted plant life. Engineers, on the other hand, define soil as any solid earth material that can be removed without blasting. Both of these definitions are important in environmental geology, and geologists must be aware of the different views concerning processes that produce soils, how soils are defined, and the role of soils in understanding environmental problems.

Consideration of soils, particularly with reference to limitations in land use, is becoming an important aspect of environmental work. In the field of land-use planning, land capability (use of land for a particular purpose) is often determined in part by the soils present, especially for such usages as urbanization, timber management, and agriculture, among others. Second, soils are critical when considering waste disposal problems because interactions between waste, water, soil, and rock often determine the suitability for a particular site to receive waste. Finally, study of soils helps in evaluating some natural hazards, including floods, landslides, and earthquakes. In the case of the first, since floodplain soils differ from upland soils, consideration of soil properties helps delineate natural floodplains for land-use planning. Soils may also influence rates of run-off and thus the potential for producing floods. Evaluation of relative ages of soils on landslide deposits may provide an estimate of the average recurrence of slides and thus assist in development of plans to minimize future impact from landslides in an area. Finally, the study of soils has been a powerful tool in establishing the chronology of earth materials deformed by faulting, which has led to better calculations of recurrence intervals of earthquakes on a fault at a particular site.

The purpose of this chapter is to introduce soils in terms of how they develop, soil chronosequences, soil fertility, water in soil, soil classification, engineering properties of soils, land use and soils, sediment pollution, rates of soil erosion, soil pollution, desertification, and, finally, the soil survey and land-use planning.

▼ DEVELOPMENT OF SOILS

Intimate interactions of processes between the rock and hydrologic cycles produce weathered rock materials that are basic ingredients of soils. **Weathering** is the physical and chemical breakdown of rocks and the first step in the soil-forming process. The more insoluble weathered material may remain essentially in place and be further modified by organic processes to form a *residual soil,* as, for example, the red soils of the Piedmont in the southeastern United States. If the weathered material is transported by water, wind, or ice (glaciers), then deposited and further modified by organic processes, a

CHAPTER FOUR

▼

▼

▼

Soils and Environment

transported soil forms. The fertile soils formed from glacial deposits in the Midwest are an example.

Soil Horizons

Soil can be considered an open system. As such, soils are a function of climate, topography, parent material (material from which the soil is formed), maturity (time), and organic activity. Many of the differences we see in soils are effects of climate and topography, but the type of parent rock, the organic processes, and the amount of time the soil-forming processes have operated are also important.

Vertical and horizontal movements of materials in soil systems provide a distinct soil layering, or *soil profile*, divided into zones, or **soil horizons**. Figure 4.1 shows the common master or prominent horizons. Highly concentrated organic material is found in the **O** and **A horizons**. Differences between the two reflect the amount of organic material present. Generally the O horizon consists of plant litter and other organic material that overlies the A horizon, which contains a good deal of organic materials as well as mineral material. The **E horizon**, if present, is a light-colored horizon underlying the O or A that is leached of iron-bearing components. The horizon color is light because it contains less organic material than the O and A horizons and little coloring material such as iron oxides.

The **B horizon** underlies the O, A, or E horizons, and several types have been recognized. Probably the most important type of B horizon is the *argillic B,* which is designated as B_t. This horizon is enriched in clay minerals that have been translocated downward by soil-forming processes. Another type of B horizon of interest to environmental geologists is the B_k horizon. This horizon is characterized by accumulation of calcium

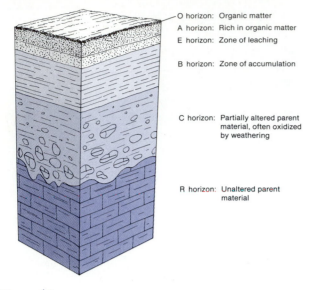

O horizon: Organic matter
A horizon: Rich in organic matter
E horizon: Zone of leaching

B horizon: Zone of accumulation

C horizon: Partially altered parent material, often oxidized by weathering

R horizon: Unaltered parent material

Figure 4.1
Schematic soil profile showing the soil horizons. Not all horizons may be present in a given soil.

carbonate. The carbonate coats individual soil particles in the soils and may fill some pore spaces but does not dominate the morphology of the horizon. A soil horizon that is so impregnated with calcium carbonate that its morphology is dominated by the carbonate is designated as a **K horizon**. Carbonate often forms laminar layers parallel to the surface in K horizons, and carbonate completely fills pore spaces between soil particles. The term *caliche* is often used for laminar layers of calcium carbonate in soils. The **C horizon** is directly over the unaltered parent material and consists of parent material partially altered by weathering processes. The **R horizon** consists of consolidated bedrock that underlies the soil. However, some of the fractures and other pore spaces might contain some clay that has been translocated downward (1).

The above discussion of soil horizons includes only the most common ones. Additional information is available from detailed soils texts (1, 2). As a final note, the term *hardpan* is often used in the literature on soils. A hardpan soil horizon is defined as a hard, compacted, or cemented soil horizon. The hardpan is often composed of clay but may be cemented with calcium carbonate, iron oxide, or silica. Hardpan horizons are nearly impermeable and thus restrict the downward movement of soil water.

Soil Color

The color of a soil or the colors of its various soil horizons are among the first observations that are made concerning a soil. O and A horizons tend to be dark in color because of abundant organic material. The E

horizon, if present, may be almost white due to leaching of iron and aluminum oxides from the horizon. The B horizon often shows the most dramatic changes in color, varying from yellow/brown to light red/brown to dark red, depending upon the presence of clay minerals and iron oxides. B_k horizons may have a lighter color because they contain carbonates, but they are sometimes mixed with redder colors due to iron oxide accumulation. If a true K horizon has developed, it may be almost white in color because of the great abundance of calcium carbonate. Although soil color can be an important diagnostic tool in looking at soils, care must be taken because the original parent material, if rich in iron, may produce a very red soil even when there has been relatively little soil profile development.

Soil color may be an important indicator of how well drained a soil is. Well-drained soils are well-aerated (oxidizing conditions), and iron oxidizes to a red color. Poorly drained soils are wet, and iron is reduced rather than oxidized. The color of such a soil is often yellow. This distinction is important because poorly drained soils are associated with environmental problems, such as lower slope stability and inability to be utilized as a disposal medium for household sewage systems (septic tank and leach field).

Soil Texture

The texture of a soil depends upon the relative proportions of sand, silt, and clay-sized particles. Clay particles have a diameter of less than 0.002 mm, silt is defined as particles with diameters ranging from 0.002 to 0.05 mm, and sand consists of particles from 0.05 to 2.0 mm in diameter. Given this, the soil textural classes are shown in Figure 4.2. A useful field technique for estimating the size of soil particles is as follows: it is sand or larger if you can see individual grains; silt if you can see the grains with a 10X hand lens; and clay if you cannot see grains with such a hand lens. Another method is to feel the soil: sand is gritty (crunches between the teeth); silt feels like baking flour; and clay is cohesive. When mixed with water, smeared on the back of the hand, and allowed to dry, clay can't easily be dusted off, whereas silt or sand can.

Soil texture is commonly identified in the field by estimation and refined in the laboratory through determination of the percentages of sand, silt, and clay on the soil fraction that is less than 2 mm in diameter. Soil texture is an important property of soils because it can be used to identify soil-forming processes and to interpret the geological history of a particular deposit. That is, the textures of over-bank stream deposits, for example, are much different from channel deposits, which in turn are different from the textures of lake deposits. These distinctions are important in trying to understand

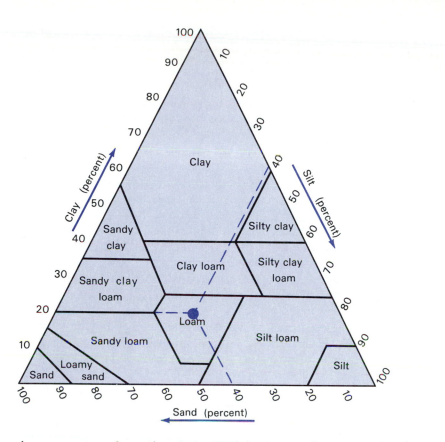

Figure 4.2
Soil textural classes. For example, the point connected by dashed lines represents a soil composed of 40% sand, 40% silt, and 20% clay, and thus is classified as loam. (U.S. Department of Agriculture standard textural triangle.)

the properties of a soil and potential limitations for a particular use.

Soil Structure

Soil particles often cling together in aggregates, called *peds,* that are classified according to shape into several types. Figure 4.3 shows some of the common structures of peds found in soils. The type of structure present is related to soil-forming processes, but some of these are poorly understood (1). For example, *spheroidal structure* is fairly common in *A* horizons, whereas *blocky* and *prismatic structures* are most likely to be found in *B* horizons. Soil structure is an important diagnostic tool in evaluating development of soil profiles. In general, as the profile matures, structure becomes more complex and may go from granular to blocky to prismatic as the clay content in the *B* horizons increases.

Figure 4.3
Chart of different soil structures (peds).

Type		Typical Size Range	Horizon Usually Found In	Comments
Granular	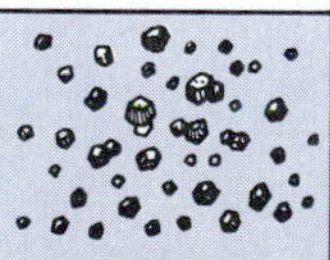	1–10 mm	A	Can also be found in B and C horizons
Blocky		5–50 mm	B_t	Are usually designated as angular or subangular
Prismatic		10–100 mm	B_t	If columns have rounded tops, structure is called columnar
Platy		1–10 mm	E	May also occur in some B horizons

Relative Profile Development

Most environmental geologists will not have occasion to make detailed soils descriptions and analyses of soils data. However, it is important for geologists to recognize differences between weak, moderate, and well-developed soils. These distinctions are useful in preliminary evaluation of soil properties and determination of whether the opinion of a soil scientist is necessary in a particular project.

A *weakly developed soil profile* is generally characterized by an *A* horizon directly over a *C* horizon. The *C* horizon may be oxidized. Such soils tend to be only a few hundred years old in most areas.

A *moderately developed soil profile* may consist of an *A* horizon overlying an argillic B_t horizon which overlies the *C* horizon. A carbonate B_k horizon may also be present but is not necessary for moderate development. These soils have a *B* horizon with a better developed texture and redder colors than those that are weakly developed. Moderately developed soils are usually at least Pleistocene (greater than 10,000 years in age).

A *strongly developed soil* is characterized by redder colors in the *B* horizon, more translocation of clay to the *B* horizon, and stronger structure. A *K* horizon may also be present but is not necessary for strong development. Strongly developed soils vary widely in age, with typical ranges between 40 thousand and several hundred thousand years and older.

▼ SOIL CHRONOSEQUENCES

A **soil chronosequence** is a series of soils, arranged from youngest to oldest in terms of relative soil profile development. Soil chronosequences help provide necessary information for understanding the recent history of a landscape. Evaluation of a site for location of critical facilities, such as a waste disposal operation or large power plant, should take into account site stability for such hazards as earthquakes, floods, and landslides.

A soil chronosequence is valuable in such hazards work because it provides a chronology in which to frame the environmental assessment. It may provide data necessary to make such inferential statements as, there is no evidence of ground rupture in the last 1000 years, or, the last mudflow was at least 30,000 years ago. It takes a lot of soils work to establish a chronosequence in a particular area. However, once such a chronosequence is developed and dated through relative soil profile development tied to absolute dates, it may be applied to a specific problem. For example, Figure 4.4 shows an offset alluvial fan along the San Andreas fault in the Indio Hills of southern California. Soil pits excavated in the alluvial fan suggest that the most probable age is about 20,000 to 30,000 years. The age was estimated based on consideration of soil chronosequence work in the nearby Mojave Desert. Although material suitable for absolute dating, such as charcoal, was not found, the soil development on the alluvial fan allowed an estimated age, from which the

(a)

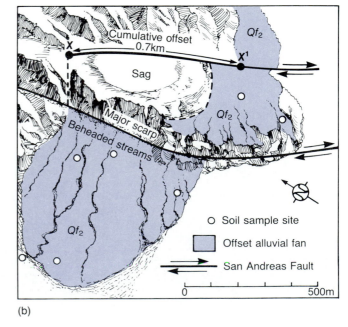

(b)

Figure 4.4
(a) Aerial photograph and (b) sketch map of off-set alluvial fan along the San Andreas fault near Indio, California.

slip rate for this part of the San Andreas fault could be estimated at 25 to 30 mm per year (3). The slip rate for this segment of the fault was not previously known, and the rate is significant because it is a necessary ingredient in eventually estimating the probability and recurrence interval of large damaging earthquakes.

▼ SOIL FERTILITY

A soil may be considered a complex ecosystem. A single cubic meter of soil may contain millions of living things, including small rodents, insects, worms, algae, fungi, and bacteria. These animals and plants are important in releasing or converting nutrients in soils into forms that are useful for plants and in the mixing and aeration of soil particles (4). *Soil fertility,* then, refers to the capacity of the soils to supply nutrients (such as nitrogen, phosphorous, and potassium) needed for plant growth when other factors are favorable (5). Soils developed on some floodplains and glacial deposits contain sufficient nutrients and organic material to be naturally fertile. Other soils, developed on highly leached bedrock or loose deposits with little organic material, may be nutrient-poor with low fertility. Soils are often manipulated to increase plant yield by applying fertilizers or other materials to improve soil texture and retention of moisture. Soil fertility can be reduced by soil erosion or leaching that removes nutrients, by interruption of natural processes (such as flooding) that supply nutrients, or by continued use of pesticides that alter or damage organisms that render nutrients into forms available to plants.

▼ WATER IN SOIL

If you analyze a block of soil in terms of what is present, you will find it is composed of bits of solid mineral material and organic matter and that this material contains pore spaces filled with either gases (mostly air), water, or other liquids. If all the pore spaces in a block of soil are filled with water, then it is in a *saturated* condition. This occurs for some soils part of the year and for some soils all of the year. For example, soils in swampy areas may be saturated year-round, whereas soils in arid regions may be saturated only occasionally.

The amount of water in a soil can be very important in determining such engineering properties as potential to shrink and swell and the strength of a soil. If you have ever built a sand castle at the beach, you know that dry sand is impossible to work with, but that moist sand will stand vertically, producing walls for your castle. Differences between wet and dry soils are also very apparent to anyone who lives in or has visited in areas with dirt roads that cross clay-rich soils. When the soil is dry, driving conditions are excellent, but following rain storms, the same roads are nearly impassable (6).

Water in soils may flow laterally or vertically through soil pores, which are the void spaces between grains or fractures produced as a result of soil structure. The flow is termed *saturated flow* if all the pore spaces are filled with water or, more commonly, *unsaturated flow* when only part of the pores are filled with water.

In unsaturated flow, movement of water is related to such processes as thinning or thickening of films of water in pores and on the surrounding soil grains (6). The water molecules closest to the surface of a particle are held the tightest, and as the films thicken, the water content increases and the outer layers of water may begin to move. Flow is fastest therefore in the center of pores and slower near the edges.

The study of soil moisture relations and water movement in the unsaturated zone, along with how to monitor water in the unsaturated zone, is an important research topic. Many employment opportunities will be available to people working in this area in the next few years.

▼ SOIL CLASSIFICATION

Terminology and classification of soils for environmental studies are a problem because in many cases we are interested in both the soil processes and the materials as they relate to human use and activity. A genetic classification that includes engineering properties is therefore most appropriate—but none exists. So, we will discuss *soil taxonomy* and the *engineering classification,* which groups soils by material types and engineering properties.

Soil scientists have developed a comprehensive and systematic classification of soils known as **soil taxonomy**. This classification is not as dependent on climate and vegetation as earlier schemes; rather, physical and chemical properties of the soil profile are emphasized by generic terminology. There is a six-fold hierarchy in the classification: Order, Suborder, Great Group, Subgroup, Family, and Series. There are ten Orders (Table 4.1) that are based on gross soil morphology, nutrient status, organic content, color, and general climatic considerations. With each step down the hierarchy, more information about a specific soil becomes known.

Soil taxonomy is especially useful for agricultural and related land-use purposes. It has been criticized for being too complex and lacking sufficient textural and engineering information to be of optimal use in specific site evaluation for engineering purposes. Nevertheless, the serious earth scientist must have knowledge of this classification because it is commonly used by soil scientists and Quaternary geologists.

Table 4.1
General properties of soil order used with soil taxonomy by soil scientists.

Order	General Properties
Entisols	No horizon development (azonal); many are recent alluvium; synthetic soils are included; are often young soils.
Vertisols	Include swelling clays (greater than 35 percent) that expand and contract with changing moisture content. Generally form in regions with a pronounced wet and dry season.
Inceptisols	One or more of horizons have developed quickly; horizons are often difficult to differentiate; most often found in young but not recent land surfaces, have appreciable accumulation of organic material; most common in humid climates but range from the Arctic to the tropics; native vegetation is most often forest.
Aridisols	Desert soils; soils of dry places; low organic accumulation; have subsoil horizon where gypsum, caliche (calcium carbonate), salt, or other materials may accumulate.
Mollisols	Soils characterized by black, organic rich A horizon (prairie soils); surface horizons are also rich in bases. Commonly found in semiarid or subhumid regions.
Spodosols	Soils characterized by ash-colored sands over subsoil, accumulations of amorphous iron-aluminum sesquioxides and humus. They are acid soils that commonly form in sandy parent materials. Are found principally under forests in humid regions.
Alfisols	Soils characterized by a brown or gray-brown surface horizon and an argillic (clay rich) subsoil accumulation with an intermediate to high base saturation (greater than 35 percent as measured by the sum of cations, such as calcium, sodium, magnesium, etc.). Commonly form under forests in humid regions of the mid-latitudes.
Ulfisols	Soils characterized by an argillic horizon with low base saturation (less than 35 percent as measured by the sum of cations); often have a red-yellow or reddish-brown color; restricted to humid climates and generally form on older landforms or younger, highly weathered parent materials.
Oxisols	Relatively featureless, often deep soils, leached of bases, hydrated, containing oxides of iron and aluminum (laterite) as well as kaolinite clay. Primarily restricted to tropical and subtropical regions.
Histosols	Organic soils (peat, muck, bog).

Source: After Soil Survey Staff, 1975, *Soil Taxonomy*. Soil Conservation Service, U.S. Department of Agriculture, Agriculture Handbook No. 436.

The **unified soil classification system** is widely used in engineering practice. Because all natural soils are mixtures of coarse particles such as gravel and sand, fine particles such as silt and clay, and organic material, this classification is divided into these three groups. Each group is based on the predominant size or the abundance of organic material (Table 4.2). Coarse soils are those in which greater than 50 percent of the particles (by weight) are larger than 0.074 millimeter in diameter. Fine soils are those with less than 50 percent of the particles greater than 0.074 millimeter (7). Organic soils have a high organic content and are identified by their black or gray color and sometimes by an odor of hydrogen sulfide, which smells like rotten egg.

▼ ENGINEERING PROPERTIES OF SOILS

All soils above the water table have three distinct phases: the solid material, the liquid (mostly water), and the gas (mostly air). The usefulness of a soil is greatly affected by the variations in the proportions and structure of the three phases. The types of solid materials, the size of soil particles, and the water content are probably the most significant variables that determine engineering properties. For fine-grained soils, properties such as the *liquid limit* (LL) and the *plastic limit* (PL) are used to classify soils for engineering purposes. The liquid limit is defined by the water content at which the soil behaves as a liquid; and the plastic limit is defined by the water content below which the soil no longer behaves as a

One of the many clay minerals. When present in soils, clays may exhibit undesirable properties such as low strength, high water content, poor drainage, and high shrink-swell potential.

Mica minerals, biotite (black) and muscovite (light), both of which are important rock-forming minerals. It is unusual to find both together in one sample.

Clear and impure varieties of halite (salt). Halite is the main mineral in rocksalt, a potential host rock for high-level radioactive wastes.

Several blue and/or green copper minerals are present in this rock.

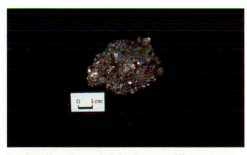

Pyrite (fool's gold) is iron sulfide, a common mineral associated with coal, which reacts with water and oxygen to form sulfuric acid.

Two varieties of feldspar, the most common rock-forming mineral in the earth's crust.

Two of the many varieties of quartz.

Calcite, the abundant mineral in limestone and marble. Limestone terrain is associated with caverns, sinkholes, subsidence, and potential water pollution and construction problems.

PLATE 1   Minerals

(Photographs by Ricardo Zepeda.)

Three samples of granite (intrusive igneous rock). White and pink mineral grains are feldspar, dark grains are ferromagnesian minerals, and quartz is clear or gray.

Basalt (extensive igneous rock). Note the fine-grain size compared to granite.

Two specimens of limestone (biochemical or chemical sedimentary rock). The sample on the left contains numerous fragments of shell material; the sample on the right is composed mostly of massive impure calcite.

Two samples of sandstone (detrital sedimentary rock). Red banding (left sample) and red color of the relatively coarse-grained sandstone (right sample) indicate the presence of iron oxides.

Serpentine, a weak rock that causes slope stability problems when encountered in excavations for roads or other structures.

Schist, or foliated metamorphic rock. Parallel alignment of mineral grains, in this case mica, demonstrates foliation.

Gneiss, a foliated metamorphic rock, with minerals segregated into white bands of feldspar and dark bands of ferromagnesian.

This iron-cemented conglomerate (detrital sedimentary rock) obviously formed quite recently.

PLATE 2   Rocks

(Photographs by Ricardo Zepeda.)

Table 4.2
Unified soil classification system used by engineers.

Major Divisions				Group Symbols	Soil Group Names
COARSE-GRAINED SOILS (Over half of material larger than 0.074 mm)	Gravels	Clean Gravels	Less than 5% fines	GW	Well-graded gravel
				GP	Poorly graded gravel
		Dirty Gravels	More than 12% fines	GM	Silty gravel
				GC	Clayey gravel
	Sands	Clean Sands	Less than 5% fines	SW	Well-graded sand
				SP	Poorly graded sand
		Dirty Sands	More than 12% fines	SM	Silty sand
				SC	Clayey sand
FINE-GRAINED SOILS (Over half of material smaller than 0.074 mm)	Silts	Non-plastic		ML	Silt
				MH	Micaceous silt
				OL	Organic silt
	Clays	Plastic		CL	Silty clay
				CH	High plastic clay
				OH	Organic clay
Predominantly Organics				PT	Peat and Muck

plastic material. The numerical difference between the liquid and plastic limits is the *plasticity index* (PI), the range in moisture content within which a soil behaves as a plastic material. Soils with a very low plasticity index (5%) may cause problems because only a small change in water content can change the soil from a solid to a liquid state. On the other hand, a large PI (greater than 35%) is suggestive of a soil likely to have excessive potential to expand and contract on wetting and drying.

The more important engineering properties of soils for planning are strength, sensitivity, compressibility, erodability, permeability, corrosion potential, ease of excavation, and shrink-swell potential.

Strength of soils is difficult to generalize, and numerical averages are often misleading, because earth materials are often composed of several different mixtures, zones, or layers of materials with different physical and chemical properties. The strength of a particular soil type is a function of cohesive and frictional forces. *Cohesion* is a measure of the ability of soil particles to stick together. The attraction of particles to each other in fine-grained soils is primarily the result of the presence of molecular and electrostatic forces acting between the particles, and is the significant factor determining the strength of the soil. The presence of moisture films between coarse grains of partly saturated soils may cause an apparent cohesion, explaining the ability of wet sand (which is cohesionless when dry) to stand in vertical walls in children's sand castles on the beach (Figure 4.5) (8). *Frictional forces* that contribute to the strength of a soil are the result of interactions among the soil's individual grains. The total frictional force is a function of the density, size, and shape of the soil particles as well as the weight of overlying particles forcing the grains together. Frictional forces are most significant in coarse-

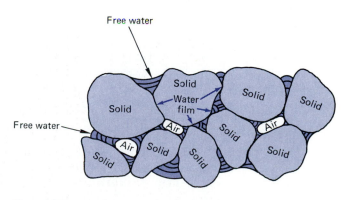

Figure 4.5
Partly saturated soil showing particle-water-air relationships. Particle size is greatly magnified. Attraction (apparent cohesion) between the water and soil particles (surface tension) develops a stress that holds the grains together. The cohesion is destroyed if the soil dries out or becomes completely saturated. (After R. Pestrong, *Slope Stability,* American Geological Institute, 1974.)

grained soils rich in sand and gravel. Since most soils are a mixture of coarse and fine particles, the strength is usually the result of both cohesion and internal friction. Although it is dangerous to generalize, clay and organic-rich soils tend to have lower strengths than coarser soils.

Vegetation may also play an important role in soil strength. For example, tree roots may provide considerable apparent cohesion through the binding characteristics of a continuous root mat or by anchoring individual roots to bedrock beneath thin soils on steep slopes.

Sensitivity of soils reflects changes in the soil strength resulting from disturbances such as vibrations or excavations. Sand and gravel soils with no clay are the least sensitive. As fine material becomes abundant, soils become more and more sensitive. Some clay soils may lose 75 percent or more of their strength following disturbance (7).

Compressibility of soils is a measure of the tendency to consolidate, or decrease in volume. It is partly a function of the elastic nature of the soil particles and is directly related to settlement of structures. Excessive settlement will crack foundations and walls. Coarse materials such as gravels and sands have a low compressibility, and settlement will be considerably less than that of fine-grained or organic soils which have a high compressibility.

Erodibility of soils refers to the ease with which soil materials can be removed by wind or water. Easily eroded materials (soils with a high erosion factor) include unprotected silts, sands, and other loosely consolidated materials. Cohesive soils, greater than 20 percent clay or naturally cemented soils, are not easily moved and therefore have a low erosion factor.

Permeability is a measure of the ease with which water moves through a material. Clean gravels or sands have the highest permeabilities. As fine particles in a mixture of clean gravel and sand increase, permeability decreases. Clays generally have a very low permeability.

Corrosion is a slow weathering or chemical decomposition that proceeds from the surface into the ground. All objects buried in the ground—pipes, cables, anchors, fenceposts—are subject to corrosion. The corrosiveness of a particular soil depends upon the chemistry of both the soil and the buried material, and on the amount of water available. It has been observed that the more easily a soil carries an electrical current (low resistivity due to more water in the soil), the higher its corrosion potential. Therefore, measurement of soil resistivity is one way to estimate the corrosion hazard (9).

Ease of excavation pertains to the procedures, and hence equipment, required to remove soils during construction. There are three general categories of excavation techniques. *Common excavation* is accomplished with an earth mover, backhoe, or dragline. This equipment essentially removes the soil without having to scrape it first. *Rippable excavation* requires breaking the soil up with a special ripping tooth or teeth before it can be removed. *Blasting* or *rock cutting* is the third, and often most expensive, category.

Shrink-swell potential in soil is its tendency to gain or lose water. Soils that tend to increase or decrease in volume with water content are *expansive*. The swelling is caused by chemical attraction and addition of layers of water molecules between the flat submicroscopic clay plates of certain clay minerals with high plasticity indices. The plates are primarily composed of silica, aluminum, and oxygen atoms, and layers of water are added between the plates as the clay expands or swells (Figure 4.6a) (10). Expansive soils often have a high plasticity index reflecting the tendency to take up a lot of water while in the plastic state.

Expansive soils in the United States cause significant environmental problems and, as one of our most costly natural hazards, are responsible for more than $2 billion in damages annually to highways, buildings, and other structures. Every year more than 250,000 new houses are constructed on expansive soils. Of these, about 60 percent will experience some minor damage (cracks in foundation, walls, or walkways) (Figure 4.6b), but 10 percent will be seriously damaged—some beyond repair (11, 12).

Montmorillonite is the common clay mineral associated with most expansive soils. With sufficient water, pure montmorillonite may expand up to 15 times its original volume, but fortunately most soils contain limited amounts of the clay, so it is unusual for an expansive soil to swell beyond 25 to 50 percent. However, an increase in volume of greater than 3 percent is considered potentially hazardous (10).

Damages to structures on expansive soils are caused by volume changes in the soil in response to changes in moisture content. Factors that tend to determine the moisture content of an expansive soil at a site include climate, vegetation, topography, drainage, site control, and quality of construction (11). Regions in which most of the rainfall occurs in a pronounced wet season followed by a dry season (for example, the southwestern United States) are more likely to experience an expansive soil problem than regions where precipitation is more evenly distributed throughout the year. This results because the wet and dry seasonal pattern allows for a regular shrink-swell sequence. Vegetation can cause changes in the moisture content of soils. Especially during the dry season, large trees draw and use a lot of local soil moisture, facilitating soil shrinkage (Figure 4.6b). Therefore, in areas with expansive soil, trees should not be planted close to foundations of light structures (such as homes). Topography and drainage are significant because adverse topographic and drainage conditions cause ponding of water around or near structures, increasing the swelling of expansive clays. Site control and construction procedures are aspects of the

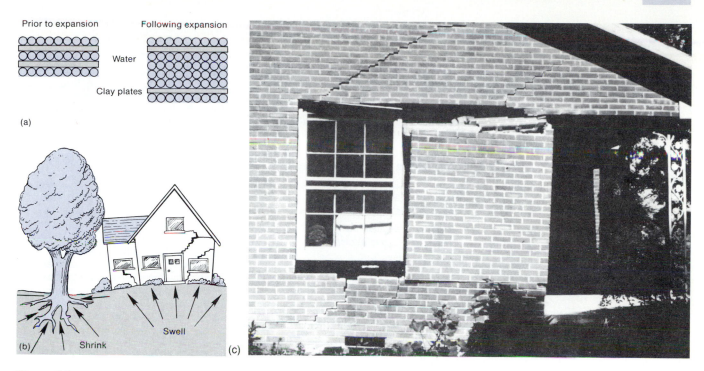

Figure 4.6
Expansive soils. Diagram of (a) expansive clay (montmorillonite) as layers of water molecules are incorporated between clay plates on a microscopic scale and (b) effects of soil's shrinking and swelling at a home site. (After Mathewson and Castleberry, *Expansive Soils: Their Engineering Geology,* Texas A & M University.) (c) Cracked wall resulting from expansion of clay soil under the foundation. (Photo courtesy of U.S. Department of Agriculture.)

expansive soil problem that the owner and contractor can advantageously manipulate. Proper design of subsurface drains, rain gutters, and foundations can minimize damages associated with expansive soils by improving drainage and allowing the foundation to accommodate some shrinking and swelling of the soil (11).

It should be apparent that some soils are more desirable than others for specific uses. Planners concerned with land use will not make soil tests to evaluate engineering properties of soils, but they will be better prepared to design with nature and take advantage of geologic conditions if they understand the basic terminology and principles of earth materials. Our discussion of engineering properties established two general principles. First, because of their low strength, high sensitivity, high compressibility, low permeability, and variable shrink-swell potential, clay soils should be avoided in projects involving heavy structures, structures with minimal allowable settling, or projects needing well-drained soils. Second, soils that have high corrosive potentials or that require other than common excavation should be avoided if possible. If such soils cannot be avoided, extra care, special materials and/or techniques, and higher-than-average initial costs (planning, design, and construction) must be expected. The second costs (operation and maintenance) may also be greater. Tables 4.3 and 4.4

further summarize engineering properties of soils in terms of the unified soil classification.

▼ LAND USE AND SOILS

The principal human influences affecting the pattern, amount, and intensity of surface-water runoff, erosion, and sedimentation are the varied land uses and manipulations of our surface water, such as artificial drainage and construction of detention and recharge ponds and dams. Such works are undertaken for several reasons and are justified most frequently as "improvements to the environment" or "for the betterment of civilization." Figure 4.7 (p. 54) indicates the changes in water and sediment processes that might occur as a landscape is modified from natural forest to clear-cut timber harvesting, urbanization, or agriculture. Following timber harvesting by clear-cutting, one can expect a decline in interception of precipitation and evapotranspiration, resulting in an increased surface runoff and production of sediment. As tree roots decay, landsliding may also increase, further delivering sediment to local stream channels. If the same land were converted to farmland, the result would also be less interception and evapotranspiration as well as increased surface runoff and soil erosion, but to a lesser degree than following clear-cut timber harvesting. The

Table 4.3
Generalized sizes, descriptions, and properties of soils.

Soil	Soil Component	Symbol	Grain Size Range and Description	Significant Properties
Coarse-grained components	Boulder	None	Rounded to angular, bulky, hard, rock particle; average diameter more than 25.6 cm.	Boulders and cobbles are very stable components used for fills, ballast, and riprap. Because of size and weight, their occurrence in natural deposits tends to improve the stability of foundations. Angularity of particles increases stability.
	Cobble	None	Rounded to angular, bulky, hard, rock particle; average diameter 6.5–25.6 cm.	
	Gravel	G	Rounded to angular, bulky, hard, rock particles greater than 2 mm in diameter.	Gravel and sand have essentially the same engineering properties, differing mainly in degree. They are easy to compact, little affected by moisture, and not subject to frost action. Gravels are generally more pervious and more stable and resistant to erosion and piping than are sands. The well-graded* sands and gravels are generally less pervious and more stable than those which are poorly graded and of uniform gradation.
	Sand	S	Rounded to angular, bulky, hard, rock particles 0.074–2 mm in diameter.	
Fine-grained components	Silt	M	Particles 0.004–0.074 mm in diameter; slightly plastic or nonplastic regardless of moisture; exhibits little or no strength when air dried.	Silt is inherently unstable, particularly with increased moisture, and has a tendency to become quick when saturated. It is relatively impervious, difficult to compact, highly susceptible to frost heave, easily erodible, and subject to piping and boiling. Bulky grains reduce compressibility, whereas flaky grains (such as mica) increase compressibility, producing an elastic silt.
	Clay	C	Particles smaller than 0.004 mm in diameter; exhibits plastic properties within a certain range of moisture; exhibits considerable strength when air dried.	The distinguishing characteristic of clay is cohesion or cohesive strength, which increases with decreasing moisture. The permeability of clay is very low; it is difficult to compact when wet and impossible to drain by ordinary means; when compacted is resistant to erosion and piping; not susceptible to frost heave; and subject to expansion and shrinkage with changes in moisture. The properties are influenced not only by the size and shape (flat or platelike), but also by their mineral composition. In general, the montmorillonite clay mineral has the greatest and kaolinite the least adverse effect on the properties.
	Organic matter	O	Organic matter in various sizes and stages of decomposition.	Organic matter present even in moderate amounts increases the compressibility and reduces the stability of the fine-grained components. It may decay, causing voids, or change the properties of a soil by chemical alteration; hence, organic soils are not desirable for engineering purposes.

*The term *well-graded* indicates an even distribution of sizes and is equivalent to *poorly sorted*.
Note: The Unified Soil Classification does not recognize cobbles and boulders with symbols. The size range for these as well as the upper limit for clay are according to Wentworth (1922).
Source: After Wagner, "The Use of the Unified Soil Classification System by the Bureau of Reclamation," International Conference on Soil Mechanics and Foundation Engineering, Proceedings [London], 1957.

Table 4.4
Selected engineering properties of soils

	Unified Soil Classification Symbol	Properties					General Use As Construction Material
		Strength	Frost Susceptibility	Permeability (after compaction)	Compressibility (after compaction and saturation)	Erodibility (compacted and saturated)	
Coarse Soils — Gravels	GW: well-graded gravel	Very high	Negligible	High	Insignificant	Low-High	Excellent
	GP: poorly graded gravel	High	Very low	Very high	Insignificant	Low-High	Good
	GM: silty gravel	High	Medium-High	Low	Insignificant	Medium-High	Good
	GC: clayey gravel	High-Medium	Low	Very low	Very low	Medium-High	Good
Coarse Soils — Sands	SW: well-graded sand	Very high	Negligible	High	Insignificant	Very high	Excellent
	SP: poorly graded sand	High	Low	High	Very low	High	Fair
	SM: silty sand	High	Medium-High	Low	Low	High	Fair
	SC: clayey sand	High-Medium	Low-Medium	Very low	Low	Medium-High	Good
Fine Soils — Silts	ML: silt	Medium	High	Low	Medium	High	Fair
	MH: micaceous silt	Medium-Low	Very high	Low	High	High	Poor
	OL: organic silt	Low	High	Low	Medium	High	Fair
Fine Soils — Clays	CL: silty clay	Medium	Medium-High	Very low	Medium	Medium	Good-Fair
	CH: high plastic clay	Low	Medium	Very low	High	Low	Poor
	OH: organic clay	Low	Medium	Very low	High	Low	Poor

Source: Modified after Wagner, 1957, and Detwyler and Marcus, 1972.

53

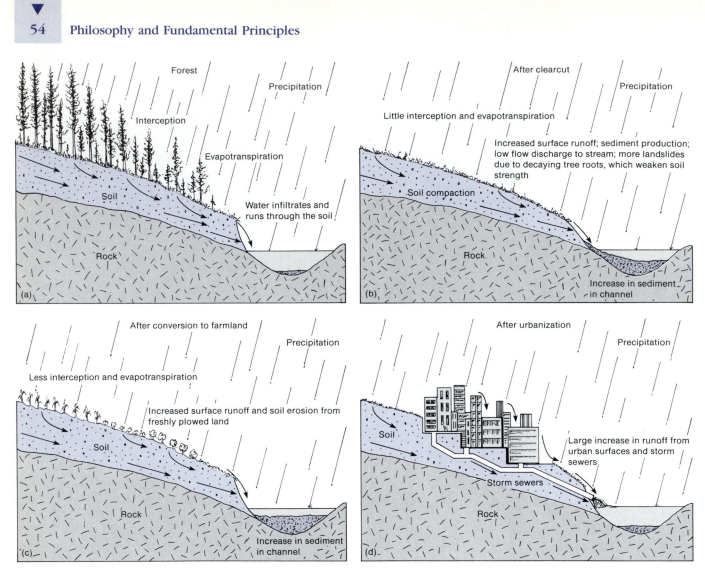

Figure 4.7
Water relations for (a) a natural forested slope, and following several land-use changes: (b) after clear-cut, (c) after conversion to farmland, and (d) after urbanization.

most intensive change in land use would be urbanization, which would bring large increases in runoff from urban surfaces and storm-water sewers.

Figure 4.8 summarizes estimated and observed variation in sediment yield under various historical changes in land use in the Piedmont region of the eastern United States from 1800 to the present. Notice the sharp peak in sediment production during the construction phase of urbanization. These data suggest that the effects of land-use change on the drainage basin, its streams, and sediment production may be quite dramatic. Streams and naturally forested areas are assumed to be relatively stable, that is, without excessive erosion or deposition. As noted above, a land-use change that converts forested land for agricultural purposes generally increases the runoff and sediment yield (erosion) of the land. As a result, the streams become muddy and may not be able to transport all the sediment delivered to the channels. Therefore, the channels will aggrade (partially fill with

sediment), possibly increasing the magnitude and frequency of flooding.

Urbanization

The change from agricultural, forested, or rural land to highly urbanized land causes dramatic changes. First, during the construction phase, there is a tremendous increase in sediment production, which may be accompanied by a moderate increase in runoff. The response of streams in the area is complex and may include both channel erosion (widening) and deposition, resulting in wide, shallow channels. The combination of increased runoff and shallow channels increases the flood hazard. Following the construction phase, the land is mostly covered with buildings, parking lots, and streets, and thus the sediment yield drops to a low level. However, the runoff increases further because of the large impervious areas and use of storm sewers, again increasing the

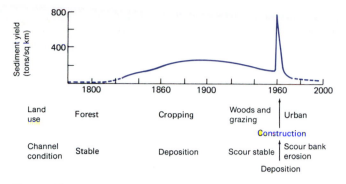

Figure 4.8
Schematic sequence of land use, sediment yield, and channel response from a fixed area. Graph shows the effects of land-use changes on sediment yield in the Piedmont region of the eastern United States before the beginning of extensive farming and continuing through a period of construction and urbanization. (200 tons/sq km = 500 tons/sq mi) (After M. Gordon Wolman, *Geografiska Annaler* 49A, 1966.)

magnitude and frequency of flooding. The streams respond to the lower sediment yield and higher runoff by eroding (deepening) their channels.

The process of urbanization directly affects soils in several ways:

▼ Soil may be scraped off and lost.
▼ Once sensitive soils are disturbed, they may have lower strengths when they are remolded.
▼ Materials may be brought in from outside areas to fill a depression prior to construction, leading to a much different soil than was previously there.
▼ Draining soils and pumping them to remove water may cause dessication (drying out) and other changes in soil properties.
▼ Soils in urban areas are susceptible to soil pollution resulting from deliberate or inadvertent addition of chemicals to soils. This problem is particularly serious if hazardous chemicals have been applied.

Off-Road Vehicles

Urbanization is not the only land-use change that causes increased soil erosion and hydrologic changes. In recent years, the popularity and numbers of off-road vehicles have increased enormously, and demand for recreational areas to pursue this interest has led to serious environmental problems and conflicts among various users of public lands.

Widespread use of off-road vehicles (ORVs) is causing significant environmental impact. There are now more than 12 million ORVs, many of which are invading the deserts, coastal dunes, and forested mountains of many parts of the United States. Problems associated with ORVs (including snowmobiles) are commonly reported from the shores of North Carolina and New York to sand

dunes in Michigan and Indiana to deserts and beaches in the western U.S. The major areas of concern are soil erosion, changes in hydrology, and damage to plants and animals. The problem is not small. A single motorcycle need travel only 7.9 kilometers to have an impact on 1,000 square meters, and a four-wheel-drive vehicle has an impact over the same area by traveling only 2.4 kilometers. In some desert areas, the tracks produce scars that may remain part of the landscape for hundreds of years (13,14).

ORVs cause direct mechanical erosion and facilitate wind and water erosion of materials loosened by their passing. Runoff from ORV sites is as much as eight times greater than for adjacent unused areas, and sediment yields are comparable to those found on construction sites in urbanizing areas (14). Figure 4.9 shows an ORV site in the Mojave Desert. Motorcycles and dune buggies nearly destroyed the vegetative cover on the sandy soil, thus contributing to a 1973 dust storm that was visible on satellite imagery. Figure 4.10 shows eroded ORV trails in the San Gabriel Mountains of California. The accumulation of eroded sediment in the dry pond at the base of the

Figure 4.9
South end of the Shadow Mountains, Mojave Desert. Dune buggies and motorcycles have destroyed the vegetative cover, allowing the sandy soil to be removed easily by wind and running water. This area contributed sediment to a dust storm, recognized on satellite imagery. (Photo by H. G. Wilshire, courtesy of U.S. Geological Survey.)

Figure 4.10
Soil erosion along off-road vehicle trails at Tejon, San Gabriel Mountains, California. Some of the eroded materials have filled in the dry pond (15 × 50 meters) at the base of the slope. (Photo by H. G. Wilshire, courtesy of U.S. Geological Survey.)

hill is approximately 15 by 50 meters. In areas of intensive use, ORVs are the dominant agents of erosion, turning vegetated hills into eroding wasteland.

Hydrologic changes from ORV activity are primarily the result of near-surface soil compaction that reduces the ability of the soil to absorb water. Furthermore, what water is in the soil becomes more tightly held and thus less available to plants and animals. In the Mojave Desert, the tank tracks produced 50 years ago are still visible and the compacted soils have not recovered (14). Compaction also changes the variability of the soil temperature. This is especially apparent near the surface, where daytime soil temperatures become hotter and nighttime soil temperatures lower.

Animals are killed or displaced and vegetation damaged or destroyed by intensive ORV activity. The damage is a result of a combination of soil erosion, compaction, temperature change, and moisture content change (14).

Management of ORVs is a difficult task. In 1972, the president issued an Executive Order requiring agencies in charge of managing public lands to adopt specific policies for ORV traffic. The object was to promote safety and minimize conflicts among the various uses of public lands (14).

There is little doubt that some land must continue to be set aside for ORV use. The problems are how much land should be involved and how environmental damage can be minimized. Sites should be chosen in closed basins with minimal soil and vegetation variation. The possible effects of airborne removal (by wind) must be evaluated carefully, as must the sacrifice of nonrenewable cultural, biological, and geological resources (13). A major problem remains: intensive ORV use is incompatible with nearly all other land use, and it is very difficult to restrict damages to a specific site. Material removed by mechanical, water, and wind erosion will always have an impact on other areas and activities (13,14).

In recent years the demand for self-propelled vehicles has been increasing. Use of off-road, all-terrain bicycles has grown dramatically, and is having an impact on the environment. Bicycles have damaged mountain meadows in Yosemite National Park, and their intensive use also contributes to trail erosion. Those who use all-terrain bicycles respond by stating that bicycles cause less erosion than horses and are lobbying to gain entry into even more locations in the national forests, parks, and wilderness areas. The problem is that although bikes potentially cause less erosion than horses, all-terrain bicycles are cheaper and easier to maintain than horses; as a result, a larger number of people may own them. Thus, the cumulative effect of many bicycles on trails may be greater than that of horses. Also, hikers and other visitors may not mind seeing animals such as horses in wilderness areas but may be less receptive to bicycles, which are fast and almost silent. This is a sensitive area for managers in the wilderness because many people who enjoy riding all-terrain bicycles come from environmentally concerned groups. As with the case of motorized off-road vehicles, management plans will have to be developed to ensure that over-enthusiastic people do not damage sensitive environments.

▼ SEDIMENT POLLUTION

Sediment is probably our greatest pollutant. In many areas, it chokes the streams, fills in lakes, reservoirs, ponds, canals, drainage ditches, and harbors, buries vegetation, and generally creates a nuisance that is difficult to remove. It is truly a resource out of place in that it depletes a land resource (soil) at its site of origin (Figure 4.11), reduces the quality of the water resource it enters, and may deposit sterile materials on productive croplands or other useful land (Figure 4.12) (15).

Most natural pollutional sediments consist of rock and mineral fragments, ranging in size from sand particles less than 2 millimeters in diameter to silt, clay, or very fine colloidal particles. Rates of storage depletion caused by sediment filling in ponds and reservoirs are shown in Table 4.5. These data suggest that small reservoirs will fill up with sediment in a few decades, while large reservoirs will take hundreds of years to fill.

construction, farming, deforestation, and channelization works. In short, a great deal of sediment pollution is the result of human use of the environment and civilization's continued change of plans and direction. It cannot be eliminated, but only ameliorated.

In this century in the United States, emphasis on soil and water conservation has grown considerably. The first programs emphasized stabilization of areas of excessive wind and water erosion, as well as flood control and irrigation water for arid and semiarid land. As the programs progressed, adverse environmental effects from some of the work, such as accelerated stream erosion and rapid reservoir siltation, were discovered. Simultaneously with implementation of new conservation programs, additional needs produced demands for broader and expanded use of completed improvements for recreation, water supply, and navigation, which necessitated research and development to find solutions to new or rediscovered problems. Design changes were developed and applied to correct the erosion and sedimentation problems. These changes included contour plowing, changes in farming practices, and construction of small dams and ponds to hold runoff or trap sediments.

Currently, the same basic problems—erosion control and sediment pollution—occupy a significant portion of the public's attention. Sources of the sediment have expanded, however, to include those from land development, highways, mining, and production of new, unusual chemical compounds and products, all in the interest of a better life. Sediment pollution affects rivers, streams, the Great Lakes, and even the oceans, and the problem promises to be with us indefinitely.

Figure 4.11
Severe soil erosion and loss of a valuable resource. (Photo by F. M. Roadman, courtesy of U.S. Department of Agriculture.)

Some pollutional sediments are debris resulting from the disposal of industrial, manufacturing, and public wastes. Such sediments include trash directly or indirectly discharged into surface waters. (Litter is not confined to roadsides and parks). Most of these sediments are very fine-grained and difficult to distinguish from the naturally occurring sediments unless they contain unusual minerals or other particular characteristics.

The principal sources of artificially induced pollutional sediments are disruption of the land surface for

Figure 4.12
Deposition of unwanted sediment—an example of sediment pollution. (Photo by W. B. King, courtesy of U.S. Department of Agriculture.)

Table 4.5
Summary of reservoir capacity and storage depletion of the nation's reservoirs.

Reservoir Capacity (acre-ft)[a]	Number of Reservoirs	Total Initial Storage Capacity (acre-ft)[a]	Total Storage Depletion		Individual Reservoir Storage Depletion		Average Period of Record yr
			(acre-ft)[a]	%	Average %/yr	Median %/yr	
0–10	161	685	180	26.3	3.41	2.20	11.0
10–100	228	8,199	1,711	20.9	3.17	1.32	14.7
100–1,000	251	97,044	16,224	16.7	1.02	.61	23.6
1,000—10,000	155	488,374	51,096	10.5	.78	.50	20.5
10,000–100,000	99	4,213,330	368,786	8.8	.45	.26	21.4
100,000–1,000,000	56	18,269,832	634,247	3.5	.26	.13	16.9
Over 1,000,000	18	38,161,556	1,338,222	3.5	.16	.10	17.1
Total or average	968	61,239,020	2,410,466	3.9	1.77	.72	18.2[b]

Source: F. E. Dendy, "Sedimentation in the Nation's Reservoirs." *Journal of Soil and Water Conservation* 23,1968.

[a]1 acre-ft = approximately 1234 m³.

[b]The capacity-weighted period of record for all reservoirs was 16.1 years.

Figure 4.13
Example of a sediment and erosion control plan for a commercial development. (Courtesy of Braxton Williams, Soil Conservation Service.) (384 to 400 ft = 117 to 122 m)

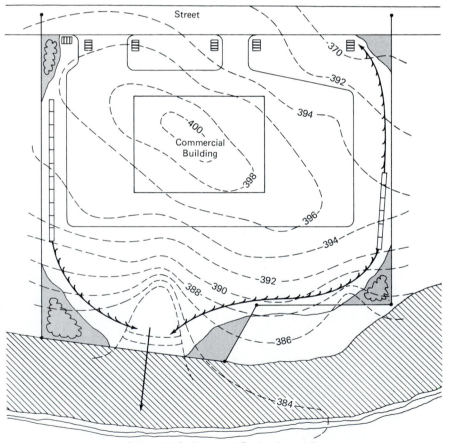

Commercial Sediment and Erosion Control Plan

Sediment basin
Storm sewer catch basin
Diversion
Drainage pattern
Buffer zone

Undisturbed area
Perennial stream
Straw bale diversion
Contour showing elevation   - - 392 - -

The solution to sediment pollution requires sound conservation practices, particularly in urbanizing areas where tremendous quantities of sediment are produced during construction. Figure 4.13 shows a typical sediment and erosion control plan for a commercial development. The plan calls for diversions to collect runoff and a sediment basin to collect sediment and keep it on the site, thus preventing stream pollution. A generalized cross section of a sediment control basin is shown in Figure 4.14.

A study in Maryland demonstrates that sediment control measures can reduce sediment pollution in an urbanizing area (16). The suspended sediment transported by the northwest branch of the Anacostia River near Colesville, Maryland, with a drainage area of 54.6 square kilometers, was measured over a ten-year period. During that time, urbanization (construction) within the basin involved about 3 percent of the area each year. The total urban land area in the basin was about 20 percent at the end of the ten-year study.

Sediment pollution was a problem because the soils are highly susceptible to erosion and there is sufficient precipitation to ensure their erosion when not protected by a vegetative cover. Most of the sediment is transported during spring and summer rainstorms (16). A sediment-control program was initiated, and the sediment yield was reduced by about 35 percent. The basic sediment control principles were to tailor development to the natural topography, expose a minimum amount of land, provide protection for exposed soil, minimize surface runoff from critical areas, and trap eroded sediment on the construction site. Specific measures included scheduled grading to minimize the time of soil exposure, mulch protection and temporary vegetation to protect exposed soils, sediment diversion berms, stabilized waterways (channels), and sediment basins. This Maryland study concluded that even further sediment control can be achieved if major grading is scheduled during periods of low erosion potential and if better sediment traps are designed to control runoff during storms (16).

▼ RATES OF SOIL EROSION

Measurements of soil erosion rates have generally taken several different approaches. The most direct way of estimating soil erosion is to make actual measurements on slopes over a period of years and use these values from that particular region as representative of what is happening in a more general sense. This approach is not often taken, however, because data from individual slopes and drainage basins are very difficult to obtain. A second approach is to use data obtained from resurveying reservoirs and calculating the volume change in stored sediment. This can be done for a region to obtain a sediment yield rate curve. Figure 4.15 is an example of such a curve for drainages in the southwestern United States. This approach has merit but is unlikely to be reliable for small drainage basins where sediment yield can vary dramatically. A final approach is to use an equation to calculate rates of sediment eroded from a particular site. Probably the most commonly used equation is the *Universal Soil Loss Equation* (17), which predicts the amount of soil moved from its original position (18). The equation is written as:

$$A = RKLSCP$$

where A = the long-term average annual soil loss for the site being considered
R = the long-term rainfall runoff erosion factor
K = the soil erodability index
L = the hillslope/length factor
S = the hillslope/gradient factor
C = the soil cover factor
P = the erosion-control practice factor

The Universal Soil Loss Equation has certain advantages in that once the various factors have been determined and multiplied together to produce predicted soil loss, then conservation practices may be applied through factors C and P to reduce the soil loss to the desired level. In the case of slopes that are amenable to shaping, factors of $K, L,$ and S may also be manipulated to achieve desired results in terms of sediment loss. This is particularly valuable when evaluating construction sites such as those for the development of shopping centers and along corridors such as pipelines and highways. For construction sites, the Universal Soil Loss Equation can be used to predict rates of sediment loss and their impact on local streams and other resources. Predicted rates can also be used to develop management strategies for minimizing sediment pollution from a construction site (17, 18).

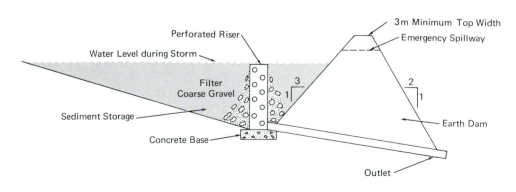

Figure 4.14
Cross section of a sediment basin. Storm water carrying sediment runs into the sediment basin where the sediment settles out and the water filters through loose gravel and into a pipe outlet. Accumulated sediment is periodically removed mechanically. (After *Erosion and Sediment Control,* Soil Conservation Service, 1974.)

Figure 4.15
Sediment yield for the southwestern United States, based on rates of sediment accumulation in reservoirs. (Data from R. I. Strand, "Present and Prospective Technology for Predicting Sediment Yield and Sources," *Proceedings of the Sediment-Yield Workshop, 1972,* U.S. Department of Agriculture Publication ARS-5-40, 1975, p. 13.)

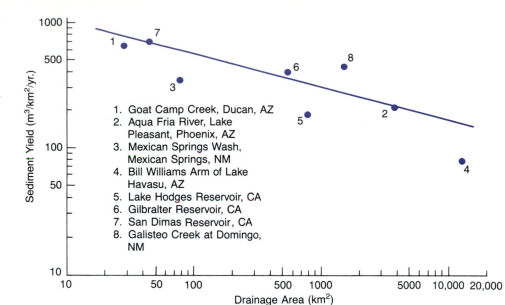

1. Goat Camp Creek, Ducan, AZ
2. Aqua Fria River, Lake Pleasant, Phoenix, AZ
3. Mexican Springs Wash, Mexican Springs, NM
4. Bill Williams Arm of Lake Havasu, AZ
5. Lake Hodges Reservoir, CA
6. Gilbralter Reservoir, CA
7. San Dimas Reservoir, CA
8. Galisteo Creek at Domingo, NM

▼ SOIL POLLUTION

Pollution of soils results when materials detrimental to people and other living things are inadvertently or deliberately applied to soils. Soil pollution came to the forefront of public attention in 1983 when the town of Times Beach, Missouri, with a population of about 2,400, was evacuated and purchased by the government, becoming a ghost town. Evacuation and purchase occurred after it was discovered that oil sprayed on the town's roads to control dust contained *dioxin,* a colorless compound composed of oxygen, hydrogen, carbon, and chlorine, which is known to be extremely toxic to mammals. Although it has never been proven that dioxin has killed any person and the dose necessary to cause adverse health effects to people is not known, dioxin is suspected of being a **carcinogen** (cancer-causing material).

Many different types of materials, including chemicals and heavy metals, may contaminate soils. Clay particles in soils may selectively attract and hold contaminants and thus serve as filters. Soils may also contain organisms that may break certain contaminants down into less harmful materials. Problems result when soils intended for uses other than waste disposal are contaminated, or when people discover that soils have been contaminated by previous uses. Of particular concern is the building of houses and other structures, such as schools, over sites where soils may have been previously contaminated. Many sites contaminated from old waste disposal facilities or inadvertent dumping of chemicals are now being discovered; some of these are being treated. Treatment of soils to remove contaminants, however, can be a very costly endeavor.

In recent years, contamination of soil and water by leaking underground tanks has become a significant environmental concern. Businesses are now adding systems to monitor storage tanks so that leaks may be detected before significant environmental damage occurs. Many environmental geology firms get a good deal of their business from monitoring and working with hazardous wastes that may affect soil and water resources.

▼ DESERTIFICATION

Desertification may be defined as the conversion of land from some productive state to that more resembling a desert. The term was probably first used to describe the advance of the Sahara Desert in Algeria and Tunisia. Driving forces of desertification include, among others, overgrazing, deforestation, adverse soil erosion, poor drainage of irrigated land, and overdraft of water supplies.

Desertification, considered a major problem today, is most pronounced during times of drought stress. In recent years, desertification has been associated with great human misery, characterized by malnutrition and starvation of millions of people, particularly in India and Africa. The effects of desertification, particularly around highly populated areas, are greatly aggravated by political stress associated with dislocated people concentrated in large encampments. When this situation occurs, the surrounding countryside may be picked clean of all natural vegetation. Livestock overgrazes the available food sources to produce a human-induced desert.

Figure 4.16 shows the degree of desertification as a hazard on a global scale. While such a map is useful in identifying general areas that may have problems with desertification, it is of little value in particular case histories or in specifying local reasons for desertification. Of particular importance in evaluating desertification is

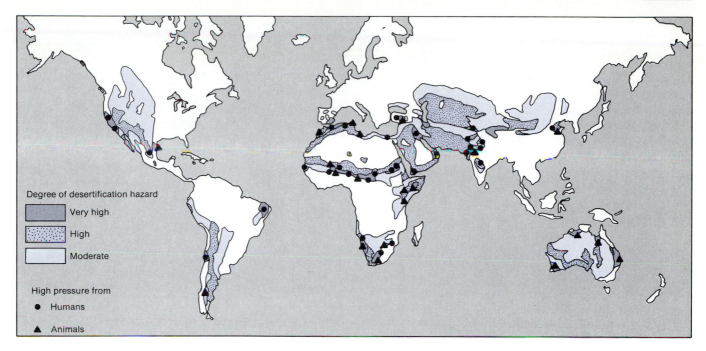

Figure 4.16
Degree of desertification hazards on a global scale. (Council on Environmental Quality, 1978, Annual Report.)

consideration of ecological principles and linkages between people, animals, and the physical environment, including long-term climatic cycles that affect hydrologic processes and soil conditions.

Desertification of North America

The process of desertification in North America has not received as much attention as other parts of the world because the effects have not been nearly so severe in terms of damage to human systems. Nevertheless, desertification is significantly reducing the productivity of land in North America, particularly as it relates to the ability of the land to support life. Major symptoms of desertification in North America are (19): declining groundwater table; salinization of soil and near-surface-soil water; reduction in aerial extent of surface water in streams, ponds, and lakes; unnaturally high rates of soil erosion; and damage to native vegetation. An arid area undergoing desertification may have any or all of the above symptoms to a lesser or greater extent. Furthermore, the symptoms are obviously interrelated. For example, salinization of topsoil may lead to loss of vegetation, which then leads to accelerated soil erosion. The process of desertification is not ordinarily characterized by a continuous change in land along some advancing front, rather it is a patchy

process that proceeds over a wide area with degraded land areas being related to local conditions concerning water, geology, soil, and land use (19).

As an example of desertification in the United States, consider the San Joaquin Valley in California. The major forces of desertification there include, among other factors (19): poor drainage of irrigated land, overgrazing, overdraft of groundwater, cultivation of highly erodable soils, and damage resulting from off-road vehicles (particularly in foothill areas). Figure 4.17 shows the San Joaquin Valley and the area with either present or potential drainage problems. The problem is primarily confined to the west side and southern end of the valley. The precipitation in the northern part of the valley is approximately 35 cm per year, while in the southern end it is closer to 12 cm per year. The San Joaquin Valley is one of the most productive agricultural areas in the world, and yet in many areas the soils are becoming more saline and salty water is rising closer to the surface. Today as much as 16,000 hectares are affected by a high salty water table; by the year 2080, as many as 460,000 hectares may be unproductive (19).

The foothill area in the southern part of the San Joaquin Valley is particularly vulnerable to erosion due to the combined effects of overgrazing, lack of windbreaks in agricultural lands, and off-road-vehicle activities. This

Figure 4.17
San Joaquin Valley, California. (a) Parts with present or potential drainage problems; (b) cross-section showing surface land-forms and drainage conditions. (Modified from Sheridan, *Desertification of the United States,* Council on Environmental Quality, 1981.)

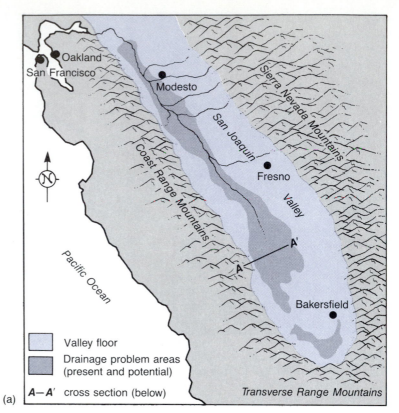

(a)

(b)

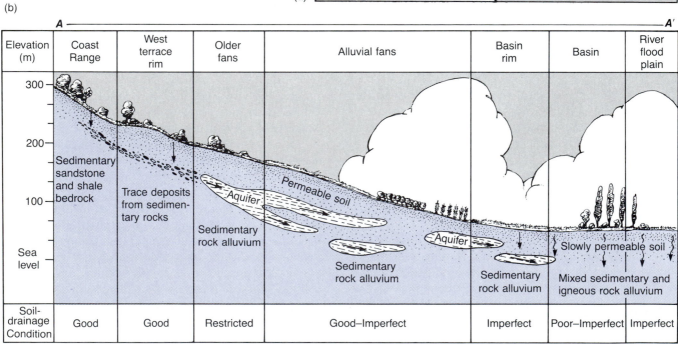

Elevation (m)	Coast Range	West terrace rim	Older fans	Alluvial fans	Basin rim	Basin	River flood plain
Soil-drainage Condition	Good	Good	Restricted	Good–Imperfect	Imperfect	Poor–Imperfect	Imperfect

was dramatically shown during and following a high magnitude windstorm on December 20, 1977. Winds that may have locally reached velocities as high as 300 kilometers per hour caused moderate to heavy damage to buildings, crops, automobiles, wildlife, and soil over an area of 2,000 square kilometers. Five people were killed in automobile accidents caused by impaired visibility. Soil erosion released valley-fever spores that caused significant increase in the incidence of the disease in distant downwind locations. The winds mobilized about 42,000 tons per square kilometer of soil over 600 square kilometers of grazing lands in a 24-hour period. The wind-scoured land was vulnerable to erosion, and subsequent rainstorms the following February caused accelerated erosion and flooding (20). Along the flanks of the mountains at the southern end of the valley, soil erosion from the wind was severe. In some locations, the entire soil was stripped, bedrock was eroded, and soil losses as high as 60 cm were observed. In some cases animals were literally excavated from their burrows by the wind erosion and killed. Figure 4.18 shows a fence post wind-scoured entirely by the December 20 storm. Soil erosion on this site was approximately 20 cm. In Figure 4.19, photographs taken before and after the windstorm indicate the extent of soil erosion as shown by removal of bulldozer-tread scars and loss of vegetation (20). Subsequent rainstorms caused accelerated runoff and flooding, creating numerous gullies that will produce erosion problems for years to come.

▼ SOIL SURVEYS AND LAND-USE PLANNING

Soils greatly affect the determination of the best use of land, and a soil survey is an important part of planning for nearly all engineering projects. A soil survey should include a soil description; soil maps showing the horizontal and vertical extent of soils; and tests to determine grain size, moisture content, and strength. The purpose of the survey is to provide necessary information for identifying potential problem areas before construction.

When properly used the information from detailed soil maps can be extremely helpful in land-use planning. Soils can be rated according to their limitations for land uses, such as housing, light industry, septic-tank systems, roads, recreation, agriculture, and forestry. Soil characteristics that help determine possible limitations for particular land use include slope, water content, permeability, depth to rock, susceptibility to erosion, shrink-swell potential, bearing strength (ability to support a load, such as a building), and corrosion potential.

The limitations can be mapped from detailed soil maps and the accompanying description of the individual soil types. As an example, a soil evaluation to determine limitations for buildings in recreation areas will emphasize the value of a soil survey (21). Figure 4.20a is a detailed soil map and brief description of soils for 4.1 square kilometers in Aroostock County, Maine. Table 4.6 (p. 66) lists soil limitations for buildings in recreational areas. Information from Figure 4.20a and Table 4.6 is combined to produce the map in Figure 4.20b showing where buildings could be located. The standard limitation classes are defined as follows: *none to slight*— soils are relatively free of limitations that affect the planned use, or have limitations that can readily be overcome; *moderate*— soils with moderate limitations resulting from effects of soil properties (these limitations can normally be overcome with correct planning, careful design, and good construction); and *severe*— soils with

Figure 4.18
Fence post scoured to a depth of 10.5 cm as a direct result of the December 20, 1977, windstorm. Total soil erosion on slopes in this area was about 20 cm. (Photograph courtesy of H. G. Wilshire and U.S. Geological Survey.)

(a)

(b)

Figure 4.19
Photographs of area 30 km east of Bakersfield, California. (a) Taken April 1977 during drought, shows bare soil at top right due in part to drought and grazing pressure. Note bulldozer tracks. (b) Taken in March 1978 following windstorm of December 1977, shows extensive soil erosion and loss of vegetation. Bulldozer scars were partly removed by wind erosion, and gullies in middle-ground were eroded by post windstorm runoff from precipitation. (Photographs courtesy of H. G. Wilshire and U.S. Geological Survey.)

severe limitations that make the proposed use doubtful. Careful planning and more extensive design and construction measures are required, possibly including major soil reclamation work (21).

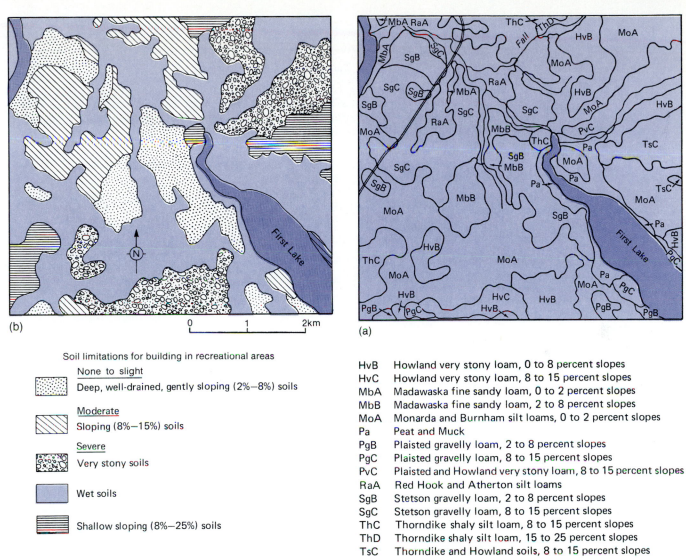

Soil limitations for building in recreational areas

None to slight
Deep, well-drained, gently sloping (2%—8%) soils

Moderate
Sloping (8%—15%) soils

Severe
Very stony soils

Wet soils

Shallow sloping (8%—25%) soils

HvB	Howland very stony loam, 0 to 8 percent slopes
HvC	Howland very stony loam, 8 to 15 percent slopes
MbA	Madawaska fine sandy loam, 0 to 2 percent slopes
MbB	Madawaska fine sandy loam, 2 to 8 percent slopes
MoA	Monarda and Burnham silt loams, 0 to 2 percent slopes
Pa	Peat and Muck
PgB	Plaisted gravelly loam, 2 to 8 percent slopes
PgC	Plaisted gravelly loam, 8 to 15 percent slopes
PvC	Plaisted and Howland very stony loam, 8 to 15 percent slopes
RaA	Red Hook and Atherton silt loams
SgB	Stetson gravelly loam, 2 to 8 percent slopes
SgC	Stetson gravelly loam, 8 to 15 percent slopes
ThC	Thorndike shaly silt loam, 8 to 15 percent slopes
ThD	Thorndike shaly silt loam, 15 to 25 percent slopes
TsC	Thorndike and Howland soils, 8 to 15 percent slopes

Figure 4.20
(a) Detailed soil map of a 4-square-kilometer tract of land, Aroostock County, Maine. (b) Soil limitations for buildings in recreational areas for the same tract. (Reproduced from *Soil Surveys and Land Use Planning*, 1966, by permission of the Soil Science Society of America.)

Table 4.6
Soil limitations for buildings in recreational areas.

Soil Items Affecting Use	Degree of Soil Limitation[a]		
	None to Slight	Moderate	Severe[b]
Wetness	Well to moderately well drained soils not subject to ponding or seepage. Over 1.2 meters to seasonal water table.	Well and moderately well drained soils subject to occasional ponding or seepage. Somewhat poorly drained, not subject to ponding. Seasonal water table 0.6–1.2 meters[c]	Somewhat poorly drained soils subject to ponding. Poorly and very poorly drained soils.
Flooding	Not subject to flooding	Not subject to flooding	Subject to flooding
Slope	0%–8%	8%–15%	15% +
Rockiness[d]	None	Few	Moderate to many
Stoniness[d]	None to few	Moderate	Moderate to many
Depth to hard bedrock	more than 1.5 meters	0.9–1.5 meters[c]	Less than 1 meter

[a]Soil limitations for septic-tank filter fields; hillside slippage, frost heave, piping, loose sand, and low bearing capacity when wet are not included in this rating, but must be considered. Soil ratings for these items have been developed.

[b]Soils rated as having severe soil limitations for individual cottage sites may be best from an aesthetic or use standpoint, but they do require more preparation or maintenance for such use.

[c]These items are limitations only where basements and underground utilities are planned.

[d]Rockiness refers to the abundance of stones or rock outcrops greater than 25 cm in diameter. Stoniness refers to the abundance of stones 8 to 25 cm in diameter.

Source: Reproduced from Soil Surveys and Land Use Planning (1966) by permission of the Soil Science Society of America.

▼ ▼ ▼ SUMMARY AND CONCLUSIONS

A basic understanding of soils and their properties is becoming crucial in several areas of environmental geology, especially when considering land capability for particular land uses such as urbanization, timber harvesting, agriculture, or siting of waste-disposal facilities. Soils are also helpful in evaluating the recurrence of natural hazards such as flooding, landslides, and earthquakes.

Soils result from intimate interactions of the rock and hydrologic cycles with the biologic environment. Soils may be defined in several ways, depending upon the interest of the persons involved. For example, engineers define soil as any earth material that may be removed without blasting, whereas to a soil scientist, a soil is solid earth material that has been sufficiently altered by physical, chemical, and organic processes so that it can support rooted plant life. A soil may be considered as a complex ecosystem consisting of many different types of living things mixed with solid, gas, and liquid materials. Soil fertility refers to the capacity of the soil to supply nutrients needed for plant growth.

Soils may be considered as a function of several variables including climate, topography, parent material, time, and organic activity. Soil-forming processes tend to produce distinctive soil layering, or horizons. Each horizon is defined on the basis of the process of formation or the type of materials present. Of particular importance are processes of leaching, oxidation, and accumulation of materials in various soil horizons. Development of the argillic B horizon, for example, depends on the translocation of clay minerals from upper to lower horizons, where they accumulate as oriented thin skins or plates of clay minerals surrounding soil grains and filling pore spaces between grains. Three properties of importance in describing soils are the color, texture, and structure.

An important concept in studying soils is relative profile development. Whereas very young soils are poorly developed, soils that are older than 10,000 years tend to show moderate development characterized by stronger development of soil structure, redder soil color, and more translocated clay in the B horizon. Strongly developed soils are similar to those of moderate development, but the properties of the B soil horizon tend to be better developed. Such soils may range in age from several tens of thousands of years to several hundred thousand years or older. Distinction between weakly developed soils and strongly developed soils is fairly easy, but distinction between moderate development and continued development is more difficult, often requiring careful evaluation.

A soil chronosequence is a series of soils arranged from youngest to oldest in terms of relative soil profile development. Establishment of a soil chronosequence in a region is useful in evaluating rates of processes and recurrence of hazardous events such as earthquakes and landslides.

Water may enter a soil mass through surface infiltration or from lateral or upward migration. Water content or, as it is sometimes called, moisture content, is an important property of soils and may significantly affect the strength and the shrink-swell potential of a soil.

Water in soils may flow vertically or laterally through soil pores that consist of void spaces between grains or fractures related to soil structure. Flow is termed either saturated (all pore spaces are filled with water) or, more commonly, unsaturated (only a portion of the pores are filled with water). Study of soil moisture and how water moves through soils is becoming a very important topic in environmental geology.

Several types of soil classification exist. Unfortunately, no one classification integrates both engineering properties and soil process. As a result, the environmental geologist must be aware of both the agricultural classification (soil taxonomy) and the unified soil classification (widely used in engineering practice).

Basic understanding of engineering properties of soils is crucial in many environmental problems. These properties include soil strength, sensitivity, compressibility, erodibility, permeability, corrosion potential, ease of excavation, and shrink-swell potential. Shrink-swell potential is particularly important because expansive soils in the United States today cause significant environmental problems and is one of our most costly natural hazards.

Varied land uses and manipulations of our surface waters and groundwaters significantly affect the pattern, amount, and intensity of surface water runoff, soil erosion, and sediment pollution. Of particular concern in recent years have been the effects of urbanization and use of off-road vehicles. Urbanization often involves loss of soil, change of soil properties, adverse soil erosion during construction, and pollution of soils. Use of motorized and nonmotorized off-road vehicles continues to cause significant impacts on the environment. Major areas of concern are soil erosion, changes in hydrology, and damage to plants and animals.

Sediment may be one of our greatest pollutants. In many locations it reduces water quality, and chokes streams, lakes, reservoirs, and harbors. With good conservation practice, sediment pollution can be much reduced.

Measurement of rates of soil erosion has involved several different approaches. These have included direct observation of erosion of soil from slopes, measurement of accumulation of sediment in reservoirs, and development of theoretical equations to calculate rates of soil erosion. The most commonly used equation is the Universal Soil Loss Equation, which predicts the amount of soil moved from its original position. One advantage of the equation is that once all the variables have been identified and measured, predicted rates can be put into a management strategy for minimizing sediment pollution from a particular site.

Soil pollution has also become a public concern. In many locations hazardous materials have been inadvertently or deliberately added to soils, sometimes polluting them and limiting their future usefulness or even rendering them hazardous to life.

Desertification may be considered the conversion of land from some productive state to that more resembling a desert. Desertification, considered a major problem today, is associated with malnutrition and starvation of people, particularly in Africa and India. Driving forces of desertification include, among others, overgrazing, deforestation, adverse soil erosion, poor drainage of irrigated land, overdraft of water supplies, and damage from off-road vehicles.

Information from detailed soil maps can be extremely useful in land-use planning because soils can be rated according to their limitations for such land uses as housing, roads, agriculture, and forestry. Information from detailed soils maps and descriptions are combined to produce simplified maps indicating limitations that range from none to slight to moderate to severe.

▼ ▼ ▼ REFERENCES

1. BIRKLAND, P. W. 1984. *Soils and geomorphology.* New York: Oxford University Press.
2. BUOL, S. W.; HOLE, F. D.; and MCCRACKEN, R. J. 1973. *Soil genesis and classification.* Ames, Iowa: Iowa State University Press.
3. KELLER, E. A.; BONKOWSKI, M. S.; KORSCH, R. J.; and SHLEMON, R. J. 1982. Tectonic geomorphology of the San Andreas fault zone in the southern Indio hills, Coachella Valley, California. *Geological Society of America Bulletin* 93:46–56.
4. ANONYMOUS, 1979. *Environmentally sound small scale agricultural projects.* Mt. Rainier, Maryland: Mohonk Trust, Vita Publications.
5. OLSON, G. W. 1981. *Soils and the environment.* New York: Chapman and Hall.
6. SINGER, M. J., and MUNNS, D. N. 1987. *Soils.* New York: Macmillan.
7. KRYNINE, D. P., and JUDD, W. R. 1957. *Principles of engineering geology and geotechnics.* New York: McGraw-Hill.
8. PESTRONG, R. 1974. *Slope stability.* American Geological Institute. New York: McGraw-Hill.
9. FLAWN, P. T. 1970. *Environmental geology.* New York: Harper & Row.
10. HART, S. S. 1974. Potentially swelling soil and rock in the Front Range Urban Corridor. *Environmental Geology* 7. Colorado Geological Survey.
11. MATHEWSON, C. C.; CASTLEBERRY, J. P., II; and LYTTON, R. L. 1975. Analysis and modeling of the performance of home foundations on expansive soils in central Texas. *Bulletin of the Association of Engineering Geologists* 17, no. 4:275–302.
12. JONES, D. E., Jr., and HOLTZ, W. G. 1973. Expansive soils: The hidden disaster. *Civil Engineering,* August:49–51.
13. WILSHIRE, H. G., and NAKATA, J. K. 1976. Off-road vehicle effects on California's Mojave Desert. *California Geology* 29:123–32.

14. WILSHIRE, H. G., et al. 1977. *Impacts and management of off-road vehicles.* Geological Society of America. Report to the Committee on Environment and Public Policy.

15. ROBINSON, A. R. 1973. Sediment: Our greatest pollutant? In *Focus on environmental geology,* ed. R. W. Tank, 186–92. New York: Oxford University Press.

16. YORKE, T. H. 1975. Effects of sediment control on sediment transport in the northwest branch Anacostia River Basin, Montgomery County, Maryland. *U.S. Geological Survey Journal of Research* 3:487–94.

17. WISCHMEIER, W. H., and MEYER, L. D. 1973. Soil erodibility on construction areas. In *Soil erosion: Causes, mechanisms, prevention and control.* Highway Research Board Special Report 135. Washington D.C. p. 20–29.

18. DUNNE, T., and LEOPOLD, L. B. 1978. *Water in environmental planning.* San Francisco: W. H. Freeman.

19. SHERIDAN, D. 1981. *Desertification of the United States.* Council on Environmental Quality, Washington D.C.

20. WILSHIRE, H. G., and NAKATA, J. K. 1981. *Field observations of the December 1977 wind storm, San Joaquin Valley, California.* Geological Society of America Special Paper 86, p. 233–51.

21. MONTGOMERY, P. H., and EDMINSTER, F. C. 1966. Use of soil surveys in planning for recreation. In *Soil surveys and land use planning,* ed. L. J. Bartelli et al., pp. 104–12. Soil Science Society of America and American Society of Agronomy.

The earth is a dynamic, evolving system with complex interactions of internal and external processes. Internal processes are responsible for moving giant lithospheric plates and thus the continents. Interactions at plate junctions generate internal stress that causes rock deformation, resulting in earthquakes, volcanic activity, and tectonic creep (slow surface movement resulting from movement along a fault zone). These processes trigger numerous external events such as landslides, mudflows, and tsunamis (giant sea waves). Other external activities, such as running water or moving waves, are the result of the interaction among the hydrosphere, atmosphere, and lithosphere. For example, tremendous flooding may result from hurricane activity. The nature and extent of an external event, however, is affected by internal processes which produce the uplifted land surface that running water and waves erode.

People generally do not perceive the full significance of processes that periodically cause loss of life and property. The casual attitude toward natural hazards is a complex manifestation of an internal optimism people have about their lives and their homes, combined with a lack of understanding of physical relations that control the magnitude (severity) and frequency of hazardous processes.

Chapter 5 provides a brief overview of hazardous earth processes. Chapters 6 through 10 discuss our interaction with natural geological events that have damaged and destroyed, and will continue to damage and destroy, human life and property. These events include river floods, landslides, earthquakes, volcanic eruptions, and coastal processes. Especially important are the natural and artificial aspects of the magnitude and frequency of hazardous earth processes and how they relate to human use and interest. These will be discussed in detail for each hazard.

PART TWO

▼

▼

▼

Hazardous Earth Processes

Photo courtesy of Tom Hanks and U.S. Geological Survey.

CHAPTER FIVE

▼
▼
▼

Natural Hazards: An Overview

There have always been natural earth processes that are hazardous to people. Natural disasters such as earthquakes, floods, and hurricanes killed approximately 3 million people on this planet during the 1970s and 1980s. The financial loss probably exceeded $100 billion and does not include social losses such as loss of employment, mental anguish, and reduced productivity (1). The natural processes that produce natural disasters must be recognized. They must be avoided where possible, and their threat to human life and property must be minimized.

We established from our discussion of uniformitarianism in Chapter 2 that present physical and biological processes have been operating a good deal longer than people have been on the earth. Therefore, people have always had to contend with processes that make their lives difficult. Surprisingly, however, *Homo sapiens* appears to have evolved during recent ice ages—a period of harsh climates and environments.

Many physical processes continue to cause loss of life and property damage, including earthquakes, landslides, and flooding of coastal or floodplain areas. In addition, biological processes often mix with physical events to produce hazards. For example, after earthquakes and floods, water may be contaminated by bacteria and the rate of the spread of diseases increased. The magnitude (intensity of energy released) and frequency (recurrence interval) of natural, hazardous processes depend on such factors as the region's climate, geology, and vegetation. In general, there is an inverse relationship between the magnitude of an event and its frequency. That is, the larger the flood, the less frequently it occurs. Studies have demonstrated generally that much of the work in forming the earth's surface is done by processes of moderate magnitude and frequency rather than by

processes with low magnitude and high frequency or by extreme events of high magnitude and low frequency. However, there are many exceptions. For example, in arid regions much of the sediment in normally dry channels may be transported by rare high-magnitude flows produced by intense (but infrequent) rainstorms. Similarly, inlets that cause major changes in the pattern and flow of sediment along barrier island coasts of the eastern United States often are cut by high-magnitude storms. As an analogy to the magnitude-frequency concept, consider the work in logging a forest done by termites, people, and elephants. The termites are small but work quite steadily; the people work less often than termites but are stronger. Given enough time, the people are able to fell most of the trees in the forest and therefore do a great deal of work. Imagine several elephants that rarely visit the forests, but when they do are capable of knocking down many trees. We can see from this analogy that most of the work is done by people, who work at a rather moderate expenditure of energy and time, rather than by the termites' frequent but low expenditure of energy or the elephants' infrequent high expenditure of energy.

Natural hazards are nothing more than natural processes. They become hazards only when people live or work in areas where these processes occur naturally. The naturalness of these hazards is a philosophical barrier that we encounter when we try to minimize their adverse effects. It is the environmental geologist's job, therefore, to identify potentially hazardous processes and make this information available to planners and decision makers so that they can formulate various alternatives to avoid or minimize the threat to human life or property.

The purpose of this chapter is to provide a brief overview of natural processes that cause loss of life or property damage. The following chapters will discuss many of these processes in detail.

▼ SERVICE FUNCTIONS OF NATURAL PROCESSES

It is ironic that natural processes or hazards, while taking human life and destroying property, also perform important service functions. For example, flooding supplies nutrients to the floodplains, as in the case of the Mississippi River or the Nile Delta prior to the building of the Aswan Dam. Flooding also causes erosion on mountain slopes, delivering sediment to beaches from rivers and flushing pollutants from estuaries in the coastal environment.

Landslides also may perform some natural service functions, particularly in the formation of lakes. Landslide debris may form dams, making lakes in mountainous areas. These lakes provide valuable water storage and are an important aesthetic resource.

Volcanic eruptions, while having the potential to produce real catastrophes, perform numerous public service functions. New land can be created, as in the Hawaiian Islands, which are completely volcanic in origin. In addition, nutrient-rich volcanic ash may settle on soils and quickly become incorporated in them. Earthquakes also perform a number of natural service functions. For example, groundwater barriers may be created when rocks are pulverized to form a clay zone known as **fault gouge**. There are numerous cases where groundwater has been dammed upslope from a fault, producing a water resource. Earthquakes are also important in mountain building and thus are directly responsible for many of the scenic resources of the western United States.

▼ NATIONAL AND REGIONAL OVERVIEW

Table 5.1 summarizes selected information about natural hazards or processes for the United States. The largest loss of life every year is associated with tornadoes and windstorms, but other processes such as lightning strikes, floods, and hurricanes also take a heavy toll in human life. The loss of lives as shown in Table 5.1 is difficult to estimate for hazards such as earthquakes, as one great event can cause tremendous loss. For example, it is estimated that a great earthquake in a densely populated part of California could inflict $100 billion in damages while killing several thousand people (1). Property damage from individual hazards is considerable. Floods, landslides, expansive soils, and frost each cause mean annual damages in the United States in excess of $1.5 billion. Surprisingly, expansive soils are one of the most costly hazards, causing over $3 billion in damages each year.

An important aspect of all natural hazards and processes is the potential to produce a **catastrophe**, defined as any situation in which the damages to people, property, or society in general are sufficient that recovery and/or rehabilitation is a long, involved process (2). Those processes most likely to produce a catastrophe include floods, hurricanes, tornadoes, earthquakes, volcanoes, and large fires (Table 5.1). Other processes, such as landslides, generally cover a smaller area and may have only a moderate catastrophe potential. Drought, which may cover a wide area but generally involves plenty of warning time, also has a moderate catastrophe potential. Processes with a low catastrophe potential include coastal erosion, frost, lightning strikes, and expansive soils (2).

Loss of life and property damage in the United States from natural hazards shift with time because of changes in land-use patterns, which influence people to develop on marginal lands; urbanization, which changes the physical properties of earth materials; and increasing population. Damage from most hazards in the United States is increasing, but the number of deaths from many are decreasing due to better warning, forecasting, and prediction of hazards.

▼ PREDICTION OF HAZARDS

Learning how to predict hazards so we can minimize human loss and property damage is an important endeavor. For each particular hazard or process, we have a certain amount of information; in some cases it is

Table 5.1
Effects of selected natural hazards/processes in the United States.

Hazard	Deaths per Year	Occurrence Influenced by Human Use	Catastrophe Potential[b]
Flood	86	Yes	H
Earthquake[a]	50+?	Yes	H
Landslide	25	Yes	M
Volcano[a]	<1	No	H
Coastal erosion	0	Yes	L
Expansive soils	0	No	L
Hurricane	55	Perhaps	H
Tornado and windstorm	218	Perhaps	H
Lightning	120	Perhaps	L
Drought	0	Perhaps	M
Frost and freeze	0	Yes	L

[a]Estimate based on recent or predicted loss over 150-year period. Actual loss of life and/or property could be much greater.

[b]Catastrophe potential: high (H), medium (M), low (L).

Source: Modified after White and Haas, 1975.

sufficient to predict or forecast events accurately. When there is insufficient information to make accurate forecasts or predictions, the best we may be able to do is simply locate areas where hazardous events have occurred and infer where and when similar future events might take place. Thus, reducing the effects of hazards involves the following aspects: identified locations where a hazard occurs, probability of occurrence, precursor events, size of event, forecast, and warning.

For the most part, we know where a particular hazard is likely to occur. For example, the major zones for earthquakes and volcanic eruptions have been delineated satisfactorily on a global scale by mapping earthquake epicenters and recent volcanic rocks and volcanoes. On a regional scale, based on the past record of activity, areas likely to have a significant hazard from a large mudflow or ash eruptions associated with a volcanic eruption have been delineated for several Cascade volcanoes, including Mt. Rainier, as well as for specific volcanoes in Japan, Italy, Colombia, and other places. On a local scale, detailed work with soils, rocks, and hydrology may identify slopes that are likely to fail (landslide) or where expansive soils exist. Certainly we can predict where flooding is likely to occur on the basis of location of the floodplain and evidence from recent floods, such as flood debris and the high-water line.

We can determine the probability of occurrence of a particular event, such as a flood, hurricane, or drought, as part of a hazard prediction. For many rivers we have sufficiently long records of flow to develop probability models that will accurately predict the floods. However, this probability is similar to the chances of throwing a particular number on a die or drawing to an inside straight in poker; thus it is possible for several floods of the same magnitude to occur in any one year, just as it is possible to throw two straight sixes with a die. Likewise, droughts may be assigned a probability on the basis of past occurrence of rainfall in a particular region.

Many hazardous earth processes have precursor events. For example, the surface of the ground may creep or move slowly for a long period prior to an actual landslide. Often the rate of creep increases up to the final failure and landslide. Volcanoes have been noticed to swell or bulge before an eruption, and there often is a significant increase in local seismic activity in the area surrounding the volcano.

Precursor events associated with earthquakes are not particularly well known nor understood, but increases in emission of radon gas from wells, foreshock activity, unusual tilt or uplift of the land, and perhaps even strange animal activity may be precursor events. Anomalous tilt or uplift may begin months or even years prior to the earthquake, whereas unusual animal activity may occur close to the time of the event (use of anomalous animal behavior is very speculative, but is being studied seriously). Seismic gaps (areas where earthquakes are expected but have not occurred) at both regional and local scales have also been valuable in predicting some earthquakes.

With some natural processes it is possible to accurately forecast when the event will arrive. For example, Mississippi River flooding, which occurs in the spring in response to snow melt or very large regional storm systems, is fairly predictable, and we can sometimes forecast when the river will reach a particular flood stage. When hurricanes are spotted far out to sea and tracked toward the shore, we can forecast when and where they will likely strike land. Tsunamis, or seismic sea waves, generated by disturbance of ocean waters by earthquakes or submarine volcanoes, may also be forecast. The tsunami warning system has been fairly successful in the Pacific Basin, and in some instances the time of arrival of the waves has been forecast precisely.

After a prediction for a hazardous event has been made and verified, the public must be warned. The flow of information leading to the warning of a possible hazard such as a large earthquake or flood should move along a path similar to that shown in Figure 5.1. When a prediction or advisory warning has been issued by scientists, the public may not welcome that prediction. For example, in 1982, when geologists advised that a volcanic eruption near Mammoth Lakes, California, was quite likely, the advisory caused loss of tourist business and apprehension on the part of the residents. The eruption did not occur and the advisory was eventually lifted. Similarly, in July 1986 scientists issued an advisory that a large earthquake was likely to occur in the Bishop, California, area. The advisory was issued following a series of earthquakes, over a four-day period, that began with an earthquake of magnitude 3 and culminated in an earthquake of magnitude 6.1, which caused significant damage. Local business owners stated that the advisory in their opinion was irresponsible and would chase potential tourists away from the eastern Sierra Nevada during the summer months when tourism is generally high. Scientists justified their advisory by pointing out that they were trying to make a responsible statement concerning the hazard. Some residents stated that they believed that government scientists were simply trying to scare them and that such advisories ought not to be issued. One resident was reported to have stated that if there was going to be a big earthquake, he wished it would just occur and get it over with! Clearly, issuing the advisory was producing anxiety among the local inhabitants north of Bishop, where the larger earthquake was thought likely to occur. Although we are not yet able to accurately predict earthquakes, it does seem that scientists have a responsibility to make informed judgments even if their predictions do not come to pass. An informed public is probably better able to act responsibly than an uninformed public, even if the subject leads to uncomfortable feelings. Captains of ships regularly review weather

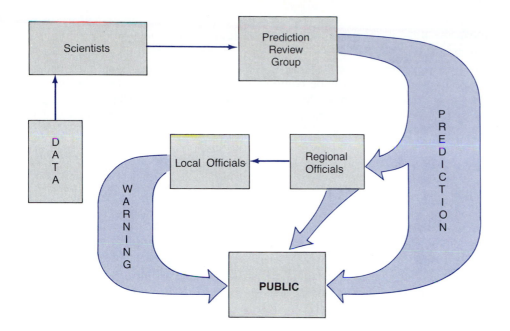

Figure 5.1
Possible flow path for issuance of a prediction or warning for natural hazards.

advisories and warnings, knowing that conditions can change. Weather warnings have proven to be very useful for planning and the careful skipper takes them seriously. Likewise, official warning of hazardous events such as earthquakes, landslides, and floods will also be useful to people who must make decisions concerning where they live, work, and travel. Considering the potential for volcanic eruption and possible loss of life in the Mammoth Lake area, it would have been irresponsible not to issue the advisory. Scientists knew that the seismic data suggested that molten rock was moving toward the surface. Issuing the advisory, even though the eruption did not occur, led to development of evacuation routes and consideration of disaster preparedness. Furthermore, it is very likely that a volcanic eruption will occur in the Mammoth Lake area in the future. The most recent event occurred only six hundred years ago! Certainly the community is better informed today than it was several years ago concerning the volcanic hazard. When an event does occur, they should be better prepared to handle the situation.

▼ SCIENTISTS, THE MEDIA, AND HAZARDS

People today learn about events happening in the world by watching television, listening to the radio, or reading newspapers and magazines. The people who report for the media are generally more interested in the impact of a particular event on people than in the scientific aspects of an event. Even major volcanic eruptions or earthquakes in unpopulated areas may receive little media attention, whereas moderate or even small events in populated areas are reported in great detail. Reporters want to sell stories and what sells is spectacular events that impact people and their possessions (3).

In a perfect world we would like to see good relations between scientists and the news media, but this lofty ideal may be difficult to achieve. This results because on the one hand are scientists, who tend to be conservative, critical people afraid of being misquoted by what they perceive to be pushy, aggressive reporters who tend to report only half-truths while seeking out differences of scientific opinion to embellish a story. On the other hand, those doing the gathering of material for stories perceive scientists as speaking a lot of jargon, while being generally uncooperative, aloof, and unappreciative of the deadlines that the reporters face (3). These statements about scientists and communicators are obviously stereotypic. Both groups have high ethical and professional standards; nevertheless, communication problems and conflicts of interest often occur.

Since scientists do have an obligation to provide information to the general public concerning natural hazards, it is a good policy for a research team to pick one spokesperson to talk to the media and public so that the information is consistent with what is known concerning the problem at hand. For example, if scientists are studying a swarm of earthquakes near Los Angeles and there is speculation among them as to what the swarm means (as there usually is), then it is best to report a consensus to the media rather than a variety of opinions—the general public may be led to believe that the scientists don't know what they are talking about. The development of several working hypotheses is the general rule for the earth scientist—a group of scientists may talk for long periods of time concerning various possibilities and scenarios for the future. When dealing with the news media, however, on a topic that may concern people, their lives, and their property, it is best to be conservative and provide information with as little jargon as possible. Reporters, on the other hand, should strive to provide their readers or viewers or listeners with factual information that has been verified as correct by the scientists. Embarrassing scientists by misquoting them will only lead to more mistrust and poor communication between scientists and journalists (3).

▼ RISK ASSESSMENT

Before rational people can discuss and consider adjustments to hazards, they must have a good idea of the risk that they face under various scenarios. The field of risk assessment is a rapidly growing one in the analysis of hazards, and its use and application should probably be expanded. The *risk* of a particular event is defined as the product of the probability of that event occurring times the consequences should it occur (4). When considering a particular risk-assessment problem, such as, for example, damage to a nuclear reactor by an earthquake, risks for a variety of different combinations of possible events should be calculated. Researchers may wish to know the risk involved with a large versus a small earthquake or perhaps the risk of a moderate event. The large or rare event has a lower probability of occurrence but the consequences may be greater. Consequences could be expressed in a variety of scales. In the nuclear reactor example, the consequences may be evaluated in terms of radiation released, which then can be related to damages to people and other living things.

The risk that society or people are willing to take depends upon the situation involved. For example, even though driving an automobile is fairly risky, most of us accept that risk as part of living in a modern world. On the other hand, acceptable risk of radiation poisoning from a nuclear power plant is very low because such events are considered unacceptable. The reason there is so much controversy over nuclear power plants today is that many people perceive them as the source of a very high risk. Even though the probability of an accident may be relatively low, the consequences are very high, resulting in a high risk. This was dramatically pointed up during and after the nuclear accident at a power plant near Chernobyl, U.S.S.R. Millions of people were exposed to elevated levels of radioactivity that killed approximately 20 people. The exact toll from the event will not be known for many years until it is determined how many cancers may have been caused by the accident.

One problem with risk analysis today has to do with the reliability of geologic data necessary for the analysis. It is often very difficult to assign probabilities to a series of events. This results because, for many geologic processes such as earthquakes and volcanic eruptions, the known chronology for past events may be very inadequate or short. Thus the probability calculations are limited at best (4). Similarly, it may be very difficult to determine the consequences of a particular event or series of events. For example, if we are concerned with the release of radiation into the environment, then we need to have a lot of information about the local biology, geology, hydrology, and meteorology, some of which may be complex and difficult to understand. In spite of these limitations, risk analysis is a step in the right direction, and as we learn more concerning the calculation of the probability of an event and its consequences, we should be able to provide the better analysis necessary for decision making.

▼ AVOIDING AND ADJUSTING TO HAZARDS

Regardless of the strategy we choose to minimize or avoid hazards, it is imperative that we understand and anticipate hazards and their physical, biological, economic, and social impacts. Unfortunately, the ways we choose to adjust to hazards are too often primarily reactive: search and rescue; evacuation; firefighting; providing emergency food, water, and shelter. There is no denying that these activities reduce loss of life and property and need to be continued. However, the move to a higher level of hazard reduction will require increased efforts to anticipate hazards and their impacts. Land-use planning to avoid hazardous locations, hazard-resistant construction, and hazard modification or control (such as flood control channels) are some of the adjustments that anticipate future events and may reduce our vulnerability to hazardous processes (1). Table 5.2 summarizes the physical and social adjustments needed for avoiding and minimizing impacts of natural hazards. Which options are chosen by an individual or society depends upon a number of factors, including hazard perception.

In recent years a good deal of work has been done to try to understand how people perceive various natural hazards. This is important because the success of hazard reduction programs depends on the attitudes of the people likely to be affected by the hazard. While there may be an adequate perception of hazards at the institutional level, this may not filter down to the general population. This is particularly true for those hazards that occur infrequently.

People are more aware of hazards such as brush or forest fires, which may occur every few years. There may even be institutionalized as well as local ordinances to control damages resulting from these events. For example, homes in some areas of southern California are roofed with shingles that will not burn readily and may even have sprinkler systems, and the lots are often cleared of brush. Such safety measures are often noticeable during the rebuilding phase following a fire.

One of the most environmentally sound adjustments to hazards involves land-use planning. That is, people can avoid building on floodplains, in areas where there are active landslides or active fault traces, and in areas where

Table 5.2
Physical and social adjustments for minimizing or avoiding natural hazards.

Physical Adjustments
Identifying and avoiding locations where hazards are likely to occur.
Building homes, offices, and other structures to better resist damage from hazardous processes.
Predicting hazards.
Preventing hazards or altering their effects.

Social Adjustments
Developing land-use planning and controls to avoid hazardous areas and conditions.
Implementing emergency preparedness programs, including evacuation plans to protect life and property once a warning is issued or an event occurs.
Spreading economic loss to a larger population through insurance, taxation, grants, and loans.
Initiating public education campaigns to raise community consciousness concerning natural hazards.
Following a disaster, planning reconstruction so that the community is less vulnerable to future hazards.

Source: Modified after Advisory Committee on the International Decade for Natural Hazard Reduction, *Reducing Disaster's Toll* (Washington, D.C: National Academy Press, 1989).

coastal erosion is likely to occur. In many cities, floodplains have been delineated and zoned for a particular land use. Zoning associated with active and potentially active faults is also commonplace in California. With respect to landslides, legal requirements for soils engineering and engineering geology studies at building sites may greatly reduce potential damages. Although it may be possible to control physical processes in specific instances, certainly land-use planning to accommodate natural processes is often preferable to a technological fix that may or may not work.

Insurance is another option that people may exercise in dealing with natural hazards. Flood insurance is relatively common in many areas, and earthquake insurance is also available. However, other than fire insurance or enforced insurance against flood, few people purchase extra policies. Only a small percentage of people in southern California, for example, have earthquake insurance.

Evacuation is an important option or adjustment to the hurricane hazard in the Gulf States and along the eastern coast of the United States. Often there will be sufficient time for people to evacuate provided they heed the predictions and warnings. However, if people do not react quickly and the affected area is a large urban region,

then evacuation routes may be blocked by people leaving in a last minute panic.

Disaster preparedness is an option that individuals, families, cities, states, or even entire nations can implement. Of particular importance here is training individuals and institutions to handle large numbers of injured people or people attempting to evacuate an area after a warning is issued.

An option that all too often is chosen is bearing the loss caused by a natural hazard. Many people are optimistic about their chances of making it through any sort of natural hazard and therefore will take little action in their own defense. This is particularly true for those hazards—such as volcanic eruptions and earthquakes—that may occur only rarely in a particular area.

▼ IMPACT OF AND RECOVERY FROM DISASTERS

The impact of a disaster upon a population may be either direct or indirect. Direct effects include people killed, injured, dislocated, or otherwise damaged by a particular event; indirect impacts are generally responses to the disaster and include responses of people who are generally bothered or disturbed by the event and of people who donate money or goods, as well as the taxing of people to help pay for emergency services, restoration, and eventual reconstruction. These concepts are summarized in Figure 5.2, which shows that direct effects cost more but affect fewer people, while indirect effects cost less but affect more people (5, 6).

The stages following a disaster are emergency work, restoration of services and communication lines, and reconstruction. Figure 5.3 shows an idealized model of recovery. This model can be compared to actual recovery activities following the 1964 earthquake in Anchorage, Alaska, and the 1972 flash flood in Rapid City, South Dakota. Restoration following the earthquake in Anchorage began almost immediately in response to a tremendous influx of dollars from federal programs, insurance companies, and other sources approximately one month after the earthquake. As a result, reconstruction was a hectic process, with everyone trying to obtain as much of the available funds as possible. In Rapid City, the restoration did not peak until approximately 10 weeks after the flood, and the community took time to carefully think through the best alternatives. As a result, Rapid City today has an entirely different land use on the floodplain, and the flood hazard is much reduced. On the other hand, in Anchorage the rapid restoration and reconstruction were accompanied by little land-use planning. Apartments and other buildings were hurriedly constructed across areas that had suffered ground rupture and were simply filled in and regraded. In ignoring the potential benefits of careful land-use planning, Anchorage is vulnerable to the same type of earthquake that

Figure 5.2
Impact of disaster in terms of the continuum of effects. (Adapted from Bowden and Kates, 1974.)

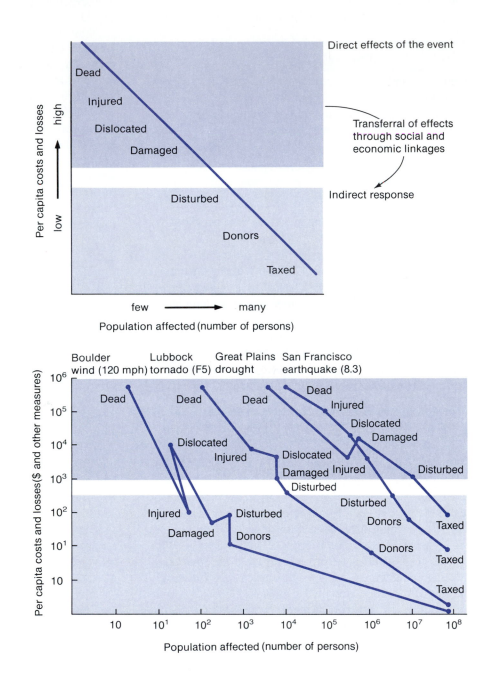

struck in 1964. In Rapid City, the floodplain is now a green belt with golf courses and other such activities—a change that has reduced the flood hazard (2, 5, 6).

▼ ARTIFICIAL CONTROLS OF NATURAL PROCESSES

Attempts to artificially control natural processes such as landslides, floods, and lava flows have had mixed success, and even the best designed structures cannot be ex-

pected to always successfully defend against an extreme event.

Retaining walls and other structures to defend slopes from failure by landslide have generally been successful when well designed. Even the casual observer has probably noticed the variety of such structures along highways and urban land in hilly areas. Structures to defend slopes have limited impact on the environment and are necessary where construction demands that artificial cuts be excavated or where unstable slopes impinge on human structures.

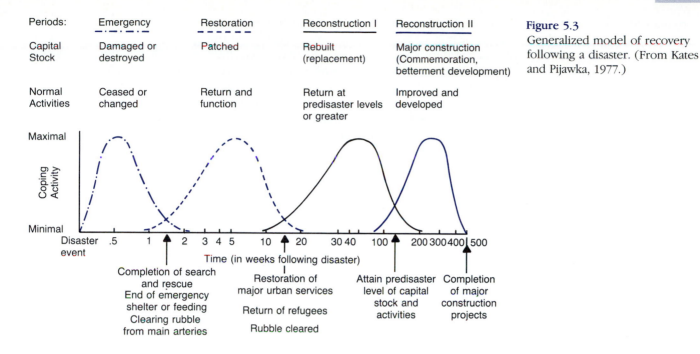

Periods:	Emergency	Restoration	Reconstruction I	Reconstruction II
Capital Stock	Damaged or destroyed	Patched	Rebuilt (replacement)	Major construction (Commemoration, betterment development)
Normal Activities	Ceased or changed	Return and function	Return at predisaster levels or greater	Improved and developed

Figure 5.3
Generalized model of recovery following a disaster. (From Kates and Pijawka, 1977.)

Completion of search and rescue
End of emergency shelter or feeding
Clearing rubble from main arteries

Restoration of major urban services
Return of refugees
Rubble cleared

Attain predisaster level of capital stock and activities

Completion of major construction projects

Common methods of flood control are construction of dams and levees and channelization. Unfortunately, flood control projects tend to provide floodplain residents with a false sense of security because no method can be expected to protect people and their property absolutely from high-magnitude floods. We will return to this discussion in Chapter 6.

▼ GLOBAL CLIMATE AND HAZARDS

Global and regional climatic change can significantly affect the incidence of hazardous natural processes such as storm damage (floods and erosion), landslides, drought, and fires. This was dramatically illustrated during the 1982–1983 El Niño event (in which the trade winds weaken or even reverse, the eastern equatorial Pacific becomes anomalously warm, and the equatorial ocean current that usually moves west weakens or even reverses), which may have brought with it a tremendous increase in hazards on a nearly global scale by putting an unusual amount of heat energy into the atmosphere. Figure 5.4 shows the effects of the hazardous events in 1982–83 that killed several thousand people while causing billions of dollars in damages to crops, structures, utilities, and so forth. Particularly hard hit were Australia, the Americas, and Africa. Drought struck Australia, where a single brush fire in February 1983 burned over 400,000 hectares, killing 74 people and destroying

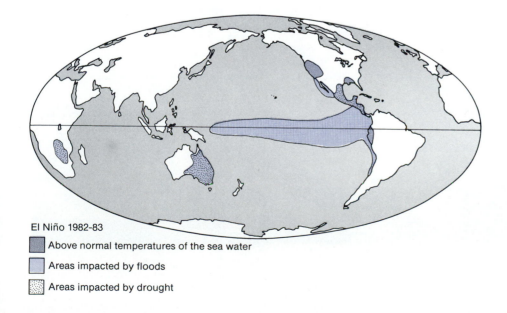

El Niño 1982-83

◼ Above normal temperatures of the sea water

◻ Areas impacted by floods

▦ Areas impacted by drought

Figure 5.4
The 1982–1983 El Niño event. Map shows the general extent of the El Niño effects and the regions damaged by floods or drought. (Data in part from Dennis, 1984).

more than 2000 houses (7). Total damage in Australia was about $3 billion. In normally arid areas of Bolivia, Peru, and Ecuador on the west coast of South America, over 350 centimeters of rain fell compared to a normal 10–12 centimeters, resulting in catastrophic floods and landslides that killed over 600 people while causing damages of about $1 billion (Figure 5.5). Floods and drought in other parts of South America caused another $3 billion in damages and killed over 150 people. In North America, droughts struck Mexico and Central America, and storms in the United States killed about 100 people and inflicted over $2 billion in damages. The Pacific Coast and mountain states were particularly hard hit; coastal erosion destroyed numerous homes and businesses and the flooding of inland rivers caused severe damage to crops and structures (8, 9).

The impact on Africa, due in part to El Niño, was drought that destroyed crops, resulting in people starving. The drought was particularly cruel in Africa because previous drought was already a problem in some areas; political problems and desertification in some areas make planting and harvesting crops difficult even in good times; and overpopulation is a growing problem in much of the troubled parts of Africa. Although there is disagreement on how much of the damage and loss of life in 1982–83 due to natural hazards is directly attributable to El Niño—some researchers believe the effects were confined to tropical regions, causing about $2 billion in damages (10)—it certainly was a year to remember when it comes to hazardous natural processes.

El Niño lasted a short time. It naturally simulated in a compressed time frame some of the effects of the long-range global warming trend we are likely precipitating by increasing the concentration of atmospheric

CO_2 through burning fossil fuels. Carbon dioxide heats the atmosphere by trapping or absorbing infrared radiation emitted from the earth. This is known as the *greenhouse effect*. What might be the effect of a prolonged climatic change on the magnitude and frequency of hazardous earth processes? If a warming trend occurs, climatic patterns will change and coastal erosion will increase as sea level rises in response to melting of glacial ice and thermal expansion of warming ocean waters. Present food production areas will change as some areas receive more precipitation and others less. Deserts and semiarid areas would likely expand and more northern latitudes would become productive. We may not be able to do anything about El Niño (although some venture that human activity contributes to such events), but we may be able to do something about atmospheric CO_2 through managing the burning of fossil fuels.

▼ THE NATURAL ENVIRONMENT AND EXTREME EVENTS

Many parts of our natural environment have evolved over a long period of time and are adjusted to a variety of natural processes of variable magnitude and frequency. The important principle here is that relatively extreme rare events of high magnitude, such as hurricanes and wildland fires, are probably as significant to long-term ecosystem stability and evolution as lesser magnitude more common events. For example, wildland fires can either maintain, advance, or retard the stage of ecological succession in forest or brushlands. Burning of plants (trees, brush, or grass) releases important nutrients for recycling through new plant growth. Effects of fire and

Figure 5.5
Flood damage in Ecuador caused by the 1982–1983 El Niño event. (Photo by David Perry.)

nutrient release can release fire-resistant seeds and therefore reduce competition for specific plants (7). In many brushlands, grasslands, and forests of the world, periodic fires are important in maintaining the ecosystem through cycling of plant growth and nutrients. In forests, high-magnitude storms are the events that often cause tree mortality and litter fall, which are important in nutrient cycling, soil formation, and introduction of new trees in open spots created by fallen trees. Trees that fall on the forest floor or in streams may reside there for decades to centuries, providing significant and variable forest and stream habitat for animals and fish. In fact, in many forested areas, pools in streams are created or enhanced by large, fallen organic debris (tree trunks), and these pools provide important aquatic habitat.

▼ POPULATION INCREASE AND NATURAL HAZARDS

Population increase is a major environmental problem. As our population continues to increase, putting greater demands on our land and resources, the need for planning to minimize losses from natural hazards/processes also increases. Specifically, an increase in population puts a greater number of people at risk from a natural event and forces the use of marginal lands, creating additional risks. This is dramatically illustrated by the recent loss of thousands of lives in Mexico and Colombia. In the autumn of 1985, Mexico endured a magnitude 7.8 earthquake that killed about 10,000 people in Mexico City alone. Two months later, mudflows

following the eruption of a volcano killed approximately 25,000 people in Colombia.

Mexico City is the center of the world's most populous urban area. Approximately 23 million people are concentrated in an area of about 2300 square kilometers, and about one-third of the families (which average five members) live in a single room. The city is built on ancient lake beds, which accentuate earthquake shaking, and parts of the city have been sinking at the rate of a few centimeters per year owing in part to ground-water withdrawal. The subsidence has not been uniform, so the buildings tilt and are even more vulnerable to the shaking of earthquakes (11).

The Colombian volcano Nevado del Ruiz erupted in February of 1845, producing a mudflow that roared down the east slope of the mountain. The mudflow killed approximately 1000 people in the town of Ambalema, located on the banks of the Lagunilla River 80 kilometers from the volcano's summit. Deposits from that event produced rich soils at a site 32 kilometers up the river valley, and an agricultural center developed there. The town that the area supported was known as Armero and, by 1985, had a population of about 22,500 people.

On November 13, 1985, another mudflow associated with a volcanic eruption buried Armero, leaving about 20,000 people dead or missing. A matter of 140 years multiplied the volcano's mudflow toll twenty times because of population increase over that period. Ironically, the same force (volcanic eruption and mudflow) that produced productive soils which stimulated development and population growth in the area later decimated it (12).

▼ ▼ ▼ SUMMARY AND CONCLUSIONS

One of the fundamental principles of environmental geology is that there have always been earth processes that are hazardous to people. The emphasis is on the term "earth process." That is, most natural hazards are simply natural processes that become a problem when people live close to a potential danger or modify processes in such a way as to increase the hazard.

Many processes will continue to cause loss of life and property damage, including flooding of coastal or flood-plain areas, landslides, earthquakes, volcanic activity, wind, expansive soils, drought, fire, and coastal erosion. However, the magnitude and frequency of these processes or events depend on such diverse factors as climate, geology, vegetation, and human use of the land. Once a process has been identified and the potentially hazardous as-

pects studied, this information must be made available to planners and decision makers so that they may avoid these hazards or processes or minimize their threat to human life and property. Of particular significance are how a warning is issued, how scientists communicate with the media and public, and how we calculate risks associated with hazards.

Major adjustments to natural hazards and processes include land-use planning, artificial control, insurance, evacuation, disaster preparedness, and bearing the loss. Which of these options is chosen by an individual or segment of society depends upon a number of factors, the most important of which may be hazard perception. Regardless of how we choose to minimize or avoid natural hazards we must increase our understanding of hazards

and do a better job of anticipating them. Increased anticipation will reduce but not replace reactive measures such as search and rescue and providing emergency aid.

The impact of a hazardous process (disaster) upon a population may be either direct or indirect. Direct effects include people killed, dislocated, or otherwise damaged by a particular event. Indirect impacts involve people generally bothered or disturbed by the event, people who donate money or goods, and taxing people to help pay for emergency services, restoration, and eventual reconstruction following a disaster. The reconstruction phase following a disaster often takes place through several stages, including emergency work, restoration of services and communication lines, and, finally, rebuilding.

Attempts to artificially control natural processes have had mixed success and usually cannot be expected to defend against extreme events.

Global climatic change, even if it is a short-term event, can affect the incidence of hazardous natural processes. The 1982–83 El Niño event is a good example.

Many natural systems have evolved with and are dependent upon high-magnitude events such as fires or storms for stability. Examples include brushland fires and storms in forests.

As the world's population increases, there will be greater demand on all land resources. This will force more people to live on marginal lands and in more hazardous locations. Therefore, as population increases, better planning at all levels will be necessary if we are to minimize losses from natural hazards/processes.

The discussion of natural processes suggests a view of nature as dynamic and changing. Our modern understanding of natural processes therefore tells us that we cannot view our environment as fixed in time. A landscape without natural hazards would also have less variety; it would be safer but less interesting and probably less aesthetically pleasing. The jury is still out on how much natural hazards should be controlled and how much they should be allowed to occur. However, we should remember that disturbance is a natural occurrence, therefore management of natural resources must manage for and with disturbances such as fires, storms, and floods.

▼ ▼ ▼ REFERENCES

1. ADVISORY COMMITTEE ON THE INTERNATIONAL DECADE FOR NATURAL HAZARD REDUCTION. 1989. *Reducing disaster's toll.* Washington, D.C.: National Academy Press.
2. WHITE, G. F., and HAAS, J. E. 1975. *Assessment of research on natural hazards.* Cambridge, Mass.: The MIT Press.
3. PETERSON, D. W. 1986. Volcanoes—Tectonic setting and impact on society. In *Studies in geophysics: Active tectonics,* pp. 231–46. Washington, D. C.: National Academy Press.
4. CROWE, B. W. 1986. Volcanic hazard assessment for disposal of high-level radioactive waste. In *Studies in geophysics: Active tectonics,* pp. 247–60. Washington, D. C.: National Academy Press.
5. KATES, R. W., and PIJAWKA, D. 1977. Reconstruction following disaster. In *From rubble to monument: The pace of reconstruction,* eds. J. E. Haas, R. W. Kates, and M. J. Bowden. Cambridge, Mass.: The MIT Press.
6. COSTA, J. E., and BAKER, V. R. 1981. *Surficial geology: Building with the earth.* New York: Wiley.
7. ALBINI, F. A. 1984. Wildland fires. *American Scientist* 72: 590–97.
8. SIEGEL, B. 1983. El Niño: The world turns topsy-turvy. *Los Angeles Times,* Aug. 17.
9. CANBY, T. Y. 1984. El Niño's ill winds. *National Geographic* 165: 144-81.
10. DENNIS, R. E. 1984. A revised assessment of worldwide economic impacts, 1982–1984 El Niño/Southern Oscillation Event. *EOS,* Transactions of the American Geophysical Union. 65(45): 910.
11. MAGNUSON, E. 1985. A noise like thunder. *Time* 126(13): 35–43.
12. RUSSELL, G. 1985. Colombia's mortal agony. *Time* 126(21): 46–52.

Streams and rivers are the basic transportation systems of the part of the rock cycle involved with erosion and deposition of sediments. They are also a primary erosion agent in the sculpture of our landscape.

Rivers carry three types of load: suspended, bed, and dissolved. The sum of the three is the *total load*. The *suspended load* is carried above the stream bed by the flowing water. It is often the largest part of the total load, and makes rivers look muddy. The *bed load* moves along the bottom of the channel by bouncing, rolling, or skipping. The *dissolved load* is carried in chemical suspension and is derived from chemical weathering of rocks in the drainage basin. It is the dissolved load that may make stream water taste salty (if the dissolved load contains large amounts of chloride and sodium), and may make the stream water hard (if the dissolved load contains high concentrations of calcium and magnesium). It is the suspended and bed load of streams that, when deposited in undesirable locations, produces the sediment pollution discussed in Chapter 4.

Streams and rivers are open systems that generally maintain a dynamic equilibrium between the work done (sediment transported by the stream) and the load imposed (sediment delivered to the stream from tributaries and hillslopes) (1). To accomplish this, the stream must maintain a delicate balance between the flow of water and movement of sediment. Since the stream cannot increase or decrease the amount of water or sediment it receives, adjustments are made in terms of channel slope and cross-sectional shape, which effectively change the velocity of the water. The change of velocity may, in turn, increase or decrease the amount of sediment carried in the system. The stream tends to have a slope to provide just the velocity of water necessary to do the work of moving the sediment load (2). Since this is a delicate balance, any change in the sediment load or discharge will initiate slope changes to bring this system into balance again. Take, for example, a land-use change from forest to agricultural row crops. This change will cause increased soil erosion and an increase in the load supplied to the stream. The stream will be initially unable to transport the entire load, and sediment will be deposited, increasing the slope, which will in turn increase the velocity of water and allow the stream to move more sediment. Slope will increase by deposition in the channel until the velocity increases sufficiently to carry the new load. A new dynamic equilibrium may be reached, provided the rate of sediment increase levels out and the channel can adjust the slope and channel shape before another land-use change. If the reverse situation occurs, that is, if farmland is converted to forest, the sediment load will decrease and the stream will react by eroding the channel to lower the slope, which in turn will lower the velocity of the water. This will continue until an equilibrium between the load imposed and work done is achieved again.

CHAPTER SIX

This sequence of change is occurring in parts of the Piedmont of the southeastern United States. There, land that was forest in the early history of the country was cleared for farming, producing accelerated soil erosion and subsequent deposition of sediment in the stream (Figure 6.1). The land is now reverting to pine forests, and this, in conjunction with soil conservation measures, has reduced the sediment load delivered to streams. Thus, formerly muddy streams choked with sediment are now clearing slightly and eroding their channels. Whether this trend continues depends on future conservation measures and land use.

Consider now the effect of constructing a dam on a stream. Considerable changes will take place both upstream and downstream of the reservoir. Upstream, the effect will be to slow down the stream, causing deposition of sediment. Downstream, the water coming out below the dam will have little sediment. (Most sediment is trapped in the reservoir.) As a result, the stream will erode its channel, picking up sediment, thus decreasing the slope until new equilibrium conditions are reached (Figure 6.2). We will return to the topic of dams on rivers in Chapter 11, with a more detailed discussion.

Channel pattern is the configuration of the channel in plan view, as from an airplane. Two main patterns (Figure 6.3) are **braided** channels, characterized by numerous bars and islands that divide and reunite the channel, and those that do not braid. Because long, straight reaches are relatively rare, channels that do not braid are designated as **sinuous**; however, sinuous channels often contain relatively short straight reaches. Individual bends in sinuous channels are called **meanders**, which migrate back and forth across the stream valley, depositing sediment on the inside of bends, forming point bars, and eroding the outside of bends.

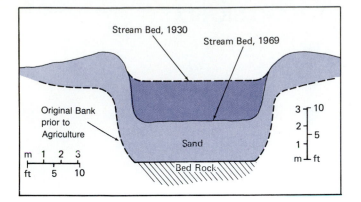

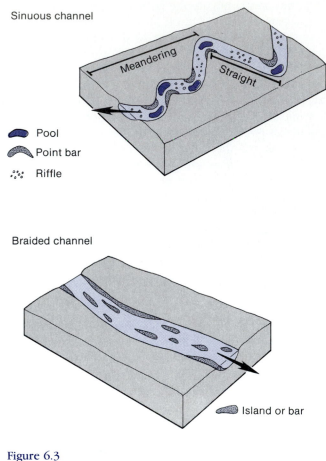

Figure 6.1

Accelerated sedimentation and subsequent erosion resulting from land-use changes (natural forest to agriculture and back to forest) at the Mauldin Millsite on the Piedmont of middle Georgia. (After S. W. Trimble, "Culturally Accelerated Sedimentation on the Middle Georgia Piedmont," Master's thesis [Athens: University of Georgia, 1969.] Reproduced by permission.)

This process is prominent in constructing and maintaining floodplains. Overbank deposition during floods causes vertical accretion that is also important in development of floodplains.

Sinuous channels often contain a series of regularly spaced pools and riffles (Figure 6.4). **Pools** at low flow are deep areas characterized by relatively slow movement of water. They are produced by scour at high flow. **Riffles** at low flow are shallow areas recognized by relatively fast water. Riffles are produced by depositional processes at high flow. With respect to erosion and deposition of sediment, we therefore conclude that pools scour at high flow and fill at low flow, whereas riffles fill at high flow and scour at low flow (Figure 6.5). It is this pattern of scour and fill that maintains pools and riffles. The pool-riffle sequence is repeated approximately every five to seven times the channel width. Streams with well-developed pools and riffles tend to have considerable gravel in the streambed and a relatively low slope. Streams with finer bed material or steep slopes tend to lack regularly spaced pools and riffles. Bedrock channels also may develop pools and riffles. These forms are

Figure 6.3

Channel patterns.

important in environmental science because they provide a variety of flow conditions characterized by deep, slow-moving water alternating with shallow, fast-moving water which facilitates desirable biologic activity. They also provide visual and other sensory variety that increases the aesthetic amenity as well as the recreational potential by providing better fishing and boating conditions. In some instances, pools and riffles also help to stabilize the channel. For example, a straight reach with pools and riffles in which the deep part of the channel alternates from bank to bank may be morphologically more stable than a channel that lacks these forms.

Figure 6.2

Upstream deposition and downstream erosion from construction of a dam and a reservoir.

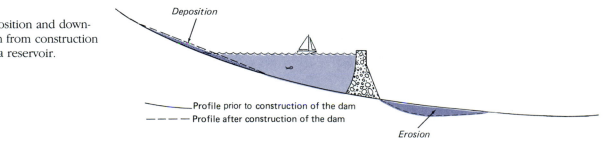

Figure 6.4
Well-developed pool-riffle sequence in Sims Creek near Blowing Rock, North Carolina. A deep pool is apparent in the middle distance and shallow riffles can be seen in the far distance and the foreground.

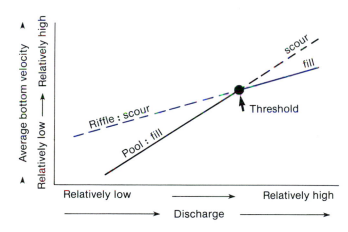

Figure 6.5
Scour-fill pattern characteristic of a pool-riffle sequence. The threshold is a critical discharge at which change in process (scour-to-fill or fill-to-scour) occurs.

▼ FLOODING: AN OVERVIEW

Much of the sediment transported in rivers is periodically stored by deposition in the channel and on the adjacent floodplain (Figure 6.6). These areas, collectively called the **riverine environment**, are the natural domain of the river. Lateral migration of bends of rivers and overbank flow combine to produce the **floodplain**, which is periodically inundated by water and sediment. This natural process of overbank flow is termed *flooding*. Channel discharge (the volume of water per unit time flowing past a particular location; cubic meters per second, m^3/s) at the point where water overflows the channel is called the *flood discharge* and may or may not coincide with property damage. The term *flood stage* frequently connotes that the elevation of the water surface has reached a high-water condition likely to cause damage to personal property on the floodplain. This

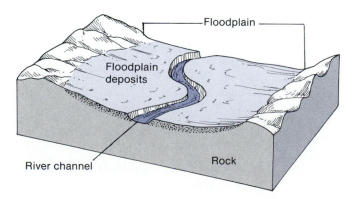

Figure 6.6
Block diagram showing floodplain and river channel.

Table 6.1
Selected severe river floods in
the United States.

Year	Month	Location	Lives Lost	Property Damage (Millions of dollars)
1937	Jan.–Feb.	Ohio and lower Mississippi river basins	137	417.7
1938	March	Southern California	79	24.5
1940	Aug.	Southern Virginia and Carolinas, and Eastern Tennessee	40	12.0
1947	May–July	Lower Missouri and middle Mississippi river basins	29	235.0
1951	June–July	Kansas and Missouri	28	923.2
1955	Dec.	West Coast	61	154.5
1963	March	Ohio River basin	26	97.6
1964	June	Montana	31	54.3
1964	Dec.	California and Oregon	40	415.8
1965	June	Sanderson, Texas (flash flood)	26	2.7
1969	Jan.–Feb.	California	60	399.2
1969	Aug.	James River basin, Virginia	154	116.0
1971	Aug.	New Jersey	3	138.5
1972	June	Black Hills, South Dakota (flash flood)	242	163.0
1972	June	Eastern United States	113	3,000.0
1973	March–June	Mississippi River	—	1,200.0
1976	July	Big Thompson River, Colorado (flash flood)	139	35.0
1977	July	Johnstown, Pennsylvania	76	330.0
1977	Sept.	Kansas City, Missouri and Kansas	25	80.0
1979	April	Mississippi and Alabama	10	500.0
1983	Sept.	Arizona	13	416.0
1986		Western states, especially California	17	270.0
1990	Jan.–May	Trinity River, Texas	—	1,000.0
1990	June	Eastern Ohio (flash flood)	21	several

Sources: NOAA, *Climatological Data, National Summary*, 1970, 1972, 1973, and 1977; U.S. Geol. Survey. Updated by author in 1990.

is obviously based on human perception of the event and therefore changes or varies as human use of the floodplain changes (3).

Flooding is one of the most universally experienced natural hazards. In the United States, the lives lost to river flooding number about 100 per year, with property damage of about $3 billion per year. Table 6.1 lists several severe U.S. floods. The loss of life is relatively low compared to loss in preindustrial societies that lack monitoring and warning systems before a flood and disaster relief afterward. Although preindustrial societies with dense populations on floodplains lose a larger proportion of lives, they have a relatively lower amount of property damage than industrial societies (3, 4).

Most river flooding is a function of the total amount and distribution of precipitation and the rate at which it infiltrates the rock or soil and the topography; however, some floods result from rapid melting of ice and snow in the spring or, on rare occasions, from the failure of a dam. Finally, land use can greatly affect flooding in small drainage basins.

▼ MAGNITUDE AND FREQUENCY
OF FLOODS

Flooding is intimately related to the amount and intensity of precipitation and runoff. Catastrophic floods reported on television and in newspapers often are produced by

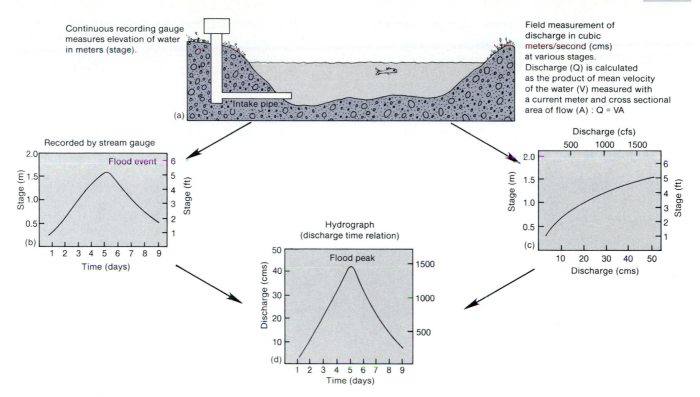

Figure 6.7
Field data (a) consist of a continuous recording of the water level (stage), which is used to produce a stage-time graph (b). Field measurements at various flows also produce a stage-discharge graph (c). Then graphs (b) and (c) are combined to produce the final hydrograph (d).

infrequent, large, intense storms. Smaller floods or *flows* may be produced by less intense storms, which occur more frequently. All flow events that can be measured or estimated from stream-gauging stations (Figure 6.7) can be arranged in order of their magnitude of discharge, generally measured in cubic meters per second (m³/s). The list of flows so arranged is called an *array* and can be plotted on a discharge-frequency curve (Figure 6.8) by deriving the recurrence interval R for each flow from the relationship

$$R = \frac{N+1}{M}$$

where R is a recurrence interval in years, N is the number of years of record, and M is the rank of the individual flow in the array (5). The highest flow for nine years of data for the stream shown in Figure 6.8 is almost 283 m³/s and so has a rank M equal to one (6). Therefore, the recurrence interval of this flood is

$$R = \frac{N+1}{M} = \frac{9+1}{1} = 10$$

which means that a flood with a magnitude equal to or exceeding 283 m³/s can be expected about every ten years. Studies of many streams and rivers show that

channels are formed and maintained by bank-full discharge with a recurrence interval of 1.5 to 2 years (28 m³/s on Figure 6.8). Therefore, we can expect a stream to emerge from its banks and cover part of the floodplain with water and sediment once every year or so.

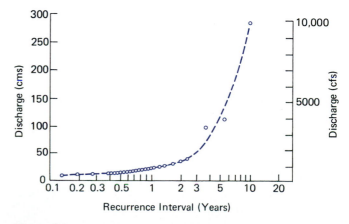

Figure 6.8
Example of a flood frequency curve. Each circle represents a flow event with recurrence interval plotted on probability paper. (After L. B. Leopold, U.S. Geological Survey Circular 559, 1968.)

The longer that flow records are collected, the more accurate the prediction of floods is likely to be. However, designing structures for a 10-year, 25-year, 50-year, or even 100-year flood, or in fact any flow below possible maximum, is itself a calculated risk because predicting a flood of a certain magnitude is based on probability, and therefore has an element of chance. Theoretically, a 25-year flood should happen on the average of once every 25 years, but two 25-year floods could occur in any given year (7). We can thus conclude that as long as we continue to build dams, highways, bridges, homes, and other structures on flood-prone areas, we can expect continued loss of lives and property.

In discussing the concepts of magnitude and frequency of floods, we should distinguish between *upstream* floods and *downstream* floods (Figure 6.9). Upstream floods are in the upper parts of drainage areas and are generally produced by intense rainfall of short duration over a relatively small area. Although these floods can be severe over a relatively small area, they generally do not cause floods in the larger streams they join downstream. Downstream floods, on the other hand, cover a wide area. They are usually produced by storms of long duration that saturate the soil and produce increased runoff. Flooding on small tributary basins is limited, but the contribution of increased runoff from thousands of tributary basins may cause a large flood downstream, which can be recognized by the downstream migration of an ever-increasing flood wave with large rise and fall of discharge (8). The example of the Chattooga-Savannah River system in Figure 6.10a shows the downstream migration with time of a flood crest. It illustrates that a progressively longer time is necessary for the rise and fall of water as the flood wave proceeds downstream, and shows dramatically the tremendous increase in discharge from low-flow conditions to over 1,700 m^3/s in five days and 257 kilometers of downstream flow (9). It is the large downstream floods that usually make television and newspaper headlines. Figure 6.10b illustrates the same flood in terms of discharge per unit

Figure 6.9
Idealized diagram comparing upstream flood (a) to downstream flood (b). Upstream floods generally cover relatively small areas and are caused by intense local storms, whereas downstream floods cover wide areas and are caused by regional storms or spring runoff. (Modified after U.S. Department of Agriculture drawing.)

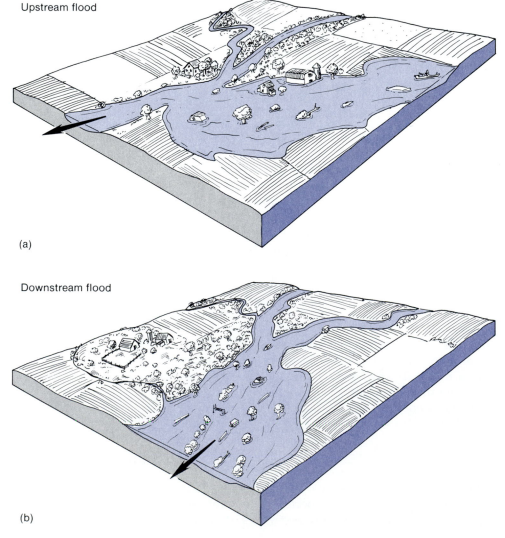

Upstream flood

(a)

Downstream flood

(b)

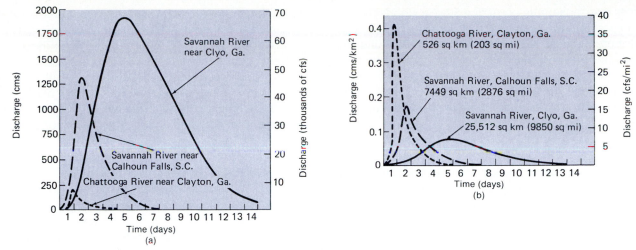

Figure 6.10
Downstream movement of a flood wave on the Savannah River, South Carolina and Georgia. (After William G. Hoyt and Walter B. Langbein, *Floods* [copyright 1955 by Princeton University Press] Fig. 8, p. 39. Reprinted by permission of Princeton University Press.)

area, thus eliminating the effect of downstream increase in discharge. This better illustrates the shape and form (sharpness of peaking) of the flood wave as it moves downstream (9).

A few upstream floods of very high magnitude have been caused directly by structural failure. For example, the most destructive flood in West Virginia's history was caused by the failure of a coal-waste dam on the middle fork of Buffalo Creek. On the morning of February 26, 1972, that flood, which lasted only three hours, cost at

least 118 lives, destroyed 500 homes, left 4,000 persons homeless, and caused more than $50 million of property damage. A wall of water 3 to 6 meters high swept through the Buffalo Creek Valley at an average rate of about 2.1 meters per second, or 8 kilometers per hour. For three days before the disaster, nearly 10 centimeters of rain fell in the area, producing a 10-year flood in local streams similar to Buffalo Creek. The volume of water released by the collapse of the coal-waste dam was approximately 500 thousand cubic meters, producing a flood about 40 times

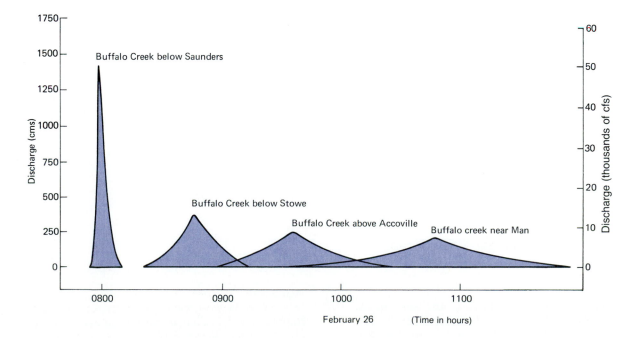

Figure 6.11
Estimated flood hydrographs for Buffalo Creek, West Virginia, on February 26, 1972. (After Davies, Bailey, and Donovan, U.S. Geological Survey Circular 667, 1972.)

greater than a naturally occurring 50-year flood. The U.S. Geological Survey concluded that the failure of the dam contributed almost all of the peak flood flow and that direct runoff from other sources was not significant. It was further concluded that there were several causes for the dam failure. First, the dam was neither designed nor constructed to withstand the amount and depth of water it impounded. It was, in fact, a waste pile that grew as more and more material was dumped. Second, there was no spillway or other adequate water-level control in the dam to provide a means of removing water. Third, sludge waste from the coal mining operation was an inadequate foundation material for the dam, and seepage through the foundation caused openings to form through which water flowed, greatly reducing the dam's stability. Fourth, stability of the dam was further reduced because of its great thickness, and the dam became saturated and somewhat buoyant. Fifth, the dam itself was constructed of coal waste, including fine coal, shale, clay, and mine rubbish, all of which disintegrate rapidly and are very

unstable. A safe, economical dam cannot be constructed of this material alone (10).

After the failure of the dam, the impounded water emptied into Buffalo Creek in 15 minutes. At the time, the flow in the creek was well below bank-full stage. The flood wave traveled three hours before reaching the mouth of Buffalo Creek at the town of Man. Along the way, the peak flattened and the flood took longer to pass a particular point (Figure 6.11). When the wave hit the Guyandotte River at 11:00 A.M., it produced a sudden high peak in the already rising river (Figure 6.12).

Because the Buffalo Creek flood did not occur naturally, it is not strictly valid to compare it to other natural floods. Nevertheless, this flood was a serious warning of the possible effects of human activities on even small streams (10).

▼ URBANIZATION AND FLOODING

Human use of land in the urban environment has increased both the magnitude and frequency of floods in small drainage basins of a few square kilometers. The rate of increase is a function of the percentage of the land that is covered with roofs, pavement, and cement (referred to as *impervious cover*) and the percentage of area served by storm sewers. Storm sewers are important because they allow urban runoff from impervious surfaces to quickly reach stream channels. Therefore, impervious cover and storm sewers are collectively a measure of the degree of urbanization. The graph in Figure 6.13 shows that an urban area with 40 percent impervious surface and 40 percent served by storm sewers can

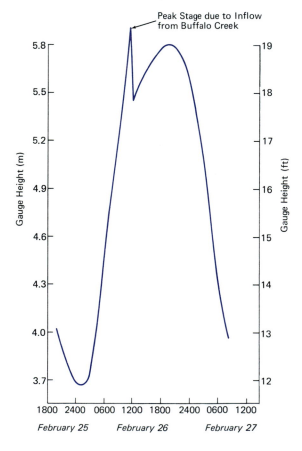

Figure 6.12
Hydrograph of the Guyandotte River at Man, West Virginia, during the period February 25 to February 27, 1972. The flood inflow from Buffalo Creek produced a sudden high peak of 5.89 meters at the gaging station. Peak discharge was 2,950 cubic meters per second. (After Davies, Bailey, and Donovan, U.S. Geological Survey Circular 667, 1972.)

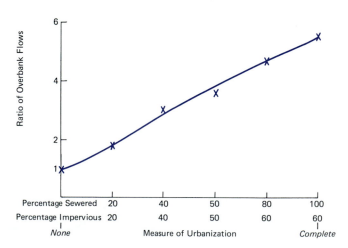

Figure 6.13
Relationship between the ratio of overbank flows (after urbanization to before urbanization) and measure of urbanization. This figure shows that as the degree of urbanization increases, the number of overbank flows per year also increases. (From L. B. Leopold, U.S. Geological Survey Circular 559, 1968.)

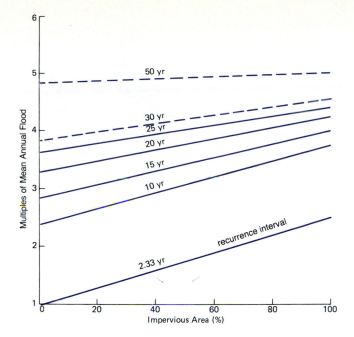

Figure 6.14
Graph showing the variation of flood frequency with percentage of impervious area. The mean annual flood is the average (over a period of years) of the largest flow that occurs each year. The mean annual flood in a natural river basin with no urbanization has a recurrence interval of 2.33 years. Note that the smaller floods with recurrence intervals of just a few years are much more affected by urbanization than the larger floods. The 50-year flood is little affected by the amount of area that is rendered impervious. (From L.A. Martens, U.S. Geological Survey Water Supply Paper 1591C, 1968.)

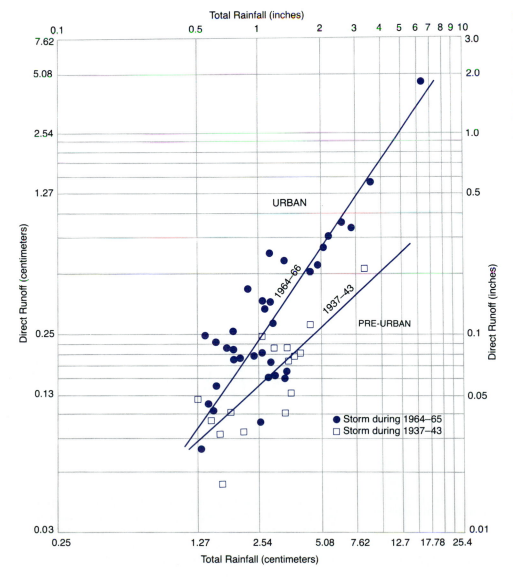

Figure 6.15
Comparison of the rainfall–runoff relationships for preurban and urban conditions. Data are for individual storms in one subarea of Nassau County, New York. (After G. E. Seaburn, U.S. Geological Survey Professional Paper 627B, 1969.)

Figure 6.16

Flood frequency curve for a 2.6-square-kilometer (one-square-mile) basin in various states of urbanization. Note 100–60 means basin is 100 percent sewered and 60 percent of surface area is impervious. Dashed line shows increase in mean annual flood with increasing urbanization. (After L. B. Leopold, U.S. Geological Survey Circular 559, 1968.)

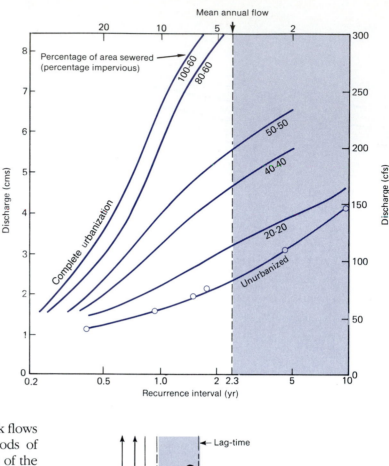

expect to have about three times as many overbank flows as before urbanization. This ratio holds for floods of small and intermediate frequency, but as the size of the drainage basin increases, floods of high magnitude with frequencies of 50 years or so are not much affected by urbanization (Figure 6.14).

Floods are a function of rainfall-runoff relations, and urbanization causes a tremendous number of changes in these relations. One study showed that urban runoff is 1.1 to 4.6 times preurban runoff (7). (See Figure 6.15.) The greatest differences correspond to larger storms. Estimates of discharge for different recurrence intervals at different degrees of urbanization are shown in Figure 6.16. The estimates dramatically indicate the tremendous increase of runoff with increasing impervious areas and storm sewer coverage.

The increase of runoff with urbanization occurs because less water infiltrates the ground, as suggested by the significant reduction in time between the majority of rainfall and the flood peak (lag-time) for urban versus rural conditions (Figure 6.17). Short lag-times, referred to as *flashy discharge,* are characterized by rapid rise and fall of flood water. Since little water infiltrates the soil, the low water or dry season flow in urban streams, sustained by groundwater seepage into the channel, is greatly reduced. This effectively concentrates any pollutants present and generally lowers the aesthetic amenities of the stream (6).

Urbanization generally increases runoff and thus flooding; however, in specific instances, relationships

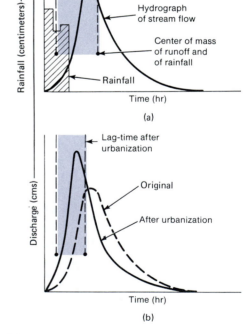

Figure 6.17

Generalized hydrographs. Hydrograph (a) shows the typical lag between the time when most of the rainfall occurs and the time when the stream floods. Hydrograph (b) shows the decrease in lag-time because of urbanization. (After L. B. Leopold, U.S. Geological Survey Circular 559, 1968.)

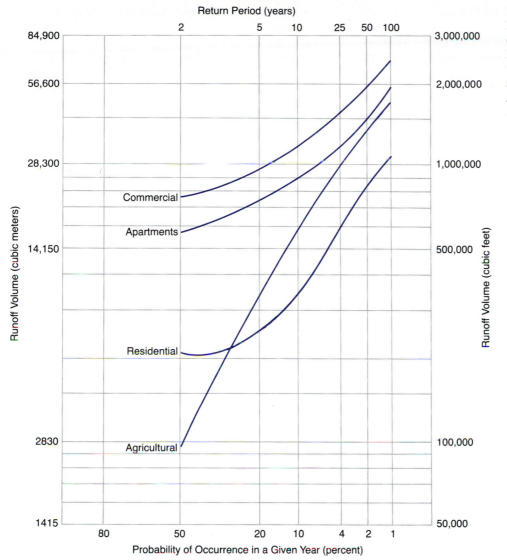

Return Period (years)

Probability of Occurrence in a Given Year (percent)

Figure 6.18
Runoff volume-frequency curves of a small tributary of the Fox River, Illinois, for different land uses. (After Terstriep, et al., *Conventional Urbanization and Its Effect on Storm Runoff,* Illinois State Water Survey Publication, 1976.)

between land use and flooding for small drainage basins may be quite complex. One study concludes that not all types of urbanization increase all runoff and flood events (11). When row crops (corn and soybeans) are replaced by low-density residential development, the predicted runoff and flood peaks for low magnitude events with recurrence intervals of two to four years increase (as expected). For events with recurrence intervals exceeding four years, however, the predicted runoff and flood peaks for the agricultural land may exceed that for residential development (Figure 6.18). The reason for this change in hydrologic character is that, as row crops are replaced by paved areas and grass, the runoff from the paved areas is greater, but the grass produces less runoff than the agricultural land. Therefore, the effect of the land-use change on runoff and flooding depends on the nature and extent of urbanization and in particular on the proportion of paved and grass-covered areas. Thus, for the land uses specified in Table 6.2 and shown in

Figure 6.18, the predicted runoff and flood from a 2-year storm are greater for residential areas than for agricultural land. This results because, for low magnitude (light) rainfall events, the runoff from the small paved areas under residential development is a significant contribution to the total runoff, whereas the runoff from the row

Table 6.2.
Relative cover (pavement or grass) as a function of different types of urbanization for small basins in Illinois.

Type of Urbanization	Paved Area (%)	Grass (%)
Residential	23	77
Apartments	60	40
Commercial	75	25

Source: Data from Terstriep, et al., "Conventional Urbanization and Its Effect on Storm Runoff." Illinois State Water Survey, 1976.

crop areas is low because of infiltration. With higher magnitude (more intense) rainfall events, however, the runoff from agricultural lands increases, and in comparison, the runoff from the small paved areas in residential developments is not very significant. Furthermore, grass areas have higher infiltration rates than do row crop areas. When urbanization consists of apartments or commercial development, the amount of paved area increases rapidly and the grassed areas decrease (Table 6.2); thus, the predicted runoff and floods are larger (at all recurrence intervals) than for agricultural land (Figure 6.18) (11). Therefore we conclude that, because most urbanization (except for very small drainage basins) tends to be a mixture of various types of development, the concept that urbanization increases runoff and flooding remains valid in most situations.

▼ NATURE AND EXTENT OF FLOOD HAZARD

For more than 200 years, Americans have lived and worked on floodplains. Advantages and enticement to do so stem from the abundance of rich alluvial soil found on the floodplain, conveniences such as abundant water supply and ease of waste disposal, and proximity to communications, transportation, and commerce that developed along the rivers. Of course, building houses, industry, public buildings, and farms on the floodplain invites disaster, but floodplain residents have refused to recognize the natural floodway of the river for what it is: part of the natural river system. As a result, flood control and drainage of wetlands became prime concerns. It is not too great an oversimplification to infer that, as the pioneers moved west, they had a rather set procedure for modifying the land: first, clear the land by cutting and burning the trees, then modify the natural drainage.

From that history came two parallel trends: an accelerating program to control floods, matched by an even greater growth of flood damages (12).

Historically, the response to flooding has been to attempt to control the water by constructing dams, modify the stream by constructing levees, or even rebuild the entire stream to more efficiently remove the water from the land. Every new project has the effect of luring more people to the floodplain in the false hope that the flood hazard is no longer significant. However, we have yet to build a dam or channel capable of controlling the heaviest rainwaters, and when the water finally exceeds the capacity of the structure, flooding may be extensive (12).

Factors that control the damage caused by floods include

▼ land use on the floodplain
▼ magnitude (depth and velocity of the water and frequency of flooding)
▼ rate of rise and duration of flooding
▼ season (for example: crops on floodplain)
▼ sediment load deposited
▼ effectiveness of forecasting, warning, and emergency systems.

Effects of the flooding may be *primary,* that is, caused by the flood, or *secondary,* caused by disruption and malfunction of services and systems associated with the flood (11). Primary effects include injury and loss of life, and damage caused by swift currents, debris, and sediment to farms (Figure 6.19), homes, buildings, railroads (Figure 6.20), bridges (Figure 6.21), roads, and communication systems. Erosion and deposition of sediment in the rural and urban landscape may also involve a loss of considerable soil and vegetation (Figure 6.22). Secondary effects may include short-term pollution of

Figure 6.19
Flood-damaged corn near Ottawa, Ohio, in June 1981 after high water caused a dike to fail. (Photo courtesy of U.S. Department of Agriculture.)

Figure 6.20
The 1952 flooding of the Missouri River in western Iowa caused disruption of transportation systems (railroad) and damage to government-stored grain. (Photo by Forsythe, courtesy of U.S. Department of Agriculture.)

Figure 6.21
Collapse of a section of the New York State Thruway bridge over Schoharie Creek, near Amsterdam, April 5, 1987. (Photo by Sid Brown, Schenectady Gazette; AP/Wide World Photos.)

Figure 6.22
Aerial view of flooding from heavy spring rains and overflow from the Obion River in Dyer County, Tennessee, May 1983. (Photo by Tim McCabe, courtesy of U.S. Department of Agriculture.)

rivers, hunger and disease, and displacement of persons who have lost their homes. In addition, fires may be caused by short circuits or broken gas mains (13).

▼ PREVENTIVE AND ADJUSTMENT MEASURES

Preventive and adjustment measures for floods involve different approaches. *Prevention* includes engineering structures and projects such as levees and flood walls to serve as physical barriers against high water; reservoirs to store water for later release at safe rates; on-site storm water detention systems or retention basins (Figure 6.23); channel improvements *(channelization)* to increase channel size and move water off the land quickly; and channel diversions to route flood waters around areas requiring protection. A major problem with a structural approach to flood hazard is that upstream reservoirs, levees, and flood walls tend to produce a false sense of security that leads to floodplain development. *Adjustment* measures include floodplain regulation and flood insurance (13).

From an environmental view, the best approach to minimizing flood damage in urban areas is floodplain regulation. This is not to say that physical barriers, reservoirs, and channel works are not desirable. In areas developed on floodplains, they will be necessary to protect lives and property (Figure 6.24). We need to recognize, however, that the floodplain belongs to the river system, and any encroachment that reduces the cross-sectional area of the floodplain increases flooding (Figure 6.25). Therefore, an ideal solution would be to discontinue development of areas subject to flooding that necessitates new physical barriers. In other words, we must "design with nature." Realistically, in most cases, the most effective and practical solution is a combination of physical barriers and floodplain regulations that will result in less physical modification of the river system.

For example, reasonable floodplain zoning in conjunction with a diversion channel project or upstream reservoir may result in a smaller diversion channel or reservoir than would be necessary if no floodplain regulations were used.

The purpose of floodplain regulation is to obtain the most beneficial use of floodplains while minimizing flood damage and cost of flood protection (14). It is a compromise between indiscriminate use of floodplains, which results in loss of life and tremendous property damage, and complete abandonment of floodplains, which gives up a valuable natural resource. A preliminary step to floodplain regulation is flood-hazard mapping, which is a means of providing floodplain information for land-use planning (15). These maps may delineate past floods or floods of a particular frequency, say, the 100-year flood, and are useful in deriving regulations for private development, purchasing of land for public use as parks and recreational facilities, and providing guidelines for future land use on floodplains.

Developing flood-hazard maps for a particular drainage basin can be difficult and expensive. The maps are generally prepared by analyzing stream flow data from gaging stations over a period of years; however, flow data are not available in many cases, especially for small streams, so alternative data sources must be used to assess the flood hazard. Methods of upstream flood hazard evaluation may involve estimations of flood peak discharges based on physical properties of the drainage basin. A study of streams in central Texas characterized by periodic intense upstream flooding produced a preliminary empirical model to estimate flood peak discharges by measuring the number of source streams *(stream magnitude)* of a drainage basin and the *drainage density* (total length of all streams in a basin divided by the area of the basin) (16). In other words, the Texas work produced a statistically valid relationship between

Figure 6.23

Comparison of runoff from a paved area through a storm drain with runoff from a paved area through a temporary storage site (retention pond). Notice that the paved area drained by way of the retention pond produces a lesser peak discharge and therefore is less likely to contribute to flooding of the stream. (Modified after U.S. Geological Survey Prof. Paper 950.)

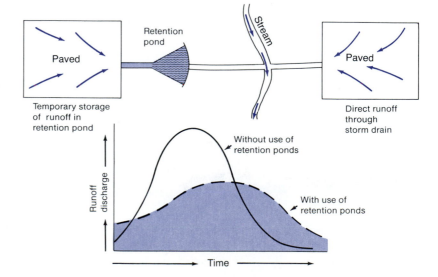

Figure 6.24
Dike on the righthand side of the river protects commercial property from flooding of the Missouri River in 1973. (Photo courtesy of Department of the Army, U.S. Corps of Engineers.)

measured flood peak discharge and two measured physical parameters, stream magnitude and drainage density, that can be used to predict floods in basins where hydrologic information is unavailable or insufficient for detailed evaluation.

Flood hazard evaluation for downstream areas may also be accomplished in a general way by direct observation and measurement of physical parameters. For example, extensive flooding of the Mississippi River

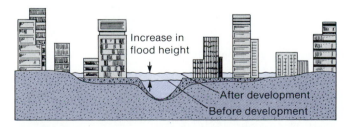

Figure 6.25
Development that encroaches on the floodplain can increase the heights of subsequent floods. (From Water Resources Council, *Regulation of Flood Hazard Areas,* vol. 1, 1971.)

Valley during the spring of 1973 shows clearly on images produced from satellite-collected data (Figure 6.26).

Floods can also be mapped from aerial photographs taken during flood events or can be estimated from high water lines, flood deposits, scour marks, and trapped debris on the floodplain, measured in the field after the water has receded (Figure 6.27).

Careful mapping of soils and vegetation can also help evaluate the downstream flood hazard. Soils on floodplains are often different from upland soils, and, with favorable conditions, certain soils can be correlated with flooding of known frequency. Figure 6.28 shows a fair correlation between soils and the 100-year flood (16). Vegetation type may facilitate flood hazard assessment because there is often a rough zonation of vegetation in river valleys that may correlate with flood zones. Some types of trees with shallow roots require an abundant supply of water and benefit from frequent submergence. These trees are often found near the banks of perennial streams that frequently flood. Other species of trees are more restricted to well-drained soils that are not subjected to prolonged or frequent flooding. While zonation of vegetation is helpful in evaluating flood-prone areas,

Figure 6.26
The March 31, 1973, flood of the Mississippi River as interpreted from satellite (Landsat) images. Point A is the confluence of the Missouri and Mississippi Rivers, and Point B is the confluence of the Illinois and Mississippi Rivers. Notice that, during the flood, the river is still very narrow through St. Louis. This is a result of the city's flood protection works. Points labelled C are large areas inundated by the flood waters, in part resulting from the flood control protection works that constrict the flow at St. Louis, causing upstream ponding of water. (After N. M. Short, et al., *Mission to Earth: Landsat Views the World*, NASA, 1976.)

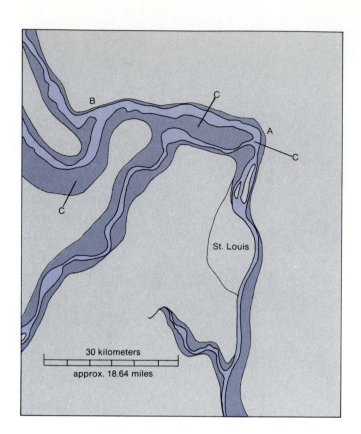

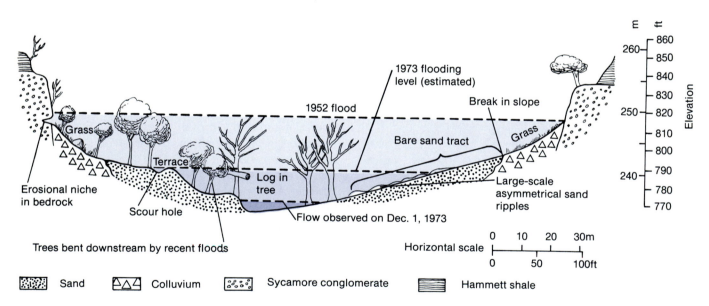

Figure 6.27
Schematic cross section of the Pedernales River Valley, Texas, illustrating floodplain features useful in estimating floods. (After V. R. Baker, "Hydrogeomorphic Methods for the Regional Evaluation of Flood Hazards," *Environmental Geology*, vol. 1, pp. 261–81, 1976.)

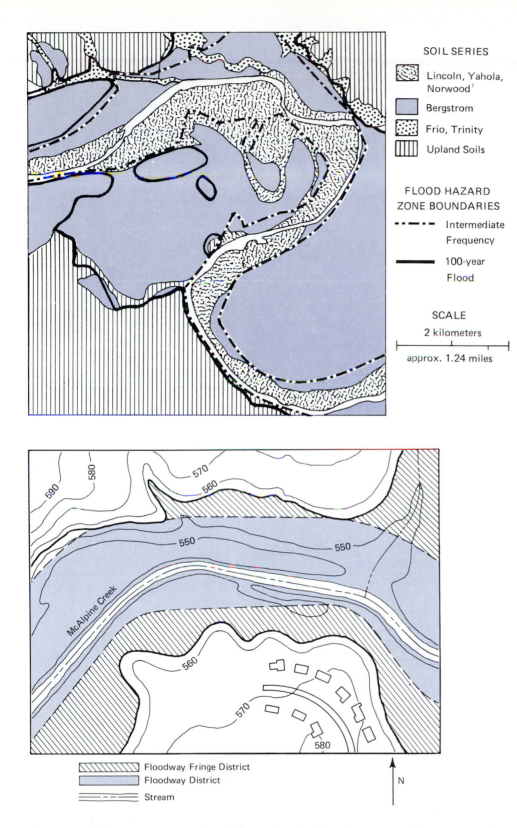

SOIL SERIES

Lincoln, Yahola, Norwood

Bergstrom

Frio, Trinity

Upland Soils

FLOOD HAZARD ZONE BOUNDARIES

— · · — · · Intermediate Frequency

————— 100-year Flood

SCALE

2 kilometers

approx. 1.24 miles

Figure 6.28
Relationship between soils and flooding for a reach of the Colorado River near Austin, Texas. The intermediate frequency floods with a recurrence interval of 10 to 30 years are primarily associated with the Lincoln, Yahola, Norwood, and Bergstrom soils (on the floodplain). The 100-year flood tends to roughly correlate with the lower topographic boundary of the upland soils. (After V. R. Baker, "Hydrogeomorphic Methods for the Regional Evaluation of Flood Hazards," *Environmental Geology*, vol. 1, pp. 261–81, 1976.)

Floodway Fringe District
Floodway District
Stream

N

Figure 6.29
Example of a flood hazard map showing the floodway fringe district and the floodway district. (From County Engineer, Mecklenburg County, North Carolina.) (550 to 580 ft = 168 to 177 m)

the cause of the zones is complex and not directly caused by flooding. Therefore, use of vegetation, as with soils, should be combined with other methods of flood hazard evaluation such as satellite data, aerial photographs, historical records, and floodplain features (16).

The primary advantages of evaluating both upstream and downstream flood hazards from direct observation or properties of the drainage basin and river valley are expediency and cost. A recent study of the Colorado River Valley near Austin, Texas, showed that these methods

Figure 6.30
Idealized diagram showing a valley cross section and its flood hazard area, floodway district, and floodway fringe district. (From Water Resources Council, *Regulation of Flood Hazard Areas,* vol. 1, 1971.)

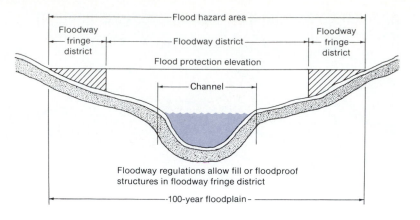

could easily distinguish areas with a frequent flood hazard (1- to 4-year recurrence interval) from areas with an intermediate (10- to 30-year) or infrequent (greater than 100-year) hazard (16). The only disadvantage is that of relative accuracy. Flood hazard evaluation based on sufficient hydrologic (stream flow) data generally provides more accurate prediction of flood events.

In urbanizing areas, the accuracy of flood-hazard mapping based entirely on stream flow data is questionable. A better map may be produced by assuming projected urban conditions with estimated percentage of impervious areas. A theoretical 100-year flood map can then be produced. The U.S. Geological Survey in Charlotte, North Carolina, is producing such maps for use with the Mecklenburg County Floodway Regulations (Figure 6.29, p. 99). Two districts are mapped in the flood hazard area, the floodway district and the floodway fringe district (Figure 6.30).

The **floodway district** is that portion of the channel and floodplain of a stream designated to provide passage of the 100-year regulatory flood without increasing the elevation of the flood by more than 0.3 meters. On this land, permitted uses include farming, pasture, outdoor plant nurseries, wildlife sanctuaries and game farms; loading areas, parking areas, rotary aircraft ports and similar uses, provided they are further than 8 meters from the stream channel; golf courses, tennis courts, archery ranges, and hiking and riding trails; streets, bridges, overhead utility lines, and storm drainage facilities; temporary facilities for certain specified activities such as circuses, carnivals, and plays; boat docks, ramps, and piers; and dams, if they are constructed in accordance with approved specifications. Other uses of the floodway district, such as open storage of equipment or material, structures designed for human habitation, or underground storage of fuel or flammable liquids, require special permits.

The second district is the **floodway fringe district,** which consists of the land located between the floodway district and the maximum elevation subject to flooding by the 100-year regulatory flood. Permitted uses in this area include any uses permitted in the floodway district;

residential accessory structures, provided they are firmly anchored to prevent flotation; fill material that is graded to a minimum of 1 percent grade and protected against erosion; and structural foundations if firmly anchored to prevent flotation. Aboveground storage or processing of any material that is flammable or explosive or that could

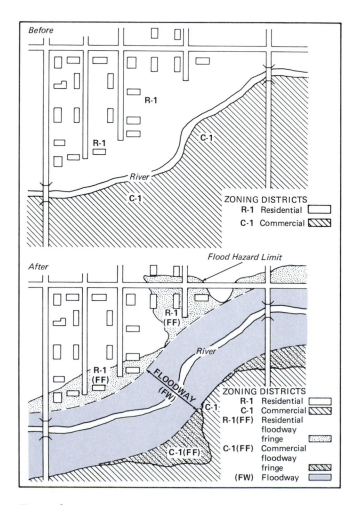

Figure 6.31
Typical zoning map before and after the addition of flood regulations. (From Water Resources Council, *Regulation of Flood Hazard Areas,* vol. 1, 1971.)

cause injury to human, animal, or plant life in time of flooding is prohibited on the floodway fringe district.

Figure 6.31 shows a typical zoning map before and after establishment of floodplain regulations.

▼ FLOODS OF THE 1970S AND 1980S

Floods recorded in the United States from 1972 through 1990 took hundreds of lives and caused several billion dollars of property damage. Floodplain residents will certainly long remember these years both for the high and dangerous waters and for the incalculable suffering of the survivors. Two floods in June of 1972 rank among the worst flood disasters in the history of the United States (4). The first occurred on June 9 and 10 in the Black Hills area of South Dakota, and the second occurred in the eastern United States from June 21 through 24.

The South Dakota event claimed 242 lives in a catastrophic flash flood. Damages were estimated at $163 million. Most of the deaths and property losses were along Rapid Creek near Rapid City, South Dakota. The flood resulted from torrential rains that struck the Black Hills, delivering, at places, 38 centimeters of rain in five hours. Total rainfall from the 7-hour storm nearly equaled the yearly average. A storm of this magnitude is likely to strike only every several thousand years (4). Following the flood, the people of Rapid City initiated a floodplain management program in which the floodway was delineated and all homes, motels, and most industrial facilities were removed from the floodplain. Today the floodplain is a green belt extensively used for recreational activities (17). As a result, the inhabitants need not fear another flood. The previous floodplain development at Rapid City had been encouraged due to an upstream reservoir that was thought capable of controlling floods. The dam provided a false sense of security (as many dams do), since the 1972 floodwaters resulted from precipitation that fell on drainage areas downstream from the reservoir.

The 1972 floods in the eastern United States were produced by tremendous rainstorms associated with Hurricane Agnes. In terms of property damage (exceeding $3 billion), this was the most devastating flood ever to strike this country. One hundred thirteen persons lost their lives. Torrential rains and floods were recorded everywhere east of the Appalachians from North Carolina to New York State (4).

Total loss of property from flood damage in 1973 was second only to that in 1972. The largest flood of the year was the spring flooding of the Mississippi River (Figure 6.26), which resulted in approximately $1.2 billion in property damage. Although tens of thousands of persons had to be evacuated and suffered appreciably as thousands of square kilometers of farmland were inundated throughout the Mississippi River Valley, fortunately,

few deaths were caused directly by the flooding (18). The 1973 flooding of the Mississippi River occurred despite a tremendous investment in flood-control dams upstream on the Missouri River. Reservoirs behind these dams inundated some of the most valuable farmland in the Dakotas, and in spite of these structures the downstream floods near St. Louis were of record-breaking magnitude (18).

Violent flash floods, nourished by a complex system of thunderstorms delivering up to 25 centimeters of rain, swept through several canyons west of Loveland, Colorado, on the night of July 31–August 1, 1976, killing 139 people and inflicting over $35 million in damages to highways, roads, bridges, homes, and small businesses. Most of the damage and loss of life was in the Big Thompson Canyon, where hundreds of residents, campers, and tourists were caught in the flood with little or no warning. Although the storm and flood were rare events (several times the magnitude of the 100-year flood), comparable floods have occurred in the past and others can be expected in the future for similar canyons along the Front Range of Colorado. This prediction is possible because many streams draining the Front Range, including Boulder Creek with substantial urban development in Boulder, Colorado, contain geologically very young flood deposits analogous to those transported and deposited by the Big Thompson Canyon flood (19, 20, 21).

The September 1983 floods in Arizona killed at least 13 people and inflicted more than $416 million in damages to homes (over 1,300 destroyed or damaged), highways, roads, and bridges. Interestingly, officials complained that donated food, clothing, and money were so abundant that at times they actually made disaster relief efforts more difficult. Disaster planners should learn to live with such difficulties—congratulations are in order to Arizona's "good neighbor policy."

Important questions were raised by the floods in 1983 at Tucson, Arizona. During the flood event, extensive bank erosion occurred as banks receded at active meander bends and other sites. Figures 6.32 and 6.33 show typical damage resulting from bank erosion on the Rillito, a tributary of Santa Cruz River in Tucson. The damage clearly suggested that those responsible for flood planning were not prepared for such massive bank erosion. Rather than adopting an overall river management plan to retard erosion, Tucson had piecemeal bank protection that generated greater channel instability than would have occurred without the structures. That is, although existing bank protection structures protected short reaches of the channel, they enhanced the bank erosion in the unprotected reaches immediately downstream (22).

The winter of 1986 brought tremendous storms and flooding to the western states, particularly California, Nevada, and Utah. In all, damages exceeded $270 million, and at least 17 people died (23). During one of the

Figure 6.32
Tucson flood damage on the Rillito in 1983. Flow is from bottom to top in the photograph. Note damage to townhouses (center) due to lateral migration of the upstream meander bend that eroded behind and through bank protection (soil cement) emplaced to protect the houses. (Photograph courtesy of V. R. Baker.)

storms and floods, a levee broke on the Yuba River in California, causing more than 20,000 people to flee their homes. An important lesson learned during this flood is that many of the levees constructed many years ago along California's and other states' rivers are in poor condition and subject to future failure. Construction of levees encourages overconfidence. No matter how large the levees or dams we construct, there will always be high magnitude storms likely to cause flooding. When construction of homes and businesses on floodplains is stimulated by the false sense of security provided by levees and dams, then flood losses may eventually be all the greater. Thus, overconfidence in levees and dams is all too often a key factor in extensive flood damage. Unfortunately, the potential benefits of flood control measures from engineering structures are often lost because of development on floodplains supposedly

protected by these structures. Structural control for floods must go hand in hand with floodplain regulations if the hazard is to be truly minimized (17).

1990 Flash Floods in Eastern Ohio

On Friday, June 15, 1990, over 14 centimeters of precipitation fell in approximately three and a half hours in some areas of eastern Ohio. Two tributaries of the Ohio River, Wegee and Pipe Creeks, generated flash floods near the small town of Shadyside that killed 21 people and left 13 missing and presumed dead. The floods were described as walls of water as high as 5 meters that rushed through the valley. In all, approximately 70 houses were destroyed and another 40 were damaged. Trailers and houses were seen washing down the creeks, bobbing like corks in the torrent.

The waves or walls of water that rushed down the creek were apparently related to failure of debris dams that developed upstream of bridges across the creeks. That is, the runoff from rainfall washed debris into the channel from sideslopes and the debris (tree trunks and other material) became lodged upstream of bridges. When the bridges could no longer contain the weight of the debris, the dams broke loose, sending surges of water downstream. This scenario of debris dams forming upstream of bridges has been played and replayed in many flash floods around the world. The problem is that in many places we simply have not built the bridges large enough to allow debris to be swept downstream. All too often, the supports for bridges are not spaced sufficiently far enough apart to allow the passage of large debris, and as a result it often piles up against the upstream side of the bridge during a flood event.

Flooding in Las Vegas, Nevada: A Case History

Las Vegas, Nevada, has a flooding problem that dates back to the early 1900s when people began developing the area. The city is vulnerable to flash floods that generally occur in July and August in response to high magnitude thunderstorm activity. Las Vegas is surrounded by mountains that drain to alluvial fans with channels known as *washes* (Figure 6.34). The largest of these is the Las Vegas Wash, which drains from the northwest through Las Vegas to Lake Mead on the Colorado River. Other washes, most notably from the west, join the Las Vegas Wash in the city. Of these, the Flamingo Wash has a notorious flood history. Damage from flooding is severe because in recent years several major developments have completely obstructed the natural washes. This construction has unfortunately coincided with increased thunderstorm activity from 1975 through the mid 1980s.

A flood that came down the Flamingo Wash in July of 1975 caused particular concern. At Caesar's Palace (a prominent casino), hundreds of cars in a parking lot that

Figure 6.33
Bridge damaged by the September 1983 floods in Tucson, Arizona. The channel of the Rillito enlarged, eroding the north abutment of the bridge. (Photograph courtesy of V. R. Baker.)

covered part of the Flamingo Wash were damaged, many of them moved downstream by floodwaters. Damage to vehicles and from lateral bank erosion along the wash was reported to be as high as $5 million. Following the flood, the Flamingo Wash was routed by way of a tunnel

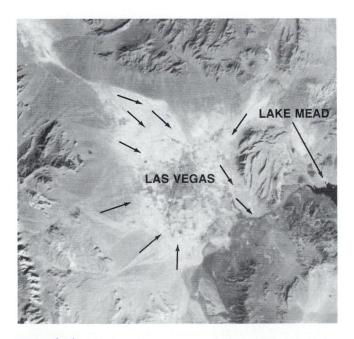

Figure 6.34
Aerial view of the Las Vegas, Nevada, area. The city is in a natural basin and many structures have been built across natural drainage channels (washes). Arrows show the direction of flow into the urban area and out to Lake Mead. (Image courtesy of Map and Image Library, University of California.)

beneath Caesar's Palace. Where the tunnel emerges, the wash flows at the surface again and often down city streets! Flooding again occurred in 1983 and 1984 in the months of July and August. In particular, eight major storms from July through September 1984 resulted in tens of millions of dollars of damage in Clark County, Nevada.

An interesting aspect of the flood hazard in the Las Vegas area is that a small area of a particular drainage basin may generate a large volume of runoff in a relatively short time. For example, in 1983 a small area of approximately 12 square kilometers of the Flamingo Wash was the center of a thunderstorm. Rainfall in amounts of approximately 10 cm over a three-hour period produced a runoff of approximately 66 cubic meters per second (cms). The runoff was produced from alluvial surfaces just upstream from the Las Vegas area rather than in the surrounding mountains. The reasons for the high runoff are two-fold: first, the high intensity storm favored rapid runoff; second, soils developed on the alluvial surfaces are characterized by slow surface infiltration with resulting high runoff potential. The reason for the slow infiltration rate is related to the development of calcic soils horizons at shallow depths. These horizons are characterized by development of calcium-carbonate-cemented hardpan layers that retard infiltration of water. These hardpan layers often are less than a few cm from the surface. In soils description, these horizons are designated as *K* horizons. In some instances, these calcium carbonate horizons are several meters thick.

Las Vegas is located in a basin at the foot of alluvial fan surfaces. In many instances, alluvial fans are very porous and surface runoff infiltrates rapidly. In the Las Vegas area, however, the calcium-carbonate-rich soil horizons cement alluvial deposits together and retard surface infiltration. As a result, the Las Vegas area is more susceptible to flood hazard than otherwise might be expected. Nonetheless, whenever urban development is built directly across active washes, trouble can be expected.

Clark County, Nevada, is currently developing a master plan that addresses the flood hazard in Las Vegas. If fully implemented the plan would cost nearly $1 billion; it is designed to protect the area from the 100-year peak flow resulting from a three-hour thunderstorm. Of particular importance in the master plan are alternatives consisting of a mix of activities: first, construction of channels, pipelines, and culverts to convey floodwaters away from developed areas; second, construction of storm-water detention basins that will release the flow over a longer period of time; and third, a combination of the first two approaches. Phase 1 of the master plan will cost just under $500 million and should help alleviate the flood problem in Las Vegas and other parts of Clark County. Because there has been so much development on the washes in Las Vegas, however, it seems unlikely that any master plan is likely to completely alleviate the flood problem. That would require spending considerably more money to completely channel the washes and route the floodwaters around all the existing development. But even this precaution would not free the Las Vegas area from flood hazard resulting from those larger storms that will occasionally occur. Implementation of a flood management plan, along with land-use planning that discourages development on natural washes, will certainly help alleviate their flood problems. However, it is a gamble as to how much flood-control measures can be expected to help, given the rapid development that is occurring in Las Vegas.

▼ PERCEPTION OF FLOODING

At the institutional level, perception of flooding appears to be adequately realistic; however, on the individual level, the situation is not as clear. Several conclusions have been drawn from perception studies (3):

▼ Accurate knowledge of flood hazard does not inhibit all persons from moving onto a floodplain.

▼ Flood hazard maps are not effective as a form of communication.

▼ Upstream development is frequently the scapegoat for downstream floodplain residents threatened by floods.

▼ People are tremendously variable in their knowledge of flooding, anticipation of future flooding, and willingness to accept adjustments caused by the hazard.

Recent progress at the institutional level includes mapping of flood-prone areas (thousands of maps have been prepared), areas with a flash flood potential downstream from dams, and areas where urbanization is likely to cause problems in the near future. In addition, the U.S. Flood Insurance Administration program has encouraged states and local communities to adopt floodplain management plans and has sponsored preparation of flood insurance plans in more than 6,000 of the country's nearly 21,000 communities with a history of flood problems (24).

▼ CHANNELIZATION: THE CONTROVERSY

Channelization of streams consists of straightening, deepening, widening, clearing, or lining existing stream channels. Basically, it is an engineering technique, with the objectives of controlling floods, draining wetlands, controlling erosion, and improving navigation (12). Of the four objectives, flood control and drainage improvement are the two most often cited in channel improvement projects.

Thousands of kilometers of streams in the United States have been modified, and more thousands of kilometers of channelization are being planned or constructed. Federal agencies alone carried out several thousand kilometers of channel modification in the 1950s and 1960s (25). Considering that there are approximately 5.6 million kilometers of streams in the United States, and that the average yearly rate of channel modification is only a very small fraction of one percent of the total, the impact may appear small. Nevertheless, nearly 1,600 kilometers of channel per year is a significant amount of environmental disruption and, therefore, merits evaluation.

Channelization is not necessarily a bad practice; however, inadequate consideration has been given to its adverse environmental effects. The picture emerging is that not enough is known about these effects, and that too little is being done to evaluate them (12).

Adverse Effects of Channelization

Opponents of channelizing natural streams emphasize that the practice is completely antithetical to the production of fish and wetland wildlife, and that, furthermore, the stream suffers from extensive aesthetic degradation. The argument is as follows: Drainage of wetlands adversely affects plants and animals by eliminating habitats necessary for the survival of certain species. Cutting trees along the stream eliminates shading and cover for fish, while exposing the stream to the sun, which results in damage to plant life and heat-sensitive aquatic organisms. Cutting of bottomland (floodplain) hardwood trees eliminates the habitats of many animals and birds and also facilitates erosion and siltation of the stream. Straightening and modifying the streambed destroys the diversity of flow patterns, changes peak flow, and destroys feeding and breeding areas for aquatic life. Finally, the conversion of wetlands from a meandering stream to a straight, open ditch seriously degrades the aesthetic value of a natural area (12). Figure 6.35 summarizes some of the differences between natural streams and those modified by people.

It is commonly believed that channelization increases the flood hazard downstream from the modified channel. Although in many cases this is true, it is far from the entire story. One study concluded that, on the contrary, increases of normal and peak flows from channelization are not particularly significant, and in some cases the flood peaks are actually reduced (25).

This results partly because the contribution of runoff from the modified channels tends to be a small fraction of the total basin runoff; peak runoff from channelized streams may not coincide with basinwide runoff, and thus the quicker flow from a modified stream may pass prior to the natural flood flow from the entire basin, thereby reducing the normal aggregated peak flow; and many streams that are modified have gradients so low that no amount of straightening could significantly increase the downstream flow. However, we emphasize that channel modification, especially straightening, can increase flooding directly downstream from the project, and the problem may be compounded if there are a number of projects in the same basin.

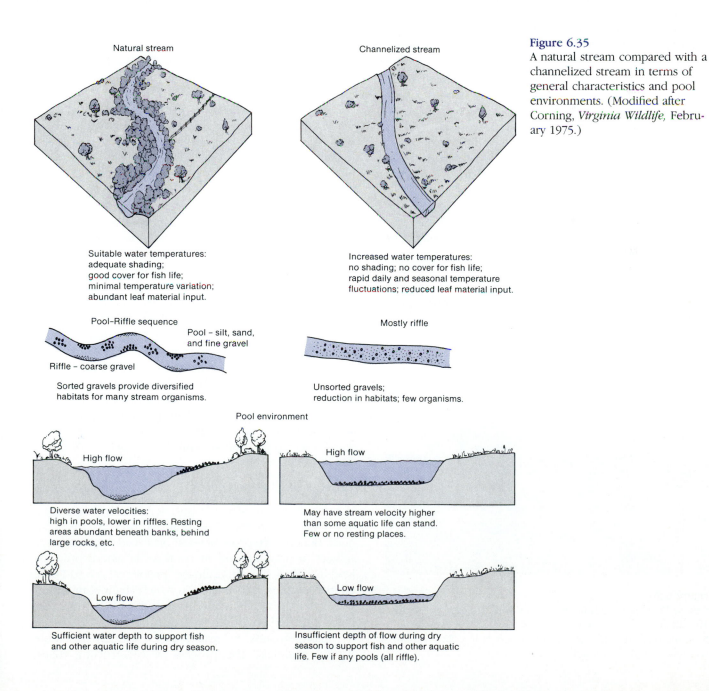

Figure 6.35
A natural stream compared with a channelized stream in terms of general characteristics and pool environments. (Modified after Corning, *Virginia Wildlife,* February 1975.)

Natural stream

Channelized stream

Suitable water temperatures: adequate shading; good cover for fish life; minimal temperature variation; abundant leaf material input.

Increased water temperatures: no shading; no cover for fish life; rapid daily and seasonal temperature fluctuations; reduced leaf material input.

Pool–Riffle sequence

Pool – silt, sand, and fine gravel

Riffle – coarse gravel

Sorted gravels provide diversified habitats for many stream organisms.

Mostly riffle

Unsorted gravels; reduction in habitats; few organisms.

Pool environment

High flow

High flow

Diverse water velocities: high in pools, lower in riffles. Resting areas abundant beneath banks, behind large rocks, etc.

May have stream velocity higher than some aquatic life can stand. Few or no resting places.

Low flow

Low flow

Sufficient water depth to support fish and other aquatic life during dry season.

Insufficient depth of flow during dry season to support fish and other aquatic life. Few if any pools (all riffle).

Examples of channel work projects that have adversely affected the environment are well known. The Willow River in Iowa and Blackwater River in Missouri illustrate some of the possible adverse impacts on streams that are channelized.

The channelization of the Willow River in Iowa has been extensively studied (26, 27). As early as 1851, the river was known for frequent flooding that caused extensive damage to crops. As a result, the valley was generally considered too risky for agriculture. At that time, the river was confined at low flow to a meandering channel about 1 meter below the floodplain. To alleviate flooding, channelization in three stages from 1906 to 1920 converted about 45 kilometers of the Willow River into a straight drainage ditch. This had the effect of producing a much shorter channel with a steeper gradient. Changes in the size and shape of the drainage ditch in the Willow River valley since construction ended in 1920 show that there has been progressive widening and deepening of the channel. In 1958, the channel was characteristically U-shaped with nearly vertical walls (Figure 6.36). In 1920, the ditch in this vicinity was only about 5 meters deep and less than 15 meters wide. In 1958, it was over 11 meters deep and 30 meters wide; and this excessive bank erosion is not unique along the ditch. The history is one of continuous enlargement since the project was completed.

The channelization designed for the Willow River was inadequate and resulted in excessive channel erosion. The drainage ditch response is a classic example of adverse channel change caused by artificial reshaping of a natural channel (27).

Channelization of the Blackwater River in Missouri has also resulted in environmental degradation, including channel enlargement, reduced biological productiv-

ity, and increase in downstream flooding (28). The natural meandering channel was dredged and shortened in 1910, nearly doubling the natural gradient. This change evidently initiated a cycle of channel erosion that is still in progress, and the channel has shown no tendency to resume meandering (28). Maximum increase in channel cross-sectional area exceeds one thousand percent, and, as a result, many bridges have collapsed and have had to be replaced. One particular bridge over the Blackwater collapsed because of bank erosion in 1930 and was replaced by a new bridge 27 meters wide. This bridge had to be replaced in 1942 and again in 1947. The 1947 bridge was 70 meters wide, but it too collapsed from bank erosion (28).

Biological productivity in the Blackwater River was reduced because the channel was excavated in easily erodible shale, sandstone, and limestone, producing a smooth streambed that does not easily support bottom fauna after it has been disturbed. As a result, the channelized reaches contain far fewer fish than the natural reaches (28).

The downstream end of the channelization project of the Blackwater terminates where the rock type changes to a more resistant limestone, making dredging financially unfeasible. Since the channelized portion has a larger channel, it can carry more water without flooding than the natural, downstream channel. As a result, when heavy rains fall, runoff carried by the upstream channelization section is discharged into the unchannelized downstream section; and because the channel cross section is much less there, frequent flooding results (28).

The Blackwater River situation is a good example of possible trade-offs and unforeseen degradations caused by channelization. The upper reaches benefited from the use of more floodplain land; but this benefit must be weighed against loss of farmland to erosion, cost of bridge repair, loss of biological life in the stream, and downstream flooding. A reasonable trade-off might have been to straighten the channel less, still providing a measure of flood protection without causing such rapid environmental degradation. The question is, how much straightening can be done before inducing unacceptable damage to the riverine system?

Benefits of Channelization

Not all channelization causes serious environmental degradation; in many cases, drainage projects are beneficial. Benefits are probably best observed in urban areas subject to flooding and in rural areas where previous land use has caused drainage problems. In addition, other examples can be cited in which channel modification has improved navigation or reduced flooding and has not caused environmental disruption.

Many streams in urban areas scarcely resemble natural channels. The process of constructing roads,

Figure 6.36
Willow Drainage Ditch near the mouth of Thompson Creek. The ditch here is about 11 meters deep and 30 meters wide. Note incipient development of meanders at low flow. (Photo courtesy of R. B. Daniels.)

utilities, and buildings with the associated sediment production is sufficient to disrupt most small streams. Channel restoration is often needed to clean urban waste from the channel, allowing the stream to flow freely, to protect the banks by not removing existing trees and, where necessary, planting additional trees and other vegetation. The channel should be made as attractive as possible by allowing the stream to meander and, where possible, by providing for variable, low water flow conditions—fast and shallow (**riffle**) alternating with slow and deep (**pool**). Where lateral bank erosion must be absolutely controlled, the outsides of bends may be defended with large stones known as **riprap** (Figure 6.37).

An attempt to improve the aesthetic appeal of an open cut ditch—Whiskey Creek near Grand Rapids, Michigan—emphasizes that engineers are beginning to consider channel restoration seriously. The idea was to make the open drain look like a country stream meandering through substantial residential and commercial properties. Several of the channel's meander bends are shown during construction in Figure 6.38. Although it may be argued that open cut drains are unlikely to be considered pleasing sights under any circumstances,

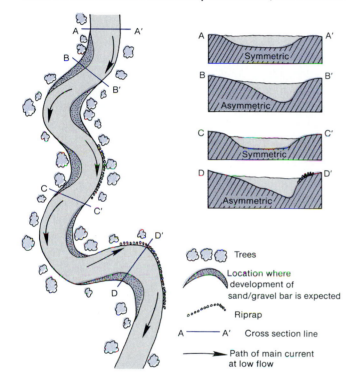

Figure 6.37
Channel-restoration design criteria for urban streams, using a variable cross-channel profile to induce scour and deposition at desired locations. (Modified after Keller and Hoffman, 1977, *Journal of Soil and Water Conservation,* vol. 32, no. 5.)

Figure 6.38
Channelization of Whiskey Creek near Grand Rapids, Michigan. This is an attempt to improve the aesthetic appeal of a stream project by constructing a series of meandering bends. (Photo courtesy of Williams and Works.)

certainly the Whiskey Creek project was something out of the ordinary and is an improvement over a straight ditch designed only for flood control.

Another example of channel restoration is found in the coastal plain of North Carolina where many streams have been channelized. As part of a mosquito-abatement program in Onslow County, the state is attempting to drain some areas and thus destroy the insects' habitat. Examination of some of the streams before improvement reveals that they are nearly blocked by debris and logs. Evidence such as buried, hand-cut cypress logs and handmade cypress shingles along the stream bank suggests that streams that were once free-flowing have been choked with debris left from logging practices years ago. The channel restoration involves excavating the old prelogging channel. Engineers on the project say the old channel can be traced by the high organic soils that form along the channel bank. The resulting restoration is a meandering stream that is allowed to flow freely again. The material dredged in restoring the channel is left on the bank and quickly supports new vegetation. Small cypress seedlings are present only a few months after the work has been completed. Although this particular practice of channel work needs further study and evaluation, it seems to be a step in the right direction. It also appears to be popular with landowners who request that the work be done rather than having to be convinced that it should be, as is often the case with drainage work.

River restoration of the Kissimmee River in Florida may become the most ambitious restoration project ever attempted in the United States. Channelization of the river began in 1962; after nine years of construction at a cost of $24 million, the meandering river had been turned into an 83-kilometer-long straight ditch. Now, at what will probably exceed the original cost of channelization, the river may be returned to its original sinuous path. Work will begin on a 16 km experimental stretch of river, at a cost of about $1 million; if successful, restoration may be undertaken in other sections of the river. The restoration is being seriously considered because the channelization failed to provide the expected flood protection, degraded valuable wildlife habitat, contributed to water quality problems associated with land drainage, and caused aesthetic degradation. In Los Angeles, California, a group called "Friends of the River" has suggested that the Los Angeles river be restored. This will be a difficult, if not impossible, task as most of the river bed and banks are concrete lined.

Channel modification on the Missouri River is an interesting example of river training in which the power of running water is used to maintain the desired channel. The project involves engineering improvements to produce a series of relatively narrow, gentle meanders rather than the previously existing succession of wide, straight reaches with a steep gradient and abrupt changes in flow

direction. That is, before modification, the channel was semibraided in places and contained numerous islands and bars that formed as a result of deposition of large amounts of sediment. This was especially true downstream from the point where the Platte River enters the Missouri River (27).

The engineering project that was started in the 1930s to induce the Missouri River to narrow and meander rather than braid involves construction of permeable pile dikes (Figure 6.39) that reduce the stream's velocity, causing deposition of sediment in desired locations. The water passing around the dikes, on the other hand, is constricted, causing an increase in velocity and scour in desired areas. As erosion and deposition progress, the dikes silt in, and natural growth of willows is established. However, the dikes have caused a reduction in aquatic habitat by filling in sloughs and marshland downstream from the dikes (Figure 6.40). As a result, the Corps of Engineers has designed a new type of dike, with a notch at the top, that will allow more water to pass through and, it is hoped, will minimize the loss of valuable wildlife and fish habitat. Furthermore, there is concern that the river training has reduced the total channel capacity and is thus facilitating more frequent flooding.

As a final example of river control, consider the tremendous undertaking by the Corps of Engineers to keep the lower Mississippi River from shifting its course by New Orleans to a shorter route along the Atchafalaya River, approximately 180 km to the west. The Atchafalaya River via the Red River provides a shorter path to the Gulf

Figure 6.39
Upstream view of the Treville Bend of the Missouri River. The permeable pile dikes on the inside of the bend effectively converge the flow of water toward the outside, where the faster water scours the channel. Thus, the power of running water is used to maintain the channel. (Photo courtesy of U.S. Army Corps of Engineers, Omaha District.)

Figure 6.40
Upstream view of Browers Bend on the Missouri River. Notice the Iowa Public Service power plant on the right bank. A new design of permeable pile dikes is now under experimentation. It will allow water to flow through notches in the dikes, thus causing less filling-in of aquatic habitat (marsh and slough), shown on the left bank behind the dikes. The dikes shown here are of the conventional type, and filling of the marsh is clearly reducing wildlife and aquatic habitat. (Photo courtesy of U.S. Army Corps of Engineers, Omaha District.)

of Mexico, and if natural processes were left alone the Mississippi would shift, leaving Baton Rouge and New Orleans along an abandoned and nearly dry channel. Economic and social costs of such a shift are enormous and so the Corps has spent several hundred million dollars in damlike river control structures to keep the Mississippi where it is. Because the potential new path is much shorter and steeper it may be only a matter of time before the shift takes place, regardless of what structures are built. This view is shared by some geologists, but does not reflect the opinion of the Corps of Engineers. The river nearly shifted during the 1973 flood when major damage to one of the control structures occurred. Larger floods are likely, and because the two rivers are close together, the capture of the Mississippi River by a shorter route to the Gulf seems very probable, even if the structures don't fail. That is, overflow and channel cutting could occur at other locations, causing the shift (29).

The Future of Channelization

Although considerable controversy and justifiable anxieties surround channelization, it is expected that channel modification will remain a necessity as long as land-use changes, such as urbanization or conversion of wetlands to farmland, continue. Therefore, we must strive to design channels that reduce adverse effects. Effective channelization can be accomplished if project objectives of flood control or drainage improvement are carefully tailored to specific needs.

If the primary objective is drainage improvement in areas where natural flooding is not a hazard, then there is no need to convert a meandering stream into a straight ditch. Rather, design might involve cleaning the channel

and maintaining a sinuous stream. In addition, for gravel-bed streams with a low gradient, a series of relatively deep areas (pools) and shallow areas (riffles) might be constructed by changing cross-sectional shape, asymmetrically to form pools and symmetrically to form riffles, as found in natural streams. Where this is possible, the resulting channel would tend to duplicate nature rather than be alien to it. In addition, cutting trees along the channel bank would be minimized and new growth would be encouraged. This plan would tend to minimize adverse effects by producing a channel that is more biologically productive and aesthetically pleasing.

Channel design for flood control is more complicated because natural channels are maintained by flows with a recurrence interval of one to two years, whereas flood-control projects may need to carry the 25- or even 100-year flood. A 100-year channel cannot be expected to be maintained by a 2-year flood. Such a channel would probably be braided and choked with migrating sandbars. Therefore, it is suggested that a pilot channel be constructed where possible. A meandering channel designed to be maintained by the 2-year flood is superimposed on the larger flood-control channel. Addition of pools and riffles in the pilot channel, when possible, would help provide fish habitat and better low flow conditions. The large channel might be vegetated, and the untrained observer might not easily recognize it. Obviously this plan will not work in urbanizing areas where sediment production is high and property is not available to be purchased for the larger channel. However, if sediment is reduced by good conservation practice and the property is available to be purchased, the urban area will benefit from a more aesthetically pleasing and more useful stream.

▼ ▼ ▼ SUMMARY AND CONCLUSIONS

Rivers are one of the basic transport systems of the rock cycle, and generally maintain a dynamic equilibrium between the work done and the load imposed. River flooding is the most universally experienced natural hazard. Loss of life is relatively low in developed countries, with adequate monitoring and warning systems, compared to that in preindustrial societies; but property damage is much greater in industrial societies because floodplains are often extensively developed.

The magnitude and frequency of flooding are a function of the intensity and distribution of precipitation and the rate of infiltration of water into the soil, rock, and topography. Human use and interest in urban areas have increased the flood hazard in small drainage basins by increasing the amount of ground covered with buildings, roads, parking lots, and so on, which force storm water to run off the land rather than infiltrate it.

Two types of floods are recognized: *upstream* floods produced by intense rainfall of short duration over a relatively small area; and *downstream* floods produced by storms of long duration over a large area that saturate the soil, causing increased runoff from thousands of tributary basins and resulting in large floods in the major rivers.

Factors that control damage caused by flooding include land use on the floodplain, magnitude and frequency of the flooding, amount of sediment deposited, and effectiveness of forecasting, warning, and emergency systems. From an environmental view, the best solution to minimizing flood damage is floodplain regulation. In highly urban areas, however, it will remain necessary to use physical barriers, reservoirs, and channel works to protect existing development. The realistic solution to minimizing flood damage involves a combination of floodplain regulation and engineering techniques. The inclusion of floodplain regulation is critical because engineered structures alone, such as dams and levees, tend to lure people into development of floodplains by producing a false sense of security.

An adequate perception of flood hazards exists at the institutional level; however, on the individual level, more public-awareness programs are needed to help people perceive the hazard of living in flood-prone areas.

Channelization is the straightening, deepening, widening, cleaning, or lining of existing streams. The most commonly cited objectives of channel modification are flood control and drainage improvement. Although many cases of environmental degradation resulting from channelization can be cited, it need not necessarily be a bad practice. In fact, in several cases, channel modification has clearly been beneficial. Nevertheless, the present state of channel modification technology is not sufficient for us to be reasonably sure that environmental degradation will not result. Therefore, until new design criteria compatible with natural stream processes are developed and tested, channelization should be considered only as a solution to a particular flood or drainage problem where the impact and alternatives are carefully studied to minimize environmental disruption.

▼ ▼ ▼ REFERENCES

1. LEOPOLD, L. B., and MADDOCK, T., Jr. 1953. *The hydraulic geometry of stream channels and some physiographic implications.* U.S. Geological Survey Professional Paper 252.

2. MACKIN, J. H. 1948. Concept of the graded river. *Geological Society of America Bulletin* 59:463–512.

3. BEYER, J. L. 1974. Global response to natural hazards: Floods. In *Natural hazards,* ed. G. F. White, pp. 265–74. New York: Oxford University Press.

4. U.S. DEPARTMENT OF COMMERCE. 1972. *Climatological data, national summary* 23, no. 13.

5. LINSLEY, R. K., Jr.; KOHLER, M. A.; and PAULHUS, J. L. 1958. *Hydrology for engineers.* New York: McGraw-Hill.

6. LEOPOLD, L. B. 1968. *Hydrology for urban land planning.* U.S. Geological Survey Circular 559.

7. SEABURN, G. E. 1969. *Effects of urban development on direct runoff to East Meadow Brook, Nassau County, Long Island, New York.* U.S. Geological Survey Professional Paper 627B.

8. AGRICULTURAL RESEARCH SERVICE. 1969. *Water intake by soils.* Miscellaneous Publication No. 925.

9. STRAHLER, A. N., and STRAHLER, A. H. 1973. *Environmental geoscience.* Santa Barbara, California: Hamilton Publishing.

10. DAVIES, W. E.; BAILEY, J. F.; and KELLY, D. B. 1972. *West Virginia's Buffalo Creek flood: A study of the hydrology and engineering geology.* U.S. Geological Survey Circular 667.

11. TERSTRIEP, M. L.; VOORHEES, M. L.; and BENDER, G. M. 1976. *Conventional urbanization and its effect on storm runoff.* Illinois State Water Survey Publication.

12. HOUSE REPORT No. 93–530. 1973. *Stream channelization: What federally financed draglines and bulldozers do to our nation's streams.* Washington, D.C.: U.S. Government Printing Office.

13. OFFICE OF EMERGENCY PREPAREDNESS. 1972. *Disaster preparedness* 1, 3.

14. BUE, C. D. 1967. *Flood information for floodplain planning.* U.S. Geological Survey Circular 539.

15. SCHAEFFER, J. R.; ELLIS, D. W.; and SPIEKER, A. M. 1970. *Flood-hazard mapping in metropolitan Chicago.* U.S. Geological Survey Circular 601C.

16. BAKER, V. R. 1976. Hydrogeomorphic methods for the regional evaluation of flood hazards. *Environmental Geology* 1:261–81.

17. RAHN, P. H. 1984. Flood-plain management program in Rapid City, South Dakota. *Geological Society of America Bulletin* 95: 838–43.

18. U.S. DEPARTMENT OF COMMERCE. 1973. *Climatological data, national summary* 24, no. 13.

19. McCAIN, J. F.; HOXIT, L. R.; MADDOX, R. A.; CHAPPELL, C. F.; and CARACENA, F. 1979. *Storm and flood of July 31–August 1, 1976, in the Big Thompson River and Cache la Poudre River Basins, Larimer and Weld Counties, Colorado.* U.S. Geological Survey Professional Paper 1115A.

20. SHROBA, R. R.; SCHMIDT, P. W.; CROSBY, E. J.; and HANSEN, W. R. 1979. *Storm and flood of July 31–August 1, 1976, in the Big Thompson River and Cache la Poudre River Basins, Larimer and Weld Counties, Colorado.* U.S. Geological Survey Professional Paper 1115B.

21. BRADLEY, W. C., and MEARS, A. I. 1980. Calculations of flows needed to transport coarse fraction of Boulder Creek alluvium at Boulder, Colorado. *Geological Society of America Bulletin* Part II, v. 91:1057–90.

22. BAKER, V. R. 1984. Questions raised by the Tucson flood of 1983. *Proceedings* of the 1984 meetings of the American Water Resources Association and the Hydrology Section of the Arizona-Nevada Academy of Science, pp. 211–19.

23. *U.S. News & World Report.* 1986. The angry waters of winter. March 3, p. 9.

24. EDELEN, G. W., Jr., 1981. Hazards from floods. In *Facing geological and hydrologic hazards, earth-science considerations,* ed. W. W. Hays. U.S. Geological Survey Professional Paper 1240-B:39–52.

25. ARTHUR D. LITTLE, INC. 1972. *Channel modifications: An environmental, economic and financial assessment.* Report to the Council on Environmental Quality.

26. DANIELS, R. B. 1960. Entrenchment of the Willow drainage ditch, Harrison County, Iowa. *American Journal of Science* 225:161–76.

27. RUHE, R. V. 1971. Stream regimen and man's manipulation. In *Environmental geomorphology,* ed. D. R. Coates, pp. 9–23. Binghamton, New York: Publications in Geomorphology, State University of New York.

28. EMERSON, J. W. 1971. Channelization: A case study. *Science* 173: 325–26.

29. McPHEE, J. 1989. *The control of nature.* New York: Farrar, Straus & Giroux.

CHAPTER SEVEN

▼
▼
▼

Landslides and Related Phenomena

Slopes are the most common landforms, and although most slopes appear stable and static, they are actually dynamic, evolving systems. Slope processes are one significant reason that stream valleys are much wider than the stream channel and adjacent floodplain.

On examining slope processes it is useful to identify slope elements as shown in Figure 7.1. The slope in Figure 7.1a shows four distinct elements consisting of a convex crest, a nearly vertical free-face, a debris slope at approximately 30°–35°, and a lower concave or wash slope. All slopes are composed of one or more of these elements. Furthermore, distinct processes are associated with each element. The convex slope at the crest is associated with slow downslope movement of rock and soil known as *creep*. The free-face is usually associated with processes such as rockfall, and the debris slope is where the material from the free-face accumulates. The angle of the debris slope is called the angle of repose, which is the steepest angle that a given loose material can maintain. Finally, the concave or wash slope is produced by processes associated with running water. Slopes such as that shown in Figure 7.1a are commonly found in semiarid regions or regions where resistant rocks are found. In the more subhumid regions or in areas with relatively soft rocks, a simpler form of slope exists, as illustrated in Figure 7.1b. There the slope elements are an upper convex slope and a lower concave slope. These gentler slope profiles are often associated with a relatively thick soil cover and often are underlain by rocks of relatively low resistance. Thus, in a general sense, we see that the form of a slope is controlled by climate and rock resistance. However, other processes such as stream erosion or wave erosion may produce a prominent free-face, as may tectonic processes such as faulting.

The two types of slope shown in Figure 7.1 may also be distinguished by the depth of soil cover. Soil of Figure 7.1a is found at the convex or crest portion of the slope, where weathered bedrock exists, and on the lower concave or wash slope, where weathering can again occur. Soils are not found on the free-face, where removal processes are rapid. In Figure 7.1b the soil is thicker in the upper part of the convex slope segment and thinner in the central portion of the slope where it is steepest, and thick again in the lower concave segment. The thinning of the soil in the central part of the slope occurs because the downslope processes are more rapid there, owing to the steeper slope. As a result, removal of material in the steeper central part of the slope nearly matches accumulation there.

Material on most slopes is constantly moving down the slope at rates that vary from imperceptible creep of soil and rock to thundering avalanches and rockfalls that move at tremendous speeds. As with floods, it may not be the largest (rare and infrequent) nor the smallest (most common) event that moves the greatest amount of material down slopes as valleys develop. Rather, events of moderate magnitude and frequency may play the most important role.

Landslides and related phenomena such as mud-flows, earthflows, rockfalls, snow or debris avalanches, and subsidence are natural events that would occur with or without human activity. However, human use and interest has led to both an *increase* in some of these events, such as landslides on hillside developments because of oversteeping of slopes, and a *reduction* of landslides by means of stabilizing structures or techniques on naturally sensitive slopes likely to experience landslides.

Loss of life and damages from landslides in the United States are substantial. Each year about 25 people are killed by landslides, and this number increases to between 100 and 150 if we include collapses of trenches and other excavations. Total annual cost of damages exceeds $1 billion (1).

Earth materials on slopes may fail and move or deform in several ways, including flowage, sliding, falling, and subsidence. Figure 7.2 illustrates these failure mechanisms and the processes involved with each. Table 7.1 classifies the common types of downslope movements and reflects the terminology used in this discussion and in other chapters. Important variables in classifying downslope movements are: type of movement, slope material type, amount of water present, and rate of movement. In general, a landslide is considered to move rapidly if the motion can be discerned with the naked eye; otherwise it is classified as slow. Actual rates vary from a slow creep of a few millimeters or centimeters per year to very rapid, at 1.5 meters per day, to extremely rapid, at several tens of meters per second (2).

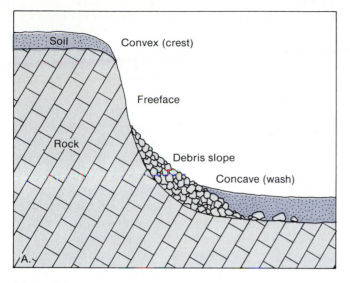
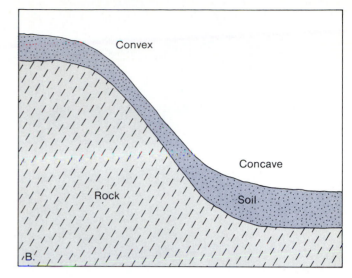

Figure 7.1
Common slope elements. (a) Slope elements common in semiarid regions or on rocks resistant to weathering and erosion. (b) A convex-concave slope more common to semihumid regions or in areas with relatively soft rocks.

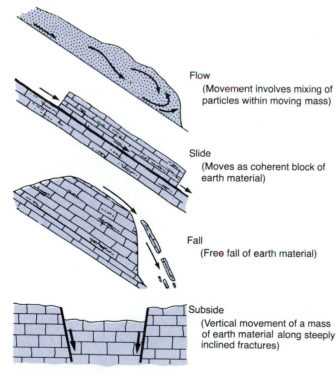

Figure 7.2
Common ways that earth materials fail and move in landslides and related phenomena.

Table 7.1
Classification of landslides and other downslope movements.

Type of Movement	Materials	
	Rock	**Soil**
Slides (variable water content and rate of movement)	Slump blocks Translation slide	Slump blocks (Rotational slide) Soil slip (planar)
Falls	Rock fall	Soil fall
slow	Rock creep	Soil creep
	Unconsolidated materials (saturated)	
Flows	Earthflow Mudflow Debris flow	
rapid	Debris avalanche	
Complex	Combinations of slides and flows	

Landslides are commonly complex, as shown in Figure 7.3, which is a failure consisting of an upper slump with rotational movement that is transformed to a flow in the lower part of the slide. Such complex landslides may form when water-saturated earth materials flow from the lower part of the slope, which undermines the upper part, causing slumping of blocks of earth materials with back rotations.

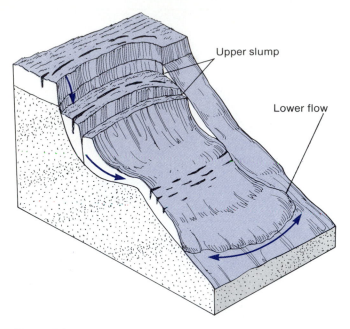

Figure 7.3
Block diagram of a common type of landslide consisting of an upper slump motion and a lower flow. (After R. Pestrong, *Slope Stability,* American Geological Institute, 1974.)

▼ SLOPE STABILITY

The nature and extent of slope stability for the various types of downslope movements generally can be explained by examining forces on slopes in terms of the role of these significant, and somewhat interrelated, variables: types of earth materials, slope (topography), climate, vegetation, water, and time.

Forces on Slopes

The causes of many landslides and related types of downslope movement can be examined by studying relations between **driving forces**, those which tend to move earth materials down a slope, and **resisting forces**, those which tend to oppose such movement. The most common driving force is the downslope component of the weight of the slope material, including anything superimposed on the slope, such as vegetation, fill material, or buildings. The most common resisting force is the shear strength of the slope material acting along potential slip planes. Recall from our discussion of soils in Chapter 4 that the shear strength is a function of cohesion and internal friction.

Slope stability is evaluated by computing a safety factor (SF) defined as the ratio of the resisting forces to the driving forces. If the safety factor is greater than one, the resisting forces must exceed the driving forces and the slope is considered stable. If, on the other hand, the safety factor is less than one, then the driving forces exceed the resisting forces and a slope failure can be

expected. Of course, anything that reduces the resisting forces or increases the driving forces will lower the safety factor and thus increase the chances of a landslide or other type of downslope movement. Driving and resisting forces are not static; rather, they tend to change with time. Thus, depending on changes in local conditions, the safety factor may increase or decrease. For example, consider construction of a road cut in the toe of a slope with a potential slip plane (Figure 7.4). The effect of the road cut in the critical toe of the slope will reduce the driving forces because some of the slope material is removed, and also reduce the resisting forces because the length of the slip plane is reduced, and it is this plane along which the resisting force (shear strength) acts. The overall effect of the road cut will be to lower the safety factor, because the reduction of the driving force is small compared to the reduction of the resisting force. That is, only a small portion of the potential slide mass is removed, but a relatively large portion of the total length along which resisting forces act is removed. Thus, the safety factor must be reduced. Safety factors are commonly computed for natural slopes or slopes constructed as part of site development or highway construction.

Role of Earth Material Type

The type of downslope movement is partly a function of slope material type. For landslides, two basic patterns of movement are recognized: **rotational** and **translational** (Figure 7.5). Rotational slides are most common for soil slopes, but are also associated with slumping in rock slopes, especially where a permeable rock such as sandstone overlies a weak rock such as shale. If, however, a very resistant rock overlies a rock of very low resis-

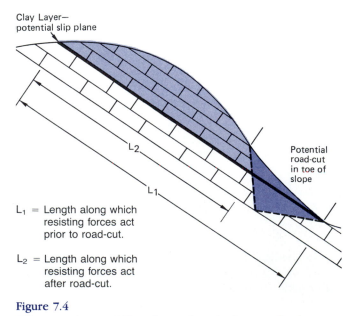

L$_1$ = Length along which resisting forces act prior to road-cut.

L$_2$ = Length along which resisting forces act after road-cut.

Figure 7.4
Effects on slope stability of a road-cut in the toe of a slope.

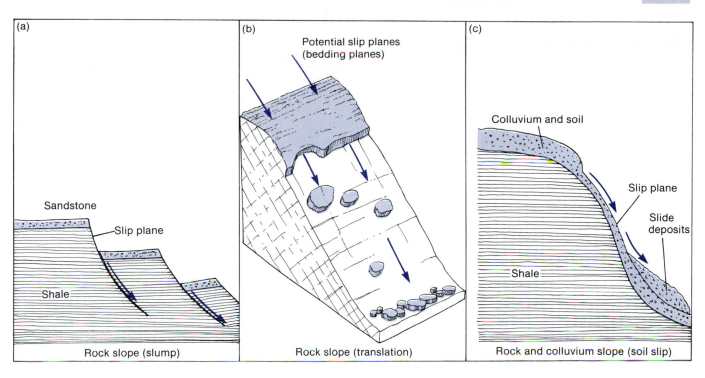

Figure 7.5
Patterns of downslope movement: (a) rotational, (b and c) translational.

tance, then rapid undercutting of the resistant rock may cause a slab failure (rock fall) (Figure 7.6).

Translation slides occur along geologic planes of weakness within slopes. Common translation slip planes in rock slopes include fractures, bedding planes, clay partings, and foliation planes. In some areas, very shallow slides parallel to the slope, known as *soil slips*, also occur. For soil slips, the slip plane is usually above the bedrock but below the soil within slope material known as **colluvium**, a mixture of weathered rock and other material.

The strength of the slope materials may greatly influence the magnitude and frequency of landslides and related events. For example, creep, earthflows, slumps, and soil slips are much more common on shale slopes or slopes on weak pyroclastic (volcanic) materials than on slopes on more resistant rock such as well-cemented sandstone, limestone, or granite. In fact, shales are so notorious for landslide activity that what is called "shale terrain" is characterized by irregular, hummocky topography produced by a variety of downslope movement processes. For all types of land use, from agricultural to urban, on shale or other weak rock slopes, one must carefully consider the possible landslide hazard prior to development.

Role of Slope and Topography

Slope angle greatly affects the relative magnitude of driving forces on slopes. Figure 7.7 illustrates that, as the

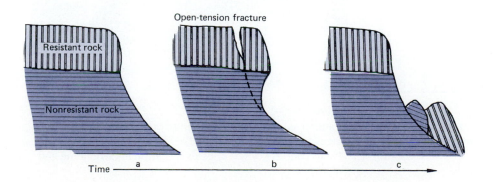

Figure 7.6
Development of a slab slide (type of rock fall). Undercutting of resistant "cap" rock at time (b) causes the development of tension fractures and eventual failure by rock fall at time (c).

Figure 7.7

Effect of slope on the driving force. Notice that, for a hypothetical slide mass, the driving force increases as the slope increases.

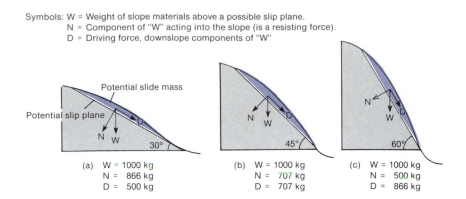

Symbols: W = Weight of slope materials above a possible slip plane.
N = Component of "W" acting into the slope (is a resisting force).
D = Driving force, downslope components of "W"

Potential slide mass

Potential slip plane

(a) W = 1000 kg
N = 866 kg
D = 500 kg

(b) W = 1000 kg
N = 707 kg
D = 707 kg

(c) W = 1000 kg
N = 500 kg
D = 866 kg

angle of a potential slip plane increases, the driving force also increases. Thus, everything else being equal, landslides should be more frequent on steep slopes. In general, this is true; for example, a study of landslides that occurred during two rainy seasons in California's San Francisco Bay area established that 75 to 85 percent of landslide activity is closely associated with urban areas on slopes greater than 15 percent, or 8.5 degrees (Figure 7.8) (3). On a national scale, the coastal mountains of California and Oregon, the Rocky Mountains, and the Appalachian Mountains have the greatest frequency of landslides. All the types of landslides listed in Table 7.1 occur in those locations.

To some extent, the type of landslide activity is also a function of slope and topography. For example, rock

falls and debris avalanches are associated with very steep slopes, and shallow soil slips in southern California are common on steep saturated slopes. These soil slips are often transformed downslope into debris flows that are extremely hazardous (Figure 7.9). At the other extreme, earthflows may occur on moderate slopes, and the effect of creep may be observed on slopes of only a few degrees.

Role of Climate and Vegetation

Climate and vegetation can influence the type of landslide or other downslope movement that occurs on a particular slope. Climate is significant in influencing downslope movement because it controls the nature and

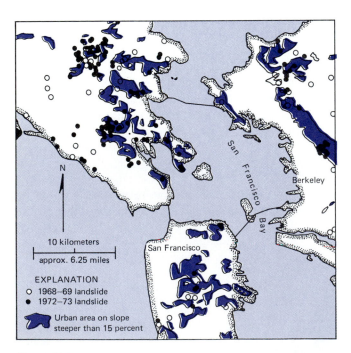

Figure 7.8

Relationship between landslide frequency and urban areas on slopes in the San Francisco Bay area exceeding 15 percent. (After T. H. Nilsen, F. A. Tayler, and R. M. Dean, U.S. Geological Survey Bulletin 1424, 1976.)

San Francisco Bay

Berkeley

N

10 kilometers
approx. 6.25 miles

San Francisco

EXPLANATION
○ 1968–69 landslide
● 1972–73 landslide

Urban area on slope steeper than 15 percent

Figure 7.9

Home in southern California destroyed by debris flow that originated as a soil slip. This 1969 event claimed two lives. (Photo courtesy of Los Angeles Department of Building and Safety.)

extent of the precipitation and thus the moisture content of slope materials. For example, both earthflows (which usually occur on slopes) and mudflows (which initially may be confined to channels) involve downslope movement of water-saturated earth materials; but earthflows are common and mudflows relatively rare in humid regions. This results because a good deal of water infiltrates into slopes in the humid areas, facilitating earthflows. Furthermore, humid regions have many perennial streams that continuously transport the materials delivered from slopes out of the drainage basin. Thus, little debris builds up in the basin, so little is available for mobilization in a mudflow.

The role of vegetation in landslides and related phenomena is complex because the vegetation in an area is a function of several factors, including climate, soil type, topography, and fire history, each of which also influences what happens on slopes. Vegetation is a significant factor in slope stability for three reasons. First, vegetation provides a cover that cushions the impact of rain falling on slopes, thus facilitating infiltration of water into the slope while retarding grain-by-grain erosion on the surface. Second, vegetation has root systems that tend to provide an apparent cohesion to the slope materials. And third, vegetation adds weight to the slope.

Most problems concerning slope stability and vegetation result from disturbance or removal of vegetation from slopes. In some cases, however, vegetation increases the probability of a landslide, especially for specific types of shallow soil slips on steep slopes. For example, one type of soil slip in southern California coastal areas occurs on steep-cut slopes covered with ice plant (Figure 7.10). During especially wet winter months, the shallow-rooted ice plants take up water, adding considerable weight to steep slopes and increasing the driving forces. The plants also cause an increase in the infiltration of water into the slope, which decreases the resisting forces. When failure occurs, the plants and

several centimeters of roots and soil slide to the base of the slope.

Soil slips on natural steep slopes in southern California are a more serious problem (Figure 7.9), and again vegetation, in this case chaparral (dense shrubs or brush), facilitates an increase in water infiltrating into the slope which tends to lower the safety factor. One study concluded that, in some instances, susceptibility to soil-slip may be greater on vegetated slopes than on slopes where vegetation had been recently removed by fire. This should not be interpreted to mean that burning reduces the landslide hazard. Even though soil slips may sometimes be reduced by removal of vegetation, they are not eliminated; in addition, the grain-by-grain erosion caused by rain splash and sheet wash on the surface greatly increases. The eroded sediment tends to fill up the ravines (steep stream valleys) with a meter or more of debris that may be mobilized into debris flows and mudflows during wet winters (4).

The type of vegetation also affects the frequency and morphology of shallow soil slips in southern California. To compare grass-covered with chaparral-covered slopes, soil slips on grassy slopes are three to five times more frequent, have a lower minimum angle at which failure occurs, and tend to be shorter and wider than those on chaparral-covered slopes (4).

Fire may facilitate development of hydrophobic soil horizons that retard infiltration of water and establishment of vegetation after burning. The "nonwettability" or hydrophobic properties are produced when waxy organic coatings on soil particles are mobilized by the fire and accumulate in a soil horizon (4). Hydrophobic soils have been recognized in areas with chaparral vegetation as well as forest cover, and the problem is more serious than generally realized because the behavior of water in the near-surface environment is critical to the establishment of vegetation and retardation of soil erosion. Unfortunately, our often necessary practice of not allowing periodic natural burns has increased the chances of formation of hydrophobic soils because the less frequent fires have more fuel and burn very hot, increasing the likelihood of hydrophobic properties developing in the soils.

Disturbance or removal of vegetation by logging has also been associated with an increase in landslides. Clearcutting, or removal of all trees, has caused several problems. First, the rate of transpiration (loss of soil water through the trees) is greatly reduced; thus soil moisture increases, tending to reduce the stability of the slopes. Second, in specific instances, infiltration of water into a slope may be increased. This is especially likely if a permeable soil on relatively low-gradient slopes is covered in winter with thick snow pack that slowly melts in the spring; as the slope becomes saturated, resisting forces are lowered. Third, with the exception of redwood trees that regenerate after logging, the roots of cut trees

Figure 7.10
Shallow soil slip on steep slopes covered with shallow rooted ice plants near Santa Barbara, California.

decay with time, reducing their strength and thus the apparent cohesion of the soil. This tends to reduce resisting forces within the slope, helping to explain the increased frequency in shallow landslides several years following timber harvesting (5).

Role of Water

Water is almost always directly or indirectly involved with landslides, and its role is thus particularly important. Water is the universal solvent, and much weathering of rocks, which slowly reduces their shear strength, takes place near the surface of the earth. The weathering contributes to the instability of rocks and is especially significant in areas with limestone, which is very susceptible to chemical weathering. It has been argued that water has a lubricating effect on individual soil grains and potential slide planes. This notion is incorrect; water is not a lubricant (6). On the other hand, the amount of water in a soil is very significant. For example, dry sand will form a cone when dumped, whereas moist sand will stand nearly vertical because surface tension in the water holds the grains together, and with all its pore spaces filled, saturated sand will flow like mud.

The effects of water on slopes and landslides are quite variable. First, saturation of earth materials causes a rise in pore water pressure. In general, as the water pressure in slopes increases, the shear strength (resisting force) of the slope decreases and the weight (driving force) increases; thus, the net effect is to lower the safety factor. This is thought to be a significant factor in the development of soil slips and debris avalanches, as well as other types of landslides. Soil slips generally occur during heavy rainfall when near-surface temporary or *perched* water table conditions may be present. During a rainstorm, the rate of surface infiltration in the unsaturated zone of the soil or colluvium exceeds the rate of deep percolation in the rock below the colluvium, and even though some water moves as seepage parallel to the slope, a temporary (perched) water table develops (Figure 7.11). Failure of the slope occurs when resisting forces are reduced sufficiently, that is, when the safety factor becomes less than one. The safety factor will be at a minimum when the perched water table rises to the surface, indicating that the potential slide mass is entirely saturated.

A rise in water pressure is present prior to many landslides, and it is safe to say that most landslides are caused by an abnormal increase of the water pressure in the slope-forming materials (6).

Rapid draw-down, when the water level in a reservoir or river is lowered quickly at a rate of at least a meter per day, is a second way that water may affect the stability of slopes. When the water is at a relatively high level, a large amount may enter the banks, producing a bank storage phenomenon. Then, when the water level suddenly drops, the water stored in the banks is left unsupported. This produces an abnormal distribution of pore water pressures that reduces the resisting forces; simultaneously, the weight of the stored water increases the driving forces. For this reason, bank failures (slumps) tend to occur along streams after flood water has receded.

A third way that water affects landslides is by contributing to spontaneous liquefaction of clay-rich sediment or **quick clay**. When disturbed, some clays lose their shear strength, behave as a liquid, and flow. The shaking of clay below Anchorage, Alaska, during the 1964 earthquake produced this effect and was extremely destructive (Figures 7.12 and 7.13). Other examples of

Figure 7.11
Idealized diagram showing development of a perched water table in colluvial material during heavy rainfall and relationship to increased instability of the slope. (After R. H. Campbell, U.S. Geological Survey Professional Paper 851, 1975.)

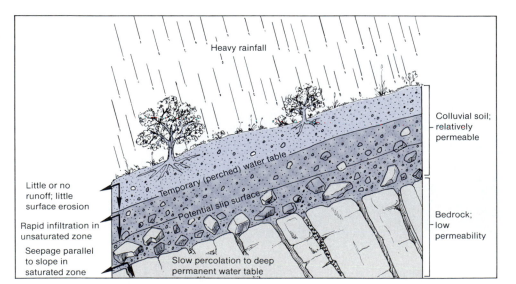

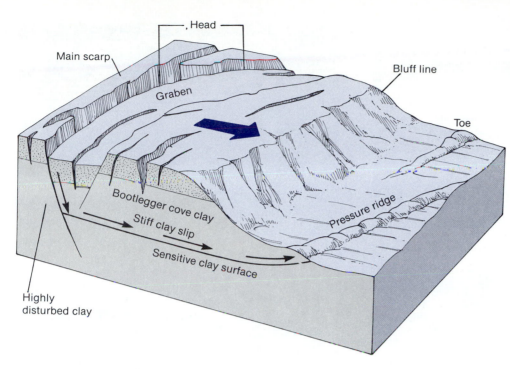

Main scarp

Head

Graben

Bluff line

Toe

Bootlegger cove clay

Stiff clay slip

Sensitive clay surface

Pressure ridge

Highly
disturbed clay

Figure 7.12
Block diagram showing how land-slides developed in response to the 1964 Alaskan earthquake in Anchorage. The graben is a down-drop block of earth mate-rial. (From W. R. Hansen, U.S. Geological Survey Professional Paper 542A, 1965.)

slides associated with sensitive clays are found in Quebec, Canada, where several large earth movements have destroyed numerous homes and killed about seventy people in recent years. The slides occur on river valley slopes in initially solid material that is converted into a liquid mud as movement begins (7). These slides are often initiated by river erosion at the toe of the slope and, although they start in a small area, may develop into large events. Since they often involve the reactivation of an older slide, future problems may be avoided by restricting development. Fortunately, older slides, even though masked from ground view by vegetation, are often visible on aerial photographs. The Quebec slides/flows are especially interesting because the liquefaction of clays occurs without earthquake shaking, unlike those in Anchorage.

Seepage of water from artificial sources, such as reservoirs, septic systems, and unlined canals, into adjacent slopes may also affect slope stability. Seepage can lower the cohesion and thus the shear strength by removing cementing materials, as was the case in the St. Francis Dam failure. Seepage may also cause an increase of water pressure in adjacent slopes, causing a reduction in the resisting force.

The ability of water to erode also affects the stability of slopes. Stream or wave erosion may remove and steepen the toe area of a slope, thus reducing the safety factor. This problem is particularly critical if the toe of the slope is an old, inactive landslide that is likely to move again if stability is reduced. Therefore, it is important to recognize old landslides along potential road cuts and other excavations prior to construction, to isolate and correct potential problems.

Role of Time

The forces on slopes often change with time. For example, both driving and resisting forces may change seasonally as the moisture content or water table position alters. Furthermore, these changes are greater in especially wet years, as reflected by the increased frequency of landslides in wet years. In other slopes, a continuous reduction in resisting forces may occur with time, perhaps due to weathering, which reduces the cohesion in slope materials, or to a regular increase in water pressure from natural or artificial conditions. A slope that is becoming less stable with time may have an increasing rate of creep until failure occurs (Figure 7.14a). A slope's safety factor might also decrease with time because of progressive wetting, which causes disarrangement of the soil particles in the slope, lowering internal friction and thus the strength of the slope materials. Figure 7.14b illustrates this phenomenon as a result of road building and culvert installation that periodically dumps water downslope. This situation emphasizes an important point: we may design slopes that are stable when constructed, but stability has a way of changing with time. Therefore, slope design should include provisions to minimize processes that might progressively weaken slope materials.

(a)

(b)

(c)

Figure 7.13
Landslides in Anchorage, Alaska, caused by the 1964 earthquake. The slides are similar to those diagrammed in Figure 7.12: (a) Government Hill School slide, (b) aerial view of slide near the hospital on Fourth Avenue, and (c) slides that caused the collapse of Fourth Avenue near C Street. (Photos [a] and [b] by W. R. Hansen, courtesy of U.S. Geological Survey. Photo [c] courtesy of U.S. Army.)

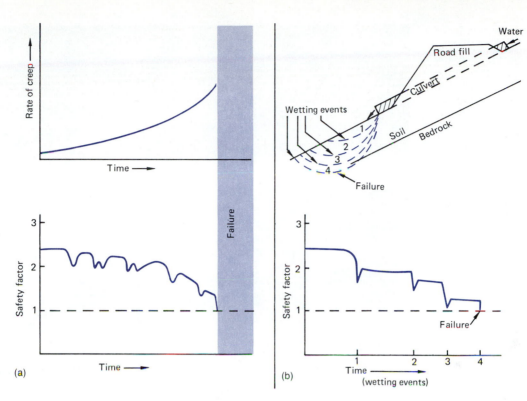

Figure 7.14
Idealized diagrams showing the influence of time on the development of a landslide: progressive creep (a) and progressive wetting (b). Progressive creep is symptomatic of a decreasing safety factor, whereas progressive wetting may cause a reduction in the resisting forces and thus produce a lower safety factor. (Diagram [b] after C. S. Yee and D. R. Harr, *Environmental Geology,* vol. 1, p. 374, 1977.)

▼ CAUSES OF LANDSLIDES

In many cases, the real causes of landslides—increase in driving force or decrease in resisting force—are masked by immediate causes, such as earthquake shocks, vibrations, or a sudden increase in the amount of moisture in slope material. The distinction between real and immediate causes is very important in court when a landslide case is heard and a definitive statement concerning the cause of a landslide is expected from an earth scientist (8). For example, a translation slide may have as an *immediate* cause heavy rains that saturated the earth material, while the *real* cause is the potential to slide upon long, weak, clay layers. A similar hypothetical example might be given for the slide of an artificial slope in a housing development in which the immediate cause is an earthquake shock, but the real cause is a poorly designed slope.

Causes of landslides may also be grouped according to whether they are *external* or *internal*. External causes produce an increase in the shear stress (driving force per unit area) at relatively constant shear strength (resisting force per unit area). Examples of external causes include loading of a slope, steepening of a slope by erosion or excavation, and earthquake shocks. Internal causes

produce landslides without any recognized external changes, and include processes that reduce the shear strength. Examples include such changes as increase in water pressure or decrease in cohesion of the slope materials. In addition, some causes of landslides are intermediate, having some attributes of both external and internal causes. For example, rapid draw-down involves an increase in the shear stress accompanied by decrease in shear strength. Other intermediate causes include spontaneous liquefaction, and subsurface weathering and erosion (6).

▼ HUMAN USE AND LANDSLIDES

Effects of human use and interest on the magnitude and frequency of landslides vary from nearly insignificant to very significant. In cases where our activities have little to do with the magnitude and frequency of landslides, we need to learn all we can about where, when, and why they occur to avoid hazardous areas and minimize damage. In cases where human use has increased the number and severity of landslides, we need to learn how to recognize, control, and minimize their occurrence wherever possible.

Figure 7.15
Floresta Creek Valley mudflow at the base of the Sierra Das Araras Escarpment in Brazil. Before the mudflow, the valley bottom contained a village and a highway construction camp. Several hundred people lost their lives when the mudflow, which was about 4 meters deep, moved through the area. (Photo by F. O. Jones, courtesy of U.S. Geological Survey.)

Mixtures of adverse geologic conditions such as weak soil or rock and potential slip planes on steep slopes with torrential rains, heavy snowfall, or seasonably frozen ground (permafrost) will continue to produce landslides, mudflows, and avalanches regardless of human activities. These are natural processes reacting to natural conditions. For example, the night of January 22, 1967, produced a landslide disaster of tremendous magnitude in Brazil. Following a three-hour electrical storm and cloudburst, an area of about 194 square kilometers was devastated by landslides and erosion (floods) that claimed the lives of about 1,700 people. Damage to property and effects on industry were inestimable, as tens of thousands of landslides turned green hills into wastelands and valleys into mud baths (Figure 7.15). All the slides were debris avalanches and mudflows (9). The slopes in disaster areas characteristically had only a thin veneer of residual soil over hard rock, resulting in a contact of relatively easy parting (Figure 7.16). More avalanches cut areas of high vegetation than grass-covered slopes, and many destroyed forests left in their natural state for many years. The combination of high-frequency vibrations of the electrical storm, pressure of the falling water, rising pore water pressure, groundwater movement, vegetation, and erosion were responsible for producing the slope failures. The magnitude of this landslide activity, while unique in recorded history, has probably been relatively frequent along the mid-southern coast of Brazil during the geologic evolution of the slopes (9).

Figure 7.16
Debris avalanche in thin residual soils in Brazil. (Photo by F. O. Jones, courtesy of U.S. Geological Survey.)

The experience in Brazil is similar to a widespread episode of debris avalanches that occurred in August of 1969. Remnants of Hurricane Camille, moving eastward from Kentucky and the Appalachian Mountain ridges, mixed with a mass of saturated air to produce thunderstorms of catastrophic proportions. These storms locally produced 71 centimeters of rain in eight hours and triggered a great many debris avalanches in central Virginia (Figure 7.17). The storms claimed 150 lives. The greatest loss of life was the result of flooding, although most people died from broken bones and blunt-force injuries rather than drowning. It is impossible to estimate how many died as a result of the debris avalanches, but the amount of debris delivered to channels in floods certainly was significant. The avalanches generally followed preexisting depressions, moved a layer of soil and vegetation 0.3 to 1 meter thick, and left a pronounced linear scar of exposed bedrock. The average amount of rock and soil debris moved was 2,500 cubic meters, or

Figure 7.18
Oblique aerial view of the debris avalanche that destroyed Yungay and Ranrahirca, Peru, killing about 20,000 people. (Photo by George Plafker, courtesy of U.S. Geological Survey.)

Figure 7.17
Debris avalanche scar in Virginia. (Photo by G. P. Williams, courtesy of U.S. Geological Survey.)

about 36,000 metric tons (10). Although the loss of life in Virginia was terrible, it was relatively low because of the sparse population. In 1970, inhabitants of Yungay and Ranrahirca, Peru, were not so fortunate when a debris avalanche triggered by an earthquake roared 3,660 meters down Mt. Huascaran at a speed in excess of 300 kilometers per hour, killing about 20,000 persons, depositing many meters of mud and boulders, and leaving only scars where the villages had been (Figure 7.18) (11).

Many landslides have been caused by interactions of adverse geologic conditions, excess moisture, and artificial changes in the landscape and slope material. Examples include the Vaiont Reservoir slide of 1963 in Italy, the Handlova, Czechoslovakia, slide in 1960, landslides associated with timber harvesting, and numerous slides in urban areas such as Rio de Janeiro, Brazil; Los Angeles,

California; Hamilton County, Ohio; and Allegheny County, Pennsylvania.

The world's worst dam disaster occurred on October 9, 1963, when approximately 2,600 lives were lost at the Vaiont Dam in Italy. As reported by George Kiersch, the disaster involved the world's highest thin-arch dam (267 meters at the crest), yet, strangely, no damage was sustained by the main shell of the dam or the abutments (12). The tragedy was caused by a huge landslide in which more than 238,000,000 cubic meters of rock and other debris moved at speeds of about 95 kilometers per hour down the north face of the mountain above the reservoir, and completely filled it with slide material for 1.8 kilometers along the axis of the valley to heights of nearly 152 meters above the reservoir level (Figure 7.19). The rapid movement created a tremendous updraft of air and propelled rocks and water up the north side of the valley, higher than 250 meters above the reservoir level. The slide, and accompanying blasts of air and water and rock, produced strong earthquakes recorded many kilometers away. It blew the roof off one man's house well over 250 meters above the reservoir and pelted the man with rocks and debris. The filling of the reservoir produced waves of water over 90 meters high that swept over the abutments of the dam. More than 1.5 kilometers downstream, the waves were still over 70 meters high, and everything for kilometers downstream was completely destroyed. The entire event, slide and flood, was over in less than seven minutes. The landslide followed a

three-year period of monitoring the rate of creep on the slope which varied from less than 1 to as many as 30 centimeters per week, until September 1963, when it increased to 25 centimeters per day. Finally, on the day before the slide, it was about 100 centimeters per day. Engineers expected a landslide, but one of much smaller magnitude, and they did not realize until October 8, one day before the slide, that a large area was moving as a uniform, unstable mass. Animals grazing on the slope had sensed danger and moved away on October 1.

The slide was caused by a combination of factors. First, adverse geologic conditions, including weak rocks and limestone with open fractures, sinkholes, and clay partings which were inclined toward the reservoir, produced unstable blocks (Figure 7.20), and very steep topography created a strong driving (gravitational) force. Second, water pressure was increased in the valley rocks because of the impounded water in the reservoir. Groundwater migration into bank storage raised the water pressure and reduced the resisting forces in the slope. The rate of creep before the slide increased as the water table rose in response to higher reservoir levels. Third, heavy rains from late September until the day of the disaster further increased the weight of the slope materials, raised the water pressure in the rocks, and produced runoff that continued to fill the reservoir even after engineers tried to lower the reservoir level. It was concluded that the immediate cause of the disaster was an increase in the driving force accompanied by

Figure 7.19

Sketch map of the Vaiont Reservoir showing the 1963 landslide which displaced water that overtopped the dam and caused severe flooding and destruction over large areas downstream. A-A′ and B-B′ are section lines shown on Figure 7.20 (After Kiersch, *Civil Engineering* 34, 1964.)

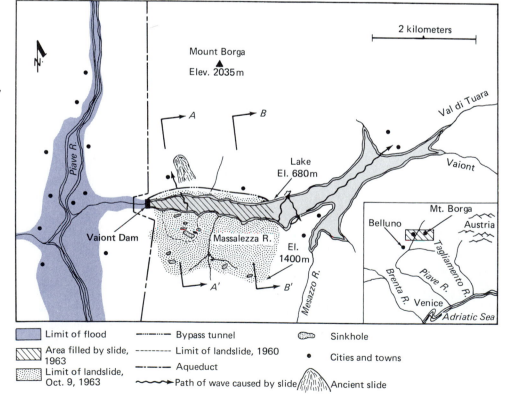

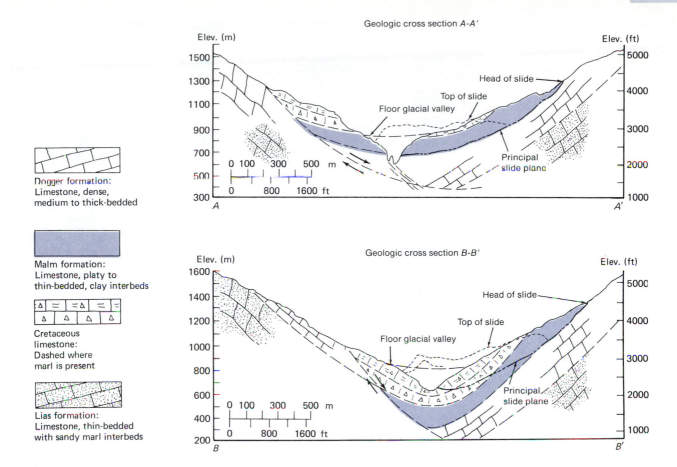

Geologic cross section A-A'

Dogger formation:
Limestone, dense,
medium to thick-bedded

Malm formation:
Limestone, platy to
thin-bedded, clay interbeds

Cretaceous
limestone:
Dashed where
marl is present

Lias formation:
Limestone, thin-bedded
with sandy marl interbeds

Geologic cross section B-B'

Figure 7.20
Generalized geologic cross sections through the slide area of the Vaiont River Valley. Location of sections are shown on Figure 7.19. (After Kiersch, *Civil Engineering* 34, 1964.)

great decrease in the resisting force; but the real cause was excess groundwater that raised the water pressure, in turn changing the weight of the slope material and producing a buoyancy effect along planes of weakness in the rock (12).*

The Handlova Slide in Czechoslovakia is an interesting example of an extremely hazardous landslide caused by a mixture of human use and adverse geologic conditions (Figure 7.21). A large coal-burning power plant in Handlova used a soft coal which emitted a large amount of fly ash that prevailing winds carried and deposited to the south. The land was changed by the accumulation of ash to such an extent that some land formerly used for grazing purposes had to be plowed. The plowing allowed rains to infiltrate at a greater rate and disturb an already delicate groundwater situation. After a heavy rainfall in 1960, which further raised the water table, a large landslide involving approximately 20,000,000 cubic meters of earth material developed. The slide moved nearly 152 meters in one month and

threatened to severely damage the town. A well-organized program to control the slide and stop the movement by draining the slide material was successful within 60 days. Even so, 150 homes were destroyed (7).

The possible cause and effect relationship between timber harvesting and erosion in northern California, Oregon, and Washington is a controversial topic. Landslides become important in the discussion because there is good reason to believe that landslides, especially shallow debris avalanches and more deeply seated earthflows, are responsible for much of the erosion. In fact, one study in the western Cascade Range of Oregon concluded that shallow slides are the dominant erosion process in the area. Furthermore, whereas timber-harvesting activities (clearcutting and road building), over approximately a twenty-year observation period on geologically stable land, did not greatly increase landslide-related erosion, over the same period of time, logging on weak, unstable slopes did increase landslide erosion by several times that which occurred on forested land (13).

There is no doubt that clearcutting and construction of logging roads can increase erosion by activating old

*G. A. Kiersch, *Civil Engineering* 34 (1964), pp. 32–39.

Figure 7.21
Part of the Handlova landslide.
(Photo courtesy of Professor M.
Matula.)

landslides and causing new ones. Figure 7.22 shows before and after photographs of an area in Humboldt County, California, that was logged. One large landslide and two smaller ones are clearly contributing a good deal of sediment to the Mattole River. Although slides A and D were both present before logging, their size increased dramatically following timber-harvesting activities. It might be argued that Figure 7.22 depicts a nightmare from the past, because new regulations on logging practices would prevent this from happening now. To some extent this is true; however, even with stringent regulations, logging on unstable slopes is still likely to influence, initiate, or increase landslide erosion. For example, one of the largest landslides in 25 years in the Oregon Coast Ranges occurred in 1975 during logging operations. The slide was a rather fast-moving, deep-seated earth flow that completely blocked the headwater of Drift Creek, producing a small lake (Figure 7.23). Although there is a fair chance the slide would have occurred even in the absence of logging, it is likely that building roads and removing trees reduced the stability of the slope.

The construction of roads in areas to be logged is an especially serious problem because roads may interrupt surface drainage, alter subsurface movement of water, and adversely change the distribution of mass on a slope by cut-and-fill (grading) operations (13). As we learn more about erosional processes in forested areas, we should be able to develop improved management procedures to minimize the adverse effects of timber harvesting. Until then, we will not be out of the woods with respect to landslide erosion problems.

Human use and interest in the landscape are most likely to cause landslides in urban areas where there are high densities of people and supporting structures such as roads, homes, and industries. Examples from Rio de Janeiro, Brazil, and Los Angeles, California, illustrate the situation.

Rio de Janeiro, with a population of nearly 6 million people, may have more slope-stability problems than any other city its size (9). The city is noted for the beautiful granite peaks that spectacularly frame the urban area. Combinations of steep slopes and fractured rock mantled with surficial deposits contribute to the problem. In earlier times, many such slopes were logged for lumber and fuel and to clear space for agriculture. This early activity was followed by landslides associated with heavy rainfall. Recently, lack of room on flat ground has led to increased urban development on slopes. Vegetation cover has been removed, and roads leading to development sites at progressively higher areas are being built. Excavations have cut the toe of many slopes and severed the soil mantle at critical points. In addition, placing slope fill material below excavation areas has increased the load (driving force) on slopes already unstable before the fill. Furthermore, this area periodically experiences tremendous rainstorms. Thus, it is easy to understand why Rio de Janeiro has a serious problem. In 1966, following heavy rains, numerous landslides occurred (Figure 7.24). The terrible storms of 1967 that we discussed earlier missed the city; if they had not, the results would have been catastrophic (9).

The city was not so fortunate in February of 1988 when an intense rainstorm dumped over 12 centimeters of rain in four hours. The storm caused flooding and mudslides that killed about 90 people, leaving about 3,000 people homeless. Restoration costs may exceed $100 million. Many of the landslides were initiated on

(a)

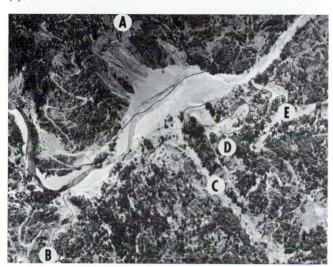

(b)

Figure 7.22
Mattole River area, Humboldt County, California. Note the contrast between the area prior to logging in 1948 (a) and after logging in 1972 (b). Landslides A and D were present in 1948, but were much smaller than in 1972; slide E did not exist in 1948. Streams B and C were nearly hidden by vegetation in 1948, but were open scars in 1972. (After M. E. Huffman, "Geology for timber harvest planning," *California Geology,* vol. 30, no. 9, pp. 195–201, 1977. Photos courtesy of California Department of Water Resources [a] and California Division of Mines Geology [b].)

steep slopes where housing is precarious and control of stormwater runoff nonexistent. It was in these hill-hugging shantytown areas that most of the deaths from mudslides occurred. However, one wing of a three-story nursing home in a more affluent mountainside area was also knocked down by a landslide, killing about 25 patients and members of the staff. Rio de Janeiro is in dire need of measures to control storm runoff and increase slope stability if future disasters are to be avoided.

Los Angeles, and more generally Southern California, has experienced a remarkable frequency of landslides

associated with hillside development. For example, in one storm period, landslides, soilslides, and mudflows in the Los Angeles area claimed two lives and forced evacuation of more than 100 homes. Millions of dollars in property damage occurred to structures such as homes, swimming pools, patios, and utilities (Figure 7.25) (14). Landslides in Southern California result from complex physical conditions, including great contrasts in topography, rock and soil types, climate, and vegetation. Interactions between the natural environment and human activity are complex and notoriously unpredictable. For this reason, the area has the sometimes dubious honor of showing the ever-increasing value of urban geology (14). As a result, more consulting geologists in Southern California are employed in the analysis of slope stability than in any other field, and Los Angeles has led the nation in developing codes concerning grading (artificial excavation and filling) for development.

As computed from geologic mapping, slides affect 60 percent of the length of sea cliffs in Southern California, and the retreat of the sea cliff is probably controlled by landslides (14). Similar estimates for slopes are not available, but the complex geology and terrain features, as well as evidence from old landslide scars and landslide deposits, suggest that slopes historically have been active. But, human activity has tremendously increased the magnitude and especially the frequency of landslides.

Figure 7.23
Large landslide in the Coast Ranges of Oregon that was associated with timber harvesting activities.

Figure 7.24
Landslide that demolished several houses and two apartment buildings in Rio de Janeiro. More than 132 people died. The large slide was evidently facilitated by a smaller landslide caused by a highway cut that overloaded the slope. (Photo by F. O. Jones, courtesy of U.S. Geological Survey.)

Figure 7.25
Damage to property in Los Angeles caused by a mudflow. Note the enormous size of material that mudflows can carry. (Photo courtesy of *Los Angeles Times.*)

The grading process in Southern California has been responsible for many landslides. It took natural processes many thousands, if not millions, of years to produce valleys, ridges, and hills. In this century, our technology has developed the machines to grade them. Leighton writes: "With modern engineering and grading practices and appropriate financial incentive, no hillside appears too rugged for future development" (14). No earth material can withstand the serious assault of modern technology; therefore, human activity is a geological agent capable of carving the landscape as do glaciers and rivers, but at a tremendously faster pace. We can, for example, convert steep hills almost overnight into a series of flat lots and roads, and such conversions have led to numerous artificially-induced landslides. As shown in Figure 7.26, oversteepened slopes mixed with increased water from sprinkled lawns or septic systems, as well as the additional weight of fill material and a house, make formerly stable slopes unstable. Any project that steepens or saturates a slope, increases its height, or places an extra load on it, may cause a landslide (Figure 7.27) (14).

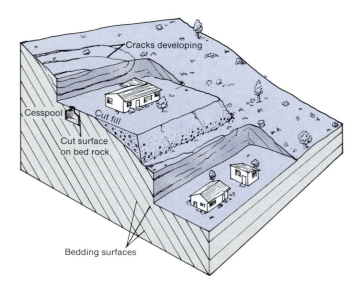

Figure 7.26
Development of artificial translation landslides. Stable slopes may be made unstable by removing support from the bedding plane surfaces. The cracks shown in the upper part of the diagram are one of the early signs that a landslide is likely to occur soon. (Reprinted, by permission, from F. B. Leighton, "Landslides and Urban Development," *Engineering Geology in Southern California* [Whittier, California: Association of Engineering Geologists], 1966.)

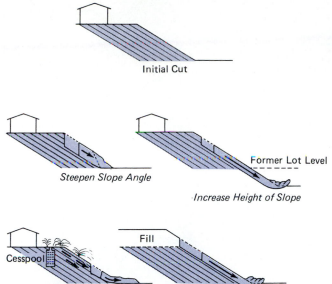

Figure 7.27
Four ways in which a stable cut may be made unstable by human activity. (Reprinted, by permission, from F. B. Leighton, "Landslides and Urban Development," *Engineering Geology in Southern California* [Whittier, California: Association of Engineering Geologists], 1966.)

Landslides on both private and public land in Hamilton County, Ohio, have been a serious problem. The slides occur in glacial deposits (mostly clay, lakebeds, and till) and colluvium and soil developed on shale; the average cost of damage exceeds $5 million per year. In 1974, a major landslide in Cincinnati, Ohio, damaged a highway under construction as well as several private structures. The landslide may be one of the most expensive in the history of the United States. Estimates in 1979 for permanent repair were $22 million (1).

Modification of sensitive slopes associated with urbanization in Allegheny County, Pennsylvania, is estimated to be responsible for 90 percent of the landslides, which produce an average of about $2 million in damages each year. Most of the landslides are slow-moving, but one rockfall, a number of years ago in an adjacent county, crushed a bus and killed 22 passengers. Most of the landslides in Allegheny County result from construction activity that loads the top of a slope, cuts into a sensitive location such as the toe of a slope, or alters water conditions on or in a slope (15).

▼ **IDENTIFICATION, PREVENTION, AND CORRECTION OF LANDSLIDES**

To minimize the landslide hazard, it is necessary to *identify* areas in which landslides are likely to occur, *design* slopes or engineering structures to *prevent*

landslides, and *control* and *stop* slides after they have started moving.

Identification of Landslides

Identifying areas with a high potential for landslides is a first step in developing a plan to avoid landslide hazards. Slide tendency can be recognized by examining geologic conditions and identifying previous slides. This information can then be used to produce slope stability maps (Figure 7.28). The utility of these maps in predicting a possible hazard is well illustrated in the before-and-after sequence of photographs of San Clemente, California (Figure 7.29). The house in the upper left corner of photograph (a) was mapped as sitting on unstable earth material (see the arrow on Figure 7.28), and, as shown in the oblique air photograph (b), a landslide eventually destroyed the home. Of course, we need this type of mapping before homes are constructed, but this example certainly verifies the validity and worth of such mapping.

In general, the procedure for evaluating the landslide risk is to first make a landslide inventory, which may be a reconnaissance map showing areas that apparently have experienced slope failure. This may be done by aerial photographic interpretation with field check. At a more detailed level, the landslide inventory may be a map that shows definite landslide deposits in terms of their relative activity as being active; inactive, geologically young; or inactive, geologically old. An example of such a map for part of Santa Clara County, California, is shown in Figure 7.30a. Information concerning landslide activity may then be combined with land-use considerations to develop a landslide risk map with recommended land uses as shown in Figure 7.30b. The latter map is of most use to planners, whereas the former supplies useful information for the engineering geologist. It is emphasized, however, that these maps do not take the place of detailed field work to evaluate a specific site but serve only as a general guideline for land-use planning and more detailed geologic evaluation.

The individual homeowner, buyer, or builder can evaluate the landslide hazard on hillside property by looking for specific physical evidence that may indicate a potential or real landslide problem. Signs include cracks in buildings or walls around yards; doors and windows that stick or jam; retaining walls, fences, or posts that are not aligned in a normal way; breakage of underground pipes or other utilities; leaks in swimming pools; tilted trees and utility poles with taut or sagging wires; cracks in the ground; hummocky or steplike ground features; and seeping water from the base or toe of a slope (15). This list was derived for a landslide-prone area in western Pennsylvania, but is applicable to many areas. When applying this information, however, keep in mind that the presence of one or more of the features is not absolute

Figure 7.28
Relative slope stability map of a portion of the San Clemente area in California. (After Ian Campbell, *Environmental Planning and Geology,* U.S. Department of Housing and Urban Development, 1971.)

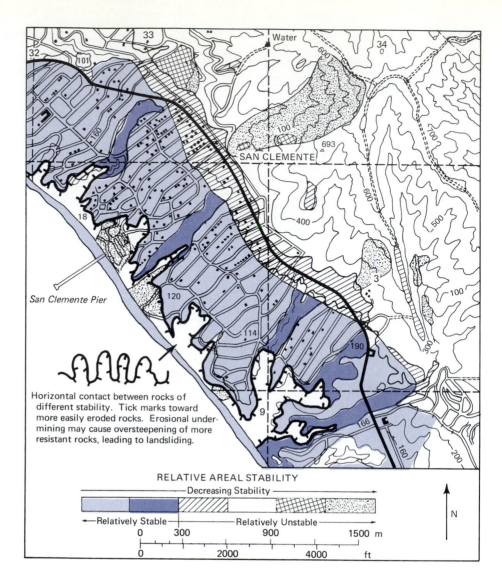

Horizontal contact between rocks of different stability. Tick marks toward more easily eroded rocks. Erosional undermining may cause oversteepening of more resistant rocks, leading to landsliding.

RELATIVE AREAL STABILITY

— Decreasing Stability —

← Relatively Stable — Relatively Unstable →

0   300   900   1500 m
0   2000   4000   ft

N

San Clemente Pier

(a)

(b)

Figure 7.29
Before (a) and after (b) photographs of an area mapped as unstable (see arrow in Figure 7.28) that subsequently failed. (Photos courtesy of George B. Cleveland, California Division of Mines and Geology; Ian Campbell, California Academy of Sciences; and U.S. Geological Survey at Menlo Park.)

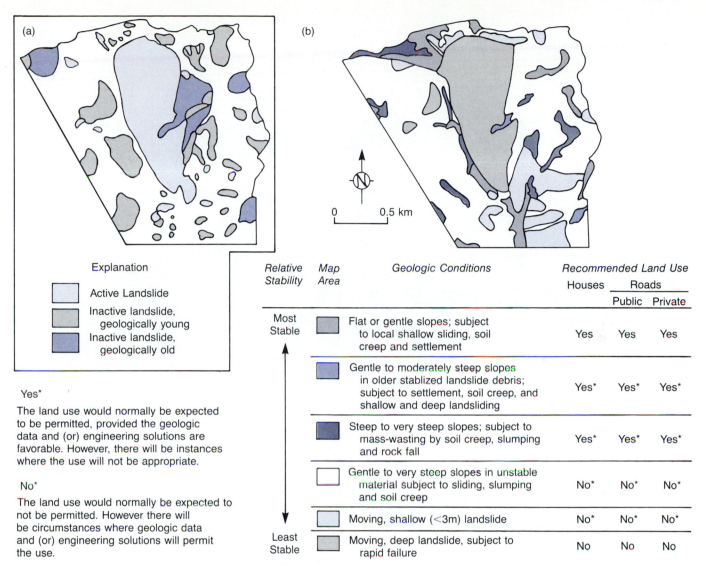

(a)

(b)

0   0.5 km

Explanation

- ☐ Active Landslide
- ▨ Inactive landslide, geologically young
- ▨ Inactive landslide, geologically old

Yes*

The land use would normally be expected to be permitted, provided the geologic data and (or) engineering solutions are favorable. However, there will be instances where the use will not be appropriate.

No*

The land use would normally be expected to not be permitted. However there will be circumstances where geologic data and (or) engineering solutions will permit the use.

Relative Stability	Map Area	Geologic Conditions	Recommended Land Use		
			Houses	Roads	
				Public	Private
Most Stable	▨	Flat or gentle slopes; subject to local shallow sliding, soil creep and settlement	Yes	Yes	Yes
	▨	Gentle to moderately steep slopes in older stablized landslide debris; subject to settlement, soil creep, and shallow and deep landsliding	Yes*	Yes*	Yes*
	▨	Steep to very steep slopes; subject to mass-wasting by soil creep, slumping and rock fall	Yes*	Yes*	Yes*
	☐	Gentle to very steep slopes in unstable material subject to sliding, slumping and soil creep	No*	No*	No*
	▨	Moving, shallow (<3m) landslide	No*	No*	No*
Least Stable	▨	Moving, deep landslide, subject to rapid failure	No	No	No

Figure 7.30
Landslide inventory map (a) and landslide risk and land-use map (b) for part of Santa Clara County, California. (After U.S. Geological Survey Circular 880, 1982, Goals and tasks of the landslide part of a ground-failure hazards reduction program.)

proof that a landslide is likely. For example, cracks in walls may also be caused by expansive soils or creep. Other features, such as hummocky or steplike ground on moderately steep slopes (greater than 15 percent, or 15 meters fall in 100 meters horizontal distance), probably do represent a potential landslide hazard that should be evaluated by a geologist. Furthermore, it is advisable not to limit your inspection only to the property in which you are interested; landslides are often larger than individual lots. Inspect adjacent areas, especially those that are upslope and downslope from your property (15).

Motivation to initiate a plan to reduce the landslide hazard through grading codes in the Los Angeles area came in the aftermath of very high losses of lives and property, to landsliding in the 1950s and 1960s. During

this period, the Portuguese Bend landslide (Figure 7.31) was responsible for damaging or destroying more than 150 homes. This slide is really part of an older, larger slide that was reactivated partly by road-building activities and partly by alteration of a delicate subsurface water situation by urban development. Movement first started in 1956 during construction of a county road that placed approximately 23 meters of fill over the upper area of the landslide, increasing the driving forces. During subsequent litigation, the county of Los Angeles was found responsible for the landslide. Movement of the Portuguese Bend slide was continuous from 1956 to 1978, averaging approximately 0.3 to 1.3 centimeters per day. The rate accelerated to more than 2.5 centimeters per day in the late 1970s and early 1980s following several years

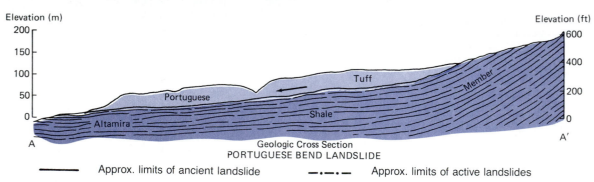

Approx. limits of ancient landslide   —.—.—   Approx. limits of active landslides

Figure 7.31
Aerial view of Palos Verdes showing the approximate known limits of a large, ancient landslide and the outline of the active Portuguese Bend and Abalone Cove landslides. Arrows
show the direction of movement. Geologic cross section shows that volcanic tuff (consolidated ash) is sliding over shale. (Photo and cross section courtesy of Los Angeles County,
Department of County Engineer.)

of above-normal precipitation. Total displacement near
the coast has been over 200 meters. Also shown in Figure
7.31 is the Abalone Cove slide, which began to move
during the above-mentioned wet period. The new slide
prompted additional geologic investigation, and a landslide control program was initiated. The program consisted of several dewatering wells installed in 1980 to
remove groundwater from the slide mass. By 1985 the
slide had apparently been stabilized (16). Based in part
on the successful stabilization of the Abalone Cove slide,
a program has been suggested to help control the
Portuguese Bend landslide. The plan involves drainage of
the slide mass by pumping groundwater from wells,

combined with improvement of surface drainage and
remolding of surface topography to reduce driving forces
and inhibit infiltration of runoff (16).

During the last 20 years of activity, one home on the
Portuguese Bend landslide has moved about 25 meters
and is constantly shifting in position. Other homes have
moved up to 50 meters in the same time, and people
living in homes about one kilometer from the ocean
seem to have adjusted to the slow movement and the
ever-changing view! Homes remaining in the active slide
area have to be adjusted every year or so with hydraulic
jacks, and utilities are on the surface. With one exception,
no new homes have been constructed since the landslide

began to move. The remaining occupants have elected to adjust to the landslide rather than bear total loss of their property. Nevertheless, few geologists would probably choose to live there now.

Grading codes in the Los Angeles area have been effective in reducing damage to hillside homes from landslides and floods. Table 7.2 illustrates this well; since detailed engineering geology studies have been required, the percentage of hillside homes damaged by landslides and floods has been greatly reduced. Although original costs are greater because of the strict codes, they are more than balanced by the reduction of losses in subsequent wet years. Therefore, we conclude that, even though landslide disasters during extremely wet years will continue to plague us, the application of geologic and engineering information prior to hillside development can help minimize the hazard. We will now discuss specific ways to prevent or control landslides.

Prevention of Landslides

Prevention of large, natural landslides is difficult, but common sense and good engineering practice can do much to minimize the hazard. For example, loading the top of slopes, cutting into sensitive slopes, placing fills on slopes, or changing water conditions on slopes should be avoided or done very cautiously (15).

Common engineering techniques for landslide prevention include provisions for surface and subsurface drainage, removal of unstable slope materials (grading), construction of retaining walls, or some combination of these (2, 11).

Drainage control is usually an effective way to stabilize a slope. The basic idea is to keep water from running across or infiltrating the slope. Surface water may be diverted around the slope by a series of gutters.

This practice is common for roadcuts. The amount of water infiltrating a slope may also be controlled by covering the slope with an impermeable layer such as soil-cement, asphalt, or even plastic.

Groundwater may be inhibited from entering a slope by excavating a cutoff trench. The trench is filled with gravel or crushed rock and positioned so as to intercept and divert groundwater away from a potentially unstable slope (2).

Grading slopes is another way to increase slope stability. Two common techniques are reducing the gradient of a slope by a single cut-and-fill operation, and benching. In the first case, material from the upper part of a slope is removed and placed near the base. The overall gradient is thus reduced and material is removed from an area where it contributes to the driving force, and placed at the toe of the slope, where it increases the resisting forces. This method is not practical on very steep, high slopes. As an alternative, the slope may be cut into a series of benches or steps. The benches are designed with surface drains to divert runoff. The benches do reduce the slope and, in addition, are good collection sites for falling rock and small slides (2).

Retaining walls, constructed from concrete cribbing, gabions (stone-filled wire baskets), or piles (long concrete, steel or wooden beams driven into the ground), are designed to provide support at the base of a slope. They should be keyed in well below the base of the slope, backfilled with permeable gravel or crushed rock, and provided with drain holes to reduce the chances of water pressure building up in the slope (Figure 7.32).

A less common method of increasing slope stability involves insertion of heavy bolts (rock bolts) through holes drilled through potentially unstable rocks into stable rocks. This technique was used to secure the slopes at the Glen Canyon Dam on the Colorado River

Table 7.2
Landslide and flood damages to hillside homes in Los Angeles County, California, from before 1952 to 1969.

Construction Dates and Legal Requirements	Number of Homes Built on Hill-side Sites	Damaged Homes		Total Damage	Average Cost Prorated for Total Number of Homes Built
		Number	Percent of Total (%)		
Pre−1952 No legal requirement for soils engineering or engineering geology studies	10,000	1,040	10	$3,300,000	$300
1952−1963 Soils engineering studies required. Minimum engineering geology studies	27,000	350	1.3	$2,767,000	$100
1963−1969 Extensive engineering geology and soils engineering studies required	11,000	17	0.15	$ 80,000	$ 7

Source: After Slosson, J. E., and Krohn, J. P., 1977, "Effective Building Codes," *California Geology* 30, no. 6, pp. 136−319. Data from Los Angeles Department of Building and Safety.

Figure 7.32
Retaining wall (concrete cribbing) used to help stabilize a roadcut.

small rock falls on roads and other areas can be noted for quick removal. Having people monitor the hazard tends to have advantages of reliability and flexibility but becomes disadvantageous during adverse weather and in hazardous locations (19). Other warning methods include electrical systems, tilt meters, and geophones that pick up vibrations from moving rocks. Shallow wells can be monitored to signal when slopes contain a dangerous amount of water and, in some regions, monitoring rainfall is useful for detecting when a threshold precipitation has been exceeded and shallow soil slips become much more probable.

and the Hanson Dam on the Green River in Washington (17).

Preventing the occurrence of landslides can be expensive, but the rewards can be even greater. It has been estimated that the benefits to cost ratio for landslide prevention ranges from approximately 10 to 2000. That is, for every dollar spent on landslide prevention, the savings will vary from $10 to $2000 (18). For example, in April 1983 a massive landslide in Utah, known as the Thistle slide, moved across a canyon creating a natural dam about 60 meters high, flooding the community of Thistle, the Denver-Rio Grande railroad and its switchyard, and a major U.S. highway (18). The landslide, which involved a reactivation of an older slide, caused approximately $200 million in damages. It had been known for many years that the slide was occasionally active in response to high precipitation. Therefore it could have been recognized that the extremely high amounts of precipitation in 1983 would cause a problem. In fact, a review of the landslide history suggests that the Thistle landslide was recognizable, predictable, and preventable! Analysis of the pertinent data suggests that implacement of subsurface drains and control of surface runoff would have lowered the water table in the slide mass enough to have prevented failure. Cost of preventing the landslide was estimated to be between $300,000 and $500,000, a small amount compared to the damages (18). Because the benefit to cost ratio in landslide prevention is so favorable, it seems prudent to evaluate active and potentially active landslides in areas where considerable damage may be expected and possibly prevented.

Landslide Warning Systems

Landslide warning systems do not prevent landslides, but they can provide time to evacuate people and their possessions, and to stop trains or reroute traffic. Surveillance provides the simplest type of warning. Hazardous areas can be visually inspected for apparent changes, and

Landslide Correction

After movement of a slide has begun, the best way to stop it is to attack the process that started the slide. In most cases, the cause of the slide is an increase in water pressure, and in such cases, an effective drainage program must be initiated. This may include surface drains at the head of the slide to keep additional surface water from infiltrating and subsurface drainpipes or wells to remove water and lower the water pressure. Draining will tend to increase the resisting force of the slope material and therefore stabilize the slope.

The tremendous success of drainage is demonstrated by this description from Karl Terzaghi (6). After a high-magnitude rainstorm, movement on a 30-degree slope of deeply weathered metamorphic rock was noted. The slide plane was approximately 40 meters below the surface, and the slide area was about 150 meters wide by 300 meters long. The slide was close to a hydroelectric power station, so immediate action was deemed necessary. Field work established that if the water level could be lowered approximately 5 meters, then the increase in resisting force would be sufficient to stabilize the slide. Drainage was accomplished by trenches and horizontal drill holes extending into the water-bearing zones of the rock. After drainage, the movement stopped, and even though the next rainy season brought record rainfall, no new movement was observed (6).

When slope materials are fine-grained and relatively impermeable, drainage is difficult at best and may even be ineffective. In this situation, common correction methods include reducing the slope angle and constructing artificial barriers such as retaining walls. Unusual methods have also been used; for example, tunnels have been excavated in the slope and hot air circulated to dry out the unstable material. In other cases, unstable soil material has been frozen during construction to stabilize the slope. Drying the material with hot air was apparently successful in the Pacific Palisades area of Southern California, and the freezing technique was used when a slide threatened to impede construction of the Grand Coulee Dam. Freezing was accomplished by using 377 freezing points. The refrigeration system provided 73,000 kilograms of ice per day, and a frozen dam 12 meters high, 6 meters thick, and 30 meters long formed to stabilize the slope material (8).

▼ SNOW AVALANCHE

Approximately 10,000 snow avalanches occur each year in the mountains of the western United States, and about one percent of these cause loss of human life or property damage, killing an average of 7 people and inflicting $300,000 in damage (20). Loss of life is increasing, however, as more people venture into mountain areas for recreation during the winter.

Avalanches may occur in dry or wet snow and are of two general types: *loose-snow avalanches* that occur in cohesionless snow and tend to be relatively small and shallow failures; and *slab avalanches* that may initially vary from about 100 to 10,000 square meters in area and 0.1 to 10 meters in thickness (20). It is the large slab

avalanches that are most dangerous, releasing tremendous energy by mobilizing up to a million tons of snow and ice and moving downslope at velocities of 5 to 30 meters per second (18 to 100 km per hour) or more. Horizontal thrust (or impact) from such events tends to vary from 5 to 50 tons per square meter, but may in extreme cases exceed 100 tons per square meter. By comparison, a thrust of only about 3 tons per square meter is necessary to collapse a frame house, and 100 tons per square meter will move reinforced concrete structures (20).

Avalanches are initiated when a mass of snow and ice on a slope fails because of the overload of a large volume of new snow, or when internal changes in a snowpack produce zones of weakness (low shear strength) along which failure occurs. When conditions are very unstable, even the weight of a single skier can start an avalanche.

Avalanches tend to follow certain paths, chutes, or tracks (see Figure 7.33) that are often well-channeled, but

Figure 7.33
Avalanche chutes or tracks in the Swiss Alps. Left photo shows two chutes, one of which is quite close to the small village; right photo shows a closeup of an avalanche chute devoid of vegetation.

unconfined tracks on open slopes also occur. Avalanche tracks often have several branches near the top that coalesce downslope; thus, it is possible for several avalanches to move through the main track in a short period of time as snowpacks in upper branches fail. Failure to recognize this possibility has caused loss of several lives when workers clearing debris from a first avalanche have been struck by a second (20).

The avalanche hazard can be reduced by avoiding dangerous areas; stabilizing slopes by clearing them with carefully placed explosions; building structures to divert or retard avalanches; and reforesting avalanche paths, since large avalanches are seldom initiated on densely forested slopes (20).

Avalanches are primarily a threat to skiers on high, steep mountain slopes, but they also threaten mountain resorts, villages, railways, highways, and even sections of some cities. For example, Alaska's capital, Juneau, has a significant avalanche hazard. In the last 100 years, a major avalanche chute above Juneau has released snow and ice six times, events that reached the sea. There have been no damaging avalanches in the last 20 years, however, so an entire subdivision has been constructed across the chute. If another large event occurs, it will destroy about 30 homes, part of a school, and a motel, and eventually roar into the harbor where several hundred boats are docked. It has been estimated that a home in the chute area, with a 40-year life span, has a 96 percent probability of being struck by an avalanche, yet the people who live there have been almost nonchalant about the hazard (21).

▼ SUBSIDENCE

Various interactions among geologic conditions and human activity have also been factors in numerous incidents of subsidence. Most subsidence is caused by withdrawal of fluids from subsurface reservoirs or by the collapse of surface and near-surface soil and rocks over subterranean voids.

Withdrawals of such fluids as oil with associated gas and water, groundwater, and mixtures of steam and water for geothermal power have caused subsidence (22). In all cases, the general principles are the same. Fluids in earth materials below the earth's surface have a high fluid pressure that tends to support the material above. This is why a large rock at the bottom of a swimming pool seems lighter: buoyancy produced by the liquid tends to lift the rock. If support or buoyancy is removed from earth materials by pumping out the fluid, the support is reduced, and surface subsidence may result. The actual subsidence mechanism involves compaction of individual grains of the earth material as the grain-to-grain load increases because of a lowering of fluid pressure. Subsidence of oil fields generally involves considerable reduction of fluid pressure, up to 280 kilograms per square centimeter, at great depth (thousands of meters) over a relatively small area, less than 150 square kilometers. On the other hand, subsidence resulting from withdrawal of groundwater generally involves a relatively low reduction of fluid pressure, often less than 14 kilograms per square centimeter, at relatively shallow depths (less than 600 meters), over a large area, sometimes many hundreds of square kilometers (22).

The Wilmington oil field in the harbor area of Long Beach near Los Angeles, California, is a spectacular example of land subsidence that has caused over $100 million in damage to structures, several hundred ruptured oil wells, and localized flooding (Figure 7.34). Subsidence was first noted there in 1940, and by 1974, had increased to 9 meters in the central area (Figure 7.35). Predictions of ultimate subsidence as great as 13.7 meters promoted a joint effort to stop it. A repressuring program was begun in 1958. Tremendous quantities of water were injected into the rocks from which the oil was pumped out, raising the fluid pressure. This procedure actually reversed the subsidence, and by 1963, there had been appreciable rebound amounting to as much as 15 percent of the initial subsidence in some parts of the oil field (22). Figure 7.36 shows the vertical movement at Pier A from 1952 to 1974, where the rebound was significant.

Thousands of square kilometers of the central valley of California have subsided. More than 5,000 square kilometers in the Los Banos–Kettleman City area alone have subsided more than 0.3 meters, and within this area, one 113-kilometer stretch has subsided an average of more than 3 meters, with a maximum of 8.5 meters (Figure 7.37). The cause was overpumping of groundwater from a deep confined aquifer. As the water was mined, the fluid pressure was reduced and the grains were compacted (23). The effect at the surface was subsidence. Similar examples of subsidence caused by overpumping are documented near Phoenix, Las Vegas, and Houston–Galveston. In some instances, the subsidence is accompanied by surface faulting, producing linear scarps and fissures which, in turn, erode to form gullies. In other instances where a great volume of groundwater has been removed, a few centimeters' uplift has actually been recorded. This anomalous uplift is apparently due to elastic rebound when the load of groundwater on near-surface rocks is removed relatively suddenly.

Another example of subsidence caused by overpumping is found in Mexico City. The population in Mexico City increased from fewer than one-half million in 1895 to 1 million in 1920, 5 million in 1960, and 23 million in 1991. As the population grew, so did the demand for groundwater, which is pumped from sand and gravel aquifers separated by clay and silt deposits. These conditions are continuous at depths of about 60 meters to over 500 meters. The discharge of thousands of

Figure 7.34
Flooding at Long Beach, California, caused by subsidence resulting from withdrawal of oil. (Photo courtesy of City of Long Beach.)

TOTAL SUBSIDENCE 1928 TO 1974

Figure 7.35
Subsidence in the Long Beach Harbor area, California, from 1928 to 1974. (Photo courtesy of Port of Long Beach.) (2 to 29 ft = 0.6 to 8.8 m)

private and municipal wells far exceeds the natural recharge of the aquifers (22). Reduction of fluid pressure as the water table lowers has resulted in as much as 7 meters of subsidence, most of which has occurred since 1940. The continuous sinking has produced many prob-lems in drainage and building construction (22). The Palace of Fine Arts, constructed of marble in 1934, has subsided approximately 5 meters, and, whereas steps once went up to the first floor, the present steps go down to that floor.

Figure 7.36
Vertical movement of bench mark
(point of known elevation) on
Pier A, Long Beach, from 1952 to
1974. Location is just right of center of Figure 7.35. (After data,
courtesy of Port of Long Beach.)

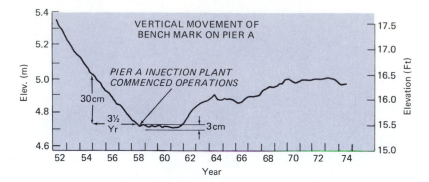

Sinkholes

Subsidence is also caused by removal of subterranean
earth materials by natural or artificial processes. Subterranean voids, for example, often form in certain soluble
rocks, such as limestone and dolomite. Lack of support
for overlying rock may lead to collapse and the formation
of large sinkholes, some of which are over 30 meters
across and 15 meters deep. One near Tampa, Florida,
collapsed suddenly in 1973, swallowing part of an orange
orchard.

What may be the largest sinkhole in the United States
formed in 1972 near Montevallo, Alabama (24). A massive
hole 120 meters wide and 45 meters deep, named the
December Giant by the press, developed suddenly when
topsoil and subsurface clay collapsed into an underlying
limestone cavern (Figure 7.38). The collapse was caused
by loss of support to the clay and soil over the cavern. A
nearby resident reported hearing a roaring noise accompanied by breaking timber and earth tremors that shook
his house. Sinkholes of this type have caused considerable damage to highways, homes, sewage facilities, and
other structures. Natural or artificial fluctuations in the
water table are probably the trigger mechanism. High
water table conditions favor solutional enlarging of the
cavern, and the buoyancy of the water helps support the
overburden. Lowering of the water table eliminates some
of the buoyant support and facilitates collapse. This was
dramatically illustrated on May 8, 1981, in Winter Park,
Florida, when a large sinkhole began developing. The
sink grew rapidly for three days, swallowing part of a
community swimming pool, parts of two businesses,
several automobiles, and a house (Figure 7.39). Damage
caused by the sinkhole exceeded $2 million. Sinkholes
form nearly every year in central Florida when the
groundwater level is lowest. The Winter Park sinkhole
formed during a drought, when groundwater levels were
at a record low. Although exact positions cannot be
predicted, their occurrence is greater during droughts; in
fact, several smaller sinks developed about the same time
as the Winter Park event.

On June 23, 1986, a large subsidence pit developed at
the site of an unrecognized, filled sinkhole in Lehigh
Valley, near Allentown, Pennsylvania. Within a period of

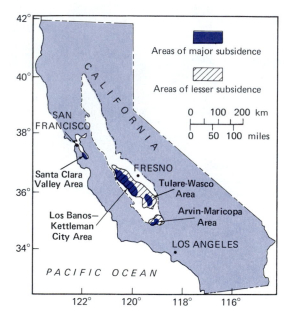

Figure 7.37
Principal areas of land subsidence in California resulting from
groundwater withdrawal. (After W. B. Bull, Geological Society
of America Bulletin 84, 1973. Reprinted by permission.)

Figure 7.38
A very large sinkhole known as the *December Giant* near
Montevallo, Alabama. (Photo courtesy of Bill Warren and
Geological Survey of Alabama.)

Figure 7.39
The Winter Park, Florida, sinkhole of May 1981. The aerial view (upper) shows the extent of the damage, and the lower photograph is a closer view of it. (Upper photo by George Remaine; lower photo by Barbara Vitaliano; both of *Orlando Sentinel Star.*)

only a few minutes, the collapse left a pit approximately 30 meters in diameter and 14 meters deep. Fortunately, the damage was confined to a street, parking lots, sidewalks, sewer lines, water lines, and utilities. Seventeen residences adjacent to the sinkhole narrowly escaped damage or loss; subsequent stabilization and repair cost was nearly one-half million dollars. Allentown is located in eastern Pennsylvania; Figure 7.40 shows the generalized geology of the valley. The northern part of the valley is underlain by shale, whereas limestone comprises the southern portion. The valley is bounded by resistant sandstone rocks to the north and resistant Precambrian granitic and gneissic rocks to the south (Figure 7.40) (25).

Evidence from a series of photographs from the 1940s to 1969 suggest that in the 1940s the sinkhole was delineated by a pond of approximately 65 meters in diameter, and during that time the sinkhole was used as a dump site. By 1958 the pond had dried up and the sinkhole was covered by vegetation and the surrounding area planted in crops. Ground photographs in 1960 suggest people were using the sinkhole as a site to dump tree stumps, blocks of asphalt, and other trash. By 1969 there was no surface expression of the sinkhole; it evidently was completely filled and corn was planted over it. Even though the sinkhole was completely filled with trash and other debris, it still received runoff water that was later increased in volume by urbanization.

Figure 7.40
Generalized geologic map of the Lehigh Valley in eastern Pennsylvania. (Modified from P. H. Dougherty and M. Perlow, Jr., *Environmental Geology and Water Science* 12 [1987], no. 2:89–98.

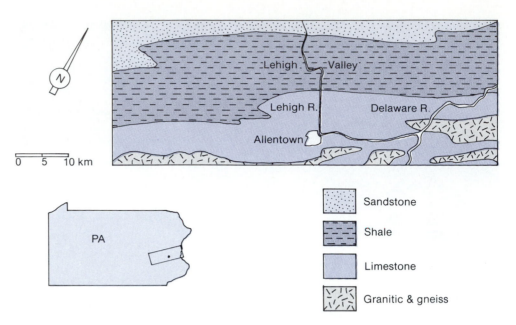

Sources of water included storm runoff from adjacent apartments and townhouses as well as streets and parking lots. It is also suspected that an old, leaking water line contributed to runoff into the sinkhole area. Urbanization also placed increased demand on local groundwater resources and resulted in the lowering of the water table. It is believed that hydrologic conditions contributed to the sudden failure. That is, the increased urban runoff facilitated the loosening or removal of the plug (soil, clay, and trash) that filled the sinkhole while the lowering of the groundwater table reduced the overlying support, as was the case with the Winter Park sinkhole (25).

The lesson learned from the Allentown sinkhole is that the cumulative effect of not recognizing the sinkhole, filling it with urban debris, and subsequently developing the site was probably responsible for the sudden failure.

Serious subsidence events have been associated with mining of salt, coal, and other minerals. Salt is often mined by solution methods in which water is injected through wells into salt deposits. The salt dissolves, and water supersaturated with salt is pumped out. The removal of salt leaves a cavity in the rock and weakens support for the overlying rock, which may lead to large-scale subsidence. In 1970, one event near Detroit produced a subsidence pit 120 meters across and 90 meters deep. Another near Saltville, Virginia, produced 75 meters of subsidence relatively quickly (Figure 7.41). Two homes went down with the Saltville subsidence. According to local residents, one family moved out the day before the event because a family member had dreamed the mountain was falling. As of 1975, large and still actively growing fractures in the rocks surrounding the subsidence area indicated that the area involved in the subsidence was considerably larger than what actually collapsed into the subterranean cavity.

Large sinks associated with bedded salt may also occur without solution mining. For example, in June of 1980, a large depression southwest of Kermit, Texas, known as the Wink Sink, developed over a period of about forty-eight hours. At the end of that time, the sinkhole was approximately 110 meters across and 34 meters deep. The Wink Sink and other similar features evidently form by natural processes when groundwater, even if it is salty (brine), slowly dissolves caverns in bedded salt that underlies less soluble rock such as sandstone. When caverns reach critical size, the overlying rocks can no longer be supported, and collapse occurs. Because this is a natural process and other caverns undoubtedly exist, future sinks probably will develop in the area without warning (26).

Even though the Wink Sink developed in an oil field, the impact on operations was minimal. Damage consisted of a broken pipeline carrying crude oil to tanks located approximately 300 meters from the sinkhole (Figure 7.42). In addition, two water disposal lines were broken by the expanding sinkhole, reducing production by about 150 barrels of oil per day (26).

Lake Peigneur

A bizarre example of subsidence associated with a salt mine occurred on November 21, 1980, when shallow Lake Peigneur (with an average depth of 1 meter) in southern Louisiana drained following collapse of the salt mine below. The collapse occurred after an oil-drilling operation apparently punched a hole into an abandoned mine shaft of a still active multimillion-dollar salt mine (at a depth of about 430 meters) in the Jefferson Island salt dome. As the hole enlarged because of water entering the mine, scouring and dissolving pillars of salt, the roof of

Figure 7.41
Large subsidence pit resulting from subsurface solutional mining of salt near Saltville, Virginia.

the mine collapsed, producing a large subsidence pit (Figure 7.43). The lake drained so fast that ten barges, a tugboat, and an oil drilling barge disappeared in a whirlpool of water into the mine, which in places has tunnels as wide as four-lane freeways and 25 meters high. (Mining is done with the aid of trucks and bulldozers.) The subsidence also claimed more than 25 hectares of Jefferson Island including historic botanical gardens, greenhouses, and a $500,000 private home. What is left of

the gardens is disrupted by large fractures that roughly step the land down to the new edge of the lake. These fractures are tensional in origin and are commonly found on the margins of large subsidence pits. Future movement may take place on these fractures, enlarging the pit.

Lake Peigneur refilled as water from a canal to the Gulf of Mexico was drawn in, and nine of the barges popped to the surface two days later. There was fear that potentially larger subsidence would take place as pillars of

(a)

(b)

Figure 7.42
Photographs of the Wink Sink. Depth to water in (a) is about 10 meters. Note the road and automobiles for scale. Photograph (b) is a view from the ground and shows some of the oil storage tanks near this sinkhole. (Photographs courtesy of Robert W. Baumgardner, Jr. and the Texas Bureau of Economic Geology.)

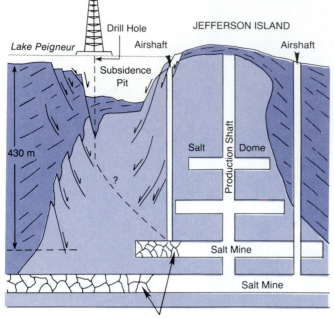

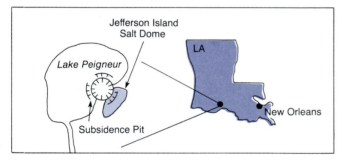

Figure 7.43
Idealized diagram showing the Jefferson Island Salt Dome collapse that caused a large subsidence pit to form in the bottom of Lake Peigneur, Louisiana.

salt holding up the roof of the salt dome dissolved. Fortunately, debris in the form of soil and lake sediment that was pulled into the mine apparently sealed the hole, and it is hoped that further major collapses will not occur.

Approximately 15 million cubic meters of water entered the salt dome and the mine is a total loss. Fortunately, the 50 miners and 7 people on the oil rig escaped. The previous shallow lake now has a large deep hole in the bottom which undoubtedly will change the aquatic ecology. In a 1983 out-of-court settlement, the salt mining company reportedly was compensated $30 million by the oil company involved. The owners of the botanical garden and private home apparently were compensated $13 million by the oil company, drilling company, and mining company.

The flooding of the mine raises important questions concerning the structural integrity of salt mines. The federal strategic petroleum reserve program is planning to store 75 million barrels of crude oil in an old salt mine of the Weeks Island salt dome about 19 kilometers from

Jefferson Island. On the other hand, while the role of the draining lake in the collapse is very significant, few salt domes have lakes above them. The Jefferson Island subsidence is thus a very rare type of event.

Coal mining

In coal mining, the practice of full recovery (removing all the coal) in subsurface mines has produced subsidence problems. The Pittsburgh area is a good example. Mining has been going on there for more than 100 years. In early years, companies purchased mining rights permitting removal of the coal with no responsibility for surface damage. The results were not so serious when mining was conducted under farmland, but as recent rapid urbanization has progressed faster than coal can be extracted, problems have resulted. If all the coal is removed, the chance of subsidence and damage to homes is high; however, if about 50 percent of the coal is left, it will usually provide sufficient support. The Bituminous Mine Subsidence and Land Conservation Act of 1966 provided for protection of public health, welfare, and safety by regulating coal mining, but this act will cause hundreds of millions of tons of coal to remain in the ground, attesting to the nature of trade-offs when there is conflict in surface and subsurface human use of the land (27).

Subsidence incidents have also been reported over coal mines that have not been worked for more than 50 years (27). On a January morning in 1973, a few residents of Wales were driving over a section of the road that suddenly collapsed into a pit 10 meters deep. Their car tottered on the brink while they scrambled to safety. The collapse was over an air shaft of a lost mine. The subsidence disrupted some utility service. Other similar subsidences have happened in the past and are likely to occur in the future.

▼ PERCEPTION OF THE LANDSLIDE HAZARD

The common reaction of homeowners in Southern California is, "It could happen on other hillsides, but never this one" (14). As with flooding, landslide hazard maps will probably not prevent people's moving into hazardous areas, and prospective hillside occupants who are initially unaware of the hazards probably cannot be swayed by technical information. Furthermore, the infrequency of large slides tends to reduce awareness of the hazard, where evidence of past events is not readily visible. Unfortunately, it often takes catastrophic events, such as the recent massive landslide in the Laguna Hills area of California, which claimed numerous expensive homes, to bring the problem to the attention of many people. In the meantime, people in many parts of the Rocky Mountains, Appalachian Mountains, and other areas continue to build homes in areas subject to future landslides.

▼ ▼ ▼ SUMMARY AND CONCLUSIONS

The most common landforms are slopes—dynamic, evolving systems in which surficial material constantly is moving downslope at rates varying from imperceptible creep to thundering avalanches.

All slopes are composed of one or more slope elements, including the crest, free-face, debris slope, and wash slope. The presence of particular slope elements on a specific slope is related to climate and rock type, which affect slope processes.

Important aspects of landslides are the type of earth material on the slope, topography, climate, vegetation, water, and time. The cause of most landslides can be determined by examining the relations between the forces that tend to make earth materials slide (*driving* forces) and forces that tend to oppose movement (*resisting* forces). The most common driving force is the weight of the slope materials, and the most common resisting force is the shear strength of the slope materials.

The role of water in landslides is especially significant and is nearly always directly or indirectly involved. Water in streams, lakes, or oceans erodes the toe area of slopes, increasing the driving forces. Excess water increases the weight of the slope materials while raising the water pressure, which in turn decreases the resisting forces in the slope materials. A rise in water pressure occurs before many landslides, and, in fact, most landslides are a result of an abnormal increase in water pressure in the slope-forming materials.

Effects of human use on the magnitude and frequency of landslides vary from insignificant to very significant. Where human activities have little effect on landslides, we need to learn all we can about where, when, and why landslides occur so that we can avoid development in hazardous areas or, if necessary, provide protective measures. In cases in which human use has increased the number and severity of landslides, we need to learn how to recognize, control, and minimize these occurrences.

To minimize landslide hazard, it is necessary to establish identification, prevention, and correction procedures. Monitoring and mapping techniques facilitate identification. Prevention of large natural slides is very difficult, but good engineering practices can do much to minimize the hazard when it cannot be avoided. Correction of landslides must attack the processes that started the slide.

Snow avalanches present a serious hazard on snow-covered, steep slopes. Loss of human life because of avalanches is increasing as more people venture into mountain areas for winter recreation.

Withdrawal of fluids such as oil and water and subsurface mining of salt, coal, and other minerals has led to subsidence hazards. In the case of fluid withdrawal, the cause of subsidence is a reduction of fluid pressures that tend to support overlying earth materials. In the case of solid material removal, subsidence may result from loss of support for the overlying material.

Perception of the landslide hazard by most people, unless they have prior experience, is negligible. Furthermore, hillside residents, like floodplain occupants, are not easily swayed by technical information.

▼ ▼ ▼ REFERENCES

1. FLEMING, R. W., and TAYLOR, F. A. 1980. *Estimating the cost of landslide damage in the United States.* U.S. Geological Survey Circular 832.
2. PESTRONG, R. 1974. *Slope stability.* American Geological Institute. New York: McGraw-Hill.
3. NILSEN, T. H.; TAYLOR, F. A.; and DEAN, R. M. 1976. *Natural conditions that control landsliding in the San Francisco Bay Region.* U.S. Geological Survey Bulletin 1424.
4. CAMPBELL, R. H. 1975. *Soil slips, debris flows, and rainstorms in the Santa Monica Mountains and vicinity, southern California.* U.S. Geological Survey Professional Paper 851.
5. BURROUGHS, E. R., Jr., and THOMAS, B. R. 1977. *Declining root strength in Douglas fir after felling as a factor in slope stability.* USDA Forest Service Research Paper INT-190.
6. TERZAGHI, K. 1950. *Mechanisms of landslides.* The Geological Society of America: Application of Geology to Engineering Practice, Berken volume: 83–123.
7. LEGGETT, R. F. 1973. *Cities and geology.* New York: McGraw-Hill.
8. KRYNINE, D. P., and JUDD, W. R. 1957. *Principles of engineering geology and geotechnics.* New York: McGraw-Hill.
9. JONES, F. O. 1973. *Landslides of Rio de Janeiro and the Sierra das Araras Escarpment, Brazil.* U.S. Geological Survey Professional Paper 697.
10. WILLIAMS, G. P., and GUY, H. P. 1973. *Erosional and depositional aspects of Hurricane Camille in Virginia, 1969.* U.S. Geological Survey Professional Paper 804.
11. OFFICE OF EMERGENCY PREPAREDNESS. 1972. *Disaster preparedness* 1, 3.
12. KIERSCH, G. A. 1964. Vaiont Reservoir disaster. *Civil Engineering* 34:32–39.
13. SWANSON, F. J., and DRYNESS, C. T. 1975. Impact of clear-cutting and road construction on soil erosion by landslides in the Western Cascade Range, Oregon. *Geology* 3, no. 7:393–96.
14. LEIGHTON, F. B. 1966. Landslides and urban development. In *Engineering geology in southern California,* ed. R. Lung and R. Proctor, pp. 149–97, a special publication of the Los Angeles Section of the Association of Engineering Geology.
15. BRIGGS, R. P.; POMEROY, J. S.; and DAVIES, W. E. 1975. *Landsliding in Allegheny County, Pennsylvania.* U.S. Geological Survey Circular 728.
16. EHLEY, P. L. 1986. The Portuguese Bend landslide: Its mechanics and a plan for its stabilization. In *Landslides and landslide mitigation in*

southern California, ed. P. L. Ehley, pp. 181–90, guidebook for field-trip, Cordilleran Section of the Geological Society of America meeting, Los Angeles, California.

17. MORTON, D. M., and STREITZ, R. 1975. Mass movement. In *Man and his physical environment,* ed. G. D. McKenzie and R. O. Utgard, pp. 61–70. Minneapolis: Burgess.

18. SLOSSON, J. E.; YOAKUM, D. E.; and SHUIRAN, G. 1986. Thistle, Utah, landslide: Could it have been prevented? *Proceedings of the 22nd Symposium on Engineering Geology and Soils Engineering,* pp. 281–303.

19. PITEAU, D. R., and PECKOVER, F. L. 1978. Engineering of rock slopes.

In *Landslides,* ed. R. Schuster and R. J. Krizek. Transportation Research Board, Special Report 176:192–228.

20. PERLA, R. I., and MARTINELLI, M., Jr. 1976. *Avalanche handbook.* U.S. Department of Agriculture, Forest Service, Agriculture Handbook 489.

21. CUPP, D. 1982. Battling the juggernaut. *National Geographic* 162:290–305.

22. POLAND, J. F., and DAVIS, G. H. 1969. Land subsidence due to withdrawal of fluids. In *Reviews in engineering geology,* ed. D. J. Varnes and G. Kiersch, pp. 187–269, The Geological Society of America.

23. BULL, W. B. 1974. Geologic factors affecting compaction of deposits in a landsubsidence area. *Geological*

Society of America Bulletin 84:3783–3802.

24. CORNELL, J., ed. 1974. *It happened last year—earth events—1973.* New York: Macmillan.

25. DOUGHERTY, P. H., and PERLOW, M., Jr. 1987. The Macungie sinkhole, Lehigh Valley, Pennsylvania: Cause and repair. *Environmental Geology and Water Science* 12, no. 2:89–98.

26. BAUMGARDNER, R. W.; GUSTAVSON, T. C.; and HOADLEY, A. D. 1980. Salt blamed for new sink in W. Texas, *Geotimes,* 25, no. 9: 16–17.

27. VANDALE, A. E. 1967. Subsidence: A real or imaginary problem. *Mining Engineering* 19, no. 9: 86–88.

The earth is a dynamic, evolving system. Earthquakes are a natural consequence of the dynamic processes that form ocean basins, continents, and mountain ranges. On a global scale (Figure 8.1), the locations of earthquakes generally coincide with the boundaries of lithospheric plates that make up the outer layer of the earth.

As new lithosphere is produced at oceanic ridge systems and older lithosphere is either consumed at subduction zones or slides past another plate, stress is produced and strain builds up in the rocks. When the stress exceeds the strength of the rocks, the rocks fail, and energy is released in the form of an earthquake. This action can be compared to pushing together and sliding two rough boards past one another. Friction along the boundary slows their motion, but rough edges break off and motion occurs at various places along the boundary. This process is similar to what happens at plate boundaries where one plate slides past another or one overrides another. The rocks undergo strain, and if the stress continues, they eventually break along a **fault**—a fracture or fracture system along which the sides move in relation to each other.

Because most natural earthquakes are initiated near plate boundaries, there tend to be linear or curvilinear continuous zones along which most seismic activity will take place (Figure 8.2). Figure 8.3 (p. 148) shows the locations of damaging historical earthquakes in the U.S. Most are in the western part of the country near the boundary between the North American and Pacific plates. However, large earthquakes have occurred far from plate boundaries. For example, during the winter of 1811–1812, a series of strong earthquakes struck the central Mississippi River valley, nearly destroying the town of New Madrid, Missouri, and killing an unknown number of people. The earthquakes, which caused church bells to ring in Boston over 1,600 km away, produced intensive surface deformation over a wide area from Memphis, Tennessee, north to the confluence of the Mississippi and Ohio Rivers; forests were flattened; fractures opened so wide that people had to cut down trees to cross them; the land sank several meters in some areas, causing flooding; and the Mississippi River supposedly reversed its flow during the shaking (1). The earthquakes occurred along a seismically active structure known as the Mississippi Embayment. The embayment (a downwarped trough or rift, at depth) is an area where the crust and lithosphere are relatively weak, and they break repeatedly as stress associated with plate motion (even though plate boundaries are far away) strains the rocks. The recurrence interval for large earthquakes along the embayment is estimated at 600–700 years; therefore, the inevitability of future damage demands that earthquake hazard be considered in design and construction of critical facilities such as nuclear power plants. Another example of a large intraplate earthquake is the 1886 event that devastated

CHAPTER EIGHT

Earthquakes and Related Phenomena

Charleston, South Carolina, claiming 60 lives and causing $23 million in property damage.

A great earthquake ranks as one of nature's most catastrophic and devastating events. Earthquakes and their related hazards have destroyed large cities and taken thousands of lives in a matter of seconds. One sixteenth century earthquake in China reportedly claimed 850,000 lives; in 1923, an earthquake near Tokyo claimed 143,000 lives; and in 1976, a catastrophic earthquake in China claimed several hundred thousand lives. It has been estimated that if a great earthquake were to hit San Francisco again, as it did in 1906, thousands of people might die in that one event. The October 17, 1989, earthquake that struck the San Francisco Bay region was a large but not great event. Nevertheless, it caused more than $5 billion of damage and killed 62 persons. Table 8.1 (p. 148) lists some of the major historical earthquakes that have occurred in the United States.

▼ SEISMIC WAVES

Breaking of rocks and resulting movement along faults produces seismic waves that cause the ground to vibrate. Some seismic waves travel within the earth (body waves), whereas others travel along the surface (Figure 8.4, p. 149). The two types of body waves generated by earthquakes are the primary, or *P*, waves and the secondary, or *S*, waves. **P waves** are the fastest of the seismic waves and may travel through both solid and liquid mediums. The rate of propagation through granite mountains for *P* waves is approximately 5.5 kilometers per second (km/sec); the rate is much reduced through liquid materials. For example, *P* waves travel at about 1.5 km/sec through water. **S waves** can only travel through

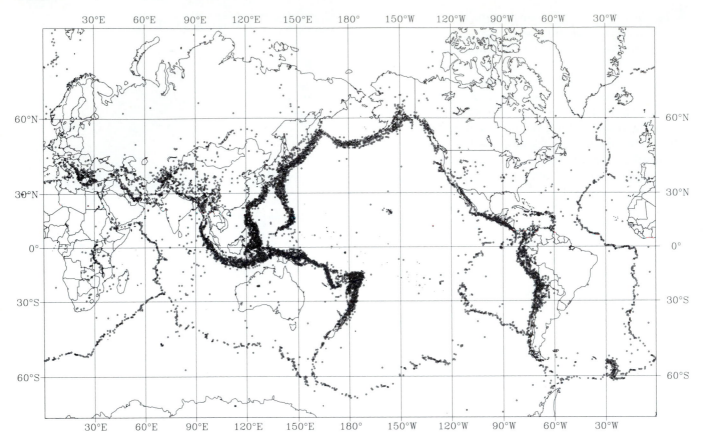

Figure 8.1
Map of global seismicity (1963–1988, Richter magnitude (M) = 5+) delineating plate boundaries and earthquake belts shown in Figure 8.2. (Courtesy of National Earthquake Information Center.)

solid earth materials, and their speed through granite is approximately 3 km/sec. Surface waves include Love and Rayleigh waves. Both travel slower than body waves, but in general, the Love waves travel faster than Rayleigh waves. Because different types of waves and waves of different frequencies travel at different speeds away from an earthquake source area, they become organized into groups of waves traveling at similar velocities. However, near the source of a large earthquake, there is not time for this segregation of the waves to take place and the shaking may be severe and complex. Waves traveling through rocks are reflected and refracted across boundaries between different earth material types and at the surface of the earth, producing amplification that may enhance shaking and damage buildings. This is further complicated by the fact that, as an earthquake takes place, the propagation of the waves is also affected by the rupturing along the fault, which may take place in a particular direction and thus tend to focus some of the earthquake energy. Finally, complexity of surface shaking may be further accentuated because both near-surface soil conditions and topography may increase or decrease the size of seismic waves at a particular site (2).

Body waves, compressional (P) and shear (S), have a wide range of frequencies but because of attenuation the most pronounced are relatively high frequencies, 0.5 to 20 Hertz (cycles per second), whereas the more complex surface waves (Love and Rayleigh waves) have lower frequencies (less than 1 Hertz). Buildings and other structures often have natural frequencies of vibration in the same range as earthquakes. This is unfortunate because shaking of buildings and other structures is facilitated when the frequency of vibration of earthquake waves is close to the natural frequency of buildings or other structures. Low buildings have higher natural frequencies than tall buildings. As a result, compressional and shear waves, with relatively high frequencies, tend to cause low buildings and other low structures to vibrate, whereas surface waves with lower frequencies tend to cause tall buildings to vibrate. This relationship between frequency of surface waves and damage was tragically demonstrated during the 1989 Loma Prieta (San Francisco) earthquake when the upper tier of the Nimitz Freeway in Oakland, California, collapsed, killing at least 41 people. It is believed that the tiered freeway had a natural frequency close to that of the bay mud on which

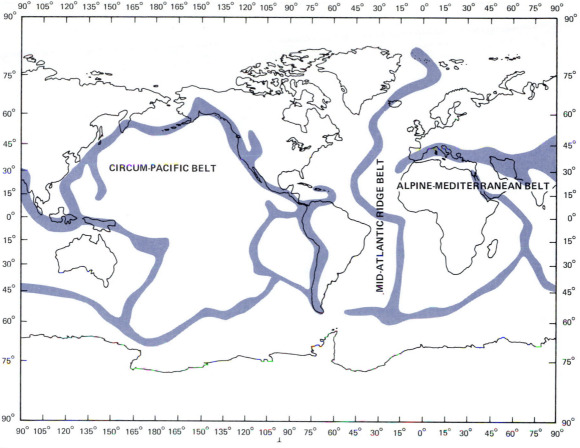

Figure 8.2

Map of the world showing the major earthquake belts as shaded areas. (Base map from NOAA.)

it was constructed and this contributed in part to the collapse (4). High frequency waves decay much more quickly with distance from a generating earthquake than do low frequency waves. Thus tall buildings may be damaged at relatively long distances (up to several hundred km) from large earthquakes (2, 3), whereas low buildings are more sensitive to shaking when they are near where an earthquake is generated.

The point or area within the earth where the first motion along a fault takes place is called the focus or **hypocenter** of the earthquake. The **epicenter** is the point on the surface of the earth directly above the hypocenter (Figure 8.4), although in some cases, as for an earthquake originating deep in a subduction zone, the epicenter may not be directly over the hypocenter. The location of an earthquake as reported by the news media is the epicenter.

▼ MAGNITUDE, INTENSITY, AND FREQUENCY OF EARTHQUAKES

The **Richter magnitude** of an earthquake is a measure of the amount of energy released and is useful in comparing

earthquakes quantitatively. It is determined by the largest amplitude of the shear wave recorded on a **seismograph**, an instrument for recording earthquake waves. The amplitude is converted to a Richter magnitude using logarithms. Thus a magnitude 6 earthquake, for example, produces a displacement on the seismograph ten times larger than does a magnitude of 5. However, the energy released, which is proportional to magnitude, may be about 30 times greater. Thus, between 27,000 and 216,000 shocks of magnitude 5 are required to release as much energy as a single earthquake of magnitude 8. On a global scale, there are probably about one million earthquakes a year that are felt by people. However, of these only a small percentage can be felt any considerable distance from their source. Nevertheless, in any given year, there is a good chance that an event of magnitude 7 to 8 or greater will occur at some location.

In a very general sense, the Richter magnitude (M) can also be related to the damage expected from an earthquake. For example, an earthquake with a magnitude of 4.0 can in rare instances cause some local damage near the source of an earthquake. An earthquake of magnitude 5 can cause moderate damage and one of

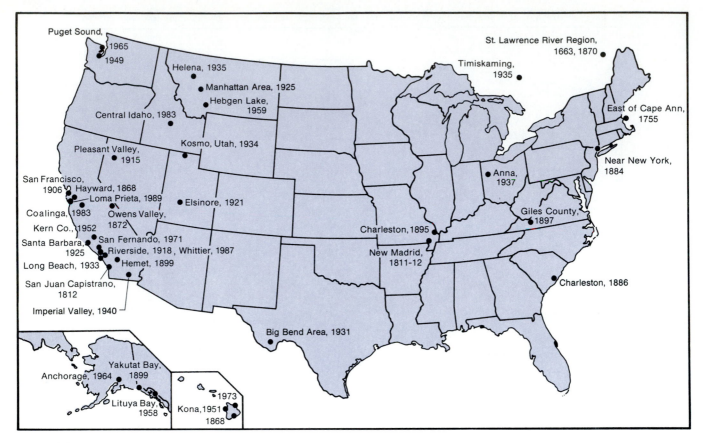

Figure 8.3
Locations of historical earthquakes in the United States that have caused damages. (From
W. W. Hays, 1981, *Hazards from Earthquakes.* In *Facing Geologic and Hydrologic Hazards,*
ed. W. W. Hays. U.S. Geological Survey Professional Paper 1240-B.)

TABLE 8.1
Selected major earthquakes in the United States.

Year	Locality	Damage $ million	Lives Lost
1811-12	New Madrid, Missouri	Unknown	
1886	Charleston, South Carolina	23	60
1906	San Francisco, California	524	700
1925	Santa Barbarta, California	8	13
1933	Long Beach, California	40	115
1940	Imperial Valley, California	6	9
1952	Kern County, California	60	14
1959	Hebgen Lake, Mont. (damage to timber and roads)	11	28
1964	Alaska and U.S. West Coast (includes tsunami damage from earthquake near Anchorage)	500	131
1965	Puget Sound, Washington	13	7
1971	San Fernando, California	553	65
1983	Coalinga, California	31	—
1983	Central Idaho	15	2
1987	Whittier, California	358	8
1989	Loma Prieta (San Francisco), California	5,000	62

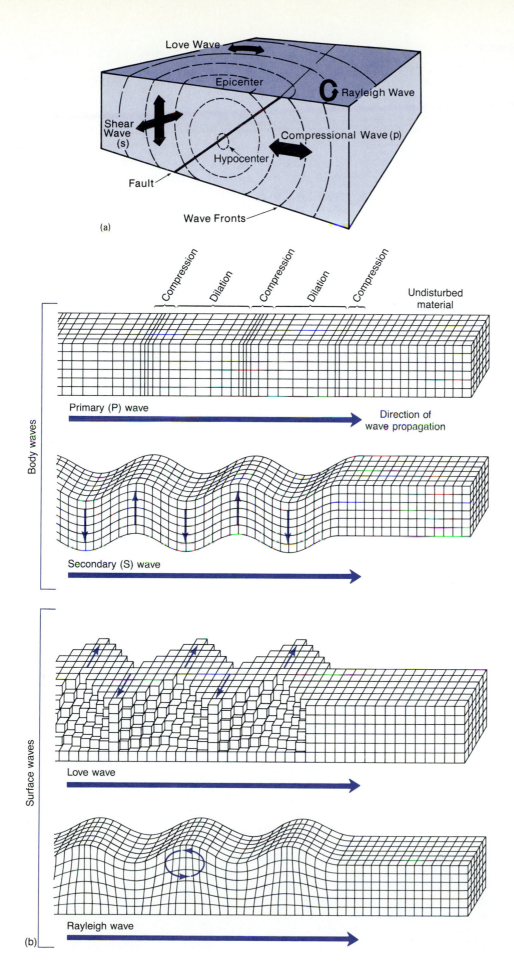

(a)

(b)

Figure 8.4
(a) Diagram of directions of vibration of body (P and S) and surface waves (Love and Rayleigh) generated by an earthquake associated with the illustrated fault. Also shown are the hypocenter and epicenter of the earthquake event. (b) Propagation of body and surface waves. ([a] from W. W. Hays, 1981, Hazards from Earthquakes. In *Facing Geologic and Hydrologic Hazards,* ed. W. W. Hays. U.S. Geological Survey Professional Paper 1240-B. [b] modified after B. A. Bolt, 1978, *Earthquakes: A Primer,* W. H. Freeman.)

magnitude 6 considerable damage. Severe damage may be expected from an earthquake with magnitude greater than 6, and an event with magnitude greater than 7 is a major earthquake capable of causing widespread damage. An earthquake with magnitude 8 or above is considered a great earthquake capable of causing catastrophic damage. However, it is important to recognize that there are exceptions to this discussion. For example, the M = 5.4 earthquake on October 10, 1986, in San Salvador did far more than moderate damage, killing about 1000 people while inflicting about $2 billion in damages. Large buildings failed as a result of the shaking and strong ground acceleration.

In practice, the Richter magnitude of an earthquake can be obtained as illustrated in Figure 8.5. The maximum amplitude and difference in arrival time of P and S waves (P waves travel faster than S waves) from a distant earthquake are measured from a seismograph. For example, Figure 8.5a shows the seismic record of an earthquake with amplitude of 85 mm and difference in arrival time of 34 seconds. The line connecting the amplitude and difference in arrival time indicates the magnitude is 6 and the distance from the epicenter is about 300 km. Records from several seismographs in the region are necessary to precisely locate the epicenter (Figure 8.5b).

Recently there has been a move to change from the Richter magnitude to the *moment magnitude* scale. The moment magnitude is considered to be a natural progression to a more quantitative and accurate scale that

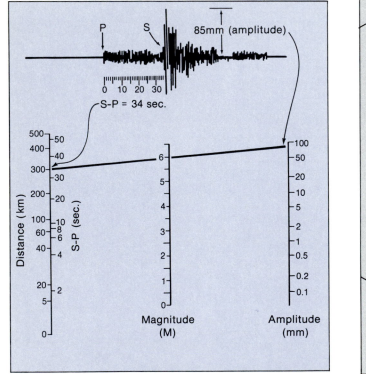

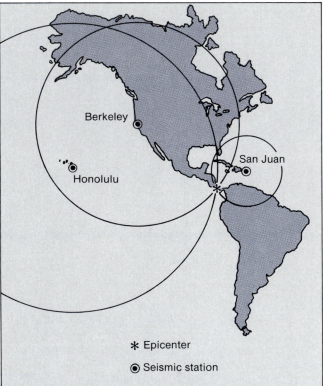

Figure 8.5

(a) Idealized diagram showing the procedure for calculating the Richter magnitude (M) of an earthquake. For our example, the maximum amplitude (85 mm) is measured from the seismic record; the difference in arrival time between the S and P waves (34 seconds) is also taken from the seismic record; and the approximate magnitude of the earthquake as well as distance from the recording station is obtained by placing a straight line between the amplitude in millimeters and difference in arrival time in seconds, as shown on the diagram. Here, the magnitude is 6 and the distance is approximately 300 km. (b) Generalized concept of how the epicenter of an earthquake is located. Distance to event from at least three seismic stations is determined (Figure 8.5a) and plotted. The intersection of the arc distances (circle radii) defines the epicenter. For the diagram the epicenter in Central America is located from data supplied by three seismic stations. Accurate location of the epicenter is not always as simple as the above hypothetical example. ([a] Modified after B. A. Bolt, 1978, *Earthquakes: A primer.* San Francisco: W. H. Freeman and Company.)

has more physical meaning. The scale is based on the seismic *moment,* defined as the product of the average amount of slip on the fault that produced the earthquake, the area that actually ruptured, and the shear modulus (strength) of the rocks that failed (5). Figure 8.6 illustrates the concepts associated with the seismic moment. In actual practice the seismic moment may be estimated for an earthquake by examining the records from seismographs, measuring the amount and length of rupture, and estimating the strength of the rocks involved. The moment magnitude scale is then determined from the mathematical relationship:

$$M = 2/3 \log M_0 - 10.7.$$

where M is the moment magnitude and M_0 is the seismic moment (5). Because the moment magnitude scale has a sounder physical base and is applicable over a wider range of ground motions than the Richter magnitude, its use is being encouraged in reporting earthquake statistics. The scale itself provides numbers that are very close to the Richter magnitude in most instances.

An earthquake's *intensity* is based on personal observations concerning the severity of its shaking. Table 8.2 is the Modified Mercalli Scale (abridged), with 12 divisions of intensity. A particular earthquake has only one magnitude, but depending on proximity to the epicenter and local geologic and engineering features, many levels of intensity can be recorded. Table 8.3 shows approximate relationships among magnitude, intensity, and average peak horizontal acceleration of the ground for earthquakes, with the 1971 (magnitude 6.6) San Fernando Valley earthquake as a specific example. Maps produced from questionnaires collected after an earthquake showing the variability of intensity (see Table 8.3) are valuable as a crude index of ground shaking and provide a detailed picture of the earthquake, including how and where it was felt. Thus, while the magnitude of

an earthquake provides information concerning the amount of energy released, the intensity reflects how people perceived the shaking.

▼ ACTIVE FAULT ZONES

As mentioned previously, a *fault* is a fracture or, more accurately, a set of fractures along which rocks have been displaced; that is, rocks on one side of the fracture have moved relative to rocks on the other side. Figure 8.7 shows the three main types of faults based on sense of relative movement.

A *fault zone* is a group of related faults or fault traces that usually form a braided pattern subparallel to the direction of the zone. The fault zone may have considerable width, ranging from a meter or so to several kilometers. Most long faults, such as the San Andreas Fault in California, are segmented, with each segment having an individual history of activity. Rupture during an earthquake is thought to stop at the boundaries between segments. Considerable research is being conducted to better understand the geology and processes that govern fault segmentation and earthquake generation on individual segments.

In assessing fault hazard, faults are commonly classified according to their recency of movement with the implicit assumption that the younger the earth materials offset by the fault, the more hazardous the fault is likely to be.

There is no real agreement as to what always constitutes an active fault. Many geologists would label a fault *active* if it can be demonstrated that there has been movement during the last 10,000 years (Holocene time). The Quaternary (last 2 million years) is the most recent period of geologic time; most of our landscape has been produced in that time. Therefore, it seems reasonable to classify any fault that has moved during the Quaternary as

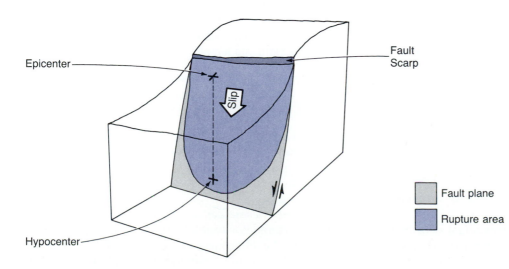

Figure 8.6
Block diagram of a fault plane and rupture area associated with an earthquake. Also shown are the hypocenter, epicenter, and surface rupture that produce a fault scarp.

Table 8.2
Modified Mercalli Intensity Scale
(Abridged)

Intensity	Effects
I	Not felt except by a very few under especially favorable circumstances.
II	Felt only by a few persons at rest, especially on upper floors of buildings. Delicately suspended objects may swing.
III	Felt quite noticeably indoors, especially on upper floors of buildings, but many people do not recognize it as an earthquake. Standing motor cars may rock slightly. Vibration like passing of truck. Duration estimated.
IV	During the day felt indoors by many, outdoors by few. At night some awakened. Dishes, windows, doors disturbed; walls make cracking sound. Sensation like heavy truck striking building; standing motor cars rocked noticeably.
V	Felt by nearly everyone; many awakened. Some dishes, windows, etc., broken; a few instances of cracked plaster; unstable objects overturned. Disturbance of trees, poles and other tall objects sometimes noticed. Pendulum clocks may stop.
VI	Felt by all; many frightened and run outdoors. Some heavy furniture moved; a few instances of fallen plaster or damaged chimneys. Damage slight.
VII	Everybody runs outdoors. Damage negligible in buildings of good design and construction; slight to moderate in well-built ordinary structures; considerable in poorly built or badly designed structures; some chimneys broken. Noticed by persons driving motor cars.
VIII	Damage slight in specially designed structures; considerable in ordinary substantial buildings with partial collapse; great in poorly built structures. Panel walls thrown out of frame structures. Fall of chimneys, factory stacks. columns, monuments, walls. Heavy furniture overturned. Sand and mud ejected in small amounts. Changes in well water. Disturbs persons driving motor cars.
IX	Damage considerable in specially designed structures; well-designed frame structures thrown out of plumb; great in substantial buildings, with partial collapse. Buildings shifted off foundations. Ground cracked conspicuously. Underground pipes broken.
X	Some well-built wooden structures destroyed; most masonry and frame structures with foundations destroyed; ground badly cracked. Rails bent. Landslides considerable from river banks and steep slopes. Shifted sand and mud. Water splashed (slopped) over banks.
XI	Few, if any (masonry) structures remain standing. Bridges destroyed. Broad fissures in ground. Underground pipe lines completely out of service. Earth slumps and land slips in soft ground. Rails bent greatly.
XII	Damage total. Waves seen on ground surfaces. Lines of sight and level distorted. Objects thrown upward into the air.

From Wood and Neuman, 1931, by U.S. Geological Survey, 1974, Earthquake Information Bulletin, v. 6 no. 5, p. 28.

potentially active (Table 8.4). Faults that have been inactive for the last 2 million years are generally classified as *inactive,* but, whereas faults accompanying recorded earthquakes certainly are active, it is often difficult to prove the activity of a fault unrelated to such easily measured phenomena.

Prehistoric earthquakes and general fault activity can be estimated by investigating fault-related landform assemblages and soils. Figure 8.8 shows the typical landforms found along active strike-slip faults character-

ized by dominant horizontal motion. Features such as offset streams, sags, linear ridges, and fault scarps may indicate recent faulting, but care must be exercised in interpreting them because some of these forms may have a complex origin.

The study of soils can also be useful in estimating the activity of a fault. For example, soils on opposite sides of a fault may be of similar or quite dissimilar age, thus establishing a relative age for the displacement. This method, in conjunction with other data, was used along

Table 8.3
Approximate relationships between the magnitude and intensity of an earthquake, with the 1971 San Fernando Valley earthquake as an example.

Magnitude	Area Felt Over (square kilometers)	Distance Felt (kilometers)	Intensity (maximum expected Modified Mercalli)	Ground Motion: (Average peak horizontal acceleration g = gravity = 9.8 meters per second per second)
3.0−3.9	1,950	25	II−III	Less than 0.15 g
4.0−4.9	7,800	50	IV−V	0.15−0.04g
5.0−5.9	39,000	110	VI−VII	0.06−0.15g
6.0−6.9	130,000	200	VII−VIII	0.15−0.30g
7.0−7.9	520,000	400	IX−X	0.50−0.60g
8.0−8.9	2,080,000	720	XI−XII	Greater than 0.60g

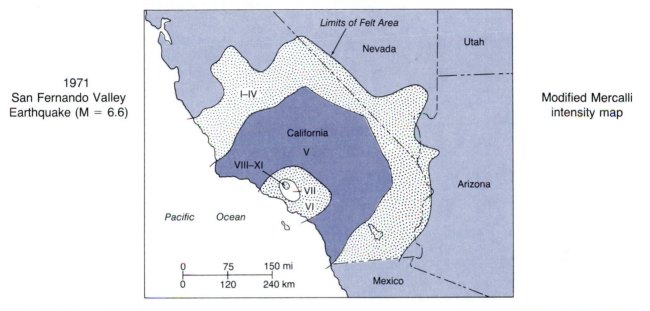

1971 San Fernando Valley Earthquake (M = 6.6)

Modified Mercalli intensity map

Modified after U.S. Geological Survey, *Earthquake Information Bulletin,* v. 6, n. 5, 1974.

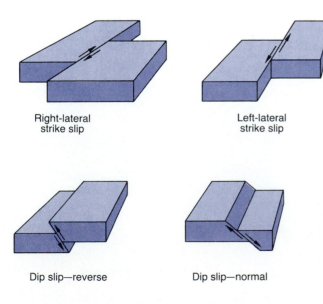

Right-lateral strike slip

Left-lateral strike slip

Dip slip—reverse

Dip slip—normal

Figure 8.7
Types of fault movement based on the sense of motion relative to the fault. (From R. L. Wesson, E. J. Helley, K. R. Lajoie, and C. M. Wentworth, U.S. Geological Survey Professional Paper 941A, 1975.)

Table 8.4
Terminology related to recovery
of fault activity.

Geologic age			Years before present	Fault activity
Era	Period	Epoch		
Cenozoic	Quaternary	Historic (Calif.) Holocene	— 200 —	Active
			—10,000—	
		Pleistocene		Potentially active
			—2,000,000—	
	Tertiary	Pre-Pleistocene	—65,000,000—	
	Pre-Cenozoic time			Inactive
Age of the earth			—4,500,000,000—	

Source: After California State Mining and Geology Board Classification 1973.

parts of the Ventura fault in California (Figure 8.9) to show its recent activity.

Prehistoric earthquake activity sometimes can be dated radiometrically (absolutely) if suitable materials can be found. For example, radiocarbon dates from faulted sediments have been obtained from the Coyote Creek fault zone and San Andreas Fault zone in southern California.

The Coyote Creek fault is characterized by predominant horizontal movement, but there is a vertical component of motion, and a trench across the fault after a 1968 earthquake revealed three older displacements of the flat-lying sediments that were dated radiometrically (carbon 14) (Figure 8.10). These data suggest that the recurrence interval for earthquakes the size of the 1968 event is about every 200 years (6).

A trench across the right-lateral, strike-slip San Andreas Fault at Pallett Creek, California, has revealed significant data concerning the earthquake history (7). During the last 1,800 years, deposition of several meters of sediment, consisting of alternating peat and gravel/sand (flood) deposits, occurred at Pallett Creek. A careful study of the peat deposits involving carbon 14 dating by Kerry Sieh suggests that over the period of record there have been 12 large earthquakes recorded at Pallett Creek. The average interval between the most recent 10 events is about 132 years, but the average apparently has little meaning because the events tend to be clustered into four groups of two or three events each. The time interval between clusters is about two to three centuries compared to a much shorter period of several decades between events within a cluster. The last large event recorded at Pallett Creek was in 1857, suggesting that the Pallett Creek is in a segment of the San Andreas Fault that currently is experiencing between-cluster activity. Thus, it is unlikely that a large earthquake will occur there in

Figure 8.8
Assemblage of fault-related land-forms along a large strike-slip fault such as the San Andreas Fault. (After R. L. Wesson, et al., U.S. Geological Survey Professional Paper 941 A, 1975.)

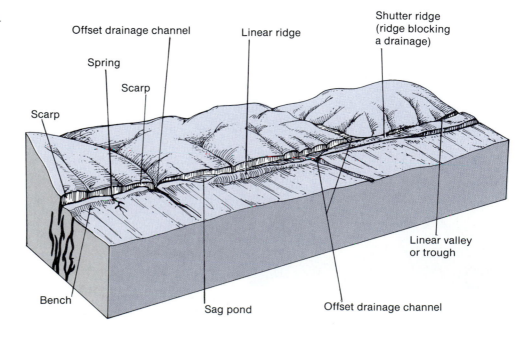

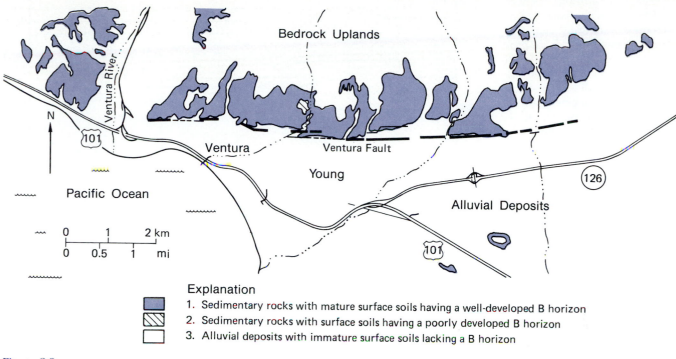

Figure 8.9
Map illustrating the difference in soils on opposite sides of the Ventura Fault in southern
California. (After Wojcicki, et al., U.S. Geological Survey Map MF-781.)

the next several decades. That is, since the time interval
between clusters is at least 200 years, then it should be
another 70 years or so before the next large earthquake
occurs at Pallett Creek. However, it is emphasized that a
record of 12 earthquakes over 1,800 years is a small
sample. Although Pallett Creek has the most complete
record for large earthquakes that has been discovered
anywhere in the world, the San Andreas Fault may yet
fool us. Figure 8.11 shows the deformation of sedimentary units associated with an earthquake that occurred at
Pallett Creek about 800 years ago (7).

▼ ESTIMATION OF SEISMIC RISK

Seismic risk maps for the United States are based on
either relative hazard or probability. Assignment of
seismic risk based on past earthquake history and on an
arbitrary relative numerical scale has serious problems
because it may assign the same risk to several areas, even
though there is not an equal probability of earthquake
occurrence. Therefore, a probabilistic approach to seismic risk (Figure 8.12), while more complicated, is
preferable. This map shows contours of maximum values
of horizontal ground acceleration caused by seismic
shaking that, with a 90 percent probability, are not likely
to be exceeded in 50 years. Another way of looking at this
is that the darkest area on Figure 8.12 represents the area
of greatest seismic hazard, interpreted as an area where
horizontal acceleration from seismic shaking of greater

than 4 meters per second (40 percent of the acceleration
of gravity) can be expected, with odds of only one in ten
to be exceeded in a 50-year period. Thus, when one uses
the probabilistic method, the total number of earthquakes as well as other information becomes important
in the analysis. Although regional earthquake hazard
maps are valuable, considerably more data are necessary
to identify hazardous areas more precisely to assist in
developing building codes and determining insurance
rates.

At a regional scale, geologists in California have
estimated the conditional probability of major earthquakes along segments of the San Andreas Fault for the
30-year period from 1988 to 2018 (Figure 8.13). The
probabilities are assigned based on synthesis of historical
records and geologic evaluation of prehistoric earthquakes (8). Results suggest that the Parkfield segment is
almost certain to rupture by the year 2018. This estimate
is based in part on the observation that moderate
earthquakes have occurred on the Parkfield segment in
1857, 1881, 1901, 1922, 1934, and 1966, or, on the
average, every 21 to 22 years. In fact, the specific
prediction is that a moderate earthquake will occur with
a probability of 95 percent before the year 1993. The
southern segment of the fault has nearly a 50 percent
probability of producing a major earthquake by the year
2019. The study assigned a probability of about 30
percent for a major event on the San Andreas Fault in the
Santa Cruz mountains where the magnitude 7.1 Loma

Figure 8.10
Generalized sketch of trench wall showing vertical displacement of flat-lying sediments associated with the Coyote Creek fault, California (a). Older offsets are progressively greater. Graph of vertical deformation against time (b). Derived from radiocarbon dates of the faulted sediments. (After M. M. Clark, A. Grantz, and R. Meyer, U.S. Geological Survey Professional Paper 787, 1972.)

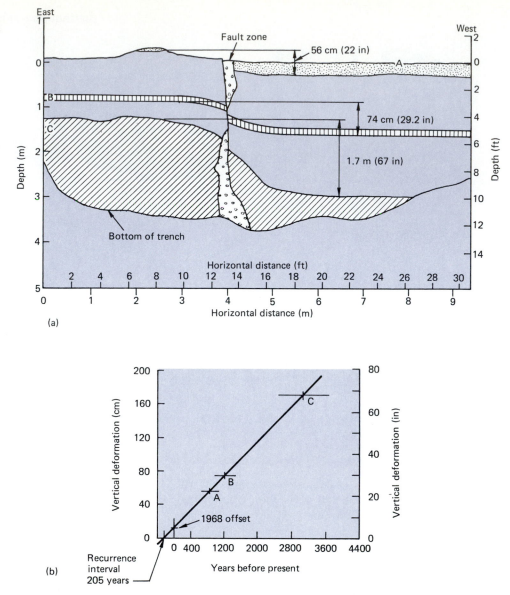

(a)

(b)

Figure 8.11
Part of the trench wall crossing the San Andreas fault at Pallett Creek showing a 30-centimeter offset. Notice that the strata above the 22-centimeter scale (central part of photograph) are not off-set. The offset sedimentary layers (white beds) and the dark organic layers were produced by an earthquake that occurred about 800 years ago. (Photo courtesy of Kerry Sieh.)

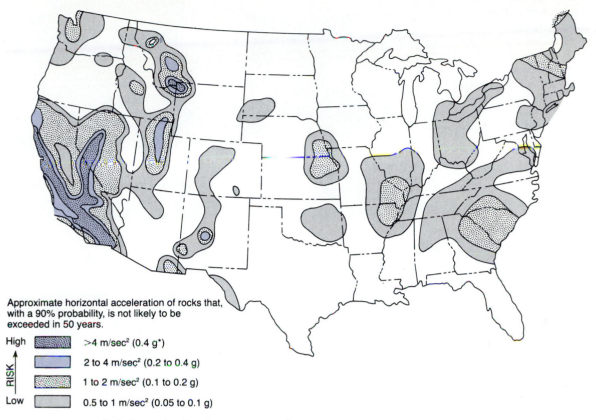

Figure 8.12

A probabilistic approach to the seismic hazard in the United States. The darker the area on the map, the greater the hazard. (From S. T. Algermissen and D. M. Perkins, 1976. U.S. Geological Survey Open File Report 76–416.)

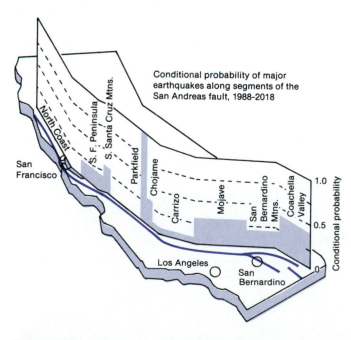

Figure 8.13

Conditional probability of a major earthquake along segments of the San Andreas fault (1988–2018). (After Heaton, T. H., et al., 1989. U.S. Geological Survey Circular 1031.)

Prieta earthquake occurred on October 17, 1989. This is strong support for the validity of the probability approach.

▼ EFFECTS OF EARTHQUAKES

Like other natural hazards, earthquakes often produce both primary and secondary effects.

Primary Effects

Primary effects of earthquakes are caused directly by the phenomena. They include violent ground motion accompanied by surface rupture and, perhaps, permanent displacement of a meter or so; for example, the 1906 earthquake in San Francisco produced 5 meters of horizontal displacement. Such violent motions produce sudden surface accelerations of the ground which can snap and uproot large trees and knock people to the ground. This motion may shear or collapse large buildings, bridges, dams, tunnels, pipelines, and other rigid structures (9). For example, the great 1964 Alaskan earthquake caused extensive damage to transportation

Figure 8.14
Damage to Penney's department store in Anchorage, Alaska. The building failed after sustained seismic shaking. At the time of the photograph, most of the rubble had been cleared from the streets. (Photo by George Plafker, courtesy of U.S. Geological Survey.)

Figure 8.15
Damage to homes in San Fernando, California, caused by the 1971 earthquake. (Photo [a] by R. Castle, courtesy of U.S. Geological Survey. Photo [b] courtesy of Los Angeles City Department of Building and Safety.)

(a)

(b)

Figure 8.16
Rescue operations at Veterans Hospital in San Fernando, California, after the 1971 earthquake. (Photo courtesy of Los Angeles City Department of Building and Safety.)

Figure 8.17
Possible failure of the lower Van Norman Dam, severely damaged as a result of the 1971 earthquake, threatened the lives of thousands of people. (Photo by R. E. Wallace, courtesy of U.S. Geological Survey.)

systems—railroads, airports, and buildings (Figure 8.14). Fortunately, Alaska has a relatively low population density. What would be the effect of a large earthquake in a highly populated area? A hint was given by the 1971 earthquake that occurred on the fringe of a highly populated area of the San Fernando Valley in California. The earthquake with magnitude of 6.6 released less than 1 percent of the energy of the Alaskan earthquake, yet it claimed one-half as many lives and did more property damage than the Alaskan event of much greater magnitude. Damage to homes and larger buildings was extensive in the city of San Fernando (Figures 8.15 and 8.16), and the lower Van Norman Dam above the highly populated valley was severely damaged (Figure 8.17) and on the brink of catastrophic failure, threatening the lives of 80,000 people who evacuated their homes (10, 11). They returned safely to their homes four days later, after the water in the reservoir had been lowered to a safe level. The 1989 Loma Prieta (San Francisco) earthquake, with a magnitude of 7.1, was also much smaller than the Alaskan event and yet caused about $5 billion of damage, which rates it one of the most expensive hazardous events ever in the United States.

Secondary Effects

Secondary effects of earthquakes include a variety of short-range events, such as liquefaction, landslides, fires, tsunamis, and floods; and long-range effects, including such phenomena as regional subsidence or emergence of land-masses and regional changes in groundwater levels.

Liquefaction.   Water-saturated granular material may be transformed from a solid state to a liquid state in a process called **liquefaction**, which results from an increase in the pore-water pressure caused by intense shaking. Earthquakes may cause liquefaction of near-surface, water-saturated silts and sands, causing the materials to lose their shear strength and flow. Three types of failure in response to liquefaction are recognized: flow landslides (earthflows) may occur on even moderate slopes; laterally spreading landslides may occur on gentle or nearly flat slopes, in which case the ground pulsates with the quaking and is accompanied by cracks, fissures, and differential settling; and quick condition failure may occur, characterized by a complete loss of shear strength (bearing capacity), in which case buildings may tilt and sink into the liquefied sediments while buried tanks may rise buoyantly (12).

Landslides.   Other varieties of landslides, in addition to those associated with liquefaction, are triggered or directly caused by earthquakes. They can be extremely destructive and can cause great loss of life, as demonstrated by the 1970 Peru earthquake. In that earthquake, more than 70,000 people died, of whom 20,000 were killed by a gigantic avalanche. The 1964 Alaskan earthquake produced thousands of landslides and avalanches, some of which devastated Anchorage (13, 14). Numerous landslides associated with the 1989 Loma Prieta earthquake caused extensive damage to buildings and roads in the Santa Cruz Mountains south of San Francisco.

Fires.   Disruption of electrical power lines and broken gas lines can start fires that are difficult to control because firefighting equipment may be damaged and essential water mains may be broken. Access to fires is often hampered by blocked and damaged roads. Earthquakes in Japan and the United States have been accompanied by terrible fires. The San Francisco earthquake of 1906 has repeatedly been referred to as the "San Francisco fire," and, in fact, 80 percent of the damage was caused by

Table 8.5
Casualties and damage in the United States caused by tsunamis from 1900 to 1971.

Year	Dead	Injured	Estimated Damage ($ × 1000)	Area
1906	—	—	5	Hawaii
1917	—	—	*	American Samoa
1918	—	—	100	Hawaii
1918	40	—	250	Puerto Rico
1922	—	—	50	Hawaii, California, American Samoa
1923	1	—	4,000	Hawaii
1933	—	—	200	Hawaii
1946	173	163	25,000	Hawaii, Alaska, West Coast
1952	—	—	1,200	Midway Island, Hawaii
1957	—	—	4,000	Hawaii, West Coast
1960	61	282	25,500	Hawaii, West Coast, American Samoa
1964	122	200	104,000	Alaska, West Coast, Hawaii
1965	—	—	10	Alaska

*Damage reported but no estimates available.
Source: Eskite, "Analysis of ESSA Activities Related to Tsunami Warning," NOAA.

terrible fires that ravaged the city for several days. The 1989 earthquake that struck San Francisco also caused large fires in the Marina District of the city. The 1923 earthquake in Japan killed 143,000 people, 40 percent of whom died in a fire storm that engulfed an open space where people had gathered in an unsuccessful attempt to reach safety (9).

Tsunamis. Often incorrectly called *tidal waves, tsunamis* or seismic sea waves are extremely destructive and present a serious natural hazard. Fortunately, they are relatively rare and are usually confined to the Pacific basin. The frequency of these events in the United States is about one every eight years (15). Table 8.5 lists the casualties and damage in the United States from tsunamis during the period from 1900 to 1971.

Tsunamis apparently originate when ocean water is vertically displaced during large earthquakes or other phenomena. In open water, the waves may travel at speeds as great as 800 kilometers per hour, and the distance between successive crests may exceed 100 kilometers. Wave heights in deep water may be less than one meter, but when the waves enter shallow coastal waters, they slow to less than 60 kilometers per hour, and wave heights may increase to more than 15 meters. Figure 8.18 shows a tsunami that struck Hawaii.

Figure 8.18
Sandy Beach on the Island of Oahu (a) moments before a tsunami generated by an earthquake in the vicinity of the Aleutian Trench struck the beach. The arrow on photograph (a) shows the exact location of photograph (b), taken a few minutes later after the tsunami struck. Notice the man circled on the lower photograph fleeing from the wave and the several automobiles, also circled, that were swept off the highway by the surging wave. (Photo by Y. Ishii, courtesy of *Honolulu Advertiser.*)

(a)

(b)

Figure 8.19
Tsunami warning system. Map shows reporting stations and tsunami travel times to Honolulu, Hawaii. (From NOAA).

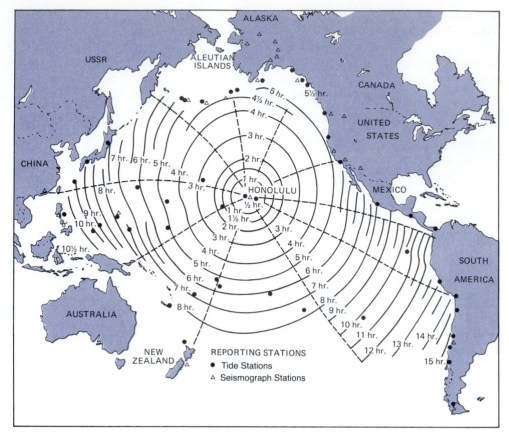

Tsunamis claimed most of the lives lost in the 1964 Alaskan earthquake and can cause catastrophic events thousands of kilometers from where they are generated, as exemplified by the 1960 seismic waves that killed 61 people in Hawaii. The waves were caused by an earthquake in Chile and reached Hawaii in only 15 hours.

Although there is no way to predict tsunamis, it is possible to detect them and warn coastal communities that lie in their path. After an earthquake or another tsunami-generating disturbance has been recognized, the arrival time of a sea wave can be predicted to within plus or minus 1.5 minutes per hour of travel time. This

Figure 8.20
Tsunami damage to railyards in Seward, Alaska, caused by the 1964 earthquake. (Photo courtesy of NOAA.)

Figure 8.21
Tsunami damage to fishing boats at Kodiak, Alaska, caused by the 1964 earthquake. (Photo courtesy of NOAA.)

information can be used to produce a tsunami warning-system map, as shown for Hawaii in Figure 8.19.

Damage caused by tsunamis is most severe at the water's edge, where boats, harbors, buildings, transportation systems, and utilities may be destroyed (Figures 8.20 and 8.21). The waves may also be disastrous to aquatic life and plants in the nearshore environment (15).

Regional changes in land elevations. Vertical deformations (uplift and subsidence) are another secondary effect of some large earthquakes. The Alaskan earthquake of 1964 caused vertical deformation of more than 250,000 square kilometers (13). The deformation includes two major zones of warping, each about 1,000 kilometers long and more than 210 kilometers wide (Figure 8.22). The seaward zone is one of uplift as great as 10 meters; the landward zone is characterized by subsidence as great as 2.4 meters centered in the vicinity of Portage. Effects of this deformation vary from severely disturbed coastal marine life to changes in groundwater levels. In addition, flood damage occurred in some communities that experienced subsidence (Figure 8.23), whereas in areas of uplift, canneries and fishermen's homes were placed above the reach of most high tides. Before the earthquake, these structures and their docks were on the water (Figure 8.24).

▼ TECTONIC CREEP

Slow, nearly continuous movement along a fault zone not accompanied by felt earthquakes is known as **tectonic creep.** This process can slowly damage roads, sidewalks, building foundations, and other structures (Figure 8.25,

p. 166). It has damaged culverts under the football stadium at the University of California at Berkeley (Figure 8.26, p. 166). Periodic repairs have been necessary as the cracks develop, and movement of 3.2 centimeters in eleven years was measured (16). Faster rates of tectonic creep have been recorded on the Calaveras fault zone near Hollister, California. There, a winery located on the fault is slowly being pulled apart at about 1 centimeter per year (17). Damage from tectonic creep generally occurs along narrow zones that delineate a fault along which movement is slow and continuous.

▼ EARTHQUAKES CAUSED BY HUMAN USE OF THE LAND

Various human activities have increased or caused earthquake activity. The damage is regrettable, but the lessons we have learned may help in the future to control or stop large catastrophic earthquakes. Three ways people have caused earthquakes are by loading the earth's crust, disposing of waste into deep wells, and setting off underground nuclear explosions (18).

About 600 local tremors occurred during the first ten years following completion of Hoover Dam, which supplies Lake Mead in Arizona and Nevada. Most of the tremors were very small, but one had a magnitude of about 5 and two had magnitudes of about 4 (18). An earthquake in India that killed 200 people and one in Zambia, both with magnitudes of about 6, were also associated with reservoirs. Evidently, fracture zones (faults) are activated by the increased load of water on the land and by increased water pressure in the rocks below the reservoir.

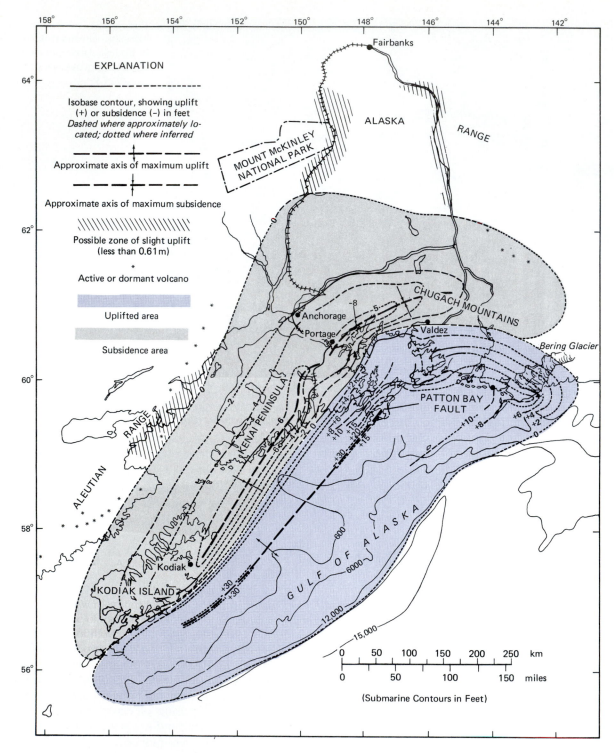

Figure 8.22
Map showing the distribution of tectonic uplift and subsidence in south central Alaska
caused by the Alaskan earthquake of 1964. (2 to 30 ft = 0.6 to 9.2 m). (From E. B. Eckel,
U.S. Geological Survey Professional Paper 546, 1970.)

Figure 8.23
Flooding of the Portage area of Alaska caused by regional subsidence resulting from tectonic land changes accompanying the 1964 earthquake. (Photo courtesy of U.S. Geological Survey.)

Figure 8.24
Canneries and fishing homes along Orca Inlet in Prince William Sound after the 1964 Alaskan earthquake. The photograph was taken at a 3-meter tidal stage, which would have reached beneath the docks prior to the earthquake. (Photo courtesy of U.S. Geological Survey.)

Events in 1966 dramatically suggested that deep-well disposal of liquid chemical waste activated fracture zones and caused more than 600 earthquakes in the Denver, Colorado, area from April 1962 to November 1965 (19). The greatest magnitude was 4.3, large enough to knock bottles off shelves in stores. The chemical waste was a by-product of the Rocky Mountain Arsenal's manufacturing of materials for chemical warfare. The arsenal started production in 1942, and evaporation from surface ponds disposed of the waste water until 1961, when groundwater pollution became a problem and engineers decided to pump the waste into a 3,600-meter-deep disposal well. The rock accepting the waste was highly fractured metamorphic rock. The new liquid increased the fluid pressure, facilitating slippage along fractures and precip-

itating earthquakes. The well-known correlation between rate of injection of waste and frequency of earthquakes is shown in Figure 8.27. When the injection stopped, the earthquakes stopped. Scientists speculate as to whether this principle of starting and stopping earthquakes by varying the fluid pressure in rocks can be used to prevent large earthquakes.

Numerous earthquakes have been triggered by nuclear explosions at the Nevada test site (18). After underground explosion of a 1.1-megaton nuclear device, thousands of aftershocks were recorded. The depth of the shocks ranged from the surface down to 7 kilometers. Magnitudes did not exceed 5.0, which is considerably less than the magnitude of 6.3 from the initial explosion. Analysis of the aftershocks suggested they were triggered

Figure 8.25
Curb displaced by fault creep on the Calaveras Fault, central California. This type of movement is generally very slow and not accompanied by earthquakes noticeable to people. (Photo courtesy of U.S. Geological Survey.)

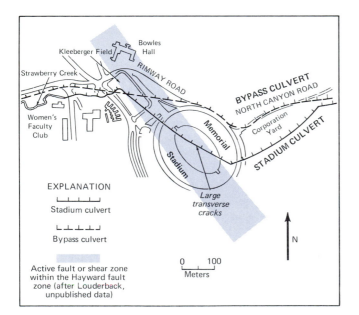

Figure 8.26
Map showing the location of the University of California Memorial Stadium, Berkeley; the active fault or shear zone within the Hayward fault zone; and the stadium culvert where major cracking has taken place. (After Radbruch, et al., U.S. Geological Survey Circular 525, 1966.)

primarily by release of natural tectonic strain. Again, scientists wonder if this principle might be used to prevent a large earthquake.

▼ EARTHQUAKE HAZARD REDUCTION

Like landslides and floods, earthquakes are natural processes, and even though human use and interest may affect, to a lesser or greater extent, some aspects of the magnitude and frequency of these processes, they cannot be completely eliminated. Therefore, we must learn to minimize the hazards.

A comprehensive earthquake hazard reduction program must be multifaceted and include recognition of active fault zones and earth materials sensitive to shaking, continued research to better predict and possibly control earthquakes, and development and improvement of possible adjustments to earthquake activity.

Seismic Shaking and Ground Rupture

Active faults present two basic hazards to people and their structures: *seismic shaking* and *surface rupture*. Seismic shaking occurs during sudden movement along a deep portion of a fault and may or may not be accompanied by surface rupture. On the other hand, rupture of the earth's surface along a fault or fault zone is characterized by fault motion that may be instantaneous or a slow creep, and may or may not be accompanied by an earthquake (20).

Two types of active faults exist that can be defined in terms of the type of hazard each presents: faults with both ground rupture and seismic shaking hazard; and faults with seismic shaking but slight or no ground rupture. It is argued that recognition of the type of fault activity associated with a particular fault is important and should be considered in land-use planning and programs to minimize potential damages from earthquakes or other activities associated with faults or fault zones.

The San Andreas fault in California is an example of a fault with both a seismic shaking and a ground rupture hazard. Historic earthquakes and fault scarps in alluvial fan and other geologically young material demonstrate the potential of the fault to produce large earthquakes and ground rupture.

Faults that can produce damaging earthquakes but do not break surficial earth materials are called *buried faults*. Several damaging earthquakes in California (for example, the 1952 Kern County, 1983 Coalinga, and 1987 Whittier events) were on *buried reverse faults*, which do cause surficial uplift and folding. These faults are being studied to learn more about the subsurface earthquake hazard.

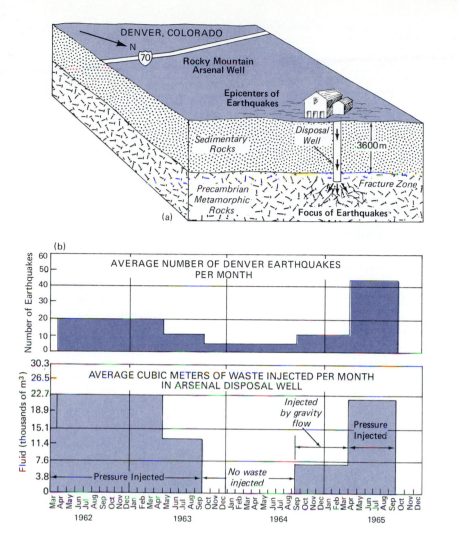

Figure 8.27
Generalized block diagram showing the Rocky Mountain Arsenal
well (a) and graph showing the
relationship between earthquake
frequency and rate of injection of
liquid waste for five characteristic
time periods (b). (Graph [b] after
D. M. Evans, *Geotimes* 10, 1966.
Reprinted by permission.)

Earth Materials and Seismic Activity

Surficial earth materials such as mud, alluvium, and bedrock behave differently in response to seismic shaking of various frequencies. For example, the intensity of shaking may be much more severe for unconsolidated sediment than for bedrock (Figure 8.28), and, therefore, shaking may be stronger for sites underlain by mud and alluvium than for bedrock sites much closer to the fault (21). This was dramatically and tragically shown during the 1985 earthquake that struck Mexico City, killing approximately 10,000 people. Severe seismic shaking caused the collapse of approximately 1,000 buildings in Mexico City, located several hundred kilometers from the source of the earthquake in the floor of the Pacific Ocean off Mexico. Coastal cities much closer to the earthquake suffered considerably less damage; the reasons are clearly related to the geology of the Mexico City area.

Figure 8.29a shows the generalized geology of Mexico City and the areas where the greatest damage occurred. Mexico City is built on ancient lake beds consisting of clays, silts, and sands which greatly intensi-

fied the seismic shaking. Seismic waves from the 8.1-magnitude earthquake offshore of Mexico initially contained many different frequencies, but the seismic waves that survived the several-hundred-kilometer journey inland to Mexico City were those with a relative long period of perhaps one to two seconds (frequencies 1.0 to 0.5 Hertz). It is speculated that when these waves struck the lake beds beneath Mexico City, the amplitude of shaking may have increased at the surface by a factor 4 to 5 times. The intense regular shaking caused the buildings to sway back and forth in time with the shaking and eventually many of them collapsed or pancaked as upper stories collapsed onto lower ones (Figure 8.29b). Most of the damage was done to buildings with six to sixteen stories because these buildings have a natural frequency that nearly matched that of the arriving seismic waves (21).

The major lesson from the Mexico City earthquake is clear—buildings constructed on materials likely to accentuate or increase seismic shaking are extremely vulnerable to earthquakes, even if the event is centered

Figure 8.28
Generalized relationship between near-surface earth material and amplification of shaking during a seismic event.

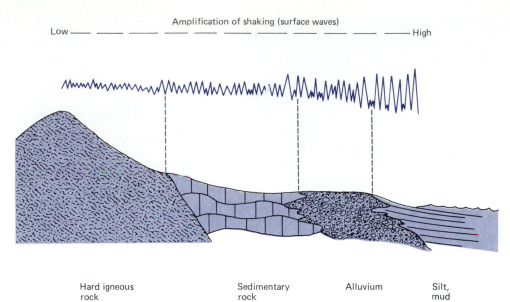

Hard igneous rock   Sedimentary rock   Alluvium   Silt, mud

several hundred kilometers away. For example, a large earthquake on the San Andreas fault near San Francisco could conceivably heavily damage the city of Sacramento, 130 km away, where the substrata are composed of alluvial deposits likely to accentuate seismic shaking. The U.S. building codes are roughly equivalent to those in Mexico City for the newer buildings (21)!

Loma Prieta (San Francisco) earthquake.   The largest earthquake to occur in the San Francisco Bay region since 1906 struck on October 17, 1989, with a magnitude of 7.1. The epicenter was approximately 80 kilometers southeast of the city of San Francisco and caused extensive damage over a wide area. The number of fatalities associated with the earthquake was 62 and, as mentioned previously, the cost of property damage— about $5 billion—greatly exceeded that of previous earthquakes in the United States. To put the event in perspective, the Loma Prieta earthquake was approximately the same magnitude as the earthquake of December 1988 that killed approximately 25,000 people in Armenia. A big reason for the large difference in fatalities is that design and construction of buildings is far better in California than in Armenia.

Prior to the Loma Prieta earthquake, it was predicted that, should a large earthquake occur in the San Francisco Bay region, damage would be most severe in areas of bay fill and mud. Figure 8.30 shows the San Francisco Bay area, the earthquake epicenter on the San Andreas fault, and the distribution of bay fill and mud. Extensive damage occurred over a wide area, including near the epicenter in the Santa Cruz Mountains and in the San Francisco area. Particularly heavy damage occurred in the Marina district in San Francisco, where seismic shaking, liquefaction, and fire destroyed many structures (Figure

8.31). Figure 8.32 shows the Oakland area where the Nimitz Freeway collapsed, killing about 41 people. Structural damage and destruction in the Marina district occurred primarily in areas of construction on bay fill and mud. Much of the fill is debris dumped in the bay during the cleanup following the 1906 earthquake. Collapse of the tiered freeway occurred on the section of the roadway constructed on bay fill and mud. Where the freeway was constructed on older alluvium, less shaking occurred and the structure survived the earthquake. Investigation following the earthquake concluded that ground motion amplification of seismic waves was a significant factor in the failure of the tiered freeway, and reconstruction of the highway must consider future events that would cause similar shaking (4).

Minimizing damage associated with shaking.   The obvious procedure for minimizing damage associated with shaking are, first, to map areas that are especially sensitive, and, second, to make this information available to those who are responsible for land-use decisions. In fact, this has now been done for the Los Angeles and San Francisco Bay areas in California.

As part of earthquake disaster preparation, it is important to predict possible geologic effects of an earthquake of a particular magnitude. This has been done for the San Francisco Bay region (22). Figure 8.33 shows a generalized geological cross-section of the southern part of the bay through Menlo Park as well as expected effects of a moderate to large earthquake on the San Andreas fault. Notice that, although surface faulting and landsliding are almost entirely restricted to the fault zone and strong bedrock shaking decreases with distance from the fault, the bay, at 16 kilometers from the fault, is very vulnerable to liquefaction and severe shaking because of

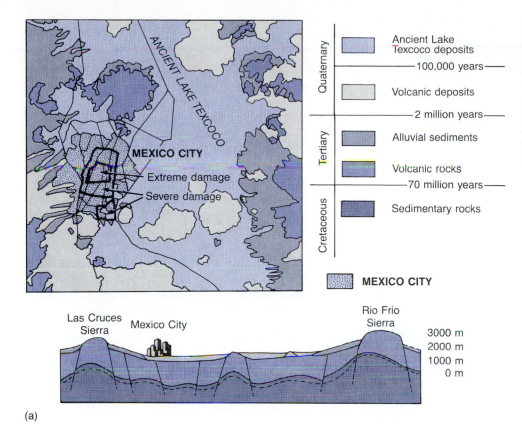

(a)

		Ancient Lake Texcoco deposits
Quaternary		
		──100,000 years──
		Volcanic deposits
		──2 million years──
Tertiary		Alluvial sediments
		Volcanic rocks
		──70 million years──
Cretaceous		Sedimentary rocks

MEXICO CITY

Las Cruces Sierra   Mexico City   Rio Frio Sierra

3000 m
2000 m
1000 m
0 m

Figure 8.29
Earthquake damage, Mexico City, 1985. (a) Generalized geologic map of Mexico City showing ancient lake deposits where greatest damage occurred. (b) Highrise building, one of many that collapsed. (Map and photo courtesy of U.S. Geological Survey, T. C. Hanks, and Darrell Herd.)

(b)

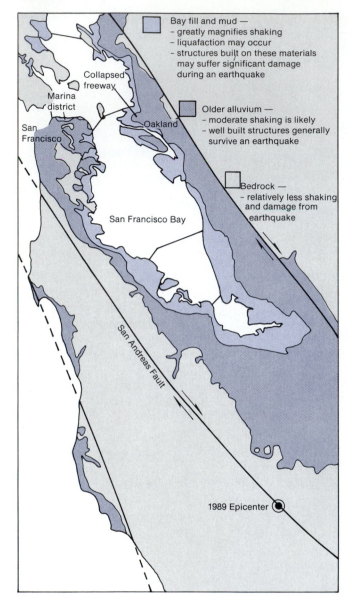

Figure 8.30
San Francisco Bay region, showing San Andreas fault and epicenter of 1989 earthquake, which had a magnitude of 7.1. (Modified after T. Hall. Data from U.S. Geological Survey.)

the amplification of surface seismic waves in the bay mud. Damage from the 1989 earthquake in the San Francisco Bay area was predictable based on the study of the distribution of earth materials and their response to seismic shaking.

▼ EARTHQUAKE PREDICTION

Earthquake prediction is an area of serious research. The Japanese made the first attempts, with some success, using as data the frequency of microearthquakes, repetitive level surveys, water-tube tilt meters, and geomagnetic observations. They found that earthquakes in the

area they studied were always accompanied by swarms of microearthquakes that occurred several months before the major shocks. Furthermore, ground tilt correlated strongly with earthquake activity: anomalous tilt occurred shortly before some shocks of a magnitude near 5. Anomalous magnetic fluctuations were also reported (18).

Chinese scientists in 1975 made the first successful prediction of a major earthquake. The Haicheng earthquake of February 4, 1975, had a magnitude of 7.3 and destroyed or damaged about 90 percent of the structures in a city of 90,000 people. The lives of thousands of people undoubtedly were saved by the massive evacuation from unsafe housing just before the earthquake. The short-term prediction was based primarily on a series of foreshocks that began four days prior to the main shock. On February 1 and 2 there were several shocks with a magnitude less than 1. On February 3, less than 24 hours before the main quake, a foreshock of a magnitude of 2.4 occurred, and in the next 17 hours, 8 shocks with magnitude greater than 3 occurred. Then, as suddenly as it began, the foreshock activity became relatively quiet for 6 hours prior to the main earthquake (23, 24).

Unfortunately, foreshocks do not always precede large earthquakes, and in 1976, a catastrophic earthquake, one of the deadliest in recorded history, struck near the mining town of T'anshan, China, killing several hundred thousand people. There were no foreshocks!

Optimistic scientists in the United States, Japan, the USSR, and China believe that eventually we consistently will be able to make long-range prediction (10,000 years to about 100 years), medium-range prediction (a few years to a few months), and short-range prediction (a few days or hours) for general locations and magnitudes of earthquakes. Earthquake prediction is a complex problem, however, and it will probably be many years before dependable short-range prediction is possible. Such prediction will most likely be based on these factors: previous patterns and frequency of earthquakes, anomalous changes in the stored strain of rocks, magnetic properties of rocks, vertical or horizontal motions (tilting) of rocks, seismic gaps along active faults, changes in the electrical resistivity of the earth, changes in the amount of dissolved radioactive gas (radon) in groundwater, and, perhaps, anomalous behavior of animals.

Rates of uplift and subsidence, especially rapid or anomalous change, may be significant in predicting shocks. For example, for more than 10 years before the 1964 earthquake near Niigata, Japan (magnitude 7.5), there was anomalous uplift of the earth's crust (Figure 8.34) (25).

Seismic gaps are defined as areas along active fault zones that are capable of producing large earthquakes but have not recently produced a large earthquake. These areas are thought to store tectonic strain and thus are candidates for future large earthquakes (26).

Figure 8.31
Damage to buildings in the Marina district of San Francisco resulting from the 1989 earthquake. (Photograph by John K. Nakata, courtesy of U.S. Geological Survey.)

Seismic gaps have been extremely useful in medium-range earthquake prediction. At least ten large plate-boundary earthquakes have been successfully forecast since 1965, including one in Alaska, three in Mexico, one in South America, and three in Japan. In the United States, seismic gaps are recognized along the San Andreas fault near Tejon—the Big Bend section of the fault northwest of Los Angeles that last ruptured in 1857—and on the southern part of the fault in the Coachella Valley that hasn't produced a great earthquake for several hundred years. Both sections are likely candidates to produce a great earthquake in the next few decades (5, 26).

As earth scientists examine patterns of relative seismicity, two ideas are emerging. First, there are sometimes reductions in small or moderate earthquakes prior to a larger event; and second, small earthquakes may tend to ring an area where a larger event eventually occurs. Such a ring (or doughnut) was noticed to have developed in the 16 months prior to the 1983, M = 6.2 earthquake in Coalinga, California. Unfortunately, this was not noticed until after the event, and quantitative

criteria to identify seismic doughnuts, although the subject of serious research, are not yet refined.

Electrical resistivity is a measure of a material's ability to retard an electrical current. Conductors such as copper and aluminum have a very low resistance, whereas other materials such as quartz (an insulator) have a very high resistance to electric current. In general, the earth is a good conductor, but its resistivity varies with the amount of groundwater present and many other factors. If a current is fed into the earth between two points separated by several kilometers, changes in the voltages will be noted if any change in electrical resistivity takes place in the rocks. Changes before earthquakes have been reported in the United States, the USSR, and China (Figure 8.35) (25).

The amount of the radioactive gas radon that is dissolved in the water of deep wells may apparently increase significantly before an earthquake (Figure 8.36). The technique of monitoring radon appears promising and is being studied carefully in the USSR, China, and the United States.

Figure 8.32
(a) Generalized geologic map of part of the San Francisco Bay showing bay fill and mud and older alluvium. (b) Collapsed freeway. (Map [a] modified from Hough, S. E., et al., 26 April, 1990. *Nature* Vol. 344, No. 6269, pp. 853–855. Copyright © Macmillan Magazines Ltd., 1990. Used by permission of the author. Photo [b] courtesy of John K. Nakata, U.S. Geological Survey.)

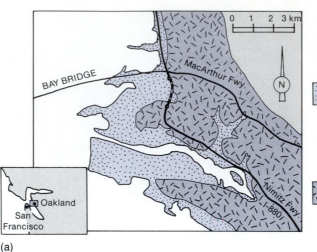

Collapse of two-tier section of Nimitz Freeway.

Bay fill and mud. Greatly magnifies shaking — liquafaction may occur. Structures built on these materials may suffer significant damage during an earthquake.

Older alluvium. Moderate shaking is likely. Well-built structures generally survive in an earthquake.

(a)

(b)

Anomalous animal behavior is often reported prior to large earthquakes: everything from unusual barking of dogs, chickens that refuse to lay eggs, snakes that crawl out of the ground in winter and freeze, and horses or cattle that run in circles, to rats emerging from the ground and acting as if drugged or dazed. Anomalous behavior was apparently common before the Haicheng earthquake (23). Unusual animal behavior was also observed prior to the 1971 San Fernando earthquake. Unfortunately, the significance and reliability of animal behavior is difficult to evaluate. Nevertheless, there is considerable interest in the topic and research is being conducted.

Toward a Prediction Method

We are a long way from a working, practical methodology to predict earthquakes reliably. On the other hand, a good deal of data is currently being gathered concerning possible precursor phenomena associated with earthquakes.

Although progress on short-range earthquake prediction has not matched expectations, medium to long-range prediction, including hazard evaluation and probabilistic analysis of areas along active faults, has progressed faster than expected (27). Techniques based on field work have been developed to calculate slip rates

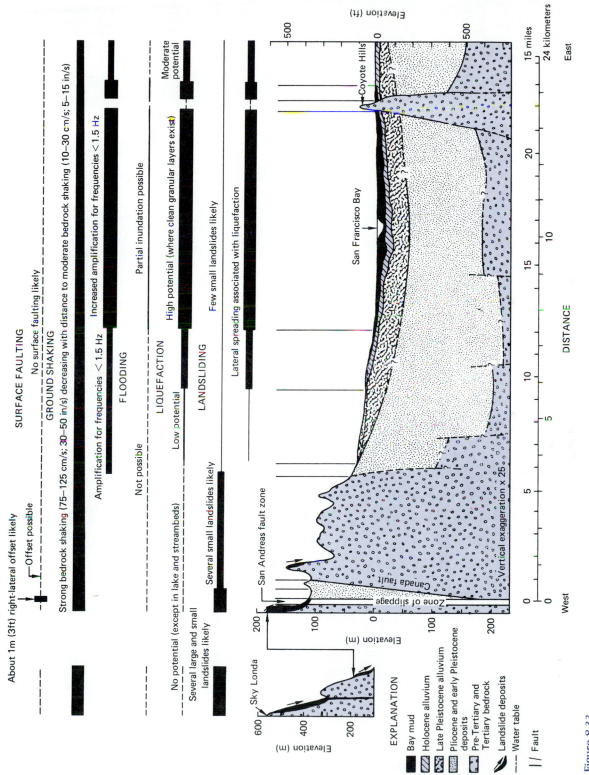

Figure 8.33
Predicted geologic effects of a magnitude 6.5 earthquake along the San Andreas fault. (After R. D. Borcherdt, et al., U.S. Geological Survey Professional Paper 941A, 1975.)

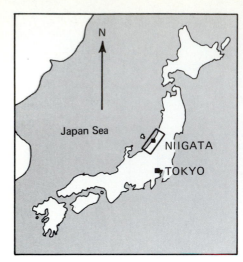

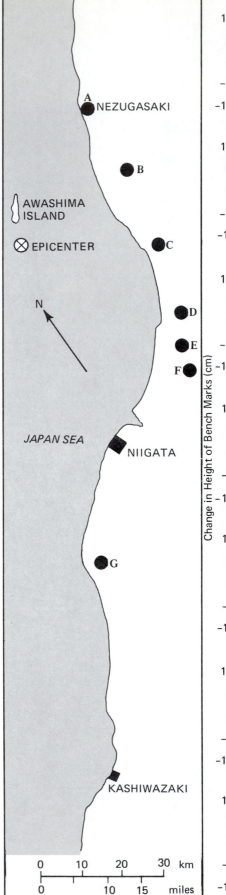

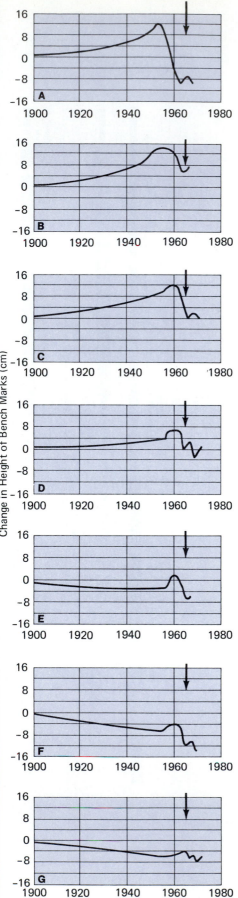

Figure 8.34
Anomalous uplift of the crust of the earth was observed for approximately ten years before the 7.5-magnitude earthquake which struck Niigata, Japan, in 1964. The uplift was measured by plotting changes in the elevation of bench marks (points of known elevation) with time. The black dots are the location of the bench marks, and the graphs correspond to these. (From "Earthquake Prediction," Frank Press. Copyright © May 1975 by Scientific American, Inc. All rights reserved.)

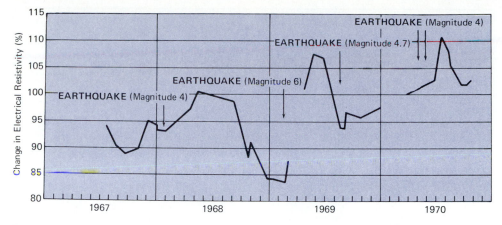

Figure 8.35
Changes in electrical resistivity of the earth's crust before earthquakes were measured in the USSR, China, and the United States. The data for this graph were obtained for earthquakes monitored in the USSR between 1967 and 1970. The measurements are made by feeding an electric current into the ground and recording voltage changes a few kilometers away. In general, it has been observed that earthquakes are preceded by a decrease in crustal resistivity. (From "Earthquake Prediction," Frank Press. Copyright © May 1975 by Scientific American, Inc. All rights reserved.)

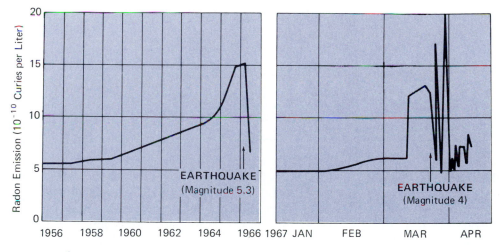

Figure 8.36
The amount of radioactive gas (radon) dissolved in the water of deep wells may increase significantly before an earthquake. The two examples shown here were recorded before earthquakes in the USSR. The 1966 event (left) had a magnitude of 5.3; the 1967 aftershock (right) had a magnitude of 4. The technique is used in both the USSR and China, and is being tried in the United States. (From "Earthquake Prediction," Frank Press. Copyright © May 1975 by Scientific American, Inc. All rights reserved.)

on faults and average recurrence intervals of large earthquakes, and geologists are beginning to make statements as to the probability of a large earthquake's occurring on a particular fault segment in the next 20 to 30 years. It has even been argued that determining recurrence of earthquakes on faults such as the San Andreas fault in California and the Wasatch fault in Utah is providing more valuable information for earthquake hazard reduction programs through development of appropriate land-use planning, building codes, and engi-

neering design than would short-term earthquake prediction capability (27).

As an example of a medium- to long-term earthquake prediction, consider California's San Andreas fault. Figure 8.37 shows four main segments of the fault, differentiated by historic and prehistoric activity derived from seismicity, strain measurement, and field observations. The area with the highest probability for a great earthquake (M = 8) is the southern segment, where there have been no historic earthquakes and no evidence for large earth-

Figure 8.37

Four segments of the San Andreas fault in California. Shown are ground rupture zones associated with the 1857 and 1906 great earthquakes as well as minimum stored strain (deficit fault slip) since the last great earthquake. Estimates of probability for earthquake events in the next several decades are also listed.

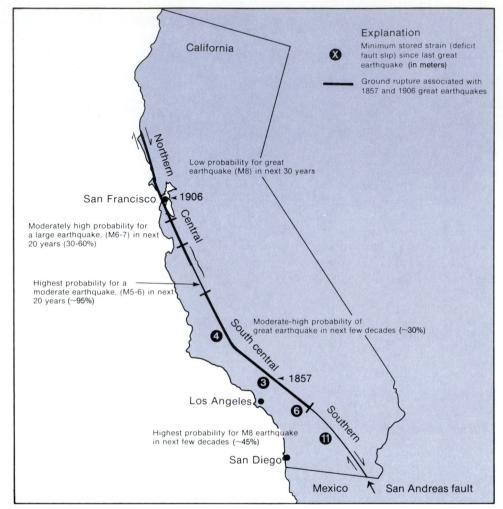

quakes in the last several hundred years. The recent geologic record suggests, however, that large earthquakes should be expected. At one location near Indio, California, several small streams are offset several meters and a late Pleistocene alluvial fan is offset nearly 700 m. The slip rate is estimated to be a few cm/yr for this segment of the fault and the strain rate is about 2 cm/yr across the fault (28). Thus, the southern segment of the San Andreas fault may be a nearly mature seismic gap with as much as 11 m of accumulated strain (deficit fault slip stored in rocks). For these reasons, geologists believe this portion of the fault is a candidate for a large earthquake in the relatively near future. Other general probabilistic statements concerning other sections of the fault are shown in Figure 8.37 (28, 29). One of the most interesting areas is near Parkfield, California, located just north of the 1857 ground rupture where magnitude 5.5 to 6 earthquakes have recurred at surprisingly regular intervals. Scientists with the U.S. Geological Survey have predicted with a high degree of certainty that another such event will occur by the year 1995 (5).

Most probabilities are only ball-park estimates and *not* certainties—we may yet be fooled again by the complex behavior of the San Andreas fault. However, the 1989 event on the central segment of the fault is close to what was predicted and suggests we are headed in the right direction with earthquake prediction.

The Borah Peak earthquake of October 28, 1983, in central Idaho, is another success story for medium-range earthquake hazard evaluation. Previous evaluation of the Lost River fault suggested that the fault was active (30). The earthquake, which was about a magnitude 7 event, killed two people and did about $15 million in damage, producing fault scarps up to several meters high and numerous ground cracks along 36 km of the fault. The important fact was that the scarps and faults produced *during* the earthquake were *superimposed* on pre-existing fault scarps, validating the usefulness of carefully mapping scarps produced from prehistoric earthquakes—the ground may break there again!

Even though we are a long way from a working methodology to predict earthquakes, speculating on a possible model is interesting. Assume that we can develop a set of relationships from which an earthquake magnitude and date (or time) of occurrence are predicted. The first step is to identify empirical relationships among precursor events (such as foreshock activity, anomalous tilt, radon gas emission) and probable earthquake magnitude. The precursor events may be measured and evaluated for a specific area or by their intensity, and we would expect to find a relationship with earthquake magnitude. In other words, after observing many earthquakes, we would empirically derive Figure 8.38a. Then, given a set of precursor observations (point *A* of Figure 8.38a), we could predict the magnitude of a possible earthquake (point *B* of the same graph). We might also assume that there is a relationship between the precursor time interval (time between the start of precursor events and the occurrence of the quake) and the magnitude of an expected earthquake. This relationship is also derived empirically from direct observation of known earthquake activity. Now take the predicted earthquake magnitude from Figure 8.38a, point *B*, and use this in Figure 8.38b to predict the precursor time interval (point *C*). Thus, we now know the magnitude, and, if we know when the precursor events started, we can calculate when the expected earthquake should occur. It sounds simple, but it is pure fiction: none of these relationships have been derived with sufficient precision to use in predicting real earthquakes. Furthermore, the potential for errors of measurement with this kind of work is considerable. Nevertheless, it is valuable to explore one example of how earthquake predictions may eventually come about.

Earthquakes and Critical Facilities

Critical facilities are those which if damaged or destroyed might cause catastrophic loss of life, property damage, or disruption of society. Examples include large dams, nuclear power plants, and liquid natural gas (LNG) facilities. There are three aspects of the decision process concerning critical facilities and earthquake hazard (31): first, evaluation of the hazard; second, evaluation of whether the facility may be designed or modified to accommodate the hazard; and third, subjective evaluation of an "acceptable risk." The first two factors have a strong scientific component, while risk assessment is a public safety issue since no facility can be rendered absolutely safe.

Consider seismic safety for a nuclear power plant. Evaluation of the earthquake hazard requires estimating *fault capability* and the *maximum credible earthquake* that might be expected along faults in the vicinity of a particular site. A *capable fault,* as defined by the U.S. Nuclear Regulatory Commission, is a fault that has exhibited movement at least once in the last 35,000 years or multiple movements in the last 500,000 years. These are more conservative criteria with a greater safety factor than the definition of an active fault (Table 8.4), reflecting the concern for siting nuclear power plants.

Estimation of the maximum credible earthquake can sometimes be accomplished by field work to evaluate effects of historical seismicity and prehistoric activity. Length of surface rupture and displacement per event are both related to earthquake magnitude, and equations are available to estimate magnitude if expected surface rupture length or displacement per event is known or assumed (32). Average recurrence of earthquakes of a particular magnitude can also be estimated if a slip rate (shear rate) is known or assumed (see Figure 8.39).

Unfortunately, the data base concerning fault rupture length and displacement is limited. Therefore, evaluation of the earthquake hazard for many critical facilities will continue to be subjective at best. All geologists can do is gather as much pertinent geologic data as possible to ensure that our decisions are based on adequate information. We are presently arguing over what is adequate. Certainly, for a specific fault, we would like to know the length of the fault, style of faulting, total displacement, age of the most recent movement, approximate recurrence interval, and magnitude of the maximum credible earthquake.

In most cases regarding siting of critical facilities, the main problem concerning seismic risk is estimating the activity of a particular fault system. For example, the seismic evaluation for the Auburn Dam site near Sacramento, California, has been controversial. Millions of

Figure 8.38
Hypothetical method of how earthquake magnitude and time of event might be predicted. Graph (a) is used to predict the magnitude based on given precursor events, and graph (b) is used to predict the precursor time interval from the earlier predicted magnitude. (See text for full explanation.)

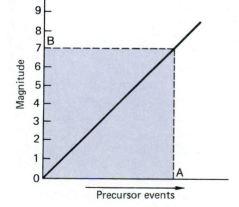
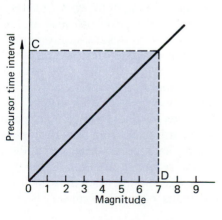

Figure 8.39
Relationships between recurrence interval, slip rate, and earthquake magnitude. (Modified after B. Slemmons and C. M. Depolo, 1986, Evaluation of Active Faulting and Associated Hazards, in *Active Tectonics,* National Academy Press, pp. 45–63.)

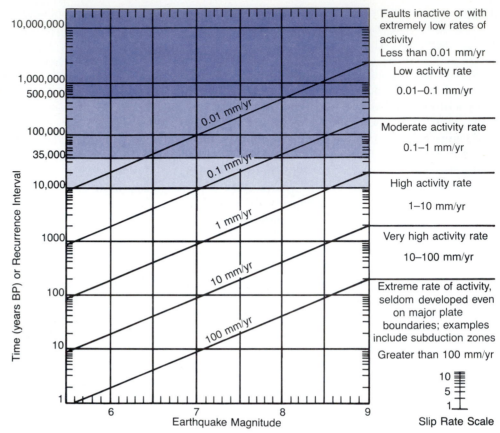

dollars have been spent and hundreds of meters of trenches excavated during the geologic studies, and there is still no agreement as to the hazard associated with the system of possibly active faults in the vicinity of the foundation area for the dam. The general area was considered to have relatively low seismic risk until a magnitude 5.7 earthquake occurred in 1975 near the Oroville Dam about 80 kilometers away from the Auburn site. This earthquake rekindled concern, initiated a new round of seismic risk evaluation, and led to the virtual abandonment of the site.

Geological evaluation of the Point Conception, California, site for a liquid natural gas (LNG) terminal (1979–1980) has also been controversial as well as time consuming. One of the major points of contention is the existence of numerous small faults that probably are associated with flexural-slip (faulting parallel to bedding planes of rocks) during folding of the rocks. Many large trenches (Figure 8.40) were excavated to study these faults, emphasizing the detailed nature of the geologic investigation required for critical facilities. The flexural-slip faults may not be capable of producing moderate to large earthquakes, but they do present a ground-rupture hazard that must be dealt with in the engineering design for the terminal. Another problem at the LNG site has been determining when these faults were active. Some of the faults cut various materials of differing ages, and

obtaining reliable dates has been a problem. The site is actually on a marine platform that may be as old as 100,000 years or more, but subsequent erosion and deposition have taken place so that younger materials are deposited on that older surface. All this makes the geological evaluation of the hazard concerning the faults difficult and somewhat speculative.

The problem remains: it is very difficult to absolutely date past earthquake events. Furthermore, the problem will only intensify as more critical facilities are planned in areas with potential seismic risk. Considerable research by many scientists is now being conducted in hopes that we will improve our knowledge of how to identify active faults and estimate the maximum credible earthquake for a particular area.

▼ EARTHQUAKE CONTROL

To aid in possible earthquake control, any system of earthquake prediction must assume that the true nature and causes of earthquakes are fairly well understood. Observations of the 1906 San Francisco earthquake led to a hypothesis known as the *earthquake cycle.* The salient features of the hypothesis are related to drop in elastic strain following an earthquake and reaccumulation of strain prior to the next event. That is, following an earthquake the elastic strain drops, and it takes time for

Figure 8.40
Large trench excavated as part of the geologic seismic safety study for the proposed liquid natural gas terminal at Point Conception. Note the people and ladders on the walls of the trench for scale. Seismic studies in southern California for critical facilities are taken very seriously and very detailed work is necessary, resulting in large excavations.

sufficient strain to accumulate again to produce another earthquake. If a particular fault produces a "characteristic earthquake," then the time necessary for elastic strain to accumulate will be fairly constant between events—which would produce a fairly constant recurrence interval for earthquakes (5).

It has been speculated that a typical earthquake cycle has two to three stages. The first stage is a long period of seismic inactivity following a major earthquake and associated immediate aftershocks. This is followed by a second period characterized by increased seismicity as the accumulated elastic strain approaches and locally exceeds strain levels to initiate faulting that produces small earthquakes. The third and final stage of the cycle, which occurs only hours or days prior to the next large earthquake, consists of foreshocks. In some cases this third stage may not even occur. Following the major earthquake, the cycle starts over again (5). Although this cycle is hypothetical, the major stages have been recognized as associated with the pattern and frequency of large earthquakes.

Unfortunately, even though we have a tremendous quantity of empirical observations concerning physical changes in earth materials before, during, and after earthquakes, there is no general agreement on a physical model to explain the observations. One model, known as the *dilatancy-diffusion model* (2), assumes that the first stage in earthquake development is an increase of elastic strain in rocks that causes them to undergo a **dilatancy** state, which is an inelastic increase in volume that starts

after the stress on a rock reaches one-half its breaking strength. During dilatancy, open fractures develop in the rocks, so it is in this stage that the first physical changes take place that might indicate a future earthquake. The model assumes that the dilatancy and fracturing of the rocks are first associated with a relatively low water content in the dilated rocks (Stage 2, Figure 8.41), which helps to produce lower seismic velocity, more earth movement, higher radon emission, lower electrical resistivity, and fewer minor seismic events. An influx of water then enters the open fractures, causing the pore water pressure to increase (which increases the seismic velocity while further lowering electrical resistivity), weakening the rocks, and facilitating movement along the fractures which is recorded as an earthquake. After the movement and release of stress, the rocks resume many of their original characteristics (25).

Scientists realized that earthquakes could be turned on and off by human activity when they noted how effectively the rates of injection of liquid waste into disposal wells controlled earthquake activity in Denver, Colorado. A U.S. Geological Survey project was initiated near Rangely, Colorado, to study the feasibility of actually *starting* earthquakes by raising the fluid pressure in rocks by injecting water into fracture zones, and *stopping* earthquakes by decreasing the fluid pressure by pumping water out of fracture zones. The project demonstrated that earthquake frequency can be controlled by manipulating the fluid pressure (Figure 8.42) (33). It is hoped that results of this research will provide the information

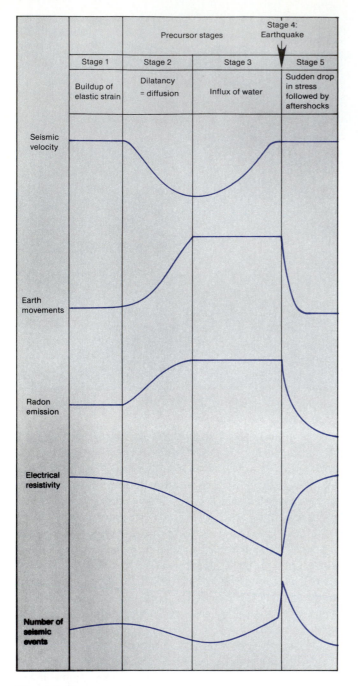

	Precursor stages			Stage 4: Earthquake
Stage 1	Stage 2	Stage 3		Stage 5
Buildup of elastic strain	Dilatancy = diffusion	Influx of water		Sudden drop in stress followed by aftershocks

Seismic velocity

Earth movements

Radon emission

Electrical resistivity

Number of seismic events

Figure 8.41
Dilatancy-diffusion model to explain the mechanism responsible for earthquakes. The curves show the expected precursory signals. (Modified from "Earthquake Prediction," Frank Press. Copyright © May 1975 by Scientific American, Inc. All rights reserved.)

necessary to reduce the earthquake hazard by releasing strain in rocks slowly through numerous controlled small shocks rather than having the strain released quickly in a large earthquake. The objective is to drill down into the fault zone (some are accessible to drilling) and then to manipulate the fluid pressure by either pumping out or injecting fluids, initiating a series of small shocks that release the stored strain in the rocks.

Proper use of nuclear explosions also may be effective in earthquake control. We have noted that such explosions probably do release natural tectonic strain. Since it is expected that the longer strain builds up, the larger the eventual earthquake will be, it is prudent to try to reduce that strain. Some scientists believe that properly spaced underground nuclear explosions could be used in conjunction with fluid-injection methods to release strain in rock and thereby limit the severity of earthquakes (34).

▼ ADJUSTMENTS TO EARTHQUAKE ACTIVITY

The mechanism of earthquakes is still poorly understood, and, therefore, such adjustments as warning systems and earthquake prevention are not yet reliable alternatives. There are, however, reliable protective measures we can take. First is structural protection, the construction of large buildings able to accommodate at least a moderate shock. Second is land-use planning—large structures, schools, hospitals, disaster relief facilities, and communication systems should not be placed on or near active faults or on sensitive earth material such as landfills, unconsolidated clay deposits, or water-saturated silts or sand. These areas are likely to accentuate earthquake waves and, therefore, be areas of high damage. Third are insurance and relief measures. Insurance, while not overly expensive in areas with a high earthquake risk, often has a high deductible before it pays benefits, and few people purchase it (34). The United States has federal disaster systems to provide relief and rehabilitation funds, and emergency systems are also provided by such organizations as the Salvation Army and the Red Cross. A fourth alternative is to take no action and pay the consequences. This philosophy is not what we espouse, but it is, in fact, what we often do. At the same time that we are developing ways to predict earthquakes and prepare for relief and rehabilitation, we are increasing the need for prediction and relief by creating new hazards where none existed. When we place any new building or structure on sensitive earth materials or on active faults in areas with an earthquake hazard, we are, in effect, seeking new problems where there were none. We know from studying earthquake damage that building on unconsolidated deposits, such as channel fill, that become unstable (subject to liquefaction) or shake easily (amplify surface waves) during an earthquake is more hazardous than building on bedrock. We also know that building on active faults (Figure 8.43) is unwise.

It is encouraging that southern California has a cooperative state-federal project to develop a comprehensive preparedness program to respond to a predicted or unexpected catastrophic earthquake. Considering that

Figure 8.42
Earthquake control at the Rangely oil field, Colorado. This is the first example in which earthquakes were intentionally started and stopped by controlling the reservoir pore pressure by controlled pumping. It was found that a particular threshold pressure was needed to initiate earthquakes. (After R. E. Wallace, U.S. Geological Survey Circular 701, 1974.)

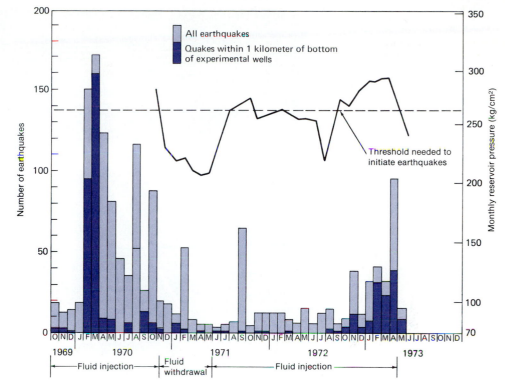

a magnitude 8.3 earthquake on the San Andreas fault north of Los Angeles may kill thousands of people, cause nearly $20 billion in damages, sever two major aqueducts that serve the region, reduce electrical power to the area by about 40 percent, close all main highways and airports, cause widespread loss of telephone communication, and damage other facilities such as natural gas lines, sewage lines, and fresh water supply lines, it is obviously prudent to evaluate the earthquake hazard and develop emergency plans. Considerable work has been accomplished in the Los Angeles region: maps have been prepared that show areas that are particularly vulnerable to shaking, ground rupture, and liquefaction (35).

We hope eventually to be able to predict earthquakes. The federal plan for issuing prediction and warning is shown in Figure 8.44. The general flow of information is from scientists to a prediction council for verification. A prediction that a damaging earthquake of a specified magnitude will occur at a particular location over a specific time span may then be issued to state and local officials, who will be responsible for issuing a warning to the public to take defensive action (that has, one hopes, been planned in advance). Potential response to a prediction depends on lead time (Table 8.6), but even a short time—as little as a few days—would be sufficient to mobilize emergency services, shut down reactors, and evacuate particularly hazardous areas.

Development of an Earthquake Warning System

It is technically feasible to develop an earthquake warning system (EWS) that would provide up to about

one minute warning to the Los Angeles area prior to the arrival of damaging earthquake waves. The Japanese have had a warning system in place for nearly twenty years that provides warning for their high-speed trains, which if derailed during or shortly after an earthquake could result in the loss of hundreds of lives. What is being talked about for California would be a more sophisticated system along the San Andreas fault that would first sense motion associated with a large earthquake and send a warning to the city, which would relay it to critical facilities, schools, and the general population. Depending upon where the earthquake is initiated, the warning time would vary from 15 to 30 seconds to as long as about one minute. The warning could provide time for people to shut down machinery and computers and take cover.

The warning system would consist of a set of sensors (seismometers) along the fault that would relay a warning to the Los Angeles area when triggered by seismic waves of a particular magnitude (Figure 8.45). One potential problem with the system is the chance of false alarms. Using the Japanese system for comparison, the number of false alarms is probably going to be less than 5 percent. However, because the length of warning time is so short, some have expressed concern as to whether much evasive action could really be taken. There is also concern for liability issues resulting from false alarms, warning system failures, and damage and suffering resulting from actions taken as a result of the early warning.

It is emphasized that the earthquake warning system is not a prediction tool, as it provides only a warning that

Figure 8.43
Houses should not have been constructed astride the 1906 break of the San Andreas fault (white line) or in areas of potential landslides. The dotted line is an older buried trace of the fault. (Photo by R. E. Wallace, courtesy of U.S. Geological Survey.)

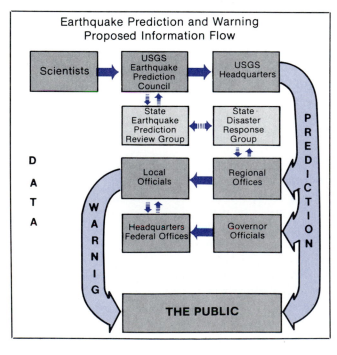

Figure 8.44
A federal plan for issuance of earthquake predictions and warning: the flow of information. (From: V. E. McKelvey, 1976, *Earthquake Prediction—Opportunity to Avert Disaster.* U.S. Geological Survey Circular 729.)

an earthquake has occurred. A recent study of the potential EWS in California concluded that the system is technically feasible, but based on a benefits to costs ratio is not economically feasible. That is, the study concluded that the potential economic benefits are not sufficient to justify the construction and operating costs. Total cost for construction is estimated to be about $3.3 to $5.8 million, with annual operating costs of $1.6 to $2.4 million (36). For comparison, damages from the 1989 Loma Prieta (San Francisco) earthquake with a magnitude of 7.1 were about $5,000 million. Larger, more damaging events can be expected. However, it is difficult to place an economic value on all issues related to public safety, especially as they relate to potential injuries or loss of life. Rejection of technology that could provide warning of approaching, potentially dangerous, seismic waves based mainly on economics may be unwise.

▼ PERCEPTION OF THE EARTHQUAKE HAZARD

That terra firma is not so firm in places is disconcerting to people who have experienced even a moderate earthquake. The relatively large number of people, especially children, who suffered mental distress following the San

Table 8.6
Potential response to an earthquake prediction with given lead time.

Lead Time	Buildings	Contents	Lifelines	Special Structures
3 Days	Evacuate previously identified hazards	Remove selected contents	Deploy emergency materials	Shut down reactors, petroleum products pipelines
30 Days	Inspect and identify potential hazards	Selectively harden (brace and strengthen) contents	Shift hospital patients; alter use of facilities	Draw down reservoirs, remove toxic materials
300 Days	Selectively reinforce		Develop response capability	Replace hazardous storage
3,000 Days		Revise building codes and land-use regulations: enforce condemnation and reinforcement		Remove hazardous dams from service

Source: C. C. Thiel, 1976, U.S. Geological Survey Circular 729.

Fernando earthquake attests to the emotional and psychological effects. This one experience was sufficient to influence several hundred Los Angeles families formerly from the Midwest to move back to that area.

Regardless of their mental distress after earthquakes, people in areas of potential disaster generally do not really understand the earthquake hazard. In addition, there appears to be a considerable difference between what people say and what they do. Near Tokyo, six new earthquake antidisaster centers and a steel mill are being constructed. While the centers are to aid victims of earthquakes, the steel mill is being placed near sea level on landfill that is part of a reclamation project for Tokyo Bay. The site is subject to severe earthquake and tsunami hazard (37). At the same time, across the Pacific Ocean, people in San Francisco continue their casual attitude

Figure 8.45
Idealized diagram showing an earthquake warning system. (After Holden, R., Lee, R., and Reichle, M. 1989. California Division of Mines and Geology Bulletin 101.)

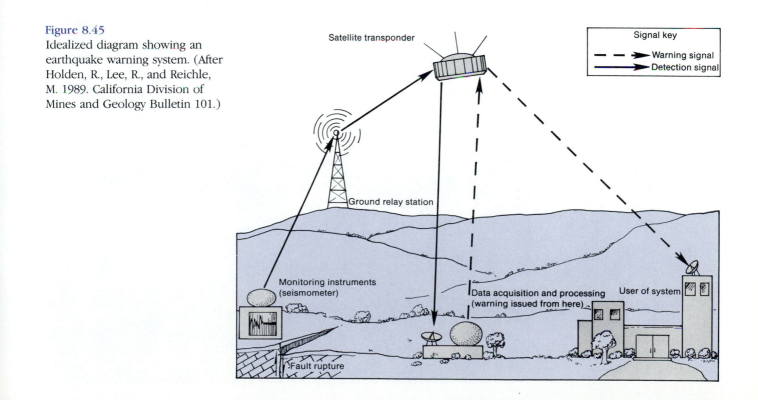

toward a possible earthquake, which promises to be a tremendous disaster.

People in hazardous areas could remove dangerous structures on old buildings that overhang the streets.

They could practice better planning for siting buildings and stop filling in bays and other sensitive areas. But they probably will not, because a hazardous event, even a tremendously catastrophic one that comes only once

Figure 8.46
Example of easements required for building setbacks from active fault traces in Portola Valley, California. Where the location of the fault trace is well known, no new buildings are allowed to be constructed within 15 meters of each side of the fault. Structures with occupancies greater than single family dwellings are required to be 38 meters from the fault. Where the precise location of the fault trace is less well known, more conservative setbacks of 30 meters for single family residences and 53 meters for higher occupancies are required. (After Mader, et al., and U.S. Geological Survey Circular 690, 1972.)

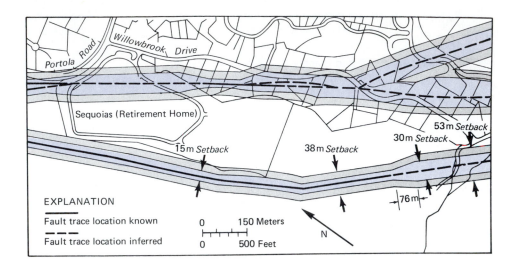

Figure 8.47
Oblique aerial photograph showing approximate location of fault traces within a part of the Hayward fault zone, some of which experienced movement during the 1868 earthquake. Present development compatible with a possible earthquake hazard include the undeveloped land, the plant nursery, the cemetery, and the highway. Other such uses might include golf courses, riding stables, drive-in theaters, or other recreational activities. Locating schools, police departments, hospitals, and other similar public facilities along active fault traces is unwise land-use planning. (Photo by R. E. Wallace, courtesy of U.S. Geological Survey.)

every few generations, is simply not perceived by the general population as a real threat. It is encouraging, however, that some communities are developing ordinances to limit development along active faults (Figure 8.46) and are considering land-use alternatives (Figure 8.47). In fact, the state of California has passed legislation that requires detailed geologic evaluation prior to construction of structures for human use that are sited on or near active or potentially active faults. Areas requiring the evaluation are known as *special studies zones* and are detailed on official maps. The number of areas requiring detailed evaluation has increased as the geologic data base has improved.

▼ ▼ ▼ SUMMARY AND CONCLUSIONS

Large earthquakes rank among nature's most catastrophic and devastating events. Most earthquakes are located in tectonically active areas where lithospheric plates, on which the continents and ocean basins are superposed, interact.

The Richter magnitude of an earthquake is a measure of the amount of energy released. It is determined by the amplitude of the largest horizontal trace recorded on a standard seismograph. A newer scale known as the moment magnitude, based on the seismic moment, has been developed. Resulting magnitudes are similar to those of the Richter scale but are more soundly based on physical principles.

The intensity of an earthquake is based on the severity of shaking as reported by observers, and varies with proximity to the epicenter and local geologic and engineering features.

Primary effects of earthquakes are violent ground motion accompanied by fracturing which may shear or collapse large buildings, bridges, dams, tunnels, and other rigid structures. Secondary effects include short-range events, such as fires, landslides, tsunamis, and floods, and long-range effects, such as regional subsidence and uplift of landmasses and regional changes in groundwater levels.

Tsunamis, or seismic sea waves, are produced by earthquakes or other phenomena that displace oceanic water and generate long waves that travel at speeds as great as 800 kilometers per hour. The waves, less than 0.5 meters high in deep water, may grow to a height of 15 meters or more on reaching the coastline. The majority of the lives lost in the 1964 Alaskan earthquake was attributed to tsunamis.

Tectonic creep resulting from slow movement along fault zones is a less hazardous process associated with earthquake activity. Nevertheless, it can cause considerable damage to roads, sidewalks, building foundations, and other structures.

Human activity has caused increasing earthquake activity in three ways: first, by loading the earth's crust through construction of large reservoirs; second, by disposing of liquid waste in deep disposal wells, which raises fluid pressures in rocks and facilitates movement along fractures; and third, by setting off underground nuclear explosions. The accidental damage caused by the first two activities is regrettable, but what we learn from all the ways we have caused earthquakes may eventually help us to control or stop large earthquakes.

Reduction of earthquake hazard will be a multifaceted program, including recognition of active faults and earth materials sensitive to shaking, and development of improved ways to predict, control, and adjust to earthquakes.

Prediction and control of earthquakes are now subjects of serious research. Optimistic scientists believe that we will eventually be able to make long-, medium-, and short-range predictions of earthquakes. Many believe such predictions may be based on previous patterns and frequency of earthquakes, seismic gaps, anomalous uplift and subsidence, and monitoring of ground tilt, strain in rocks, microearthquake activity, magnetic activity, radon concentration, electrical resistivity changes in earth materials, and perhaps anomalous animal behavior. Scientists hope to control earthquakes by manipulating fluid pressure in rocks along faults, thereby causing a series of small earthquakes, rather than let a natural catastrophic event occur. It has also been suggested that nuclear explosions, in conjunction with manipulation of fluid pressure, might be used to release natural tectonic strain slowly before a large earthquake.

To date, long-term and medium-term earthquake prediction and probabilistic analysis have been much more successful than short-term prediction, and supply important information for land-use planning, developing building codes, and engineering design of critical facilities.

Warning systems and earthquake prevention are not yet reliable alternatives; but more communities are developing emergency plans to respond to a predicted or unexpected catastrophic earthquake. Seismic zoning and other methods of hazard reduction are also being practiced.

The earthquake hazard in potential disaster areas remains poorly perceived by inhabitants of the areas. Their lack of concern probably results because a process that produces even catastrophic damage may not be perceived as a real threat if it strikes the same area only every few generations. Education programs and legislation are making more people aware of the earthquake hazard.

▼ ▼ ▼ REFERENCES

1. HAMILTON, R. M. 1980. Quakes along the Mississippi. *Natural History* 89:70–75.

2. BOLT, B. A. 1977. *Earthquakes: A Primer.* San Francisco: W. H. Freeman.

3. HAYS, W. W. 1981. *Facing geologic and hydrologic hazards.* U.S. Geological Survey Professional Paper 1240 B.

4. HOUGH, S. E.; FRIBERG, P. A.; BUSBY, R.; FIELD, E. F.; JACOB, K. H.; and BORCHERDT, R. D. 1989. Did mud cause freeway collapse? *EOS. Transactions, American Geophysical Union* 70, no. 47:1497, 1504.

5. HANKS, T. C. 1985. The national earthquake hazards reduction program: Scientific status. *U.S. Geological Survey Bulletin* 1659.

6. CLARK, M. M.; GRANTZ, A.; and MEYER, R. 1972. Holocene activity of the Coyote Creek fault as recorded in the sediments of Lake Cahuilla. In *The Borrego Mountain earthquake of April 9, 1968,* pp. 1112–30. U.S. Geological Survey Professional Paper 787.

7. SIEH, K.; STUIVER, M.; and BRILLINGER, D. 1989. A more precise chronology of earthquakes produced by the San Andreas fault in southern California. *Journal of Geophysical Research* 94:603–23.

8. HEATON, T. H.; ANDERSON, D. L.; ARABASZ, W. J.; BULAND, R.; ELLSWORTH, W. L.; HARTZELL, S. H.; LAY, T.; and SPUDICH, P. 1989. *National seismic system science plan.* U.S. Geological Survey Circular 1031.

9. OFFICE OF EMERGENCY PREPAREDNESS. 1972. *Disaster preparedness* 1, 3.

10. U.S. GEOLOGICAL SURVEY and the NATIONAL OCEANIC AND ATMOSPHERIC ADMINISTRATION. 1971. *The San Fernando, California, earthquake of February 9, 1971.* U.S. Geological Survey Professional Paper 733.

11. NATIONAL ACADEMY OF SCIENCES and NATIONAL ACADEMY OF ENGINEERING. 1973. The San Fernando earthquake of February 9, 1971: Lessons from a moderate earthquake on the fringe of a densely populated region. In *Focus on environmental geology,* ed. R. W. Tank, pp. 66–76. New York: Oxford University Press.

12. YOUD, T. L.; NICHOLS, D. R.; HELLEY, E. J.; and LAJOIE, K. R. 1975. Liquefaction potential. In *Studies for seismic zonation of the San Francisco Bay region,* ed. R. D. Borcherdt, pp. 68–74. U.S. Geological Survey Professional Paper.

13. HANSEN, W. R. 1965. *The Alaskan earthquake, March 27, 1964: Effects on communities.* U.S. Geological Survey Professional Paper 542A.

14. ECKEL, E. B. 1970. *The Alaskan earthquake, March 27, 1964: Lessons and conclusions.* U.S. Geological Survey Professional Paper 546.

15. OFFICE OF EMERGENCY PREPAREDNESS. 1972. *Disaster preparedness* 1, 2.

16. RADBRUCH, D. H., et al. 1966. *Tectonic creep in the Hayward fault zone, California.* U.S. Geological Survey Circular 525.

17. STEINBRUGGE, K. V., and ZACHER, E. G. 1960. Creep on the San Andreas fault. In *Focus on environmental geology,* ed. R. W. Tank, pp. 132–37. New York: Oxford University Press.

18. PAKISER, L. C.; EATON, J. P.; HEALY, J. H.; and RALEIGH, C. B. 1969. Earthquake prediction and control. *Science* 166: 1467–74.

19. EVANS, D. M. 1966. Man-made earthquakes in Denver. *Geotimes* 10: 11–18.

20. YEATS, R. S.; CLARK, M.; KELLER, E. A.; and ROCKWELL, T. K. 1981. Active fault hazard in southern California: Ground rupture vs. seismic shaking. *Geological Society of America Bulletin* 92(I): 189–96.

21. JONES, R. A. 1986. New lessons from quake in Mexico. *Los Angeles Times,* September 26.

22. BORCHERDT, R. D., et al. 1975. Predicted geologic effects of a postulated earthquake. In *Studies for seismic zonation of the San Francisco Bay region,* ed. R. D. Borcherdt, pp. 88–95. U.S. Geological Survey Professional Paper 941A.

23. RALEIGH, B., et al. 1977. Prediction of the Haicheng earthquake. *Transactions of the American Geophysical Union* 58, no. 5: 236–72.

24. SIMONS, R. S. 1977. Earthquake prediction, prevention and San Diego. In *Geologic hazards in San Diego,* ed. A. L. Patrick. San Diego Society of Natural History.

25. PRESS, F. 1975. Earthquake prediction. *Scientific American* 232: 14–23.

26. RIKITAKR, TSUNEJI. 1983. *Earthquake forecasting and warning.* London: D. Reidel.

27. ALLEN, C. R. 1983. Earthquake prediction. *Geology* 11: 682.

28. RALEIGH, C. B.; SYKES, L. R.; SIEH, K. E.; and ANDERSON, D. L. 1983. Forecasting southern California earthquakes. *California Geology* March: 54–63.

29. SIMON, C. 1983. California's quakes: Narrower odds. *Science News* 124–404.

30. HAIT, M. H. 1978. Holocene faulting, Lost River Range, Idaho. *Geological Society of America Abstracts with Programs* 10(5): 217.

31. CLUFF, L. S. 1983. The impact of tectonics on the siting of critical facilities. *EOS* 64 no. 45: 860.

32. SLEMMONS, D. B. 1977. *State-of-the-art for assessing earthquake hazards in the United States* (series): Determination of design earthquake magnitude from fault length and maximum displacement data. U.S. Army Engineers Waterways Experimental Station, Vicksburg, Miss.

33. WALLACE, R. E. 1974. *Goals, strategy, and tasks of the earthquake hazard reduction program.* U.S. Geological Survey Circular 701.

34. EMILIANI, C.; HARRISON, C.; and SWANSON, M. 1969. Underground nuclear explosions and the control of earthquakes. *Science* 165: 1255–56.

35. ZIONY, J. I., ed. 1985. *Evaluating earthquake hazards in the Los Angeles region: An earth-science perspective.* U.S. Geological Survey Professional Paper 1360.

36. HOLDEN, R.; LEE, R.; and REICHLE, M. 1989. *Technical and economic feasibility of an earthquake warning system in California.* California Division of Mines and Geology Special Publication 101.

37. NICHOLAS, T.C., Jr. 1974. Global summary of human response to natural hazards: Earthquakes. In *Natural hazards,* ed. G. F. White, pp. 274–84. New York: Oxford University Press.

One of the approximately 40 people killed when Mt. Unzen in Japan erupted in June, 1991, was Harry Glicken. Harry was a brave and dedicated young scientist who, through good fortune, had escaped death in the May 18, 1980, eruption of Mt. St. Helens. This chapter on volcanic processes is dedicated to Harry, who loved volcanoes and who was my friend.

On a worldwide basis, volcanic activity usually affects sparsely populated areas. However, volcanic eruptions can be tremendously destructive, and when such an event occurs near a densely populated area, a catastrophe may be recorded. Japan, Indonesia, and the U.S. Pacific Northwest, among other areas, have high population densities close to active or potentially active volcanoes. On the global scale, 5 to 15 volcanoes are active every month.

Causes of volcanic activity are directly related to plate tectonics, and most active volcanoes are located near plate junctions where **magma** (hot, partially liquid rock) is produced as spreading or sinking lithospheric plates interact with other earth materials. Approximately 80 percent of all active volcanoes are located in the "ring of fire" which circumscribes the Pacific Ocean (Figure 9.1). This area essentially corresponds to the Pacific plate (Table 9.1). In the United States, several volcanoes in Hawaii, about 25 in Alaska, and several in the Cascade Range of the Pacific Northwest can be considered active.

The Hawaiian volcanoes are interesting because they are located well within the Pacific plate rather than near a plate boundary. It is currently believed that there is a **hot spot** below the Pacific plate where magma is generated. Magma moves upward through the plate and produces a volcano on the bottom of the sea that eventually may become an island. Because at the Hawaiian Islands the plate is moving roughly northwest over the stationary hot spot, a chain of volcanoes, older to the northwest, is formed. The Big Island, Hawaii, is presently near the hot spot and is experiencing active volcanism and growth. Islands to the northwest, such as Maui, Molokai, and Oahu, are evidently off the hot spot and volcanoes on those islands are no longer active. Another hot spot is probably located below the Yellowstone National Park area, and is responsible for the famous hot springs and geysers found there.

Volcanoes in the Cascade Range of Washington, Oregon, and California continue to pose potential danger to the population and agricultural centers of the Pacific Northwest (Figure 9.2). Table 9.2 summarizes recent activity of major volcanoes in the Cascade Range.

Three common types of volcanoes are the shield and composite volcanoes and the volcanic dome (Figure 9.3). Each type of volcano has a characteristic style of activity that is primarily a result of the viscosity of the magma, which is related to its silica content.

CHAPTER NINE

▼

▼

▼

Volcanic Activity

Shield volcanoes, by far the largest volcanoes, are characterized by relatively nonexplosive activity, resulting from the low silica content (about 50 percent) of the magma. Shield volcanoes are built up almost entirely from numerous lava flows, but can also produce a lot of volcanic ash. They are common in the Hawaiian Islands and are also found in some areas of the Pacific Northwest and Iceland.

Composite volcanoes, known for their beautiful cone shape, are associated with a magma of intermediate silica content (about 60 percent) which is more viscous than the low-silica magma of shield volcanoes. Volcanic activity characteristic of composite volcanoes is a mixture of explosive activity and lava flows. Examples in the United States include Mt. St. Helens and Mt. Rainier (Figure 9.2), and as the 1980 eruption of Mt. St. Helens demonstrated, these volcanoes can be violent and dangerous.

Volcanic domes are characterized by viscous magma with a high silica content (about 70 percent). Activity is generally explosive, making volcanic domes very dangerous. Mt. Lassen in northeastern California, which last erupted from 1914 to 1917 (Figure 9.4), is a good example of a volcanic dome. One eruption included a tremendous horizontal blast that destroyed a large area.

A group of volcanic domes known as the Chaos Crags, located just north of Lassen Peak, has a recent history of explosive activity and large, rockfall debris avalanches. Fear of another rockfall or volcanic eruption from the Chaos Crags recently caused closing of the Manzanita Lake campground, lodge, and visitors' center in Lassen National Park.

Although relatively rare, volcanic activity, from a historical standpoint, has destroyed considerable prop-

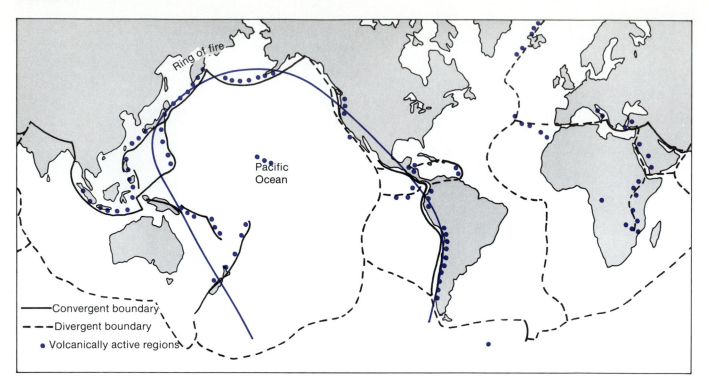

Figure 9.1

Volcanically active regions of the world. (Modified from Costa, J. E., and Baker, V. R. 1981. *Surficial Geology*. Originally published by John Wiley & Sons, New York and republished by TechBooks, Fairfax, Virginia.)

Table 9.1

Distribution of the world's active volcanoes.

Area		Percentage of Active Volcanoes
Pacific		79
Western Pacific Islands	45	
North and South America	17	
Indonesian Islands	14	
Central Pacific Islands (Hawaii, Samoa)	3	
Indian Ocean Islands		1
Atlantic		13
Mediterranean and Asia Minor		4
Other Areas		3
		100

Source: A. Rittman, *Disaster Preparedness*, Office of Emergency Preparedness, 1962.

erty and taken many thousands of human lives. Table 9.3 lists some of the more catastrophic events. The loss of life, though tragic, is relatively small compared to the number of deaths caused by floods or earthquakes during roughly the last 2,000 years. Unfortunately, volcanic activity cannot be prevented; therefore, programs to minimize this hazard involve prediction, warning, preparation, and experience gained from past events.

▼ EFFECTS OF VOLCANIC ACTIVITY

Primary effects of volcanic activity include lava flows, ash flows (also known as hot avalanches, nuées ardentes, and ignimbrites), lateral blasts, ash falls, and the release of gases (mostly water as steam, but sometimes corrosive or poisonous). Secondary effects include mudflows, floods, and fires (1). Table 9.4 briefly describes these hazards for the U.S., including predictability, frequency of occurrence, and degree of risk.

Lava Flows

Lava flows are the most familiar product of volcanic activity (Figure 9.5, p. 195). They result when magma reaches the surface and overflows the crater or volcanic vent along the flanks of the volcano. Lava flows can be quite fluid and move quite rapidly or relatively viscous and move slowly. Most lava flows are slow enough that people can easily move out of the way as they approach (2). Lava flows from Kilauea volcano, Hawaii, (Figure 9.5) have been active for the past several years. By May of 1990, over 50 homes in the village of Kalapana were destroyed by slow-moving flows and by August 1990 lava flowed across part of the famous Kaimu Black Sand Beach and into the ocean. The village of Kalapana has virtually disappeared and it will be many decades before much of the land is productive again. On the other hand, the

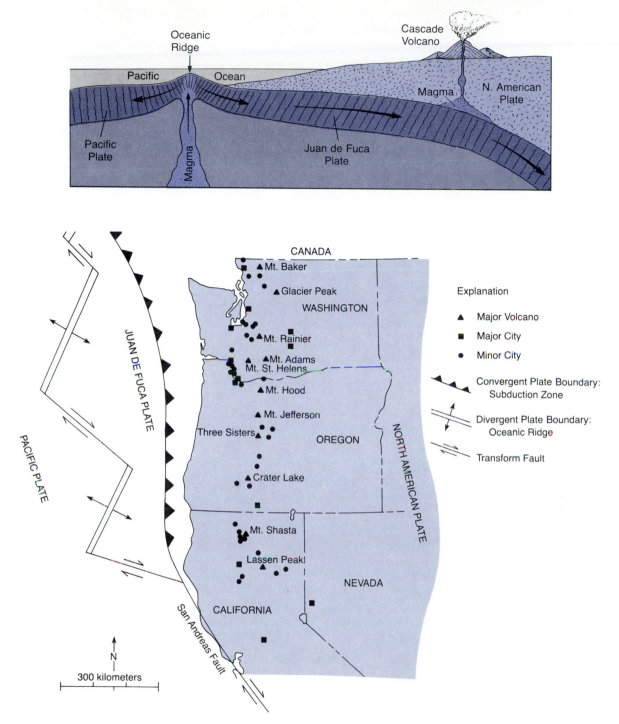

Figure 9.2
Map and plate tectonic setting of the Cascade Range showing major volcanoes and cities in their vicinity. See Figure 3.2 for regional tectonic environment. (Map modified after Crandall and Waldron, *Disaster Preparedness,* Office of Emergency Preparedness, 1969.)

eruptions, in concert with beach processes, have produced new and larger black sand beaches. The sand is produced when the molten lava enters the ocean and fragments into sand-sized particles.

Several methods, such as bombing, hydraulic chilling, and constructing walls, have been employed to deflect lava flows away from populated or otherwise valuable areas. These methods have had mixed success. They cannot be expected to block large flows, and need further evaluation.

Observation of lava movement in Japan suggests that there are two types of pressure associated with lava flows:

Table 9.2
Summary of recent activity of major volcanoes in the Cascade Range.

	Mt. Baker	Glacier Peak	Mt. Rainier	Mt. St. Helens	Mt. Adams	Mt. Hood	Mt. Jefferson	Three Sisters	Mt. Mazama (Crater Lake)	Newberry Volcano (Newberry Crater)	Mt. Shasta	Glass Mountain area	Lassen Peak—Chaos Crags area
Active in historic time	X		X	X		X					?	?	X
Known products of eruptions, last 12,000 years Lava flow	X		X	X	X	X	X	X	X	X	X	X	X
Tephra (airborne rock debris)	X	X	X	X		?		X	X	X	X	X	X
Pyroclastic flow		X	X	X					X		X		X
Mudflow	X	X	X	X	X	X					X		X
Estimated population at risk More than 1,000	X		X	X		?			X		X		?
Less than 1,000			?		X		X	X		X		X	

Source: Crandell and Mullineaux, "Technique and Rationale of Volcanic Hazards." *Environmental Geology* 1 (New York: Springer-Verlag, 1975).

pressure related to the vertical height of the lava and pressure from the momentum of the flow. Both of these must be considered when building walls to deflect lava. Although even walls of rubble and loosely laid stone have fortuitously withstood lava flows in Italy and Hawaii, experience suggests that proper planning is necessary for best results (3). The following criteria are suggested by A. C. Mason and H. L. Foster. First, walls about 3 meters high are usually sufficient. Second, the upslope side should be steeper to prevent the lava overriding the wall. Third, the wall should be located diagonally to the slope so that the lava will be deflected to an area where little damage will occur. Fourth, guide channels should be located upslope to facilitate the direction of lava to the proper location. Fifth, walls should be located below likely vents and as far upslope as possible.

Bombing of the lava flows has been attempted to stop their advance. It has proven most effective against lava flows in which fluid lava is confined to a channel by congealing lava on the margin of the flow. The objective is to partly block the channel by bombing, thereby causing the lava to pile up, facilitating an upstream break through which the lava escapes to a less damaging route. The established procedure is successive bombing at higher and higher points as necessary to control the threat. Bombing has some merit for future research, but

it probably provides little protection, is unpredictable, and, in any event, cannot be expected to affect large flows. Poor weather conditions and the abundance of smoke and falling ash can also reduce the effectiveness of bombing (3).

Mt. Helgafell.   The world's most ambitious program to control lava flows was initiated in 1973 on the Icelandic island of Heimaey, when lava flows with low silica content nearly closed the harbor to the island's main town and thereby threatened continued use of the island as Iceland's main fishing port. The situation prompted immediate action. Experiments on the island of Surtsey in 1967 showed that water could be used to stop the advance of lava, and favorable conditions existed on Heimaey to try it on a large scale. First, the main flows were viscous and slow moving, allowing the necessary time to initiate a control program. Second, transportation by sea and the local road system was adequate to move the necessary pumps, pipes, and heavy equipment. Third, water was readily available.

The procedure was to first cool the margin and surface of the flow with numerous fire hoses fed from a 13-centimeter pipe. Then bulldozers were moved up on the slowly advancing (at less than 1 meter per hour) flow, making a track or road on which the plastic pipe was

(a) Kilauea, Hawaii

(b) Mt. St. Helens, Washington, before May 18, 1980, eruption

Figure 9.3
(a) Shield volcano has low-silica (approximately 50 percent) magma and relatively nonviolent activity. The largest type of volcano, it may measure 200 kilometers in width and is constructed of many lava flows. (b) Composite volcano has magma of intermediate silica content (approximately 60 percent) and activity alternating from explosive to lava flows. The cone is constructed of alternating layers of pyroclastic material (mostly ash) and lava flows. (c) Volcanic dome has high-silica magma (approximately 70 percent). It is one of the most violent and explosive of all volcanoes. (Photos [a] and [b] courtesy of U.S. Geological Survey. Photo [c] by Mary R. Hill, courtesy of California Division of Mines and Geology.)

(c) Mt. Lassen, California

Figure 9.4
Eruption of Lassen Peak in June 1914. (Photo by B. F. Loomis, courtesy of Loomis Museum Association.)

Table 9.3
Selected historic volcanic events.

Volcano or City	Effect
Vesuvius, Italy, A.D. 79	Destroyed Pompeii and killed 16,000 people. City was buried by volcanic activity and rediscovered in 1595.
Skaptar Jokull, Iceland, 1783	Killed 10,000 people (many died from famine) and most of the island's livestock. Also killed some crops as far away as Scotland.
Tambora, Indonesia, 1815	Global cooling; produced "year without a summer."
Krakatoa, Indonesia, 1883	Tremendous explosion; 36,000 deaths from tsunami.
Mt. Pelée, Martinique, 1902	Ash flow killed 30,000 people in a matter of minutes.
La Soufrière St. Vincent, 1902	Killed 2,000 people and caused the extinction of the Carib Indians.
Mt. Lamington, Papua New Guinea, 1951	Killed 6,000 people.
Villarica, Chile, 1963–64	Forced 30,000 people to evacuate their homes.
Mt. Helgafell, Heimaey Island, Iceland, 1973	Forced 5,200 people to evaculate their homes.
Mt. St. Helens, Washington, USA, 1980	Debris avalanche, lateral blast, and mudflows killed 54 people, destroyed over 100 homes.
Nevado del Ruiz, Colombia, 1985	Eruption generated mudflows that killed at least 22,000 people.
Mt. Unzen, Japan, 1991	Ash flows and other activity killed 41 people and burned over 125 homes. Over ten thousand people evacuated.
Mt. Pinatubo, Philippines, 1991	Tremendous explosions, ash flows, and mud flows combined with typhoon killed over 300 people; several hundred thousand people evacuated.

Source: Data partially derived from C. Ollier, *Volcanoes* (Cambridge, Mass.: MIT Press, 1969).

placed. The pipe did not melt as long as water was flowing in it, and small holes in the pipe also helped cool particularly hot spots along various parts of the flow; at each place, approximately 0.03 to 0.3 cubic meters of water per second was delivered at a distance of about 50 meters in back of the edge of the flow (Figure 9.6, p. 196). The cooling was done in conjunction with diversion barriers constructed by bulldozers mounding up loosened material in front of the advancing flow. Lava tended to pile up against these barriers. Watering at each location lasted about two weeks, or until the steam stopped coming out of the lava in that particular area. Watering had little effect the first day, but then that part of the flow began to slow down. The program undoubtedly had an important effect on lava flows. It tended to restrict their movement and thus reduced property damage. After the outpouring of lava stopped in June 1973, the harbor was still usable (4). In fact, by fortuitous circumstances, the shape of the harbor was actually improved, since the new rock provides additional protection from the sea.

Pyroclastic Activity: Ash Flow, Ash Fall, and Lateral Blasts

Pyroclastic activity is characterized by explosive volcanism in which **tephra** (all types of volcanic debris from ash to very large particles) is physically blown from a volcanic vent into the atmosphere. Several types of activity are recognized: *volcanic ash eruptions,* in which a tremendous quantity of rock fragments, natural glass fragments, and gas are blown high into the air by explosions from the volcano; *lateral blasts of gas and ash,* which are explosive and may travel at tremendous velocity (perhaps exceeding the speed of sound) away from the volcano; and *volcanic ash flows,* which are hot avalanches of ash, rock, and volcanic glass fragments mixed with gas, which are blown out a vent and move very rapidly down the sides of the volcano.

Volcanic ash eruptions, or **ash fall,** can cover hundreds or even thousands of square kilometers with a carpet of volcanic ash. An eruption at Crater Lake about

Table 9.4
Volcanic hazards for the United States: Events, effects, predictability, frequency, and risk.
Note: Ash flows and hot avalanches are also known as ignimbrites, and *nuées ardentes*
(French for "glowing clouds"). Volcanic mudflows are also called lahars.

	Lava Flows	Ash Flow, Hot Avalanches, Mudflows, and Floods	Ash Fall, Lateral Blast, and Gases
Origin and characteristics	Result from nonexplosive eruptions of molten lava. Flows are erupted slowly and move relatively slowly; usually no faster than a person can walk.	Hot avalanches can be caused directly by eruption of fragments of molten or hot solid rock; mudflows and floods commonly result from eruption of hot material onto snow and ice and eruptive displacement of crater lakes. Mudflows also commonly caused by avalanches of unstable rock from volcano. Hot avalanches and mudflows commonly occur suddenly and move rapidly, at tens of kilometers per hour.	Produced by explosion or high-speed expulsion of vertical to low-angle columns or lateral blasts of fragments and gas into the air; materials can then be carried great distances by wind. Gases alone may issue nonexplosively from vents. Commonly produced suddenly and move away from vents at speeds of tens of kilometers per hour.
Location	Flows are restricted to areas downslope from vents; most reach distances of less than 10 kilometers. Distribution is controlled by topography. Flows occur repeatedly at central-vent volcanoes, but successive eruptions may affect different flanks. Elsewhere, flows occur at widely scattered sites, mostly within volcanic "fields."	Distribution nearly completely controlled by topography. Beyond volcano flanks, effects of these events are confined mostly to floors of valleys and basins that head on volcanoes. Large snow-covered volcanoes and those that erupt explosively are principal sources of these hazards.	Distribution controlled by wind directions and speeds, and all areas toward which wind blows from potentially active volcanoes are susceptible. Zones around volcanoes are defined in terms of whether they have been repeatedly and explosively active in the last 10,000 years.
Size of area affected by single event	Most lava flows cover no more than a few square kilometers. Relatively large and rare flows probably would cover only hundreds of square kilometers.	Deposits generally cover a few square kilometers to a few hundreds of square kilometers. Mudflows and floods may extend downvalley from volcanoes many tens of kilometers.	An eruption of "very large" volume could affect tens of thousands of square kilometers, spread over several states. Even an eruption of "moderate" volume could significantly affect thousands of square kilometers.
Effects	Land and objects in affected areas subject to burial and generally total destruction. Flows that extend into areas of snow may melt it and cause potentially dangerous and destructive floods and mudflows. May start fires.	Land and objects subject to burning, burial, dislodgement, impact damage, and inundation by water.	Land and objects near an erupting vent subject to blast effects, burial, and infiltration by abrasive rock particles, accompanied by corrosive gases, into structures and equipment. Blanketing and infiltration effects can reach hundreds of kilometers downwind. Odor, "haze," and acid effects may reach even farther.

continued

Table 9.4
continued.

	Lava Flows	Ash Flow, Hot Avalanches, Mudflows, and Floods	Ash Fall, Lateral Blast, and Gases
Predictability of location of areas endangered by future eruptions	Relatively predictable near large, central-vent volcanoes. Elsewhere, only general locations predictable.	Relatively predictable, because most originate at central-vent volcanoes and are restricted to flanks of volcanoes and valleys leading from them.	Moderately predictable. Voluminous ash originates mostly at central-vent volcanoes; its distribution depends mainly on winds. Can be carried in any direction; probability of dispersal in various directions can be judged from wind records.
Frequency, in contiguous United States	Probably one to several small flows per century that individually cover less than 10 square kilometers. Flows that cover tens to hundreds of square kilometers probably occur at an average rate of about one every 1,000 years. (In Hawaii, eruption of many flows per decade would be expected.)	Probably one to several events per century caused directly by eruptions. Probably only about one event per 1,000 years caused directly by eruption at "relatively inactive" volcanoes.	Probably one to a few eruptions of "small" volume every 100 years. Eruption of "large" volume may occur about once every 1,000 to 5,000 years. Eruption of "very large" volume, probably no more than once every 10,000 years.
Degree of risk in affected area	To people, low. To property, high.	Moderate to high for both people and property near erupting volcano. Risk relatively high to people because of possible sudden origin and high speeds. Risk decreases gradually downvalley and more abruptly with increasing height above valley floor.	Moderate risk to both people and property near erupting volcano; decreases gradually downwind to very low.

Source: D. R. Mullineaux, 1981. U.S. Geological Survey Professional Paper 1240B.

7,000 years ago blanketed an area of several hundreds of thousands of square kilometers in the northwestern United States with ash. Approximately 2,100 square kilometers were covered with more than 15 centimeters of ash (Figure 9.7). Several volcanoes in the Cascade Range could have similar eruptions in the future (2).

Volcanic ash eruptions create several hazards. First, vegetation, including crops and trees, may be destroyed (Figure 9.8). Second, surface water may be contaminated by sediment, resulting in temporary increase in acidity of the water. The increase in acidity generally lasts only a few hours after the eruption ceases. Third, structural damage to buildings may occur, caused by the increased load on roofs. A depth of one centimeter of ash places an extra 2.5 tons of weight on a roof with a surface area of about 140 square meters. Weight of the ash was a major problem during the 1973 Icelandic eruption. Houses,

public buildings, and businesses were literally buried in ash (Figure 9.9). Many buildings collapsed from the excess load, but inhabitants did save many by diligently shoveling the debris from the roofs (4). Fourth, health hazards such as irritation of the respiratory system and eyes are caused by contact with the ash and associated caustic fumes (2).

Volcanic **ash flows**, which may move as fast as 100 kilometers per hour down the sides of a volcano, can be catastrophic if a populated area is in the path of the flow. On the morning of May 8, 1902, a flow of hot, incandescent ash, steam, and other gases called a *nuée ardente*, roared down Mt. Pelée through the West Indies town of St. Pierre, killing 30,000 people. A prisoner in jail was one of the two survivors, and he was severely burned and horribly scarred. Reportedly, he spent the rest of his life touring circus sideshows as the "Prisoner of St. Pierre."

Figure 9.5
Lava flows from Kilauea volcano, Hawaii, December 1986: (a) lava flow entering road intersection in Kalapana subdivision; (b) at this orchid farm, only the metal roofing remains from a house destroyed by lava flow. (Photos by J. D. Griggs, U.S. Geological Survey.)

(a)

(b)

Ash flows are often exploding avalanches that may be as hot as hundreds of degrees Celsius and incinerate everything in their path. They are one of the most lethal aspects of volcanic eruptions. Flows like these have occurred on volcanoes of the Pacific Northwest in the past and can probably be expected in the future. Fortunately, they seldom occur in populated areas.

Another type of ash flow is a *base surge,* which forms when ascending magma comes in contact with water on or near the earth's surface in a violent steam explosion. Such an eruption occurred in 1911 on an island in Lake Taal, Philippines, killing about 1,300 inhabitants on the island and lake shore as a tremendous blast swept across

the water. A similar event occurred at the same location in 1965, this time claiming perhaps as many as two hundred lives. Base surge eruptions are commonly associated with small volcanoes with bowl-shaped craters like that of Diamond Head, Hawaii. Many such extinct volcanoes can be found in the Christmas Lake valley region, an ancient lake in south-central Oregon, and in the Tule Lake region of northern California.

Poisonous Gases

A number of gases, including water vapor, carbon dioxide, carbon monoxide, sulfur dioxide, and hydrogen

Figure 9.6
Close-up of an attempt to control the movement of a lava flow with water on the island of Heimaey. (Photo courtesy of Icelandic Airlines.)

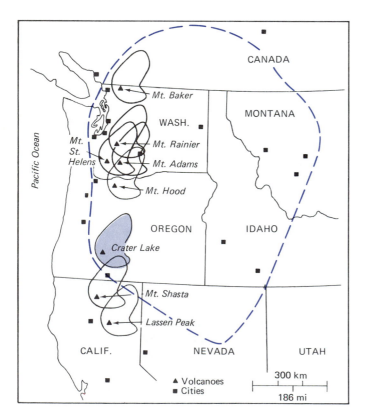

Figure 9.7
Map showing area covered by a volcanic ash eruption at Crater Lake about 7,000 years ago. The outer line shows the approximate limit of the ash fall. The shaded pattern shows the area covered by 15 centimeters or more of volcanic ash. This same area is superimposed on other major volcanoes of the Cascade Range. (After Crandell and Waldron, *Disaster Preparedness,* Office of Emergency Preparedness, 1969.)

Figure 9.8
Volcanic ash damaged or killed the trees shown here during the 1959–60 Kilauea volcano, Hawaii. (Photo by J. P. Eaton, courtesy of U.S. Geological Survey.)

sulfide, among others, are emitted during volcanic activity. In addition, dormant volcanoes may emit gases for long periods following eruptions. This was dramatically and tragically illustrated on the night of August 21, 1986, when Lake Nios in Cameroon, Africa, vented a poisonous gas, probably consisting mostly of carbon dioxide. The gas was heavier than air and settled in nearby villages, killing approximately 2,000 people by asphyxiation. It is speculated that the carbon dioxide and other gases were slowly released into the bottom waters of the lake, which occupies part of the crater of a dormant volcano. As long as such gas is dissolved in the lake water and kept there by the pressure of the overlying water, there is no problem. However, if the lake water is suddenly disturbed by a subaqueous landslide or small earthquake that brings the water from the bottom of the lake suddenly to the surface (as was probably the case with Lake Nios), the gases may enter the atmosphere. The Cameroon event is not unique, for in 1984 an underwater landslide in a nearby lake evidently initiated the release of carbon dioxide gas that took the lives of 37 people.

In Japan, volcanoes are monitored to detect releases of poisonous gas such as hydrogen sulfide. When releases are detected sirens are sounded to advise people to evacuate to high ground to escape the gas. Scientists speculate that a warning system in Cameroon would not have been effective because the release of gas was very sudden. On the other hand the event is evidently quite rare and unlikely to occur in other parts of the world with any great regularity (5).

Caldera-Forming Eruptions

Giant volcanic craters, or **calderas**, are produced by very rare but extremely violent eruptions. None have oc-

(a)

(b)

Figure 9.9
(a) Volcanic eruption in 1973, on the island of Heimaey, about 16 kilometers southwest of Iceland. This eruption forced the evacuation of 5,200 residents and covered more than half the buildings with ash and lava. (b) House nearly buried in ash. (Photos courtesy of Icelandic Airlines.)

curred anywhere on earth in the last few thousand years, but at least 10 have occurred in the last million years, three in North America. A large caldera-forming eruption may explosively extrude up to 1,000 cubic kilometers of pyroclastic debris, mostly ash, which is about 1,000 times that ejected by the 1980 eruption of Mt. St. Helens; produce a crater more than 10 km in diameter; and blanket an area of several tens of thousands of square kilometers with ash flow and fall deposits that vary from up to 100 m thick near the crater rim to a meter or so up to 100 km away from the source (6).

The most recent caldera-forming eruptions in North America occurred about 600,000 and 700,000 years ago at Yellowstone National Park in Wyoming and in the Owens Valley, California (Figure 9.10), respectively. The main

events in these eruptions can be over very quickly, in a few days to a few weeks, but intermittent, lesser-magnitude volcanic activity can linger on for a million years. Thus, the Yellowstone event has left us hot springs and geysers, including Old Faithful, while the Owens Valley event (Long Valley caldera) is producing a potential hazard. Actually, both sites are still capable of producing future volcanic activity because magma is still present at variable depths beneath the caldera floors. Both calderas are *resurgent calderas* because the floors of the calderas have slowly domed upward since the explosive eruptions that formed them. The problem at Long Valley is that around 1980, the rate of uplift accelerated up to 25 cm in only about two years and was accompanied by earthquake swarms. Some of these

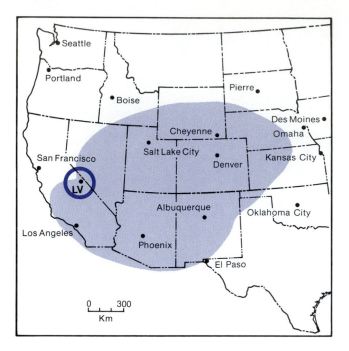

Figure 9.10
Area covered by ash from the Long Valley caldera eruption approximately 700,000 years ago. Circle near Long Valley (LV) has a radius of 120 km and encloses the area subject to at least 1 m of downwind ash accumulation if a similar eruption were to occur again. Also within this circle, hot pyroclastic flows (ash flows) are likely to occur and, in fact, could extend farther than shown by the circle. (From C. D. Miller, D. R. Mullineaux, D. R. Crandell, and R. A. Bailey, 1982. *Potential Hazards from Future Volcanic Eruptions in the Long Valley-Mono Lake Area, East-Central California and Southwest Nevada—A Preliminary Assessment.* U.S. Geological Survey Circular 877.)

earthquakes had a magnitude of 5 to 6, became harmonic, which is symptomatic of moving magma, and were moving closer to the surface (from an 8-km depth in 1980 to 3.2 km in 1982). Concern over a possible volcanic eruption prompted the U.S. Geological Survey to issue a notice of potential volcanic hazard in 1982. A notice of hazard from continued, severe earthquakes was issued in 1980. Swarms of earthquakes continued into 1983, and although the notices have been lifted, the future situation is uncertain (7).

A map and cross-section of the Long Valley caldera and Mammoth Lakes, a popular ski area, is shown in Figure 9.11. The last major volcanic eruption in the caldera since its explosive beginning about 700,000 years ago occurred about 100,000 years ago, but smaller eruptions, including steam explosions and explosive eruptions of rhyolite, occurred as recently as 400 to 500 years ago at the site of the Inyo craters (see Figure 9.11a). The magnitude of potential future eruptions at Long Valley is uncertain, but range from the unlikely very large catastrophic event to more probable smaller eruptions.

Geologists now believe that magma is moving up as a shallow intrusion, as shown in Figure 9.11b, but it could reach the surface. Thus, the monitoring of uplift, tilt, earthquakes, and hot springs activity continues. Figure 9.12 shows the potential ash fall and ash flow hazard zones from a possible future Long Valley explosive eruption that ejects about 1 cubic kilometer of material. Citizens of Mammoth Lakes and other nearby communities are taking the warning seriously, "preparing for the worst and hoping for the best," while pursuing their business in providing recreation to visiting urbanites (7, 8). Given the potentially catastrophic loss of life and property from a major eruption in this area, concern is certainly warranted.

Mudflows and Fires

Two serious secondary effects of volcanic activity are **mudflows** (also called *lahars*), produced when a large volume of loose volcanic ash and other ejecta becomes saturated and unstable and moves suddenly downslope, and *fires*. Large mudflows are especially likely to be generated from volcanoes with an abundance of snow and/or ice that can be melted rapidly. A small mudflow resulting from the 1915 eruption of Mt. Lassen in California is shown in Figure 9.13. Figure 9.14 shows the effects of mudflows and hot fiery blasts to forested land. What if there had been a town there?

Gigantic mudflows have originated on the flanks of volcanoes in the Pacific Northwest (1). The paths of two mudflows, the Osceola and Electron, that originated on Mt. Rainier are shown in Figure 9.15. Deposits of the Osceola mudflow are 5,000 years old. This mudflow moved over 80 kilometers from the volcano and involved over 1.9 billion cubic meters of debris, equivalent to 13 square kilometers piled to a depth of more than 150 meters. Deposits of the younger, 500-year-old Electron mudflow reached about 56 kilometers from the volcano and involved in excess of 150 million cubic meters of mud. More than 300,000 people now live on the area covered by these old flows, and there is no guarantee that similar flows will not occur again. Figure 9.16 shows the potential risk of mudflows and tephra accumulation for Mt. Rainier. Someone in the valley in view of such a flow would describe it as a wall of mud a few meters high moving at about 30 kilometers per hour, see it at a distance of perhaps 1.6 kilometers, and would need a car headed in the right direction toward high ground to escape being buried alive (2).

Crandell and Waldron emphasize the potential hazard of volcanic mudflows as compared to flood hazards. Floods are usually preceded by heavy rains that cause a gradual rise in water level. People in flood-prone areas generally have time to escape, and when the flood recedes, the water and danger are essentially gone. However, catastrophic mudflows can occur with little or

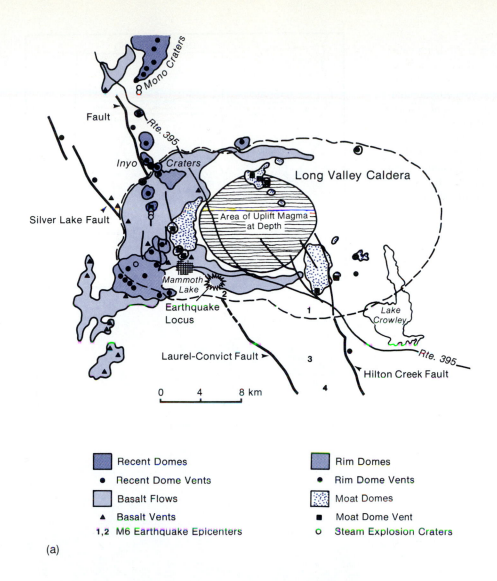

Figure 9.11
Map (a) and diagram (b) illustrating the volcanic hazard near Mammoth Lakes, California. The map (a) shows the location of past volcanic events, the area of uplift where magma seems to be moving up, and finally the area where earthquake swarms have occurred near Mammoth Lakes. The geologic cross-section (b) shows a section (northeast-southwest) through the Long Valley caldera. Shown are geologic relations inferred for the 1980 magma rise that produced the uplift and swarms of earthquakes. (From R. A. Bailey, 1983. Mammoth Lakes Earthquakes and Ground Uplift: Precursor to Possible Volcanic Activity? In *U.S. Geological Survey Yearbook, Fiscal Year 1982.*)

Legend:

- Recent Domes
- Recent Dome Vents
- Basalt Flows
- Basalt Vents
- 1,2 M6 Earthquake Epicenters
- Rim Domes
- Rim Dome Vents
- Moat Domes
- Moat Dome Vent
- Steam Explosion Craters

(a)

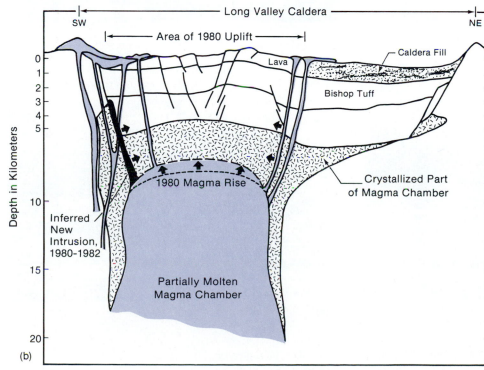

(b)

Figure 9.12
Potential hazards from a volcanic eruption at the Long Valley caldera near Mammoth Lakes, California. Colored area and diagonal lines show the hazard from flowage events out to a distance of approximately 20 km from recognized potential vents. Lines surrounding the hazard areas represent potential ash thicknesses of 20 cm at 35 km (dashed line), 5 cm at 85 km (dotted line), and 1 cm at 300 km (solid line). These estimates of potential hazards assume an explosive eruption of approximately 1 cubic km from the vicinity of recently active vents. (From C. D. Miller, D. R. Mullineaux, D. R. Crandell, and R. A. Bailey, 1982. *Potential Hazards from Future Volcanic Eruptions in the Long Valley-Mono Lake Area, East-Central California and Southwest Nevada—A Preliminary Assessment.* U.S. Geological Survey Circular 877.)

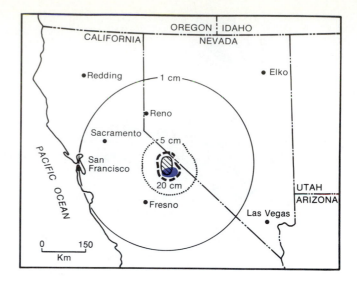

Figure 9.13
Several small mudflows on the west side of Lassen Peak. Four such mudflows reached Manzita Lake at the base of the mountain in 1915. Some of the dark area at the summit is new lava. (Photo by B. F. Loomis, courtesy of Loomis Museum Association.)

(a)

(b)

Figure 9.14
Jessen Meadow with Mt. Lassen in the background. Photograph (a) was taken in August of 1910. Photograph (b) was taken in 1925, from the same viewpoint but after a mudflow and hot blast from the volcano. Notice that all the timber in the 1910 photograph was destroyed. (Photo by B. F. Loomis, courtesy of Loomis Museum Association.)

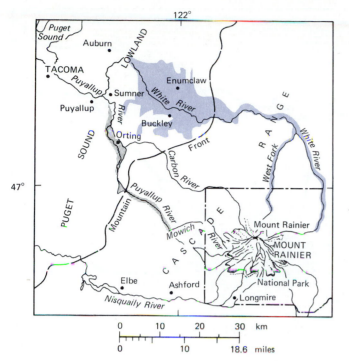

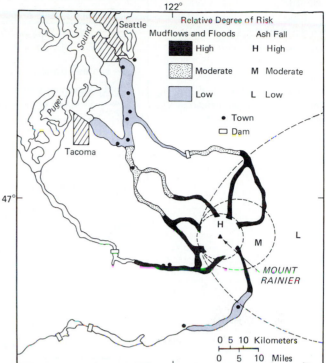

Figure 9.15

Map of Mt. Rainier and vicinity showing the extent of the Osceola mudflow in the White River Valley (shaded) and the Electron mudflow (dot pattern) in the Puyallup River Valley. (From Crandell and Mullineaux, U.S. Geological Survey Bulletin 1238, 1967.)

Figure 9.16

Sketch map of Mt. Rainier and vicinity showing relative degrees of potential hazards from ash fall (tephra) and from mudflows and floods that might result from an eruption. (From Crandell and Mullineaux, "Techniques and Rationale of Volcanic Hazards," *Environmental Geology,* vol. 1 [New York: Springer-Verlag, 1975].)

no warning. They are likely to start when the volcano is hidden by clouds of smoke, and after the event, the mud, perhaps a few meters thick, remains. Since mudflows are confined to valleys, there is another possible hazard when the valley is artificially dammed to produce hydroelectric power. A large mudflow could fill a reservoir, pushing the water over the spillway and causing a severe flood downstream. On the other hand, used wisely, reservoirs might be a safety factor for all except the really large mudflows. The water level in reservoirs might be drawn down during an upstream volcanic event, and the storage basin behind the dam could be used to contain a possible mudflow. This is not the intended function of the dam, but it is a fortunate safety mechanism (2).

Nevado del Ruiz

The volcano in Colombia known as Nevado del Ruiz erupted on November 13, 1985. The eruption triggered catastrophic mudflows that killed at least 22,000 people while inflicting more than $200 million in property damage. The eruption followed a year of precursory activity, including earthquakes and hot-spring activity. Monitoring of the volcano began in July of 1985, and in October a risk assessment and hazards map was com-

pleted that correctly identified events that actually occurred on November 13. The report accompanying the map gave a 100 percent probability that potentially damaging mudflows would be produced by an eruption, as had been the case with previous eruptions.

The November 13 event included two large explosions—the eruptions produced pyroclastic flows that scoured and melted glacial ice on the mountain, generating the mudflows that raced down river valleys. Figure 9.17 shows the areas impacted by base surges and pyroclastic flows as well as the location of glacial ice that contributed the water necessary to produce the mudflows. Of particular significance was the mudflow that raced down the river Lagunillas and destroyed part of the town of Armero, where most of the deaths occurred. Figure 9.18 shows the volcano and the town of Armero. The mudflows buried the southern half of the town, sweeping buildings completely off their foundations. The rectilinear pattern visible beneath the mud shows where some of the buildings and streets and other structures were located before being inundated by the flow (9).

The real tragedy of the catastrophe was that the outcome was predicted; in fact, there were several attempts to warn the town and evacuate it. Despite these

Figure 9.17
Map of the volcano Nevado del Ruiz area showing some of the features associated with the eruption of November 13, 1985. (Modified after D. G. Herd, 1986, "The Ruiz Volcano disaster." *EOS, Transactions of the American Geophysical Union,* May 13: 457–60.)

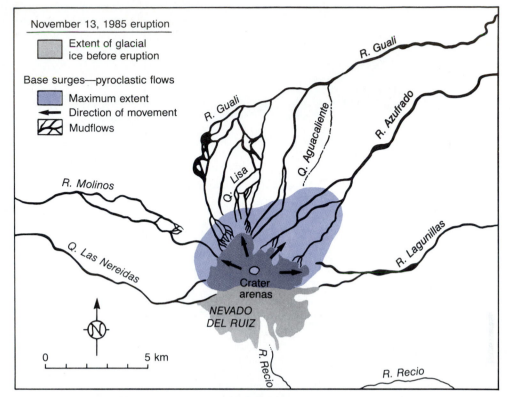

warnings there was little response and, as a result, approximately 21,000 people in the town were killed. Early in 1986 a permanent volcano observatory center was established in Colombia to continue monitoring the Ruiz volcano as well as others in South America. As a result South America should be better prepared to deal with future volcanic eruptions. Had there been better communication lines from civil defense headquarters to local towns, and a better appreciation of potential volcanic hazards even 40 km from the volcano, evacuation would have been possible for Armero. Hopefully, the lessons learned from this event will help minimize future loss of life associated with volcanic eruptions and other natural disasters.

Mt. Unzen

Japan has nineteen active volcanoes, and nearly 200 years ago Mt. Unzen in southwestern Japan erupted, killing about 15,000 people. The mountain violently erupted again in June of 1991, and authorities ordered the evacuation of thousands of people. Nevertheless, about 40 people, including journalists, police, firemen, and scientists, were killed by ash flows and other volcanic activity. Many homes were ignited and damage was extensive.

Mt. Pinatubo

The June 15–16, 1991, eruptions of Mt. Pinatubo in the Philippines may be the largest of the century—several times larger than the 1980 Mt. St. Helens eruption. Combined effects of ash fall, ash flows, mud flows, and a typhoon resulted in the deaths of more than 300 people. Evacuation of about 200,000 people, including an entire U.S. military base, undoubtedly saved many lives. The eruption column from the tremendous explosions sent ash to elevations of 30 kilometers. As with similar past events of this magnitude, the aerosol cloud of ash, including sulfur dioxide, will remain in the atmosphere, circling the earth, for as long as one year. The ash particles and sulfur dioxide scatter incoming sunlight, and perhaps will slightly alter (cool) the global climate during the next year, as well as produce spectacular sunsets (10).

Mt. St. Helens

The May 18, 1980, eruption of Mt. St. Helens in the southwest corner of Washington (Figure 9.2) exemplifies the many types of volcanic events expected from a Cascade volcano. Although the eruption, like many natural events, was unique and complex, making generalizations somewhat difficult, we have learned a great deal from Mt. St. Helens, and the entire story is not yet complete.

Mt. St. Helens awoke in March of 1980, after 120 years of dormancy, with seismic activity and small explosions as groundwater came in contact with hot rock. At 8:32 A.M. on May 18, 1980, a magnitude 5.0 earthquake was registered on the volcano. That earthquake triggered a

(a)

(b)

Figure 9.18
(a) Summit of Nevado del Ruiz as viewed from the northeast on December 10, 1985. The plume is rising from Arenas crater at the head of Rio Azufrado valley, where large mudflows were generated. (b) Remains of Armero, where about 21,000 people died on November 13, 1985. The rectilinear pattern of building foundations is visible through the thin volcanic debris flow (mudflow) deposits. (Photos courtesy of Thomas C. Pierson and U. S. Geological Survey.)

large landslide/debris avalanche (approximately 2.5 cubic kilometers) that involved the entire northern flank of the mountain on which a bulge had been growing at a rate of about 1.5 meters per day for at least several weeks and probably since activity started in March. The landslide/debris avalanche shot down the north flank of the mountain, displacing water in nearby Spirit Lake, struck and overrode a ridge 8 km to the north, then made an abrupt turn and moved for a distance of 18 km down the Toutle River. The initial failure of the bulge released internal pressure and Mt. St. Helens erupted with a lateral blast directly from the area previously occupied by the bulge. After the lateral blast, a large vertical cloud rose quickly to an altitude of approximately 19 km (Figures 9.19 and 9.20). Eruption of the vertical column continued for more than nine hours, and large volumes of volcanic ash fell on a wide area of Washington, northern Idaho,

and western and central Montana (Figure 9.20). The total amount of volcanic ash ejected was about one cubic kilometer, and during the nine hours of eruption a number of ash flows swept down the northern slope of the volcano. The entire northern slope of the volcano, the upper part of the north fork of the Toutle River basin, was devastated as forested slopes were transformed into a gray, hummocky landscape consisting of volcanic ash, rocks, blocks of melting glacial ice, narrow gullies, and hot steaming pits (Figure 9.21) (11).

The first of several mudflows, consisting of a mixture of water, volcanic ash, rock, and organic debris (such as logs), occurred minutes after the start of the eruption. Some time later, a young couple fishing on the Toutle River, approximately 36 km downstream from Spirit Lake, were awakened by a loud rumbling noise from the river, which was covered by felled trees. The pair attempted to

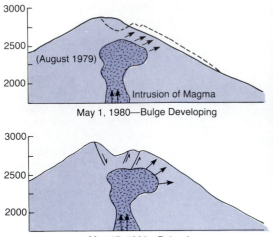

(August 1979)

Intrusion of Magma

May 1, 1980—Bulge Developing

May 17, 1980—Bulge Area

(a)

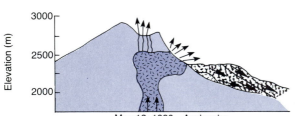

Elevation (m)

May 18, 1980—Avalanche
8:32 Eruption Starts

(b)

Avalanche

Seconds Later—Strong Lateral Blast

(c)

Figure 9.19

Diagrams and photographs showing the sequence of events for the May 18, 1980, eruption of Mt. St. Helens. (Photographs [a], [b], and [c] © 1980 by Keith Ronnholm [the Geophysics Program, Univ. of Washington, Seattle], and photograph [d] by Robert Krimmel, U.S. Geological Survey. Drawings inspired from lecture by James Moore, U.S. Geological Survey.)

(d)

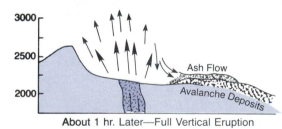

About 1 hr. Later—Full Vertical Eruption

Figure 9.19, *continued.*

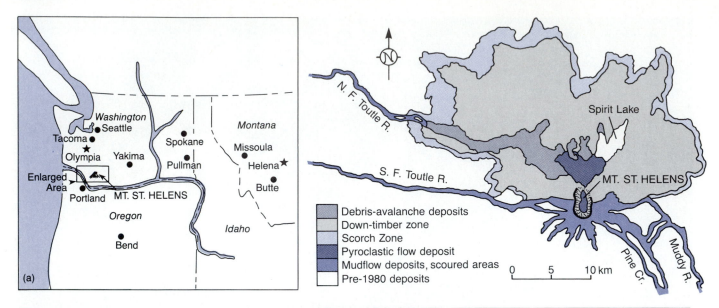

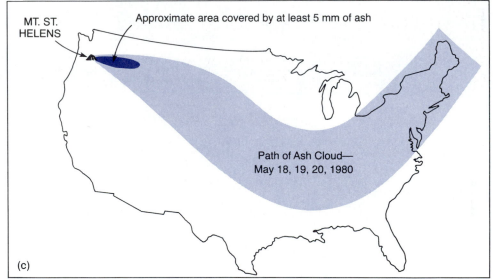

	Debris-avalanche deposits
	Down-timber zone
	Scorch Zone
	Pyroclastic flow deposit
	Mudflow deposits, scoured areas
	Pre-1980 deposits

0   5   10 km

Figure 9.20

Mt. St. Helens: (a) location; (b) debris-avalanche deposits, tree blowdown, and mudflows associated with the May 18, 1980, eruption; (c) path of the ash cloud from the 1980 eruption. (Data from various U.S. Geological Survey publications.)

run to their car, but water from the rising river poured over the road, preventing their escape. A mass of mud then crashed through the forest toward the car, and the couple climbed on top of the roof to escape the mud. They were safe only momentarily, as the mud pushed the car over the bank and into the river. They leaped off the roof and fell into the river, which was by now a rolling mass of mud, logs, collapsed train trestles, and other debris. The water also was increasing in temperature. One of them got trapped between logs and disappeared several times beneath the flow but was lucky enough to emerge again. The couple were carried downstream for approximately 1½ km before another family of campers spotted and rescued them. The flows and accompanying floods raced down the valleys of the north and south forks of the Toutle River at estimated speeds of 29–55 km per hour. Water levels in the river reached at least 4 meters above flood stage and nearly all bridges along the

river were destroyed. The hot mud quickly raised the temperature of the Toutle River to as high as 38°C. Mud, logs, and boulders were carried 70 km downstream into the Cowlitz River and eventually 28 km further downstream into the Columbia River. Nearly 40 million cubic meters of material was dumped into the Columbia River, reducing the depth of the shipping channel from a normal 12 m to 4.3 m for a distance of 6 km (11).

When the volcano could be viewed again following the eruption, it was observed that the maximum altitude of the volcano was reduced by about 450 m and the original symmetrical mountain was now a huge, steep-walled amphitheater facing northward (Figure 9.22). The debris avalanche, horizontal blast, pyroclastic flows, and mudflows devastated an area of nearly 400 square kilometers, killing 54 people. More than 100 homes were destroyed by the flooding, and approximately 800 million board feet of timber was flattened by the blast (Figure

(a)

(b)

Figure 9.21
Aerial view (a) and ground view (b) of deposits from the May 18, 1980, landslide/debris avalanche associated with the eruption of Mt. St. Helens. Notice the hummocky nature of the topography. The person in photograph (b) provides a scale by which the size of some of the large blocks of the debris can be estimated. The total volume of debris is in excess of 2.5 cubic kilometers. (Photographs courtesy of U.S. Geological Survey, Robert Krimmel [a], Harry Glicken [b].)

9.23). The total damage was estimated to be near three billion dollars, but long-range damage to fisheries and other resources have been difficult to estimate.

Following the catastrophic eruption of Mt. St. Helens, an extensive program was established to monitor volcanic activity and in particular the construction (by lava flows) of the dome that occupies part of the crater produced by the May 18, 1980, eruption. The movement of water and sediment in the disturbed land, particularly on the northern flank of the volcano, is also being monitored. The purpose of the extensive monitoring is to provide advance warnings of potential eruptions, floods, or mudflows likely to present a hazard to people and facilities in the Mt. St. Helens area (12).

During the first three years following the May 18 eruption, at least eleven other, smaller events occurred that contributed to the building of a lava dome approximately 250 meters high within the crater. Each of these

(a)

(b)

Figure 9.22
Mt. St. Helens (a) before and (b) after the May 18, 1980, eruption. As a result of the eruption, much of the northern side of the volcano was blown away and the altitude of the summit was reduced by approximately 450 meters. (Photographs courtesy of U.S. Geological Survey and Harry Glicken.)

(a)

(b)

Figure 9.23
The lateral blast from the 1980 eruption of Mt. St. Helens downed approximately 800,000,000 board feet of timber. Shown here (a) are some of those trees. Photograph (b) is a close-up of one splintered tree trunk where it was severed by the blast. (Photographs courtesy of U.S. Geologial Survey and Harry Glicken.)

PLATE 3A
In June of 1991, Mt. Unzen in Japan erupted. This firefighter runs from sure death as the ash flow moves down the valley. (Photo courtesy of The Yomiuri Shimbun)

PLATE 3B
Thousands of people were evacuated, including over a thousand from U.S. Clark Air Base, during this large ash eruption and explosion of Mt. Pinatubo in the Philippines on June 12, 1991. (Photo courtesy of Claro Cortes/Reuters/Bettman Archive)

PLATE 4
Two extremes of channelization. Concrete ditch, Carpinteria Creek, California (top) and channel restoration of Briar Creek near Charlotte, North Carolina (bottom).

PLATE 5A
Oil fields were deliberately set on fire during the Iraqi occupation of Kuwait. Soot and smoke from hundreds of oil fires entered the lower atmosphere, causing local and regional environmental degradation. (Photo courtesy of L. Van der Stockt/Gamma Liaison)

PLATE 5B
Coastal erosion and associated landsliding claimed this house at Cove Beach, Oregon, in 1988, proving that building close to the ocean is hazardous. (Photo courtesy of Gary Braasch/AllStock)

PLATE 6
Large storm waves attacking a beach house in Ventura, California during the winter of 1982–1983 (top). Note large rocks thrown onto the street by wave action (bottom). (Photograph courtesy of C. G. Marshall.)

eruptions added to the dome as lava was extruded near the top of the dome and slowly flowed toward the base. Monitoring the growth of the dome and the deformation of the crater floor, along with such other techniques as monitoring the gases emitted and general growth of the dome, was useful in predicting eruptions. Each eruption added a few million cubic meters to the size of the dome; if the rate of accumulation were to continue for several hundred years, Mt. St. Helens would be rebuilt to its former height. However, some scientists believe that this may be unlikely and the future is uncharted (12).

An extensive system of stream gages and lake gages has been established in the Mt. St. Helens area (Figure 9.24). Monitoring is most extensive on the northern flank of the volcano, particularly in the area covered by the debris avalanche deposits. The deposits raised the level of Spirit Lake by over 60 meters and produced an even higher natural dam. In addition, the avalanche deposits dammed up several tributaries, forming lakes such as Castle Lake and Coldwater Lake shown on Figure 9.24. It was recognized that failure of the dams produced by the avalanche deposits could lead to catastrophic flooding and generation of mudflows similar to those that occurred during the 1980 eruption. To defuse this hazard, outlets were constructed for the lakes and a series of stream gaging stations installed to provide information concerning movement of flood waters. Sediment data collected at the stream gages also provide long-term information concerning the movement and routing of sediment down the river system toward the Columbia River (12).

▼ PREDICTION OF VOLCANIC ACTIVITY

It is unlikely that we will be able in the near future to predict accurately all volcanic activity, but valuable information is being gathered about phenomena that occur prior to eruptions. One problem is that most prediction techniques require experience with actual eruptions before the mechanism is understood. Thus, we are better able to predict eruptions in the Hawaiian Islands because we have had so much experience there.

Possible methods of predicting volcanic eruptions include geophysical observations of thermal and magnetic properties, topographic monitoring of tilting or swelling of the volcano, monitoring of seismic activity, monitoring of gas emissions, and studying the geologic history of a particular volcano or volcanic center (13).

Geophysical monitoring of volcanoes is based on the fact that, prior to an eruption, a large volume of magma moves up into some sort of holding reservoir beneath the volcano. The hot material changes the local magnetic, thermal, hydrologic, and geochemical conditions. As the surrounding rocks heat, the rise in temperature of the surficial rock may be detected by infrared aerial photography. Thus, periodic remote sensing of a volcanic chain may detect new hot points that could indicate possible future volcanic activity. This method was used with some success at Mt. St. Helens prior to the main eruption on May 18, 1980.

When older volcanic rocks are heated by new magma, magnetic properties, originally imprinted when the rocks cooled and crystallized, may change. These changes might be detailed by ground or aerial monitoring. Unfortunately, this method has not been sufficiently researched and tested to be applied to actual prediction of volcanic events (13).

Monitoring of topographic changes and seismic behavior of volcanoes has been useful in predicting some volcanic eruptions. The Hawaiian volcanoes, especially Kilauea, have supplied most of the data. The summit of Kilauea actually tilts and swells prior to an eruption and

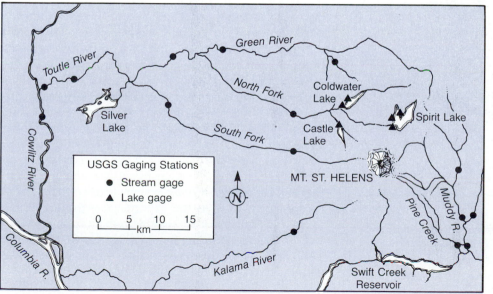

Figure 9.24
Map of Mt. St. Helens area showing locations of U.S. Geological Survey gaging stations where water and sediment are monitored. (After S. Brantley and L. Topinka, 1984, *Earthquake Information Bulletin*, v. 6, no. 2.)

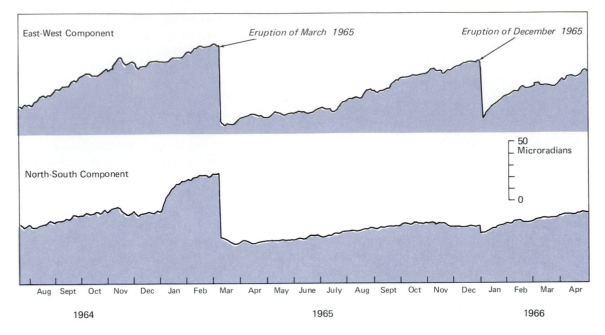

East-West Component   Eruption of March 1965   Eruption of December 1965

North-South Component

50 Microradians

0

Aug Sept Oct Nov Dec Jan Feb Mar Apr May June July Aug Sept Oct Nov Dec Jan Feb Mar Apr

1964   1965   1966

Figure 9.25
Graph showing east-west component and north-south component of ground tilt recorded from 1964 to 1966 on Kilauea Volcano, Hawaii. Notice the slow change in ground tilt before eruption and rapid subsidence during eruption. (From R. S. Fiske and R. Y. Koyanagi, U.S. Geological Survey Professional Paper 607.)

subsides during the actual outbreak (Figure 9.25). This movement, in conjunction with earthquake swarms that reflect the moving subsurface magma and announce a coming eruption, was used to predict volcanic activity in the vicinity of the farming community of Kapoho on the flank of the volcano (Figure 9.26), 45 kilometers from the summit. The inhabitants were evacuated before the activity, which overran, thrust aside, and floated away lava barriers (walls) and eventually destroyed most of the

village (Figure 9.27) (14). Scientists expect that, because of characteristic swelling and earthquake activity before eruptions, the Hawaiian volcanoes will be predictable. Predicting eruptions of less active volcanoes, such as in the Cascade Range, is much more difficult, as we learned from Mt. St. Helens.

Our understanding of volcanic hazards has come a long way, however. The Mt. St. Helens eruption was predicted from a long-range standpoint (geologists in the 1960s predicted it would very likely erupt before the end of the century), and we were fairly successful in making intermediate-range predictions—the increased seismic activity and bulge in the volcano were predicted to lead to some sort of eruption. Unfortunately, the very short-range prediction of a few hours or even a few days for the large explosion on May 18 was not made. Hazard zones were established and entry was restricted before the May 18 eruption. After the eruption the designated red zone within 30 km of the volcano and a blue zone within an additional 15 km to the east and southeast were closed to entry. This paid off on August 7, when a prediction for an explosive event, based on change in volcanic gas chemistry and earthquake activity, was made just hours prior to the eruption. Thus, as the experience with the volcano increased, chances for successful prediction increased. In fact, 13 eruptions, from June 1980 through December 1982, were predicted by monitoring seismicity, deformation of the crater floor, and gas emissions (15).

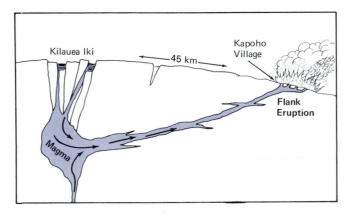

Kilauea Iki   Kapoho Village

← 45 km →

Flank Eruption

Magma

Figure 9.26
Idealized diagram showing the 1960 flank eruption on Kilauea which destroyed the village of Kapoho. (After Eaton and Schmidt, *Atlas of Volcanic Phenomena*, U.S. Geological Survey.)

Cascade Volcanoes

The Cascade volcanoes are complex and do not fit as easily into a pattern of eruption as do the shield volcanoes of the Hawaiian Islands. Despite this limitation, it is advantageous to attempt to generalize and make models for the Cascade volcanoes. As we learn more from direct observation and experience, the models should improve our ability to predict eruptions.

Detailed geologic investigations of a volcano or volcanic center may yield valuable information concerning possible future volcanism. For example, there is abundant evidence that the Cascade volcanoes have a potential for future eruptions, information that was used to predict that Mt. St. Helens would erupt before the end of the twentieth century.

Stages of development. Although each volcano or volcanic center in the Cascades is somewhat individual with a unique history, we can make some generalizations. Many of the Cascade volcanoes have developed or evolved through several stages. In the first stage, the main composite cone is constructed of lava flows and pyroclastic materials with an intermediate silica content (about 60%), known as *andesite*. The second stage is characterized by less frequent but divergent activity: the sporadic development of satellite cones composed of relatively low-silica volcanic rock (approximately 50%) known as *basalt;* and the development of domes composed of relatively high-silica (approximately 70%) volcanic rock known as *rhyolite*. The trend toward volcanic domes may be a dangerous sign, as their eruption can be very explosive (16).

The amount of the materials and the composition in the evolutionary stages of Cascade volcanoes vary considerably. While several volcanoes reflect the various stages, no two are exactly alike and no one volcano follows exactly the evolutionary path described above. Mts. Adams, Baker, and Rainier are still in the main cone-building stage, and all three are composed of a rather uniform andesite (Figure 9.2). Mt. St. Helens may still be in the cone-building stage, as (prior to the May 18, 1980, explosion) the entire upper part of the volcano was constructed during the last 2,500 years, and much of that in the last few hundred years. It is possible that the explosive activity is part of a general cone-building cycle or episode of unknown time length that involves both explosive and constructive eruptions. Thus, in the future, the volcano may again build a cone characteristic of a young composite volcano. On the other hand, the eruptive products, including domes, and their chemical compositions have varied considerably during the last few thousand years at Mt. St. Helens. It is therefore difficult to place the volcano into a general model, emphasizing our earlier point that each volcano may have a unique history.

Mts. Jefferson and Hood are evidently in the divergent stage, with the development of basalt satellite vents and rhyolite domes. For these two volcanoes, the second stage developed following at least one long period of dormancy (16).

At the end of the second stage, a volcano may become extinct, reenter the first stage, or enter a third stage known as *maturity*. Mt. Mazama (Crater Lake in Oregon) is the only known example (Figure 9.2) of this stage. After the second stage, Mazama was dormant for several thousand years, resumed divergent activity, was dormant for another 5,000 years, and, in a spectacular climax, erupted violently about 7,000 years ago, ejecting approximately 60 cubic kilometers of tephra (mostly ash) over an area in excess of 12 million square kilometers. When the eruption was over, the summit of Mt. Mazama had disappeared, and the present site of Crater Lake, a caldera nearly 10 kilometers wide and 120 meters deep, was in its place. We conclude that the third, and, presumably, final stage may be very explosive and, thus, very dangerous (16).

The environmental significance of our discussion concerning the evolutionary history of the Cascade volcanoes is that we may be able to estimate the chance of catastrophic eruption of a particular volcano by examining its past activity. The trend toward divergent activity (satellite cones and domes) is a clear danger sign, since this increases the chances of explosive eruptions.

▼ ADJUSTMENT TO AND PERCEPTION OF THE VOLCANIC HAZARD

Apart from the psychological adjustment to losses, the only major human adjustment to volcanic activity is evacuation (17). The small number of adjustments probably reflects the fact that few eruptions occur near populated areas. The experience of about 5,000 Icelandic people on the island of Heimaey offers an example of a hardy people's response to a serious destructive hazard. In January 1973, the dormant Mt. Helgafell came alive, and subsequent eruptions nearly buried the town of Vestmannaeyjar in ash and lava flows. The harbor, a major fishing port, was nearly blocked. With the exception of about 300 town officials, fire fighters, and police, the people were evacuated. They returned six months later, in July 1973, to survey the damage and estimate the chances of rebuilding their town and lives. The situation was grim. Ash had drifted up to 4 meters thick and covered much of the island and town. Molten lava was still steaming near the volcano. Their first task was to dig out their homes and shops, salvaging what they could. Then they used the same volcanic debris that had buried their town to pave new roads and an airport to allow materials to be moved in and distributed. They also decided to use the volcano to advantage, and by January

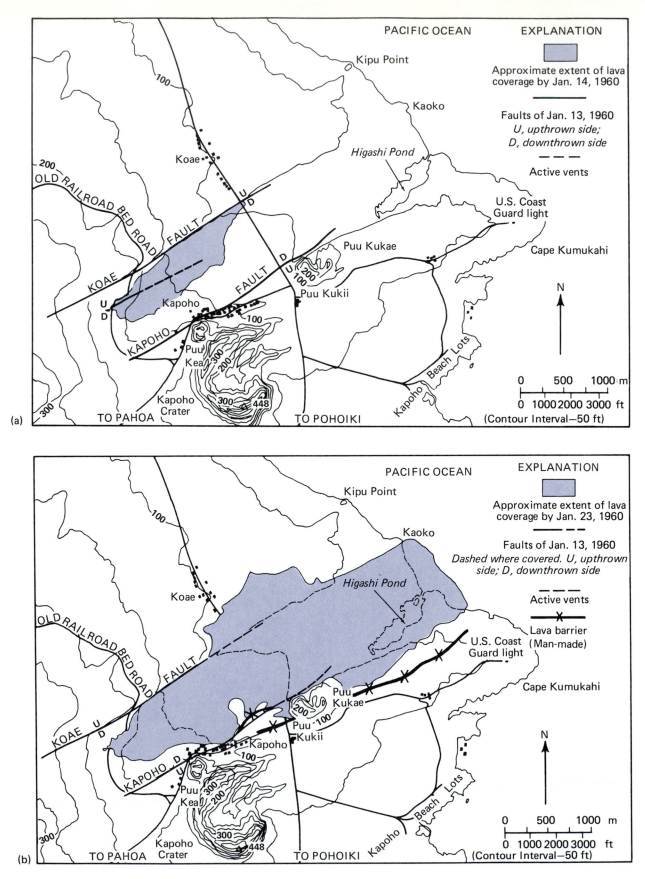

Figure 9.27
Volcanic eruption on the flank of Kilauea from January 23 to February 5, 1960. Notice the lava barriers that failed. (100 ft = 30.5 m) (After Richter, et al., U.S. Geological Survey Professional Paper 537E, 1970.)

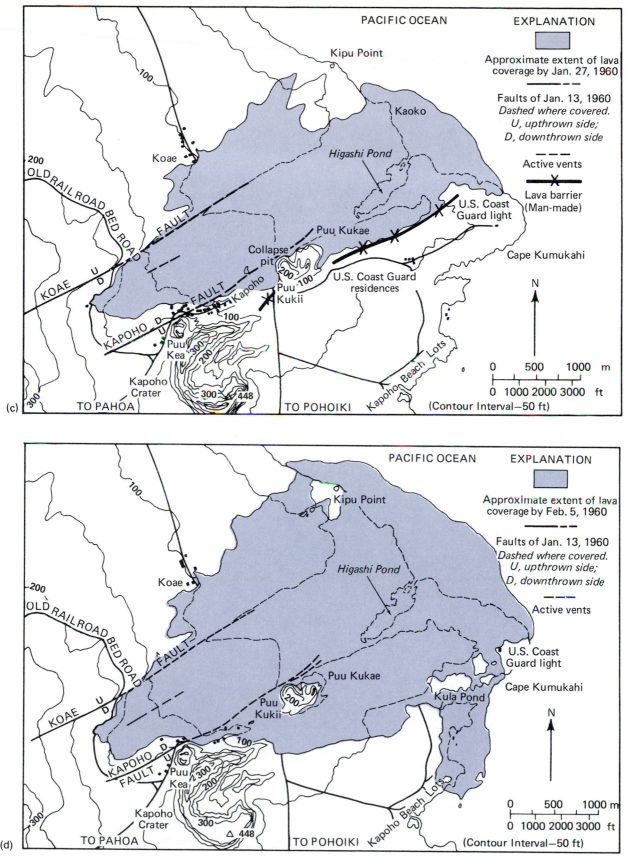

1974, the first home was heated by lava heat. By 1980, 15 percent of the homes were heated this way, and the system was being expanded to the whole town. The heat source is expected to last at least 15 years. More significantly, two fish-meal factories and three freezing plants were reopened, revitalizing the island's fishing industry (18, and B. Voight, personal communication).

Since eruptions seldom occur near populated areas, little information is available concerning how people perceive the volcanic hazard. One study of perception evaluated volcanic activity in Hawaii; the study makes two suggestions. First, perception of volcanic activity and what people will do in case of an eruption has little to do with proximity to the hazard, home ownership, knowledge of necessary adjustments, and income level. Second, a person's age and length of residence near the hazard are significant in that person's knowledge of volcanic activity and possible adjustments (17).

Mt. St. Helens provided a valuable example of what kind of human difficulties can be expected during and after a high magnitude physical event that disrupts a large area. The experience should help in devising emergency plans for possible future volcanic eruption or other events such as large earthquakes. For example, the political pressures that arose during the early days of the awakening of Mt. St. Helens are quite interesting. On May 17, just one day before the eruption, a caravan of local cabin owners entered the Spirit Lake area at the base of the north flank of the volcano to inspect their properties. They had been requesting admittance for some time, and it was finally granted. A second trip was planned for May 18. If the eruption had not taken place on the 18th, but, say, one week later, one wonders how many people would have returned to the mountain. As we learn more about the magnitude and frequency of natural processes, we are doing a better job of short-range prediction, which should help to save lives in the future.

▼ ▼ ▼ SUMMARY AND CONCLUSIONS

Volcanic activity is a process generally related to plate tectonics. Most volcanoes are located at plate junctions where magma is produced as spreading or sinking lithospheric plates interact with other earth material. Volcanic eruptions have destroyed property, taken many human lives, and have a high potential to produce catastrophes.

Primary effects of volcanic activity include lava flows and pyroclastic activity. Secondary effects include mudflows and fires. All these effects have occurred in the recent history of the Cascade Range of the Pacific Northwest, and there is no reason to believe that further activity will not occur there in the future.

Several methods, including construction of walls, bombing, and hydraulic chilling, have been used in attempts to control lava flows. These methods have had varying degrees of success and require further evaluation.

Pyroclastic activity includes volcanic ash eruptions, which may cover large areas with a carpet of ash; volcanic ash flows, which move as fast as 100 kilometers per hour down the side of a volcano; and lateral blast, which can be very destructive.

Giant caldera-forming eruptions are violent but rare geologic events. Following their explosive beginning, however, they often resurge and may present a volcanic hazard for a million years or longer. Recent uplift and earthquakes at the Long Valley caldera in California are reminders of the potential hazard.

Mudflows are generated when melting snow and ice mix with volcanic ash. These flows are a serious hazard and can devastate an area many miles from the volcano.

Sufficient monitoring of geophysical properties, topographic changes, seismic activity, and recent geologic

history of volcanoes may eventually result in reliable prediction of volcanic activity. Prediction based on ground tilting, earthquake activity, deformation of the crater floor, and gas emissions is being attempted with some success in Hawaii and at Mt. St. Helens. Applying these techniques to other areas may lead to better prediction of volcanic activity. On a worldwide scale, however, it is unlikely that we will be able to predict accurately all volcanic activity in the near future.

Apart from psychological adjustment to losses, the only major human adjustment to volcanic activity is evacuation.

Perception of the volcanic hazard is apparently a function of one's age and length of residency near the hazard. Direct proximity to a volcano, home ownership, knowledge of necessary adjustments, and income level have little effect on the perception.

▼ ▼ ▼ REFERENCES

1. OFFICE OF EMERGENCY PREPAREDNESS. 1972. *Disaster preparedness* 1, 3.
2. CRANDELL, D. R., and WALDRON, H. H. 1969. Volcanic hazards in the Cascade Range. In *Geologic hazards and public problems, conference proceedings,* ed. R. Olsen and M.

Wallace, pp. 5–18. Office of Emergency Preparedness Region 7.
3. MASON, A. C., and FOSTER, H. L. 1953. Diversion of lava flows at Oshima, Japan. *American Journal of Science* 251:249–58.
4. WILLIAMS, R. S., Jr., and MOORE, J. G. 1973. Iceland chills a lava flow.

Geotimes 18: 14–18.
5. SINOLOWE, J. 1986. The lake of death. *Time,* Sept. 8: 34–37.
6. FRANCIS, P. 1983. Giant volcanic calderas. *Scientific American* 248, no. 6:60–70.
7. BAILEY, R. A. 1983. Mammoth Lakes earthquakes and ground uplift: Pre-

cursor to possible volcanic activity? *U.S. Geological Survey Yearbook, 1982*:5–13.

8. MILLER, C. D.; MULLINEAUX, D. R.; CRANDELL, D. R.; and BAILEY, R. A. 1982. *Potential hazards from future volcanic eruptions in the Long Valley–Mono Lake Area, east central California and southwest Nevada—a preliminary assessment.* U.S. Geological Survey Circular 877.

9. HERD, D. G. 1986. The 1985 Ruiz Volcano disaster. *EOS.* Transactions, American Geophysical Union, May 13: 457–60.

10. AMERICAN GEOPHYSICAL UNION. 1991. Pinatubo cloud measured. *EOS.* Transactions, American Geophysical Union, 72:29, 305–306.

11. HAMMOND, P. E. 1980. Mt. St. Helens blasts 400 meters off its peak. *Geotimes* 25: 14–15.

12. BRANTLEY, S., and TOPINKA, L. 1984. *Earthquake Information Bulletin* 16, no. 2.

13. FRANCIS, P. 1976. *Volcanoes.* England: Pelican Books.

14. RICHTER, D. H.; EATON, J. P.; MURATA, K. J.; AULT, W. U.; and KRIVOY, H. L. 1970. *Chronological narrative of the 1959–60 eruption of Kilauea Volcano, Hawaii.* U.S. Geological Survey Professional Paper 537E.

15. SWANSON, D. A.; CASADEVALL, T. J.; and DZURISIN, D. 1983. Predicting eruptions at Mount St. Helens, June 1980 through December 1982. *Science* 221:1369–76.

16. WISE, W. S., for SHANNON & WILSON, INC. 1976. *Volcanic hazard study.* Report to Portland General Electric Company. Portland, Oregon.

17. MURTON, BRIAN J., and SHIMABUKURO, SHINZO. 1974. Human response to volcanic hazard in Puna District, Hawaii. In *Natural hazards,* ed. G. F. White, pp. 151–59. New York: Oxford University Press.

18. CORNELL, JAMES, ed. 1974. *It happened last year—earth events—1973.* New York: Macmillan.

CHAPTER TEN

▼

▼

▼

Coastal Hazards

Coastal areas are varied in topography, climate, and vegetation, and they are generally dynamic environments. Continental and oceanic processes converge along coasts to produce landscapes that are characteristically capable of rapid change. The impact of hazardous coastal processes is considerable, because many populated areas are located near the coast. This is especially true in the United States where it is expected that most of the population will eventually be concentrated along the nation's 150,000 kilometers of shoreline, including the Great Lakes. Today, the nation's largest cities lie in the coastal zone, and approximately 75 percent of the population lives in coastal states (1).

The most serious coastal hazards are *tropical cyclones,* which claim many lives and cause enormous amounts of property damage every year; *tidal floods* caused by combination of a high tide with a storm surge; *tsunamis,* or seismic sea waves (discussed in Chapter 8), which are particularly hazardous to coastal areas of the Pacific Ocean; and *coastal erosion,* which continues to produce considerable property damage that requires human adjustment.

▼ TROPICAL CYCLONES

Tropical cyclones have taken hundreds of thousands of lives in a single storm. A tropical cyclone that struck the northern Bay of Bengal in Bangladesh in November of 1970 produced a 6-meter rise in the sea. Flooding killed approximately 300,000 people, caused $63 million in crop losses, and destroyed 65 percent of the total fishing capacity of the coastal region (2). Another devastating cyclone hit Bangladesh in the spring of 1991, killing over 100,000 people while inflicting more than $1 billion in damage.

Tropical cyclones, known as *typhoons* in most of the Pacific Ocean and *hurricanes* in the Western Hemisphere, cause damage and destruction from high winds; river flooding that results from intense precipitation and usually causes more deaths and destruction than the wind; and storm surges (wind-driven oceanic waters) that are the most lethal aspect of tropical cyclones (Figure 10.1). Most deaths in these storms result from drowning (3). Property damage from hurricanes can be staggering. For example, several hurricanes that struck the United States in the last ten years each caused over $1 billion in property damage. Hurricane Hugo in September of 1989 was the costliest in U.S. history with estimated damages to the U.S. mainland, Puerto Rico, and the U.S. Virgin Islands of $5 to $7 billion (4). Despite the increasing population along the Atlantic and Gulf coasts, the loss of lives from hurricanes has decreased significantly because of more effective detection and warning; however, the amount of property damage has greatly increased. Concern is growing that continued population increase accompanied by unsatisfactory evacuation routes, building codes, and refuge sites may contribute to hurricane catastrophes along the Atlantic and Gulf coasts (3).

A secondary effect of tropical cyclones is flash flooding caused by intense rainfall as the storm moves inland. Hurricane Camille, in 1969, first damaged the Louisiana and Mississippi coastlines and, two days later, caused record amounts of precipitation in the mountains of western Virginia, resulting in more than $100 million in damage (3).

Most hurricanes form in a belt between 8 degrees north and 15 degrees south of the equator, and the areas most likely to experience cyclones in this zone are those with warm surface-water temperatures. The storms are generated as tropical disturbances and dissipate as they move over the land. Wind speeds in these storms are greater than 100 kilometers per hour, and the winds blow in a large spiral around a relatively calm center called the *eye* of the hurricane. Winds of 100 kilometers per hour or greater are generally recorded over an area about 160 kilometers in diameter, while gale-force winds greater than 60 kilometers per hour are experienced over an area about 640 kilometers in diameter. During an average year, about five hurricanes will develop that might threaten the Atlantic and Gulf coasts. Figure 10.2 shows the storm tracks of and loss of life caused by the several devastating hurricanes that struck the United States from 1964 to 1989 (3). Three general storm tracks are possible: (1) a storm heads toward land and then veers back into the Atlantic; (2) a storm travels over Cuba and into the Gulf of Mexico to strike the Gulf Coast; and (3) a storm skirts along the East Coast and may strike land from Florida to New York. Figure 10.3 shows the probability (as a percent chance) of a hurricane striking a particular segment of coast in any one year. Notice that

Figure 10.1
Storm surge from Hurricane Camille, 1969. (Photo courtesy of NOAA.)

the areas most likely to experience a hit are the Texas coast and southern Florida.

A fully developed hurricane is an awesome spectacle. Figure 10.4 shows Hurricane Allen on Friday, August 8, 1980. The storm, with winds of nearly 300 kilometers per hour and six-meter waves, nearly covered the entire Gulf of Mexico and was one of the largest hurricanes to strike the United States in this century. Approximately 200,000 people heeded the warning and evacuated coastal areas of southern Texas—only a few stubborn people remained. The storm killed more than 100 people before reaching Texas, but fortunately stalled for several hours, losing energy, off the coast of Texas before striking the land in a sparsely populated area. Loss of life in the United States was light (13 people were killed in a helicopter accident while evacuating offshore oil platforms), but damage to crops from torrential rains that exceeded 25 centimeters was at least several hundred million dollars. On the positive side, the rains ended a month-long drought and thus were welcomed in some areas. In the final analysis, we were lucky with Hurricane Allen: the storm stalled offshore and lost energy before striking land; evacuation involved several hundred thousand rather than several million people; and its final path over land was in rural range land.

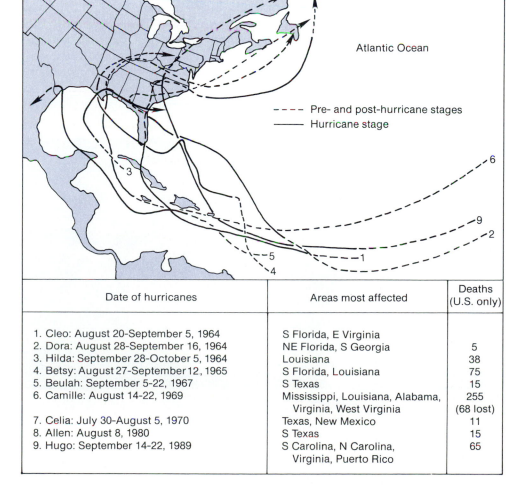

Atlantic Ocean

- - - - Pre- and post-hurricane stages
——— Hurricane stage

Figure 10.2
Catastrophic hurricanes affecting the United States from 1964 to 1989. The tracks for Celia and Allen are not shown. (Modified after *Some Devastating North Atlantic Hurricanes of the 20th Century,* U.S. Department of Commerce, 1970. Updated by NOAA.)

Date of hurricanes	Areas most affected	Deaths (U.S. only)
1. Cleo: August 20–September 5, 1964	S Florida, E Virginia	
2. Dora: August 28–September 16, 1964	NE Florida, S Georgia	5
3. Hilda: September 28–October 5, 1964	Louisiana	38
4. Betsy: August 27–September 12, 1965	S Florida, Louisiana	75
5. Beulah: September 5–22, 1967	S Texas	15
6. Camille: August 14–22, 1969	Mississippi, Louisiana, Alabama, Virginia, West Virginia	255 (68 lost)
7. Celia: July 30–August 5, 1970	Texas, New Mexico	11
8. Allen: August 8, 1980	S Texas	15
9. Hugo: September 14–22, 1989	S Carolina, N Carolina, Virginia, Puerto Rico	65

Figure 10.3
Probability that a hurricane will strike a particular 80-km south Atlantic coastal segment in a given year. (From Council on Environmental Quality, *Environmental Trends*, 1981.)

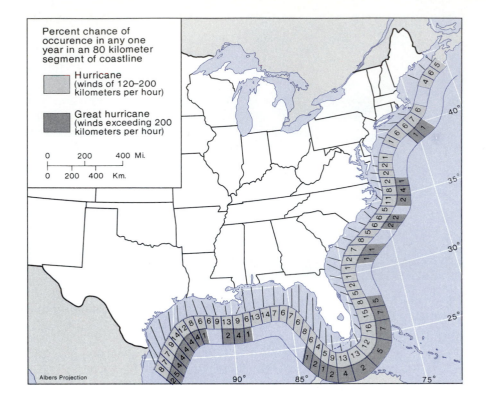

Hurricane Hugo was first detected by satellite imagery on September 9, 1989, as a cluster of thunderstorms off the coast of Africa (Figure 10.2). The tropical disturbance became a tropical storm on September 11, and a hurricane on September 13, with wind speeds of about 120 kilometers per hour. On Monday, September 17, in the early afternoon, Hugo struck the northeast tip of Puerto Rico, killing 12 people and leaving approxi-

mately 30,000 homeless. In the early hours of September 22, the hurricane ploughed into the United States mainland near Charleston, South Carolina, with winds as high as 220 kilometers per hour at the coast and 170 kilometers per hour at the historic town of Charleston. The storm surge at the coast was about 2.4 meters above the predicted normal tide, with waves reported as high as 6 meters above mean sea level. Low-lying coastal resorts

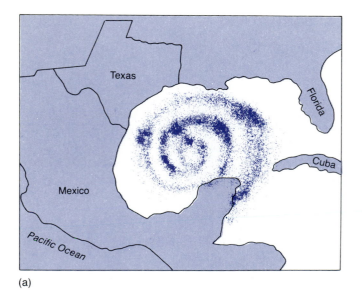

(a)

(b)

Figure 10.4
Hurricane Allen on Friday, August 8, 1980. Drawing (a) shows that the hurricane covered the entire Gulf of Mexico. More detail is shown in the actual satellite photograph (b). (Photograph courtesy of NOAA and Jeff Dozier.)

were devastated. Garden City beach, about 112 kilometers northeast of Charleston, was nearly destroyed. Heavy damage, with loss of several beach homes, was reported at Myrtle Beach, a popular resort area. In all, over $3 billion in property damage and 34 deaths were reported in South and North Carolina. Total property loss in the United States was about $5–7 billion, making Hugo the most costly hurricane to ever strike the United States. After landfall, Hugo weakened somewhat, but still had sufficient strength to knock out power at Charlotte, North Carolina, several hundred kilometers inland. The storm then moved northward into the Appalachian Mountains, where it weakened to become a tropical storm. The storm moved inland across West Virginia, Virginia, eastern Ohio, just east of Lake Erie, and across eastern Canada before finally moving out to sea into the North Atlantic (Figure 10.2) (4).

Although Hugo was extremely costly in terms of property damage, loss of life was minimal due to the long warning time that allowed for evacuation of low-lying coastal areas. Before striking land at South Carolina, the storm slowed somewhat, regaining its strength after passing through the Puerto Rico area. This gave authorities ample time to complete the evacuation; even so, traffic jams occurred in some areas and unfortunately, as is often the case, some people refused to evacuate. Had the storm moved more quickly, providing a shorter warning time, loss of life could have been much higher. Thus, we should not be overconfident and believe that all low-lying areas on the Gulf and Atlantic coasts are adequately prepared for future large hurricanes.

▼ TIDAL FLOODS

Hurricanes are not the only coastal storms that inflict damage. For example, a storm surge from a lesser tropical disturbance, when combined with a high tide, may produce a "flash" tidal flood. Such an event occurred in Bangor, Maine, on February 2, 1976. The city, located 32 kilometers inland from Penobscot Bay at the confluence of the Kenduskeag Stream and the Penobscot River, was flooded with 3.7 meters of water. The flood resulted when a tidal storm surge, caused by strong south-southeasterly winds up to about 100 kilometers per hour off the coast of New England, moved up the funnel-shaped Penobscot Bay (Figure 10.5). When the surge reached Bangor, flood waters rose very quickly, reaching the maximum depth in less than 15 minutes. Approximately 200 motor vehicles parked in lots along the Kenduskeag Stream were submerged, and when the flood receded about one hour later from this first documented tidal flood at Bangor, damages in the downtown area exceeded $2 million (5).

In the first century A.D., the Romans founded the city of Londinium on the banks of the River Thames, and until only recently, London has been at the mercy of tidal

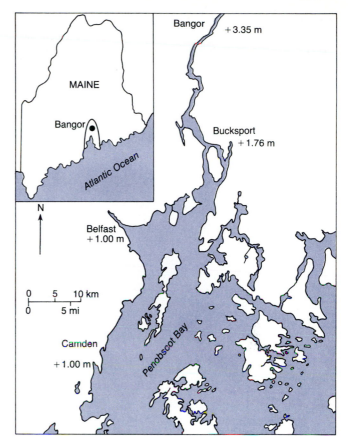

Figure 10.5
Bangor, Maine, and the funnel-shaped Penobscot Bay that modified the tidal-storm surge of February 2, 1976. The numbers next to the various locations are the height of the water above that predicted. Notice the dramatic increase in water height from Bucksport to Bangor. (Data from U.S. Geological Survey, Professional Paper 1087, 1979.)

floods caused by storm surges. There have been at least seven disastrous floods since the thirteenth century, two of the more recent in 1928 and 1953. It was not a question of whether London would flood, but *when* the city would be inundated again. Estimates that a great storm surge could do catastrophic damage, perhaps $6 billion worth, threaten the lives of many people, flood buildings, disrupt government, and close the rail system and underground transit, led to construction of the Thames barrier, which was commissioned and tested in 1983. The great barrier cost about $750 million and is constructed of ten huge steel gates separated by nine boat-shaped ribs extending over 500 meters from bank to bank (Figure 10.6). When not in use, the barrier gates rest on concrete sills in the chalk (limestone) bed of the river and therefore do not present a barrier to navigation. When needed, the gates, powered by electrically driven hydraulics, rotate 90 degrees to stand nearly 15 meters above the bed of the river. Along with downstream embankments and smaller barriers, they protect the city of London against even the 1,000-year tidal flood (6).

Figure 10.6
Map and diagrams of the storm-surge tidal barrier constructed across the River Thames near London, England. The site of the barrier (a) is just downstream from the main city. A plan view of the river (b) shows the piers and submerged gates across the river where the width is approximately 500 m. The steel gate rests in a concrete sill in the chalk bed of the river (c) and is rotated into place to provide protection from North Sea storm surges as needed.

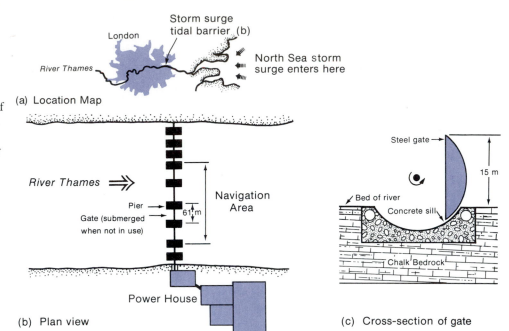

(a) Location Map

(b) Plan view

(c) Cross-section of gate

▼ COASTAL EROSION

As a result of global rise in sea level and unwise development in the coastal zone, coastal erosion is becoming recognized as a serious national and world-wide problem. Compared to other natural hazards such as earthquakes, tropical cyclones, or floods, erosion of coasts is generally a more continuous, predictable process, and large sums of money are spent in attempts to control it. If extensive development of coastal areas for vacation and recreational living continues, problems of coastal erosion are certain to become more serious.

Geologic conditions for the East and West coasts of the United States are strikingly different. This difference is in part related to plate tectonics. The West Coast is on the leading edge of the North American plate and experiences active uplift and deformation. As a result, the coastline is characterized by rugged sea cliffs, a locally irregular coastline, and narrow continental shelf. Such a coast, where rocks project out into the ocean (headlands) is conducive to vigorous wave action and coastal erosion (Figure 10.7). Upon striking a shore, waves expend more

Figure 10.7
Breaking waves are powerful agents of erosion. (Photo courtesy of California Division of Beaches and Parks.)

Figure 10.8
Oblique aerial photograph showing breaking waves and surf zone. Notice that most of the wave energy is expended on the headland protruding into the water. If rock resistance is constant, which it often is not, this area erodes considerably faster than the bay areas. (Photo courtesy of California Division of Beaches and Parks.)

energy on the protruding areas than on the embayed areas (Figure 10.8). Therefore, the effect of wave erosion is to straighten the shoreline—eroding the headlands and depositing the eroded material in the intervening bays. The East Coast, on the other hand, is on the trailing edge of the North American plate. It is a tectonically quiet coast with a broad continental shelf. For this reason and others, the major coastline features are chains of barrier islands (linear, relatively narrow islands composed of sediment transported and deposited in the coastal environment) separated from the mainland by bays, sounds, and lagoons. Exchange of water between lagoons and the open sea takes place through inlets that form when storm waves breach the barrier islands (7).

Beach Form and Process

The sand on beaches is not static. Wave action constantly keeps the sand moving in the surf and swash zones (Figure 10.9), and when waves strike the coast at an angle, the net result is a longshore current and beach drift, which collectively move beach material along a coast in a process called littoral transport (Figure 10.10). Along both the East and West coasts of the United States, the direction of littoral transport (although it can be quite variable) is most often to the south, and sediment transport is on the order of 200,000 to 300,000 cubic meters per year.

Figure 10.9 shows the basic terminology of an idealized near-shore environment. The landward extension of the beach terminates at a natural topographic and morphologic change, such as a seacliff or dune line. The *berms* are flat backshore areas on beaches (where people sunbathe). Berms are formed by deposition of sediment as waves rush up and expend the last of their energy. The *beach face* is the sloping portion of the beach below the berm, part of which is exposed by the uprush and backwash of waves *(swash zone)*. It is in the swash zone that *beach drift*, the up and back movement of beach material along the beach face as idealized in Figure 10.10,

occurs. The *surf zone* is that portion of the seashore environment where borelike waves of translation occur after the waves break. It is in the surf zone that *longshore current* occurs. The *breaker zone* is the area where the incoming waves become unstable, peak, and break. The *longshore trough* and *bar* are an elongated depression and adjacent ridge of sand produced by wave action. A particular beach, especially if it is wide and gently sloping, may have a series of long shore bars, troughs, and breaker zones (8).

Erosion Factors

The sand on many West Coast beaches is supplied to the coastal areas by rivers that transport it from areas upstream where it has been produced by weathering of quartz- and feldspar-rich rocks. Our technology has interfered with this material flow of sand from mountains to beach by building dams that effectively trap the sand. As a result, some West Coast beaches are deprived of sediment.

Where a seacliff is present along a coastline, there may be added erosion problems because the seacliff is exposed to both marine and land processes. These processes may work together to erode the cliff at a greater rate than either process could without the other. The problem is further compounded when people interfere with the seacliff environment through inappropriate development.

Figure 10.11 shows a typical southern Californian seacliff environment at low tide. The rocks of the cliff are steeply inclined and folded shale. A thin veneer of sand and coarser material (pebbles and boulders) near the base of the cliff mantles the wave-cut platform. A mantle of sand approximately one meter thick covers the beach during the summer, when long gentle waves construct a wide berm while protecting the seacliff from wave erosion. During the winter, storm waves, which have a high potential to erode beaches, remove the mantle of sand, exposing the base of the seacliff. Thus, it is not

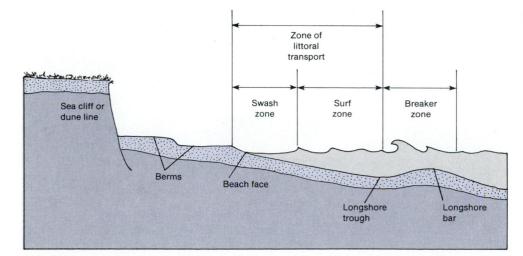

Figure 10.9
Basic terminology for landforms and wave action in the beach and near-shore environment.

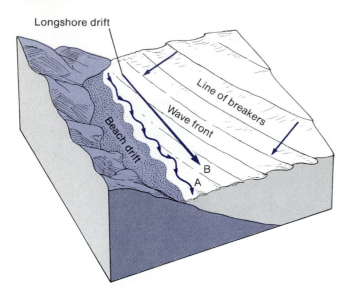

Figure 10.10
Block diagram showing the processes of beach drift and long-shore drift, which collectively move sand along the coast (littoral transport). Sediments in the swash zone (A) and surf zone (B) follow paths shown by the arrows.

surprising that most erosion of the seacliffs in southern California takes place during the winter months.

In addition to wave erosion, processes that attack the seacliff include biological erosion, weathering, rain wash, landslides, and artificially induced erosion (9). Biological processes facilitate and directly cause some erosion of the seacliff; for example, boring mollusks, marine worms, and some sponges can destroy rock. Weathering is significant in weakening the rocks of the seacliff and acts as an aid to erosion: trees on the top of the seacliff may have roots that penetrate the rock and wedge them apart; salt spray may enter small holes and fractures and, as the water evaporates, the salt crystallizes, exerting pressure on the rock that weakens it and can break off small pieces. Rain wash can cause a considerable amount of seacliff erosion; however, the amount of erosion depends upon the nature and extent of the rainfall and the erodability of the rocks that make up the seacliff.

A variety of human activities can induce seacliff erosion. Urbanization, for example, results in increased runoff which—if not controlled, carefully collected, and diverted away from the seacliff—can result in serious

Figure 10.11
Generalized cross section (a) and photograph (b) of seacliff, beach, and wave-cut platform, Santa Barbara, California. (Photo courtesy of Donald Weaver.)

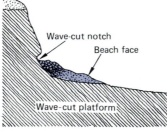

(a)

(b)

(a)   (b)

Figure 10.12
Erosion of seacliff composed of soft compaction shale near Isla Vista, California. Uncontrolled runoff from storm drains on the seacliff results in serious erosion (a). Controlled runoff released at the base of the seacliff on the beach causes much less erosion (b).

erosion. Figure 10.12a shows a drain pipe through which water is simply dumped on the seacliff. The result is active erosion. On the other hand, a short distance away, another drain pipe is routed to the base of the seacliff and the water runs out on the face of the beach (Figure 10.12b), resulting in much less erosion. Watering lawns and gardens on top of a seacliff also adds a good deal of water to the slope. This water tends to migrate toward the base of the seacliff, where it may emerge as small seeps or springs, effectively reducing the stability of the seacliff and facilitating landslides. Figure 10.11 shows a section of seacliff with a park on top. Recreational use of the seacliff is certainly superior to residential use and should be encouraged. The watering of the large lawn, however, has encouraged the continuous flow of several small seeps along the seacliff. Over a period of years, this may lower the resisting forces of the rocks in the cliff and facilitate an increase in the frequency of landslides in an area where they are already all too common.

Structures such as walls, buildings, swimming pools, and patios may also decrease the stability of the seacliff by increasing the driving forces (Figure 10.13). Strict regulation of development in many areas of the coastal zone now forbids most unsafe construction, but we must continue to live with our past mistakes.

The rate of seacliff erosion is variable, and few measurements are available. Along parts of the southern California coast near Santa Barbara, the rate averages 15 to 30 centimeters per year, depending on the resistance of the rocks and the height of the seacliff (9). These erosion rates are moderate compared to other parts of the world. Along the Norfolk coast of England, for example, erosion rates in some areas are about 2 meters per year. More remarkable, in the North Sea off Germany, one small island with soft erodable sandstone seacliffs had a perimeter of about 200 kilometers in A.D. 800, and by 1900, the perimeter had been reduced to only about 3 kilometers by seacliff retreat. At that time, a concrete sea wall was constructed around the entire island to control the erosion (9).

The main conclusion concerning seacliff erosion and retreat is that it is a natural process that cannot be

Figure 10.13
Development on the top of a seacliff at Isla Vista, California. The cantilevered deck has since been removed; such structures are no longer allowed. (Photo courtesy of Robert Norris.)

completely controlled unless large amounts of time and money are invested—and even then there is no guarantee. Therefore, it seems we must learn to live with some erosion. It can be minimized, however, by applying sound conservation practices, such as controlling the water on and in the cliff and not placing homes, walls, large trees, or other loading on top of the cliff.

Sea walls of concrete or riprap (large stones) may help retard erosion, but are sometimes ineffective because considerable erosion may occur at the top of the cliff above the protective structure. Furthermore, sea walls tend to produce a narrower beach with less sand, particularly if waves are strongly reflected, and, unless carefully designed to complement existing land use, may cause aesthetic degradation. Design and construction of sea walls must thus be tailored to specific sites.

Wave erosion on beaches that do not have a seacliff is also a problem in some West Coast areas. Dams on the coastal rivers decrease the sand supply to the beaches, but that isn't the whole story. During winters of exceptionally large storms and high rainfall, large waves up to five meters high may strike a beach, quickly eroding it.

Also, some areas that have wide beaches may erode over a period of several years; for example, one area in Ventura County, California, experienced about 60 meters of beach erosion in the early 1960s, damaging many homes and roads (Figure 10.14).

Beach erosion along the East Coast is a result of tropical cyclones and severe storms; a rise in sea level; and interference of human use and interest with natural shore processes (7). Tropical cyclones and severe storms can greatly alter a coastline by causing beach erosion and erosion of the foreshore sand dune system and by opening new inlets through the barrier islands. In addition, storm surges cut channels through the foreshore dunes, and eroded sediment is deposited in back of the beach as washover deltas (7).

Recent rise in sea level is eustatic (worldwide) and independent of continental movement, at the rate of about 2 to 3 millimeters per year. Evidence suggests that the rate of rise has increased during the last 60 years as a result of melting of the polar ice caps and thermal expansion of the upper ocean waters, triggered by global warming, hypothetically related to increased atmospheric carbon dioxide produced by burning fossil fuels. Sea levels could rise by 700 mm over the next century, ensuring that coastal erosion would become an even greater problem than it is today. There is evidence for submergence of the coast in many areas along the East Coast. Along the Outer Banks of North Carolina (Figure 10.15), the Shackleford Banks that were low-lying forests in 1850 are today salt marshes (10).

Human interference with natural shore processes has caused considerable coastal erosion. Most problems have arisen in areas that are highly populated and developed. For example, in some areas, artificial barriers

Figure 10.14
Coastal erosion at Oxnard Shores, Ventura County, California, required that this house be placed on a raised foundation. Notice the fireplace that originally was at ground level. Several homes in this location were destroyed by wave action.

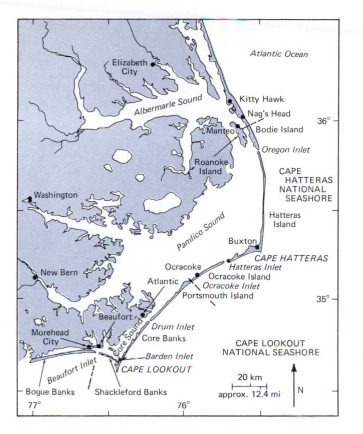

Figure 10.15
The Outer Banks of North Carolina showing the locations of Cape Hatteras and Cape Lookout National Seashores. (From Godfrey and Godfrey, *Coastal Geomorphology,* ed. D. R. Coates [Binghamton, New York: Publications in Geomorphology, State University of New York, 1973].)

retard the movement of sand, causing beaches to grow in some areas and erode in others, resulting in damage to valuable property.

Littoral Cell, Beach Budget, and Wave Climate

The concepts of the littoral cell, beach budget, and wave climate are basic to understanding coastal problems. A *littoral cell* is a segment of coastline that includes an entire cycle of sediment delivery to the coast (usually by rivers), longshore littoral transport, and eventual loss of sediment from the near-shore environment (11). Figure 10.16 shows five littoral cells in southern California. Each cell involves the erosion, transport, and deposition of more than 200,000 cubic meters per year of sand. Furthermore, each cell has an annual *budget* of sediment, including sources and losses to the beach environment. Whenever more sediment is transported out of a particular area in a cell than is delivered to that site, erosion results. Negative beach budgets are common, partly because dams on rivers store sediments that would otherwise be delivered to coastal areas, and because

coastal structures interfere with longshore transport of sand. The ultimate sink for sand in most of the California littoral cells is transport through submarine canyons to deep sea basins. As an example, consider the Santa Barbara littoral cell of Figure 10.16. The sand on the beaches from Point Conception to the sink at Hueneme and Magu canyons is supplied from streams and rivers that transport sand from the inland mountains to the coastal zone. Beach drift and longshore drift move the sand to the east and south along the coast. When the sand reaches the heads of the submarine canyons it is funneled down the canyons and the beaches physically end. Beaches are again present further east in the Santa Monica cell where streams again deliver sand to the coastal environment. *Wave climate* is a statistical characterization on an annual basis of wave height, period (time in seconds between arrival of successive waves), and direction, for calculating wave energy at a particular site. Figure 10.17 shows three aspects of wave climate (period, direction, and season of arrival) for the coastal area near Oxnard Shores in Ventura County, California (see also Figure 10.14). The data for Oxnard Shores suggest that the coast there is vulnerable to northwesterly winter waves that move most of the sediment in the Santa Barbara littoral cell (see Figure 10.16).

The beach budget and wave climate provide the basic data necessary for formulating and evaluating beach sediment supply and coastal erosion plans. Without this basic information, comprehensive coastal planning is significantly handicapped. Unfortunately, coastal erosion is usually defended against in a piecemeal manner, community by community, rather than from a littoral cell and beach budget approach that involves planning for a more extensive length of shoreline. As a result, there is little coordination of effort and the problem of beach erosion persists—winter storms in 1983 did over $100 million damage to beach property in California alone. If communities in the same littoral cell worked together, they would be much more successful in fighting coastal erosion.

Examples of Human Activity and Erosion

The Atlantic coast. The Atlantic coast from northern Florida to New York is characterized by barrier islands, many of which have been altered to a lesser or greater extent by human use and interest. Two examples, the barrier island coast of Maryland and the Outer Banks of North Carolina, illustrate the spectrum of human activity.

Demand for the 50 kilometers of Atlantic oceanfront beach in Maryland is very high, and the limited resource is used seasonally by residents of the Washington, D.C., and Baltimore, Maryland, urban centers (Figure 10.18). Ocean City on Fenwick Island has, since the early 1970s, promoted high-rise condominium and hotel development on the waterfront of the narrow island. As a result,

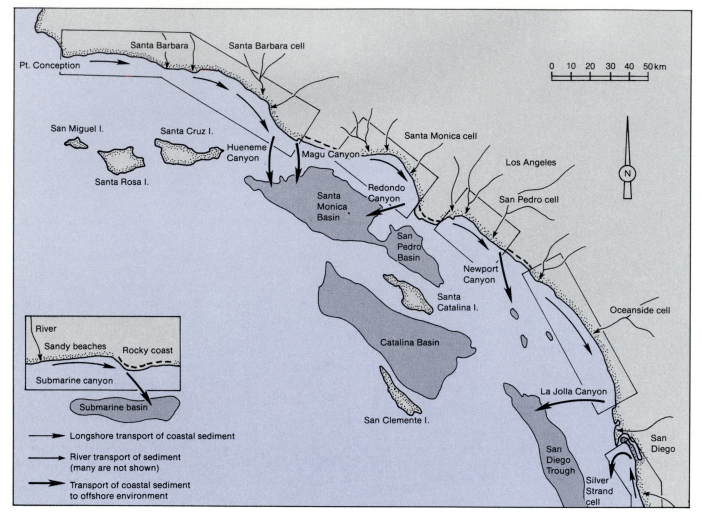

Figure 10.16
Littoral cells in southern California. Shown are five of the major cells, four of which deliver sediment transported along the coast to the submarine basin by way of a submarine canyon. (From D. L. Inman, 1976, *Man's impact on the California coastal zone*. State of California, Department of Navigation and Ocean Development.)

the natural frontal dune system has been removed in many locations, resulting in a serious beach erosion problem. Winter storms in 1977 and 1978 removed a good deal of sand, further narrowing the beaches. Attempts by the city to broaden the beaches by moving sand with bulldozers have not been successful, and expensive beach nourishment (bringing in sand from other areas) is under consideration, as loss of the beaches presents a serious economic problem to the resort area (12). More ominous is the almost certain possibility that a future hurricane will cause serious damage to Fenwick Island. The inlet south of Ocean City formed during a hurricane in 1933, and there is no guarantee, despite attempts to stabilize the inlet by coastal engineering, that a new inlet will not form at the present site of Ocean City some time in the future.

Across the Ocean City inlet to the south is Assateague Island, which encompasses two-thirds of the Maryland coastline. In contrast to the highly urbanized Fenwick

Island, Assateague Island is in a much more natural state. The island is used for passive recreation such as sunbathing, swimming, walking, and wildlife observation. However, both islands are in the same littoral cell and so are connected by common supply of sand. At least, that was the case until the jetties at the Ocean City inlet were constructed in 1935. Since construction of the jetties, coastal erosion in the northern few kilometers of Assateague Island has averaged about 11 meters per year, which is nearly twenty times the long-term rate of shoreline retreat for the Maryland coastline. During this same time, beaches immediately north of the inlet became considerably wider, requiring the lengthening of a recreational pier (13).

Observed changes in the Atlantic coast of Maryland are clearly related to the pattern of longshore drift of sand and human interference. Longshore drift is to the south at an average annual volume of about 150,000 cubic yards. Construction of the Ocean City inlet jetties, in an

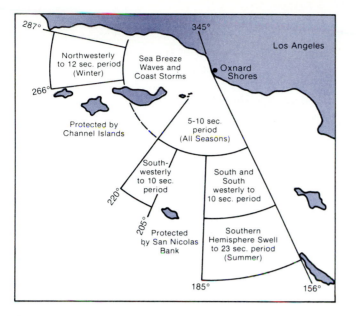

Figure 10.17
Part of the wave climate for Oxnard Shores, west of Los Angeles, California. Shown are the directions and periods of dominant waves that strike the area during particular seasons of the year. (From D. L. Inman, 1976, *Man's impact on the California coastal zone.* State of California, Department of Navigation and Ocean Development.)

attempt to stabilize the inlet, interfered with the natural southward flow of sand and diverted it offshore rather than allowing it to continue southward to nourish the beaches on Assateague Island. Starved of sand, the northern portions of the island have experienced serious shoreline erosion during the past 50 years. This example has been cited as the most severe case of beach erosion associated with engineering structures that block long-shore transport of sediment that can be found anywhere along the U.S. coastline (13). Thus our fundamental principle involving environmental unity holds: we cannot do only one thing, because everything is connected to everything else.

Another good example of our influence on coastal processes is found in the Outer Banks of North Carolina (Figure 10.15). The northern section near Cape Hatteras (Figure 10.19) has been progressively stabilized and developed (particularly near the beach), whereas the southern section near Cape Lookout is nearly unaltered by human activity.

The northern section is characterized by a continuous artificial dune system that was built to control erosion. The desire to protect roads and buildings behind the dunes led to an attempt at total stabilization. The idea was to hold everything in place and prevent the

Figure 10.18
The barrier island coast of Maryland. Fenwick Island is experiencing rapid urban development and there is concern for potential hurricane damage. What if a new inlet forms (during a hurricane) at the site of Ocean City? Inset shows details of the Ocean City inlet and effects of jetty construction.

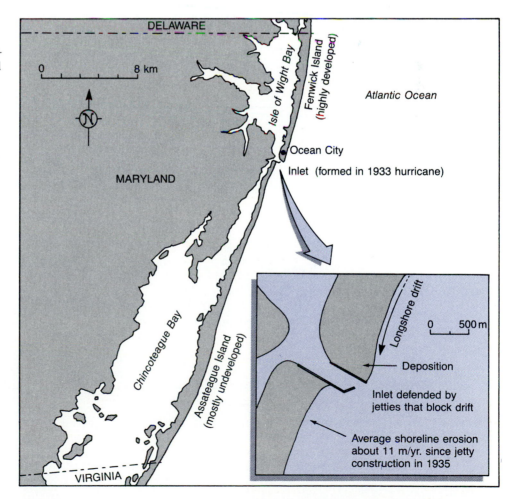

Figure 10.19
The dune line and the beach at Cape Hatteras. (Photo by R. Anderson, courtesy of National Park Service.)

sea from washing over the island. In contrast, the southern section is characterized by a natural zone of dunes rather than a line. The zone is frequently broken by overwash passes that allow water to flood across the island (Figure 10.20) (10).

Several studies have shown that building an artificial dune line has resulted in erosion and narrowing of the beach and has necessitated spending millions of dollars to stabilize artificial dunes that tend to erode rapidly (10, 14). Even with this effort, the structures and roads behind the dune line will eventually have to be relocated as the shoreline continues to erode and recede. In contrast, where the barrier islands have been left in a natural state, severe storms cause little erosion. As a result, the dunes are not washing away but are moving back by natural processes responsible for the origin and maintenance of the barrier island system. We emphasize, however, that there is considerable controversy as to whether the islands identified in the studies as natural are "typical" barrier islands, since overgrazing with subsequent wind erosion of the frontal dune line may have facilitated the overwash. If this is true, then it may be necessary to reevaluate the benefits of frequent overwash.

The Gulf coast. Coastal erosion is also a serious problem along the Gulf of Mexico. One study in the Texas coastal zone suggests that, in specific areas where long-term coastal recession rates can be determined from geologic data, human modification of the coastal zone in the last 100 years has accelerated coastal erosion by 30 to 40 percent over prehistoric rates (15). The human modifications that appear most responsible for the accelerated erosion are construction of coastal engineering structures, subsidence as a result of groundwater withdrawal, and damming of rivers that supply sand to the beaches. Erosion along the Gulf of Mexico in

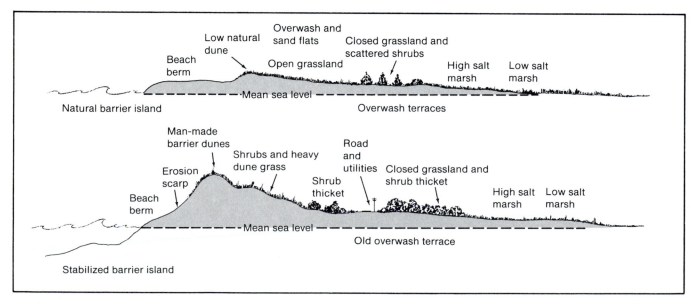

Figure 10.20
Cross sections of the Barrier Islands of North Carolina. The upper diagram is typical of the more natural systems and the lower typical of the artificially stabilized Barrier Islands. (From Dolan, *Coastal Geomorphology,* ed. D. R. Coates [Binghamton, New York: Publications in Geomorphology, State University of New York, 1973].)

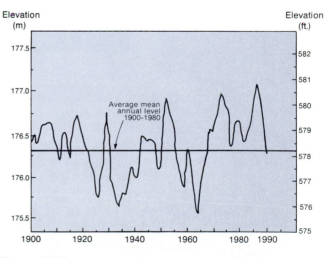

Figure 10.21
Estimated rates of coastal erosion for the southwestern Matagorda Peninsula, Texas. (After Wilkinson and McGowen, *Environmental Geology*, vol. 1 [New York: Springer-Verlag, 1977].)

Texas has apparently been going on for several thousand years as a natural process; therefore, it is often difficult to determine the exact extent to which human interference has increased or decreased the erosion. Nevertheless, it appears that the natural rate of landward migration of barrier islands and other coastal beach features in many areas has definitely increased in recent years (historic time). For example, Figure 10.21 shows estimated erosion rates for the southwestern Matagorda Peninsula in Texas. The recent (1856 to 1956) erosion rates are 34 percent greater than the prehistoric rates. The situation is quite complex, however, because, as the peninsula has retreated nearly two kilometers in the last 1,500 years, the nearby Matagorda Island has grown (accreted) by nearly the same amount. Most of the erosion is in response to high magnitude storms (hurricanes) that produce a net bayward migration of the entire peninsula. The recent situation (post-1965) was further complicated by construction of a ship channel and accompaning jetties to improve navigation that have produced highly variable erosion rates (15).

The Great Lakes. Erosion is a periodic problem along the coasts of the Great Lakes and has been particularly troublesome along the Lake Michigan shoreline. Damage is most severe during prolonged periods of high lake levels that occur following extended periods of above-normal precipitation. The relationship between precipitation and lake level has been documented by the United States Corps of Engineers since 1860. The data show that the lake level has fluctuated about 2 meters during this time. Figure 10.22 shows the level of Lake Michigan from 1900 to 1990. Lake level has been rising in recent years from a low stage in 1964, and in 1973, it again reached the high level of 1952. In the mid-1980s the lake was at a record high level. During a highwater stage there is considerable coastal erosion and many buildings, roads, retaining walls, and other structures are destroyed by wave erosion (Figure 10.23) (16). For example, damages

from fall storms in 1985 alone caused an estimated $15 million to $20 million in damages.

During periods of below-average lake level, wide beaches develop that dissipate energy from storm waves and protect the shore. With rising lake-level conditions, however, the beaches become narrow, and storm waves exert considerable energy against coastal areas. Even a small lake-level rise on a gently sloping shore will inundate a surprisingly wide section of beach (16).

Long-term rates of bluff-top, shoreline recession at many Lake Michigan sites average about 0.4 meters per year (17). Severity of erosion at a particular site depends on such factors as the presence or absence of a frontal dune system (dune-protected bluffs erode at a slower rate); orientation of the coastline (sites exposed to high-energy storm winds erode faster); groundwater seepage (seeps along the base of a coastal bluff cause slope instability, increasing the erosion rate); and existence of protective structures (structures may be locally

Figure 10.22
Lake levels for Lake Michigan from 1900 through 1991. (Data from U.S. Department of Commerce and U.S. Army Corps of Engineers. Modified after Buckler and Winters, 1983.)

Figure 10.23
Gale-force winds combined with record high levels of Lake Michigan caused significant damage to Chicago's north shore early in 1987. (a) Workmen attempt to reinforce a seawall on Febuary 8. (b) Photo taken the following day shows the new wall washed out and sandbags piled at apartment doors and windows to prevent further damage. (AP/Wide World photos.)

(a)

(b)

beneficial but often accelerate coastal erosion in adjacent areas) (16, 17). A few construction options and the relocation alternative are summarized, with respective advantages and disadvantages, in Figure 10.24. In recent years beach nourishment (artificial placement of sand on a beach) has been attempted for some Great Lakes beaches. The added sands are deliberately sized much coarser than the natural sands that had eroded, which, it is hoped, will reduce the erosion potential.

Engineering Structures

Engineering structures in the coastal environment are primarily designed to improve navigation or retard erosion. They include groins, breakwaters, and jetties.

Because they tend to interfere with the littoral transport of sediment along the beach, these structures all too often cause undesirable deposition and erosion in their vicinity.

Groins are linear structures placed perpendicular to the shore. They are usually constructed in groups called *groin fields* (Figure 10.25). The basic idea is that each groin will trap a portion of the sand that is moving in the littoral transport system. A small accumulation of sand will develop updrift of each groin, thus building an irregular but wider beach. The problem is that, while deposition occurs updrift of a groin, erosion tends to occur in the downdrift direction. Thus, a groin or groin field results in a wider, more protected beach in a desired area, but may cause a zone of erosion to develop

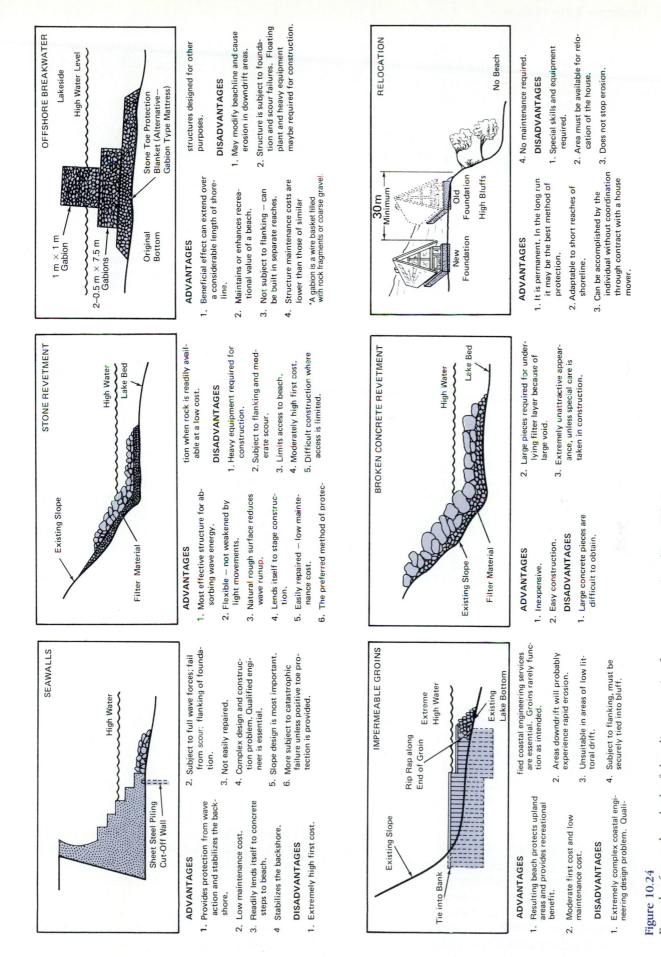

OFFSHORE BREAKWATER

Lakeside
High Water Level
1 m × 1 m Gabion
2–0.5 m × 7.5 m Gabions
Stone Toe Protection Blanket (Alternative— Gabion Type Mattress)
Original Bottom

structures designed for other purposes.

ADVANTAGES

1. Beneficial effect can extend over a considerable length of shoreline.
2. Maintains or enhances recreational value of a beach.
3. Not subject to flanking — can be built in separate reaches.
4. Structure maintenance costs are lower than those of similar

DISADVANTAGES

1. May modify beachline and cause erosion in downdrift areas.
2. Structure is subject to foundation and scour failures. Floating plant and heavy equipment maybe required for construction.

*A gabion is a wire basket filled with rock fragments or coarse gravel.

STONE REVETMENT

High Water
Lake Bed
Existing Slope
Filter Material

tion when rock is readily available at a low cost.

ADVANTAGES

1. Most effective structure for absorbing wave energy.
2. Flexible — not weakened by light movements.
3. Natural rough surface reduces wave runup.
4. Lends itself to stage construction.
5. Easily repaired — low maintenance cost.
6. The preferred method of protec-

DISADVANTAGES

1. Heavy equipment required for construction.
2. Subject to flanking and moderate scour.
3. Limits access to beach.
4. Moderately high first cost.
5. Difficult construction where access is limited.

SEAWALLS

High Water
Sheet Steel Piling Cut-Off Wall

ADVANTAGES

1. Provides protection from wave action and stabilizes the backshore.
2. Low maintenance cost.
3. Readily lends itself to concrete steps to beach.
4. Stabilizes the backshore.

DISADVANTAGES

1. Extremely high first cost.

2. Subject to full wave forces; fail from scour; flanking of foundation.
3. Not easily repaired.
4. Complex design and construction problem. Qualified engineer is essential.
5. Slope design is most important.
6. More subject to catastrophic failure unless positive toe protection is provided.

RELOCATION

30 m Minimum
New Foundation
Old Foundation
High Bluffs
No Beach

ADVANTAGES

1. It is permanent. In the long run it may be the best method of protection.
2. Adaptable to short reaches of shoreline.
3. Can be accomplished by the individual without coordination through contract with a house mover.
4. No maintenance required.

DISADVANTAGES

1. Special skills and equipment required.
2. Area must be available for relocation of the house.
3. Does not stop erosion.

BROKEN CONCRETE REVETMENT

High Water
Lake Bed
Existing Slope
Filter Material

ADVANTAGES

1. Inexpensive.
2. Easy construction.

DISADVANTAGES

1. Large concrete pieces are difficult to obtain.

2. Large pieces required for underlying filter layer because of large void.
3. Extremely unattractive appearance, unless special care is taken in construction.

IMPERMEABLE GROINS

Rip Rap along End of Groin
Extreme High Water
Existing Lake Bottom
Existing Slope
Tie into Bank

ADVANTAGES

1. Resulting beach protects upland areas and provides recreational benefit.
2. Moderate first cost and low maintenance cost.

DISADVANTAGES

1. Extremely complex coastal engineering design problem. Quali-

fied coastal engineering services are essential. Groins rarely function as intended.
2. Areas downdrift will probably experience rapid erosion.
3. Unsuitable in areas of low littoral drift.
4. Subject to flanking, must be securely tied into bluff.

Figure 10.24

Example of several methods of shoreline protection from wave erosion. (From U.S. Army Corps of Engineers.)

Figure 10.25
Diagram of two beach groins. Deposited sediment builds a wide beach in the updrift direction; in the downdrift direction, the sparsity of sediment for transport can cause erosion to occur.

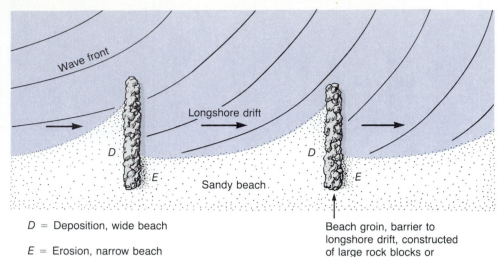

Wave front

Longshore drift

D

E

Sandy beach

D

E

D = Deposition, wide beach

E = Erosion, narrow beach

Beach groin, barrier to longshore drift, constructed of large rock blocks or other materials

in the adjacent downcoast shoreline. The erosion results primarily as a groin or groin field becomes filled with trapped sediment. Once a groin is filled, sand is transported around its offshore end to continue its journey along the beach. Therefore, erosion may be minimized by artificially filling each groin; this is known as *beach nourishment,* and requires trucking sand out onto the beach. Thus nourished, the groins will draw less sand from the natural littoral transport system, and the downdrift erosion will be reduced (8). Even with beach nourishment and other precautions, however, groins may cause undesirable erosion; therefore, their use should be carefully evaluated.

Breakwaters are designed to intercept waves and provide a protected area (harbor) for boat moorings, and may be attached to the beach or be separated. In either case, they block the natural littoral transport of beach sediment and, thus, locally change the configuration of the coast as new areas of deposition and erosion develop (Figure 10.26). Breakwaters have caused serious erosion problems downdrift of the structures. In addition, they produce a sand trap that continues to accumulate sand in the updrift direction. Eventually the trapped sand may fill or block the entrance to the harbor as a sand spit or bar develops. As a result, a dredging program, or artificial bypass, is often necessary to keep the harbor open and clear of sediment. The sediment that is removed by dredging should be transported and released on the beach downdrift of the breakwater to rejoin the natural littoral transport system, thus reducing the erosion problem.

Jetties are often constructed in pairs at the mouth of a river or inlet to a lagoon, estuary, or bay (Figures 10.18 and 10.21). They are designed to stabilize the channel, prevent or minimize deposition of sediment in the channel, and generally protect it from large waves (8). Jetties tend to block the littoral transport of beach sediment, thus causing the updrift beach adjacent to the

jetty to widen while downdrift beaches erode. The deposition at jetties may eventually fill the channel, making it useless, while downcoast erosion damages coastal development. Mechanically bypassing (dredging) the sediment minimizes but does not eliminate all undesirable deposition and erosion.

There is no way to build a breakwater or jetty along a coast with an active littoral transport system so that it will not interfere with and partially or almost totally block the longshore movement of beach sediment. These structures must therefore be carefully planned and protective measures taken early to eliminate or at least minimize adverse effects. This may include installation of a dredging and artificial sediment-bypass system, beach nourishment program, seawalls, riprap, or some combination of these measures (8).

Beach Nourishment

In the above discussion we introduced the topic of beach nourishment as it pertains to engineering structures in the coastal zone. Beach nourishment can also be an alternative to engineering structures. In its purest form, beach nourishment involves artificially placing sand on beaches in the hope of constructing a positive beach budget. The procedure has distinct advantages in that it is aesthetically preferable to many engineering structures and it provides a recreation beach as well as providing some protection from shoreline erosion.

The city of Miami Beach, Florida, and the U.S. Army Corps of Engineers began an ambitious beach nourishment program in the mid-1970s to reverse a serious beach erosion problem that had plagued the area since the 1950s and to provide protection from storms. The natural beach had nearly disappeared by the 1950s, and only small pockets of sand could be found associated with various shoreline protection structures, including seawalls and groins. As the beach disappeared, coastal

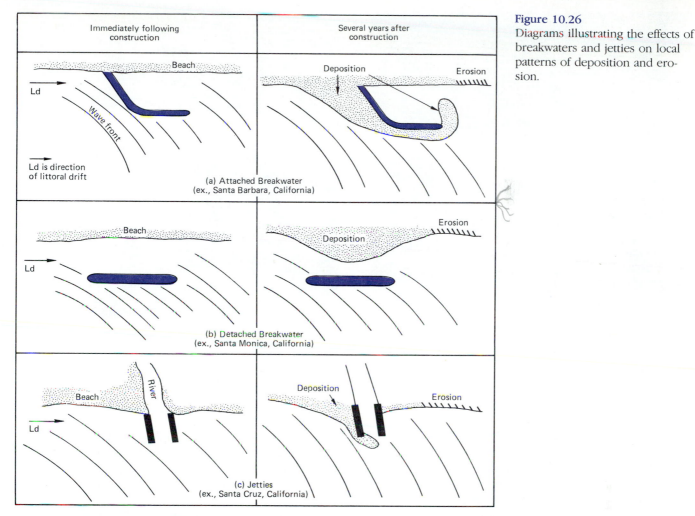

Figure 10.26
Diagrams illustrating the effects of breakwaters and jetties on local patterns of deposition and erosion.

resort areas, including high-rise hotels, became vulnerable to storm erosion (18).

The purpose of the nourishment program for Miami Beach was to produce a positive beach budget, and thus a wide beach, and provide additional protection from storm damage. The project will cost about $62 million over 10 years and involve nourishment of about 160,000 cubic meters of sand per year to replenish erosion losses. By 1980, about 18 million cubic meters of sand had been dredged and pumped from an offshore site onto the beach, producing a 200-meter-wide beach (18). Figure 10.27 shows Miami Beach before and after the nourishment—the change is dramatic. Figure 10.28 shows the cross-section design for the project which includes two components (a wide berm and frontal dune system) to function as a buffer to wave erosion and storm surge. The Miami project was expanded in the mid- to late 1980s to include dune restoration, which involves establishing native vegetation on the dune shown in Figure 10.28. Public access through the dunes is along special wooden walkways. Other areas of the dunes are protected. Although only time will show how well the nourishment project protects against a really large storm or hurricane, and although the project is expensive, it has the advan-

tage of providing aesthetic amenities and a degree of protection to valuable resort property. Beach nourishment is certainly preferable to the fragmented erosion control methods that preceded it.

▼ PERCEPTION OF AND ADJUSTMENT TO COASTAL HAZARDS

People adjust to the tropical cyclone hazard either by doing nothing and bearing the loss or by taking some kind of action to modify potential loss. Bearing the loss is probably the most common individual adjustment. Attempts to modify potential loss include *strengthening* the environment with protective structures and land stabilization, and *adapting* behavior by better land-use zoning, evacuation, and warning (19).

Inhabitants in areas with a serious tropical cyclone hazard, such as coastal Bangladesh and the Gulf Coast of the United States, are very much aware of the hazard (20, 21). In Bangladesh, perception of the cyclone hazard was influenced by people's economic and cultural backgrounds. It was found that the people are not particularly concerned about the hazard, and the three socioeconomic classes perceive the cyclones in much the same

(a)   (b)

Figure 10.27
Miami Beach (a) before and (b) after beach nourishment. (Photographs courtesy of U.S. Army Corps of Engineers.)

way; however, the upper class tends to favor earning a living in a less hazardous area.

Although repeatedly devastated by tropical cyclones, coastal areas in Bangladesh do not have an integrated system of private or public adjustment. Although people there are aware of the hazard, the adoption of adjustments is not a function of the frequency of cyclones. Even when shelter is available, people are reluctant to leave their homes. During a 1970 cyclone, approximately 38 percent of the people saved their lives by climbing trees, 5 percent took shelter in a two-story community center, and 8 percent remained on top of an embankment designed to protect the land against saltwater intrusion.

The embankment was incomplete when the storm hit, and was severely eroded and overlapped by storm-surge waves. It is hard to believe that 22 people in a family died only 400 meters from the community center that offered safety (20).

People's attitudes toward hurricane damage and the possibility of preventing or modifying damage are related to their educational level and their hurricane experience. Public awareness programs in areas likely to experience tropical cyclones should thus be matched to the educational level of the target group. Furthermore, since better-educated people tend to be more positive toward damage prevention, increased attention should be given

Figure 10.28
Cross section of the Miami Beach nourishment project. The dune and beach berm system provide protection against storm attack. (From U.S. Army Corps of Engineers.)

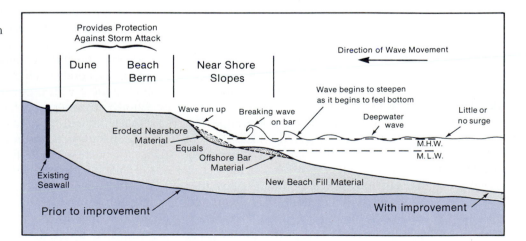

to the less-educated population to counteract their tendency to lack confidence in damage-prevention measures. Awareness programs also should be designed differently depending on the area's history of hurricane experience (21).

Adjustments to coastal erosion fall into one of several categories: beach nourishment that tends to imitate natural processes; modification of the wave energy (wave climate) through construction of near-shore structures designed to dissipate wave energy; shoreline stabilization through structures such as groins and seawalls; and land-use change that attempts to avoid the problem.

We are at a crossroads today with respect to adjustment to coastal erosion. One road leads to ever-increasing coastal defenses in an attempt to control the processes of erosion, and the second path involves learning to live with coastal erosion through flexible environmental planning and wise land-use in the coastal zone. In the second path, all structures in the coastal zone (with such exceptions as critical facilities in certain recreational areas) are considered temporary and expendable. Any development in the coastal zone must be in the best interests of the general public rather than a few who develop the ocean front. Accepting this philosophy requires an appreciation of the following five principles (22):

1. *Coastal erosion is a natural process rather than a natural hazard; erosion problems occur when people build structures in the coastal zone.* The coastal zone is an area where natural processes associated with waves and moving sediment occur. Because such an environment will have a certain amount of natural erosion, the best land uses are those compatible with change. These include recreational activities such as swimming and fishing. When we build in the coastal zone, problems develop.

2. *Any shoreline construction causes change.* The beach environment is dynamic. Any interference with natural processes produces a variety of secondary and tertiary changes, many of which may have adverse consequences. This is particularly true for engineering structures such as groins and seawalls that affect the storage and flow of sediment along a coastal area.

3. *Stabilization of the coastal zone through engineering structures protects the property of relatively few people at a larger general expense to the public.* With few exceptions such as recreational shorelines visited by thousands of people during the tourist season, shoreline protection generally benefits a relatively small number of property owners. It has been argued that the interests of those who own property in the coastal zone are not compatible with the public interest, and it is unwise to expend large amounts of public funds to protect the property of a few. It must be kept in mind that engineering structures along the shoreline are often meant to protect developed property, not the beach itself.

4. *Engineering structures designed to protect a beach may eventually destroy it.* Engineering structures often modify the coastal environment to such an extent that it may scarcely resemble a beach. For example, construction of large seawalls causes reflection of waves and turbulence that eventually removes the beach.

5. *Once constructed, shoreline engineering structures produce a trend in coastal development that is difficult, if not impossible, to reverse.* Engineering structures often lead to additional repairs and larger structures at spiraling costs. In some areas the cost of the structures eventually exceeds the value of the beach property itself. For these and other reasons, several states have recently imposed severe limitations on future engineering construction intended to stabilize the coastline. As sea levels continue to rise and coastal erosion becomes more widespread, nonstructural alternatives to the problem should continue to receive favorable attention, because of both financial necessity and a recognition that the amenities of the coastal zone should be kept intact for future generations to enjoy.

Perception of coastal erosion as a natural hazard depends primarily on an individual's past experience, proximity to the coastline, and the probability of suffering property damage. One study of coastal erosion of seacliffs near Bolinas, California, 24 kilometers north of the entrance to the San Francisco Bay, established that people living close to the coast in an area likely to experience damage in the near future are generally very well informed and see the erosion as a direct and serious threat (23). People living a few hundred meters from a possible hazard, although aware of the hazard, know little about its frequency of occurrence, severity, and predictability. Still further inland, people are aware that coastal erosion exists but have little perception of the hazard.

The Bolinas study is especially interesting because of the controversy that surrounded various alternatives the community considered. Before 1971, coastal-erosion damages and expenditures at Bolinas exceeded $300,000 (Table 10.1). The rate of cliff retreat is approximately 0.5 meters per year under natural conditions and somewhat less when seawalls and other protective measures are used. The community considered four alternatives. The first was zoning to prevent development in the hazard zone. No structures to slow the erosion of the cliffs would be provided. This, however, would be effective only for undeveloped land, and in several areas, homes and roads are close enough to the retreating cliffs to be damaged eventually regardless of zoning. Second, the community considered public land acquisition in the hazard zone. The land would be purchased for open space and

Table 10.1
Coastal erosion damages and expenditures at Bolinas, California.

	Value ($)[a]
Damages	
Homes (three)	75,000
Real property (the equivalent of 15 lots at $5,000 each)	75,000
Public utilities (roads, pipelines)	50,000
Total	200,000
Expenditures	
Seawall construction	75,000
Cliff drainage and planting	25,000
Riprap of road area at top of cliffs	5,000
Other road repairs	10,000
Moving of homes (two)	10,000
Problem studies[b]	10,000
Total	135,000

[a] 1971 dollar equivalent.

[b] U.S. Army Corps of Engineers, Bolinas Beach and Erosion Study, $5,000 per year for two years.

Source: *Natural Hazards: Local, National, Global* edited by Gilbert F. White. Copyright © 1974 by Oxford University Press, Inc. Reprinted by permission.

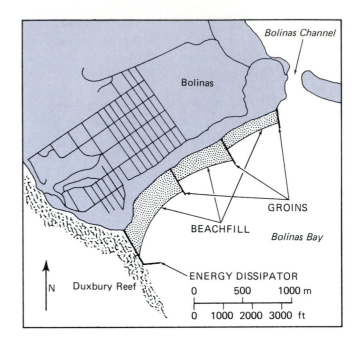

Figure 10.29
Proposed adjustments for cliff stabilization at Bolinas, California. (After R. A. Rowntree, *Natural Hazards: Local, National, Global,* ed. G. F. White [New York: Oxford University Press, 1974].)

recreation. Natural erosion of the cliffs would be allowed, and damage would be restricted to low-cost recreation facilities. Third, construction of a seawall of broken rock at the base of the cliffs to stabilize erosion was considered. This would cost $3 to $4 million, and there would still be up to 20 meters of erosion at the top of the cliffs unless other protective measures were taken. Fourth, the community considered a combination of groins, beachfill, and an energy dissipator (Figure 10.29). The groins and energy dissipator would be constructed of broken rock at an initial cost of $4 to $6 million. This would stabilize the base of the cliff, but erosion on the top would continue until a stable angle of repose was reached, which, as in the case of the seawall, would be up to 20 meters unless protective measures were taken (23).

The people of Bolinas include weekend and summer residents, retired persons who are permanent residents, long-time agricultural residents, and activist environmentalists. Each, depending on background and proximity to the hazard, had a different perception of the problem. The town formed a planning group to consider the alternatives and a conservation zone was recommended along with a policy to prevent construction within 50 meters of retreating seacliff. Expensive stabilization plans that would be an inducement to growth and tourism, something the community fears more than cliff erosion, were rejected. The community has, in effect, chosen to accommodate itself to the hazard, taking the position that

coastal erosion is a natural process rather than a natural hazard. It is likely that the community as a whole will maintain present cliff stabilization measures, but eventually the cliffs will retreat through the first row of houses regardless of interim measures.*

Cape Hatteras Lighthouse Controversy

In North Carolina a dramatic collision of opinions concerning beach erosion is now being played out. North Carolina is a leader in the philosophy that beach erosion is a natural process that can be lived with. But now erosion threatens the historic Cape Hatteras lighthouse located near Buxton on the outer banks (Figures 10.15 and 10.19). When the lighthouse was originally constructed in the late 19th century it was approximately 1/2 kilometer from the sea. Today, it is closer to 100 meters, and a major storm could take it out. The options are:

▼ Artificially control coastal erosion at the site, and reverse state policy of yielding to erosion;
▼ Do nothing and eventually lose the lighthouse and thus, an important bit of American history;
▼ Move the lighthouse inland.

*From *Natural Hazards: Local, National, Global* edited by Gilbert F. White, Copyright © 1974 by Oxford University Press, Inc. Reprinted by permission.

Relocating the lighthouse would reinforce the policy of allowing natural processes to continue. Which choice is made will have important implications for future decisions in the coastal zone. Although it is difficult to predict the outcome, I expect that the lighthouse will be relocated. However, time is of the essence because if it is not done soon, erosion may claim it.

▼ ▼ ▼ SUMMARY AND CONCLUSIONS

The coastal environment is one of the most dynamic areas on earth, and rapid change is a regular occurrence.

Migration of people to coastal areas is a continuing trend, and approximately 75 percent of the population in the United States now lives in coastal states.

The most serious coastal hazard is tropical cyclones. These violent storms bring high winds, storm surges (wind-driven oceanic waves), and river flooding. They continue to take thousands of lives and cause billions of dollars in property damage.

The combination of high tides with a storm surge can cause serious flooding. The threat of such a flood in London, England, precipitated the construction of a storm surge (tidal) barrier across the River Thames.

Although it causes a relatively small amount of damage compared to other natural hazards such as river flooding, earthquakes, and tropical cyclones, coastal erosion is a serious problem along the East, West, Gulf of Mexico, and Great Lakes coasts of the United States.

Human interference with natural coastal processes such as the building of groins, jetties, breakwaters, artificial dunes, and other structures is occasionally successful, but in many cases it has caused considerable coastal erosion. Most problems occur in areas with high population density, but sparsely populated areas along the Outer Banks in North Carolina are also having trouble with coastal erosion.

Perception of and adjustment to coastal hazards depend on such factors as the magnitude and frequency of the hazard, and the individual's educational level, experience, and proximity to the hazard.

With tropical cyclones, the most common individual adjustment is to do nothing and bear the loss. Community adjustments in developed countries generally attempt to modify the environment by building protective structures designed to lessen potential damage, or encourage change in people's behavior by better land-use zoning, evacuation, and warning.

Adjustment to coastal erosion in developed areas is often the "techno-logical fix": building seawalls, groins, and other structures to stabilize the beach and prevent erosion. These have had mixed success, and often cause additional problems in adjacent areas. In addition, engineering structures are very expensive, require maintenance, and once in place are difficult to remove. The cost of engineering structures may often eventually exceed the value of the properties they protect; such structures may even destroy the beaches they were intended to save. More recently, beach nourishment is being used to replenish beach material lost by erosion. One community in California takes the position that coastal erosion of seacliffs is a natural process rather than a natural hazard, and has rejected engineering solutions to the problem. The trend toward land-use planning as an alternative to coastal management is encouraging and is expected to continue.

▼ ▼ ▼ REFERENCES

1. COATES, D. R., ed. 1973. *Coastal geomorphology.* Binghamton, New York: Publications in Geomorphology, State University of New York.
2. WHITE, A. U. 1974. Global summary of human response to natural hazards: Tropical cyclones. In *Natural hazards: Local, national, global,* ed. G. F. White, pp. 255–65. New York: Oxford University Press.
3. OFFICE OF EMERGENCY PREPAREDNESS. 1972. *Disaster preparedness* 1, 2.
4. NATIONAL HURRICANE CENTER. 1989. *Preliminary report, Hurricane Hugo, 10–22 September, 1989.*
5. MORRILL, R. A.; CHIN, E. H.; and RICHARDSON, W. S. 1979. *Maine coastal storm and flood of February 2, 1976.* U.S. Geological Survey Professional Paper 1087.
6. SHAW, D. P. 1983. A barrier to tame the Thames. *Geographical Magazine* LV(3): 129–31.
7. EL-ASHRY, M. T. 1971. Causes of recent increased erosion along United States shorelines. *Geological Society of America Bulletin* 82: 2033–38.
8. KOMAR, P. D. 1976. *Beach processes and sedimentation.* New Jersey: Prentice-Hall.
9. NORRIS, R. M. 1977. Erosion of sea cliffs. In *Geologic hazards in San Diego,* eds. P. L. Abbott and J. K. Victoris. San Diego Society of Natural History.
10. GODFREY, P. M., and GODFREY, M. M. 1973. Comparison of ecological and geomorphic interactions between altered and unaltered barrier island systems in North Carolina. In *Coastal geomorphology,* ed. D. R. Coates, pp. 239–58. Binghamton, New York: Publications in Geomorphology, State University of New York.
11. INMAN, D. L. 1976. *Man's impact on the California coastal zone.* State of California Dept. of Navigation and Ocean Development.
12. U.S. DEPARTMENT OF COMMERCE. 1978. *State of Maryland coastal management program and final environmental impact statement.*
13. LEATHERMAN, S. P. 1984. Shoreline evolution of North Assateague Is-

land, Maryland. *Shore and Beach,* July: 3–10.

14. DOLAN, R. 1973. Barrier islands: Natural and controlled. In *Coastal geomorphology,* ed. D. R. Coates, pp. 263–78. Binghamton, New York: Publications in Geomorphology, State University of New York.

15. WILKINSON, B. H., and McGOWEN, J. H. 1977. Geologic approaches to the determination of long-term coastal recession rates, Matagordo Peninsula, Texas. *Environmental Geology:* 359–65.

16. LARSEN, J. I. 1973. *Geology for planning in Lake County, Illinois.* Illinois State Geological Survey Circular 481.

17. BUCKLER, W. R., and WINTERS, H. A. 1983. Lake Michigan bluff recession. *Annals of the Association of American Geographers* 73(1): 89–110.

18. CARTER, R. W. G., and OXFORD, J. D. 1982. When hurricanes sweep Miami Beach. *Geographical Magazine* LIV, no. 8: 442–48.

19. BAUMANN, D. D., and SIMS, J. H. 1974. Human response to the hurricane. In *Natural hazards: Local, national, global,* ed. G. F. White, pp. 25–30. New York: Oxford University Press.

20. ISLAM, M. A. 1974. Tropical cyclones: Coastal Bangladesh. In *National hazards: Local, national, global,* ed. G. F. White, pp. 19–25. New York: Oxford University Press.

21. BAKER, E. J. 1974. Attitudes toward hurricane hazards on the Gulf Coast. In *Natural hazards: Local, national, global,* ed. G. F. White, pp. 30–36. New York: Oxford University Press.

22. NEAL, W. J.; BLAKENEY, W. C., Jr.; PILKEY, D. H., Jr.; and PILKEY, O. H., Sr. 1984. *Living with the South Carolina shore.* Durham, N.C.: Duke University Press.

23. ROWNTREE, R. A. 1974. Coastal erosion: The meaning of a natural hazard in the cultural and ecological context. In *Natural hazards: Local, national, global,* ed. G. F. White, pp. 70–79. New York: Oxford University Press.

In this latter part of the twentieth century, Americans are a metropolitan people. Today, nearly 75 percent of the American people live in or near metropolitan areas that have populations greater than 50,000. It is expected that by the year 2000, 85 percent of Americans will be urban residents. Therefore, metropolitan population growth is now a basic feature of the transformation from agrarian to industrial life in the United States.

The evolution of urban areas has been from farm to small town, to city, to large metropolitan areas, to what is now referred to as the "urban region." An *urban region* is an area with at least one million people, in which there is a zone of metropolitan areas with intervening counties never far from a city[1]. In 1920, there were 10 urban regions that contained approximately one-third of the total population of the country. By 1970, there were 16 urban regions with three-fourths of the population, and by the year 2000, there will probably be at least 23 urban regions with five-sixths of the nation's people. If these projections are correct, 54 percent of all Americans will be living in the two largest urban regions—41 percent in the metropolitan belt extending from the Atlantic seaboard and westward past Chicago, and 13 percent in the California region between San Francisco and San Diego.

It is projected that by the year 2000 urban regions will occupy one-sixth of the total land area of the continental United States; the environmental impact of this urbanization will be staggering. Urbanization renders a large portion of the soils within the urbanized area impervious, modifies rivers, and otherwise changes the previous landscape to suit human use. This process need not totally destroy the land, however. If growth is properly planned and directed, and if regular review and reevaluation of the goals and objectives are conducted, urban areas should be able to maintain at least a rough balance between acceptable environmental degradation and constructive use of the environment.

The stages of conversion from a rural to an urban area are rural, early urban, middle urban, and late urban[2]. During the *rural* stage, the landscape may consist of its virgin state and/or modest to intensive cultivation. Characteristically, there may be occasional farm buildings and dwellings which draw their water supply from wells, springs, streams, or ponds. Sewage is disposed of through outhouses, cesspools, or, perhaps, septic tanks, and garbage may be fed to domestic animals. The *early urban* stage is characterized by city-type homes built on small to large plots of land in a semiorderly to orderly manner. Schools, churches, and shopping facilities are interspersed. The water supply is generally drawn from individual wells. Rubbish is buried or burned and sewage is disposed of in septic tanks or cesspools. The *middle urban* stage is characterized by large-scale housing developments. There are more schools and shopping centers and some industrial sites. During this stage, centralized systems to provide water and sewers to dispose of wastewaters may be built. Solid wastes may be collected by trucks. This is the first stage in which environmental degradation from urbanization is easily recognizable. The *late urban* stage is characterized by a large number of homes, apartments, commercial and industrial buildings, sidewalks, and parking lots. A large part of the land surface is made impervious or nearly so. Sanitary wastewater systems and storm sewers attempt to remove human and industrial wastes and storm water runoff. In this stage, environmental degradation is obvious to many and may, or has, become a serious threat.

---

[1] Commission on Population Growth and the American Future. 1972. *Population and the American future.* Washington, D.C.: U.S. Government Printing Office.

[2] Savani, J., and Kammerer, J. C. *Urban growth and the water regimen.* U.S. Geological Survey Water Supply Paper 1591A.

Photo by Matthew Neal McVay/AllStock

PART THREE

Human Interaction with the Environment

Chapters 11 through 13 consider the relations between people and our physical environment. Because these relations are often most stressed in urban areas, special attention is given to the problems stemming from human activities and inactivities that create our special urban landscape. Several relationships between hydrology and human use are reviewed in Chapter 11. Especially important here are water supply, management, and pollution.

Chapter 12 is concerned with integrated waste management, and the treatment and disposal of wastes (particularly sanitary landfills), ocean dumping, radioactive wastes, septic systems, and hazardous chemical wastes. Recent advancements involving the relations between health and the fields of geology and geography are discussed in Chapter 13. This subject promises to become more significant as our understanding of the relations among earth processes, earth materials, and health problems increases.

The more obvious and well-known detrimental aspects of human use of the surface water, groundwater, and atmospheric water in the hydrologic cycle include fouled rivers, persistent "suds" on rivers or in wells, and unusual odors from surface water or groundwater. Problems with water supplies and wastewaters were among the first environmental issues to be recognized and attacked on a scientific basis. Traditionally, water and wastewaters have been considered and treated as separate entities. Although convenient and understandable, such segmentation of a complex subject can compound problems, obscure solutions, and confuse the general public. It has been only within the last 50 years that the interrelationships among and between surface water, groundwater, and atmospheric water have been determined in a reasonable, acceptable manner, but all too frequently, they are still considered as separate entities. Consequently, many local, state, and federal agencies split jurisdictional responsibilities along artificial lines. Fortunately, during the last two decades, there has been a distinct trend to bridge these jurisdictional splits.

The fields of engineering, geology, biology, chemistry, and meteorology each progressed through philosophical ponderings and debates on certain aspects of the "nature of things" while in its founding stage. Only within this century have the various disciplines joined together to investigate and pursue quantification of elusive segments of the hydrologic cycle. In this chapter we will consider the topics of hydrology, water supply and use, water pollution and water management, and water and ecosystems. The chapter will close with a discussion of the Colorado River which draws together many diverse issues associated with water.

▼ WATER: A BRIEF GLOBAL PERSPECTIVE

On a global scale, total water abundance is not the problem; the problem is water's availability in the right place at the right time in the right form. Water is a heterogeneous resource that can be found in either liquid, solid, or gaseous form at a number of locations at or near the earth's surface. Depending upon the specific location of water, the residence time may vary from a few days to many thousands of years (Table 11.1). Furthermore, more than 99 percent of the earth's water is unavailable or unsuitable for beneficial human use because of its salinity (sea water) or form and location (ice caps and glaciers). Thus the amount of water for which all the people on earth compete is much less than 1 percent of the total.

As the world's population and the industrial production of many goods increase, the use of water will also accelerate. Today, world per capita use of water is 710 cubic meters per year, and the total human use of water is 2600 cubic kilometers per year. It is estimated that by

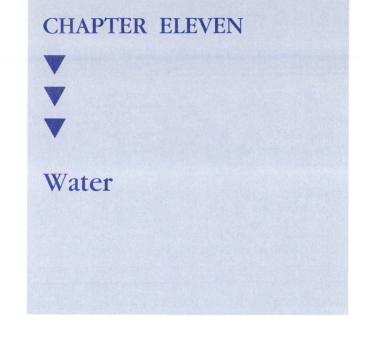

CHAPTER ELEVEN

▼

▼

▼

Water

the year 2000 world use of water will more than double to 6000 cubic kilometers per year—a significant fraction of the naturally available fresh water.

The total average annual water yield (runoff) from the earth's rivers is approximately 48,000 cubic kilometers (Table 11.2), but its distribution is far from uniform. Some runoff occurs in relatively uninhabited regions, such as Antarctica, which produces 2,300 cubic kilometers, or about 5 percent of the earth's total runoff. South America, which includes the relatively uninhabited Amazon basin, provides 16,660 cubic kilometers, or about one-third of total runoff; and in North America, total runoff is about one-half of that for South America, or 8,100 cubic kilometers per year. Unfortunately, much of the North American runoff occurs in sparsely settled or uninhabited regions, particularly in the northern parts of Canada and Alaska.

Compared with other resources, water is used in tremendous quantities. In recent years the total amount of water used on the earth each year has been approximately 1,000 times the world's total production of minerals, including petroleum, coal, metal ores, and nonmetals (1). Because of its great abundance, water is generally a very inexpensive resource. But, because the quantity and quality of water available at any particular time are highly variable, statistical statements about the cost of water on a global basis are not particularly useful. Shortages of water have occurred and will continue to occur with increasing frequency. Such shortages lead to serious economic disruption and human suffering (2).

The U.S. Water Resources Council has estimated that water use in the United States by the year 2020 may exceed surface water resources by 13 percent. As early as 1965, 100 million people in the United States used water that had already been used once before, and by the end

Table 11.1

The world's water supply (selected examples)

Location	Surface Area (km)^2	Water Volume (km)^3	Percentage of Total Water	Water: Estimated Average Residence Time
Oceans	361,000,000	1,230,000,000	97.2	Thousands of years
Atmosphere	510,000,000	12,700	0.001	9 days
Rivers and streams	—	1,200	0.0001	2 weeks
Groundwater: shallow, to depth of 0.8 km	130,000,000	4,000,000	0.31	Hundreds to many thousands of years
Lakes (fresh water)	855,000	123,000	0.009	Tens of years
Ice caps and glacers	28,200,000	28,600,000	2.15	Up to tens of thousands of years and longer

Source: Data from U.S. Geological Survey.

Table 11.2

Water budgets for the continents.

Continent	Precipitation (cm/yr)	Evaporation (cm/yr)	Runoff (cm/yr)	Runoff (km^3/yr)
Africa	69	43	26	7,700
Asia[a]	60	31	29	13,000
Australia	47	42	5	380
Europe	64	39	25	2,200
North America	66	32	34	8,100
South America	163	70	93	16,600
All continents	469	257	212	47,980

[a]Includes the USSR.

Source: Modified after Budyko, 1974.

of the century most of us will be using recycled water. How can we manage our water supply, use, and treatment to maintain adequate supplies?

▼ WATER AS A UNIQUE LIQUID

To understand water in terms of supply, use, pollution, and management, we first need a modest acquaintance with some of water's characteristics. Water is a unique liquid; without it, life as we know it would be impossible. Every water molecule contains two atoms of hydrogen and one of oxygen. The chemical bonds that hold the molecule together are *covalent,* so each hydrogen atom shares its single electron with electrons in the oxygen atom. Although the molecule is electrically neutral (having no net positive or negative charge), the hydrogen end of the molecule is more positively charged, and the oxygen end is more negatively charged. As a result, the water molecule is *dipolar*—this characteristic is responsible for many important properties of water and how it reacts in the environment. For example, water molecules are attracted to each other (more-positive ends to more-negative ends) which produces thin films or layers of water molecules between and around particles important in the movement of water in the unsaturated

(vadose) zone above the groundwater table. This process is one of *cohesion.* Water molecules may also be attracted to solid surfaces (*adhesion*); in particular, the more negative (oxygen) ends of the water molecule are attracted to positive ions such as sodium, calcium, magnesium, and potassium. Because clay particles tend to have a negative charge, they attract the more positive (hydrogen) end of water molecules and so become hydrated. Finally, the dipolar nature of the water molecule is responsible for producing surface tension because water molecules are more attracted to each other than they are to molecules of air. Surface tension is extremely important in many physical and biological processes that involve water moving through small openings and pore spaces (3).

Water is often referred to as the universal solvent. Because natural waters are slightly acidic, they dissolve a great variety of compounds from simple salts to minerals and rocks. Water is particularly important in the chemical weathering of rocks and minerals that, along with physical and biochemical processes, initiates soil formation.

Among common compounds and molecules, water is the only one whose solid form is lighter than its liquid form, which explains why ice floats. If ice were heavier than liquid water, it would sink; although this would be safer for ships traveling in the vicinity of icebergs, properties of the biosphere would be much different from what they are.

Another important feature of water is that its *triple point* (the temperature and pressure at which the three phases of water—solid [ice], liquid, and gas [water vapor]—can exist together) occurs naturally at or near the surface of the earth. The triple point for some substances, on the other hand, can only be achieved on earth under laboratory conditions.

Water has a tremendous moderating effect on the environment. This results because water has a relatively high *specific heat*, defined as the amount of heat measured in calories required to raise the temperature of one gram of a substance one Centigrade degree. The specific heat of water is 1.0 calorie per gram, as compared to the specific heats of most other organic solvents, which are about 0.5 calories per gram. Thus, compared to other common liquids, water has the greatest capacity to absorb and store heat, thus moderating the environment.

▼ THE WATER CYCLE

The global water cycle involves the flow of water from one storage "compartment" to another. In its simplest form, the global water cycle can be viewed as water flowing from the oceans to the atmosphere, falling from the atmosphere as rain onto the ocean, land, or fresh water, and then returning to the oceans as either surface runoff and subsurface flow or to the atmosphere by evaporation.

Problems related to water pollution are extremely variable. Of particular significance are the *residence times* and reservoir sizes of water in the various parts of the cycle because these factors are related to pollution potential (see Table 11.1). For example, water in rivers has a short average residence time of about two weeks. Therefore, a one-time pollution event (one that does not involve the pollutant's attaching to sediment on the river bed, which would result in a much longer residence time) will be relatively short-lived because the water will soon leave the river environment. (On the other hand, the pollutant will enter lakes or an ocean, where residence times are longer and pollutants more difficult to deal with.) Many circumstances, such as sewage spills or pollutant-carrying truck or train crashes, can produce one-time pollution events. News of such events is frequent these days. However, pollution is more likely to result from chronic processes, which discharge pollutants directly into rivers. Groundwaters, unlike river waters, have long residence times (from hundreds to thousands of years). Therefore natural removal of pollutants from groundwater is a very slow process, and correction is very costly and difficult.

Surface Runoff and Sediment Yield

Surface runoff has important effects on erosion and the transport of materials. Water moves materials either in a dissolved state or as suspended particles, and surface water can dislodge soil and rock particles on impact (Figure 11.1). The size of particles and the amount of suspended particles moved by surface waters depend in part on the volume and depth of the water and the velocity of flow. That is, the faster a stream or river flows, the larger the particles it can move and the more material is transported. Therefore, the factors that affect runoff also affect sediment erosion, transport, and deposition.

The flow of water on land is divided by watersheds. A **watershed**, or **drainage basin** (Figure 11.2), is an area of ground in which any drop of water falling anywhere in it will leave in the same stream or river (assuming, that is, that the drop is not consumed by the biosphere, evaporated, stored, or transported out of the watershed by subsurface flow). Two drops of rain separated by only centimeters along the boundary of a major continental divide may end up a few weeks later in different oceans thousands of kilometers away.

The amount of surface-water runoff from a watershed or drainage basin and the amount of sediment carried vary significantly. The variation results from geologic, physiographic, climatic, biologic, and land-use characteristics of the particular drainage basin together with variations of these factors with time. For instance, even the most casual observer is aware of the difference in the amount of sediment carried by streams in flood state (they are often more muddy) compared with what appears to be carried at low-flow conditions.

Geologic factors. The principal geologic factors affecting surface-water runoff and sedimentation include rock and soil type, mineralogy, degree of weathering, and structural characteristics of the soil and rock. Fine-grained, dense, clay soils and exposed rock types with few fractures generally allow little water to move downward and become part of the subsurface flows. The runoff from precipitation falling on such materials is comparatively rapid. Conversely, sandy and gravelly soils, well-fractured rocks, and soluble rocks absorb a larger amount of precipitation and have less surface runoff. These principles are illustrated in Figure 11.2. The upper parts of basins A and B are underlain by shale, and the lower parts are underlain by sandstone. The shale has a greater potential to produce runoff than the more porous sandstone. Therefore the *drainage density* (length of channel per unit area) is much greater in the shale areas than in the sandstone areas.

Physiographic and climatic factors. Physiographic factors that affect runoff and sediment transport include climate, shape of the drainage basin, relief and slope characteristics, the basin's orientation to prevailing

Figure 11.1
Raindrops strike the earth with enough force to tear apart unprotected soil and separate soil particles from one another. They wash the soil and generally move soil particles downslope. (Photos courtesy of U.S. Department of Agriculture.)

storms, and drainage pattern (spatial arrangement of the natural stream channels).

Climatic factors that influence runoff and stream sedimentation include the type of precipitation that

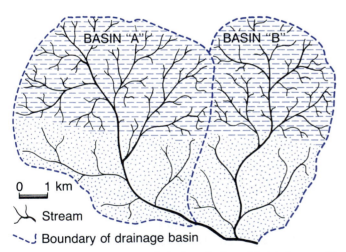

0   1 km

⅄ Stream

〰⌡ Boundary of drainage basin

▤ Part of basin underlain by shale with high runoff potential and greater drainage density

▦ Part of basin underlain by sandstone with lower runoff potential and lower drainage density

Figure 11.2
Two drainage basins. Water falling on one side of the central boundary will drain into Basin A; on the other side, water will drain into Basin B. In this case, the streams from both basins eventually converge.

occurs, the intensity of the precipitation, the duration of precipitation with respect to the total annual climatic variation, and the types of storms (whether cyclonic or thunderstorm). In general, discharge of large volumes of water and sediment is associated with infrequent high-magnitude storms that occur on steep, unstable topography underlain by soil and rocks with a high erosion potential.

The *shape of a drainage basin* is greatly affected by the geologic conditions. One principal effect of basin shape on runoff and sedimentation is its role in governing the rate at which water is supplied to the main stream. Basins that are long and narrow and have short tributaries receive flow from the tributaries much more rapidly than basins that have long, sinuous tributaries.

The factors of *relief* and *slope* are interrelated: the greater the relief, the more likely the stream is to have a steep gradient and a high percentage of steep, sloping land adjacent to the channel. Relief and slope are important because they affect not only the velocity of water in a stream but also the rate at which water infiltrates the soil or rock, the rate of overland flow, and thus the rate at which surface and subsurface runoff enters a stream.

Orientation of the stream basin to prevailing storm influences the rate of flow, the peak flow, the duration of surface runoff, and the amount of transpiration and evaporation losses. The latter is a factor because of the influence of basin orientation on the amount of heat received from the sun and exposure to prevailing winds.

The *drainage net* (set of channels that make up a drainage system) is important in its effect on the efficiency of the drainage system. It is controlled by soil, rock, and topographic features. The characteristic shape of a hydrograph from a drainage basin (land area that contributes water to a drainage net) varies with the type of drainage net. For instance, in a well-drained, efficient drainage basin, surface runoff from a rainstorm concentrates quickly to a flood peak. Discharge from such an efficient stream is commonly described as *flashy*. Conversely, in a stream that has relatively small variation between high and low flows, storm waters tend to run off more slowly, and the stream is therefore considered sluggish and less efficient. Figure 11.3 illustrates the difference in the shape of hydrographs from similar storms for an efficient drainage basin and a less efficient one.

Biological factors.   The biosphere is capable of affecting stream flow in several ways. First, vegetation may decrease runoff by increasing the amount of rainfall intercepted and removed by evaporation. Rainfall that is intercepted by vegetation also falls to the ground more gently and is more likely to infiltrate the soil. Experimental clear-cutting of forested watersheds has been shown to increase the stream flow due to decreased evapotranspiration (water used by the trees and released to the atmosphere) following timber harvesting (4). Second, streamside vegetation increases the resistance to flow, which slows down the passage of flood water. Third, streamside vegetation retards stream-bank erosion because its roots bind and hold soil particles in place. Finally, in forested watersheds, large organic debris (stems and pieces of woody debris greater than ten centimeters in diameter) may profoundly affect stream-

channel form and process. In steep mountain watersheds, many of the pool environments important for fish habitat may be produced by large organic debris.

Animals affect streams by removing vegetation or burrowing. Large grazing mammals can damage streamside environments causing bank erosion problems. Animals burrowing through flood control levees can start erosion problems that eventually lead to failure of the levees.

Soil organisms alter the physical structure and sometimes allow greater percolation of water into the soil, reducing runoff and erosion. Plant roots and burrowing animals can produce macropores (large openings) in soil that can greatly increase the rate at which water moves through soil. Soils with a high organic content tend to be relatively cohesive—they reduce surface erosion and tend to hold water tenaciously—compared to sandy soils, which have low cohesion, high porosity, and high permeability.

The above principles and processes suggest that runoff can be variable depending upon geologic, physiographic, climatic, and biologic conditions. For example, under natural conditions with continuous forest cover, direct surface runoff as shown in Figure 11.1 is unusual because trees and lower vegetation intercept the precipitation and the potential for water to infiltrate the soil is high. In such cases the runoff is subsurface by way of shallow throughflow confined to the soil above the groundwater (Figure 11.4a). An exception to this may occur near streams and in hillslope depressions if the groundwater table rises to the surface. Such saturated areas can produce surface runoff even in humid climates with good vegetation cover (point 3a, Figure 11.4a). Areas that are saturated tend to expand and contract, being larger during the time of spring snow melt than in the late summer or fall, when precipitation is less. In disturbed areas, areas with sparse vegetation cover, semiarid lands, and areas with land uses such as row crops or urbanization, overland flow is often produced because the intensity of precipitation is greater than the potential for the water to infiltrate the ground (Figure 11.4b).

Thus we can identify three major paths by which water on slopes can be transported to the stream environment and exported from the drainage basin: overland flow, throughflow (a type of shallow subsurface flow above the groundwater table), and groundwater flow (Figure 11.4). Understanding potential paths of runoff for a particular site or area is critical in evaluating hydrologic impacts of projects that involve land-use changes. Loss of vegetation and soil compaction during urbanization, for example, will produce more overland flow (point 3b, Figure 11.4b), as will land-use change from forest to row crops.

Figure 11.5 shows the water resource regions of the United States, and Table 11.3 shows the variations in the

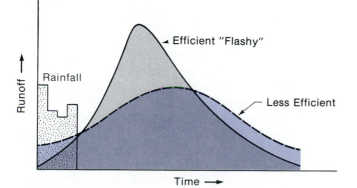

Figure 11.3
Idealized hydrographs illustrating the difference between an efficient drainage basin that removes surface runoff quickly and a less efficient one that takes a longer time to remove the water. Many natural stream basins have less efficient hydrographs, whereas artificial channels tend to be more efficient but may damage the environment.

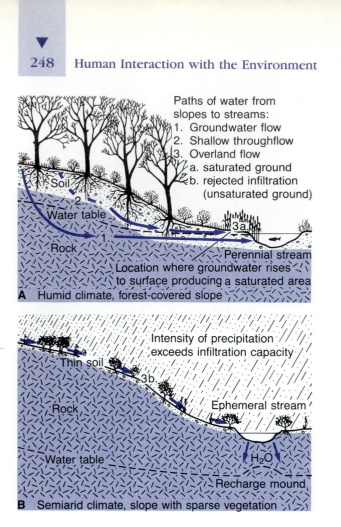

Paths of water from slopes to streams:
1. Groundwater flow
2. Shallow throughflow
3. Overland flow
   a. saturated ground
   b. rejected infiltration (unsaturated ground)

Soil

Water table

Rock

Perennial stream

Location where groundwater rises to surface producing a saturated area

A Humid climate, forest-covered slope

Intensity of precipitation exceeds infiltration capacity

Thin soil

Rock

Ephemeral stream

Water table

H_2O

Recharge mound

B Semiarid climate, slope with sparse vegetation

Figure 11.4

Paths of water from slopes to streams: (a) on a humid, forest-covered slope and (b) in a semiarid climate on a slope with sparse vegetation.

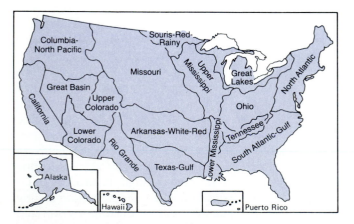

Figure 11.5

Water resources regions of the United States. (From Water Resources Council, *The Nation's Water Resources,* 1968.)

Table 11.3

Estimated ranges in sediment yields from drainage areas of 260 square kilometers or less.

Region	Estimated Sediment Yield (metric tons/km²/yr)*		
	High	Low	Average
North Atlantic	4,240	110	880
South Atlantic-Gulf	6,480	350	2,800
Great Lakes	2,800	40	350
Ohio	7,391	560	2,780
Tennessee	5,460	1,610	2,450
Upper Mississippi	13,660	40	2,800
Lower Mississippi	28,760	5,460	18,220
Souris-Red-Rainy	1,650	40	175
Missouri	23,470	40	5,250
Arkansas-White-Red	25,760	910	7,710
Texas-Gulf	8,140	320	6,310
Rio Grande	11,700	530	4,550
Upper Colorado	11,700	530	6,310
Lower Colorado	5,670	530	2,100
Great Basin	6,240	350	1,400
Columbia-North Pacific	3,850	120	1,400
California	19,510	280	4,550

*The range in high to low values reflects different years with different discharges and ability to erode and transport sediment.

Source: *The Nation's Water Resources.* Water Resources Council, 1968.

natural sediment yield for relatively small river basins. The amount of sediment carried by rivers as part of their work within the rock cycle varies with geologic, climatic, topographic, physical, vegetative, and other conditions. Hence, some rivers are consistently and noticeably different in their clarity and appearance, as can be inferred from Table 11.4. Although this table reflects varying degrees of human influence, it demonstrates the sizable variations of sediment load per unit area in various parts of the world. For instance, on the average, the Lo River of China carries nearly 200 times more suspended load than does the Nile River of Egypt. In the United States, the Mississippi is not as "muddy" as the Missouri and the Colorado rivers.

The general relationship between size of drainage basin and sediment load suggests that, as basin size increases, the sediment yield per unit area decreases (Table 11.5). This relationship results from the increase in probability of sediment storage and deposition with increased basin size, the fact that smaller basins tend to be steeper (which increases the energy available for erosion and transport of sediment), and the decreased probability of total basin coverage by a single storm event with increased basin size.

Table 11.4
Some major rivers of the world ranked by sediment yield per unit area.

River	Drainage Basin (10^3 km^2)	Sediment Load per Year (tons/km^2)
Amazon	5,776	63
Mississippi	3,222	97
Nile	2,978	37
Yangtze	1,942	257
Missouri	1,370	159
Indus	969	449
Ganges	956	1,518
Mekong	795	214
Yellow	673	2,804
Brahmaputra	666	1,090
Colorado	637	212
Irrawaddy	430	695
Red	119	1,092
Kosi	62	2,774
Ching	57	7,158
Lo	26	7,308

Source: Data from Holman, 1968.

Groundwater

The major source of groundwater is precipitation that infiltrates the surface to enter and move through the top of the vadose zone (Figure 11.6). It may then move downward through the remainder of the vadose zone, which is seldom saturated. Until recently, the vadose zone was called the unsaturated zone, but we now know that some saturated areas may exist there at times as water moves through. The vadose zone has special significance because potential pollutants applied at the surface must percolate through it before they enter the

Table 11.5
Arithmetic average of sediment-production rates for various groups of drainage areas in the United States. These data illustrate that as the size of a drainage basin (watershed) increases, the sediment production per unit area decreases

Watershed-Size Range (km^2)	Number of Measurements	Average Annual Sediment-Production Rate (m^3/km^2)
Under 25	650	1,810.3
25–250	205	762.2
250–2,500	123	481.2
Over 2,500	118	238.2

Source: From *Handbook of Applied Hydrology*, by Ven Te Chow. Copyright © 1964 by McGraw-Hill. Used with permission of McGraw-Hill Book Company.

saturated zone below the water table. Thus, in environmental subsurface monitoring the vadose zone is an area of early warning for potential pollution to groundwater resources. Water that percolates through the vadose zone may enter the groundwater system (zone of saturation where saturated flow occurs). The upper surface of this zone is the **water table**. The **capillary fringe** just above the water table is a belt of variable thickness where water is drawn up by capillarity, which is due to the attractive force of water to surfaces of earth materials, and surface tension (attraction of water molecules to each other).

In addition to precipitation, other sources of groundwater include water that infiltrates from surface waters, including lakes and rivers; artificial recharge (surface water deliberately injected into the groundwater system); storm-water retention or recharge ponds; agricultural irrigation and wastewater treatment systems such as cesspools and septic tank drain fields.

Movement of water into the zone of saturation and through earth materials is an integral part of both the rock cycle and the hydrologic cycle. For example, water may dissolve minerals from materials it moves through and deposits them elsewhere as cementing material, producing sedimentary rocks. Groundwater may transport sediment, heat, gases, and microorganisms. What actually occurs varies with the chemical and physical characteristics of the water, soil, and rock as the water infiltrates through the biological and soil-horizon environments above the water table and moves through the groundwater system below the water table.

The rate of movement of groundwater depends upon the nature and extent of the primary (or intergranular) and secondary (or fracture) openings of the soil or rocks and upon the hydraulic gradient, which for the simplest cases is approximately the slope of the water table. The percentage of void or empty space in soil and rock is called **porosity**. The ability of water to move through a particular material is called **hydraulic conductivity**, which is expressed in units such as meters per day. Some of the most porous materials, such as clay, have a very low hydraulic conductivity (Table 11.6) because while the small, flat clay particles can have a great deal of pore space, the individual openings are very small and tenaciously hold water.

Hydraulic conductivity of an earth material is a function of both the properties of the fluid (such as viscosity and density) and properties of the material (such as particle diameter, size of pores, and how interconnected the pore spaces are). In some older works *hydraulic conductivity* may be called the *coefficient of permeability*, which may lead to confusion. The term *permeability* is also used as a measure of the ability of an earth material to transmit fluid, but only in terms of the properties of that material (not the properties of the fluid). When discussing the movement of

Figure 11.6
Generalized diagram showing
zones of groundwater, capillary
fringe, and water table.

L = Length of ground water flow between
    points A and B.

h = Difference in elevation of the water
    between points A and B.

h/l = Hydraulic gradient, which is approximately
      the slope of the water table.

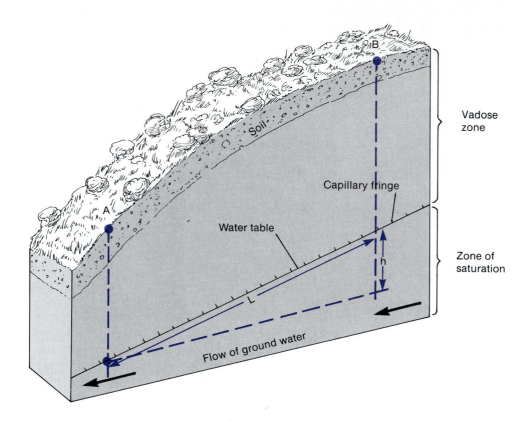

Table 11.6
Porosity and hydraulic conductivity of selected earth
materials.

Material	Porosity (%)	Hydraulic Conductivity* (m/day)
Unconsolidated		
Clay	45	0.041
Sand	35	32.8
Gravel	25	205.0
Gravel and sand	20	82.0
Rock		
Sandstone	15	28.7
Dense limestone or shale	5	0.041
Granite	1	0.0041

*In older works, may be called coefficient of permeability.

Source: Modified after Linsley, Kohler, and Paulhus, *Hydrology for
Engineers* (New York: McGraw-Hill, 1958. Copyright © 1958 by
McGraw-Hill Book Company. Used by permission of McGraw-Hill
Book Company.)

groundwater, we often will refer to a material as having a
high or low permeability or being nearly impermeable,
such as a clay liner of a waste disposal facility. Gravel and
sands have high permeabilities compared to silt and clay.
Bedrock such as granite with few fractures and most
shales are considered to have low permeability. In this
text we will use both hydraulic conductivity and permeability to describe hydraulic properties of earth materials.
However, the term *hydraulic conductivity* is usually
preferred because it is expressed in units that are easily
understood and it is commonly used in hydrogeology
today.

A zone of earth material capable of producing
groundwater at a useful rate from a well is called an
aquifer. A zone of earth material that will hold water but
not transmit it fast enough to be pumped from a well is
called an aquiclude or **aquitard**. Aquitards often form a
confining layer through which little water moves. Clay
soils, shale, and igneous or metamorphic rocks with little
interconnected porosity and/or fractures are likely to
form aquitards. On the other hand, gravel, sand, soils,

and fractured sandstone, as well as granite and metamorphic rocks with high porosity due to connected open fractures, will be good aquifers if groundwater is present.

Aquifers are said to be **unconfined** if there is no confining layer restricting the upper surface of the zone of saturation at the water table. If an aquitard (confining layer) is present, then the water below may be under pressure, forming **artesian** conditions. Water in artesian systems tends to rise to about the height of the recharge zone (the zone where precipitation infiltrates the surface to move down to the groundwater system). This condition is analogous to the function of a water tower that produces water pressure for homes (Figure 11.7). Aquifers overlain by confining layers are designated as **confined aquifers**. Both confined and unconfined aquifers may be found in the same area (Figure 11.8).

In 1856 an engineer named Henry Darcy was working on the water supply for Dijon, France. He performed a series of important experiments that demonstrated that the discharge (Q) of groundwater may be defined as the product of the cross-sectional area of flow

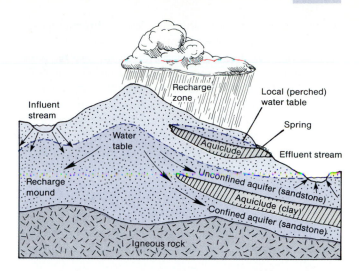

Figure 11.8
An unconfined aquifer, a local (perched) water table, and influent and effluent streams.

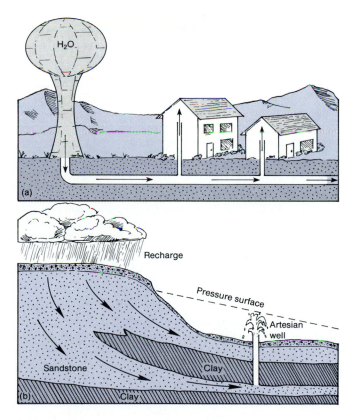

Figure 11.7
Development of an artesian well system. In (a), water rises in homes due to pressure created by water level in the tower. If friction in pipes is small, there will be little drop in pressure. As shown in (b), the pressure surface in natural systems declines away from the source because of friction in the flow system, but water may still rise above the surface of the ground if an impervious layer such as clay is present to cap the groundwater.

(A), the hydraulic gradient (I), and the hydraulic conductivity (K). Thus, $Q = KIA$. The units on both sides are a volumetric flow rate (such as cubic meters per day), and the relationship is known as Darcy's Law.

From a more theoretical point of view, the actual driving force for flow is called the fluid potential or hydraulic head, which at the point of measurement is the sum of the elevation of the water (elevation head) and the ratio of the fluid pressure to the unit weight of water (pressure head). Groundwater always moves from an area of high hydraulic head to an area of lower hydraulic head. For our simple example in Figure 11.6 the pressure head at both points A and B is atmospheric (defined as 0); thus, the hydraulic heads at A and B are their respective elevations. The difference in hydraulic heads between points A and B (h) divided by the flow length (L) is the hydraulic gradient (I). The condition shown for Figure 11.6 is an unconfined aquifer. If a confining layer is present then the fluid pressure must be considered in the calculation of the hydraulic gradient. However, Darcy's Law still applies. Because groundwater always moves from an area of higher to lower hydraulic head, it may move down, laterally, or upward depending upon local conditions. The reason that the water flows upward making an artesian well in Figure 11.7 is that the hydraulic head below the clay confining layer is greater than the hydraulic head found above the confining layer. Thus, the water moves from an area of higher hydraulic head to lower (in this case upward).

Figure 11.8 shows some of the interesting interactions between surface water and groundwater. In particular, two types of streams may be defined. **Effluent streams** tend to be perennial (flow all year). During the dry season, groundwater seeps into the channel, maintaining stream flow. **Influent streams** are everywhere above the groundwater table and flow in direct response to precipitation. Water from influent streams moves down through the vadose zone to the water table, forming a recharge mound. Influent streams may be intermittent or ephemeral in that they flow only part of the year. From an environmental standpoint influent streams are particularly important because water pollution in the stream may move downward through the stream bed and eventually pollute the groundwater found below. Dry river beds are particularly likely to experience this type of problem. For example, the Mojave River in southern California is dry near Barstow, California, almost all of the time. Cleaning solvents introduced into the dry river bed as part of a large cleaning operation for equipment have infiltrated down through the vadose zone to contaminate and threaten groundwater that is used by several localities for municipal purposes.

When water is pumped from a well, a **cone of depression** forms in the water table or artesian pressure gradient (Figure 11.9). A very large cone of depression can even alter the direction in which groundwater moves within an area. Overpumping of an aquifer causes the water level to lower continuously with time. This necessitates lowering the pump settings or drilling deeper wells. These adjustments may work, depending on the hydrologic conditions, but are not always successful and are often very costly. For instance, continued deepening to correct for overpumping of wells that tap igneous and metamorphic rocks is limited. Water from these wells is pumped from open fracture systems that tend to close or diminish in number and size with increasing depth. Also the quality of groundwater may be degraded as it is extracted from deeper sources.

The soil, sediment, and rocks that groundwater passes through act as natural filters. Under the right conditions, this filtering system cleanses the water, trapping and biodegrading disease-causing microorganisms and particulates that contain toxic compounds. The water actually exchanges materials with the soil and rocks. If the soil or rock surface is already highly contaminated or contains naturally toxic elements, the water may be rendered toxic by these natural processes. For instance, water may become toxic by dissolving sufficient amounts of a toxic element or mineral, such as arsenic. An interesting example with serious environmental implications comes from the western San Joaquin Valley in California, where **selenium** (a very toxic heavy metal) present in the soil is released by application of irrigation waters. Subsurface drainage of the selenium-rich water from fields enters the surface waters and has caused birth defects in waterfowl. The extent of the problem is only now being learned. Selenium is also toxic to people. Like many trace metals, selenium has a dual character: it is necessary for life processes at trace concentrations, but toxic at some higher concentrations. Most often, the groundwaters dissolve a mixture of minerals and some gases that can be nuisances to some human uses. Some examples are iron as ferrous hydroxide, which colors the water brown and leaves a brown discoloration on laundry and plumbing fixtures; calcium, which creates in part the so-called hardness of water; and hydrogen sulfide, which produces a "rotten egg" odor.

People's perceptions about groundwater affect the way they view our water resources. First, people tend to assume that water is available when, where, and in the amounts they want. We turn on a faucet and expect water—it is somebody else's responsibility to see that we have it. Second, because groundwater is out of sight, it is out of mind or mysterious. Third, groundwater is not as easily measured quantitatively as surface water. Therefore, precise quantitative values of groundwater reserves are not available, and we rely on estimates of the probable reserves.

The long residence time for groundwaters (Table 11.1) reflects the deep, insulated type of storage aquifers provide. Not all groundwater takes hundreds of years to rejoin the other, more rapidly moving parts of the hydrologic cycle, but most of it is well below the influence of transpiration by plants and evaporation into the atmosphere. Where it is not that deep, it is most susceptible to evapotranspiration, discharge to streams, and use or abuse by humans. The latter is of increasing concern because of the potential long-term damage to this resource, the high cost of cleaning up polluted groundwater, and the increasing need to use it as water use per capita increases.

▼ WATER SUPPLY AND USE: U.S. EXAMPLE*

The water supply at any particular point on the land surface depends upon several factors in the hydrologic

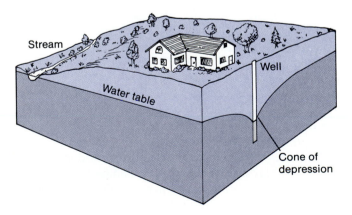

Figure 11.9
Cone of depression in water table resulting from pumping water from a well.

---

*Much of the discussion concerning water supply and use is summarized from U.S. Water Resources Council, 1978, *The nation's water resources, 1975–2000*, Vol. 1.

cycle, including the rates of precipitation, evaporation, stream flow, and subsurface flow. The various uses of water by people also significantly affect water supply. A concept useful in understanding water supply is the **water budget**—the inputs, outputs, and storage of water in a system. On a continental scale, the water budget for the conterminous United States is shown in Figure 11.10. The amount of water vapor passing over the United States every day is approximately 152,000 million cubic meters (40,000 billion gallons), and of this approximately 10 percent falls as precipitation in the form of rain, snow, hail, or sleet. Approximately two-thirds of the precipitation evaporates quickly or is transpired by vegetation. The remaining one-third, or about 5510 million cubic meters (1450 billion gallons) per day, enters the surface or groundwater storage systems, flows to the oceans or across the nation's boundaries, is used by consumption, or evaporates from reservoirs. Unfortunately, owing to natural variations in precipitation that cause either floods or droughts, only a portion of this water can be developed for intensive uses. Thus only about 2565 million cubic meters (675 billion gallons) per day are considered to be available 95 percent of the time (2).

On a regional scale, it is critical to consider annual precipitation and runoff patterns in order to develop water budgets. Figures 11.11 and 11.12 illustrate the variability of precipitation and runoff for the conterminous United States. Potential problems with water supply can be predicted in areas where average precipitation and runoff are relatively low, such as in the southwestern and Great Plains regions of the United States as well as in some of the intermontane valleys in the Rocky Mountain area. The theoretical upper limit of surface water supplies is the mean annual runoff, assuming it could be successfully stored. Unfortunately, storage of all the runoff is not possible because of evaporative losses from large reservoirs, the limited number of suitable sites for reservoirs, and need for other water uses such as river transportation and wildlife. As a result, shortages in water supply are bound to occur in areas with low precipitation and runoff. Strong conservation practices are necessary to ensure an adequate supply (2).

Because of the large annual variations in stream flow, even areas with high precipitation and runoff may periodically suffer from droughts. For example, the dry years of 1961 and 1966 in the northeastern United States and 1976–1977 and 1985–1990 in parts of the western United States produced serious water shortages. Fortunately, in the more humid eastern United States, stream flow tends to vary less than in other regions, and drought

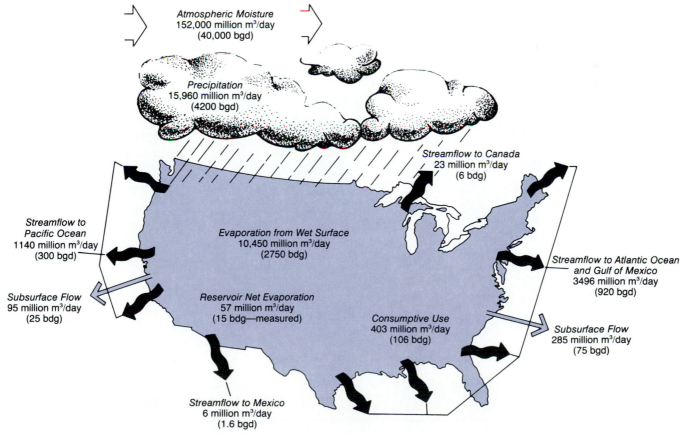

Atmospheric Moisture
152,000 million m³/day
(40,000 bgd)

Precipitation
15,960 million m³/day
(4200 bgd)

Streamflow to Canada
23 million m³/day
(6 bdg)

Streamflow to
Pacific Ocean
1140 million m³/day
(300 bgd)

Evaporation from Wet Surface
10,450 million m³/day
(2750 bdg)

Streamflow to Atlantic Ocean
and Gulf of Mexico
3496 million m³/day
(920 bgd)

Subsurface Flow
95 million m³/day
(25 bdg)

Reservoir Net Evaporation
57 million m³/day
(15 bdg—measured)

Consumptive Use
403 million m³/day
(106 bdg)

Subsurface Flow
285 million m³/day
(75 bgd)

Streamflow to Mexico
6 million m³/day
(1.6 bgd)

Figure 11.10
Water budget for the conterminous United States. (From U.S. Water Resources Council, 1978, *The Nation's Water Resources, 1975–2000.*)

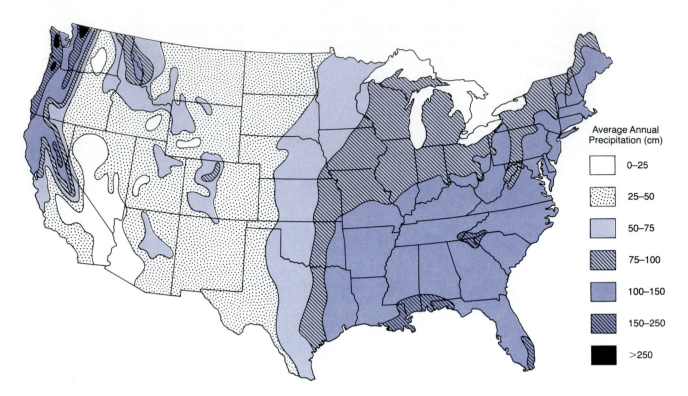

Figure 11.11
Average annual precipitation for the conterminous United States. (From U.S. Water Resources Council, 1978, *The Nation's Water Resources, 1975–2000.*)

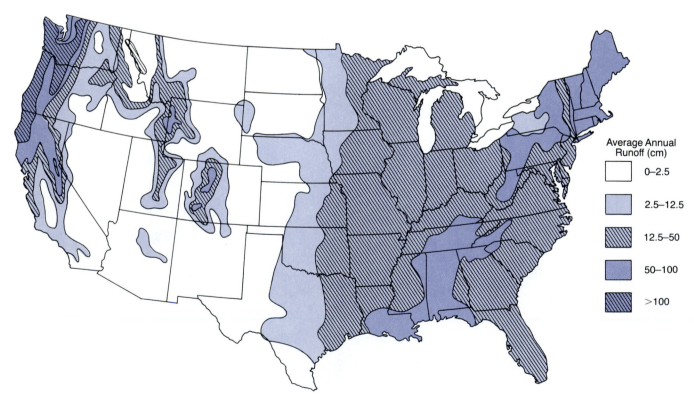

Figure 11.12
Average annual runoff for the conterminous United States. (From U.S. Water Resources Council, 1978, *The Nation's Water Resources, 1975–2000.*)

254

is less likely (2). Nevertheless, the drought of summer 1986 in the southeastern United States caused tremendous hardships and inflicted several billions of dollars in damage.

Nearly half the population of the United States uses groundwater as a primary source of drinking water. Fortunately, the total amount of groundwater available in the United States is enormous, accounting for approximately 20 percent of all water withdrawn for consumptive uses. Within the conterminous United States the amount of groundwater within 0.8 kilometer of the land surface is estimated to be between 125,000 and 224,000 cubic kilometers. To put this in perspective, the lower estimate is about equal to the total discharge of the Mississippi River during the last 200 years. Unfortunately, owing to the cost of pumping and exploration, much less than the total quantity of groundwater is available. Figure 11.13 shows the major aquifers in the United States capable of yielding more than 0.2 cubic meter (50 gallons) per minute (2).

Protecting groundwater resources is an environmental problem of particular public concern because so many people derive their domestic water supplies from groundwater. The residency time for groundwater aquifers is often measured in hundreds to thousands of years; therefore, once aquifers are damaged by pollutants, it may be difficult or impossible to reclaim them for continued use. Aquifers are also very important because approximately 30 percent of the stream flow in the United States is supplied by groundwater that emerges as springs or other seepages along the stream channel. This phenomenon, known as *base flow,* is responsible for the low flow or dry season flow of most perennial streams. Therefore, maintaining high-quality groundwater is important in maintaining good quality stream flow.

In many parts of the country, groundwater withdrawal from wells exceeds natural inflow. In such cases, water is being mined and can be considered a nonrenewable resource. Groundwater overdraft is a serious problem in the Texas–Oklahoma–High Plains area, California, Arizona, Nevada, New Mexico, and isolated areas of Louisiana, Mississippi, Arkansas, and the South Atlantic–Gulf region (Figure 11.14a). In the Texas–Oklahoma–High Plains area alone, the overdraft amount is approximately equal to the natural flow of the Colorado River (Figure 11.14b) (2). In this area lies the Ogallala aquifer, which is composed of water-bearing sands and gravel that underlie an area of about 400,000 square kilometers from South Dakota into Texas. Although the aquifer holds a tremendous amount of groundwater, it is being used in some areas at a rate that is up to 20 times that of natural recharge by infiltration of precipitation. The water level in many parts of the aquifer has declined in recent years, and eventually a significant portion of land now being irrigated may return to dry farming if the resource is used up.

To date only about 5 percent of the total groundwater resource has been depleted, but water levels have declined as much as 30–60 meters in parts of Kansas, Oklahoma, New Mexico, and Texas. As the water table becomes lower, yields from wells decrease and energy costs to pump the water increase. The most severe problems in the High Plains and the Ogallala aquifer today are in those locations where irrigation was first started in the 1940s.

In many areas, pumping of groundwater has forever changed the character of the land. For example, rivers in the Tucson, Arizona, area, prior to lowering of the water table through pumping, were perennial, with healthy populations of trout, beaver, and other animals. Today the native riparian trees have died and the rivers are dry much of the year. Ironically, these processes also increased the flood hazard in Tucson, which currently gets its entire water supply from groundwater sources! Loss of riparian trees and the root strength they provided to stream banks render the channels much more vulnerable to lateral bank erosion. During the 1983 floods in Tucson this became very apparent as roads, bridges, and buildings were damaged by the shifting channels. Tree-lined channels are much more stable, but riparian trees need a groundwater table sufficiently close to the surface for healthy growth. Unfortunately, mining of groundwater in the Tucson area has precluded restoration of trees.

▼ WATER USE

To discuss water use, we must distinguish between instream and offstream uses. *Offstream uses* remove water from its source—water is taken for drinking, for washing and sewage, and for agricultural irrigation. *Consumptive use* is an offstream use in which water does not return to the stream or groundwater resource immediately after use (2).

Instream water use includes the use of rivers for navigation, hydroelectric power generation, fish and wildlife habitat, and recreation. Multiple uses usually create controversy because each instream use requires different conditions to prevent damage or detrimental effects. Fish and wildlife require certain water levels and flow rates for maximum biological productivity; these levels and rates may differ from the requirements for hydroelectric power generation, which requires large fluctuations in discharges to match power needs. Similarly, both of these may conflict with requirements for shipping and boating. The discharge necessary to move the sediment load in a river may require yet another pattern of flow. Figure 11.15 diagrams the seasonal patterns of discharge for some of these uses. A major problem concerns how much water may be removed from a stream or river and transported to another location without damaging the stream system. This is a problem in the Pacific Northwest, where certain fish

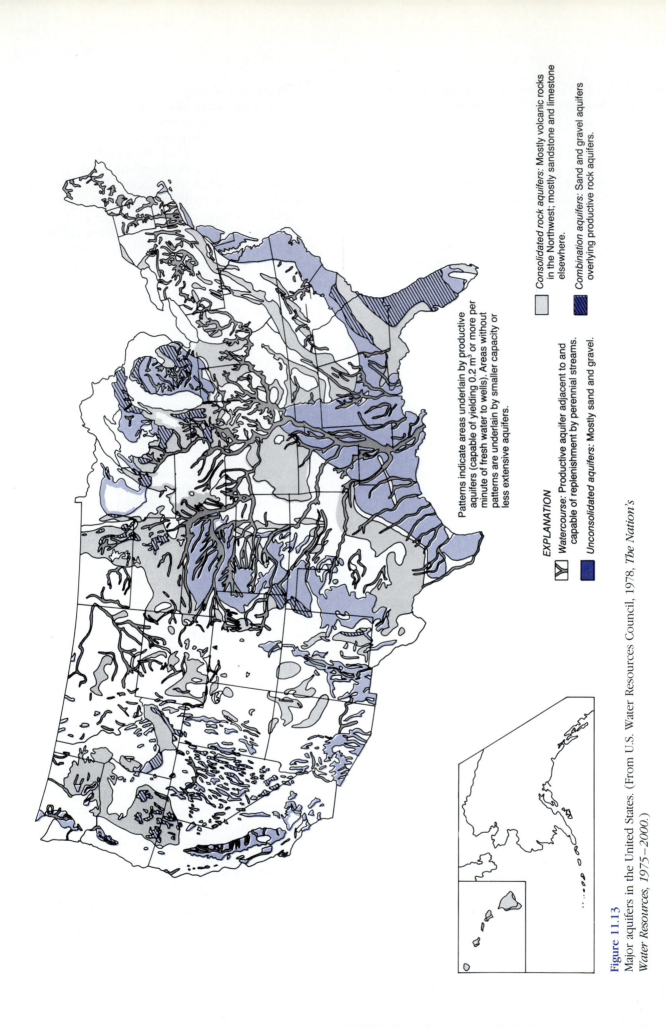

Patterns indicate areas underlain by productive aquifers (capable of yielding 0.2 m³ or more per minute of fresh water to wells). Areas without patterns are underlain by smaller capacity or less extensive aquifers.

Consolidated rock aquifers: Mostly volcanic rocks in the Northwest; mostly sandstone and limestone elsewhere.

Combination aquifers: Sand and gravel aquifers overlying productive rock aquifers.

EXPLANATION

Watercourse: Productive aquifer adjacent to and capable of replenishment by perennial streams.

Unconsolidated aquifers: Mostly sand and gravel.

Figure 11.13
Major aquifers in the United States. (From U.S. Water Resources Council, 1978, *The Nation's Water Resources, 1975–2000.*)

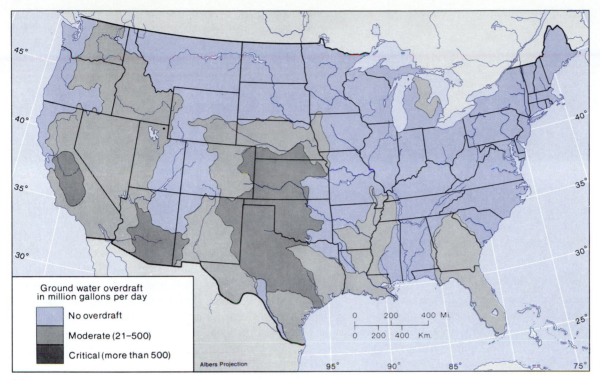

(a)

Ground water overdraft
in million gallons per day

☐ No overdraft

☐ Moderate (21–500)

■ Critical (more than 500)

Albers Projection

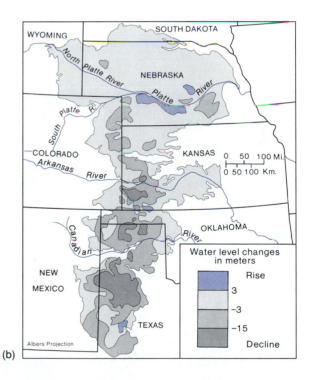

(b)

Figure 11.14

(a) Groundwater overdraft, 1975, for the conterminous United States. (b) A detail of water level changes in the Texas-Oklahoma-High Plains area. (From U.S. Geological Survey.)

including the steelhead trout and salmon are on the decline partly because people have induced alterations in land use and stream flows that have degraded fish habitats.

In California, demands are being made on northern rivers for reservoir systems to supply the cities of southern California. In our modern civilization, water is often moved vast distances from areas with abundant rainfall to areas of high usage. In California, two-thirds of the state's runoff occurs north of San Francisco, where there is a surplus of water, while two-thirds of the water use occurs south of San Francisco, where there is a deficit. In recent years, canals constructed by the California Water Project and the Central Valley Project have moved tremendous amounts of water from the northern to the southern part of the state, adversely affecting

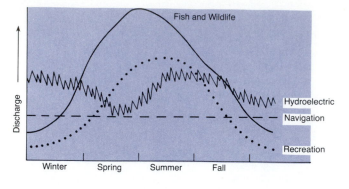

Figure 11.15
Diagram of instream water uses and the varying discharges for each use. Discharge is the amount of water passing by a particular location and is measured in cubic meters per second.

ecosystems (especially fisheries) in some northern California rivers through diversion of waters.

The major water diversion projects in California are shown in Figure 11.16. Of particular interest is the long-standing dispute between the city of Los Angeles and the people in Owens Valley on the eastern side of the Sierra Nevada. Los Angeles suffered a drought near the end of the nineteenth century and, after looking for a potential additional water supply, settled on the Owens

Valley. By various means (some of which were controversial, to say the least) the city purchased most of the water rights and constructed the Los Angeles Owens River Aqueduct, completed in 1913. Since that time groundwater has also been taken. As a result of the tremendous exportation of surface water and groundwater, the Owens Valley has suffered from desertification. Protests that were more violent in the early 1900s are now court battles; only recently have both parties come closer to a settlement that will include limits on the amount of water taken by Los Angeles and projects to halt environmental degradation.

Many large cities in the world must seek water from areas increasingly farther away. For example, New York City has imported water from nearby areas for over 100 years. Water use and supply in New York City represent a repeating pattern. Originally, local groundwater, streams, and the Hudson River itself were used. However, water needs exceeded local supply, so in 1842 the first large dam was built more than 48 kilometers north of the city. As the city expanded rapidly from Manhattan to Long Island, water needs again increased. The sandy aquifers of Long Island were at first a source of drinking water, but this water was removed faster than rainfall replenished it. Local cesspools contaminated the groundwater, and salty ocean water intruded. A larger dam was built at Croton in 1900. Further expansion of the population created the

(a)   (b)

Figure 11.16
(a) California aqueducts and irrigation canals; (b) view of the Los Angeles Owens River Aqueduct. (Photo by F. C. Whitmore, Jr., U.S. Geological Survey.)

same pattern: initial use of groundwater; pollution, salinification, and exhaustion of this resource; and subsequent building of new, larger dams farther upstate in forested areas. The pattern continues with development of tract housing in eastern Long Island, where there are now problems of pollution, salinification, and exhaustion of the resource. It is important to recognize that New York City is not unique. Many urban areas are having problems with their water supply. On the West Coast, Los Angeles faces a similar problem as a growing population demands more water which is becoming harder to obtain. One would think that eventually the cost of obtaining water from long distances would place an upper limit on growth. To some extent this may be true, but the price of water is often kept artificially low through a variety of government programs. People in urban environments could do much through water conservation to alleviate or reduce the problems related to water supply, but shortages have not yet become sufficiently acute to precipitate this.

As greater quantities of water are needed for cities and agriculture, conflicts will increase and intensive argument will center on instream water use. An important, fruitful area of research is more careful evaluation of what flows are necessary to maintain a natural river system.

Offstream water-use withdrawals in the United States today amount to more than 1700 million cubic meters per day (450 billion gallons/day)—considerably more than one-half the average flow of the Mississippi River, which drains approximately 40 percent of the conterminous United States. Consumption of water amounts to approximately 22 percent of that withdrawn, or 379 million cubic meters per day (100 billion gallons/day). Figure 11.17 lists present sources and disposition of offstream water withdrawals, amount of withdrawal by the major users (self-supplied industry, agriculture, public supply, and rural), and consumptive use of fresh water. Figure 11.18 shows recent trends in water withdrawals, consumptive use, and population. Examination

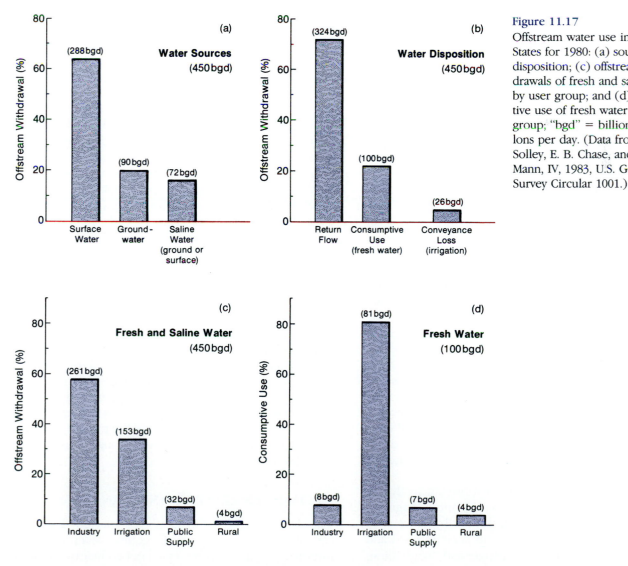

Figure 11.17
Offstream water use in the United States for 1980: (a) sources; (b) disposition; (c) offstream withdrawals of fresh and saline water by user group; and (d) consumptive use of fresh water by user group; "bgd" = billions of gallons per day. (Data from W. B. Solley, E. B. Chase, and W. B. Mann, IV, 1983, U.S. Geological Survey Circular 1001.)

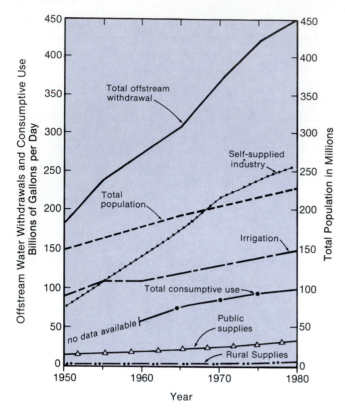

Figure 11.18
Offstream water withdrawals and consumptive use, 1950–
1980. (After W. B. Solley, E. B. Chase, and W. B. Mann, IV,
1983, U.S. Geological Survey Circular 1001.)

of the information in these figures reveals that industry (self-supplied, including thermoelectric power) and agriculture (irrigation) withdraw 58 percent and 34 percent respectively, while public supply (domestic, public, commercial, and industrial uses) and rural (domestic and livestock uses) withdraw much less (5). In terms of consumption, agriculture (irrigation) consumes about 80 percent of the total fresh water used (5), and it is agricultural use that is heavily subsidized to keep the price of water artificially low. Trends in withdrawals and consumption (Figure 11.18) suggest that self-supplied industry (especially power plants) has been responsible for much of the recent growth in water resource utilization. It is expected that withdrawals will decrease by the year 2000 while water consumption will increase slightly. It is hoped that projected reductions in offstream withdrawals will be realized from conservation endeavors, improved technology, and recycling of water. Increases in consumption from the present level will most likely reflect growth and greater water demand for industries, manufacturing, agriculture, and steam generation of electricity (2).

What can be done to use water more efficiently and reduce withdrawal and consumption? Improved agricultural irrigation could reduce withdrawals by between 20 and 30 percent. Such improvements include lined and covered canals that reduce seepage and evaporation;

computer monitoring and scheduling of water releases from canals; a more integrated use of surface waters and groundwaters; night irrigation; improved irrigation systems (sprinklers, drip irrigation); and better land preparation for water application.

Domestic use of water accounts for only about 6 percent of the total national withdrawals. Because this use is concentrated, it poses major local problems. Withdrawal of water for domestic use may be substantially reduced at a relatively small cost with more efficient bathroom and sink fixtures, night irrigation, and drip irrigation systems for domestic plants.

How water supply is perceived by people is also important in determining how much water is used. For example, people in Tucson, Arizona, perceive the area as a desert (which it is) and cultivate many native plants (cactus and other desert plants) in their yards and gardens. Tucson's water supply is all from groundwater, which is being mined (used faster than it is being naturally replenished); the water use is about 605 liters (160 gallons) per person per day. Not far away the people of Phoenix, Arizona, use about 983 liters (260 gallons) of water per person per day. Parts of Phoenix use as much as 3780 liters (1000 gallons) per person per day to water large lawns, mulberry trees, and high hedges! Phoenix has been accused of having an "oasis mentality," concerning water use. Water rates also make a difference. The people in Tucson pay about 75 percent more for water than the people in Phoenix, where the water supply is drawn from the Salt River rather than from groundwater. Water rates in Tucson are structured to encourage conservation, and some industries consider water as a cost-control measure (6). The message here is that since water in the southwestern United States and other locations will be in short supply in the future, we could all do with a little of Tucson's "desert mentality," particularly such large urban areas as Los Angeles and San Diego. Whether Tucson can maintain its desert mentality following importation of Colorado River water remains to be seen.

Considering steam generation of electricity, water removal could be reduced as much as 25–30 percent by different types of cooling towers that use less or no water. Manufacturing and industry might curb water withdrawals by increasing in-plant treatment and recycling of water or by developing new equipment and processes that require less water (2). Because the field of water conservation is changing so rapidly, it is expected that a number of innovations will reduce the total withdrawals of water for various purposes, even though consumption will continue to increase (2).

▼ DAMS, RESERVOIRS, AND CANALS

Our discussion of water supply established that many agricultural and urban areas require water delivered from nearby, and in some cases not so nearby, sources.

To accomplish this a system of water storage and routing by way of canals and aqueducts from reservoirs is needed. The parties interested in water and water development range from government agencies to local water boards and conservation groups. A good deal of controversy often surrounds water development, and the day of developing large projects in the United States without careful environmental review has passed. The resolution of development issues now involves input from a variety of groups that may have very different needs and concerns. These range from agricultural groups who see water development as critical for their livelihood to groups of people whose primary concerns are with wildlife and wilderness preservation. It is a positive sign that the various parties on water issues are now at least able to meet and communicate their desires and concerns.

Dams and Reservoirs

Dams and their accompanying reservoirs are generally designed to be multifunction structures. That is, those who propose the construction of dams and reservoirs point out that reservoirs may be used for activities such as recreation, as well as providing flood control and assuring a more stable water supply. It is important to recognize that it is often difficult to reconcile these various uses at a given site. For instance, water demands for agriculture might be high during the summer, resulting in a drawdown of the reservoir and the production of extensive mud flats. Those interested in recreation find the low water level and the mud flats to be aesthetically degrading, and these effects of high water demand may also interfere with wildlife (particularly fish) by damaging or limiting spawning opportunities. Finally, dams and reservoirs tend to instill a false sense of security to those living below the water retention structures, because dams cannot fully protect us against great floods.

There is little doubt that we will need some additional dams and reservoirs if our present practices of water use are continued. Some existing structures will also need to be heightened. As more people flock to urban areas, water demands there are going to increase and additional water storage will be required. This is particularly true in the more arid parts of the country (including the southern California belt extending eastward into Arizona), where populations are growing rapidly.

Conflicts over construction of additional dams and reservoirs are bound to occur—water developers may view a canyon dam site as a resource for water storage, while other people view it as a wilderness area and recreation site for future generations. The conflict is particularly pointed because good dam sites are often sites of high-quality scenic landscape. Unless water-use patterns change in agricultural and urban areas, however, additional water supply facilities will be a high priority for rapidly growing urban areas, perhaps taking precedence over aesthetic and environmental concerns.

Whenever a dam and reservoir are constructed on a river system, that system is changed forever. The flow of water and sediment is changed, as are the physical and biological habitats and land uses below the dam.

The Grand Canyon of the Colorado River provides a good example of a river's adjustment to the impact of a large dam. In 1963, the Glen Canyon Dam was built upstream from the Grand Canyon. Construction of the dam drastically altered the pattern of flow and channel process downstream. From a hydrologic viewpoint, the Colorado River was tamed. Before the Glen Canyon Dam, the river reached a maximum flow in May or June during the spring snow melt, then flow receded during the remainder of the year, except for occasional flash floods caused by upstream rainstorms. During periods of high discharge, the river had a tremendous capacity to transport sediment (mostly sand and silt) and vigorously scoured the channel. As the summer low flow approached, the stream was able to carry less sediment, so deposition along the channel formed large bars and terraces, known as beaches to people who rafted the river. The high floods also moved large boulders off the rapids that formed because of shallowing of the river where it flows over alluvial fan deposits delivered from tributary canyons. After the dam was built, the mean annual flood (the average of the highest flow each year for a period of years) was reduced from approximately 2,500 cubic meters per second to 800 cubic meters per second, and the ten-year flood (the flood that can be expected on an average of every ten years) was reduced from about 3,500 cubic meters per second to 860 cubic meters per second. On the other hand, the dam did control the flow of water to such an extent that the median discharge actually increased from about 210 cubic meters per second to 350 cubic meters per second. The flow is highly unstable, however, because of fluctuating needs to generate power, and the level of the river may vary by as much as 5 meters per day with a mean daily high discharge of about 570 cubic meters per second and daily low of 130 cubic meters per second. The dam also greatly affected the sediment load of the Colorado River through the Grand Canyon: the median suspended sediment concentration was reduced by a factor of about 200 times immediately downstream from the dam. A lesser reduction in sediment load occurred farther downstream because tributary channels continued to add sediment to the channel (7).

The change in hydrology of the Colorado River in the Grand Canyon has greatly changed the river's morphology. The rapids may be becoming more dangerous because large floods no longer occur—flows which had previously moved some of the large boulders farther downstream (Figure 11.19). In addition, some of the large sand bars or beaches are eroding because, since the

(a)   (b)

Figure 11.19
Crystal Rapid of the Colorado River in the Grand Canyon before and after the construction of the Glen Canyon Dam. Photograph (a) taken in 1963 shows the rapid as it existed prior to the building of the dam. (Bureau of Reclamation photograph by Al Turner.) Photograph (b) taken in 1967 shows the effect of a 1965 flash flood. The debris and rock delivered from Crystal Creek (left) forms a fan-shaped deposit in the Colorado River. This deposit has made the rapid more difficult to negotiate, and the debris will remain in the river for a relatively long time as large floods that would normally remove some of the debris are diminished by the presence of the dam and reservoir. (Bureau of Reclamation photograph by Mel Davis.)

river is deficient in sediment below the dam, it is picking up more sediment and thus causing erosion.

Changes in the river flow (mainly deleting the high flows) have also resulted in vegetational shifts. Prior to the dam, three nearly parallel belts of vegetation were present on the slopes above the river. Adjacent to the river and on sand bars grew ephemeral plants, which were scoured out by yearly spring floods. Above the high-water line were clumps of thorned trees (mesquite and catclaw acacia) mixed with cactus and Apache plume. Higher yet could be found a belt of widely spaced brittle brush and barrel cactus (8). Closing the dam in 1963 tamed the spring floods for 20 years, and a new belt of exotic plants, including tamarisk (salt cedar) and indigenous willow, became established along the river banks. In June of 1983 a record snow melt in the Rocky Mountains forced the release of about 2,500 cubic meters per second, which is about three times that normally released and about the same as an average spring flood prior to the dam. The resulting flood scoured the river bed and banks releasing stored sediment that replenished the sediment on sand bars and scoured out or broke off some of the tamarisk and willow stands. The effect of the large release of water was beneficial to the river environment and emphasizes the importance of

the larger events (floods) to maintaining the system in a more natural state. Perhaps management of rivers below some dams should call for periodic release of large flows to help cleanse the system. Such a management option would obviously require floodplain management to restrict land uses that are prone to flood damage. Unfortunately, dam construction often encourages such land uses.

One final impact is the increase in the number of people rafting through the Grand Canyon. Although rafting is now limited to 15,000 persons per year, the long-range impact on canyon resources is bound to be appreciable. Prior to 1950, fewer than 100 explorers and river runners had made the trip through the canyon. We must concede that the Colorado River of the 1970s and 1980s is a changed river and, in spite of the 1983 flood which pushed back some of the changes, it cannot be expected to return to what it was before the construction of the dam (7, 8).

Canals

Water from upstream reservoirs may be routed to downstream needs by way of natural watercourses or by canals and aqueducts.

Canals, whether lined or unlined, are often attractive nuisances to people and animals. Where they flow through urban areas, drownings are an ever-present threat. When they are unlined, canals may lose a good deal of water to the subsurface flow system. While it may be argued that this is a form of artificial groundwater recharge, it may be an inefficient one because canals may cross areas with little potential for groundwater development or areas of poor groundwater quality. In these cases, water seeping from unlined canals is essentially lost water.

The construction of canal systems, especially in developing countries, has led to serious environmental problems. For example, when the High Dam at Aswan was completed in 1964, a series of canals was needed to convey the water to agricultural sites. The canals became infested with snails that carry the dreaded disease schistosomiasis (snail fever). This disease has always been a problem in Egypt but the swift currents of Nile flood waters flushed out the snails each year. The tremendous expanse of waters in irrigation canals now provides happy homes for these snails. The disease is debilitating and so prevalent in parts of Egypt that virtually the entire population of some areas may be affected by it. The Egyptian canals are also a home for mosquitoes, one particular variety of which carries a form of malaria that killed 100,000 Egyptians in 1942.

Reservoirs and canal systems are being planned in a variety of environments around the world today. Environmental concern (and laws) in the United States ensures that important environmental review will take place. This may not be true in many developing countries, where attention to environmental concerns is not as high a priority as, for instance, the production of food. In such areas construction of large, long canals may considerably alter land use and the biologic environment by producing new and different water systems and barriers to migration of wildlife. This is not to say that water development in developing countries is not necessary, but it does emphasize the need for environmental concern at the ecosystem level when planning and developing water resources. At the very least, we can share with developing countries the mistakes we have made in the past so that they need not be repeated in the future. Water development and environmental concern are not necessarily incompatible. However, tradeoffs must be made if a quality environment is to be preserved.

▼ WATER MANAGEMENT

Management of water resources is a complex issue that will become more difficult in coming years as the demand for water increases. While this will be especially true in the southwestern United States and other arid and semiarid parts of the world, New York and Atlanta, among other U.S. cities, are also facing future water supply problems. Options open to people who want to mini-

mize potential water supply problems include locating alternative supplies, managing existing supplies better, or controlling growth. In some areas, locating new supplies is unlikely, and serious consideration is being given to ideas as original as towing icebergs to coastal areas where fresh water is needed. It seems apparent that water will become much more expensive in the future and, if the price is right, many innovative programs are possible. Recently one city in southern California began to consider a plan to import Colorado water via a private company that owns water rights there. Such thoughts and plans are causing concern among government agencies who manage water resources through regulations that control allocation and price. Cities in need of water are beginning to treat water like a commodity that can be bought and sold on the open market, like oil or gas. If cities are willing to pay for water and are allowed to somehow avoid current water regulation, then allocation and pricing as it is now known will change; if the cost rises enough, "new water" from a variety of sources may become available. For example, irrigation districts (water managers for an agricultural area) may contract with cities to supply water to urban areas. They could do this without any less water being available for crops by using conservation measures to minimize present water loss through evaporation and seepage from unlined canals. Currently most irrigation districts do not have the capital to finance expensive conservation methods, but money from cities for water could finance the projects.

Luna Leopold (a leader in the study of rivers and water resources) has suggested that a new philosophy of water management is needed—one based on geologic, geographic, and climatic factors as well as on the traditional economic, social, and political factors. He argues that the management of water resources cannot be successful so long as it is naively perceived primarily from an economic and political standpoint. However, this is how water use is approached. The term "water use" is appropriate because we seldom really "manage" water (9). The essence of Leopold's water management philosophy is summarized in this section.

Surface water and groundwater are both subject to natural flux with time. In wet years surface water is plentiful, and the near-surface groundwater resources are replenished. During these years we hope that our flood-control structures, bridges, and storm drains will withstand the excess water. Each of these structures is designed to withstand a particular flow (for example, the 20-year flood), which, if exceeded, may cause damage or flooding.

All in all, we are much better prepared to handle floods than water deficiencies. During dry years, which must be expected even though they may not be accurately predicted, we should have specific strategies to minimize hardships. For example, subsurface waters in various locations in the western United States are either too deep to be economically extracted or have marginal water quality. These waters may be isolated from the present hydrologic cycle and therefore are not subject to natural recharge. Such water might be used when the need is great. However, advance planning to drill the wells and connect them to existing water lines is necessary if they are to be ready when the need arises. Another possible emergency plan might involve the treatment of wastewater. Reuse of water on a regular basis might be too expensive, but advance planning to reuse treated water during emergencies might be wise (9).

When dealing with groundwater that is naturally replenished in wet years, we should develop plans to use surface water when available and not be afraid to use groundwater in dry years; that is, the groundwater might be pumped out at a rate exceeding the replenishment rate in dry years. During wet years natural recharge and artificial recharge (pumping excess surface water into the ground) will replenish the groundwater resources. This water management plan recognizes that excesses and deficiencies in water are natural and can be planned for.

▼ WATER POLLUTION

Water pollution refers to degradation of water quality as measured by biological, chemical, or physical criteria. Degradation of water is generally judged in terms of the intended use of the water, departure from norm, effects on public health, or ecologic impacts. From a public health or ecologic view, a pollutant is any biological, physical, or chemical substance in which an identifiable excess is known to be harmful to other desirable living organisms. Thus excessive amounts of heavy metals, certain radioactive isotopes, phosphorus, nitrogen, sodium, and other useful (even necessary) elements, as well as certain pathogenic bacteria and viruses, are all pollutants. In some instances, a material may be considered a pollutant to a particular segment of the population although not harmful to other segments. For example, excessive sodium as a salt is not generally harmful, but it is to some people on diets restricting intake of salt for medical purposes.

There are many different materials that may pollute surface water or groundwater. Our discussion here will focus on selected aspects of biochemical oxygen demand, fecal coliform bacteria, nutrients, hazardous chemicals, oil, heavy metals, radioactive materials, and sediment.

▼ SELECTED WATER POLLUTANTS

Biochemical Oxygen Demand (BOD)

Dead organic matter in streams decays. Bacteria carrying out this decay require oxygen. If there is enough bacterial activity, the oxygen in the water can be reduced to levels so low that fish and other organisms die. A stream without oxygen is a dead stream for fish and many organisms we value. The amount of oxygen required for such biochemical decomposition is called the **biochemical oxygen demand (BOD)**, a commonly used measure in water quality management. BOD is measured as milligrams per liter of oxygen consumed over five days at 20°C. Dead organic matter is contributed to streams and rivers from natural sources (such as dead leaves from a forest) as well as from agriculture and urban sewage. Approximately 33 percent of all BOD in streams results from agricultural activities, but urban areas, particularly those with sewer systems that combine sewage and storm water runoff, may add considerable BOD to streams during floods, when sewers entering treatment plants can be overloaded and overflow into streams, producing pollution events.

The Council on Environmental Quality defines the threshold for water pollution as a dissolved oxygen content of less than 5 milligrams per liter of water. The diagram in Figure 11.20 illustrates the effect of BOD on dissolved oxygen content in a stream when raw sewage is introduced as a result of an accidental spill. Three zones are recognized: the *pollution zone,* with a high BOD and reduced dissolved oxygen content as initial decomposition of the waste begins; an *active decomposition zone,* where the dissolved oxygen content is at a minimum owing to biochemical decomposition as the organic waste is transported downstream; and a *recovery zone,* in which the dissolved oxygen increases and the BOD is reduced because most oxygen-demanding organic waste from the input of sewage has decomposed and natural stream processes replenish the water with dissolved oxygen. All streams have some capability to degrade organic waste after it enters the stream. Problems result when the stream is overloaded with biochemical oxygen-demanding waste, overpowering the stream's natural cleansing function.

Fecal Coliform Bacteria

Disease-carrying microorganisms are important biological pollutants. Among the major water-borne human diseases are cholera and typhoid. Because it is often difficult to monitor the disease-carrying organisms directly, we use the count of human fecal coliform bacteria as a common measure of biological pollution and a standard measure of microbial pollution. These common, harmless forms of bacteria are normal constituents of human intestines and are found in all human waste. The threshold used by the Council of Environmental Quality for pollution is 200 cells of fecal coliform bacteria per 100 milliliters of water.

Nutrients

Nutrients released by human activity may lead to water pollution. Two important nutrients that can cause prob-

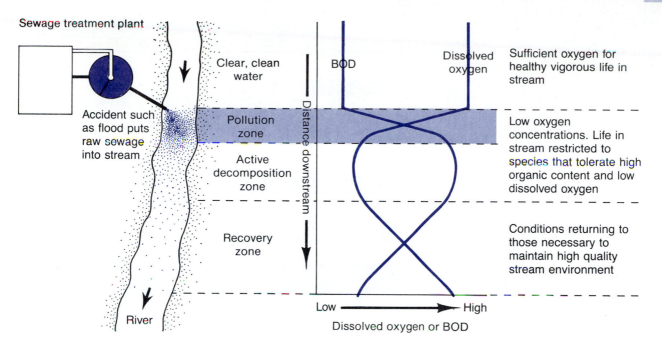

Figure 11.20
Relationship between dissolved oxygen and biochemical oxygen demand (BOD) for a stream following the input of sewage.

lems are phosphorus and nitrogen, both of which are released from a variety of sources related to land use (Figure 11.21). Forested land has the lowest concentrations of phosphorus and nitrogen in stream waters. Increase in these nutrients is observed in urban streams because of the introduction of fertilizers, detergents, and the products of sewage treatment plants. The highest concentrations of phosphorus and nitrogen are found in agricultural areas—sites of such sources as fertilized farm fields and feedlots (2).

Oil

Oil discharged into surface water (usually the ocean) has caused major pollution problems. Several large oil spills from submarine oil drilling operations have occurred in recent years, such as the 1969 oil spill in the Santa Barbara Channel and the 1979 Yucatan Peninsula spill in Mexico. The latter is one of the world's largest spills to date, spewing out about 3 million barrels of oil before being capped in 1980. Both spills were caused by an oil well blowing out, and both caused damage to beaches and marine life when the oil drifted ashore. Favorable winds averted a major disaster on Texas beaches and inland wetlands, as the oil only touched the shore briefly. Santa Barbara was not so lucky. Beaches were covered with oil and waterfowl were killed.

Oil is also released into the marine environment from oil tanker accidents. Just after midnight on March 24, 1989, the *Exxon Valdez* ran aground on Bligh Reef, 40 kilometers south of Valdez in Prince William Sound.

Crude oil delivered to Valdez via the Trans-Alaskan Pipeline poured out of the ruptured tanks of the vessel at a rate of approximately 20,000 barrels per hour. The *Exxon Valdez* was loaded with 1.2 million barrels of North Slope crude, and of this, more than 250,000 barrels (11 million gallons) gushed from the hold of the 300-meter tanker. Oil remaining in the *Exxon Valdez* was loaded into another tanker (10).

The oil spilled into what is considered one of the most pristine and ecologically rich marine environments of the world (10), and the accident is now known as the worst oil spill in the history of the United States. Many species of fish, birds, and marine mammals are present in Prince William Sound, and the long-term impact of the oil spill on the environment is difficult to ascertain. Following the spill the oil spread over a large area. Figure 11.22 shows the extent of oil sheens, tar balls, and mousse that was suspected to have come from the *Exxon Valdez* as of August 10, 1989. Mousse is a thick, weathered patch of oil with the consistency of a soft pudding that often washes up on beaches. The area covered by the spill was very large. Figure 11.22b compares the area of the spill as of August 10 to the eastern seaboard of the United States. Notice that it would extend from Massachusetts to North Carolina!

Soon after the accident the governor of Alaska declared Prince William Sound a disaster area and applied for federal assistance. Cleanup work on the coastline posed enormous problems to workers attempting the project. Photographs and videotapes of the work suggest an almost futile attempt to clean individual

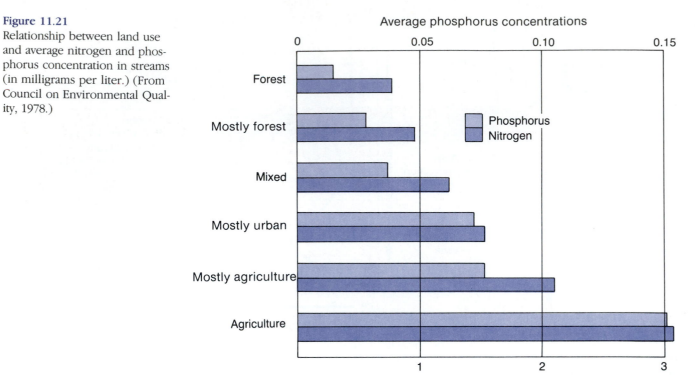

Figure 11.21
Relationship between land use and average nitrogen and phosphorus concentration in streams (in milligrams per liter.) (From Council on Environmental Quality, 1978.)

Average phosphorus concentrations

0   0.05   0.10   0.15

Forest

Mostly forest

Mixed

Mostly urban

Mostly agriculture

Agriculture

Phosphorus
Nitrogen

1   2   3

Average nitrogen concentrations

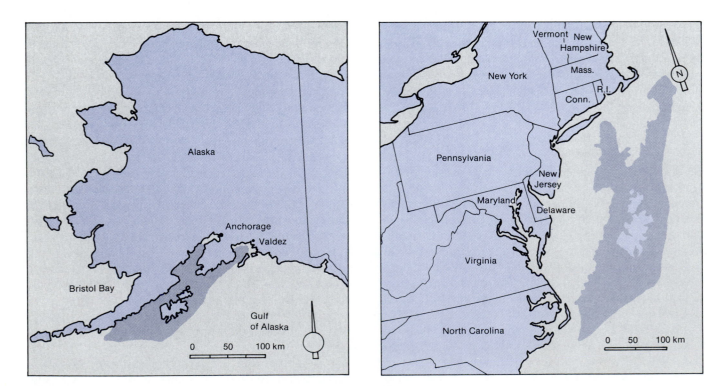

Figure 11.22
(a) Extent of Alaskan oil spill of 1989. (b) Area of 1989 Alaskan oil spill compared to the eastern coast of the United States. (From *Alaska Fish and Game,* July-August 1989, vol. 21, no. 4.)

pebbles on beaches. The spill completely disrupted the lives of people who work in the vicinity of Prince William Sound, and only after the passage of time will the true impacts of the spill be fully understood. Certainly the short-term impacts were very significant indeed. Those included disruption of the commercial fisheries, sport fisheries, and tourism as well as loss of sea birds and mammals. Interruption of the flow of North Slope crude resulted in an almost immediate increase in the price of oil to the lower 48 states. It is hoped that the lessons learned from the *Exxon Valdez* spill will be used to develop better management strategies for both the shipment of crude oil and the emergency plans to minimize environmental degradation.

On June 8, 1990, approximately 80 kilometers from Galveston, Texas, explosions and fire broke out on the supertanker *Mega Borg*. The accident released approximately 4.3 million gallons of oil into the Gulf of Mexico, making it one of the nation's largest spills. Shortly after the spill, numerous smaller boats were used to skim the oil that trailed from the stricken tanker. The oil, which was a light African crude, evaporated or burned and by June 16 less than 14,000 gallons were reported to remain in the water.

An experimental approach to oil abatement was used in the *Mega Borg* spill. The Coast Guard mixed oil-consuming bacteria with nutrient-enriched seawater and spread it on a section of the oil slick. The process, which is known as bioremediation, had never before been tried on an oil spill in the open sea. However, laboratory experiments had suggested the technique might be successful. Thus the *Mega Borg's* accident will go down in history as being the first oil spill to be treated by oil-eating microbes released onto the spilled oil.

Military activity has become another cause of pollution of the marine environment by oil. The huge oil spill in the Persian Gulf during the 1991 war released an unknown volume of oil into a fragile environment. It may be the world's largest spill, and it is certainly the largest deliberate spill.

Heavy Metals

Heavy metals such as mercury, zinc, and cadmium are dangerous pollutants and are often deposited with natural sediment in the bottoms of stream channels. If these metals are deposited on floodplains, then the heavy metals may become incorporated in plants, food crops, and animals. If they are dissolved and the water is withdrawn for agriculture or human use, heavy-metal poisoning can result. Heavy metals are discussed in detail in Chapter 13.

Thermal Pollution

A heating of waters, primarily from hot-water emission from industrial operations and power plants, causes thermal pollution. There are several problems with heated water. Even water several degrees warmer than the surrounding water holds less oxygen. Warmer water favors different species than cooler water and may increase growth rates of undesirable organisms, including certain water plants and fish. On the other hand, the warm water may attract and allow better survival of certain desirable fish species, particularly during the winter.

Hazardous Chemicals

Many synthetic organic and inorganic compounds are toxic to people and other living things. When these materials are accidentally introduced into surface or subsurface waters, serious pollution may result. The complex problem of hazardous chemicals and their management is discussed in detail in Chapter 12.

Radioactive Materials

Radioactive materials in water may be dangerous pollutants. Of particular concern are possible effects to people, other animals, and plants of long-term exposure to low doses of radioactivity. Chapters 12 and 13 discuss radiation in terms of waste disposal and environmental effects.

Sediment

Sediment consists of rock and mineral fragments ranging in size from sand particles less than 2 millimeters in diameter to silt, clay, and even finer colloidal particles. By volume, it is our greatest water pollutant. It is truly a resource out of place. It depletes a land resource (soil) at its site of origin, reduces the quality of the water resource it enters, and may deposit sterile materials on productive croplands or other useful land. Sediment pollution is discussed in detail in Chapter 4.

The above discussion of potential pollutants in water certainly is not complete. For some of the pollutants we have discussed, Table 11.7 lists the thresholds used by the Council on Environmental Quality as indicators of water quality. As we learn more about water pollution and its effects on the environment, threshold concentrations and the list of potential pollutants will certainly change.

▼ SURFACE WATER POLLUTION

Pollution of surface waters occurs when too much of an undesirable or harmful substance flows into a body of water, exceeding the natural ability of that ecosystem to utilize or remove the undesirable material or convert it to a harmless form.

Water pollutants are categorized as emitted from point or nonpoint sources. *Point sources* are discrete and confined, such as pipes that empty into streams or rivers

Table 11.7
Thresholds used by the Council on Environmental Quality in analyzing the nation's water quality.

Indicator	Abbreviation	Threshold Level
Fecal coliform bacteria	FC	200 cells/100 ml[a]
Dissolved oxygen	DO	5.0 mg/l[b]
Total phosphorus	TP	0.1 mg/l[c]
Total mercury	Hg	2.0 μg/l[d]
Total lead	Pb	50.0 μg/l[d]
Biochemical oxygen demand	BOD	5.0 mg/l[e]

1 = liter; ml = milliliter; mg = milligram; μg = microgram.

[a]Criteria level for "bathing waters" from EPA "Redbook."

[b]Criteria level for "good fish populations" from EPA "Redbook."

[c]Value discussed for "prevention of plant nuisances in streams or other flowing waters not discharging directly to lakes or impoundments" in EPA "Redbook."

[d]Criteria level for "domestic water supply (health)" from EPA "Redbook." Criteria level for preservation of aquatic life is much lower.

[e]Value chosen by CEQ.

Source: Council on Environmental Quality, 1979.

from industrial or municipal sites. In general, point-source pollutants from industries are controlled through on-site treatment or disposal and are regulated by permit. Municipal point sources are also regulated by permit. In older cities in the northeastern and Great Lakes areas of the United States, most point sources are outflows from combined sewer systems. These sewer systems combine storm-water flow with municipal waste. During heavy rains, urban storm runoff may exceed the capacity of the sewer system, causing it to back up and overflow, delivering pollutants to nearby surface waters.

Nonpoint sources are diffused and intermittent and are influenced by such factors as land use, climate, hydrology, topography, native vegetation, and geology. Common urban nonpoint sources include urban runoff from streets or fields; such runoff contains all sorts of pollutants, from heavy metals to chemicals and sediment. Rural sources of nonpoint pollution are generally associated with agriculture, mining, or forestry. Nonpoint sources are difficult to control.

Consider Lake Washington near Seattle, Washington, as a case history. When Seattle was a small city, the pollution of streams flowing into Lake Washington had little noticeable effect. Natural biological degradation and inorganic processes were sufficient to take care of the effluents. By the 1950s, however, the city had become so large that the rate at which effluents were being dumped into the lake greatly exceeded the lake's capacity to remove them or transform them into harmless forms.

Given situations such as Lake Washington, we have three ways to deal with water pollution: reduce the sources; transport the pollutants to some place where they will not do damage (as is being done at Lake Tahoe

in California); or convert the pollutants to harmless forms. In the Lake Washington case, the second method was the easiest and most practical because of a unique situation—that is, the nearness of Puget Sound with its rapid rate of flushing with the Pacific Ocean. This method was less expensive than adding improved treatment processes to remove the phosphorus. The effluent flowing from the city sewage to the lake had already been subjected to secondary treatment, meaning that 90 percent or more of the BOD had been removed; it was essentially free of disease-causing organisms and major organic compounds.

Few other cities are as fortunate as Seattle. For example, exporting and treating sewage in the Lake Tahoe basin is expensive but necessary to help maintain the high quality of water in Lake Tahoe. There are other success stories in the treatment of water pollution. One of the most notable is the cleanup of the Thames River in England. For centuries London's sewage had been dumped into that river, and few fish were to be found downstream in the estuary. In recent decades, however, improvement in water treatment has led to the return of a great number of species of fish—some not seen in the river in centuries.

Since the 1960s a serious attempt has been made in the United States to reduce water pollution and thereby increase water quality. The basic assumption is that people have a real desire for safe water to drink, to swim in, and to use in agriculture and industry. At one time water quality near major urban centers was considerably worse than it is today; in at least one instance, a U.S. river was inadvertently set on fire. In recent years the number of success stories has been very encouraging; perhaps the best known is the Detroit River. In the 1950s and the early 1960s the Detroit River was considered a dead river, having been an open dump for sewage, chemicals, garbage, and urban trash. Tons of phosphorus were discharged each day into the river, and a film of oil up to 0.5 centimeter thick was often present. Aquatic life was damaged considerably, and thousands of ducks and fish were killed. Although today the Detroit River is not a pristine stream, considerable improvement has resulted from industrial and municipal pollution control. Oil and grease emissions were reduced by 82 percent between 1963 and 1975, and the shoreline is usually clean. Phosphorus and sewage discharges have also been greatly reduced. Fish once again are found in the Detroit River. Other success stories include New Hampshire's Pemigewasset River, North Carolina's French Broad River, and the Savannah River in the southeastern United States. These examples are evidence that water pollution abatement has positive results (11).

The Hudson River assessment and cleanup of PCBs (polychlorinated biphenyls, with chemical structure similar to DDT and dioxin) is another good example of the determination of people to clean up our rivers. PCBs were used mainly in electrical capacitors and transformers; discharge of the chemicals from two outfalls on the Hudson River started about 1950 and terminated in 1977. Approximately 295,000 kilograms of PCBs are believed to be present in Hudson River sediments. Concentrations in the sediment are as high as 1000 parts per million near the outfalls, compared to less than 10 parts per million several hundred kilometers downstream at New York City (12). The U.S. Food and Drug Administration permits less than 2.5 parts per million PCBs in dairy products, while the New York State limit for drinking water is 0.1 part per billion. PCBs are carcinogenic and are known to cause disturbances of the liver, nervous system, blood, and immune response system in people. Furthermore, they are nearly indestructible in the natural environment and become concentrated in the higher parts of the food chain—thus the concern! Water samples in the 240-kilometer tidal reach of the Hudson River have yielded average PCB concentrations ranging from 0.1 to 0.4 part per billion, but PCBs are concentrated to much higher levels in some fish. As a result fishermen on the lower Hudson have suffered a significant economic impact from the contamination because nearly all commercial fishing was banned, and sport fishing was greatly reduced (12, 13).

The cleanup plan for the Hudson River will cost nearly $30 million and involves removing and treating contaminated river sediment. Dredging will be done in "hot spots," where the concentration of PCBs is greater than 50 parts per million. The dredging will greatly reduce the time necessary for the river to clean itself by the natural process of sediment transport to the ocean.

▼ GROUNDWATER POLLUTION

Approximately one-half of all people in the United States today depend on groundwater as their source of drinking water. Because most of us have long believed that groundwater is in general quite pure and safe to drink, the fact that groundwater may be quite easily polluted by any one of several sources (Table 11.8) is alarming to many people. In addition, the pollutants, even the very toxic ones, may be difficult to recognize. One of the best known examples of groundwater pollution is the Love Canal near Niagara Falls, New York, where burial of chemical wastes has caused serious water pollution and health problems.

Unfortunately, Love Canal is not an isolated case. Hazardous chemicals have been found or are suspected to be in groundwater supplies in nearly all parts of the world, developed and developing countries alike. Developed industrial countries produce thousands of chemicals; many of these, particularly pesticides, are exported to developing countries, where they protect crops that eventually are imported by the same industrial countries, completing a circle. For example, Costa Rica imports several pesticides, including DDT, aldrin, endrin, and chlordane, that are banned or heavily restricted in the United States. Most of these pesticides are used on crops destined for export to richer nations (14).

In the United States today, only a small portion of the groundwater is known to be contaminated. Nevertheless, several million people have used that water, and the problem is growing as testing of groundwater becomes more common. For example, Atlantic City, New Jersey, and Miami, Florida, are two eastern cities threatened by polluted groundwater that is slowly migrating toward their wells. It is estimated that 75 percent of the 175,000 known waste disposal sites in the country may be

Table 11.8
Common sources of groundwater pollution and/or contamination.

Leaks from storage tanks and pipes

Leaks from waste disposal sites such as landfills

Seepage from septic systems and cesspools

Accidental spills and seepage (train or truck accidents, for example)

Seepage from agricultural activities such as feedlots

Intrusion of salt water into coastal aquifers

Leaching and seepage from mine spoil piles and tailings

Seepage from spray irrigation

Improper operation of injection wells

Seepage of acid water from mines

Seepage of irrigation return flow

producing plumes of hazardous chemicals that are migrating into groundwater resources. Because many of the chemicals are toxic or suspected carcinogens, it appears we have been conducting a large-scale experiment on the effects of chronic low-level exposure of people to potentially harmful chemicals. Unfortunately the final results of the experiment won't be known for many years (15). Preliminary results suggest we had better act now before a hidden time bomb of health problems explodes.

The hazard presented by a particular groundwater pollutant depends upon several factors such as volume of pollutant discharged, concentration or toxicity of the pollutant in the environment, and degree of exposure of people or other organisms (16).

Significant differences in the physical, geologic, and biologic environments are associated with groundwater pollution as compared with those of surface-water pollution. In the latter, hydraulics of flow and availability of oxygen and sunlight (which may tend to degrade pollutants), along with the rapidity with which processes of dilution and dispersion of pollutants occur, are markedly different than for groundwater, where the opportunity for bacterial degradation of pollutants is generally confined to the soil or a meter or so below the surface. Furthermore, the channels through which groundwater moves are often very small and variable. For these reasons, the rate of movement is much reduced (except, perhaps, in large solution channels within limestones), and the opportunity for dispersion and dilution is limited. In addition, the lack of oxygen in groundwater kills aerobic types of microorganisms (that require free oxygen to live and grow) but may provide a happy home for anaerobic varieties (that prosper in an oxygen-deficient environment).

The ability of most soils and rocks physically to filter out solids, including pollution solids, is well recognized. However, this ability varies with different sizes, shapes, and arrangements of filtering particles, as evidenced in the use of selected sands and other materials in water filtration plants. Also known, but perhaps not so generally, is the ability of clays and other selected minerals to capture and exchange some elements and compounds when they are dissociated in solutions as positively or negatively charged elements or compounds. Such exchanges, along with sorption and precipitation processes, are important in the capture of pollutants. These processes, however, have definable units of capacity and are reversible. They also can be overlooked easily in designing facilities to correct pollution problems by relying on soils and rocks of the geologic environment for treatment. This oversight, which can result in possible groundwater pollution, is especially significant in land application of wastewaters.

Much conjecture, debate, and research have gone on about groundwater pollution. The principal concerns of groundwater pollution are the introduction of chemical elements, compounds, and microorganisms that do not occur naturally in aquifers that can be or are used for drinking water. Because of the degradation of the aquifer—and also because of difficulty in detection, the long-term residency, and the difficulty and expense of aquifer recovery—strong arguments can be made that no wastes or possible pollutants should be allowed to enter any part of the groundwater system. This is an impossible dream. Rather, the answer lies in knowing more about how natural processes treat wastes so that when soil and rocks are not capable of treating, storing, or recycling wastes, we can develop processes to make the pollutants treatable, storable, or recyclable.

Studies and case histories of groundwater pollution generally show that microorganisms seldom survive for distances greater than 30 meters because they either are filtered out or are killed by the hostile environment (17). Greater survival rates appear to occur in those geologic environments where porosity and permeability have high values, such as gravel, coarse sand, or fracture and cavern systems in limestone (18, 19). The spectrum of possible problems includes contamination by microorganisms from sewage disposal (20, 21), chromium pollution from industrial processes (22), contamination from oil-field brines and related industrial wastes (23), pollution caused by disposal of pharmaceutical wastes (24), pollution caused by winter road salt, pollution from waste disposal facilities, and pollution from intensive use of insecticides and pest controls for agricultural and landscaping purposes.

Aquifer pollution is not solely the result of disposal of wastes on the land surface or in the ground. Overpumping or mining of groundwater so that inferior waters migrate from adjacent aquifers or the sea also cause contamination problems. Hence, human use of public or private water supplies can accidentally result in aquifer pollution. Intrusion of salt water into freshwater supplies has caused problems in coastal areas of New York, Florida, and California, among other areas.

Figure 11.23 illustrates the general principle of salt-water intrusion. The groundwater table generally is inclined toward the ocean, while a wedge of salt water is inclined toward the land. Thus, with no confining layers, salt water near the coast may be encountered at depth. Because fresh water is slightly less dense than salt water (1.000 compared to 1.025 grams per cubic centimeter), a column of fresh water 41 centimeters high is needed to balance 40 centimeters of salt water. A more general relationship is that the depth to salt water below sea level is 40 times the height (H in Figure 11.23) of the water table above sea level. When wells are drilled, a cone of depression can develop in the freshwater table which may allow intrusion of salt water as the interface between fresh and salt water rises in response to the loss of mass of fresh water.

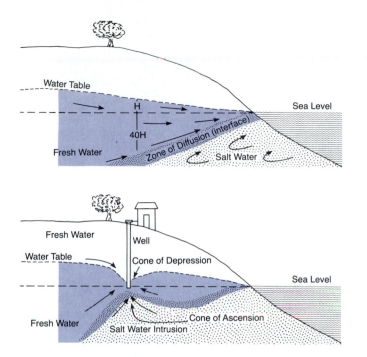

Figure 11.23
How salt water intrusion might occur. The upper drawing shows the groundwater system near the coast under natural conditions and the lower shows a well with both a cone of depression and a cone of ascension. If pumping is intensive, the cone of ascension may be drawn upward, delivering salt water to the well. The "*H*" and "40 *H*" represent the height of the fresh water table above sea level and the depth of salt water below sea level respectively.

Long Island, New York, provides a good example of an area with groundwater problems. Two counties on the island, Nassau and Suffolk, with a population of several million people, are entirely dependent on groundwater for their water supply. Two major problems associated with groundwater in Nassau County are intrusion of salt water and shallow-aquifer contamination (25). Figure 11.24 shows the general movement of groundwater under natural conditions for Nassau County. Salty groundwater is restricted from inland migration by the large wedge of fresh water moving beneath the island.

Groundwater resources of Nassau County are in two main aquifers. The upper aquifer is composed of young glacial deposits that yield large amounts of water at depths less than 30 meters. Below the glacial deposits are older marine sedimentary rocks consisting of interbedded sands, clays, and silts. Most of the fresh water in Nassau County is pumped from this lower aquifer, from sandy beds at depths below 30 meters. Most of the water-bearing sands are confined by overlying clay and silt beds of low permeability; thus the water is under artesian pressure, which causes the water in wells to rise to within 15 meters of the surface. In terms of the total volume of water and number of people who use it, the groundwater resource for Long Island is one of the world's largest (25).

In spite of the huge quantities of water in Nassau County's groundwater system, intensive pumping in recent years has caused water levels to decline as much as 15 meters in some areas. As groundwater is removed near coastal areas, the subsurface outflow to the ocean decreases, allowing salt water to migrate inland. Salt water intrusion in the deep aquifer has occurred in Nassau County. The mechanism of salt intrusion is more complex than that ideally shown in Figure 11.23. As fresh water from sandy beds in the deep aquifer is intensely pumped, salt water is drawn inland as a series of narrow wedges. Although the problem of salt water intrusion is not yet widespread, the salt water front is being carefully monitored as part of a comprehensive management program.

The most serious groundwater problem on Long Island is shallow-aquifer pollution associated with urbanization. Sources of pollution in Nassau County include urban runoff, household sewage from cesspools and septic tanks, salt used to deice highways, and industrial and solid waste. These pollutants enter surface waters and then migrate downward, especially in areas of intensive pumping and declining groundwater levels. Figure 11.25 shows the extent of high concentration of dissolved nitrate in deep groundwater zones. The greatest concentrations are located beneath densely populated urban zones where water levels have dramatically declined and where nitrates from such sources as cesspools, septic tanks, and fertilizers are routinely introduced into the hydrologic environment.

Drilling of deep wells, such as those for scientific and petroleum exploration, also can cause degradation of freshwater aquifers. Some wells allow considerable material to enter freshwater aquifers. This was particularly a problem during the early phases of petroleum and other well drilling (especially for salt), when little was known about the depths at which fresh waters could occur in many parts of the United States and throughout the world.

▼ GROUNDWATER TREATMENT

Correction of aquifer and vadose zone contamination is not impossible, though it may be a complex and expensive problem requiring careful evaluation and treatment. Important steps involved in correcting a groundwater pollution problem are:

1. *Characterizing the geology.* This is particularly important because the existence of more permeable buried channels, soil macropores, and geologic structures, such as fractured, folded, and faulted rocks, may be the dominant factors that control the direction of groundwater flow.

Figure 11.24
The general movement of fresh groundwater for Nassau County, Long Island, New York. (From U.S. Geological Survey Professional Paper 950, 1978.)

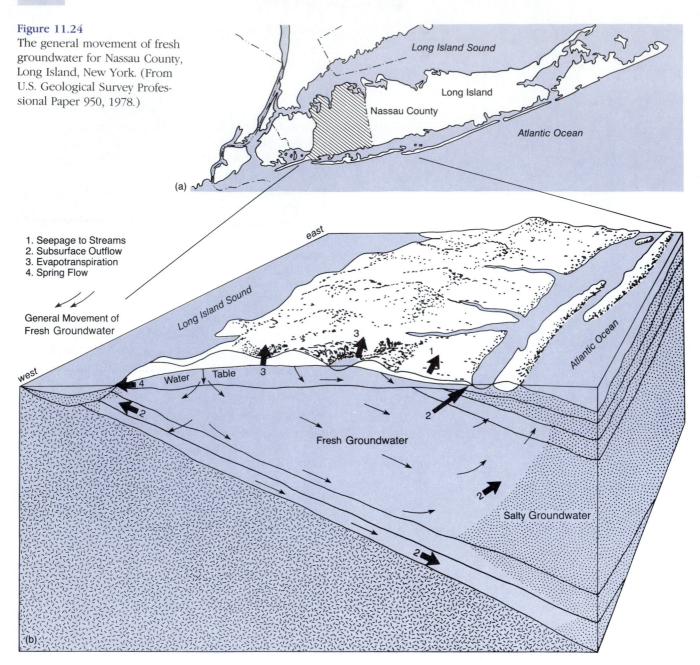

1. Seepage to Streams
2. Subsurface Outflow
3. Evapotranspiration
4. Spring Flow

General Movement of Fresh Groundwater

2. *Characterizing the hydrology.* Factors such as the depth to groundwater, its direction of flow, and rate of flow must be determined. Characterizing the hydrology also involves looking at relationships between surface water and groundwater processes that affect the site.

3. *Identifying contaminants present and the transport processes.* Contaminants are identified through careful site evaluation and gathering of samples. Some contaminants, such as gasoline, are floaters; that is, most of the gasoline will be found on top of the water table because it is lighter than water. However, some components in gasoline are soluble in water, so there will also be a dissolved phase below the water table. On the other hand, contaminants such as trichloroethylene (TCE), which is a dry-cleaning solvent that is heavier than water, will sink rather than float. Other pollutants, such as some salts, are very soluble in water and will move with the general flow of the groundwater environment.

4. *Initiating the treatment process.* Table 11.9 briefly outlines some of the methods for treating groundwater and vadose zone water. The specific treatment selected depends upon variables such as type of contaminant, method of transport, and characteristics of the local environment, such as depth to water table and geologic characteristics.

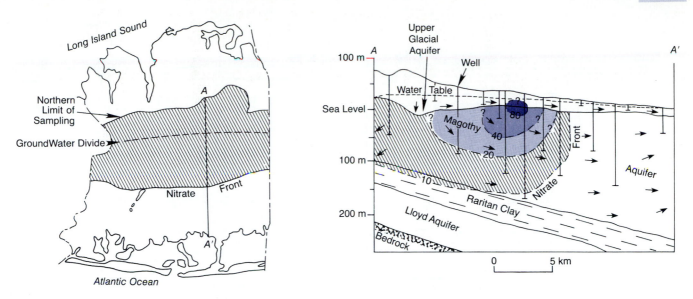

Figure 11.25
Extent of high concentration of dissolved nitrate in groundwater zone, Nassau County, Long Island, New York. The greatest concentrations are located beneath densely populated urban zones where water levels have dramatically declined and nitrates are more abundant due to urban waste disposal and horticulture practices. Contours shown in milligrams per liter of dissolved nitrate. (From U.S. Geological Survey Professional Paper 950, 1978.)

As an example of groundwater treatment consider Figure 11.26. Figure 11.26a shows a site with a service station and underground gasoline tank that is leaking. Most of the contaminant is floating on top of the water table but some is also dissolved and moves with the groundwater. The direction of the migrating vapor phase is away from the leakage plume. Figure 11.26b shows the same location after a system consisting of dewatering wells and a vapor extractor well has been installed. The dewatering wells locally lower the groundwater table and the vapor extraction well, which uses a vacuum pump, collects the contaminant in a vapor phase where it may be treated. Leaking underground gasoline tanks, such as that shown in Figure 11.26, are a very common phenomena in our urban environment today. In recent years regulation of underground tanks has been tightened. It is not uncommon to see drill rigs testing gasoline station sites, and later noticing that the tanks have been excavated and treatment for leaking gasoline has begun. In a particular pollution case it may be difficult to show

Extraction Wells	Vapor Extraction	Bioremediation	Permeable Treatment Bed
Pump out contaminated water and treat it by filtration, oxidation, or air stripping (volatilization of contaminant in an air column), or by biological processes.	Uses vapor extraction well and then treatment.	Injection of nutrients and oxygen to encourage growth of organisms that degrade the contaminant in the groundwater.	Provides contact treatment as contaminated water plume moves through a treatment bed in the path of groundwater movement. Encourages neutralization of the contaminant by chemical, physical, or biological processes.

Table 11.9
Methods of treating groundwater and vadose zone water.

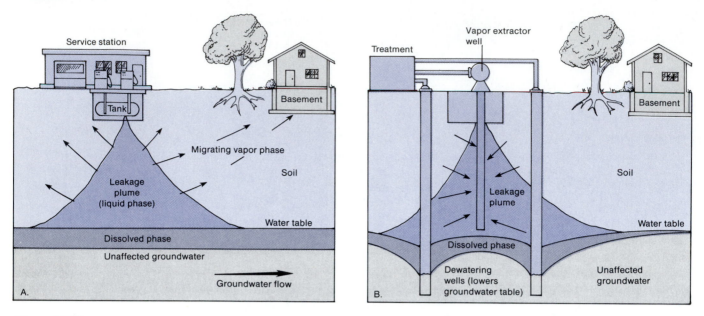

Figure 11.26
Idealized diagram of a leaking buried tank (a) and possible remediation method (b). See text for explanation. (Courtesy of University of California Santa Barbara Vadose Zone Laboratory and David Springer.)

where contaminants, such as gasoline, have come from. At many intersections there may be two or more gasoline stations that have had a series of buried tanks over a rather long period of time. Litigation over responsibilities concerning groundwater pollution from leaking underground tanks is a common process and one which may be difficult to resolve.

▼ WATER REUSE

Water reuse, which generally refers to the use of wastewater following some sort of treatment, is often discussed in terms of an emergency water supply, a long-term solution to a local water shortage, or a fringe benefit to water pollution abatement. Data have been or are being collected in many locations in the United States as part of water reuse research or implementation (26).

Water reuse can be inadvertent, indirect, or direct (Figure 11.27). *Inadvertent water reuse* results when water is withdrawn, used, treated, and returned to the environment without specific plans for further withdrawals and use, which nevertheless occur (Figure 11.27a). Such use patterns occur along many rivers and, in fact, are accepted as a common and necessary procedure for obtaining a water supply. There are several risks associated with inadvertent reuse. Inadequate treatment facilities may deliver contaminated or poor-quality water to users. Because the fate of disease-causing viruses during and after treatment is not completely known, we are uncertain about the environmental health hazards of treated water. In addition, each year many new chemi-

cals, some of which cause birth defects, genetic damage, or cancer in humans, are introduced into the environment. Unfortunately, harmful chemicals are often difficult to detect, and their effects on humans may be hidden if the chemicals are ingested in low concentrations over many years (26). In spite of these problems, inadvertent reuse of water will by necessity remain a common pattern. If we recognize the potential risks, we can plan to minimize them by using the best possible water treatment available.

Indirect water reuse is a planned endeavor (Figure 11.27b). In southern California, among other areas, several thousand cubic meters of treated wastewater per day have been applied to surface recharge areas. The treated water eventually enters into groundwater storage to be reused for agricultural and municipal purposes.

Direct water reuse refers to treated water that is piped directly to the next user (Figure 11.27c). In most cases the "user" will be industry or agricultural activity. Very little direct use of water is planned (except in emergencies) for human consumption because of cultural attitudes. Figure 11.28 summarizes these attitudes: Direct ingestion is least accepted, whereas uses in which no body contact occurs are generally much more acceptable.

▼ DESALINATION

Desalination of sea water, which contains about 3.5 percent salt (each cubic meter of sea water contains about 40 kilograms of salt), is an expensive form of water

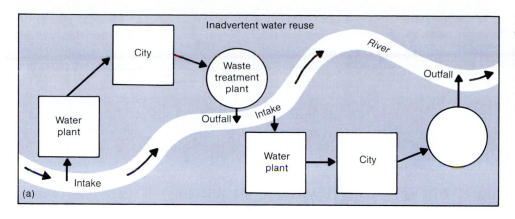

Inadvertent water reuse

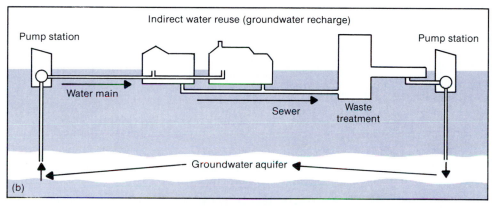

Indirect water reuse (groundwater recharge)

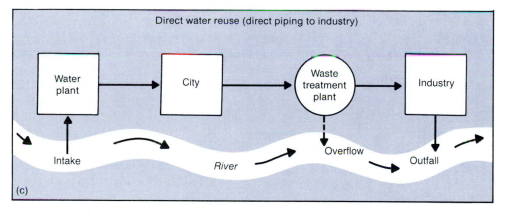

Direct water reuse (direct piping to industry)

Figure 11.27
Three types of water reuse. (From G. E. Symons, 1968, "Water Re-use: What Do We Mean?" *Water and Waste Engineering* 5: 40–41; as modified by R. E. Kasperson, 1977.)

treatment practiced at several hundred plants around the world. The salt content must be reduced to about 0.05 percent. Large desalination plants produce 20,000–30,000 cubic meters of water per day at a cost of about 10 times that paid for traditional water supplies in the United States. Desalinated water has a "place value," which means that the price increases quickly with the transport distance and elevation increase from the plant at sea level. Because the various processes that actually remove the salt require energy, the cost of the water is tied to ever-increasing energy costs. For these reasons, desalination will remain an expensive process that will be used only when alternative water sources are not available. Due to increasing population and inadequate water supply, accented by drought in the late 1980s, some

communities such as Santa Barbara, California, are seriously considering desalination.

Middle Eastern countries in particular will continue to use desalination. In many arid regions, including the Middle East, there are brackish ground and surface waters with a salinity of about 0.5 percent (one-seventh that of sea water). Obviously, desalination of this water is less expensive and plants may be located at inland sites.

▼ WATER AND ECOSYSTEMS

The major ecosystems of the world have evolved in response to physical conditions that include, among others, climate, nutrient input, soils, and hydrology. Changes in these factors will affect ecosystems; in

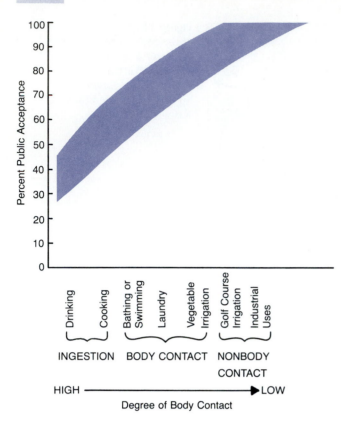

Figure 11.28
Diagram showing the public acceptance of use of treated wastewater. In general, people have a negative attitude toward direct reuse, that is, drinking of treated wastewater. (Modified after Kasperson, 1977.)

particular, changes induced by humans may cause far-reaching changes. Throughout the world today, with few exceptions, people are degrading natural ecosystems on a regional and global scale. Hydrologic conditions, particularly surface water processes and quality, are becoming limiting factors in many systems. Even the ultimate global hydrologic system, the oceans, is experiencing pollution problems. Figure 11.29 shows parts of the oceans and large lakes that are currently accumulating pollutants or have the potential for intermittent pollution. These areas often coincide with the near-shore productive areas of the marine environment.

While most coastal salt marshes are now protected in the United States, freshwater wetlands are still threatened in many areas. One percent of the nation's total wetlands is lost every two years—freshwater wetlands account for 95 percent of that loss, ranging from swamps in North Carolina to prairie potholes in the Midwest to vernal pools in California. Everyone knows wetlands are highly productive for fish and wildlife. What is needed are incentives for private land owners to preserve wetlands rather than fill them to develop the land (27) or stronger legislation to protect these areas.

Water resource development may be important to people but it can also adversely affect ecosystems.

Construction of large dams permanently changes rivers. For example, construction of the High Dam at Aswan on the Nile River in Egypt deprived the eastern Mediterranean Sea of nutrients delivered from the river, causing a one-third reduction in plankton production. As a result sardine, mackerel, lobster, and shrimp fishing have declined because these species depend on plankton for a food base. The dam, Lake Nasser, and associated canals also are breeding grounds for snails and mosquitoes. The former carry the disease schistosomiasis (snail fever) and the latter, malaria.

The Amazon River basin in South America has an average flow that amounts to 20 percent of the fresh water in the world's rivers. The amount of water reaching the ocean is so great that sailors more than 150 kilometers offshore may dip a cup over the side of their boat to obtain fresh water (28)! The river has many large tributaries, several of which surpass the Mississippi River (which drains 40 percent of the United States) in size. This great resource constitutes an immense potential for development of hydropower, and Brazil hopes to develop it. The Tucurui Dam on the Tocantins River in the basin is expected to eventually produce about 8000 megawatts of electricity—equivalent to the energy produced by about eight large nuclear power plants. Total power potential in the basin is close to 100,000 megawatts! The reservoir will flood more than 2000 square kilometers of virgin tropical rain forest. This and other similar planned projects are certain to impact upon the Amazon Basin through downstream effects on plants and wildlife. This is especially true if herbicides are necessary to control aquatic weeds, such as water hyacinth, that provide a good environment for carriers of such dangerous diseases as malaria and schistosomiasis. Herbicides used in the past include 2,4-D, which contains Agent Orange, a chemical that has been linked to cancers and birth defects. Aquatic mammals (manatees) that eat water plants are being imported and this may help reduce the need for herbicides (28).

Newly created lakes in tropical areas have historically been beset with problems—including anaerobic decay of organic materials in the deep water and filling in of the reservoir with sediment. Decay of plants produces a more acidic lake water that may corrode equipment, and sediment decreases the useful life of a reservoir for storing water and producing electricity. It is feared that continued timber harvesting upslope and upstream of the Tucurui Dam and reservoir will significantly increase soil erosion and sediment production, thus reducing the useful life of the reservoir (28).

Tropical rain forests contain over 50 percent of all species on earth. Most species have never been studied; some may contain chemicals useful in fighting disease, be useful for food crops, or have other potential utilitarian value. Large-scale experiments to change the basic hydrology in these areas should be done cautiously to

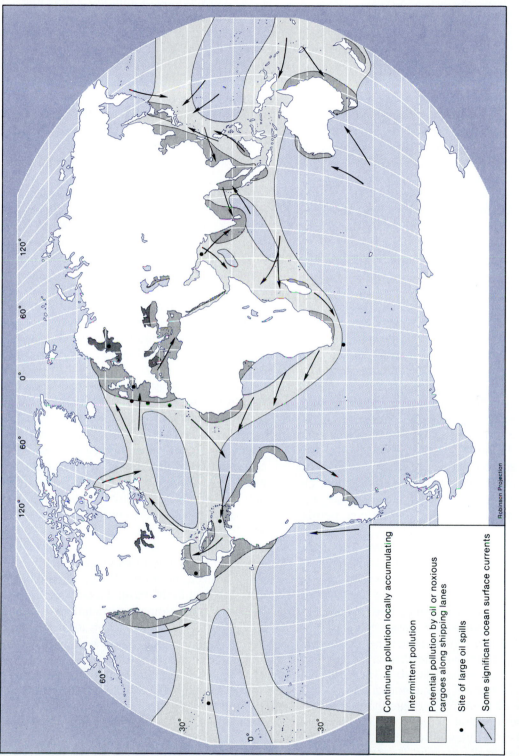

Legend:

- Continuing pollution locally accumulating
- Intermittent pollution
- Potential pollution by oil or noxious cargoes along shipping lanes
- • Site of large oil spills
- Some significant ocean surface currents

Robinson Projection

Figure 11.29
Sites of ocean pollution throughout the world. (Adapted from Council on Environmental Quality, *Environmental Trends*, 1981.)

avoid unnecessary ecologic damage. However, such caution seems unlikely considering the potential short-term benefits from development.

▼ THE COLORADO RIVER: A CASE HISTORY

No discussion of water resources and water management would be complete without a mention of the Colorado River basin and the controversy that surrounds the use of its water. People have been using the water of the Colorado River for about 800 years. Early Native Americans in the basin had a highly civilized culture with a sophisticated water distribution system. Many of their early canals were later cleared of debris and used by settlers in the 1860s (29). Given this early history, it is somewhat surprising to learn that the river was not completely explored until 1869 when John Wesley Powell, who later became director of the U.S. Geological Survey, navigated wooden boats through the Grand Canyon.

Although the waters of the Colorado River basin are distributed by canals and aqueducts to many millions of urban people, and to agricultural areas such as the Imperial Valley in California, the basin itself, with an area of approximately 632,000 square kilometers, is only sparsely populated. Yuma, Arizona, with approximately 42,000 people, is the largest city on the river, and only the cities of Las Vegas, Phoenix, and Tucson have more than 50,000 inhabitants. Rod Nash, writing about the wilderness values of the river, states that at the confluence of the Green and Colorado rivers it is 80 kilometers to the nearest video game and you are in the heart of a national park (30). Even though there are few cities in the Colorado River basin, only about 20 percent of the total population is rural! Vast areas of the basin have extremely low densities of people and over large areas measuring several thousand square kilometers there are no permanent residents (29).

The headwaters of the Colorado River are in the Wind River Mountains of Wyoming, and in its 2,300-kilometer journey to the sea the river flows through or abuts seven states (Wyoming, Colorado, Utah, New Mexico, Nevada, Arizona, California) and Mexico. Although the drainage basin is very large, encompassing much of the southwestern United States, the annual flow is only about 3 percent of that of the Mississippi River and less than a tenth of that of the Columbia. Therefore, the Colorado River for its size has only a modest flow, and yet it has become one of the most regulated, controversial, and disputed bodies of water in the world. Conflicts that have gone on for decades extend far beyond the Colorado River basin itself to involve large urban centers and developing agriculture areas of California, Colorado, New Mexico, and Arizona. The need for water in these semi-arid areas has resulted in over-use of limited supplies and deterioration of water quality. Interstate agreements, court settlements, and international agree-ments have periodically eased or intensified tensions in relations among people who use the waters along the river. The legacy of laws and court decisions, along with changing water-use patterns, continues to influence the lives and livelihood of millions of people in both Mexico and the United States (31).

Waters of the Colorado River have been appropriated among the various users including the seven states and the Republic of Mexico. This appropriation has occurred through many years of negotiation, international treaty, interstate agreements, contracts, federal legislation, and court decisions. As a whole, this body of regulation is known as the "Law of the River." Two of the more important early documents in this law were the Colorado River Compact of 1922, which divided water rights in terms of an upper and lower basin (Figure 11.30), and the treaty with Mexico in 1944, which promised an annual delivery of 1.85 cubic kilometers (1.5 million acre-feet) of Colorado River water to Mexico. More recent was a 1963 U.S. Supreme Court decision involving Arizona and California. Arizona refused to sign the 1922 compact and had a long conflict with California concerning appropriation of water. The Court decided that southern California must relinquish approximately 0.74 cubic kilometers (600,000 acre-feet) of Colorado River water when the Central Arizona Project is completed. Finally, in 1974 the Colorado River Basin Salinity Control Act was approved by Congress. The act authorized procedures to control adverse salinity of the Colorado River water, including construction of desalination plants that are scheduled for completion by the early 1990s.

Management of the Colorado River Basin and its waters has been frustrating in part because the basin is characterized by inherent instabilities (29). For example, in 1922 when the Colorado River Compact was worked out, the hypothesis was that the virgin flow of the river was approximately 20 cubic kilometers (16.2 million acre-feet) per year. That annual flow is now believed to average closer to 16.6 cubic kilometers (13.5 million acre-feet) per year (32). Even these numbers are misleading, however, because of the tremendous hydrologic instability within the basin. Flood waters in the Colorado River may come from snow melt floods, long-term winter precipitation events, or short-term summer thunderstorms; thus the total water available on a year-to-year basis is tremendously variable. Table 11.10 shows the legal water entitlements for the Colorado River Basin. Notice that the actual distribution of water is greater than the annual flow. The reason this distribution can be obtained is that the Colorado River is one of the most regulated rivers in the world. Figure 11.31 shows a profile of the river and some of the major dams and reservoirs. The 19 high dams on the river can store approximately 86.3 cubic kilometers (70 million acre-feet) of water. Of this, approximately 80 percent is stored in two reservoirs behind Hoover and Glen Canyon dams. This storage of water represents (if managed very

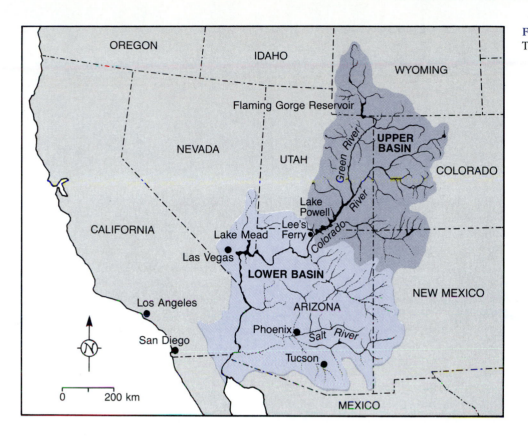

Figure 11.30
The Colorado River basin.

efficiently) a buffer of several years' water supply. However, if a severe drought of several years' duration should occur, delivery of water could become very difficult. The Colorado River was one of the nation's first major rivers to have its entire flow fully appropriated.

Balancing the future water needs of various users will continue to be a difficult and frustrating problem.

Construction of dams, reservoirs, and diversions on the Colorado River has generally been viewed as a successful venture from the viewpoint of supplying

State	Legal Entitlements (million ac ft per yr)	Actual Distribution[e] (million ac ft per yr)
California	4.400[a]	4.400[f]
Arizona	3.800[a]	2.050[f]
Nevada	0.300[a]	0.300
Lower Basin	**8.500[b]**	**6.750**
Colorado	3.881[c]	2.406
Utah	1.725[c]	1.070
Wyoming	1.050[c]	0.651
New Mexico	0.844[c]	0.523
Upper Basin	**7.500[b]**	**4.650**
Mexico	1.500[d]	1.500
Total	**17.500**	**14.500**

Table 11.10
Legal and actual distribution of Colorado River water.

[a]1928 Boulder Canyon Project Act

[b]1922 Colorado River Compact

[c]1948 Upper Colorado River Basin Compact

[d]1944 Mexico-U.S. Treaty

[e]Includes losses to evaporation of 0.6 million ac ft per year in the Upper Basin and 0.9 million ac ft per year in the Lower Basin, and 0.9 million ac ft per year inflow to Lower Basin from local streams

[f]Agreement at time of Central Arizona Project authorization by Congress. Upper Basin amounts agreed to by states as percentages.

Source: W. L. Graf, 1985, *The Colorado River,* Resource Publications in Geography, Association of American Geographers.

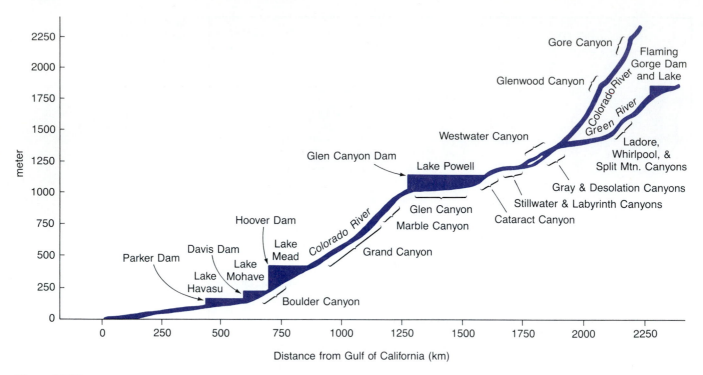

Figure 11.31
Longitudinal profiles of the Colorado and Green rivers, showing the major dams, reservoirs, and canyons. (From W. L. Graf, 1985, *The Colorado River,* Resource Publications in Geography, Association of American Geographers.)

water. However, this has not always been the case. For example, the present Salton Sea in the Imperial Valley formed in 1905 and 1906 when virtually the entire Colorado River was unintentionally diverted into the southern Imperial Valley (Salton Basin). At that time the Colorado River was completely undammed, and control works located in Mexican territory failed from flood waters in 1905 and 1906. By the time the river was controlled in 1907, the present Salton Sea had formed and was at a level higher than present. Water in the Salton Sea today is maintained through inflow from irrigation waters used to leach salts out of agricultural lands. Should this inflow stop or be reduced, the Salton Sea would soon dry up due to high evaporation rates there. Because the lake has become an important recreation (sport fishing) area, its future is controversial. If the lake waters become much saltier than they are now, the ecosystem and present fishery would be significantly damaged. However, the present lake is not unique to the Salton Basin. Others present during recent geologic history have also dried up!

While water supply is the primary problem in the Colorado River Basin, the problem of how to manage water quality is also significant. Although heavy metals and radioactive materials have become concentrated in the basin's waters and reservoirs, salt is causing the most problems. Natural salinity of the Colorado River in the headwaters is only 50 parts per million (ppm). For comparison, 550 ppm is the upper limit set for human consumption. However, as the river flows towards the sea, tributaries flow over exposed salt beds, and salt springs add salt to the river. As a result, under natural conditions, the salinity of the Lower Colorado would probably range from 250 to 380 ppm. If the salinity of irrigation water exceeds 750 parts per million, it may damage agriculture. Because of the natural salt in the river and upstream irrigation and evaporation, salinity of the Lower Colorado River averages 1,500 ppm and at times is as high as 2,700 ppm. Quality of the water is so poor that Mexican farmers have allowed it to pass their fields rather than damage their crops and soils. The United States and Mexico agreed in 1973 that the United States would deliver water to Mexico with a salinity no more than 115 parts per million greater than what occurs at the Imperial Dam, a short distance upstream from the border. The salinity there is approximately 800 ppm. To achieve this goal, a large desalination plant is currently under construction near the border, which will cost several hundred million dollars in capital expenses and over $10 million a year to operate. This is a tremendous investment in a structural effort to control the salinity of the river water. Only time will tell how effective it will be (29).

Issues of water and basin management in the Colorado River are complex but firmly illustrate some of the major problems likely to face other parts of the arid

Southwest in coming years: How are we to appropriate water? How can we best control water quality? Answers to these questions are not simple; what we have learned so far from our experiences with the Colorado River should help in future planning.

▼ ▼ ▼ SUMMARY AND CONCLUSIONS

An obvious and well-known detrimental aspect of human use of surface water, groundwater, and atmospheric water is the pollution of rivers and groundwaters. We are beginning to understand that solutions to many hydrologic problems require integrating all aspects of the water cycle.

Water is one of our most abundant and important renewable resources. However, more than 99 percent of the earth's water is unavailable or unsuitable for beneficial human use because of its salinity or location. The pattern of water supply and use at any particular point on the land surface involves interactions between the biochemical, hydrologic, and rock cycles. To evaluate a region's water resources and use patterns, a water budget is developed to define the natural variability and availability of water.

During the next several decades, the total water withdrawn from streams and groundwater in the United States is expected to decrease slightly, but consumptive use will increase because of greater demands from a growing population and industry. Water withdrawn from streams competes with other instream needs, such as maintaining fish and wildlife habitats and navigation, and may therefore cause conflicts.

Water pollution specifically refers to degradation of water quality as measured by physical, chemical, or biological criteria. These criteria take into consideration the intended use for the water, departure from the norm, effects on public health, and ecological impacts. Water pollutants have point or nonpoint sources. The major water pollutants are oxygen-demanding waste (BOD), pathogens, nutrients, synthetic organic and inorganic compounds, oil, heavy metals, radioactive materials, heat, and sediment. Since the 1960s there has been a serious attempt to improve water quality in the United States. Although the program seems to have been successful, water quality in many areas is still substandard.

The movement of water down to the water table and through aquifers is an integral part of the rock and hydrologic cycles. In moving through an aquifer, groundwater may improve in quality. However, it may also be rendered unsuitable for human use by natural or artificial contaminants.

In the case of groundwater pollution, the physical, biologic, and geologic environments are considerably different from those of surface water. The ability of many soils and rocks to physically or otherwise degrade pollutants is well known, but not so generally known is the ability of clays and other earth materials to capture and exchange certain elements and compounds.

Pollution of an aquifer may result from disposal of wastes on the land surface or in the ground. It can also result from overpumping of groundwater in coastal areas, leading to intrusion of salt water into freshwater aquifers.

Desalination of sea water in specific instances will continue, but large-scale desalination is not likely because of increasing costs of the energy and transportation required.

Water resource management needs a new philosophy that considers geologic, geographic, and climatic factors and utilizes creative alternatives.

Water is an integral part of ecosystems. It is possible to pollute even the world's largest hydrologic system, the oceans, especially near large coastal urban areas and along major shipping lanes.

Construction of dams and reservoirs on major rivers, such as the Mississippi, Nile, and Amazon, has caused or will cause significant environmental change to important regional ecosystems.

Water resources in the wetlands of the world are very important for biological productivity, especially of fish that are an important food source to people. Shellfish are particularly vulnerable to water pollution; millions of dollars have been lost to pollution of shellfish beds in the nation's wetlands.

▼ ▼ ▼ REFERENCES

1. COUNCIL ON ENVIRONMENTAL QUALITY and THE DEPARTMENT OF STATE. 1980. *The global 2000 report to the president: Entering the twenty-first century.* Vol. 2.
2. WATER RESOURCES COUNCIL. 1978. *The nation's water resources, 1975–2000.* Vol. 1.
3. SINGER, M. J., and MUNNS, D. M. 1978. *Soils: An introduction.* New York: Macmillan.
4. LIKENS, G. E.; BORMANN, F. H.; PIERCE, R. S.; EATON, J. S.; and JOHNSON, N. M. 1977. *The biogeochemistry of a forested ecosystem.* New York: Springer-Verlag.
5. SOLLEY, W. B.; CHASE, E. B.; and MANN, W. B., IV. 1983. *Estimated use of water in the United States in 1980.* U.S. Geological Survey Circular 1001.
6. ALEXANDER, G. 1984. Making do with less. *National Wildlife.* Special Report, February/March: 11–13.
7. DOLAN, R.; HOWARD, A.; and GALLENSON, A. 1974. Man's impact on the Colorado River and the Grand Canyon. *American Scientist,* v. 62: 392–401.
8. LAVENDER, D. 1984. Great news from the Grand Canyon. *Arizona Highways Magazine,* January: 33–38.
9. LEOPOLD, L. B. 1977. A reverence for rivers. *Geology* 5: 429–30.
10. *Alaska Fish and Game.* 1989. Special oil spill issue. Vol. 21, no. 4 (July/August).
11. COUNCIL ON ENVIRONMENTAL

QUALITY. 1979. *Environmental quality.*

12. ANONYMOUS. 1982. *U. S. Geological Survey Activities, Fiscal Year 1982.* U. S. Geological Survey Circular 875: 90–93.

13. GEISER, K., and WANECK, G. 1983. PCB's and Warren County. Science for the people.

14. WEIR, D., and SCHAPICO, M. 1980. The circle of poison. *The Nation,* November 15.

15. CAREY, J. 1984. Is it safe to drink? *National Wildlife.* Special Report, February/March: 19–21.

16. PYE, U. I., and PATRICK, R. 1983. Ground water contamination in the United States. *Science* 221: 713–18.

17. MALLMAN, W. L., and MACK, W. N. 1961. *A review of biological contamination of ground water.* Proceedings of Symposium on Ground Water Contamination, Taft Sanitary Engineering Center, Cincinnati, Ohio.

18. LEGRAND, H. E. 1968. Environmental framework of ground water contamination. *Journal of Ground Water* 6:14–18.

19. DEUTSCH, M. 1965. Natural controls involved in shallow aquifer contamination. *Journal of Ground Water* 3:37–40.

20. BOGAN, R. H. 1961. *Problems arising from ground water contamination by sewage lagoons at Tieton, Washington.* Public Health Service Technical Report W61-5.

21. VOGT, J. E. 1961. *Infectious hepatitis outbreak in Posen, Michigan.* Public Health Service Technical Report W61-5.

22. DEUTSCH, M. 1961. *Incidents of chromium contamination of ground water in Michigan.* Public Health Service Technical Report W61-5.

23. PETTYJOHN, W. A. 1971. Water pollution by oil-field brines and related industrial wastes in Ohio. *Ohio Journal of Science* 71:257-69.

24. BURT, E. M. 1972. The use, abuse, and recovery of a glacial aquifer. *Journal of Ground Water* 10: 65–72.

25. FOXWORTHY, G. L. 1978. Nassau County, Long Island, New York—Water problems in humid country. In *Nature to be commanded,* eds. G. D. Robinson and A. M. Spieker, pp. 55–68. U.S. Geological Survey Professional Paper 950.

26. KASPERSON, R. E. 1977. Water reuse: Need prospect. In *Water re-use and the cities,* eds. R. E. Kasperson and J. X. Kasperson, pp. 3–25. Ha-nover, N. H.: University Press of New England.

27. LEVINSON, M. 1984. Nurseries of life. *National Wildlife.* Special Report, February/March: 18–21.

28. CANFIELD, C. 1985. *In the rainforest.* New York: Alfred A. Knopf.

29. GRAF, W. L. 1985. *The Colorado River.* Resource Publications in Geography. Association of American Geographers.

30. NASH, R. 1986. Wilderness values and the Colorado River. In G. D. Weatherford and F. L. Brown (eds.) *New courses for the Colorado River.* Albuquerque, NM: Univ. of New Mexico Press, pp. 201–214.

31. HUNDLEY, N., Jr. 1986. The West against itself: The Colorado River—an institutional history. In G. D. Weatherford and F. L. Brown (eds.) *New courses for the Colorado River.* Albuquerque, NM: Univ. of New Mexico Press, pp. 9–49.

32. BALLARD, S. C.; MICHAEL, D. D.; CHARTOOK, M. A.; CLINES, M. R.; DUNN, C. E.; HOCK, C. M.; MILLER, G. D.; PARKER, L. B.; PENN, D. A.; and TAUXE, G. W. 1982. *Water and western energy: Impacts, issues, and choices.* Boulder, CO: Westview Press.

People in the United States and throughout the world are facing a tremendous solid waste disposal problem, particularly in growing urban areas. The problem boils down to the simple fact that urban areas are producing too much waste and there is far too little space for disposal. About half of the cities in the United States are estimated to be running out of landfill space. Philadelphia is essentially out of landfill space and must bargain with other states on a month by month or yearly basis to dispose of its trash; Los Angeles and New York City will run out of space before the end of this century. Costs are another limiting factor: landfill disposal cost approximately $5–$10 per ton in 1980. Today fees of $20 to $50 per ton are common, and Philadelphia pays approximately $75 per ton of trash disposed (1).

Perhaps the largest solid waste disposal site in the world is located on the 1,500-hectare site on Staten Island, New York. The waste disposal facility known as Fresh Kills accepts approximately 11,000 tons of waste every working day of the approximately 24,000 tons of municipal and commercial waste collected by the city of New York (2). Thus the site on an annual basis accepts about 2.9 million tons, which is enough waste to fill a convoy of 10-ton trucks over 4,000 kilometers long. By comparison, the total annual amount of waste collected in urban areas in the United States is about 150 million tons, and this would fill a convoy of collection trucks over 200,000 kilometers long. This distance is about 5 times around the circumference of the earth and over half the distance from the earth and the moon! (1).

At Fresh Kills refuse is piled several tens of meters above sea level, and with the prospect that the site may eventually accept up to 20,000 tons per day, it is expected that the refuse pile will eventually reach an elevation of 150–200 meters above sea level. This, in fact, is the construction of a large hill or small mountain of urban waste. As large as the Fresh Kills site is, it will be completely filled by the year 2000 (2).

A possible solution to the solid waste problem would be to develop new disposal facilities. Unfortunately, no one wants to live near a waste disposal site, be it a sanitary landfill for municipal waste, an incinerator facility that can reduce the volume of waste by 75 percent, or a disposal operation for hazardous chemical materials. This obviously creates serious siting problems even if the local geographic, geologic, and hydrologic environment is favorable. The bottom line seems to be that people have little confidence in the competency of government or industry to preserve and protect public health as it relates to waste disposal (1). The waste disposal industry in the United States represents a $20 billion sector of the economy, and it's an industry that is accustomed to the relatively simple system of collection of waste and landfill disposal (1). The rise in public consciousness concerning environmental problems and solutions is forcing the disposal industry to explore new solid-waste manage-

▼

▼

▼

CHAPTER TWELVE

Waste Management

ment systems. What is emerging is the concept known as *integrated waste management,* which is a complex set of management alternatives including source reduction, recycling, composting, landfill, and incineration (1).

▼ CONCEPTS OF WASTE MANAGEMENT

Earlier Views

During the first century of the industrial revolution, the volume of waste produced was relatively small and the concept of *"dilute and disperse"* was adequate. Factories were located near rivers because the water provided easy transport of materials by boat, ease of communication, sufficient water for processing and cooling, and easy disposal of waste into the river. With few factories and sparse population, "dilute and disperse" seemed to remove the waste from the environment (3).

Unfortunately, as industrial and urban areas expanded, the concept of "dilute and disperse" became inadequate, and a new concept known as *"concentrate and contain"* became popular. It is now apparent, however, that containment was and is not always achieved. Containers, natural or artificial, may leak or break and allow waste to escape. As a result, another concept developed—*"resource recovery."* This philosophy holds that wastes may be converted to useful materials, in which case they are no longer wastes but resources. However, even with our state-of-the-art technology large volumes of waste cannot be economically converted or are essentially indestructible. Therefore we still have waste disposal problems (3).

Modern Trends

Disposal or treatment of liquid and solid waste by federal, state, and municipal agencies costs billions of dollars every year. In fact, it is one of the most costly environmental expenditures of governments, accounting for the majority of total environmental expenditures (4).

All types of societies produce waste, but industrialization and urbanization have caused an ever-increasing effluence that has greatly compounded the problem of waste management. Although tremendous quantities of liquid and solid waste from municipal, industrial, and agricultural sources are being collected and recycled, treated or disposed of, new and innovative programs remain necessary if we are to keep ahead of what might be called a waste crisis. A popular idea is to consider the so-called waste as "resources out of place." Although we will probably never be able to recycle all waste, it seems apparent that the increasing cost of raw materials, energy, transportation, and land will make it financially feasible to recycle more resources, as well as to reuse the land where wastes not recycled are buried, creating new land resources for development. This concept is known as **sequential land use**. For example, the city of Denver, Colorado, used abandoned sand and gravel pits for landfill sites that today are sites of a parking lot and the Denver Coliseum. It is emphasized, however, that sequential land uses must be carefully selected. Urban housing, schools, hospitals, and other such construction should not be placed over old waste disposal sites.

Of particular importance is the growing awareness that many of our waste management programs simply involve moving waste from one site to another and not really properly disposing of it. For example, waste from urban areas may be placed in landfills, but eventually these may cause further problems from the production of methane gas or noxious liquids that leak from the site to contaminate the surrounding areas. Disposal sites are also capable of producing significant air pollution. Even some sewage treatment plants that received state and federal assistance in construction are now found to be producing air pollutants, some of which are carcinogenic. It is safe to assume that waste management is going to be a public concern for a long time. Of particular importance will be the development of new methods of waste management that will not endanger the public health, create a nuisance, or create an environmental time bomb. This is the essence of integrated waste management.

For discussion purposes, it is advantageous to break the management, treatment, and disposal of waste into several categories: reduce, recycle, and reuse; solid-waste disposal; hazardous chemical waste management; radioactive waste management; ocean dumping; septic-tank sewage disposal; and wastewater treatment.

▼ REDUCE, RECYCLE, AND REUSE

Reduce, recycle, and reuse are the three R's of waste management. They have the objective of reducing the amount of waste that must be disposed of in landfills,

incinerators, and other facilities. The study of the waste stream utilizing integrated waste management technology suggests that by the year 2000: (1) better design of packaging could result in a 10 percent (by weight) reduction in the total U.S. waste stream, (2) recycling programs could cause a further reduction of approximately 30 percent, and (3) composting might result in a further 10 percent reduction. Therefore, an integrated waste management system that incorporated the above elements could reduce the weight of urban refuse disposed of in landfills or incinerated by 50 percent, which is about the same savings as from incineration. It is estimated that the above rates for recycling are reasonable and in fact might be reached in some parts of the United States before the year 2000. The potential upper limit for recycling is even higher. "Intense recycling" could recover an estimated 80–90 percent of the U.S. waste stream (5). One pilot study involving 100 volunteer families in East Hampton, New York, achieved a level of 84 percent. "Partial recycling" that targets a limited number of items such as newsprint, glass, aluminum cans, plastic, and organic material for composting could reach a target of about a 30 percent reduction of the urban waste stream.

The most successful waste reduction and recycling program in the U.S. is probably at Seattle, Washington. Seattle, in 1988, was facing the closure of its only landfill, and the City Council proposed the construction of a large incineration facility. Citizen opposition was fierce and Seattle opted for a far-reaching waste reduction and recycling program. The 1998 objective is to reduce the amount of waste requiring landfill disposal by 60 percent. By 1989 the city had reached 37 percent, which is the highest of any large city in the country (5). The states of New Jersey and Rhode Island have recently set recycling quotas between 15 and 25 percent. Large cities such as New York and Los Angeles have initiated their own recycling programs. The environmentally aware city of Berkeley, California, has set its recycling goal at 50 percent and there is the possibility of legislation being passed in California that might require a 35 percent recycling rate for urban areas (1).

Integrated waste management is a complex problem, and if high levels of recycling are to be obtained then we must also develop markets and in many cases processing facilities for the recycled material. In fact, in some areas where recycling has been successful (many cities in Ohio, for example) glutted markets have forced stockpiling or temporary suspension of recycling some items. To be successful, recycling and development of markets will have to be developed together.

▼ SOLID-WASTE DISPOSAL

Disposal of solid waste is primarily an urban problem. In the United States alone, urban areas produce about 640 million kilograms of solid waste each day. That amount of

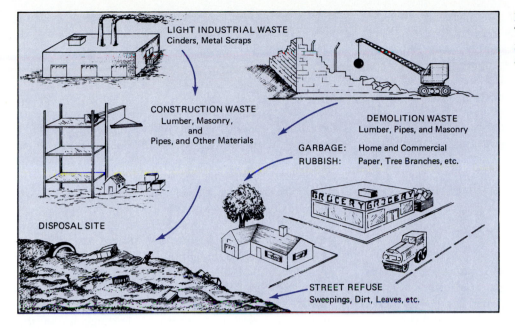

Figure 12.1
Types of materials or refuse commonly transported to a disposal site.

waste is sufficient to cover more than 1.6 square kilometers of land every day to a depth of 3 meters (6). Figure 12.1 summarizes major sources and types of solid waste, and Table 12.1 lists the generalized composition of solid waste at a disposal site in 1986 and projected for the year 2000. It is no surprise that paper is by far the most abundant of these solid wastes. We emphasize that this is only an average content, however, and considerable variation can be expected because such factors as land use, economic base, industrial activity, climate, and season of the year vary. In some areas, infectious wastes from hospitals and clinics can create problems if they are not properly sterilized before disposal (some hospitals have facilities to incinerate such wastes). In large urban areas large quantities of toxic materials may also end up at disposal sites. Urban landfills are now being considered hazardous waste sites that will require costly monitoring and cleanup.

Table 12.1
Generalized composition of urban solid waste (by weight) for 1986 and projected for the year 2000.

Material	1986 (%)	2000 (%)
Paper	36	39
Hard waste	20	19
Plastics	7	9
Metals	9	9
Food waste	9	7
Glass	8	7
Wood	4	4
Other	7	6

Source: A. M. Ujihara and M. Gough, "Managing Ash from Municipal Waste Incinerators," in *Resource for the Future* (Center for Risk Management: 1989).

The common methods of solid-waste disposal, summarized from a U.S. Geological Survey report, include on-site disposal, composting, incineration, open dumps, and sanitary landfills (6). Of these, the physical factors are most significant in planning siting of sanitary landfills. Without careful consideration of the soils, rocks, and hydrogeology, a landfill program may not function properly.

On-site Disposal

By far the most common on-site disposal method in urban areas is the mechanical grinding of kitchen food waste. Garbage disposal devices are installed in the waste-water pipe system from a kitchen sink, and the garbage is ground and flushed into the sewer system. This effectively reduces the amount of handling and quickly removes food waste, but final disposal is transferred to the sewage treatment plant where solids such as sewage sludge still must be disposed of (6). Hazardous liquid chemicals may also be inadvertently or deliberately disposed of in sewers, requiring treatment plants to handle toxic materials. Illegal dumping in urban sewers has only recently been identified as a potential major problem.

Another method of on-site disposal is small-scale incineration. This method is common in institutions and apartment houses (6). It requires constant attention and periodic maintenance to ensure proper operation. In addition, the ash and other residue must be removed periodically and transported to a final disposal site.

Composting

Composting is a biochemical process in which organic materials decompose to a humuslike material. It is rapid, partial decomposition of moist, solid, organic waste by aerobic organisms. The process is generally carried out in the controlled environment of mechanical digesters (4). Although composting is not common in the United States, it is popular in Europe and Asia, where intense farming creates a demand for the compost (4). A major drawback of composting is the necessity to separate the organic material from the other waste. Therefore, it is probably economically advantageous only when organic material is collected separately from other waste (7). Nevertheless, composting is considered part of integrated waste management, and its importance is expected to grow.

Incineration

Incineration is the reduction of combustible waste to inert residue by burning at high temperatures (900°–1000°C). These temperatures are sufficient to consume all combustible material, leaving only ash and noncombustibles. Incineration ideally reduces the volume of waste that must be disposed of by 75 to 95 percent (6). Actually, due to maintenance and waste supply problems, the realistic reduction of waste that must be disposed of is closer to 50 percent. This is about the same savings that can be gained from waste reduction and recycling (5).

Advantages of incineration of urban waste are twofold. First, it can effectively convert a large volume of combustible waste to a much smaller volume of ash to be disposed of at a landfill; and second, combustible waste can be used to supplement other fuels in generating electrical power. However, it is emphasized that burning of urban waste is certainly not a clean process. The burning produces air pollution and toxic ash that must be disposed of at landfills. Smokestacks from incinerators may emit nitrogen and sulphur oxides that are precursors of acid rain; carbon monoxide; and heavy metals such as lead, cadmium, and mercury. The smokestacks may be fitted with devices to trap some of the pollutants, but the process of pollution abatement is expensive. Other economic factors related to incinerators are also a subject of concern. The plants themselves are expensive and often need government subsidies to be established. One recent study showed that an investment of $8 billion in the United States could construct incinerators capable of burning about 25 percent of the nation's solid waste. However, a similar investment in recycling and composting facilities could handle as much as 75 percent of the nation's solid urban waste (5).

The economic viability of incinerators is dependent upon revenue from the sale of energy produced by burning waste. As a result they need to run at near capacity to remain profitable. With the increase in composting and recycling, the economics are far from certain because those processes compete directly with incineration. The main conclusion in evaluating incineration relative to waste reduction and recycling is that the latter can reduce the volume of waste that must be disposed of at a landfill by at least as much as can incineration (5).

Open Dumps

Open dumps are the oldest and most common way of disposing of solid waste. Although in recent years thousands have been closed in the United States, many still are being used in other parts of the world. In many cases, these dumps are located wherever land is available, without regard to safety, health hazards, and aesthetic degradation. The waste is often piled as high as equipment allows. In some instances, the refuse is ignited and allowed to burn. In others, the refuse is periodically leveled and compacted (6).

As a general rule, open dumps tend to create a nuisance by being unsightly, breeding pests, creating a health hazard, polluting the air, and often polluting groundwater and surface water. Fortunately, in the United States, open dumps have given way to the better planned and managed sanitary landfills that have taken their place.

Sanitary Landfill

A *sanitary landfill* as defined by the American Society of Civil Engineering is a method of solid-waste disposal that functions without creating a nuisance or hazard to public health or safety. Engineering principles are used to confine the waste to the smallest practical area, reduce it to the smallest practical volume, and cover it with a layer of compacted soil at the end of each day of operation, or more frequently if necessary. This covering of the waste with compacted soils makes the sanitary landfill "sanitary." The compacted layer effectively denies continued access to the waste by insects, rodents, and other animals. It also isolates the refuse from the air, thus minimizing the amount of surface water entering into and gas escaping from the wastes (7).

The sanitary landfill as we know it today emerged in the late 1930s; by 1960, more than 1400 cities disposed of their solid waste by this method. Two types are used: area landfill on relatively flat sites, and depression landfill in natural or artificial gullies or pits. The depths of landfills vary from about 2 to 13 meters. Normally, refuse is deposited, compacted, and covered by a minimum of 15 centimeters of compacted soil at the end of each day. The finishing cover is at least 50 centimeters of compacted soil (clay) designed to minimize infiltration of surface water (6). Compaction and subsidence can be expected for years following completion of a landfill. Therefore, any subsequent development that cannot accommodate potential subsidence should be avoided.

Leachate.   One of the most significant possible hazards from a sanitary landfill is groundwater or surface-water pollution. If waste buried in a landfill comes into contact with water percolating down from the surface, or with groundwater moving laterally through the refuse, **leachate**—obnoxious, mineralized liquid capable of transporting bacterial pollutants—is produced (8). For example, two landfills dating from the 1930s and 1940s in Long Island, New York, have produced leachate plumes that are several hundred meters wide and have migrated several kilometers from the disposal site.

The nature and strength of leachate produced at a disposal site depends on the composition of the waste, the length of time that the infiltrated water is in contact with the refuse, and the amount of water that infiltrates or moves through the waste (6). Compared to raw sewage and slaughterhouse waste, landfill leachate has a much higher concentration of pollutants. Fortunately, however, the amount of leachate produced from urban waste disposal is much less than the amount of raw sewage produced.

Another possible hazard from landfills is uncontrolled production and escape of **methane gas**, generated as organic wastes decompose. For example, gas generated in an Ohio landfill migrated several hundred meters through a sandy soil to a housing area, where one home exploded and several others had to be evacuated. Properly managed, though, methane gas (if not polluted with toxic materials) is a resource. In some cases this includes construction of barriers (plastic liner and clay) to inhibit lateral migration of gas and installation of a special well to collect the gas.

Site selection. Factors controlling the feasibility of sanitary landfills include topographic relief, location of the groundwater table, amount of precipitation, type of soil and rock, and the location of the disposal zone in the surface-water and groundwater flow system. The best sites are those in which natural conditions ensure reasonable safety in disposal of solid waste (that is, little or acceptable pollution of ground or surface waters); conditions may be safe because of climatic, hydrologic, geologic, or human-induced conditions, or combinations of these (9).

The best sites are in arid regions. Disposal conditions are relatively safe there because in a dry environment, regardless of whether the burial material is permeable or impermeable, little leachate is produced. On the other hand, in a humid environment some leachate will always be produced; therefore an acceptable level of leachate production must be established to determine the most favorable sites. What is acceptable varies with local water use, local regulations, and the ability of the natural hydrologic system to disperse, dilute, and otherwise degrade the leachate to a harmless state.

The most desirable site in a humid climate is one in which the waste is buried above the water table in clay and silt soils of low hydraulic conductivity. Any leachate produced will remain in the vicinity of the site and be degraded by natural filtering action and base exchange of some ions between the clay and the leachate. This would also hold for high water table conditions, as is often the case in humid areas, provided material with low hydraulic conductivity is present (11).

If the refuse is buried above the water table over a fractured-rock aquifer, as shown in Figure 12.2, the potential for serious pollution is low because the leachate is partly degraded by natural filtering as it moves down to the water table. Furthermore, the dispersion of contaminants is confined to the fracture zones (4). However, if the water table were higher or if the cover material were thinner with a moderate to high hydraulic conductivity, then widespread groundwater pollution of the fractured-rock aquifer might result.

If the geologic environment of a landfill site is characterized by an inclined limestone-rock aquifer, overlain by sand and gravel soils with high hydraulic conductivity, as shown in Figure 12.3, considerable contamination of the groundwater might result. This happens because the leachate moves quickly through the soil and enters the limestone, where open fractures or cavities may transport the pollutants with little degradation other than dispersion and dilution. Of course if the inclined rock is all shale, with low hydraulic conductivity, little pollution will result.

In summary, the following guidelines (10) should be followed in site selection of sanitary landfills:

▼ Limestone or highly fractured rock quarries and most sand and gravel pits make poor landfill sites because these earth materials are good aquifers.
▼ Swampy areas, unless properly drained to prevent disposal into standing water, make poor sites.
▼ Clay pits, if kept dry, may provide satisfactory sites.
▼ Flat areas are favorable sites provided an adequate layer of material with low hydraulic conductivity, such as clay and silt, is present above any aquifer.

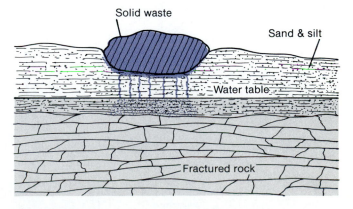

Figure 12.2
Waste-disposal site where the refuse is buried above the water table over a fractured rock aquifer. Potential for serious pollution is low because leachate is partially degraded by natural filtering as it moves down to the water table. (After W. J. Schneider, U.S. Geological Survey Circular 601F, 1970.)

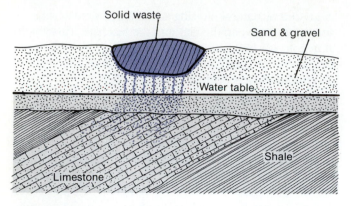

Figure 12.3
Solid-waste disposal site where the waste is buried above the water table in permeable material with high hydraulic conductivity in which leachate can migrate down to fractured bedrock (limestone). The potential for groundwater pollution may be high because of the many open and connected fractures in the rock. (After W. J. Schneider, U.S. Geological Survey Circular 601F, 1970.)

▼ Floodplains likely to be periodically inundated by surface water should not be considered as acceptable sites for refuse disposal.

▼ Any material with high hydraulic conductivity and with a high water table is probably an unfavorable site.

▼ In rough topography, the best sites are near the heads of gullies where surface water is at a minimum.

We emphasize that, while these guidelines are useful, they are not intended to preclude a hydrogeological investigation, including drilling to obtain samples, per-

meability testing to determine hydraulic conductivity, and other tests to predict the movement of leachate from the buried refuse (7).

Design of sanitary landfills. Design of modern sanitary landfills is complex and uses the multiple-barrier approach. These barriers include a compacted clay liner, leachate collection systems, and a compacted clay cap. Figure 12.4 provides an idealized diagram showing these features of a sanitary landfill. Depending upon local site conditions, landfills may also have additional synthetic liners made of plastics or other materials and a system to collect natural gas that might accumulate. Finally, sanitary landfills must have a system of monitoring wells and other devices to evaluate potential for groundwater pollution. The subject of monitoring is an important one and we will now address that issue in greater detail.

Monitoring pollution. Once a site is chosen for a sanitary landfill, monitoring the movement of groundwater should begin before filling commences. After the operation starts, continued monitoring of the movement of leachate and gases should be continued as long as there is any possibility of pollution. This is particularly important after the site is completely filled and the permanent cover material is in place because a certain amount of settlement always occurs after a landfill is completed, and if small depressions form, then surface water may collect, infiltrate, and produce leachate. Therefore, monitoring and proper maintenance of an abandoned landfill will reduce its pollution potential (7).

Hazardous waste pollutants from a solid-waste disposal site may enter the environment (11) by as many as six paths (Figure 12.5).

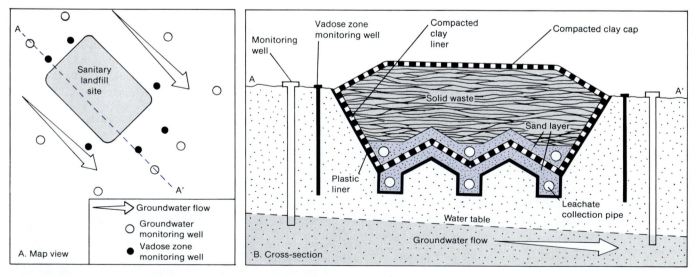

Figure 12.4
Idealized diagrams showing map view (a) and cross-section (b) of a landfill with a double liner of clay and plastic and a leachate collection system.

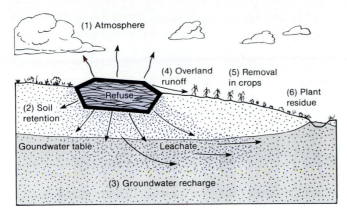

Figure 12.5
Several ways that hazardous waste pollutants from a solid-waste disposal site may enter the environment.

1. Compounds volatilized in the soil and fill after placement, such as methane, ammonia, hydrogen sulfide, and nitrogen gases, may enter the atmosphere.
2. Heavy metals such as lead, chromium, and iron are retained in the soil.
3. Soluble material such as chloride, nitrate, and sulfate readily pass through the fill and soil to the groundwater system.
4. Overland runoff may pick up leachate and transport it into the surface-water network.
5. Some crops and cover plants growing in the disposal area may selectively take up heavy metals and other toxic materials to be passed up the food chain as people and animals eat them.
6. If the plant residue left in the field contains toxic substances, it will return these materials to the environment through soil-forming and runoff processes.

A thorough monitoring program will consider all six possible paths by which pollutants enter the environment. Potential atmospheric pollution by gas from landfills is a growing concern, and a thorough monitoring program includes periodic analysis of air samples to detect toxic gas before it becomes a serious problem.

Many landfills have no surface runoff, and therefore monitoring of on site surface water is not necessary. However, thorough monitoring is required if overland runoff does occur, and monitoring of nearby downgradient streams, rivers, and lakes is necessary. Monitoring of soil and plants should include periodic chemical analysis at prescribed sampling locations. The need for monitoring groundwater quality through observation wells is minimal (but still necessary) if the landfill is in relatively impermeable soil overlying dense permeable rock. In this case, leachate and groundwater movement may be less than 30 centimeters per year. However, if permeable water-bearing zones exist in the soil or bedrock, then observation wells to sample the water

quality periodically and monitor the movement of the leachate are necessary (11). It is important also to monitor water in the unsaturated (vadose) zone above the water table. The purpose is to identify potential pollution problems *before* they reach and contaminate groundwater resources, where correction is very expensive (see Figure 12.4).

▼ HAZARDOUS CHEMICAL WASTE MANAGEMENT

The creation of new chemical compounds has proliferated tremendously in recent years. In the United States alone, approximately 1,000 new chemicals are marketed each year and a total of about 50,000 chemicals are currently on the market. Although many of the chemicals have been beneficial to people, approximately 35,000 of the chemicals used in the United States are classified as being either definitely or potentially hazardous to the health of people (Table 12.2). The United States currently is generating over 150 million metric tons of hazardous chemical waste per year. In the recent past, as much as half of the total volume of wastes was being indiscriminately dumped (12).

Uncontrolled Sites

In the United States, there are 32,000–50,000 uncontrolled waste disposal sites, and of these probably

Table 12.2
Examples of products we use and potentially hazardous waste they generate.

Products we use	Potentially hazardous waste
Plastics	Organic chlorine compounds
Pesticides	Organic chlorine compounds, organic phosphate compounds
Medicines	Organic solvents and residues, heavy metals (mercury and zinc, for example)
Paints	Heavy metals, pigments, solvents, organic residues
Oil, gasoline, and other petroleum products	Oil, phenols and other organic compounds, heavy metals, ammonia salts, acids, caustics
Metals	Heavy metals, fluorides, cyanides, acid and alkaline cleaners, solvents, pigments, abrasives, plating salts, oils, phenols
Leather	Heavy metals, organic solvents
Textiles	Heavy metals, dyes, organic chlorine compounds, solvents

Source: U.S. Environmental Protection Agency. SW-826, 1980.

1200–2000 contain sufficient waste to be a serious threat to public health and the environment. For this reason, a number of scientists believe that management of hazardous chemical materials may be the most serious environmental problem ever to face the United States.

Past uncontrolled dumping of chemical waste has polluted soil and groundwater resources in several ways (Figure 12.6). First, chemical waste stored in barrels either on the surface of the ground or buried at a disposal site eventually corrode and leak, polluting the surface, soil, and groundwater. Second, liquid chemical wastes dumped in unlined "lagoons," may percolate through the soil and rock and eventually to the groundwater table. Third, liquid chemical waste may be illegally dumped in deserted fields or even along dirt roads.

We must recognize that abandoned hazardous landfills and other sites for the disposal of chemical waste are not merely a potential problem. Examples from Kentucky, New Jersey, and New York illustrate that the problem is a serious one. At a location near Louisville, Kentucky, known as the "Valley of the Drums," thousands of leaking drums of waste chemicals were buried or stored on the surface to ooze toxic materials.

Located near Elizabeth, New Jersey, are the remains of about 50,000 charred drums, stacked four high in places, next to a brick and steel building once owned by a now-bankrupt chemical corporation. The drums and other containers were left to corrode for nearly ten years, and many had been either improperly labeled or had been burned so that the nature of the chemicals could not be determined simply from outside markings. Leaking barrels allowed unknown waste to seep into an adjacent stream that eventually flows into the Hudson River. The site was so bad that clean-up efforts were very difficult. Identification of some of the materials at the site showed that there were two containers of nitroglycerine, numerous barrels of biological agents, cylinders of phosgene and phosphoric gases (which are extremely volatile and ignite when exposed to air), as well as a variety of heavy metals, pesticides, and solvents, some of which are very toxic. It took months of work with a large crew of people to remove most of the material from the New Jersey site. Unfortunately, it is difficult to know if all the waste has been removed; additional material may be buried at other sites that are more difficult to locate (13).

Love Canal

The best known example of problems associated with chemical waste disposal comes from the town of Niagara Falls, New York, and the "Love Canal." In 1976, in a residential area near Niagara Falls, trees and gardens began to die. Children found the rubber on their tennis shoes and on their bicycle tires disintegrating. Dogs sniffing in a landfill area developed sores that would not heal. Puddles of toxic, noxious substances began to ooze to the soil surface; a swimming pool popped its foundations and was found to be floating on a bath of chemicals.

A study revealed that the residential area had been built on the site of a chemical dump. The area had originally been excavated in 1892 as the beginnings of the Love Canal, which was supposed to provide a transportation route between industrial centers. When that plan failed, the ditch was unused for decades and seemed a convenient dump for wastes. From the 1920s to the 1950s, more than 80 different chemicals were dumped there. Finally, in 1953, the company dumping the chemicals donated the land to the city of Niagara Falls for one dollar. Eventually several hundred homes and an elementary school were built on and near the site (Figure 12.7). Heavy rainfall and heavy snowfall during the winter of 1976–77 set off the events that made Love Canal a household word.

A study of the site identified a number of substances present there—including benzene, dioxin, dichlorethylene, and chloroform—that were suspected of being carcinogens. Although officials readily admitted that very little was known about the impact of these chemicals and others at the site, there was grave concern for the people living in the area. The concern has been well documented now because of the alleged higher-than-average rates of miscarriages, blood and liver abnormalities, birth defects, and chromosome damage. However, a study by the New York health authorities suggests that no chemically caused health effects have been absolutely established (14, 15, 16).

The cleanup of the Love Canal is an important demonstration of state-of-the-art technology in hazardous waste treatment. The objective has been to contain and stop the migration of wastes through the groundwater flow system and to remove and treat dioxin-contaminated soil and sediment from stream beds and sewers (17).

The method being used to minimize further production of contaminated water is to cover the dump site and

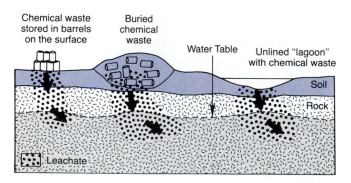

Figure 12.6
Ways that uncontrolled dumping of chemical waste may pollute soil and/or groundwater.

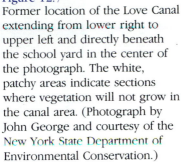

Figure 12.7
Former location of the Love Canal extending from lower right to upper left and directly beneath the school yard in the center of the photograph. The white, patchy areas indicate sections where vegetation will not grow in the canal area. (Photograph by John George and courtesy of the New York State Department of Environmental Conservation.)

adjacent contaminated area with a 1-meter-thick layer of compacted clay and a polyethylene plastic cover to reduce infiltration of surface water. Lateral movement of water is inhibited from entering the site by specially designed impervious walls. These procedures will greatly reduce subsurface seepage of water through the site, and the water that does seep out is collected and treated (14, 15, 16, 17).

The homes on and adjacent to Love Canal were abandoned and bought by the government. Approximately 200 of the homes had to be destroyed. During the 1980s approximately $275 million was spent for cleanup and relocation at Love Canal. The Environmental Protection Agency now considers the area clean and over 200 of the remaining homes were scheduled to be sold in the early 1990s. Because the price of the homes is approximately 20 percent below the market of other areas in Niagara Falls, they are expected to sell. This in spite of the reputation of the site and the adverse publicity it attracted (18).

What went wrong in Love Canal to produce a suburban ghost town? How can we avoid such disasters in the future? The real tragedy of Love Canal is that it is probably not an isolated incident. That is, there are many hidden "Love Canals" across the country, "time bombs" waiting to explode (14, 15).

Responsible Management

In the United States, the federal government moved in 1976 to begin the management of hazardous waste with the passage of the Resource Conservation and Recovery Act, which is intended to provide for "cradle-to-grave" control of hazardous waste. At the heart of the act is the identification of hazardous wastes and their life cycles. Regulations are included to ensure that stringent record keeping and reporting are done to verify that wastes do not present a public nuisance or a public health problem. The act also identifies wastes considered hazardous in terms of several categories: materials that are highly toxic to people and other living things; wastes that may explode or ignite when exposed to air; wastes that are extremely corrosive; and wastes that are otherwise unstable.

Recognizing that a great number of waste disposal sites presented hazards, Congress in 1980 passed the Environmental Response Compensation and Liability Act, which established a revolving fund (popularly called the Superfund) to clean up several hundred of the worst abandoned hazardous chemical waste disposal sites known to exist around the country. Although the fund has experienced significant management problems and is far behind schedule, a small number of sites have been

Table 12.3

Comparison of hazard reduction technologies.

	Disposal		Treatment		
	Landfills and Impoundments	Injection Wells	Incineration and Other Thermal Destruction	Emerging High-Temperature Decomposition[a]	Chemical Stabilization
Effectiveness: How well it contains or destroys hazardous characteristics	Low for volatiles, questionable for liquids; based on lab and field tests	High, based on theory but limited field data available	High, based on field tests except little data on specific constituents	Very high; commercial scale tests	High for many metals; based on lab tests
Reliability issues	Siting, construction, and operation Uncertainties: long-term integrity of cells and cover, linear life less than life of toxic waste	Site history and geology, well depth, construction and operation	Monitoring uncertainties with respect to high degree of DRE: surrogate measures, PICs, incinerability[c]	Limited experience Mobile units, on-site treatment avoids hauling risks Operational simplicity	Some inorganics still soluble Uncertain leachate test, surrogate for weathering
Environment media most affected	Surface and ground water	Surface and ground water	Air	Air	Groundwater
Least compatible wastes[b]	Liner reactive, highly toxic, mobile, persistent, and bioaccumulative	Reactive; corrosive; highly toxic, mobile, and persistent	Highly toxic and refractory organics, high heavy metals concentration	Some inorganics	Organics
Relative Costs: Low, Mod., High	L-M	L	M-H	M-H	M
Resource recovery potential	None	None	Energy and some acids	Energy and some metals	Possible building material

[a]Molten salt, high-temperature fluid well, and plasma arc treatments.

[b]Wastes for which this method may be less effective for reducing exposure, relative to other technologies. Wastes listed do not necessarily denote common usage.

[c]DRE = destruction and removal efficiency. PIC = product of incomplete combustion.

Source: Council on Environmental Quality, 1983.

treated. Unfortunately, the funds available are not sufficient to pay for decontamination of all the targeted sites. That would cost many times more, perhaps as much as $100 billion. Furthermore, because of concern that the present technology is not sufficiently advanced to treat all the abandoned waste disposal sites, the strategy may be to simply try to confine the waste to those sites until better disposal methods are developed. It seems apparent that the danger of abandoned disposal sites is likely to persist for some time to come.

Management of hazardous chemical waste includes several options, including recycling, on-site processing to recover byproducts with commercial value, microbial breakdown, chemical stabilization, high-temperature decomposition, incineration, and disposal by secure landfill or deep-well injection. A number of technological advances have been made in the field of toxic waste management, and as land disposal becomes more and more expensive, the trend toward on-site treatment that has recently started is likely to continue. However, on-site treatment will not eliminate all hazardous chemical waste; disposal will remain necessary. Table 12.3 compares hazardous waste reduction technology in terms of treatment and disposal. Notice that all of the technologies available will cause some environmental disruption. No one simple solution exists for all waste management issues.

Secure Landfill

The basic idea of the secure landfill is to confine the waste to a particular location, control the leachate that drains from the waste, collect and treat the leachate, and detect possible leaks. Figure 12.8 demonstrates these procedures. A dike and liners (made of clay and impervious material such as plastic) confine the waste, and a system of internal drains concentrates the leachate in a collection basin from which it is pumped out and transported to a waste-water treatment plant. Designs of new facilities today must include multiple barriers consisting of several impermeable layers and filters as well as impervious covers. The function of impervious liners is to ensure that the leachate does not contaminate soil and, in particular, groundwater resources. However, this type of waste-disposal procedure must have several monitor wells to alert personnel if and when leachate migrates out of the system to possibly contaminate nearby water resources.

It has recently been argued that there is no such thing as a really secure landfill, implying that they all leak to some extent. This is probably true; impervious plastic liners, filters, and clay layers can fail, even with several backups, and drains can become clogged, causing overflow. Yet landfills that are carefully sited and engineered can minimize problems. Preferable sites are those with good natural barriers such as thick clay-silt deposits, an arid climate, or a deep water table that minimize migration of leachate. Nevertheless, land disposal should be used only for specific chemicals compatible and suitable for the method.

Land Application

Application of waste materials to the surface soil horizon is referred to as land application, land spreading, or land farming. Land application of waste may be a desirable treatment method for certain biodegradable industrial wastes. Such wastes include petroleum oily waste and certain organic chemical-plant waste. A good indicator of the usefulness of land application of a particular waste is its *biopersistence* (the measure of how long a material remains in the biosphere). The greater or longer the biopersistence, the less suitable the waste is for land application procedures. Land application is not an effective treatment or disposal method for inorganic substances such as salts and heavy metals (19).

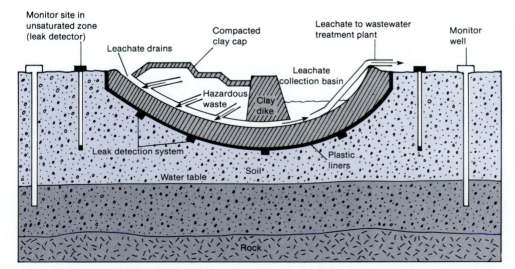

Figure 12.8

A secure landfill for hazardous chemical waste. The impervious liners and systems of drains are an integral part of the system to ensure that leachate does not escape from the disposal site. Monitoring in the unsaturated zone is important and involves periodic collection of soil water with a suction device.

Land application of biodegradable waste works because, when such materials are added to the soil, they are attacked by microflora (bacteria, molds, yeast, and other organisms) that decompose the waste material. The soil may thus be thought of as a microbial farm that constantly recycles organic and inorganic matter by breaking it down into more fundamental forms useful to other living things in the soil. Because the upper soil zone contains the largest microbial populations, land application is restricted to the uppermost 15–20 centimeters of the soil profile (19).

Land application of waste offers several potential advantages. It is relatively effective, environmentally safe, and has a reasonable cost. It is fairly simple and does not depend on high maintenance of failure-prone equipment. Finally, the method makes use of natural processes to recycle the waste (19).

Sites for land application of waste must be chosen carefully to ensure that the facility functions as designed. Of particular importance in siting are soil type, depth to groundwater, distance to streams and other surface discharges of water, rate of potential migration of waste, and topography. Because land application of waste produces a storage site in the soil for metals, salts, dusts, and other nonbiodegradable materials, care must be taken to ensure that these materials do not build up to toxic levels. In particular, wastes that are ignitable, reactive, volatile, toxic to life, or incompatible with soil should not be applied. At the present time, approximately half of all petroleum oily wastes and some pharmaceutical and organic chemical wastes are being treated by land application methods (19).

The process of land application involves a microbial conversion that is a public service function of natural ecosystems. In some cases, the microbes do not use the compounds but merely convert them from toxic to nontoxic form. However, care must be taken because the opposite may also occur. That is, microbes may convert a less toxic form to a more toxic one. For example, microbes can convert inorganic mercury to methyl mercury, which is very toxic. Nevertheless, we can and do make use of natural processes when we treat or dispose of waste into streams, marshes, and agricultural or forest soils and when we use microbial activity in sewage treatment. All these methods work as long as the concentrations of the waste do not exceed the capacities of the microbes to utilize, break down, or change the waste (19). Finally, as with other types of land disposal technology, the vadose zone and groundwater near the site must be carefully monitored to ensure the disposal system is working as planned and not polluting water resources.

Surface Impoundment

Excavations and natural topographic depressions have been used to hold hazardous liquid waste. These are primarily formed of soil or other surficial materials, but may be lined with manufactured materials such as plastic. The impoundment is designed to hold the waste; examples include aeration pits and lagoons at hazardous waste facilities. Surface impoundments have been criticized because they are especially prone to seepage, resulting in pollution of soil and groundwaters. Evaporation from surface impoundments can also produce an air pollution problem. For these reasons, some hazardous-waste facilities have been prohibited from receiving liquid waste.

Deep-well Disposal

Another method of waste disposal is by injection into deep wells. The term "deep" refers to rock (not soil) that is below and completely isolated from all freshwater aquifers, thereby assuring that injection of waste will not contaminate or pollute existing or potential water supplies. This generally means that the waste is injected into a permeable rock layer several thousand meters below the surface in geologic basins confined above by relatively impervious, fracture-resistant rock, such as shale or salt deposits (3).

Deep-well injection of oil-field brine (salt water) has been important in controlling water pollution in oil fields for many years, and huge quantities of liquid waste (brine) pumped up with oil have been injected back into the rock. Today, several billion liters per day are pumped into subsurface rocks (20). In recent years, the technique has been used more commonly for permanent storage of industrial waste deep underground. A typical well is about 700 meters deep, and wastes are pumped into a 60-meter-thick zone at a rate of about 400 liters per minute (21).

Deep-well disposal of industrial wastes should not be viewed as a quick and easy solution to industrial waste problems (22). Even where geologic conditions are favorable for deep-well disposal, natural restrictions include the limited number of suitable sites and the limited space within these sites for disposal of waste. Possible injection zones in porous rock are usually already filled with natural fluids, usually brackish or briny water. Therefore, to pump in waste, some of the natural fluid must be displaced by compression (even slight compression of the natural fluids in a large volume of permeable rock can provide considerable storage space) and by slight expansion of the reservoir rock as the waste is being injected (21).

The most favorable sites from a geologic consideration are synclinal basins and coastal plains, because they generally contain thick sequences of sedimentary rocks bearing salt water. Figure 12.9 shows geologic features that are significant in evaluating sites for deep waste-injection wells and locating some existing injection systems. The most favorable basins are tectonically stable and range from tens to hundreds of kilometers in width.

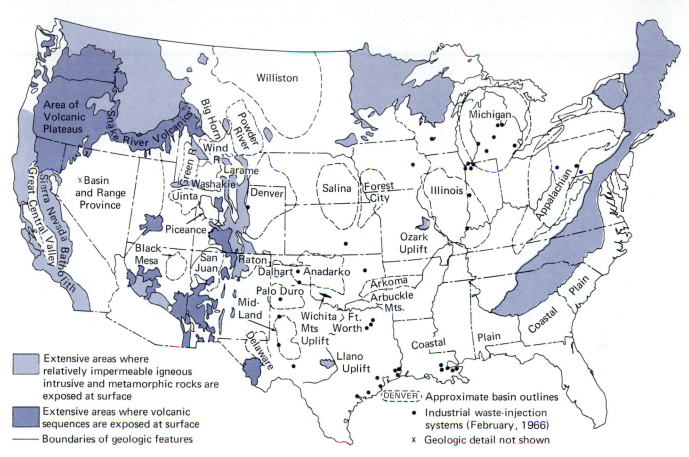

Figure 12.9
Geologic features significant in deep-well disposal. Notice that most of the industrial-waste injection systems (wells) are located in geologic basins or on coastal plains. Both of these environments are characterized by thick sequences of sedimentary rocks. (After D. L. Warner, American Association of Petroleum Geologists Memoir 10, 1968.)

Since natural flow of fluids is slow, less than about 1 meter per year, deep-well disposal should be restricted vertically and much less restricted horizontally (3).

Problems with deep-well disposal. Several problems associated with disposal of liquid waste in deep wells have been reported (21, 22). Perhaps the best known are earthquakes caused by injection of waste from the Rocky Mountain Arsenal near Denver, Colorado. These earthquakes occurred between 1962 and 1965. The injection zone was fractured gneiss at a depth of 3.6 kilometers; the increased fluid pressure evidently initiated movement along the fractures. This is not a unique case. Similar initiations of earthquakes have been reported in oil fields in western Colorado, Texas, and Utah (21). A similar activation of faults in southern California caused by injection of fluids into the Inglewood oil field for secondary recovery is thought to have contributed to the failure of the Baldwin Hills Reservoir. In a different sort of incident, in 1968, a disposal well on the shore of Lake Erie in Pennsylvania "blew out," releasing several million liters of harmful chemicals from a paper plant into the lake (22). Despite these incidents, most oil-field

brine and industrial disposal wells are functioning as designed and are not degrading the near-surface environment.

Feasibility and general site considerations. The feasibility of deep-well injection as the best solution to a disposal problem depends on four factors: the geologic and engineering suitability of the proposed site, the volume and the physical and chemical properties of the waste, economics, and legal considerations (23).

Geologic considerations for disposal wells are twofold (23). First, the injection zone must have sufficient porosity, thickness, hydraulic conductivity, and size to ensure safe injection. Sandstone and fractured limestone are the commonly used reservoir rocks (22). Second, the injection zone must be below the level of fresh-water circulation and confined by a relatively impermeable rock with low hydraulic conductivity, such as shale or salt, as shown in Figure 12.10.

Optimal use of limited underground storage space is achieved if deep-well injection is used only when more satisfactory methods of waste disposal are not available and when the volume of injected wastes is minimized by

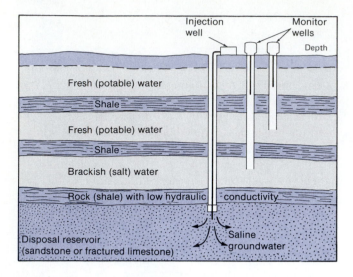

Figure 12.10
A deep-well injection system. The disposal reservoir is a sandstone or fractured limestone capped by impermeable rock and isolated from all fresh water. Monitor wells are a safety precaution to ensure that there is no undesirable migration of the liquid waste into either the formation above or fresh water. (100 m = 327 ft)

good waste management (23). Optimal use includes thorough evaluation of the physical and chemical properties of the waste to ensure that it will not adversely affect the ability of the rock reservoir to accept it. Adverse effects can be minimized in some instances by preinjection treatment of the waste to enhance compatibility with the reservoir rock and the natural fluids present there. It is also advisable to take advantage of natural buffers; for example, if the waste is acidic, use of limestone as a reservoir rock may be advantageous because the acid waste tends to increase the reservoir's permeability and

hydraulic conductivity by chemically attacking and enlarging natural fractures, thus allowing more waste to be injected. Care must be taken, however, because certain acids, such as sulfuric acid, may react to plug the porosity of limestone.

Construction and operating costs often determine the feasibility of deep-well injection as the preferred method of waste treatment. Important geologic factors pertaining to construction costs are the depth of the well and the ease with which drilling proceeds. Operating costs depend on the hydraulic conductivity, porosity, and thickness of the injection zone, and the fluid pressure in the reservoir. All of these are important in determining the rate at which the reservoir will accept liquid waste (23).

Consideration of legal aspects of deep-well disposal suggests that existing laws, regulations, and policies are probably adequate. However, regulatory agencies do need more funds and trained personnel to regulate successfully the growing number of disposal wells (20).

Monitoring disposal wells. An essential part of any disposal system is monitoring. It is very important to know exactly where the wastes are going, how stable they are, and how fast they are migrating. It is especially important in deep-well disposal where toxic or otherwise hazardous materials are involved.

Effective monitoring requires that the geology be precisely defined and mapped before initiating the disposal program. Especially important is locating all freshwater-bearing zones and old or abandoned oil or gas wells that might allow the waste to migrate up to freshwater aquifers or to the surface (Figure 12.11). A system of deep observation wells drilled into the disposal reservoir in the vicinity of the well can monitor the movement of waste, and shallow observation wells

Figure 12.11
How liquid waste might enter a freshwater aquifer through abandoned wells. This diagram illustrates why the location of all abandoned wells should be known, and it emphasizes the necessity of monitoring wells. (After Irwin and Morton, U.S. Geological Survey Circular 630, 1965.)

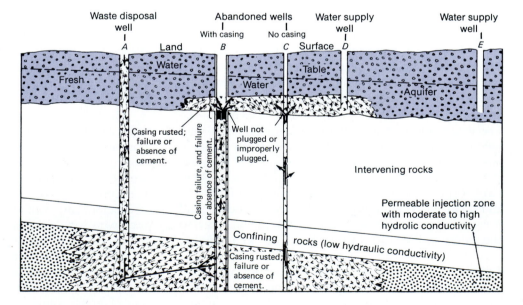

drilled into freshwater zones can monitor the water quality to quickly identify any upward migration of the waste.

Incineration of Hazardous Chemical Waste

Hazardous waste may be destroyed through high-temperature incineration. Because incineration produces an ash residue that must be buried in a landfill, it is considered an option of waste treatment. The technology used in incineration and other high-temperature decomposition or destruction is changing rapidly. Figure 12.12 diagrams one type of high-temperature incineration system that may be used to burn toxic waste. Waste—as liquid, solid, or sludge—may enter the rotating combustion chamber, where it is rolled and burned. Ash from this burning process may be collected in a water tank while remaining gaseous materials move into a secondary combustion chamber, where the process is repeated. Finally, the remaining gas and particulates move through a scrubber system that eliminates surviving particulates and acid-forming components. Carbon dioxide, water, and air then are emitted from the stack. As shown in Figure 12.12, ash particulates and wastewater are produced at various parts of the incineration process; these must be either treated or disposed of in a landfill. More advanced types of incineration and thermal decomposition of waste are being developed. One of these utilizes a molten salt bed that should be useful in destroying certain organic materials. Finally, other incineration techniques include liquid-injection incineration on land or sea, fluidized-bed systems, and multiple-hearth furnaces. Which incineration method is used for a particular waste depends upon

the nature and composition of the waste and the temperature necessary to destroy the hazardous components. For example, the generalized incineration system shown in Figure 12.12 could be used to destroy PCBs.

Summary of Management Alternatives

Direct land disposal of hazardous waste is often not the best initial alternative. Even with extensive safeguards and state-of-the-art designs, land disposal alternatives cannot guarantee that the waste is contained and will not cause environmental disruption in the future. This holds true for all land disposal facilities, including landfill, surface impoundments, land application, and injection wells. Pollution of air, land, surface water, and groundwater may result from failure of a land disposal site to contain hazardous waste. Groundwater pollution is perhaps the most significant result of such failure because it provides a convenient route for pollutants to reach humans and other living things. Figure 12.13 shows some of the paths that pollutants may take from land disposal sites to contaminate the environment. These paths include improper landfill procedures that eventually produce leakage and runoff to surface water or groundwater; seepage, runoff, or air emissions from unlined lagoons; percolation and seepage resulting from surface application of waste to soils; leaks in pipes or other equipment associated with deep-well injection; and leaks from buried drums, tanks, or other containers.

The philosophy of handling hazardous chemical waste should be multifaceted and should include such processes as source reduction, recycling and resource recovery, treatment, and incineration (24).

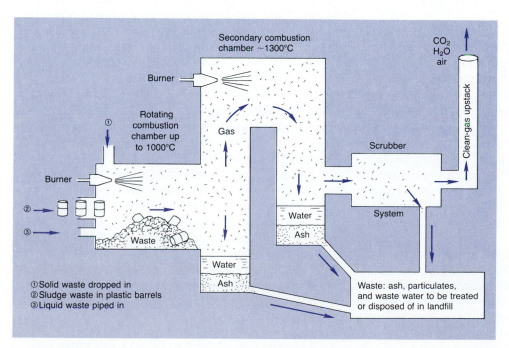

Figure 12.12
High-temperature incinerator system to burn toxic waste.

Figure 12.13
Examples of how land disposal/
treatment methods of managing
hazardous wastes may contaminate the environment. (Modified
after C. B. Cox, 1985, *The Buried
Threat*, California Senate Office of
Research, No. 115–5.)

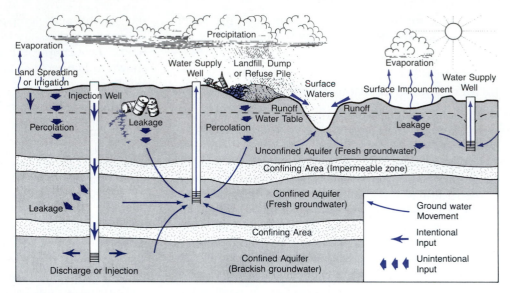

Source reduction.   Source reduction has the objective
of reducing the amount of hazardous waste that is
generated by a manufacturing or other process. For
example, changes in the chemical processes involved,
equipment used, raw materials used, or maintenance
measures may be successfully employed to reduce either the amount or toxicity of the hazardous waste
produced (24).

Recycling and resource recovery.   Hazardous chemical waste may contain materials that can be successfully
recovered for future use. For example, acids and solvents
collect contaminants when they are used in manufacturing processes. These acids and solvents can be processed
to remove the contaminants and can then be reused in
the same or different manufacturing processes (24).

Treatment.   Hazardous chemical waste may be treated
by a variety of processes to change the physical or
chemical composition of the waste in such a way as to
reduce its toxic or hazardous characteristics. Examples
include neutralizing acids, precipitation of heavy metals,
and oxidation to break up hazardous chemical compounds (2).

Incineration.   Hazardous chemical waste can be successfully destroyed by high-temperature incineration.
Incineration is considered to be a waste treatment rather
than a disposal method because the process produces
an ash residue that can be disposed of in a landfill
operation (19).

It has recently been argued that alternatives to land
disposal are not being utilized to their full potential. That
is, the volume of waste could be reduced and the
remaining waste could be recycled or treated in some
form prior to land disposal of the residues of the

treatment processes (24). Advantages to source reduction, recycling, treatment, and incineration are

▼ The actual waste that must be later disposed of is
  reduced to a much smaller volume.
▼ Useful chemicals may be reclaimed and reused.
▼ Treatment of wastes may make them less toxic and
  therefore less likely to cause problems in landfills.
▼ Because a smaller volume of hazardous waste is
  actually disposed of, there is less stress on the
  dwindling number of acceptable landfill sites.

▼ RADIOACTIVE WASTE MANAGEMENT

Radioactive wastes are byproducts that must be expected
as electricity is produced from nuclear reactors or
weapons are produced from plutonium. Considering
waste disposal procedures, radioactive waste may be
grouped into two general categories: low-level waste and
high-level waste. In addition, the tailings (materials that
are removed by mining activity but which are not
processed and which remain at the site) from uranium
mines and mills must also be considered very hazardous
(Figure 12.14). In the western United States, more than
20 million tons of abandoned tailings will produce
radiation for at least 100,000 years.

Low-level radioactive wastes are defined as radioactive materials that contain only small amounts of radioactivity and generally consist of a wide variety of items
such as residuals or solutions from chemical processing;
solid or liquid plant waste, sludges, and acids; and slightly
contaminated equipment, tools, plastic, glass, wood,
fabric, and other materials (25). Prior to disposal, liquid
low-level radioactive waste is solidified or packaged with
material capable of absorbing at least twice the volume of
liquid present (26).

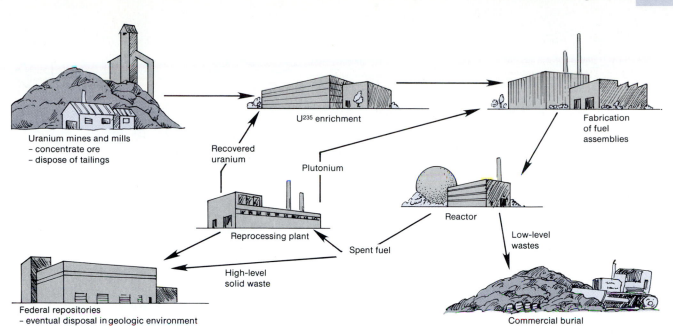

Figure 12.14
The nuclear fuel cycle. The U.S. does not now reprocess spent fuel. Disposal of tailings, which because of their large volume can be more toxic than high-level wastes, has been treated casually.

Radioactive decay of low-level waste does not generate a great deal of heat and a rule of thumb is that the material must be isolated from the environment for about 500 years to ensure that the level of radioactivity does not produce a hazard. In the United States the philosophy for management of low-level waste has been "dilute and disperse". Past experience suggests that low-level radioactive waste can be buried safely in carefully controlled and monitored near-surface burial areas in which the hydrologic and geologic conditions severely limit the migration of radioactivity (25). In the United States low-level radioactive waste has been buried at 15 main sites (Figure 12.15). Several of these sites have not provided adequate protection of the environment; this failure has been due at least in part to a poor understanding of the local hydrologic and geologic environment (26). For example, a study of the Oak Ridge National Laboratory in Tennessee suggests that the water table is less than seven meters below the ground surface in places. As a result, leachate generated from the disposal sites has a relatively short path to the groundwater. The investigation identified migration of radioactive materials from one of the burial sites and concluded that containment of the waste is difficult because of the short residence time of water in the vadose zone. That is, because the water table is shallow, it does not take long for surface waters to infiltrate and percolate down to the groundwater (26). On the other hand, the depth to the water table at the low-level radioactive waste disposal facility near Beatty, Nevada, is about 100 meters. That site

has apparently successfully contained radioactive waste, due in part to the great depth of the water table and thus low recharge rate to the groundwater. That is, because the water table is located at great depth, it takes a long time for any leachate generated to enter the groundwater environment (26).

The hydrology and geology of a particular disposal site plays a major role in containing radioactive materials at a low-level radioactive waste facility. Environmental hazards associated with low-level radioactive waste may persist for 500 years, and unfortunately most investigations of sites receiving waste have been ongoing for only about ten years. This time is obviously inadequate to properly estimate the success of particular sites. Nevertheless, several criteria have been delineated to produce an approach to disposal that is designated as a multiple-barrier concept. Figure 12.16 shows an idealized disposal site and some of the natural factors that provide for the multiple barriers likely to help confine the waste material (26). It is emphasized that no one site is likely to have all of these natural factors and so site selection must carefully consider options to minimize migration of radioactive waste from the burial site.

High-level wastes are produced as fuel assemblages in nuclear reactors become contaminated with large quantities of fission products. This spent fuel must periodically be removed and reprocessed or disposed of. Fuel assemblies will probably not be reprocessed in the near future in the United States; therefore the present waste management problems involve removal, trans-

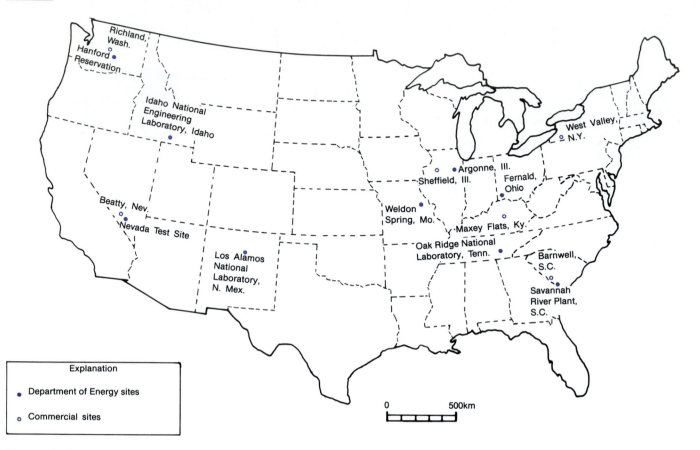

Figure 12.15
Location of low-level radioactive disposal sites. Many of these sites have been closed down.
(After J. N. Fischer, 1986, U.S. Geological Survey Circular 973.)

port, storage, and eventual disposal of spent fuel assemblies (27).

Hazardous radioactive materials produced from nuclear reactors include fission products such as krypton-85 (half-life of 10 years), strontium-90 (half-life of 28 years), and cesium-137 (half-life of 30 years). The **half-life** is the time required for the radioactivity to be reduced to one-half its original level. Generally, at least 10 half-lives (and preferably more) are minimal before a material is no longer considered a health hazard. Therefore, a mixture of the fission products mentioned above will require hundreds of years of confinement from the biosphere. Reactors also produce a small amount of plutonium-239 (half-life of 24,000 years), which is an artificial element. Because plutonium and the fission products must be isolated from the biological environment for very long periods of time (a quarter of a million years or more) their permanent disposal is a geologic problem.

Scope of Disposal Problem

High-level waste is extremely toxic and a sense of urgency surrounds its disposal as the total volume of spent fuel assemblies slowly accumulates. But there

is also conservative optimism that the waste disposal problem will be solved.

It has been projected that, if there is no disposal program, by the year 2000 several hundred thousand spent fuel elements from commercial reactors will be in storage awaiting disposal or eventual reprocessing to recover plutonium and unfissioned uranium. With reprocessing, solid high-level waste would occupy only several thousand cubic meters, a volume that would not even cover a football field to a depth of 1 meter (27).

Production of plutonium for weapons also generates high-level waste. At present, several hundred thousand cubic meters of liquid and solid high-level waste are being stored at U.S. Department of Energy repositories at Hanford, Washington; Savannah River, Georgia; and Idaho Falls, Idaho.

Serious problems with radioactive waste have occurred with liquid waste buried in underground tanks. Sixteen leaks involving 1330 cubic meters were located at Hanford from 1958 to 1973. An incident in 1973 involved a leak of 437 cubic meters of low-temperature waste. Since then, various improvements, including stronger, double-shelled storage tanks, reduction of the volume of liquid waste stored through the solidification program, and increased reserve capacity, have been made. It is

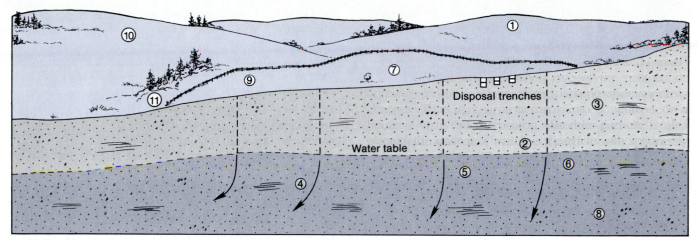

Explanation

1.	Low rainfall:	Means less surface infiltration and thus less production of leachate.
2.	Deep water table:	In general the deeper the water table the longer the time necessary for leachate to reach the groundwater.
3.	Modest soil hydrolic conductivity:	High hydrolic conductivity leads to rapid transport rates of fluids and very low hydrolic conductivity may lead to surface ponding, therefore moderate hydrolic conductivity is deemed more desirable.
4.	Slow moving groundwater:	Provides for or facilitates longer residency time and thus time of travel of contaminants from the disposal site to other locations.
5.	High adsorption and ion exchange rates:	Facilitates adhesion of radioactive molecules to the surface of other earth materials resulting in removal of contaminants.
6.	Homogeneous geology:	Makes it easier to predict likely movement of contaminants. Complex geology consisting of faults, folds, and other structures provides discontinuities along which waste may preferentially migrate.
7.	Topography and soils that minimize erosion:	Reduces the likelihood of exposure of waste by erosional processes.
8.	Absence of exploitable resources:	Waste must be isolated for hundreds of years and so obviously we do not want to pick a site with needed resources.
9.	Absence of surface-water bodies:	Reduces likelihood of pollution of water bodies.
10.	Low probability for faulting or volcanic activity:	Faulting or volcanic activity at or near a site obviously might jeopardize containment of low-level radioactive waste.
11.	Adequate buffer zone:	Recognizes that containment is not always possible and buffer zones provide additional time for radioactive decay to take place and thus are an added safety precaution.

Figure 12.16
Idealized diagram illustrating natural factors associated with suitability for burial of low-level radioactive waste. (After J. N. Fischer, 1986, U.S. Geological Survey Circular 973.)

hoped these improvements will reduce the chance of future incidents (28).

Storage of high-level waste is at best a temporary solution that allows the federal government to meet its commitments for accepting waste. Regardless of how safe any "storage" program is, it requires continuous surveillance and periodic repair or replacement of tanks or vaults. Therefore it is desirable to develop more permanent "disposal" methods in which retrievability may be possible but not absolutely necessary.

Disposal in the Geologic Environment

There is fair agreement that the geologic environment can provide the most certain safe containment of high-level radioactive waste. Because disposal of high-level waste is a certain necessity, the federal government is actively pursuing and developing possible alternative methods. Although such concepts as disposal into polar ice caps or into sediment in deep ocean basins have been explored, stable bedrock offers the most promise.

A comprehensive geologic disposal development program (27) should have a number of objectives:

▼ To identify sites that meet broad geologic criteria of tectonic stability and slow movement of groundwater with long flow paths to the surface
▼ To conduct intensive subsurface exploration of possible sites to positively determine geologic and hydrologic characteristics
▼ To predict future behavior of potential sites on the basis of present geologic and hydrologic situations and assumptions for future changes in such variables as climate, groundwater flow, erosion, and tectonics
▼ To evaluate risk associated with various predictions
▼ To make a political decision as to whether the risks are acceptable to society

The Nuclear Waste Policy Act of 1982 initiated a comprehensive federal-state, high-level nuclear waste disposal program. The Department of Energy was responsible for investigating several potential sites and the act originally called for the President to recommend a site by 1987. In December of 1987, Congress amended the act to specify that only the Yucca Mountain site in southern Nevada would be evaluated to determine if high-level radioactive waste could be disposed of there. The rock at the Yucca Mountain site where the repository would be constructed is tuff (naturally welded volcanic ash). It is believed by some scientists and others that the site was chosen, not so much for the geology, but because it is an existing nuclear reservation and therefore might draw minimal social and political opposition (28, 29).

The welded tuff at the Yucca Mountain site is a densely compacted volcanic ash. Fortunately, the site is located in an extremely dry region. Precipitation is about 15 centimeters per year and most of this runs off or evaporates. Hydrologists have estimated that less than 5 percent of the precipitation infiltrates the surface and eventually reaches the water table, which is several hundred meters below the surface. The depth to the potential repository is about 300 meters below the mountain's surface and as a result, the repository could be constructed at least 100 meters above the water table in the vadose zone (30).

During the 1990s the Department of Energy and the U.S. Geological Survey will complete extensive scientific evaluation of the Yucca Mountain site. These studies will help determine how well the geologic and hydrologic setting can isolate high-level nuclear waste from the environment. The site is attractive because there are several natural barriers present, including the arid climate, which greatly restricts downward movement of water; a strong rock (welded tuff that has a high sorption capacity for radioactive material); and a thick vadose zone (deep water table). If the studies indicate that the Yucca Mountain site could safely isolate radioactive waste, then the Department of Energy will apply to the U.S. Nuclear Regulatory Commission for a license to construct a disposal facility. If opposition from the state of Nevada is worked out and environmental and safety factors met, the repository might be ready for acceptance of high-level radioactive waste by the year 2010.

Long-term Safety

A major problem with the disposal of high-level radioactive waste remains: How credible are long-range (thousands of years to a few million) geologic predictions (27)? There is no easy answer to this question because geologic processes vary over both time and space. Climates change over long periods of time, as do areas of erosion, deposition, and groundwater activity. For example, large earthquakes hundreds or even thousands of kilometers from a site may permanently change groundwater levels. The seismic record for the western United States is only known for about the past hundred years and, therefore, estimates of future earthquake activity are tenuous at best. The bottom line is that geologists can evaluate relative stability of the geologic past, but they cannot guarantee future stability. Therefore, decision makers (not geologists) need to evaluate the uncertainty of prediction in light of pressing political, economic, and social concerns (27). These problems do not mean that the geologic environment is not suitable for safe containment of high-level radioactive waste, but care must be taken to ensure that the best possible decisions are made on this very critical and controversial issue.

▼ OCEAN DUMPING*

The oceans of the world, 363 million square kilometers of water, cover more than 70 percent of the earth. They play a part in maintaining the world's environment by providing the water necessary to maintain the hydrologic cycle, contributing to the maintenance of the oxygen–carbon dioxide balance in the atmosphere, and affecting global climate. In addition, the oceans are very valuable to people, providing such necessities as foods and minerals.

It seems reasonable that such an important resource as the ocean would receive preferential treatment, yet in 1968 alone, more than 43 million metric tons of waste were dumped at 246 sites off the coasts of the United States. Federal legislation and regulation by the Environmental Protection Agency in 1972 reduced the number of sites to 188, but ocean dumping continues to degrade the oceanic environment (4). Furthermore, if population

---

*The section on ocean dumping is summarized from *Ocean Dumping: A National Policy* (Washington, D. C.: Council on Environmental Quality, 1970), pp 1–25.

growth in coastal regions continues as expected, increased amounts of waste can be expected to adversely impact the oceans.

The types of wastes (31) that have been dumped in the oceans off the United States include the following: dredge spoils—solid materials such as sand, silt, clay, rock, and pollutants deposited from industrial and municipal discharges—removed from the bottom of water bodies generally to improve navigation; industrial waste—acids, refinery wastes, paper mill wastes, pesticide wastes, and assorted liquid wastes; sewage sludge—solid material (*sludge*) that remains after municipal wastewater treatment; construction and demolition debris—cinder block, plaster, excavation dirt, stone, tile, and other materials; solid waste—refuse, garbage, or trash; explosives; and radioactive waste.

The 1972 Marine Protection Research and Sanctuaries Act prohibits ocean dumping of radiological, chemical, and biological warfare agents and any high-level radioactive waste. Furthermore, it provides for regulation of all other waste disposal in the oceans off the United States by the Environmental Protection Agency or, in the case of dredge spoil, by the U.S. Army Corps of Engineers. In addition to materials prohibited by law from being dumped, the Environmental Protection Agency has prohibited material whose effect on marine ecosystems cannot be determined; persistent inert materials that float or remain suspended, unless they are processed to ensure that they sink and remain on the bottom; and material containing more than trace concentrations of mercury and mercury compounds, cadmium and cadmium compounds, organohalogen compounds (organic compounds of chlorine, fluorine, and iodine, etc.), and compounds that may form from such substances in the oceanic environment; as well as crude oil, fuel oil, heavy diesel oil, lubricating oils, and hydraulic fluids. The Environmental Protection Agency also requires special permits and strictly regulated dumping for arsenic, beryllium, chromium, and lead; low-level nuclear waste; organo-silicon compounds; organic chemicals; and petrochemicals. A major objective of the act is to prevent ocean dumping of wastes that, before recent federal laws were enacted, were discharged into rivers and the atmosphere (4).

Ocean Dumping and Pollution

Ocean dumping contributes to the larger problem of ocean pollution, which has seriously damaged the marine environment and caused a health hazard to people in some areas. Shellfish have been found to contain pathogens such as polio virus and hepatitis, and at least 20 percent of the nation's commercial shellfish beds have been closed because of pollution. Beaches and bays have been closed to recreational uses. Lifeless zones in the marine environment have been created. Heavy kills of fish and other organisms have occurred and profound changes in marine ecosystems have taken place (31).

Major impacts of marine pollution on oceanic life (31) include

▼ Killing or retarding growth, vitality, and reproductivity of marine organisms by toxic pollutants

▼ Reduction of dissolved oxygen necessary for marine life because of increased oxygen demand from organic decomposition of wastes

▼ Biostimulation by nutrient-rich waste, causing excessive blooms of algae in shallow waters of estuaries, bays, and parts of the continental shelf, resulting in depletion of oxygen and subsequent killing of algae that may wash up and pollute coastal areas

▼ Habitat change caused by waste-disposal practices that subtly or drastically change entire marine ecosystems

Major impacts on people and society caused by marine pollution include the following: production of a public health hazard caused by marine organisms transmitting disease to people; loss of visual and other amenities as beaches and harbors become polluted by solid waste, oil, and other materials; and economic loss. Loss of shellfish from pollution in the United States amounts to many millions of dollars per year. In addition, a great deal of money is being spent cleaning solid waste, liquid waste, and other pollutants in the coastal areas (31).

Ocean Dumping: The Conflict

It is unfortunate that the people interested in ocean dumping and those interested in harvesting marine resources such as shellfish and finned fish both prefer to do their jobs near the shore—the former because of convenience and transportation costs and the latter because of the richness of near-shore fisheries. The city of Los Angeles provides a classic example of the conflict.

For over 30 years, sewage and sewage sludge have been dumped several miles offshore into Santa Monica Bay. The rate of flow in recent years has been about 400 million gallons per day, only about 25 percent of which receive secondary treatment. Los Angeles successfully fought state and federal regulations for 14 years to avoid having to provide secondary treatment for all sewage.

It is now recognized that the bay is seriously polluted by the sewage and by earlier waste disposal going back to the 1940s, when oil refinery wastes, cyanide, and polychlorinated biphenyls (PCBs) began to be introduced into the marine environment. Concerns over potential health-related issues have convinced local government officials that all the sewage should receive secondary treatment. The Los Angeles City Council's recent decision to spend an additional $172 million for secondary treatment may eventually end a long environmental battle.

Alternatives to Ocean Dumping

The ultimate solution to the problem of ocean dumping is development of economically feasible and environmentally safe alternatives. To emphasize some of the possibilities, we will discuss some of the alternatives to ocean dumping of dredge spoils and sewage sludge.

Disposal of dredge spoils represents the vast majority of all ocean dumping. Such spoils are removed primarily to improve navigation and are usually redeposited only a few kilometers away. Approximately one-third are seriously polluted with heavy metals such as cadmium, chromium, lead, and nickel, as well as other industrial, municipal, and agricultural wastes. As a result, disposal in the marine environment can be a significant source of pollution (31).

The long-range alternative is phasing out ocean disposal of those dredge spoils that are polluted; however, this is not possible at present because of the great volume. Until land-based disposal can handle the necessary volume (which seems unlikely), interim techniques should be developed, such as determining which spoils are polluted before disposal and then hauling the polluted spoils farther from the dredging site to a safe disposal site. The main disadvantage here is the increased cost of longer hauls (31).

Sewage sludge disposed of at sea causes important environmental problems in terms of the volume, toxic and possibly pathogenic materials involved, and possible effects on marine life. Table 12.4 shows minimum, average, and maximum concentrations of three heavy metals found in sewage sludge, compared to normal concentrations in seawater and what concentrations are toxic to marine life. These data suggest that sewage sludge, like polluted dredge spoils, may be extremely harmful to marine life. Based on a very conservative estimate of 54 grams of sludge generated per person per day, by the year 2000 more than 1.8 million metric tons of sludge will be generated in the coastal regions, representing an increase of 50 percent in 30 years. Most sewage sludge is currently disposed of on land or is incinerated, but in some areas tremendous quantities are disposed of by ocean dumping (31).

Alternatives to ocean dumping of sewage sludge in coastal regions include various methods of land disposal.

Unfortunately, land disposal is considerably more expensive than near-shore marine disposal. However, if the sludge is digested (treated to control odors and pathogens) or barged long distances from shore, the cost becomes comparable, and land disposal may even be cheaper (31).

Perhaps the best long-range alternative to ocean dumping is the use of digested sludge for land reclamation, especially strip-mined land reclamation. Many strip mines in other areas are in great need of reclamation, and the nutrient value and organic material in this sludge can improve undesirable low-productivity soils by improving soil texture, composition, and nutrient levels. Unfortunately, there are limits on how much sludge can be realistically utilized and, as noted, some sludge contains toxic materials.

▼ SEPTIC-TANK SEWAGE DISPOSAL

Population movement in the United States continues to be from rural to urban, or urbanizing, areas. Although the most satisfactory method of sewage disposal is municipal sewers and sewage treatment facilities, often construction of an adequate sewage system has not kept pace with growth. As a result, the individual **septic tank** disposal system continues to be an important method of sewage disposal. Not all land, however, is suitable for installation of a septic-tank disposal system, so evaluation of individual sites is necessary and often required by law before a permit can be issued.

The basic parts of a septic-tank disposal system are shown in Figure 12.17. The sewer line from the house leads to an underground septic tank in the yard. The tank is designed to separate solids from liquid, digest and store organic matter through a period of detention, and allow the clarified liquid to discharge into the seepage bed, or absorption field, which is a system of piping through which treated sewage seeps into the surrounding soil.

Geologic factors that affect the suitability of a septic tank disposal system include type of soil, depth to the water table, depth to bedrock, and topography. These variables are generally listed with soil descriptions associated with a soil survey of a county or other area.

Table 12.4
Concentrations in parts per million (ppm) of heavy metals in sewage sludge.

Metal	Concentrations in Sewage Sludge (ppm)			Natural Concentrations in Seawater (ppm)	Concentrations Toxic to Marine Life (ppm)
	Min.	Avg.	Max.		
Copper	315	643	1,980	0.003	0.1
Zinc	1,350	2,459	3,700	0.01	10.0
Manganese	30	262	790	0.002	—

Source: *Ocean Dumping: A National Policy,* Council on Environmental Quality, 1970.

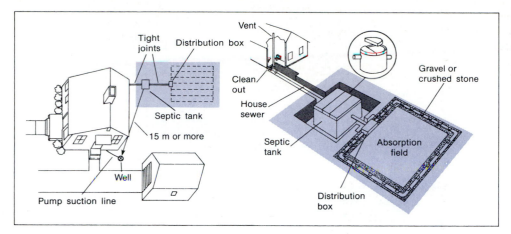

Figure 12.17
Septic-tank sewage disposal system (right) and location of the absorption field with respect to the house and well. (After Indiana State Board of Health.)

Soil surveys are published by the Soil Conservation Service and, when available, are extremely valuable in interpreting possible land use, such as suitability for a septic system. Since the reliability of a soils map for predicting limitations of soils is limited to an area larger than a few thousand square meters, and because soil types can change within a few meters, it is often desirable to have an on-site evaluation by a soil scientist. On some sites, it is also helpful to determine the rate at which water moves through the soil, which is best done by a percolation test. The test also is useful in calculating the size of the absorption field needed (Figure 12.18). Soils with a percolation rate of less than 2.5 centimeters per hour are always undesirable for a septic-tank disposal

system. Other criteria are these: the maximum seasonal elevation of the groundwater table should be at least 5 meters below the bottom of the absorption field; bedrock should be at a depth greater than 1.2 meters below the absorption field; and the surface slope generally should be less than 15 percent, or a 15-meter drop per 100-meter horizontal distance (32, 33). Figure 12.19 demonstrates some of these limitations.

Sewage absorption fields may fail for several reasons. The most common cause is poor soil drainage (Figure 12.20), which allows the effluent to rise to the surface in wet weather. Poor drainage can be expected in areas with clay or compacted soils with low hydraulic conductivity, high water table, rock with low hydraulic conductivity

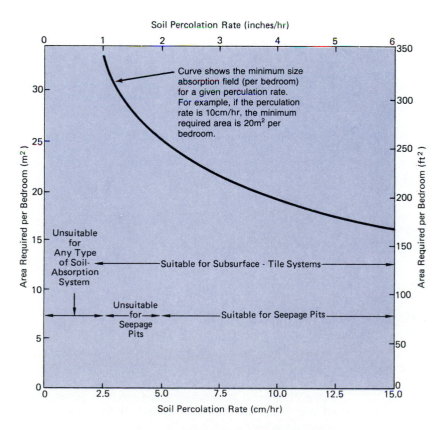

Figure 12.18
Graph used to determine the size of absorption field needed for a septic system for a private residence. (After Agriculture Information Bulletin 349, Soil Conservation Service, 1971.)

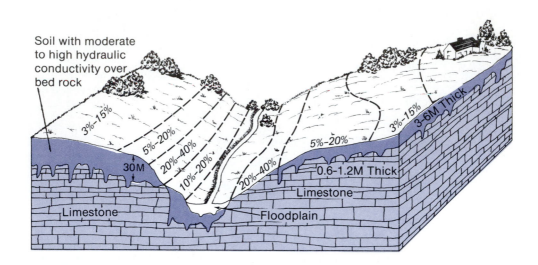

Figure 12.19
Block diagram illustrating some of the limitations of a septic system on limestone with a variable cover. The deep soil over the limestone at the left has moderate limitations, and construction may be difficult on the steeper slopes. The soil that is 3 to 6 meters thick on the extreme right may not be suitable for a septic system if the water supply is to come from a well. The shallow soil (0.6 to 1.2 meters thick) has severe limitations, and if a septic system is placed there, the stream at the bottom of the hill may become polluted. (After Agriculture Information Bulletin 349, Soil Conservation Service, 1971.)

near the surface, or frequent flooding. Steep topography also causes failure; however, if the absorption field is laid on the slope contours, a steeper slope may be accommodated (Figure 12.21) if there are no horizontal, impervious clay layers in the soil near the surface. Such layers allow the effluent to flow above them until it is discharged on the surface and runs unfiltered down the slope (32).

▼ **WASTEWATER TREATMENT**

The main purpose of wastewater treatment is to reduce the amount of suspended solids, bacteria, and oxygen-demanding materials in wastewater. In addition, new techniques are being developed to remove whatever harmful dissolved inorganic materials may be present. The new techniques, however, are more likely to reinforce rather than replace present technology (34).

Figure 12.20
Effluent from a septic-tank sewage disposal system rising to the surface in a backyard. The soil is poorly drained, and septic systems in such soils will not function well during wet weather. (Photo by W. B. Parker, courtesy of U.S. Department of Agriculture.)

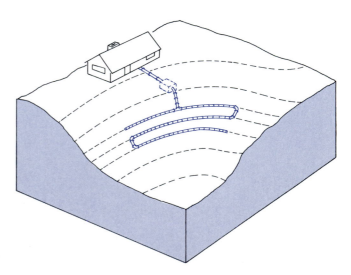

Figure 12.21
Block diagram illustrating how a septic field may be constructed on a steeper slope. The tile lines are laid out on the contour. This type of configuration is necessary on most sloping fields or in fields where there may be a change in soil type. (From Agriculture Information Bulletin 349, Soil Conservation Service, 1971.)

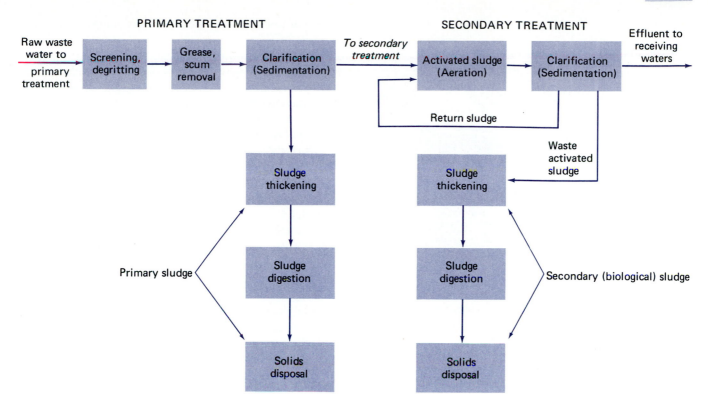

Figure 12.22

Flow chart showing various procedures in primary and secondary treatment of wastewater. (Reprinted from *Cleaning Our Environment—The Chemical Basis for Action,* a report by the Subcommittee on Environment Improvement, Committee on Chemistry and Public Affairs, American Chemical Society, 1969, p. 107. Reprinted by permission of the copyright owner.)

Existing wastewater treatment generally falls into two broad classes (Figure 12.22): *primary treatment,* involving removal of grit, screening, grinding, flocculation, and sedimentation; and *secondary treatment,* involving controlled biological assimilation and degradation processes that occur in nature by microorganisms. A third class, known as *tertiary treatment,* involves chemical treatment or other measures designed to further purify the water (34).

The most troublesome aspect of wastewater treatment is the handling and disposal of sludge produced in the treatment process. The amount of sludge produced is conservatively estimated at about 54 to 112 grams per person per day, and its disposal accounts for 25 to 50 percent of the capital and operating cost of a treatment plant.

Sludge handling and disposal have four main objectives (34):

1. To convert the organic matter to a relatively stable form
2. To reduce the volume of sludge by removing liquid
3. To destroy or control harmful organisms
4. To produce by-products whose use or sale reduces the cost of processing the sludge

Final disposal of sludge is accomplished by incineration, burying it in a landfill, using it for soil reclamation, or dumping it in the ocean. From an environmental standpoint, the best use of sludge is to improve soil texture and fertility in areas disturbed by activity such as strip mining and poor soil conservation.

Although it is unlikely that all the tremendous quantities of sludge from large metropolitan areas can ever be used for beneficial purposes, many smaller towns and many industries, institutions, and agricultural activities can take advantage of municipal and animal wastes by converting them into resources. This process of locally recycling liquid waste is called the *wastewater renovation and conservation cycle,* as shown schematically in Figure 12.23. The major processes in the cycle are: *return* of the treated wastewater to crops by sprinkler or other irrigation system; *renovation,* or natural purification by slow percolation of wastewater through soil to eventually recharge the groundwater resource with clean water; and *reuse* of the water by pumping it out of the ground for municipal, industrial, institutional, or agricultural purposes (35). Of course, not all aspects of the cycle are equally applicable to a particular waste-water problem. Waste recycling from cattle feed lots differs considerably from recycling from industrial or municipal sites. But the

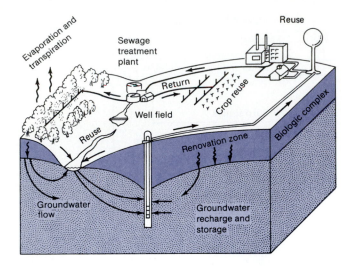

Figure 12.23
The wastewater renovation and conservation cycle. (From Parizek and Myers, American Water Resources Administration, 1968.)

general principle of recycling is valid, and the processes are similar in theory.

When recycling wastewater, the return and renovation processes are crucial, and soil and rock type, topography, climate, and vegetation play significant roles. Particularly important factors are the ability of the soil to safely assimilate waste, the ability of the selected vegetation to use the nutrients, and knowledge of how much wastewater can be applied (36).

Wastewater is recycled on a large scale near Muskegon and Whitehall, Michigan (Figure 12.24). Raw sewage from homes and industries is transported by sewers to the treatment plant where it receives primary and secondary treatment. The wastewaters are then chlorinated and pumped into a piping network that transports the effluent to a series of spray irrigation rigs. After the wastewater trickles down through the soil, it is collected in a network of tile drains and transported to the Muskegon River for final disposal. This last step is an indirect tertiary treatment, using the natural environment as a filter. The Michigan project is controversial because of concern for possible pollution of surface water and groundwater, as well as problems associated with an elevated groundwater table. However, it provides a possible alternative to direct (at a treatment plant) tertiary treatment, and experience gained from this project should be valuable in evaluating other possible sites for recycling wastewater. Table 12.5 summarizes three years of water treatment performance in Michigan. The system effectively removes most of the potential pollutants, as well as heavy metals, color, and viruses.

In Clayton County, Georgia, just south of Atlanta, a large water renovation and conservation cycle project was recently initiated. The project handles up to 760,000 cubic meters (20 million gallons) of wastewater per day, which is applied to a 972-hectare (2,400 acre) pine forest. Trees will be harvested on a 20-year rotation. The forest is part of the watershed that supplies water to the area, therefore wastewater is recycled to become part of the drinking-water supply (37).

An interesting example that served as a model project for subsequent recycling of wastewater comes from Santee, California, near San Diego. Even though the Santee project was eventually terminated, partly because of lack of customers for the reclaimed water (there is an abundant alternative water supply), it was innovative. After regular treatment of the sewage, the wastewater was stored for a month in a holding pond (Lake 1 in Figure 12.25) to allow further oxidation. Then, after chlorination, the water was pumped to a spreading area, where it trickled into the ground and flowed through a sand-gravel soil (an indirect tertiary treatment). The water surfaced again and entered a string of lakes, overflowing to form a stream used to irrigate a golf course. The geology of the Santee site in Sycamore Canyon is particularly suitable for lakes because a layer of impermeable clay with low hydraulic conductivity underlies (at a depth of 3 to 5 meters) surficial sands and gravels. The slope of the canyon and the existence of the sands and gravels overlying the clay allows for lateral movement and easy collection of the wastewater. The clay also seals the lakes, decreasing the possibility that wastewater will contaminate other water resources. Microbiological investigation suggested that the lake water was safe, and, in addition to providing wildlife habitat, the area was accepted by 12,500 residents and other people who knew where the water came from and used the lake area for a variety of recreational activities (38, 39).

A final example of an innovative system that uses naturally occurring earth materials advantageously to purify water for public consumption is reported from a Michigan community on the Lake Michigan shore. The city of Ludington, with a summer population of 11,000, uses sands and gravels below the lake bottom to prefilter and thus treat the lake water for municipal use. The procedure is shown schematically in Figure 12.26. A system of lateral intakes is buried in sand and gravel 4 to 5 meters below the lake bottom, where the water depth is at least 5 meters. The water is pumped out for municipal use, and in some cases the only additional treatment is chlorination.

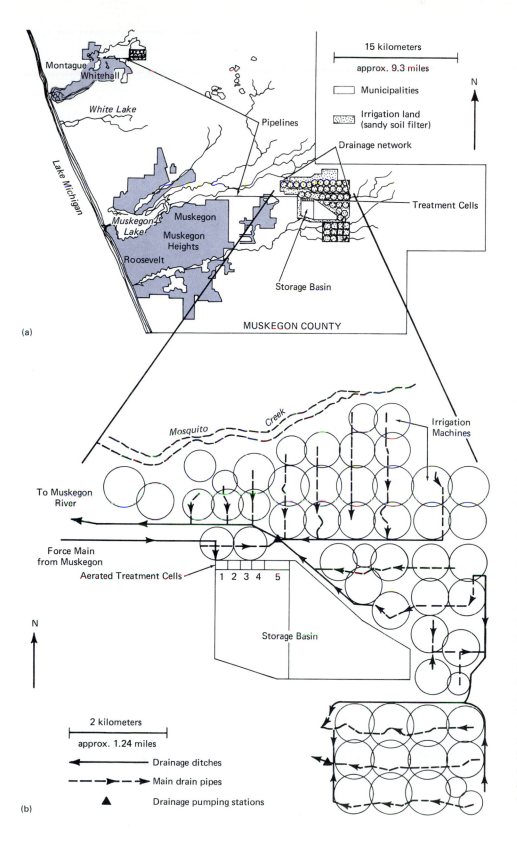

(a)

(b)

Figure 12.24
(a) Two effluent-application areas, one near Muskegon, the other near Whitehall, Michigan, eliminate the need to discharge inadequately treated industrial and municipal wastes to surface waters. These two subsystems replace four municipal treatment facilities [total flow, 11 mgd (41,500 m³/d)] and eliminate direct discharges of five industrial plants [combined flow of 19 mgd (71,800 m³/d)]. The effluent-on-land system will handle the 1992 requirements of the county, with a population of 170,000 generating a flow of 43.4 mgd (165,000 m³/d), including an industrial flow of 24 mgd (90,600 m³/d). Irrigation land consists of 6,000 acres (24.3 million m²) surrounding the treatment cells and storage basins. (b) Raw sewage from the sewers of Muskegon is transported via a pressure main to this Muskegon–Mona Lake reclamation site. There, the sewage flows through three aerated treatment lagoons (1, 2, and 3) and is then diverted either into the large storage basin or into the settling lagoon (4) before going into the outlet lagoon (5). Effluent leaving the outlet lagoon is chlorinated and then pumped through a piping system that carries it to 55 spray-irrigation rigs. The circles symbolize the radius swept out by the spray rigs. After the effluent trickles down through the soil, it is collected by a network of underground pipes and pumped to surface waters. A series of berms around the lowland edges of the irrigation circles will contain storm-water runoff. These waters will pond within the circles until handled by the drainage system. [Reprinted, by permission, from E. I. Chaiken, S. Poloncsik, and C. D. Wilson, *Civil Engineering* 42 (New York: American Society of Civil Engineers, 1973.)]

Table 12.5
Removal efficiency, Muskegon land treatment system (1976–78).

Wastewater Resource	3-Year Average Concentration (mg/liter)		Percent Removal
	Influent	**Effluent**	
BOD	300	3–4	99
Suspended solids	290	7	98
Phosphorus	3	0.08	97
Nitrogen	9	2.5	72
Potassium	10.5	3	71

Source: Y. A. Demirjian and Council on Environmental Quality, Annual Report, 1979.

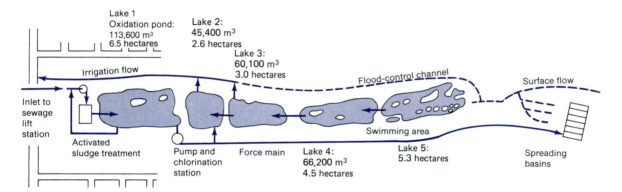

Figure 12.25
A water renovation cycle at Santee, California. After leaving the treatment plant, water is taken to spreading basins where eventually it goes into the surface-water system and through a series of lakes to be used eventually for irrigation and other purposes. Although this project was terminated in the mid 1970s, it has served as a model for other water renovation projects. (Reprinted by permission from *POWER,* June 1966.)

Figure 12.26
The water supply for Ludington, Michigan, is derived from the waters of Lake Michigan, using the natural, sandy bottom of the lake as a filter. (Courtesy of Williams and Works.)

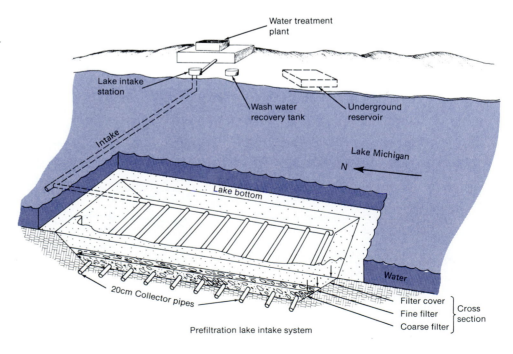

▼ ▼ ▼ SUMMARY AND CONCLUSIONS

Waste-management practices since the industrial revolution have moved from "dilute and disperse," to "concentrate and contain," to integrated waste management, which includes alternatives such as landfills, incineration, composting, recycling, reusing, and source reduction.

The most common method for disposal of urban waste is the sanitary landfill. However, around many large cities, space for landfills is hard to find and few people wish to live near any waste disposal operation. We are headed toward a disposal crisis if the new methods and ideas of integrated waste management are not acted upon soon.

Hazardous-chemical waste management may be the most serious environmental problem in the United States. Hundreds or even thousands of uncontrolled disposal sites may be time bombs that eventually will cause serious public health problems. Realistically, we will continue to produce hazardous chemical wastes. Therefore, it is imperative that safe disposal methods

be developed and used. Options in the management of hazardous chemical wastes include on-site processing to recover byproducts with commercial value, microbial breakdown, chemical stabilization, incineration and disposal of residue by secure landfill, and deep-well injection.

Radioactive-waste management presents a serious and ever-increasing problem. High-level wastes remain hazardous for thousands of years; those currently being stored will eventually have to be permanently disposed of. A promising method is disposal in a carefully and continuously monitored appropriate geologic environment. Apparently, low-level waste can be safely buried and carefully controlled and monitored at near-surface sites.

Because ocean dumping can be a significant source of marine pollution, efforts to control indiscriminate dumping are in effect. Alternatives to ocean dumping of materials such as polluted dredge spoils, sewage sludge, and other potentially hazardous materials are being developed, but in many cases such

alternatives are not yet practical or economically feasible.

With continued urban expansion, the septic-tank sewage disposal system will remain an important and common method of waste disposal. However, suitable soil and topography are necessary to reduce the chance of the system's failing, which could result in environmental pollution and possible public health problems.

The costs of wastewater treatment are expected to remain the highest environmental costs facing state and local governments. One of the largest expenses of operating and maintaining a wastewater treatment facility is that of the handling and disposal of sewage sludge.

An exciting possibility for some municipal, industrial, and agricultural wastewater is that of a wastewater renovation and conservation cycle. The basic idea is to convert treated wastewater into a resource for local recycling.

▼ ▼ ▼ REFERENCES

1. RELIS, P., and DOMINSKI, A. 1987. *Beyond the crisis: Integrated waste management.* Santa Barbara, California: Community Environmental Council.
2. MacFADYEN, J. T. 1985. Where will all the garbage go? *The Atlantic* 225, (3): 29–38.
3. GALLEY, J. E. 1968. Economic and industrial potential of geologic basins and reservoir strata. In *Subsurface disposal in geologic basins: A study of reservoir strata,* ed. J. E. Galley, pp. 1–19. American Association of Petroleum Geologists Memoir 10.
4. COUNCIL ON ENVIRONMENTAL QUALITY 1973. *Environmental quality—1973.* Washington, D.C.: U.S. Government Printing Office.
5. YOUNG, J. E. 1991. Reducing waste, saving materials. In *State of the World,* ed. L. R. Brown, pp. 39–55. New York: World Watch Institute, W. W. Norton.
6. SCHNEIDER, W. J. 1970. *Hydraulic implications of solid-waste disposal.*

U.S. Geological Survey Circular 601F.
7. TURK, L. J. 1970. Disposal of solid wastes—Acceptable practice or geological nightmare? In *Environmental geology,* pp. 1–42. Washington, D.C.: American Geological Institute Short Course, American Geological Institute.
8. HUGHES, G. M. 1972. *Hydrologic considerations in the siting and design of landfills.* Environmental Geology Notes, No. 51. Illinois State Geological Survey.
9. BERGSTROM, R. E. 1968. *Disposal of wastes: Scientific and administrative considerations.* Environmental Geology Notes, No. 20. Illinois State Geological Survey.
10. CARTWRIGHT, K., and SHERMAN, F. B. 1969. *Evaluating sanitary landfill sites in Illinois.* Environmental Geology Notes, No. 27. Illinois State Geological Survey.
11. WALKER, W. H. 1974. Monitoring toxic chemical pollution from land disposal sites in humid regions.

Ground Water 12: 213–18.
12. ENVIRONMENTAL PROTECTION AGENCY. 1980. *Everybody's problem: Hazardous waste.* SW-826.
13. MAGNUSON, E. 1980. The poisoning of America. *Time* 116(12): 58–69.
14. ELLIOT, J. 1980. Lessons from Love Canal. *Journal of the American Medical Association* 240: 2033–34, 2040.
15. KUFS, C., and TWEDWELL, C. 1980. Cleaning up hazardous landfills. *Geotimes* 25: 18–19.
16. ALBESON, P. H. 1983. Waste management. *Science* 220: 1003.
17. ESCH, M. 1985. Love Canal waste leaves ghost town. *Santa Barbara News Press.* Jan. 27.
18. Return to Love Canal. 1990. *Time,* May 28, p. 27.
19. HUDDLESTON, R. L. 1979. Solid-waste disposal: Landfarming. *Chemical Engineering* 86(5): 119–24.
20. McKENZIE, G. D., and PETTYJOHN, W. A. 1975. Subsurface waste management. In *Man and his physical*

environment, eds. G. D. McKenzie and R. O. Utgard, pp. 150–56. Minneapolis: Burgess Publishing.

21. PIPER, A. M. 1970. *Disposal of liquid wastes by injection underground: Neither myth nor millennium.* U.S. Geological Survey Circular 631.

22. COMMITTEE OF GEOLOGICAL SCIENCES. 1972. *The earth and human affairs.* San Francisco: Canfield Press.

23. WARNER, D. L. 1968. Subsurface disposal of liquid industrial wastes by deep-well injection. In *Subsurface disposal in geologic basins: A study of reservoir strata,*, ed. J. E. Galley, pp. 11–20. American Association of Petroleum Geologists Memoir 10.

24. CECELIA, C. 1985. *The buried threat.* California Senate Office of Research. No. 115-5.

25. OFFICE OF INDUSTRY RELATIONS. 1974. *The nuclear industry, 1974.* Washington, D.C.: U.S. Government Printing Office.

26. FISCHER, J. N. 1986. *Hydrologic factors in the selection of shallow land burial for the disposal of low-level radioactive waste.* U.S. Geological Survey Circular 973.

27. BREDEHOEFT, J. D.; ENGLAND, A. W.; STEWART, D. B.; TRASK, J. J.; and WINOGRAD, I. J. 1978. *Geologic disposal of high-level radioactive wastes—Earth science perspectives.* U.S. Geological Survey Circular 779.

28. HUNT, C. B. 1983. How safe are nuclear waste sites? *Geotimes* 28(7): 21–22.

29. HEIKEN, G. 1979. Pyroclastic flow deposits. *American Scientist* 67: 564–71.

30. U.S Department of Energy. 1990. *Yucca Mountain project: Technical status report.* DE90015030.

31. COUNCIL ON ENVIRONMENTAL QUALITY. 1970. *Ocean dumping: A national policy.* Washington, D.C.: U.S. Government Printing Office.

32. SOIL CONSERVATION SERVICE. 1971. *Soils and septic tanks.* Agriculture Information Bulletin 349. Washington, D.C.: U.S. Government Printing Office.

33. U.S. DEPARTMENT OF HEALTH, EDUCATION, AND WELFARE. 1967. *Manual of septic tank practice.*

Rockville, Maryland: Bureau of Community Management.

34. AMERICAN CHEMICAL SOCIETY. 1969. *Clean our environment: The chemical basis for action.* Washington, D.C.: U.S. Government Printing Office.

35. PARIZEK, R. R., and MYERS, E. A. 1968. Recharge of ground water from renovated sewage effluent by spray irrigation. *Proceedings of the Fourth American Water Resources Conference,* pp. 425–43.

36. FOGG, C. E. 1972. Waste management—nationally. *Soil Conservation,* May 1972.

37. BASTIAN, R. K., and BENFORADO, J., 1983. Waste treatment: Doing what comes naturally. *Technology Review* Feb./March: 59–66.

38. POWER. 1966. *Water: A special report.* New York.

39. McCAULEY, D. 1977. Water reclamation and re-use in six cities. In *Water re-use and the city,* ed. R. E. Kasperson and J. X. Kasperson, pp. 49–73. Hanover, New Hampshire: The University Press of New England.

The geologic cycles—tectonic, rock, geochemical, and hydrologic—are responsible on a global scale for the geographic pattern of continents, ocean basins, and climates. On a smaller scale, the cycles produce the spatial arrangement of all landforms, rocks, minerals, soils, groundwater, and surface water. As a member of the biological community, the human race has carved a niche with other members of the community—a niche highly dependent on complex interrelations among the biosphere, atmosphere, hydrosphere, and lithosphere. We are only beginning to inquire into and gain a basic understanding of the total range of factors that affect our health and well-being. As we continue our exploration of the geologic cycle—from the scale of minute quantities of elements in soil, rocks, and water to regional patterns of climate, geology, and topography—we are making startling discoveries about how these factors might influence the incidence of certain diseases and death rates. In the United States alone, the death rate varies considerably from one area to another (1).

Disease has been attributed to an imbalance resulting from a poor adjustment between an individual and the environment (2). It seldom has a one-cause, one-effect relationship. The geologist's contribution is to help isolate aspects of the geologic environment that may influence the incidence of disease. This tremendously complex task requires sound scientific inquiry coupled with interdisciplinary research with physicians and other scientists. Although the picture is now rather vague, the possible rewards of the emerging field of medical geology are exciting and may eventually play a significant role in environmental health.

▼ DETERMINING HEALTH FACTORS

To study the geologic aspects of environmental health, one must consider cultural and climatic factors associated with patterns of disease and death rates. This procedure helps to isolate the geologic influence. *Cultural aspects* of a society reflect the sum total of concepts and techniques that a group of people have developed to survive in their environment; these aspects influence the occurrence of disease by linking or separating disease and people. The nature and extent of these links depend on such factors as local customs and degree of industrialization. Primitive societies that live directly off the land and water are plagued by a different variety of health problems than urban society. Industrial societies have nearly eliminated such diseases as cholera, typhoid, hookworm, and dysentery, but are more likely to suffer from lung cancer and other diseases related to air, soil, and water pollution. Constructing homes on cement slabs that may develop cracks, and tightly insulating homes for energy conservation, have made our homes

CHAPTER THIRTEEN

▼

▼

▼

The Geologic Aspects of Environmental Health

more susceptible to indoor air pollution such as radon gas that may be linked to cancer of the lungs.

The high incidence of stomach cancer in Japanese people is an example of a relationship between culture and disease (3). The Japanese prefer rice that is polished and powdered; unfortunately, the powder contains asbestos as an impurity. Asbestos, a fibrous mineral, is either a true carcinogen (cancer-causing material) or takes a passive role as a carrier of trace-metal carcinogens.

Incidence of lead poisoning also suggests a cultural, political, and economic relationship to patterns of disease. Effects of lead poisoning can include anemia, mental retardation, and palsy. Lead is found in some moonshine whiskey and has resulted in lead poisoning among adults and even unborn or nursing infants whose mothers drank it (4). It has even been suggested that one of the reasons for the fall of the Roman Empire was widespread lead poisoning. Some estimate that the Romans produced about 55,000 metric tons of lead per year for 400 years. Lead was used in pots in which grape juice was processed into a syrup for a preservative and sweetener of wine, in cups from which the Romans drank wine, and in cosmetics and medicines. The ruling class also had water piped into their homes through lead pipes. Historians argue that gradual lead poisoning among the upper class resulted in their eventual demise through widespread stillbirths, deformities, and brain damage. This hypothesis has gained support because of the high lead content found in the bones of ancient Romans (4).

Climatic factors such as temperature, humidity, and amount of precipitation are sometimes intimately related to disease patterns. Serious health hazards exist in the tropics where two of the worst climatically controlled diseases, schistosomiasis and malaria, are found. Schisto-somiasis, called *snail fever,* is a significant cause of death among children and saps the energy of millions of people in the world. Its effects have tremendous socio-economic consequences, and some researchers consider it the world's most important disease (2).

Malaria and schistosomiasis, among other diseases, are clearly related to climatic factors because the vectors necessary to carry the disease, that is, mosquitoes and snails, are climatically controlled. Other diseases, such as Burkett's tumor—a debilitating, but seldom fatal, cancer-ous disease of the lymph system—is associated in Africa with three climatic conditions: annual rainfall greater than 51 centimeters, elevations lower than 1,500 meters, and mean annual temperature of at least 15° Celsius (3). Figure 13.1 shows the climatic conditions and the occurrence of Burkett's tumor in Africa. We emphasize that the relationship between climate and the disease may not be a cause-effect relationship; in fact, some workers believe that Burkett's tumor is caused by a virus. Nevertheless, the association of the disease with particu-lar climatic conditions is evident.

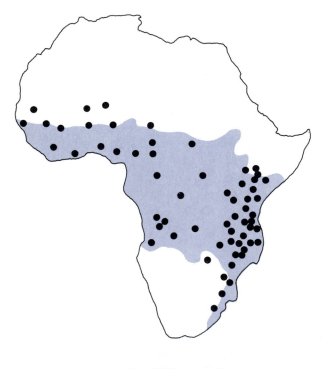

Areas below 1500 meters elevation and with mean annual temperature greater than 15°C and more than 51 centimeters of rainfall per year

● Sites of Burkett's tumor

Figure 13.1
Map of Africa showing areas below 1,500 meters in elevation with mean annual temperature greater than 15°C and more than 51 centimeters of rainfall, superimposed on sites where Burkett's tumor has been identified. (From Armed Forces Institute of Pathology.)

Assumed relations between culture and climate and the incidence of disease must be viewed with some skepticism because there is seldom a simple answer to environmental health problems. For example, if schisto-somiasis were controlled only climatically, then all areas with an appropriate climate, such as much of the Amazon River Basin, would have the disease. Fortunately, this is not the case, and in some instances the reason is geologic. The conditions for the disease in the Amazon River Basin are nearly optimal, yet it occurs only in two very limited areas, primarily because there is not a sufficient amount of calcium in the water to support the snail, the intermediate host. In other areas, the acidity of the water in the presence of copper and other heavy metals may be responsible for the absence of the necessary snails in an otherwise suitable environment for schistosomiasis (1).

From this discussion, we can see some of the complex relations between disease patterns and environ-ment. With this in mind, let us now focus on the geologic aspects of health problems by considering geologic factors of environmental health, occurrence and effects of trace elements on health, and the significance of the geologic environment to the incidence of heart disease and cancer, the leading causes of death in the United States. This chapter concludes with a discussion of radiation and radon gas, including the geologic factors involved and the potentially serious threat to human health.

▼ SOME GEOLOGIC FACTORS OF ENVIRONMENTAL HEALTH

The soil in which we cultivate plants for food, the rock on which we build our homes and industries, the water we drink, and the air we breathe all influence our chances of developing serious health problems. On the other hand, these same factors can also influence our chances of living a longer, more productive life. Surprisingly, many people still believe that soil, water, or air in a "natural," "pure," or "virgin" state must be "good" and that if human activities have changed or modified them, they have become "contaminated," "polluted," and therefore "bad." This belief is by no means the entire story (5).

Relationships between geology and health are signif-icant research and discussion topics. Although few valid cause-and-effect relationships have been isolated, we are learning more all the time about the subtle ways in which the geologic environment affects general health. Treating the various aspects of medical geology at even the introductory level requires discussion of the natural distributions of elements in the earth's crust and the ways in which natural and artificial processes concentrate or disperse those elements.

Natural Abundance of Elements

There is a very general inverse relation between atomic number and the abundance of elements in the universe. Lighter elements are encountered more frequently than most heavier ones. In general, this relation holds for both the lithosphere and the biosphere. Table 13.1, the periodic table of the elements, shows the position and atomic number of each element and indicates some of the more abundant elements in the earth's crust and environmentally important trace elements. Table 13.2 shows the most abundant elements in the average composition of the rocks of the continental crust. Note that more than 99 percent of these rocks by weight are concentrated in the first 26 elements of the periodic table. Table 13.3 shows the distribution of the more abundant elements in the average adult human body. Note that more than 99 percent of the body by weight is composed of elements in the first 20 elements of the periodic table.

Living tissue, animal or vegetable, is composed primarily of 11 elements, so-called **bulk elements**. Five of these are metals: hydrogen, sodium, magnesium, potassium, and calcium. Six are nonmetals: carbon, nitrogen, oxygen, phosphorus, sulfur, and chlorine. For those species that have hemoglobin, iron is added to the list. In addition to the bulk elements, living tissue requires several other elements to function properly. These are *trace-element metals* (present in minute quantities) that help to regulate the dynamic processes of life. Trace elements that have been studied and shown essential for nutrition include fluorine, chromium, manganese, iron, cobalt, copper, zinc, selenium, molybdenum, and iodine. This list is not complete, and it would not be surprising to learn that many more trace elements are essential or at least active in life processes (6, 7). Other elements, including nickel, arsenic, aluminum, and barium, accumulate as tissues age and are known as **age elements**. The physiological consequences of the accumulation of some elements in living tissue are known in some cases but completely unknown or poorly understood in others (7).

Concentration and Dispersion of Chemical Components

The movement of chemical compounds along various paths through the lithosphere, hydrosphere, atmosphere, and biosphere makes up the geochemical cycle. Natural processes, such as the release of gas by volcanic activity or the weathering of rock and rock debris, release chemical material into the environment. In addition, human use may result in the release of material and substances that lead to pollution or contamination of the environment. In general, but with many exceptions, concentrations of many trace elements tend to increase from rock to soil and water to plants and animals. Figure 13.2 shows some of the paths that trace elements may take to become concentrated in the human body, possibly causing health problems.

Once released by natural or artificial processes, elements and other substances are cycled and recycled by geochemical and rock-forming processes. Thus, a certain concentration of a particular trace element in igneous rocks may have quite a different concentration in sedimentary rocks formed from the weathered products of the igneous rock. Whether the concentration has increased or decreased depends on the nature of the geochemical and rock-forming processes. Table 13.4 (p. 318) lists the concentrations of selected elements in igneous and sedimentary rocks. Although this information is not detailed, it is useful because it indicates a change in the relative abundance of elements produced by rock-forming and biological processes, as, for example, the approximately tenfold increase in selenium from the original weathering of igneous rocks to the formation of shale. With the exception of coal and phosphorites (rock that is rich in calcium phosphate), other types of sedimentary rocks do not show a similar increase in selenium. This example and others suggest that geochemical and rock-forming processes such as weathering, leaching, accretion, deposition, and biological activity effectively sort, concentrate, and disperse elements and other substances throughout the environment.

Weathering is the physical and chemical breakdown of rock material and a major process in the formation of soil. Regardless of whether the parent material for soil is bedrock or rock debris transported and deposited by running water, wind, or ice, weathering is a natural process that frees trace elements to be used by the biosphere in life processes.

The artificial counterpart of weathering is pollution or contamination. These processes release trace elements into the environment; for example, lead is released into the environment when lead additives in gasoline are emitted through exhaust systems. Mercury, cadmium, nickel, zinc, and other metals are released into the atmosphere and water through industrial and mining operations.

Because some studies have shown a dramatic increase of lead in the air, water, and soil, and even the Antarctic ice, it is obvious that we need a closer examination of the possible adverse effects of chronic lead exposure. With the exception of human lead poisoning through lead-based paint and plant damage caused by exposure to lead close to highways or by heavy mineral concentrations released in mining areas, however, documented adverse health effects of lead are somewhat speculative. In fact, one author claims there is little difference in the amounts of lead in blood and urine samples from urban and rural residents (8).

Table 13.1

316

Legend:
- Atomic Number
- * Element Relatively Abundant In Earth's Crust
- Element Symbol
- ** Environmentally Important Trace Elements
- Element Name

1	2	3	4	5	6	7	8	9	10	11	12	13	14	15	16	17	18
1 H Hydrogen																	2 He Helium
3 Li ** Lithium	4 Be Beryllium											5 B * Boron	6 C * Carbon	7 N Nitrogen	8 O * Oxygen	9 F ** Fluorine	10 Ne Neon
11 Na * Sodium	12 Mg * Magnesium											13 Al * Aluminum	14 Si * Silicon	15 P Phosphorus	16 S Sulfur	17 Cl Chlorine	18 Ar Argon
19 K * Potassium	20 Ca * Calcium	21 Sc Scandium	22 Ti Titanium	23 V ** Vanadium	24 Cr ** Chromium	25 Mn ** Manganese	26 Fe ** Iron	27 Co ** Cobalt	28 Ni ** Nickel	29 Cu ** Copper	30 Zn ** Zinc	31 Ga Gallium	32 Ge Germanium	33 As Arsenic	34 Se ** Selenium	35 Br Bromine	36 Kr Krypton
37 Rb Rubidium	38 Sr Strontium	39 Y Yttrium	40 Zr Zirconium	41 Nb Niobium	42 Mo ** Molybdenum	43 Tc Technetium	44 Ru Ruthenium	45 Rh Rhodium	46 Pd Palladium	47 Ag Silver	48 Cd ** Cadmium	49 In Indium	50 Sn ** Tin	51 Sb Antimony	52 Te Tellurium	53 I ** Iodine	54 Xe Xenon
55 Cs Cesium	56 Ba Barium	57 La Lanthanum	72 Hf Hafnium	73 Ta Tantalum	74 W Wolfram	75 Re Rhenium	76 Os Osmium	77 Ir Iridium	78 Pt Platinum	79 Au Gold	80 Hg ** Mercury	81 Tl Thallium	82 Pb ** Lead	83 Bi Bismuth	84 Po ** Polonium	85 At Astatine	86 Rn ** Radon
87 Fr Francium	88 Ra ** Radium	89 Ac Actinium															

58 Ce Cerium	59 Pr Praseodymium	60 Nd Neodymium	61 Pm Promethium	62 Sm Samarium	63 Eu Europium	64 Gd Gadolinium	65 Tb Terbium	66 Dy Dysprosium	67 Ho Holmium	68 Er Erbium	69 Tm Thulium	70 Yb Ytterbium	71 Lu Lutetium
90 Th Thorium	91 Pa Protactinium	92 U ** Uranium	93 Np Neptunium	94 Pu Plutonium	95 Am Americium	96 Cm Curium	97 Bk Berkelium	98 Cf Californium	99 Es Einsteinium	100 Fm Fermium	101 Md Mendelevium	102 No Nobelium	103 Lw Lawrencium

Table 13.1
Periodic table of the elements with examples of environmentally important trace elements that are relatively abundant in the earth's crust highlighted.

Table 13.2
The relative abundance of the most common elements in the rocks of the earth's crust.

Atomic No.		Element	Weight (%)
8	O	Oxygen	46.4
14	Si	Silicon	28.15
13	Al	Aluminum	8.23
26	Fe	Iron	5.63
20	Ca	Calcium	4.15
11	Na	Sodium	2.36
12	Mg	Magnesium	2.33
19	K	Potassium	2.09
		Total	99.34

Table 13.3
Distribution of the more abundant elements in the adult human body.

Atomic No.	Element		Weight (%)
8	Oxygen	(O)	65.0
6	Carbon	(C)	18.0
1	Hydrogen	(H)	10.0
7	Nitrogen	(N)	3.0
20	Calcium	(Ca)	1.5
15	Phosphorus	(P)	1.0
16	Sulfur	(S)	0.25
19	Potassium	(K)	0.2
11	Sodium	(Na)	0.15
17	Chlorine	(Cl)	0.15
12	Magnesium	(Mg)	0.05
	Total		99.30

Leaching, accretion, deposition, biologic activity, and other processes may concentrate or disperse elements after they are released by natural and artificial processes. *Leaching* of soils is the natural removal of soluble material (in solution) from the upper to lower soil horizons. Material that leaches out of soil may enter the groundwater system and be dispersed or diluted. If the material is sufficiently abundant or toxic or otherwise harmful, it also may pollute the groundwater. Leaching from soils is most prevalent in warm, humid climates where the soil may be nutrient-poor because the nutrients are removed. Furthermore, trace elements that remain may be concentrated at undesirable levels.

Accumulation in soils refers to processes that cause or increase retention of material in soil. Examples include salts that may accumulate on the surface and the upper zones of soils through evaporation processes, and

materials that have been removed by leaching from the *A* horizon and accumulate in the *B* horizon. An example of the latter is found in semiarid regions where accumulation of calcium carbonate (caliche) is found in the *B* horizon of some soils. (See Chapter 4 for a discussion of soil horizons.)

Deposition of earth materials has two environmentally important aspects. First, materials such as heavy metals cause biologic disruptions when deposited in streams, lakes, and oceans. Second, problems develop in areas where a deficiency of needed trace elements results because the elements were not originally deposited along with other sediments.

Absorption of mercury to suspended sediments and bottom sediments may lead to high concentrations of mercury in aqueous environments. This absorption may lead to biological disruption because inorganic sub-

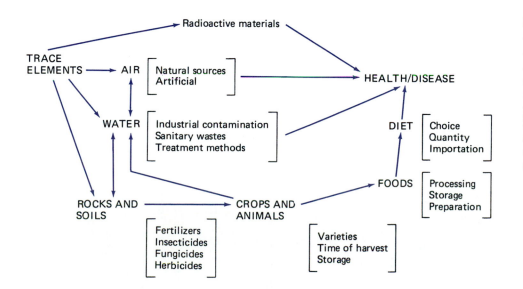

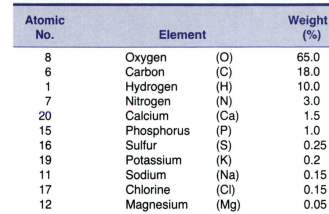

Figure 13.2
A schematic drawing showing mechanisms by which trace elements may find their way to humans and animals, thus influencing the quality of health or producing disease. (After K. E. Beeson, *Geochemistry and the Environment,* vol. 1, 1974. Reproduced with permission of the National Academy of Sciences.)

Table 13.4
Concentrations of some elements in various natural materials.

Type of Material	Concentrations by Elements (ppm)									
	Cadmium	Chromium	Copper	Fluorine	Iodine	Lead	Lithium	Molybdenum	Selenium	Zinc
Ultramafic igneous[a,b]	0–0.2	1,000–3,400	2–100	—	0.06–0.3	—	—	—	—	—
	0.05(?)	1,800	15		0.1	1	0.5	0.3	0.05	40
Basaltic igneous[a,b]	0.006–0.6	40–600	30–160	20–1,060	—	2–18	3–50	0.9–7	—	48–240
	0.2	220	90	360	0.5	6	20	1.5	0.05	110
Granitic igneous[a,b]	0.003–0.18	2–90	4–30	20–2,700	—	6–30	10–120	1–6	—	5–140
	0.15	20	15	870	0.5	18	35	1.4	0.05	40
Shales and clay[a,b,c]	0–11	30–590	18–120	10–7,600	2.2–380	16–50	4–400	—	—	18–180
	1.4	120	50	800	5(?)	20	80	2.6	0.6	90
Black shales (high C)[d]	0.3–8.4	26–1,000	20–200	—	—	7–150	—	1–300	—	34–1,500
	1.0	100	70			20		10		100(?)
Deep-sea clays[a,b]	0.1–1	—	—	—	11–50	—	—	—	—	—
	0.5	90	250		35(?)	80	57	27	0.17	165
Limestones[a,b,c,e]	—	—	—	0–1,200	0.4–29	—	5–10	—	—	—
	0.05	10	4	220	5	9	7	0.4	0.08	20
Sandstones[a,b,c]	—	—	—	10–880	—	1–31	7–90	—	—	2–41
	0.05	35	2	180	1.7	7	30	0.2	0.05	16
Phosphorites[f]	0–170	30–3,000	10–100	24,000–41,500	—	10–30	—	3–300	1–100	20–300
	30	300	30	31,000		10		30	18	50
Coals (ash)[a]	—	10–1,000	2–40	40–480	1–11	2–50	2–300	0.2–16	0.4–3.9	7–108
	2	20	15	80	4	15	50	5	2[g]	50

Note: The upper figure is the range usually reported; the lower figure, the average.

[a] Turekian and Wedepohl (1961).
[b] Parker (1967).
[c] Becker et al. (1972)
[d] Vine, J. D., and Tourtelot, E. B., *Econ. Geol.* 65, 223 (1970).
[e] Wedepohl (1970).
[f] Gulbrandsen, R. A., *Geochim. Cosmochim. Acta* 30, 769 (1966).
[g] U.S. Geological Survey (1972).

Source: M. Fleischer, U.S. Geological Survey. Reproduced with permission of the National Academy of Science.

stances in the water are basic to the life processes of low life-forms in aquatic environments; these organisms assimilate the mercury, which is then passed on at higher and higher concentrations through the food chain.

Deficiencies that result when a certain material is not deposited with sediments moved by water, ice, and wind are less well understood because of the possibility of interactions of other processes such as leaching. Erosion and deposition by wind are particularly susceptible to selective removal. One author attributes the lack of lead, iron, copper, cobalt, and other materials in the sand hills of Nebraska to the fact that these metals occur in grains smaller and heavier than quartz and therefore were not moved with the quartz grains that formed the sand hills (3).

▼ TRACE ELEMENTS AND HEALTH

Every element has a whole spectrum of possible effects on a particular plant or animal. For example, selenium is toxic in seleniferous areas, has no observable effect in most conditions, and is beneficial to animal production (raising cattle, sheep, etc.) in some areas. The apparent contradiction is resolved when we recognize that the first case is one of oversupply of selenium; the second represents a balanced state; and the third case is one of deficiency, which in some cases is rectified by supplementing the animals' food supply with the element (9).

It was recognized many years ago that the effects of a certain trace element on a particular organism depend on the dose or concentration of the element. This **dose dependency** can be represented by a dose-response curve, as shown in Figure 13.3 (9, 10). When various concentrations of an element present in a biological system are plotted against effects on the organism, three things are apparent. First, while relatively large concen-

trations are toxic, injurious, and even lethal ($D-E-F$ in Figure 13.3), trace concentrations may be beneficial or even necessary for life ($A-B$). Second, the dose-response curve has two maxima ($B-C$) forming a plateau of optimal concentration and maximum benefit to life. Third, the threshold concentration, where harmful effects to life begin, is not at the origin (zero concentration), but varies with concentrations less than at point A and greater than at point D in Figure 13.3 (9, 10).

Points A, B, C, D, E, and F in Figure 13.3 are significant threshold concentrations. Unfortunately, points E and F are known only for a few substances for a few organisms, including people, and the really important point D is all but unknown (10). The width of the maximum-benefit plateau (points B, C) for a particular life form depends on the organism's particular physiological equilibrium (10). In other words, the different phases of activity, whether beneficial, harmful, or lethal, may differ widely both quantitatively and qualitatively for different substances, and, therefore, are completely observable only under special conditions (9).

A complete discussion of the geology and environmental effects of all trace substances could be a textbook in itself. Our objective here is to discuss representative examples to show the possible effects of imbalances of trace elements. For this purpose, we have selected fluorine, iodine, zinc, and selenium. In addition, it is valuable to relate trace-element problems to human use of the land, which we will do with examples of mining activity.

Fluorine

Fluorine is fairly abundant in rocks (Table 13.4) and soils. Most of the fluorine in soils is derived from the parent rock, but it can also be added by volcanic activity, which

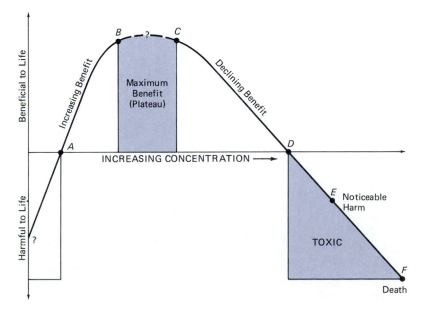

Figure 13.3
Generalized dose-response curve.

Figure 13.4
Dose-response curve for fluoride.

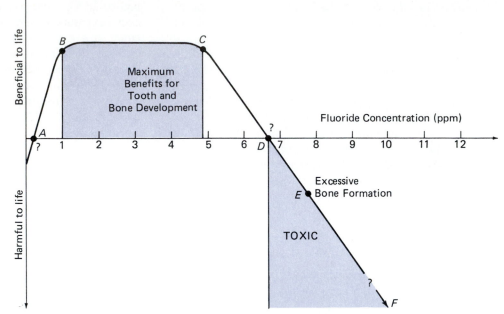

deposits fluorine-rich volcanic ash on the land. Industrial activity and application of fertilizers have also, on a limited basis, contributed locally to an increase in the concentration of fluorine in soils.

Fluorine is an important trace element that forms fluoride compounds such as calcium fluoride, which increases the crystallinity of the apatite (calcium phosphate) crystals in teeth. It helps prevent tooth decay by facilitating the growth of larger, more perfect crystals. The same processes occur in bones where fluoride assists in the development of more perfect bone structure that is less likely to fail with old age.

Relations between the concentration of fluoride (a compound of fluorine such as sodium fluoride, NaF) and health indicate a specific dose-response curve, as shown in Figure 13.4. The optimum fluoride concentration (point B) for the reduction of dental caries (DMF index: decayed, missing, and filled) is about 1 part per million (100 parts per million = 0.01 percent) (Figure 13.5). Fluoride levels greater than 1.5 parts per million do not significantly decrease the DMF index, but they do increase the occurrence and severity of mottling (discoloration of teeth) (11). In concentration of about 4 to 6 ppm, fluoride may help prevent calcification of the abdominal aorta. Finally, fluoride concentrations of 4 to 6 parts per million may also reduce the prevalence of osteoporosis, a disease characterized by reduction in bone mass (Figure 13.6) and collapsed vertebrae (12). The letters N.S. in Figure 13.6 indicate the differences that are not statistically significant. Based on the study of fluoride in osteoporosis, point C on the dose-response curve for fluoride was located (Figure 13.4). Point E on the curve was located by the fact that fluoride concentrations of 8 to 20 parts per million are associated with

excessive bone formation in the periosteum (dense, fibrous, outer layer of bone) and calcification of ligaments that usually do not calcify (7). The positions of points A, D, and F are not known precisely, but fluoride

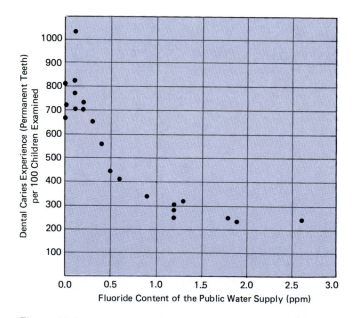

Figure 13.5
Relation between incidence of dental caries (in permanent teeth) observed in 7,257 selected 12- to 14-year-old school children in 21 cities of four states, and the fluoride content of public water supply. (Reprinted from F. J. Maier, *Water Quality and Treatment,* 3rd ed., by permission of the Association. Copyrighted 1971 by the American Water Works Association, Inc., 6666 West Quincy Avenue, Denver, Colorado 80235. Used with permission of McGraw-Hill Book Company.)

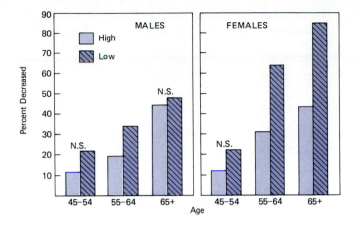

Figure 13.6
Percentage of decreased bone density in subjects from a high-fluoride area compared to those from a low-fluoride area. (Reprinted from the *Journal of the American Medical Association,* October 31, 1966, vol. 198. Copyright 1966, American Medical Association.)

in massive doses is the main ingredient of some rodent poisons.

Iodine

Thyroid disease, caused by a deficiency of iodine, is probably the best-known example of the relationship between geology and disease. The thyroid gland, located at the base of the neck, requires iodine for normal function. Lack of iodine causes goiter, a tumorous condition involving enlargement of the thyroid gland (13). Furthermore, a child born to a mother having iodine deficiency during pregnancy may suffer *cretinism,* characterized by stunted growth and mental disability. At

one time cretinism was fairly common in areas of Mexico and Switzerland that had a high goiter rate (14). The incidence of goiter is clearly related to deficiencies of iodine, as shown by the relationship between iodine-deficient areas and the occurrence of goiter in the United States, which is shown in Figure 13.7. Use of iodized salt is now common in the goiter belt, and it has been found that all adolescent and many adult goiters slowly decrease in size when iodized salt is used. In only four years (1924–28), the use of iodized salt in Michigan reduced the incidence of goiter from 38.6 to 9 percent (13). In spite of this, a recent study indicated that the use of iodized salt is not as prevalent as it was before World War II, and goiter is still endemic to some areas in the United States, especially among the poor (14).

There is considerable speculation over what processes are responsible for concentrating iodine in or removing it from surficial earth materials. The most popular hypothesis is that iodine is released by weathering of rocks. Some of it then enters the rivers and eventually the sea; therefore, the oceans have become great iodine reservoirs, containing possibly about 25 percent of the earth's total iodine (15). Because most iodine compounds are quite soluble, however, it is unlikely that much iodine is residual after long weathering processes. A more likely theory is that iodine is picked up in the ocean in a gaseous state or absorbed onto dust particles, transported by the atmosphere, and deposited by precipitation onto land areas. This idea proposes that iodine has slowly accumulated in the soil over a long period of time. Another possibility is related to the recent (geologically speaking) glaciation that ended just a few thousand years ago in the Great Lakes region. It might be argued that glaciated areas in the

Figure 13.7
Map of the United States showing relationship between goiter occurrence and iodine deficiencies. (From Armed Forces Institute of Pathology.)

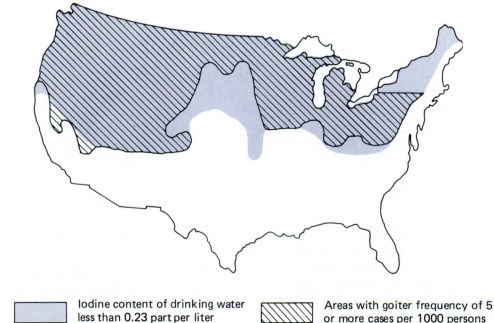

Iodine content of drinking water less than 0.23 part per liter

Areas with goiter frequency of 5 or more cases per 1000 persons

United States, which roughly coincide with the goiter belt, have low iodine concentrations in the soils because glaciation removed and destroyed the older soils that had accumulated iodine for a long period, and the young, post-glaciation soils have not had sufficient time to accumulate large concentrations of iodine from the atmosphere (15). This is unlikely, except from a relative standpoint, considering the solubility of iodine compounds.

Biologic processes also significantly affect the amount of iodine available to plants and animals in an area. Plants affect the iodine content in soils by absorbing iodine into their living tissue, and by retaining iodine in the organic-rich (humus) upper soil horizon. Several analyses have shown that iodine tends to be concentrated in the upper soil horizons. Therefore, biological processes that cycle iodine in soils and plants might be more significant in determining availability of iodine than the amount of iodine in the local bedrock (15).

Zinc

Zinc is a trace element necessary to plants, animals, and people. Although zinc is a heavy metal that in excessive amounts has been associated with disease, it is known primarily from zinc-deficiency studies. Zinc deficiencies are known in 32 states and have resulted in a variety of plant diseases that cause low yields, poor seed development, and even total crop loss (16).

Zinc deficiency in plants is related primarily to three soil conditions: low content of zinc, unavailability of zinc present in the soil, and poor soil management. The amount of zinc is low in areas that are highly leached, as in many coastal areas in the Coastal Plain of North Carolina, Georgia, and Florida. Zinc content is also low in areas such as the sand hills of northern Nebraska where wind transport of the sand that formed the hills failed to move the heavy metals along with the lighter quartz grains (16).

Zinc is recognized as essential to all animals and people, especially during early stages of development and growth. Although required concentrations are small, even slight deficiencies can cause loss of fertility, delayed healing, and disorders of bones, joints, and skin. Zinc deficiencies are not unknown in humans and may be associated with some chronic arterial disease, lung cancer, and other chronic diseases (13). One problem is that interrelationships between zinc deficiencies in hospital patients and the causes of disease are poorly understood. Do these people have the disease partially because they suffer from a zinc deficiency, or do they have a zinc deficiency because they have the disease? If the former is true, then zinc therapy with its known beneficial effects on tissue repair may be useful in treating some chronic diseases (16).

Adding zinc supplements to the soil may help prevent retardation of plant and animal growth. Care must be taken in adding zinc, however, because if the supplementary zinc is not of high quality, relatively large amounts of cadmium, which is often associated with zinc, may be inadvertently released into the environment. The result might be elimination of the growth problem, but a new hazard from the cadmium. Cadmium has been associated with bone disease, heart disease, and cancer.

Selenium

In concentrated amounts, selenium may be the most toxic element in the environment. It is a good example of why concern is increasing over the need for controlling health-related trace elements. Selenium is required in the diet of animals at a concentration of 0.04 part per million, beneficial to 0.1 part per million, and toxic above 4 parts per million. Selenium is of concern to biologists because of its toxicity, even though greater losses have resulted from selenium deficiency than from selenium toxicity (17, 18). Selenium toxicity has recently been discovered in the San Joaquin Valley in central California. The problem is threatening a large agricultural area, the environmental impact of which is discussed in Chapter 17.

The primary source of selenium is volcanic activity. It has been estimated that throughout the history of the earth, volcanoes have released about 0.1 gram of selenium for every square centimeter of the earth's surface (17). Selenium ejected from volcanoes is in a particulate form. Therefore, it is easily removed from volcanic gas by rain and is usually concentrated near the volcano. This explains why the average concentration of selenium in the crust of the earth is about 0.05 parts per million, whereas the soils of Hawaii, which are derived from volcanic material, contain 6 to 15 parts per million (17).

Selenium in soils varies from about 0.1 part per million in deficient areas to as much as 1,200 parts per million in organic-rich soils in toxic areas. Hawaii, with a selenium content of 6 to 15 parts per million in the soil, does not produce toxic seleniferous plants, whereas South Dakota and Kansas, with soils containing less than 1 part per million selenium, do produce toxic seleniferous plants. This apparent dichotomy is resolved by examining the availability of selenium. Elemental selenium is relatively stable in soils and not available to plants. Availability of the element is controlled primarily by whether the soil is alkaline or acidic. In acid soils (as in Hawaii), the selenium is in an insoluble form, whereas in alkaline soils, selenium may be oxidized to a form that is extremely soluble in water and thus readily available to plants. This explains how forage crops grown for animals in soil containing soluble selenium are toxic, whereas forage crops grown in soils containing insoluble selenium have selenium deficiencies (17).

Selenium tends to be somewhat concentrated in living organisms. Some plants, called *selenium accumulators*, may contain more than 2,000 parts per million selenium, whereas other plants in the same area may have less than 10 parts per million total selenium (18). Selenium levels in human blood range from 0.1 to 0.34 part per million, which is about 1,000 times that found in river water and several thousand times that found in seawater. Marine fish, on the other hand, contain selenium in amounts of about 2 parts per million which is many thousand times that found in seawater (17). Because selenium accumulates in organic material, it may be concentrated in organic sediment and soil. Therefore, fossil fuels such as coal that are developed from organic material also contain selenium. It has been estimated that the annual release of selenium by combustion of coal and oil in the United States is about 4,000 tons (17).

Little is known about selenium deficiency and toxicity in humans. However, one study concluded that muscular dystrophy, which is known to be related to a selenium deficiency for cows and sheep, may also be a causative factor for the disease in humans (19). A few cases of selenium poisoning have been reported in people who live on food grown in highly seleniferous soils or who drink seleniferous water in undeveloped countries. In economically developed countries where interregional shipments of food are standard and diets are high in protein, there is probably little problem with selenium (18).

Geographic patterns of selenium distribution are valuable in assessing whether forage and other feed crops are apt to contain relatively high or low selenium concentrations. Figure 13.8 shows the general distribution of selenium in plants in the United States. This map is a general guide to locating regions where the addition of selenium to animal feed might be beneficial. In some areas, selenium content cannot be accurately predicted, however, so detailed analysis and mapping remain necessary (18).

Human Use, Trace Substances, and Health

Agricultural, industrial, and mining activities have all been responsible for releasing potentially hazardous and toxic materials into the environment. This risk is part of the price we pay for our life-style. For example, unexpected and unanticipated problems have arisen from seemingly beneficial chemicals that control pests and disease. As we become more sophisticated at anticipating and correcting problems, it is expected that environmental disruption from release of trace substances will be reduced.

Geologically significant examples of unexpected problems with trace substances are found in mining processes. Ironically, we spend time, energy, and money to extract resources concentrated by the geological cycle but in doing so, sometimes concentrate and release potentially harmful and toxic trace elements into the environment. Two rather different examples illustrate this situation: in Japan, the occurrence of a serious bone disease related to mining zinc, lead, and cadmium; and in Missouri, a metabolic imbalance of cattle associated with mining clay for the ceramic industry.

A serious chronic disease known as **Itaiitai** in the Zintsu River Basin, Japan, has claimed many lives. This extremely painful disease—the name *Itaiitai* literally means "ouch, ouch"—attacks bones, causing them to become so thin and brittle that they break easily. The disease broke out near the end of World War II when the

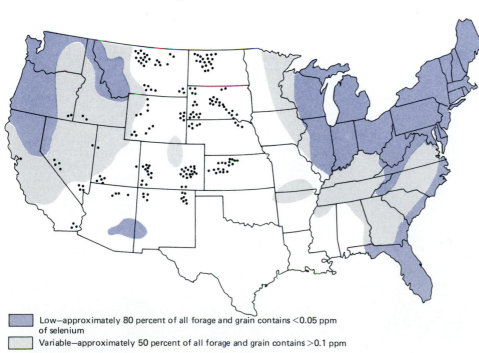

Figure 13.8
Map showing the concentration of selenium in plants in the conterminous United States. (From *Micronutrients in Agriculture*, 1972, by permission of the Soil Science Society of America, Inc.)

Low—approximately 80 percent of all forage and grain contains <0.05 ppm of selenium

Variable—approximately 50 percent of all forage and grain contains >0.1 ppm

Adequate—80 percent of all forage and grain contains >0.1 ppm of selenium
• Local areas where selenium accumulator plants contain\>50 ppm

Japanese industrial complex was damaged and good industrial-waste disposal practices were largely ignored. Mining operations for zinc, lead, and cadmium dumped mining wastes into the rivers. Farmers used the contaminated water downstream for domestic and agricultural purposes. For years, the cause of the disease was unknown. Then, in 1960, bones and tissues of the victims of Itaiitai disease were examined and found to contain large concentrations of zinc, lead, and cadmium (20).

Measurement of heavy-metal concentrations in the water, sediment, and plants of the Zintsu River Basin, and subsequent experiments on rats fed diets of heavy metal, established two facts. First, although the water samples generally contained less than 1 part per million cadmium and 50 parts per million zinc, these minerals are selectively concentrated at high rates in the sediment and higher yet in plants. One set of data for five samples shows an average of 6 parts per million cadmium in polluted soils. In plant roots, this average increased to 1,250, and in the rice it was 125. Second, rats fed a diet of 100 parts per million cadmium lost about 3 percent of their total bone tissue, and rats fed a diet containing 30 parts per million cadmium, 300 zinc, 150 lead, and 150 copper lost an equivalent of about 33 percent of their total bone tissue (20).

Although measurements of concentration of heavy metals are somewhat variable in the water, soil, and plants of the Zintsu River Basin, the general tendency is clear. Scientists are fairly certain that heavy metals, especially cadmium, in concentrations of a few parts per million in the soil and rice produce Itaiitai disease (21).

Strip mining of clay, coal, and other minerals may bring to the surface materials that contain anomalous concentrations of elements that are hazardous or toxic to an area's plants and animals. For example, a clay pit area of about 9 square kilometers in Missouri was investigated by the U.S. Geological Survey (22). Mining clay for use in the ceramic industry resulted in severe disturbances in the metabolism of beef cattle in the area; these disturbances interfered with the cattle's growth, nutrition, and reproduction.

The topography in the western part of the area is characterized by a flat upland surface generally above 244 meters in altitude. The underlying rocks are interbedded clay, shale, sandstone, and coal covered with a variable thickness of windblown silt deposits. The eastern part of the area is generally below 244 meters in altitude, and the underlying rocks are limestone. Sinkholes in the limestone often contain deposits of fire clay that is mined. Figure 13.9 shows an abandoned clay pit approximately 100 meters wide, about 20 meters deep, and nearly filled with water. The large clay pile consists of clay with numerous fragments of shale, limestone, and pyritic material (iron sulfide). The smaller pit is almost entirely clay. Part of the surface runoff drains into the clay pit, and the remainder drains into the Rocky Branch. Only the cattle located on land downstream with direct access to the Rocky Branch suffered from metabolic disorders.

Although the rock types contain normal concentrations of elements, they are relatively rich in some elements compared to the soils. The occurrence of pyrite in the rocks is a particular problem because when pyrite is exposed to water, it weathers to sulfuric acid, which increases the solubility of other compounds. This is essentially the same problem that is prevalent in coal-mining areas where acidic water from mines contaminates streams. Water that drains from clay piles into the pit and streams is very acidic and contains high concen-

Figure 13.9
Aerial view of a clay pit area, Callaway County, Missouri. Dump truck at lower right indicates the scale. (From Ebens, et al., U.S. Geological Survey Professional Paper 807, 1973.)

trations of some trace elements. This material is subsequently deposited in the bed of the stream and on the floodplains. Plants that grow on the clay piles along the floodplains contain some trace elements in concentrations toxic to animals that eat the plants (22).

Extensive analysis of more than 30 different elements in the clay, sediment, and plants of the clay pit area revealed that about 20 were present in anomalously high concentrations. These fit into one of five general groups:

1. Elements that occur in unusually high amounts in the clay or sediment or both, and are also found in anomalous amounts in many plants; examples include aluminum, copper, molybdenum, nickel, and sodium
2. Elements that occur in anomalous concentrations in the clay and sediment, but are seldom found in anomalous amounts in plants; examples include barium, beryllium, chromium, and vanadium
3. Elements that occur in relatively low amounts in the clay and sediment, but are concentrated in some plants; examples include boron, cadmium, calcium, and zinc
4. Elements that are anomalous in clay or sediment, but are not easily evaluated in plants; examples include carbon, selenium, and silicon
5. Elements whose concentrations and rates of movement through the local environment are not easily categorized at this time; examples include iron, lead, magnesium, manganese, and strontium (22).

This grouping suggests that some elements may influence metabolic imbalances in cattle because the animals eat plants with a toxic level of a particular element, as in the first and third groups. On the other hand, some elements may influence the metabolism of grazing animals because the animals directly ingest clay and sediment or drink water that contains particularly harmful elements in solution or suspension.

Evaluation of the entire geochemical environment established that, first, four elements (beryllium, copper, molybdenum, and nickel) are conspicuously concentrated in the clay, sediment, and plants; and second, some elements are highly mobile (beryllium, cobalt, copper, and nickel, among others). Of these elements, three (cobalt, copper, and molybdenum) are known to be significant in metabolic processes of animals. In trace amounts they are essential, but in high concentrations they are likely to be toxic (22).

The geologic, hydrologic, and biologic processes responsible for providing, transporting, and concentrating toxic trace elements are indeed complex. Figure 13.10 lists and describes the movement of elements through the geochemical system of the clay pit area. This example is especially valuable because it indicates the interaction of the complex, multidimensional aspects of geochemical systems. Indeed, complexity is much more common than simplicity in the study of anomalous concentrations of trace elements and their effects on the biosphere.

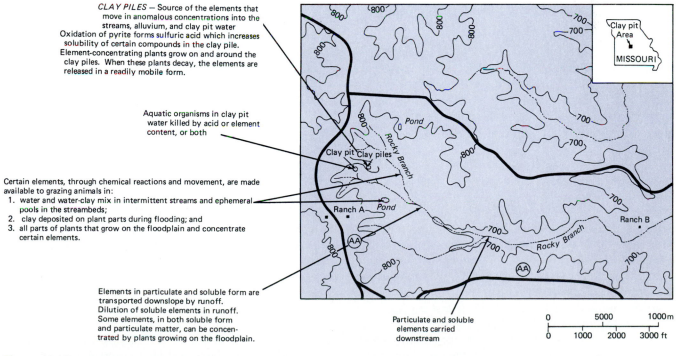

Figure 13.10
Movement of elements through the geochemical system of a clay pit area, Callaway County, Missouri. (From Ebens, et al., U.S. Geological Survey Professional Paper 807, 1973.) (700 to 800 ft = 213 to 244 m)

▼ CHRONIC DISEASE AND GEOLOGIC ENVIRONMENT

Health can be defined as an organism's state of adjustment to its own internal environment and to its external environment. The relationships between the geologic environment and regional or local variations in chronic disease such as cancer and heart disease in people have been observed for many years. Although evidence suggesting associations between the geological environment and chronic disease continues to accumulate, the real significance remains to be discovered. Reasons for the lack of conclusive results are twofold. First, hypotheses for testing relations between the geologic environment and disease have not been specific enough. Basic research and field verification need to be better coordinated. Second, many difficulties remain in obtaining reliable and comparable data for medical-geological studies (23).

Although the significance of geologic variations in contributing to disease compared to that of other environmental factors, such as climate, remains an educated guess, the benefit to humankind of learning more about these relationships is obvious. The geographic variations of the incidence of heart disease in the United States may be related to the geologic environment; and it has been estimated that two-thirds of the cancerous tumors in the Western Hemisphere result from environmental causes, while genetic and racial factors are secondary in importance (23).

Heart Disease and the Geochemical Environment

The term *heart disease* here includes coronary heart disease (CHD) and cardiovascular disease (CVD). Variations of heart disease mortality have generally shown interesting relationships with the chemistry of drinking water. For example, studies in Japan, England, Wales, Sweden, and the United States all conclude that communities with relatively soft water have a higher rate of heart disease.

A general relationship between soft water and high death rates from heart disease is also present in the United States. Figure 13.11 shows the adjusted death rate from heart disease from 1949 to 1951 and the average hardness of the water (24). The figures are consistent with more recent data which also found a significant negative correlation between hardness of water and death rates from heart disease (25). We emphasize, however, that the generally negative correlation between heart disease and hardness of water is not conclusive. A study in Indiana found a small positive correlation, suggesting that many other variables may exert considerable influence on rates of heart disease (26), and a study in Ohio suggests that sulfate (SO_4) and bicarbonate (HCO_3) concentrations possibly influence the incidence of heart disease (27). The Ohio study found a slight

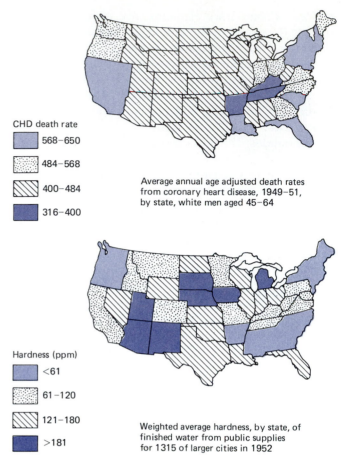

CHD death rate

☐ 568–650

⬚ 484–568

▨ 400–484

◼ 316–400

Average annual age adjusted death rates from coronary heart disease, 1949–51, by state, white men aged 45–64

Hardness (ppm)

◼ <61

⬚ 61–120

▨ 121–180

◼ >181

Weighted average hardness, by state, of finished water from public supplies for 1315 of larger cities in 1952

Figure 13.11
Maps of the conterminous United States comparing the average annual adjusted death rate from coronary heart disease (CHD) to the weighted average hardness of the public water supplies. (Reprinted from E. F. Winton and L. J. McCabe, *Journal of American Water Works Association,* Volume 62, by permission of the Association. Copyrighted 1971 by the American Water Works Association, Inc., 6666 West Quincy Avenue, Denver, Colorado 80235.)

positive relation between deaths from heart attack and sulfate concentration in drinking water. Ohio counties with sulfate-rich drinking water derived from coal-bearing rocks in the southeast part of the state tend to have a higher death rate due to heart attack (Figure 13.12). On the other hand, Ohio counties that have drinking water characterized by low-sulfate and high-bicarbonate concentration, derived from young glacial deposits, tend to have a relatively low death rate from heart attack (Figure 13.12) (27).

The meaning of relationships between water chemistry and death rates from heart disease is difficult to surmise, although there are several possibilities. First, the relationships may have nothing to do with heart disease. Second, soft water is acidic and may, through corrosion, attack pipes and release into the water trace elements that cause heart disease. Third, some substances dissolved in

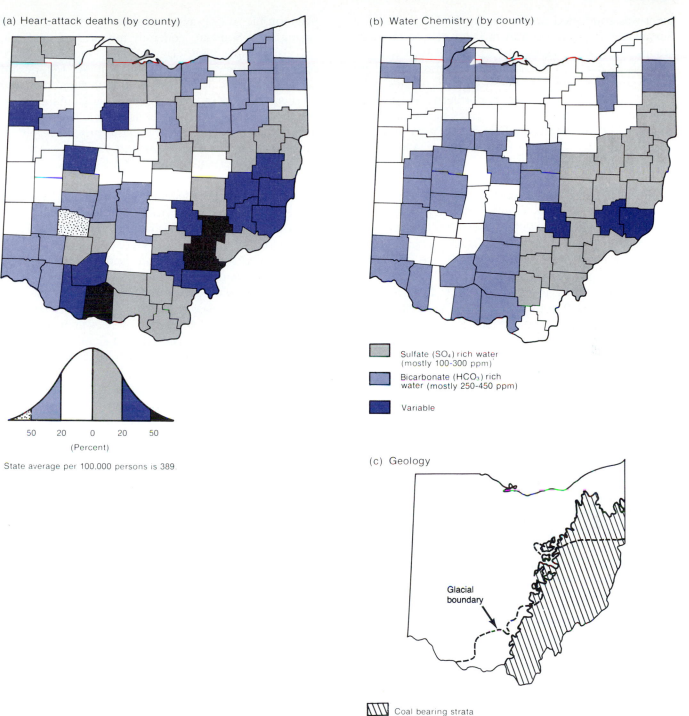

(a) Heart-attack deaths (by county)

(b) Water Chemistry (by county)

50   20   0   20   50
(Percent)

State average per 100,000 persons is 389.

Sulfate (SO₄) rich water
(mostly 100-300 ppm)

Bicarbonate (HCO₃) rich
water (mostly 250-450 ppm)

Variable

(c) Geology

Glacial
boundary

Coal bearing strata

Figure 13.12
Occurrence of heart-attack death by county and distribution of sulfate-rich and bicarbonate-rich surface water by county in Ohio. (After R. J. Bain, 1979, *Geology* 7: 7–10.)

hard water or water with low-sulfate and high-bicarbonate concentration may retard heart disease. Or, fourth, some characteristics of soft water or water with high-sulfate and low-bicarbonate concentration may enhance heart disease. Of course, some combination of the second, third, and fourth factors, along with others, is also likely, consistent with our observation that a disease

may have several causes. Additional research is needed to prove the benefit of hard water and perhaps treat soft water to reduce heart disease; for now, we can only say that heart disease may indeed be related to the chemistry of drinking water, but the explanation remains obscure.

An interesting study of patterns of heart disease mortality and possible relations to the geologic environment was conducted in Georgia by the U.S. Geological Survey (28, 29). The contrast in heart disease mortality among the 159 counties in Georgia is such that the differences between the highest and lowest rates are nearly as great as can be found in any two counties in the United States. Nine counties in northern Georgia with low rates of death caused by heart disease and nine counties in central and south central Georgia with high death rates caused by heart disease were selected for geochemical analysis.

The locations of these counties are shown in Figure 13.13. Table 13.5 lists the death rates from heart disease and other causes in the selected counties. In the nine selected counties with low rates of death from heart disease, the range is from 560 to 682 deaths per 100,000 population for males 35 to 74 years of age during the period 1950 to 1959. In the nine selected counties with high mortality rates, the range is from 1,151 to 1,446 deaths per 100,000 population for the same age group and time period (28). Mortality rates for the south central part of the state are thus about twice as high as for the northern section.

The nine counties with high heart-disease rates are primarily in southern Georgia on the Atlantic Coastal Plain (Figure 13.13). The landscape characteristically has low relief, sluggish drainage, and swamps. In general, sandy soils overlie Cenozoic marine sedimentary rocks that have undergone intensive weathering. Small-scale agricultural activity is prevalent where relief and drainage are sufficient.

The nine counties with low heart-disease rates are located in northern Georgia in the Appalachian Highlands, known as the Piedmont, the Blue Ridge, and the Ridge and Valley (Figure 13.13). The *Piedmont* extends from the Blue Ridge Mountains to the Coastal Plain. Rocks here are a mixture of Precambrian and Paleozoic

metamorphic rocks and intrusive igneous rocks of varying ages. Soils are generally mixtures of sand, silt, and clay (loam) with clay subsoils. The *Blue Ridge* topography is mountainous, with narrow valleys and turbulent streams. The rocks include metamorphic and sedimentary rocks of Precambrian to early Cambrian age and Paleozoic igneous rocks. Soils are acid with relatively high organic content; high relief is not favorable to agricultural activity. The *Ridge and Valley* is characterized by linear ridges and parallel valleys oriented northeast to southwest; rocks are folded and faulted sedimentary rock of Paleozoic age. Soils vary from rich loam soils in valleys with limestone bedrock to less fertile clay soils on shales and relatively infertile soils derived from sandstone bedrock (28).

A detailed geochemical investigation of the soils and plants of the low-death-rate area in northern Georgia and high-death-rate area in south central Georgia was conducted (28, 29). It was concluded that the geochemical variations in the soil reflected differences in the bedrock. Thirty elements were analyzed, and significant concentrations of aluminum, barium, calcium, chromium, copper, iron, potassium, magnesium, manganese, niobium, phosphorus, titanium, and vanadium were found. Zirconium was the only element that occurred in significantly larger concentrations in soils of the high-death-rate counties (29). The differences probably reflect the fact that soils in the high-death-rate counties of the Coastal Plain derive from unconsolidated sediments (sands and clays) that were previously weathered and leached during deposition.

Several trace elements are known to have beneficial effects on heart disease. Of these, manganese, chromium, vanadium, and copper are more highly concentrated in the low-death-rate areas of Georgia. Therefore, the low death rate may be a result of an abundance of beneficial

Figure 13.13

Map showing physiographic regions and Georgia counties having high and low death rates from cardiovascular disease. (After Shacklette, Sauer, and Miesch, U.S. Geological Survey Professional Paper 574C, 1970.)

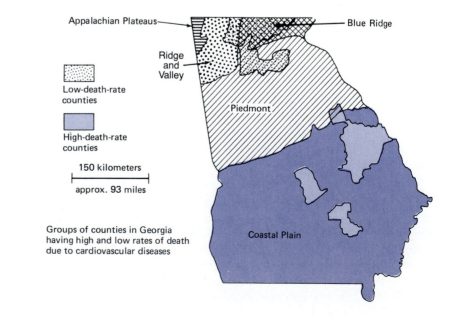

Table 13.5
The death rates from selected causes, males age 35 to 74, for selected counties of Georgia during the period 1950 to 1959.

County	Cardiovascular diseases (330−334, 400−468[a])			All Noncardiovascular Causes of Death	All Causes of Death
	Total	Coronary Heart Disease (420[a])	Other Cardiovascular Diseases		
Low-rate counties					
Cherokee	671.0	344.1	326.9	525.4	1196.4
Fannin	655.2	262.2	393.0	607.9	1263.1
Forsyth	635.7	275.1	360.6	529.6	1165.3
Gilmer	647.5	296.9	350.6	485.4	1132.9
Hall	667.7	328.6	339.1	573.3	1241.0
Murray	680.8	274.8	406.0	593.6	1274.4
Pickens	681.7	332.2	349.5	735.5	1417.2
Towns	587.6	201.2	386.4	480.0	1067.6
Union	560.1	287.1	273.0	477.5	1037.6
High-rate counties that were studied					
Bacon	1302.2	743.1	559.1	789.9	2092.1
Bleckley	1205.5	650.6	554.9	546.0	1751.5
Burke	1167.0	603.7	563.3	834.3	2001.3
Dodge	1206.3	569.4	636.9	736.1	1942.4
Emanuel	1167.8	467.4	700.4	731.5	1899.3
Jeff Davis	1446.4	701.1	745.3	568.6	2015.0
Jefferson	1151.2	679.3	471.9	677.9	1829.1
Jenkins	1330.8	530.1	800.7	802.5	2133.3
Warren	1321.7	915.6	406.1	760.2	2081.9
Other high-rate counties					
Baldwin[b]	1218.8	781.5	437.3	744.7	1963.5
Chatham	1177.2	753.7	423.5	757.2	1934.4
Johnson	1170.3	644.1	526.2	700.1	1870.4
Lee	1170.6	754.5	416.1	927.8	2098.4
Long	1162.3	614.6	547.7	638.1	1800.4
Marion	1155.8	667.2	488.6	686.6	1842.4
Randolph	1252.1	572.2	679.9	618.1	1870.2
Richmond[b]	1203.0	773.5	429.5	771.3	1974.3
Treutlen	1171.7	533.2	638.5	766.2	1937.9
All counties of Georgia	925.2	520.4	404.8	649.0	1274.2

[a]Code number of International Statistical Classification of Diseases, 7th Revision (World Health Organization, 1957).

[b]With adjustments for resident institution populations.

Note: Average annual death rates per 100,000 population, age adjusted by 10-year age groups by the direct method to the entire United States population age 35−74 in 1950. Deaths tabulated by Georgia State Department of Health, by county of usual residence.

Source: Shacklette, Sauer, and Miesch, U.S. Geological Survey Professional Paper 574C, 1970.

trace elements in the soils rather than concentrations of harmful elements. In other words, the tendency for higher mortality from heart disease in Georgia may be a causal relationship facilitated by a deficiency rather than an excess of certain trace elements (29).

Additional information is needed about other elements and their relationship to heart disease. Cadmium, fluorine, and selenium, among other elements, require further study to increase our understanding of how trace elements affect the heart and circulatory system. With respect to cadmium, we know that individuals who die from hypertensive complications generally have greater concentrations of cadmium or higher ratios of cadmium to zinc in their kidneys compared to individuals who die from other diseases. Surprisingly, however, workers who are exposed to cadmium dust and accumulate the element in their lungs do not have abnormal rates of hypertension (30).

Strokes and the Geochemical Environment in Japan

Perhaps the first report of a relationship between water and strokes came from Japan, where a prevalent cause of death is apoplexy (stroke), a sudden loss of body functions caused by the rupture or blockage of a blood vessel in the brain. The geographic variation of the disease in Japan is related to the ratio of the sulfate to bicarbonate (SO_4 to HCO_3) of river water. Water with low sulfate-to-bicarbonate ratios is relatively hard, whereas high ratios indicate soft (acid) water. Figure 13.14 shows areas in Japan where the readjusted death rates are relatively high, greater than 120 per 100,000 population, compared to areas where the sulfate-to-bicarbonate ratio is relatively high, greater than 0.6 (31).

The abundance of sulfate, especially in the northeastern part of Japan, evidently stems from the sulfur-rich volcanic rock found there. Rivers in Japan that flow through sedimentary rocks are, in contrast, low in sulfate and high in bicarbonate, like most river water in the world. In general, however, river water in Japan is relatively soft, with a hardness less than 40 parts per million, compared to water in the United States where the mean hardness of raw municipal water is 139 parts per million (25).

Cancer and Geochemical Environments

Cancer tends to be strongly related to environmental conditions. As with heart disease, however, relationships between geochemical environment and cancer cannot be proven, and undoubtedly the causes of the various types of cancers are complex and involve many variables, some of which may be the presence or absence of certain earth materials that contribute to, or help protect people from, disease.

Relationships between cancer and environment have two aspects: cancer-causing (carcinogenic) substances released into the environment by *human use* of re-sources; and cancer-causing substances that occur *naturally* in earth materials such as soil and water.

Recent information suggests that the presence of cancer-causing substances in drinking water is ubiquitous. This may or may not be true, but certainly water polluted with industrial waste containing toxic chemicals, some of which are possible carcinogens, is being released into our surficial water supplies. The Mississippi River, particularly, has pollution problems, and, ironically, even present water treatment causes problems. When combined with chlorination in water treatment, some industrial waste turns into cancer-producing material. Other carcinogens escape water treatment because antiquated treatment procedures fail to remove toxic substances.

The occurrence of certain cancers has been related to the natural environment. Examples include iodine deficiency and its relation to breast cancer (14); mineralized drinking water and its relation to stomach cancer (32); organic matter, zinc, and cobalt in soil and their relation to stomach cancer (33, 34); and soil salinity, climate, vegetation, and agriculture and their relation to esophageal cancer (35). These studies do not prove a cause-and-effect relationship between certain earth materials and cancer, but they do indicate an area for future research that promises to be significant in solving environmental health problems.

The leading cause of death for women between 40 and 44 years of age is breast cancer. It is also the leading cause of death for all types of cancer for all women between the ages 35 and 55 (14). The occurrence of breast cancer in the United States has been related to areas with an iodine deficiency (the goiter belt). Figure 13.15 shows areas with a relatively high incidence of breast cancer superimposed on areas with iodine deficiency. It has been observed that countries with iodine deficiencies and a resulting high incidence of goiter also have high incidences of breast cancer. The converse also holds; that is, countries with sufficient iodine have a low incidence of goiter and breast cancer (14). The

Figure 13.14
Maps of Japan comparing the ratio of SO_4 to CO_3 in rivers to the death rate of apoplexy in 1950. (Data from T. Kobayashi, 1957.)

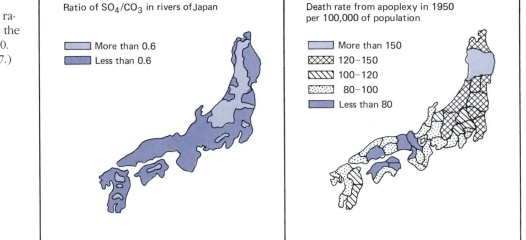

Ratio of SO_4/CO_3 in rivers of Japan

- More than 0.6
- Less than 0.6

Death rate from apoplexy in 1950 per 100,000 of population

- More than 150
- 120–150
- 100–120
- 80–100
- Less than 80

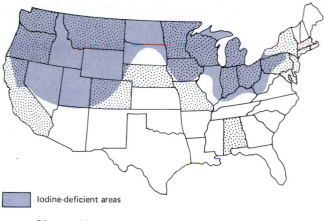

Iodine-deficient areas

24 states with greatest
incidence of breast cancer

Figure 13.15
Map of conterminous United States comparing iodine-deficient areas ("goiter belt") with the 24 states with the greatest incidence of breast cancer. (Data from Bogardus and Finley, 1961, and Spencer, *The Texas Journal of Science,* 1970).

significance of this observation is not well understood, but is apparently real.

A study into the relationship between the occurrence of cancer, especially stomach cancer, in West Devon, England, established the hypothesis that cancer in that area is connected with water supplies derived from a particular rock type. Figure 13.16 shows the geology of the area and the high incidence of cancer where drinking water is pumped from the rocks of Devonian age (345 to 400 million years old). These sedimentary rocks are highly mineralized compared to the sedimentary rocks of

Carboniferous age (280 to 345 million years old) or to the granite that is mineralized but to a lesser extent than the Devonian rocks. Research suggests the cancer occurrence is associated with the mineralization of the Devonian rocks, but no specific cancer-causing substance has been isolated (32).

Two studies in Wales and England established relationships between stomach cancer and soil characteristics. The earlier study concluded that areas with a high incidence of stomach cancer had soils with a relatively high amount of organic matter (32). The amount of organic material is measured by the percentage of weight loss of a dry soil after burning. Figure 13.17 shows the

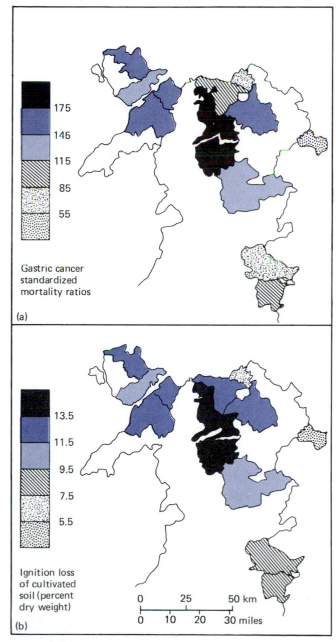

175
145
115
85
55

Gastric cancer
standardized
mortality ratios

(a)

13.5
11.5
9.5
7.5
5.5

Ignition loss
of cultivated
soil (percent
dry weight)

0   25   50 km

0   10   20   30 miles

(b)

Figure 13.17
Maps of North Wales rural districts comparing cancer mortality ratios and ignition loss of cultivated soils. (After C. D. Legon, *British Medical Journal,* September 27, 1952.)

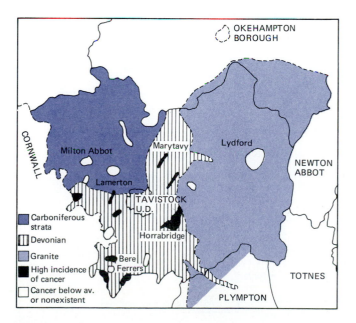

OKEHAMPTON BOROUGH

CORNWALL

Milton Abbot   Marytavy   Lydford

NEWTON ABBOT

Lamerton

TAVISTOCK U.D.

Horrabridge

Bere Ferrers

TOTNES

PLYMPTON

Carboniferous strata
Devonian
Granite
High incidence of cancer
Cancer below av. or nonexistent

Figure 13.16
Map of West Devon, England, showing generalized rock types and areas where the incidence of cancer is high or low. (From Allen-Price, *The Lancet,* vol. 1, 1960.)

correlation of areas with high mortality rates of stomach cancer with organic content of cultivated soils. This study was not able to associate disease rates with specific cancer-producing substances in the soils.

A subsequent detailed study of soils in northern Wales and Cheshire, England, combined geochemical analysis of the soil and measurement of organic content (34). It concluded that the abnormal rates of stomach cancer were associated with the amount of organic material in the soil in that abnormally high rates of the disease were related to long residence on soils with an organic content between certain limits. In addition, the study found a positive relationship between the concentration of zinc, cobalt, and chromium and cancer of the stomach. Zinc and cobalt may be especially significant for two reasons. First, zinc is an active part of some essential enzyme systems in the human body and is also active in the processes of gastric digestion. Second, cobalt is known to have carcinogenic properties and is impor-

tant in plant and animal activity. Although the high rates of cancer in this study are related to excesses of trace elements in the soil, the geographic distribution of such soils is apparently unrelated to that of stomach cancer (34).

Incidence rates of esophageal cancer in northern Iran, near the Caspian Sea, offer another opportunity to examine relations between environment and chronic disease. In this area, the rates of esophageal cancer are tremendously variable, and marked differences occur over short distances (35). Research in the area indicates a relationship between cancer and the climate, soils, vegetation, and agricultural practices. There is a great deal of interrelationship among these factors, however, as the climate obviously affects the soils, vegetation, and agricultural practices. The highest correlation between rates of esophageal cancer and environment is for soil types. Figure 13.18 shows relative rates of cancer and general soil types. It is obvious that the highest rates of

Figure 13.18
Maps of the esophageal cancer belt in Asia comparing age and standardized incidence rates of the cancer (upper) with soil types (lower). (From Kmet and Mahboubi, *Science,* vol. 175, pp. 846–53, February 1972. Copyright © 1972, American Association for the Advancement of Science.)

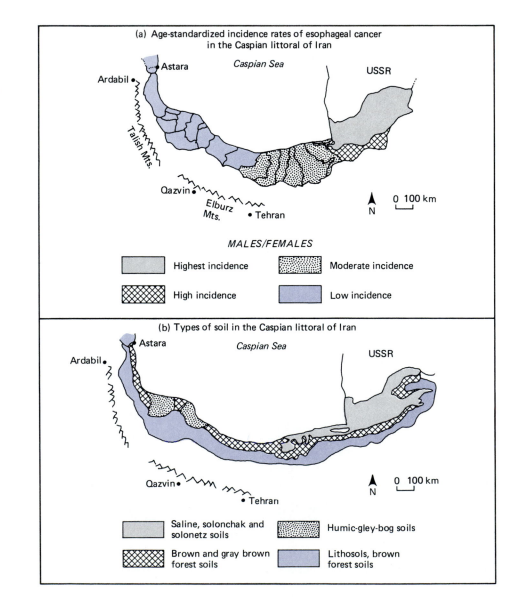

the disease are associated with the saline (salty) soils in the eastern section. These soils are in an area of little rainfall. Rainfall increases to the west by a factor of 4, and the soils become more and more leached. The lowest incidence of cancer is in the rain belt in the western section (Figure 13.18). Vegetation and agriculture of the area systematically change from grazing on sparse cover of specially adapted plants in the east, to lush forests and dry farming of rice, fruit, and tea in the west. Paralleling this change from east to west is the continuously lessening incidence of esophageal cancer (35).

The examples of relations between environment and incidence of cancer further establish the importance of recognizing multiple causes of chronic disease. Regardless of the numerous contradictions and exceptions that can be found, sufficient data are now available to establish beyond a reasonable doubt that hypothesizing a single cause for cancer is invalid. Rather, we must adopt a multifactorial approach, and variations in the environment are certainly among the significant factors.

▼ RADIATION AND RADON GAS

Radiation

A **radioisotope** is a chemical element that spontaneously undergoes radioactive decay. It changes from one isotope to another and during the process emits one or more forms of radiation. Isotopes are atoms of an element that have the same atomic number (the number of protons in the nucleus) but vary in atomic mass number (the number of protons plus neutrons in the nucleus). For example, two isotopes of uranium are $^{235}U_{92}$ and $^{238}U_{92}$. The atomic number for both atoms of uranium is 92, whereas the atomic mass numbers are 235 and 238 respectively. The two different isotopes may be written as uranium-235 and uranium-238, or U-235 and U-238.

An important characteristic of a radioisotope is its **half-life**, which is the time required for one-half of a given amount of the isotope to decay to another form. Every radioisotope has a unique characteristic half-life. Radon-222, for example, has a relatively short half-life of 3.8 days. Radioactive carbon-14 has an intermediate half-life of 5570 years. Uranium-235 has a half-life of 700 million years.

There are three major kinds of radiation emitted during radioactive decay: alpha particles (α), beta particles (β), and gamma rays. *Alpha particles* consist of two protons and two neutrons and have the greatest mass of the three types of radiation. Because of their greater mass, alpha particles do not travel very far. Alpha particles can travel approximately 5 to 8 centimeters in air before they stop, and in human tissue, which is much denser than air, they can travel only about 0.005 to 0.008 centimeters. This is a very short distance indeed and therefore if alpha particles are to cause damage to tissue

and living cells then the particle must originate very close to the cell (36).

Beta particles are electrons and are intermediate with respect to mass between alpha particles and gamma rays. Beta decay occurs when one of the protons or neutrons in the nucleus of the isotope spontaneously changes. That is, a proton turns into a neutron or a neutron turns into a proton. Another particle is also ejected, the *neutrino,* which is a particle with apparently no mass at all (36). The third type of radiation is called *gamma radiation,* in which a type of X ray is emitted from the isotope. Gamma rays are similar to X rays but are more energetic and penetrating. Gamma rays travel the longest average distance of all radiation and can penetrate thick shielding.

Each radioisotope has its own characteristic emissions; some isotopes emit only one type of radiation, and some emit a mixture. Furthermore, the three different kinds of radiation have different toxicities. In terms of human health, or the health of any organism, alpha radiation is most dangerous when inhaled or ingested. All of its radiation is stopped within a very short distance by the body's tissues; therefore, all the damaging radiation is absorbed by the body. On the other hand, when the alpha-emitting isotope is stored in a container, it is relatively harmless. Beta radiation is intermediate in its effects, although most beta radiation will be absorbed by the body when a beta-emitter is ingested. Beta particles can be stopped by moderate shielding. A gamma emitter can be dangerous outside or inside the body, but when ingested, some of the radiation passes outside the body. Protection from gamma rays requires thick shielding.

Some radioisotopes, particularly those of very heavy elements, undergo a series of radioactive decay steps, finally reaching a stable nonradioactive isotope. For example, uranium decays through a series of steps and ends up as a stable isotope of lead. The decay chain from uranium-238, with a half-life of 4 billion years, to a stable lead-206 is shown in Figure 13.19. Also shown are the half-lives of some of the isotopes and the type of radiation that occurs. The two important facts are the type of radiation emitted and the half-life. The decay from one radioisotope to another is often stated in terms of parent and daughter products. For example, radon-222, which is a gas with a half-life of 3.8 days, decays by alpha emission to its daughter polonium-218, which is a solid with a half-life of 187 seconds.

A radioisotope's degree of dangerousness, either in the environment or to human health, depends on several factors: the kind of energy of the radiation emitted, the half-life, and the chemical activity of the isotope. An isotope such as radon-222 that readily enters a gaseous phase when released through the atmosphere may be extremely dangerous if its daughter products which are solid are inhaled. This occurs even with low energies and short half-lives.

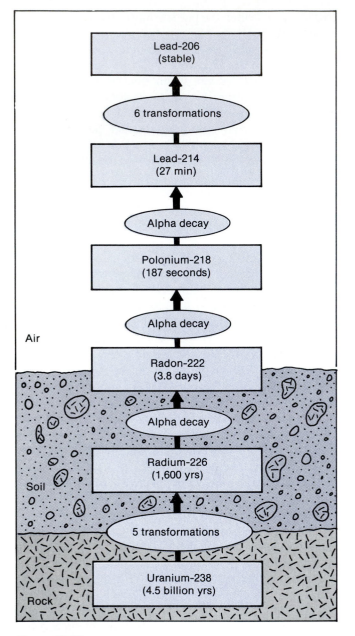

Figure 13.19
Simplified radioactive decay chain from uranium-238 to lead-206. Not all isotopes are shown. Half-lives and type of decay are shown for some.

Radiation units. A commonly used unit for measuring radioactive decay is the *curie,* which is the amount of radioactivity from 1 gram of radium-226 that undergoes about 37 billion nuclear transformations per second. Radium-226 has a half-life of 1,622 years; by the end of that time the initial gram of radium will be reduced to 0.5 gram, the nuclear transformations will be reduced to about 18.5 billion per second and the amount of radioactivity present will be 0.5 curie (37).

Radium was discovered in the 1890s by Marie Curie and her husband Pierre. They also discovered polonium, which they named after Marie Curie's homeland, Poland. Harmful effects from radiation were not known in the late 1890s and both Marie Curie and her daughter died of radiation-induced cancer (36).

In the International System (SI) of measurements the unit commonly used for radioactive decay is the *becquerel,* which corresponds to one radioactive decay per second. When working with radioactive gases such as radon-222 the unit of measurement is often becquerels per cubic meter or picocuries per liter (pCi/l). A picocurie is 1 trillionth or (10^{-12}) of a curie. Picocuries per liter are therefore a measure of the number of radioactive decays of radon every second in a liter of air.

The actual dose of radiation that is delivered by radioactivity is commonly measured in terms of *rads* and *rems.* In the International System the corresponding units are *grays* and *sieverts,* respectively. The rads or grays are the unit of absorbed dose of radiation and 1 gray is equal to 100 rads. Rems and sieverts are a unit of equivalent dose or effective-equivalent dose where 1 sievert is 100 rem (36). These units are somewhat confusing on first presentation but are important as both systems of units are commonly used. The energy retained by living tissue that has been exposed to radiation is called the radiation absorbed dose. (The initial letters of this term give us the unit rad.) However, because different types of radiation have different penetration and therefore do variable damage to living tissue, the rad is multiplied by a factor known as the relative biological effectiveness to produce the rem or sievert units. When small doses of radioactivity are being considered, the millirem (mrem) or millisievert (mSv)—one-thousandth of a rem or a sievert, respectively—are used (37, 38, 39). When measuring X rays or gamma rays the commonly used units are the *roentgen* or, in the International System of units, *coulombs* per kilogram.

Radiation doses. The natural background radiation received by Americans averages about 1.5 millisieverts per year. The range, however, is about 1.0 to 2.5 millisieverts per year (37). In general, the highest levels of radiation are in the mountain states and the lowest level is in Florida. The differences are due primarily to elevation and geology. More cosmic radiation is received at higher elevations because of the thinner atmosphere, and granitic rocks that often contain radioactive minerals are more common in mountainous areas. Florida, with a basic limestone geology and low elevation, has a relatively low level of background radiation (39). Nevertheless, there are locations in Florida where phosphate deposits occur and the background radiation there is often above average due to above-average uranium concentrations in the rocks that contain the phosphates. The precise amounts of radiation received by people from low-background sources are not completely agreed upon. However, some major sources include potassium-40 and carbon-14, which are present in our bodies and probably deliver between 0.2 and 0.25 millisieverts per year. The element potassium is an important electrolyte in our blood, and one isotope of potassium (K-40) has a very long half-life. Although potassium-40 makes up only a very small percentage of the total potassium in our bodies, it is present in all of us and so we are all slightly radioactive. Thus if you share your life with another person you are also exposed to a little bit more radiation (a small hazard compared to the many benefits of companionship). Cosmic rays deliver between 0.35 and about 1.5 millisieverts per year, depending upon elevation, and radioactive materials in rocks and soils deliver on the average about 0.35 millisieverts per year. However, the amount delivered from rocks and soils may be much larger in some areas where radon gas seeps into homes. Anthropogenic sources of low-level radiation include X rays for medical and dental purposes, which may deliver an average of 0.7–0.8 millisieverts per year; nuclear weapons testing and nuclear power plants that may be responsible for approximately 0.04 millisieverts per year; and burning of fossil fuels such as coal, oil, and natural gas, which may add another 0.03 millisieverts per year (37, 39). A person's occupation and life-style can also affect the annual dose of radiation. Every time you fly at high altitudes you receive an additional dose of radiation. If you work at a nuclear power plant or conventional coal-fired plant or in a number of industrial positions, you may also be exposed to low-level radiation. Precisely how much radiation is received on job sites is closely monitored at obvious sites such as nuclear power plants and laboratories where X rays are produced; personnel must wear badges that reflect the dose of radiation received. Figure 13.20 shows some of the sources of radiation we are commonly exposed to. Notice that the

exposure to radon gas at its high end can be about equivalent to the Chernobyl nuclear power accident that occurred in the Soviet Union in 1986. That is, in some homes the radiation is about that experienced for evacuees from Chernobyl (40).

Radiation doses and health.   A major question concerning radiation exposure to people is, When does the exposure or dose become a hazard to health? There are no easy answers to this question. A dose of about 5,000 millisieverts (5 sieverts) is considered lethal to all people exposed to it; 1,000 to 2,000 millisieverts is sufficient to cause health problems, including vomiting, fatigue, potential abortion of pregnancies of less than two months duration, and temporary sterility in males; at 500 millisieverts physiological damage is recorded; and the maximal allowable dose of radiation per year for workers in industry is 50 millisieverts, which is approximately 30 times the average whole-body natural background radiation received by people (37, 39). The maximum permissible annual dose for the general public in the United States is 5 millisieverts or about 3 times the annual background (37). Although a good deal of information is known about the effects of high doses of radiation on people, most of the information comes from a study of people who survived the atomic bomb detonations in Japan at the end of World War II, people exposed to high levels of radiation in uranium mines, workers painting watch dials with luminous paint containing radium, and people treated with radiation therapy for disease (41). Workers who are exposed to high levels of radiation in mines have been shown to suffer a significantly higher rate of lung cancer than the general population. Study of mortality suggests there is a delay of 10 to 25 years between the time of exposure and onset of disease.

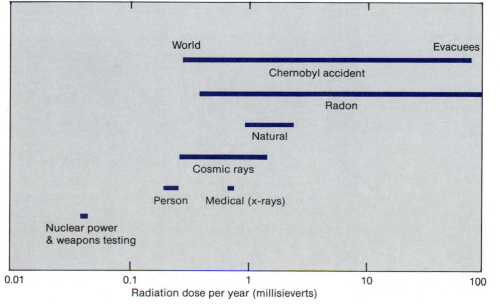

Figure 13.20
Annual radiation dose to people. (Data in part from A. V. Nero, Jr., "Controlling Indoor Air Pollution," *Scientific American* 258 [1988], 5: 42–48.)

Although there is debate about the nature and extent of the relationship between radiation exposure and cancer mortality, most scientists agree that radiation can cause cancer. Some scientists believe that there is a linear relationship such that any increase in radiation will produce an additional hazard (Figure 13.21). Others believe that the body is able to handle and recover from very low levels of radiation, but beyond some threshold health effects become more apparent. The verdict is still out on this subject, but it seems prudent to take a conservative stand and accept that there may be a linear relationship. That is, any increase in radiation is likely to be accompanied by an increase in adverse health effects.

Radon Gas

Radon is a naturally occurring radioactive gas that is colorless, odorless, and tasteless. Radioactive decay of radium-226 produces radon-222 (see Figure 13.19). Ernest Dorn, a German chemist, discovered radon in 1900 and it has had an interesting history. In the early 1900s radon water became a popular health fad and one doctor reported in a medical journal that radon had absolutely no toxic effects, and in fact was accepted harmoniously by the human system much as sunlight is accepted by plants. In 1916 the American Medical Association actually refused to endorse several radon emanators (a device for dissolving radon in water) because too little radon was produced. During this period, many products containing radium and thus radon were on the market and these included chocolate candies, bread, and toothpaste. As recently as 1953 a

contraceptive jelly that contained radium was marketed in the United States. Today the cell-killing properties of radon daughters are widely used for the treatment of cancer (36).

Why is radon dangerous? People are worried about radon gas because exposure to elevated concentrations of radon is associated with increased risk of lung cancer. The actual risk depends upon both the concentration of the radon and the length of time of exposure. It is believed that the risk increases as the level of radon concentration, the length of exposure, and a person's smoking habits increase (42).

In recent years there have been approximately 140,000 lung cancer deaths in the United States each year. The Environmental Protection Agency estimates that approximately 20,000 of these lung cancer deaths are probably related to exposure to radon gas and its daughter products, particularly polonium-218. It is believed that the combined effect of exposure to radon gas and smoking is particularly hazardous. One estimate is that the combination of exposure to tobacco and radon is approximately 10 times as hazardous as exposure to either of the pollutants by itself (42). In summary, the Environmental Protection Agency has estimated that as many as 20,000 deaths per year from lung cancer may be related to exposure to radon gas. This estimate was issued in spite of the fact that there have been no direct studies linking radon gas exposure in houses to increased incidence of lung cancer. Estimates of such linkage come from studies of people who have experienced high exposure to radiation through such activities

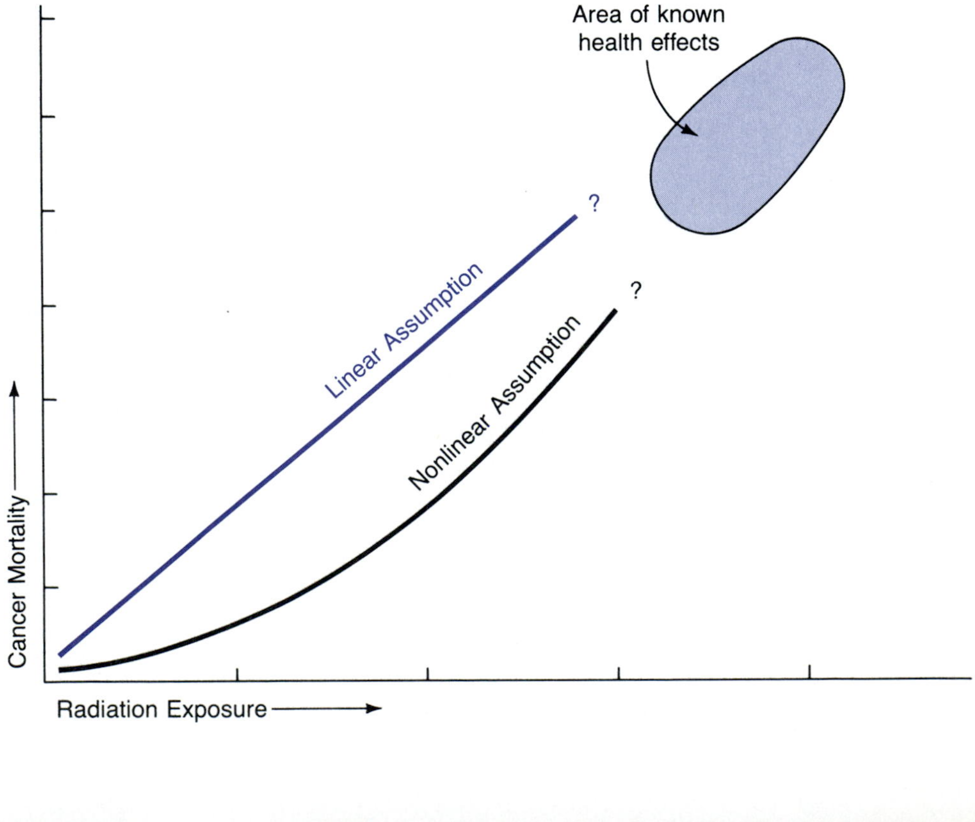

Figure 13.21
Possible relations between radiation exposure and cancer mortality. We know that high exposures increase cancer mortality: the colored area in the figure indicates this. The path there may be linear or nonlinear. (After University of Maine and the Maine Department of Human Services, 1983.)

as mining uranium. The health risk from radon gas is related primarily to daughter products such as polonium-218, which is a particle and adheres to dust. This dust may be inhaled into the lungs where alpha radiation can occur as polonium-218 decays to lead-214 with a half-life of approximately three minutes (see Figure 13.19). Inhaled aerosols that carry radon daughters such as polonium-218 are trapped in the lungs by sticky mucus on the surface of the bronchial airways of the lungs. Basal cells located just beneath the mucus may be damaged by alpha radiation. It is known that radiation can cause breaking of strands of DNA in cells, and when both strands of DNA are broken, the damage may be permanent, resulting in mutations. The mutations probably do not cause cancer by themselves but are linked to the process of initiation of the disease, which may be later promoted through exposure to chemicals such as those found in cigarette smoke (36). The United States Environmental Protection Agency has radon risk evaluation charts, one of which is shown in Figure 13.22. The chart relates exposure to radon gas in picocuries per liter to the estimated risk and estimated number of lung cancer deaths (43).

A recent study suggested radon as a causative factor in inducing myeloid leukemia, melanoma cancer of the kidney, and some childhood cancers (44). The study is based on statistical analysis of radon concentration in homes and cancer incidence, but is controversial. The controversy results in part because the data base consists of cancer rates and average radon concentrations for several countries rather than evaluating radon concentrations in homes of cancer victims. The approach, while valid, needs to be tied closely to individuals and their specific environment before more skeptical observers are convinced that radon causes cancers other than lung cancer.

The connection of radon to melanoma (a deadly form of skin cancer) is hypothetically related to cell damage to the skin by radioactive decay of radon and its daughter products that might somehow cluster on the skin. The leukemia is hypothetically related to the assumption that radon is soluble in blood and fat cells, resulting in accumulation of radon in the fatty cells of bone marrow (44). These hypotheses are interesting and have important implications that should be addressed in future research as soon as possible.

In terms of the implied risk it is known that large concentrations of radon are hazardous and furthermore that it has not been shown that small concentrations are harmless. The Environmental Protection Agency has set 4.0 pCi/l as the action level beyond which radon gas is considered to be a hazard.

The average outdoor concentration of radon gas is approximately 0.2 pCi/l, and the average indoor level about 1.0 pCi/l. The action level of 4.0 pCi/l is only an estimate of a concentration target that indoor levels should be reduced to. The comparable risk at 4.0 pCi/l is about 200 chest X rays per year, hardly insignificant.

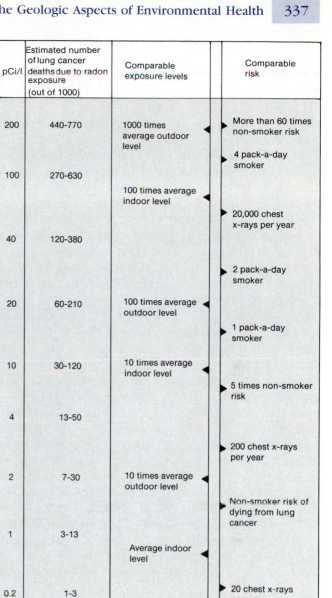

Figure 13.22
Estimated risk associated with radon. These estimates are calculated as long-term exposure risks for a person living about 70 years and spending about 75% of the time in the home with a designated level of radon. (From U.S. Environmental Protection Agency, *A Citizen's Guide to Radon,* 1986, OPA–86–004.)

If risks from radon gas are even close to EPA estimates, then the hazard is comparable to deaths from automobile accidents and hundreds of times higher than risks presented by outdoor pollutants possibly present in water and air. Pollutants in the outdoor environment are generally regulated to reduce the risk of premature death and disease to less than 0.001 percent. Assignment of risk from indoor pollutants such as organic chemicals suggest a risk of cancer is approximately 0.1 percent. These are small compared to radon. For example, people who live in homes for about 20 years with an average concentration of radon of about 25 pCi/l face a 2 to 3 percent chance of contracting lung cancer (40).

Geology of radon gas. Both rock type and structure are important in determining how much radon is likely to reach the surface of the earth. The range of uranium-238 concentration in rocks and soil can vary greatly. Rock types such as sandstone generally contain less than one part per million (ppm) U-238 but rocks such as some dark shales and some granites may contain more than 3 ppm U-238. The actual amount of radon that reaches the surface of the earth is related to the concentration of uranium in the rock and soil as well as the efficiency of the transfer processes from the rock or soil to soil-water and soil-gas.

Some regions of the United States contain bedrock with an above-average natural concentration of uranium. An area in Pennsylvania, New Jersey, and New York famous for elevated concentrations of radon gas is the Reading Prong (Figure 13.23). This region contains a large number of homes with elevated concentrations of radon gas (36). Similarly, two areas in Florida have been identified that have elevated concentrations of uranium and radioactive potassium in phosphate-rich rocks. Many other states, including Illinois, New Mexico, South Dakota, North Dakota, and Washington, have identified areas with elevated indoor radon concentrations. A dark shale in the Santa Barbara, California area has been identified as a significant producer of radon gas (Figure 13.24).

Geologic structures such as shear zones, fracture zones, and faults are commonly enriched with uranium and can produce elevated concentrations of radon gas in the overlying soils (45).One study in southern Virginia found that shear zones in granite rock are associated with as much as a ten-fold increase in the concentration of radon in overlying soils. Furthermore, one of the highest indoor levels in the United States is identified over shear zones in Boyertown, Pennsylvania (45). In Santa Barbara, California, one of the highest concentrations of in-house radon measured to date occurs at the crest of an active fold (anticline) involving superficial sand beds that probably overlie organic-rich shale at shallow depth. It is hypothesized that open tension fractures at the crest of the fold allow the radon gas to diffuse upwards to the surface.

The amount of radon gas that escapes from bedrock and soil particles is greatly influenced by water content. Relatively high moisture content favors the trapping of radon gas in the spaces between soil and fractures in rocks. This results because the presence of water greatly reduces the distance that radon can travel as a result of radioactive decay from its parent material such as radium-226. Movement of radon gas from fractures in rock and pore spaces in soil is facilitated by relatively low moisture content. That is, the process of diffusion of radon gas is much greater in unsaturated material. Thus the moisture content affects the amount of radon that may reach the surface of the earth in two opposing ways:

1. High moisture content increases the chances of radon entering and residing in the pore space in rock fractures or between soil grains.
2. High moisture content tends to inhibit radon from moving through the pore spaces in rocks and soils.

As a result of these two factors, there is an optimal water content that will result in the maximum amount of radon that actually reaches the surface of the earth. In some soils this optimal water content is approximately 20 to 30 percent (36).

Figure 13.23
The Reading Prong area in the eastern U.S. where high levels of indoor radon were first discovered in the early 1980s. (Modified after U.S. Geological Survey, *U.S.G.S. Yearbook*, 1986.)

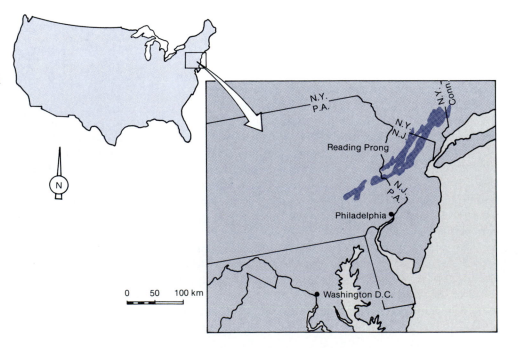

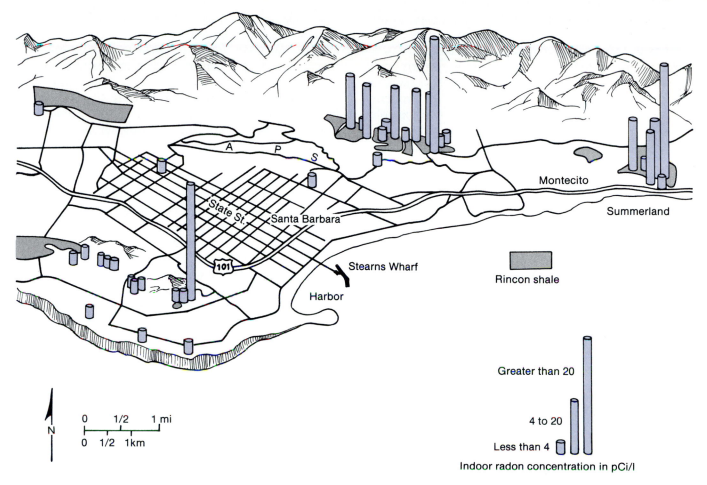

Figure 13.24

Santa Barbara, California. This seaside community has an emerging radon problem related to houses built on Rincon shale.

How radon gas enters homes. Most of the early interest concerning radon gas in homes was related to the use of using building materials that contained relatively high concentrations of radium. For example, between 1929 and 1975, materials used to manufacture concrete in Sweden contained relatively high concentrations of radium. Similarly, in the 1960s in the United States it was discovered that some homes and other buildings in the Grand Junction area of Colorado were constructed with materials contaminated by uranium mine waste. Some houses in Florida were discovered to have relatively high concentrations of radon where phosphate mining waste was used as fill dirt. Finally, some homes in New Jersey were built at a landfill site where wastes from a radium processing plant were earlier deposited (46).

In December of 1984, scientists discovered that radon gas from natural sources may enter the home and possibly present a serious health hazard. The story of Stanley Watras has been told and retold many times until it is now part of radon legend (36). Nevertheless, it is worth repeating here. Stanley Watras lives in Boyertown, Pennsylvania, and in December of 1984 had a job as technical advisor in the Limerick Nuclear Power Station. At the entrance to the plant are radiation detectors to ensure that no radioactive materials leave the facility. The reactor at the power plant had not yet been turned on when Watras, on his way into the plant, set off the alarms! Testing of his clothing suggested that the contamination could not have come from the plant, but must have come from where he lived. Power company officials checking Watras's home were astounded to find that the radiation level of the indoor air was 3,200 pCi/l. Scientists were very surprised because until then they did not believe that radon that naturally seeped into homes could be hazardous to anyone (36, 46, 47). The level of 3,200 pCi/l is 800 times higher than the level of 4 pCi/l considered as a threshold for hazard by the U.S. Environmental Protection Agency. The Watras home held the record for indoor radon concentration until the latter part of the 1980s when a home in Whispering Hills, New Jersey, was discovered with a radiation level of 3,500 pCi/l (47).

Since 1985 awareness of the radon gas problem in the United States has increased. Nevertheless, people have a hard time focusing on a problem that they can't see, smell, hear, or touch. The problem is further compounded by the fact that some property owners seem to be more concerned about potential loss of property values than the health hazard. There is a considerable hiatus between people's perception of the radon problem and the potential size of the problem itself. This is true even in states such as New Jersey where the problem has been known about for several years. One survey of homeowners concluded that only about 5 percent of the homeowners either had or were planning a radon test; about 10 percent believed that radon may be a problem in their home. The actual percentage of homes that are expected, in that part of New Jersey, to have a radon problem is approximately 30 percent (36).

There is not a good estimation of the number of homes in the United States today that may have elevated concentrations of radon gas. Looking at the tests that have been done, approximately one in twelve homes surveyed have an indoor radon level above 4 pCi/l. If this rate is a good average then approximately 7 million homes in the United States would have elevated rates, and millions more will need to be tested. On the other hand, many tests are taken in areas where radon is suspected and thus may bias the estimate. It should be mentioned that high radon concentrations are not only found in homes but also other buildings. In a recent survey of more than 100 schools, the Environmental Protection Agency found that in more than half of them the radon concentration was above the action level of 4 pCi/l, and one school in Tennessee had a level of 136 pCi/l (42). This information is alarming because some health officials believe that children, because of the small size of their lungs and the fact that they are still developing, may be more susceptible to radiation damage.

Three major sources or pathways have been identified by which radon gas may enter homes (Figure 13.25). These are:

▼ Gas that migrates up from soil and rocks into basements and other parts of houses,
▼ Groundwater pumped into wells,
▼ Construction materials such as building blocks made of substances that emit radon gas (41).

Figure 13.26 shows some of the major radon entry points into a home.

Radon is an important environmental problem that comes from natural processes. It is one of our environmental problems that is not related to any public, industrial, or government activities. The radon problem may emerge as the most significant environmental health problem in the United States today!

The occurrence of radon in homes is directly related to the fact that radon is a gas that can move through the small openings in soils and rocks. Radon migrates up with the soil gases and seeps through concrete floors and foundations, floor drains, or small cracks and pores in block walls. Radon can also enter homes through the water supply, particularly if the home is supplied by a private well. In general, approximately 10,000 pCi/l of radon in water will produce about 1 pCi/l of radon in indoor air. The radon and its daughter products are released into the air when water is used for daily functions such as showering, dish washing, and clothes washing. When water is pumped from underground sources for municipal supplies, it may also contain radon gas. However, because most of the water in these systems is stored for a few days and treated for purification, there is a delay between the time the water is pumped and when it is actually used. The storage process helps release radon from the water and the lag time allows for the remaining radon to decay away (36). As a result, public and municipal water supplies generally have a relatively low radon content.

Measurement of radon in homes. Measurement of radon concentration in homes has several inherent difficulties. First, the concentration of radon may be extremely variable in any one house depending upon the season of the year as well as other poorly understood factors related to radon emission from the soil gas. In many parts of the country higher levels of radon may be expected in the winter when the house is more closed up, compared to the summer when windows are left open for ventilation. Within the ground, emissions of radon may vary with barometric pressure, water content, and other variables that may be difficult to ascertain. Another factor concerning indoor radiation is that while one house may have a relatively high concentration, other houses in the vicinity may have much lower levels. Although the geology in a general way can be used as a diagnostic tool to identify areas that may have high concentrations of radon, only in-house testing can verify this. Even slight differences in the geology as, for example, locations of major fractures, may be very significant.

There are two major ways of measuring radon within a house. The first is the charcoal canister, which is a small tin can or paper pouch filled with activated charcoal. When the canister is opened, air is allowed to circulate and diffuse through the charcoal, and radon will be absorbed. After three to six days, the sampler is closed tight and sent to a laboratory where specialized detectors measure the radiation from the charcoal and estimate how much radon was in the air at the time of sampling.

Charcoal canister tests are good only for a short period because radon, with a half-life of 3.8 days, decays rather quickly, and any test longer than a few days will mean that some of the radon has decayed away. Advantages of the charcoal test are that it is a good screening

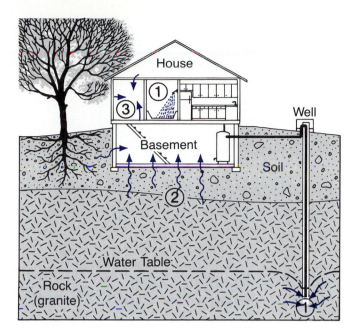

Figure 13.25
How radon may enter homes. (1) Radon in groundwater enters well and goes to house where it is used for water supply, dishwashing, showers, and other purposes. (2) Radon gas in rock and soil migrates into basement through cracks in foundation and pores in construction. (3) Radon gas is emitted from construction materials used in building the house. (Source: Environmental Protection Agency.)

device, is done quickly, and can provide data to the home owner in a time period of about two weeks. The disadvantage of the charcoal canister is that the test is a very short one and radon concentrations can vary considerably over the year. Depending upon the time of year the results could be quite different. (36).

The second type of test for measuring the concentration of radon gas in a home is the alpha-track detector. This type of test is designed to remain in a home for a much longer period of time, typically from a month or so to a year. When the container holding the alpha-track detector is opened to the air, radon atoms present undergo radioactive decay, releasing alpha particles, which hit the piece of plastic in the detector, causing microscopically small damage. This damage to the plastic is called a track, and the tracks may be counted using a microscope. From the number of tracks present the concentration of radon in the air can be estimated. Alpha-track detectors have the advantage of being a relatively long-term test and therefore are not as affected by short-term fluctuations in the concentration of radon (36).

A question commonly asked by people who test for radon in their homes is, "What should I do next?" Assuming that the first test was a screening with a charcoal canister, then the U.S. Environmental Protection Agency has provided useful guidelines in determining the urgency of the need for follow-up measurements. The following statements reflect those guidelines (43): (1) If the measurement from the initial screening is less than about 4 pCi/l, follow-up measurements are probably not necessary. If the initial measurement was made with the house closed up prior to and during the testing then there is little chance that the radon concentration will exceed 4 pCi/l on an annual basis. (2) If the initial screening measurement is about 4pCi/l to 20 pCi/l, it is advised to perform an alpha-track detector test for a period of approximately one year. (3) If the initial screening is about 20 pCi/l to 200 pCi/l, the follow-up measurements should include an alpha-track for no more than three months with the doors and windows closed as much as possible. (4) If the initial measurement is greater than about 200 pCi/l, you should perform follow-up measurements as soon as possible consisting of another charcoal canister test, and consideration must be given to taking action to reduce the exposure to radon gas. The above guidelines are likely to be revised, and it is likely that the EPA will recommend shorter alpha-track tests (one or two months) for any house with an initial charcoal test result greater than 4 pCi/l.

Methods to reduce concentrations of radon gas in homes. A successful program to reduce or limit a potential hazard from radon gas in homes centers upon three major strategies. First, the points of entry of radon may be located and sealed; second, ventilation of the home may be improved by keeping more windows open or using fans; and third, construction methods that provide a venting system may be installed (48, 49).

Radon enters homes for the same reason that smoke goes up a chimney, in a process known as the chimney effect. Houses are generally warmer than the surrounding soil and rock, so as the gases and air rise, radon is drawn into the house. Wind can also be a factor because it increases air flows into and out of buildings due to pressure differences. Radon can also enter the house through the water from a private well, but this contribution is usually a much smaller problem than the gas that seeps up through house foundations.

Probably the simplest method of reducing the radon concentration is simply to increase the ventilation of the home. Sometimes this is sufficient to solve the problem. Sealing cracks in the foundation can also reduce radon entry (Figure 13.27), particularly if it is done in specific sites where the radon is actually entering the house. Detectors are available to help identify such locations. Unfortunately, this method often fails to solve the problem, and new cracks may form.

A variety of construction options are open to the home owner. These include, among others, venting

A. Slab on grade
   A. Cracks on concrete slabs[1]
   B. Spaces behind brick veneer walls that
      rest on uncapped hollow-block foundation
   C. Loose fitting pipe penetrations
   D. Open tops of block walls
   E. Water (from some wells)
   F. Cold joint between two concrete pours
   G. Heating duct registers or sub-slab
      cold-air return pipes
   H. Hole under bathtub and under commode ring
   I. Wall/floor joint

[1]Hairline cracks probably do not contribute

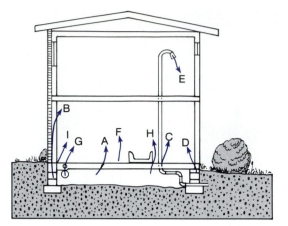

B. Raised foundation
   A. Cracks in subflooring and flooring
   B. Spaces behind stud walls and brick veneer walls
      that rest on uncapped hollow-block foundation
   C. Electrical penetrations
   D. Loose-fitting pipe penetrations
   E. Open tops of block walls
   F. Water (from some wells)
   G. Heating duct register penetrations
   H. Cold-air return ducts in crawl space

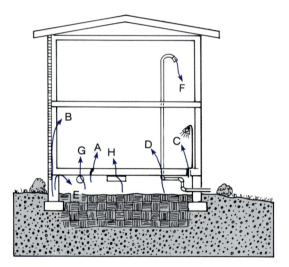

C. Basement
   A. Cracks in concrete slabs
   B. Cold joint between two concrete pours
   C. Pores and cracks in concrete blocks
   D. Floor-to-wall crack or French drain
   E. Exposed soil, as in a sump
   F. Weeping (drain) tile, if drained to open sump
   G. Mortar joints
   H. Loose fitting pipe penetrations
   I. Open tops of block walls
   J. Water (from some wells)
   K. Untrapped floor drain to a dry well or septic system

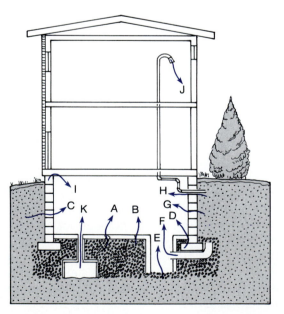

Figure 13.26
Main radon entry points for homes: (a) slab on grade; (b) raised foundation with crawl-space; and (c) basement. (Source U.S. Environmental Protection Agency, *Radon resistant residential new construction,* 1988. EPA/600-880/087.)

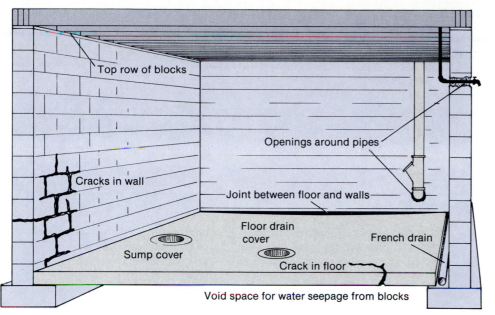

Figure 13.27
Sealing some of the main places that radon may enter a home. (From U.S. Environmental Protection Agency, *Radon Reduction Methods,* 1989, RD 681.)

systems in a basement or crawl space. If the house is built on a slab, then sub-slab ventilation systems may be installed. Figure 13.28a shows a sub-slab ventilation system for a typical home. For homes with a crawl space, vents and a suction system may be installed (Figure 13.28b).

There is good news and bad news concerning radon gas. The bad news is that many parts of the United States and other parts of the world have relatively high emissions of radon gas from soil and rocks, and the radiation is producing a hazard in homes. The good news

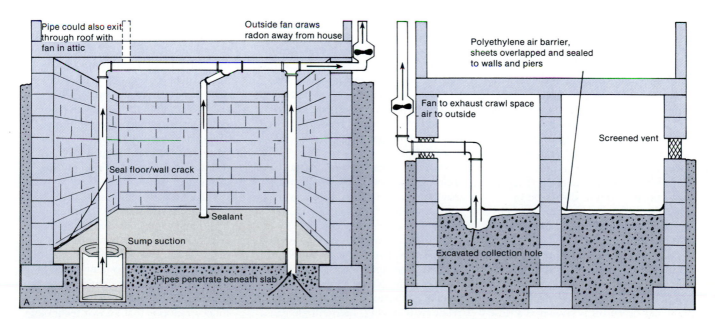

Figure 13.28
Methods to reduce radon gas in homes. (a) Sub-slab suction. (b) Crawl-space suction. (From U.S. Environmental Protection Agency, *Radon Reduction Methods,* 1989, RD 681.)

is that in most cases the problem can be fixed relatively easily. Even if a ventilation system is required it costs only a few thousand dollars, generally a small percentage of the value of the home.

Radon gas may be the most serious environmental health problem in the United States today. On the other hand, future research may show that health risks from radon gas exposure are not as great as predicted by the Environmental Protection Agency. No matter which of

these statements is true, it is important for people to become informed now concerning this potential problem. People have difficulty taking appropriate action to deal with the possibility of radon gas in their homes, and in too many cases seem to be more concerned with property values than their health. Once the entire story concerning radon is explained so that people understand it and know that the problem can be ameliorated, then the fear will be reduced.

▼ ▼ ▼ SUMMARY AND CONCLUSIONS

The death rate in some areas is certainly greater or less than in others, and a particular disease seldom has a one-cause, one-effect relationship.

Cultural factors can affect the geographic pattern of a particular disease. Examples include stomach cancer in Japan and lead poisoning in numerous areas.

Temperature, humidity, and precipitation are important factors in the climatic control of disease patterns, of which schistosomiasis and malaria are the best-known examples. In addition, other diseases are associated with a particular climate (for example, Burkett's tumor), but cause-and-effect relations are not clear.

The natural distribution of matter in the universe is such that the lighter elements are more abundant. The first 26 elements of the periodic table are sufficient to account by weight for the vast majority of the lithosphere and biosphere.

The movement of elements along numerous paths through the lithosphere, hydrosphere, biosphere, and atmosphere is known as the geochemical cycle. Along these paths, natural processes such as weathering and volcanic activity, combined with human pollution, release many types of substances into the environment. They are then concentrated or dispersed by other natural processes such as leaching, accretion, deposition, and biologic activity.

Every element has an entire spectrum of possible effects on plants and animals. The graph of concentration of a certain element against effects on a particular organism is known as a dose-response curve. For a particular element, trace concentrations may be beneficial or necessary for life. Higher

concentrations may be toxic, and still higher concentrations lethal, to a particular plant or animal. Interesting examples of environmentally important elements include iodine, fluorine, zinc, and selenium, among others.

Agricultural, industrial, and mining activities have released hazardous and toxic materials into the environment. These materials have been associated with bone disease in people, metabolic disorders in cattle, and other biological problems.

Relationships between chronic disease and geologic environment are complex and difficult to analyze. Furthermore, relationships are not proof of cause and effect. Nevertheless, considerable evidence is being gathered and studied, and preliminary results suggest that the geochemical environment is indeed a significant factor in the incidence of such serious health problems as heart disease and cancer.

Heart disease rates are apparently related to water chemistry such as total hardness and concentration of sulfate or bicarbonate ions. In addition, a study in Georgia suggests that deficiencies of trace elements may also be related to abnormal rates of heart disease.

Studies of cancer rates suggest that abnormal incidence in specific areas may be related to such factors as organic content of the soil and abundance of certain trace elements. In other areas, saline soils may be related to cancer. Although most tumors in the Western Hemisphere are assumed to be caused by environmental factors, evidence of simple cause-and-effect relations is lacking, and multiple causation of disease remains more likely.

The three major kinds of radiation are alpha particles, beta particles, and

gamma rays. Each type of radiation has a different effect and toxicity. A particular radioisotope's degree of hazard depends on several factors, including the kind of radiation emitted, the half-life, and the chemical activity. In general, low-energy, short half-lived isotopes are less dangerous than high-energy, long half-lived isotopes. However, isotopes such as radon that readily enter a gaseous phase or are readily absorbed in water become extremely dangerous when concentrated. In the United States today, it is hypothesized that up to 20,000 deaths per year from lung cancer are related to the indoor radon gas problem.

A major question concerning radiation exposure to people is, When does the exposure or dose become a hazard? There are no easy answers to this question, but two hypotheses are being tested. The first is that there is a direct linear relationship between exposure and hazard, and the second is that there is some threshold of radiation which, if exceeded, will cause damage to people and other living things. Certainly for higher levels of radiation, thresholds apparently exist. The controversy is about lower doses, where effects must be measured in terms of statistics concerning health problems.

Radon gas in homes is considered a serious environmental health problem. The factors that control the concentration of indoor radon include geology, radon concentration in the soil and rock, moisture content of soil, type of house construction, and season of the year. The good news concerning indoor radon is that known technology is available to reduce or remove the problem.

▼ ▼ ▼ REFERENCES

1. SAUER, H. I., and BRAND, F. R. 1971. Geographic patterns in the risk of dying. In *Environmental geochemistry in health,* ed. H. L. Cannon and H. C. Hopps, pp. 131–50. Geological Society of America Memoir 123.

2. HOPPS, H. C. 1971. Geographic pathology and the medical implications of environmental geochemistry. In *Environmental geochemistry in health,* ed. H. L. Cannon and H. C. Hopps, pp. 1–11. Geological Society of America Memoir 123.

3. YOUNG, K. 1975. *Geology: The paradox of earth and man.* Boston: Houghton Mifflin.

4. BYLINSKY, G. 1972. Metallic menaces. In *Man, health and environment,* ed. B. Hafen, pp. 174–85. Minneapolis: Burgess Publishing.

5. WARREN, H. V., and DELAVAULT, R. E. 1967. A geologist looks at pollution: Mineral variety. *Western Mines* 40:23–32.

6. FURST, A. 1971. Trace elements related to specific chronic diseases: Cancer. In *Environmental geochemistry in health,* ed. H. L. Cannon and H. C. Hopps, pp. 109–30. Geological Society of America Memoir 123.

7. CARGO, D. N., and MALLORY, B. F. 1974. *Man and his geologic environment.* Reading, Mass.: Addison-Wesley.

8. UNDERWOOD, E. J. 1971. Geographical and geochemical relationships of trace elements to health and disease. In *Trace substances in environmental health—IV,* ed. D. D. Hemphill, pp. 3–11. Columbia, Missouri: University of Missouri Press.

9. MERTZ, W. 1968. Problems in trace element research. In *Trace substances in environmental health—II,* ed. D. D. Hemphill, pp. 163–69. Columbia, Missouri: University of Missouri Press.

10. HORNE, R. A. 1972. Biological effects of chemical agents. *Science* 177: 1152–53.

11. MAIER, F. J. 1971. Fluorides in water. In *Water quality and treatment,* 3d ed., pp. 397–440. New York: McGraw-Hill.

12. BERNSTEIN, D. S.; SADOWSKY, N.; HEGSTED, D. M.; GURI, C. D.; and STARE, F. J. 1966. Prevalence of osteoporosis in high- and low-fluoride areas in North Dakota. *The Journal of the American Medical Association* 198: 499–504.

13. GILBERT, F. A. 1947. *Mineral nutrition and the balance of life.* Norman, Oklahoma: University of Oklahoma Press.

14. SPENCER, J. M. 1970. Geologic influence on regional health problems. *The Texas Journal of Science* 21: 459–69.

15. SHACKLETTE, H. T., and CUTHBERT, M. E. 1967. Iodine content of plant groups as influenced by variations in rock and soil type. In *Relations of geology and trace elements to nutrition,* eds. H. L. Cannon and D. F. Davidson, pp. 31–45. Geological Society of America Special Paper 90.

16. PORIES, W. J.; STRAIN, W. H.; and ROB, C. G. 1971. Zinc deficiency in delayed health and chronic disease. In *Environmental geochemistry in health and disease,* eds. H. L. Cannon and H. C. Hopps, pp. 73–95. Geological Society of America Memoir 123.

17. LAKIN, H. W. 1973. Selenium in our environment. In *Trace elements in the environment,* ed. E. L. Kothny, pp. 96–111. Advances in Chemistry Series 123, American Chemical Society.

18. ALLAWAY, W. H. 1969. Control of environmental levels of selenium. In *Trace substances in environmental health—II,* ed. D. D. Hemphill, pp. 181–206. Columbia, Missouri: University of Missouri Press.

19. ORNDAHL, G.; RIDBY, A.; and SELIN, E. 1982. Myotonic dystrophy and selenium. *Acta. Med. Scand.* 211: 493–99.

20. PETTYJOHN, W. A. 1972. No thing is without poison. In *Man and his physical environment,* ed. G. D. McKenzie and R. O. Utgard, pp. 109–10. Minneapolis: Burgess Publishing.

21. TAKAHISA, H. 1971. Discussion in *Environmental geochemistry in health and disease,* ed. H. L. Cannon and H. C. Hopps, pp. 221–22. Geological Society of America Memoir 123.

22. EBENS, R. J.; ERDMAN, J. A.; FEDER, G. L.; CASE, A. A.; and SELBY, L. A. 1973. *Geochemical anomalies of a claypit area, Callaway County, Missouri, and related metabolic imbalance in beef cattle.* U.S. Geological Survey Professional Paper 807.

23. ARMSTRONG, R. W. 1971. Medical geography and its geologic substrate. In *Environmental geochemistry in health and disease,* ed. H. L. Cannon and H. C. Hopps, pp. 211–19. Geological Society of America Memoir 123.

24. WINTON, E. F., and McCABE, L. J. 1970. Studies relating to water mineralization and health. *Journal of American Water Works Association* 62: 26–30.

25. SCHROEDER, H. A. 1966. Municipal drinking water and cardiovascular death-rates. *Journal of the American Medical Association* 195: 125–29.

26. KLUSMAN, R. W., and SAUER, H. I. 1975. *Some possible relationships of water and soil chemistry to cardiovascular diseases in Indiana.* Geological Society of America Special Paper 155.

27. BAIN, R. J. 1979. Heart disease and geologic setting in Ohio. *Geology* 7: 7–10.

28. SHACKLETTE, H. T.; SAUER, H. I.; and MIESCH, A. T. 1970. *Geochemical environments and cardiovascular mortality rates in Georgia.* U.S. Geological Survey Professional Paper 574C.

29. SHACKLETTE, H. T.; SAUER, H. I.; and MIESCH, A. T. 1972. Distribution of trace elements in the occurrence of heart disease in Georgia. *Geological Society of America Bulletin* 83: 1077–82.

30. SCHROEDER, H. A. 1965. Cadmium as a factor in hypertension. *Journal of Chronic Disease* 18: 647–56.

31. KOBAYASHI, J. 1957. On geographical relationship between the chemical nature of river water and death-rate of apoplexy. *Berichte des Ohara Institute fur landwirtschaftliche biologie* 11: 12–21.

32. ALLEN-PRICE, E. D. 1960. Uneven distribution of cancer in West Devon. *Lancet* 1: 1235–38.

33. LEGON, C. D. 1952. The aetiological significance of geographical variations in cancer mortality. *British Medical Journal,* 700–702.

34. STOCKS, P., and DAVIES, R. I. 1960. Epidemiological evidence from

chemical and spectrographic analysis that soil is concerned in the causation of cancer. *British Journal of Cancer* 14: 8–22.

35. KMET, J., and MAHBOUBI, E. 1972. Esophageal cancer in the Caspian littoral of Iran. *Science* 175: 846–53.

36. BRENNER, D. J. 1989. *Radon: Risk and remedy.* New York: W. H. Freeman.

37. EHRLICH, P. R.; EHRLICH, A. H.; and HOLDREN, J. P. 1970. *Ecoscience: Population, resources, environment.* San Francisco: W. H. Freeman.

38. WALDBOTT, G. L. 1978. *Health effects of environmental pollutants.* 2d ed. St. Louis: C. V. Mosby.

39. VAN KOEVERING, T. E., and SELL, N. J. 1986. *Energy: A conceptual approach.* Englewood Cliffs, N.J.: Prentice Hall.

40. NERO, A. V., JR. 1988. Controlling indoor air pollution. *Scientific American* 258(5): 42–48.

41. UNIVERSITY OF MAINE AND MAINE DEPARTMENT OF HUMAN SERVICES. 1983. Radon in water and air. *Resource Highlights,* February.

42. U.S. ENVIRONMENTAL PROTECTION AGENCY. 1989. *Radon measurements in schools.* Office of Radiation Programs. EPA 520/1 89–010.

43. U.S. ENVIRONMENTAL PROTECTION AGENCY. 1986. *A citizen's guide to radon.* OPA–86–004.

44. HENSHAW, D. L.; EATOUGH, J. P.; and RICHARDSON, R. B. 1990. Radon as a causative factor in induction of myeloid leukaemia and other cancers. *The Lancet,* April 28, 1008–12.

45. GATES, A. E., and GUNDERSEN, L. C. S. 1989. Role of ductile shearing in the concentration of radon in the Brookneal Zone, Virginia. *Geology* 17: 391–94.

46. HURLBURT, S. 1989. Radon: A real killer or just an unsolved mystery? *Water Well Journal,* June, 34–41.

47. EGGINTON, J. 1989. Menace of Whispering Hills. *Audubon,* January, 28–35.

48. U.S. ENVIRONMENTAL PROTECTION AGENCY. 1986. *Radon reduction techniques for detached houses.* EPA 625/5–86–019.

49. U.S. ENVIRONMENTAL PROTECTION AGENCY. 1988. *Radon-resistant residential new construction.* EPA 600/8–88/087.

One of the most fundamental concepts of environmental geology is that this earth is our only suitable habitat and its resources are limited. This belief holds even though some resources such as timber, water, air,[1] and food are renewable, whereas other resources such as oil, gas, and minerals are recycled so slowly in the geologic cycle that they are essentially nonrenewable. It is important to recognize, however, that renewable resources are renewable only as long as environmental conditions remain favorable for their natural or planned reproduction. Careless use of renewable resources, such as air, water, vegetation, and, to a lesser extent, soils, may render these renewable resources less renewable than we would wish.

Because resources are indeed limited, important questions arise. How long will a particular resource last? How much short- or long-term environmental deterioration are we willing to concede to ensure that resources are developed in a particular area? How can we make the best use of available resources? These questions have no easy answers. We are now struggling with better ways to estimate the quality and quantity of resources. Unfortunately, it is extremely hard to address the second and third questions without a satisfactory answer to the first question, but determining how long a particular resource will last is further complicated because availability will change as our technological skills and discovery techniques become more sophisticated. Furthermore, the nature and extent of people's and institutions' willingness to conserve resources and protect sensitive environments from mineral exploitation and development depends on factors such as national security, desired standard of living, environmental awareness, and economic incentives, as well as many other factors, all of which are difficult to evaluate and which change with time.

Basically, our present resource problem (or crisis) is related to a people problem—too many people. As we stated earlier, it is impossible to maintain an ever-growing population on a finite resource base. With this perspective, Chapter 14 will explore geologic and environmental aspects of mineral resources, and Chapter 15 will discuss energy resources. Although these two subjects overlap considerably, for organizational purposes we will discuss all resources associated with energy production separately. Mineral resources covers a wide variety of earth materials, including metallic and nonmetallic minerals as well as sand, gravel, crushed rock, and ornamental rock (dimension stone) that have commercial value. Some earth materials, such as those used for construction material (sand and gravel, etc.), are low-value resources and have primarily a "place value." That is, they are extracted economically because they are located near where they are to be used. Long hauls of these low-value, high-bulk materials drastically increase their price and reduce their chances of competing with other, closer supplies. On the other hand, materials such as diamonds, copper, gold, and aluminum are high-value resources. These materials are extracted wherever they are found and transported around the world to numerous markets regardless of distances.[2]

---

[1] A discussion of the air environment including: air as a renewable resource; the urban air and air pollution; the urban microclimate; and global climate change is presented in Chapter 16.

[2] FLAWN, P. T. 1970. *Environmental geology*. New York: Harper & Row.

Photo courtesy of U.S. Department of Energy

CHAPTER FOURTEEN

▼

▼

▼

Mineral Resources
and Environment

I nfinite resources, including space, could support an infinite population. Unfortunately, the earth and its resources are finite while the population grows at an ever-faster pace. Although overpopulation has been a problem in some areas for at least several hundred years, it is now apparent that it is a global problem.

When resource data are combined with population data, the conclusion is clear—it is impossible in the long run to match exponential population growth with production of useful materials based on a finite resource base. Considering that many other countries also aspire to affluence, while the world population increases, the availability of necessary resources becomes questionable.

▼ THE IMPORTANCE OF MINERALS TO SOCIETY

Modern society depends on the availability of mineral resources (1). Consider the mineral products found in a typical American home (Table 14.1). Specifically, consider your breakfast this morning. You probably drank from a glass made primarily of sand, ate food from dishes made from clay, flavored your food with salt mined from the earth, ate fruit grown with the aid of fertilizers such as potassium carbonate (potash) and phosphorus, and used utensils made from stainless steel, which comes from processing iron ore and other minerals.

Minerals are so important to people that, other things being equal, one's standard of living increases with the increased availability of minerals in useful forms. Furthermore, the availability of mineral resources is one measure of the wealth of a society. Those who have been successful in the location, extraction, or importation and use of minerals have grown and prospered. Without

mineral resources, modern technological civilization as we know it would not be possible.

▼ UNIQUE CHARACTERISTICS OF MINERALS

Minerals can be considered our nonrenewable heritage from the geologic past. Although new deposits are still forming from present earth processes, these processes are producing new mineral deposits too slowly to be of use to us today. Locations where mineral deposits are found tend to occupy a small area and tend to be hidden. They must, therefore, be discovered. Unfortunately, most of the easy-to-find deposits have been exploited, and if civilization were to vanish, our descendents would have a much harder time discovering minerals for technological advance than we did. Unlike biological resources, minerals can't be managed to produce a sustained yield—the supply is finite. Recycling and conservation will help, but eventually the supply will be exhausted.

▼ RESOURCES AND RESERVES

Mineral resources are broadly defined as elements, chemical compounds, minerals, or rocks that are concentrated in a form that can be extracted to obtain a usable commodity (2). From a practical viewpoint, this concept of a resource is unsatisfactory because, except in emergencies, a particular commodity will not be extracted unless extraction can be accomplished at a profit. Therefore, a more pragmatic definition of a **resource** is a concentration of a naturally occurring material (solid, liquid, or gas) in or on the crust of the earth in such a form that economical extraction is currently or potentially feasible (3). A **reserve**, on the other hand, is that portion of a resource which is identified and from which usable materials can be legally and economically extracted *at the time* of evaluation. The distinction between resources and reserves, therefore, is based on current geologic and economic factors (Figure 14.1). Resources include both identified reserves and other materials that are either identified but not currently economically or legally available or are undiscovered (currently hypothetical or speculative) (3).

The main point about resources and reserves is that *resources are not reserves!* An analogy from a student's personal finances will help clarify this point. A student's reserves are the liquid assets, such as money in the pocket or bank, whereas the student's resources include the total income the student can expect to earn during his or her lifetime. This distinction is often critical to the student in school because resources are "frozen" assets or next year's income and cannot be used to pay this month's bills (2).

Regardless of potential problems, it is important for planning to estimate future resources. This requires a

Table 14.1

A few of the mineral products in a typical American home.

Building materials	Sand, gravel, stone, brick (clay), cement, steel, aluminum, asphalt, glass
Plumbing and wiring materials	Iron and steel, copper, brass, lead, cement, asbestos, glass, tile, plastic
Insulating materials	Rock, wool, fiberglass, gypsum (plaster and wallboard)
Paint and wallpaper	Mineral pigments (such as iron, zinc, and titanium) and fillers (such as talc and asbestos)
Plastic floor tiles, other plastics	Mineral fillers and pigments, petroleum products
Appliances	Iron, copper, and many rare metals
Furniture	Synthetic fibers made from minerals (principally coal and petroleum products); steel springs; wood finshed with rottenstone polish and mineral varnish
Clothing	Natural fibers grown with mineral fertilizers; synthetic fibers made from minerals (principally coal and petroleum products)
Food	Grown with mineral fertilizers; processed and packaged by machines made of metals
Drugs and cosmetics	Mineral chemicals
Other items	Windows, screens, light bulbs, porcelain fixtures, china, utensils, jewelry: all made from mineral products

Source: U.S. Geological Survey Professional Paper 940, 1975.

continual reassessment of all components of a total resource by considering new technology, probability of geologic discovery, and shifts in economic and political conditions (3).

This approach to resources and reserves, developed by the U.S. Geological Survey and the U.S. Bureau of Mines, is superior to a simple periodical listing of the total amount of material available or likely to become available. Such a list is misleading when used for planning purposes. A superior method is to list or graph mineral resources in terms of the resource classification (Figure 14.1). United States domestic reserves and resources for selected materials, projected through the year 2000, are shown in Table 14.2. Figure 14.2 shows how data from an identified resource can be tabulated and how identified resources can be further subdivided. *Measure-identified* resources refer to those that are well

known and measured and for which the total tonnage or grade is well established. *Indicated-identified* resources are not so well known and measured and therefore cannot be outlined completely by tonnage or grade. *Inferred-identified* resources have quantitative estimates based on broad geologic knowledge of the deposit. The category into which a particular identified resource will fit is obviously a function of available geologic information that involves testing, drilling, and mapping, all of which become more expensive with greater depth. It is therefore not surprising that most of the information for coal is available for identified resources buried less than 300 meters from the surface, which includes 89 percent of the identified resources (Figure 14.2). With increasing depth, less information is available because most detailed mapping is still done from surface exposure (outcrop) of rocks where local relief is seldom more than 100 meters or so. Sampling of subsurface materials is less certain, and indirect geophysical observations are usually not sufficient to make accurate measurements of identified resources that are well below the earth's surface.

The example of silver will further illustrate some important points about resources and reserves. The earth's crust (to a depth of 1 kilometer) contains almost 2 million million (2×10^{12}) metric tons of silver—an amount much larger than the annual world use, which is approximately 10,000 metric tons. If this silver existed as pure metal concentrated into one large mine, it would represent a supply sufficient for several hundred million years at current levels of use. Most of this silver, however, exists in extremely low concentrations—too low to be extracted economically with current technology. The known reserves of silver, reflecting the amount we could obtain immediately with known techniques, is about 200,000 metric tons, or a 20-year supply at current use levels.

The problem with silver, as with all mineral resources, is not with its total abundance but with its concentration and relative ease of extraction. When an atom of silver is used, it is not destroyed, but remains an atom of silver. It is simply dispersed and may become unavailable. In theory, given enough energy, all mineral resources could be recycled, but this is not possible in practice. Consider lead, which is mined and was for many years used in gasoline. This lead is now scattered along highways across the world and deposited in low concentration in forests, fields, and salt marshes close to these highways. Recovery of this lead is, for all practical purposes, impossible.

▼ **AVAILABILITY OF MINERAL RESOURCES**

The earth's mineral resources can be divided into several broad categories, depending on our use of them: elements for metal production and technology, building

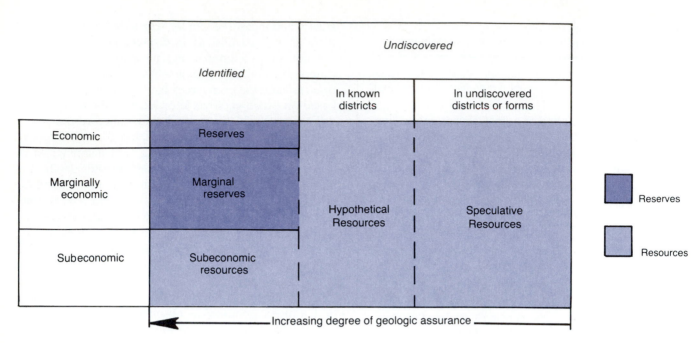

Figure 14.1
Classification of mineral resources used by the U.S. Geological Survey and the U.S. Bureau of Mines. (After U.S. Geological Survey Circular 831, 1980.)

Table 14.2
A generalized outlook of domestic reserves and resources through the year 2000.

Group 1: RESERVES in quantities adequate to fulfill projected needs well beyond the year 2000.

Coal	Silicon
Construction stone	Molybdenum
Sand and gravel	Gypsum
Nitrogen	Bromine
Chlorine	Boron
Hydrogen	Argon
Titanium (except rutile)	Diatomite
Soda	Barite[a]
Calcium	Lightweight aggregates
Clays	Helium
Potash	Peat
Magnesium	Rare earths[a]
Oxygen	Lithium
Phosphorus	

Group 2: IDENTIFIED SUBECONOMIC RESOURCES in quantities adequate to fulfill projected needs beyond the year 2000 and in quantities significantly or slightly greater than estimated UNDISCOVERED RESOURCES.

Aluminum	Vanadium
Nickel[a]	Zircon[a]
Uranium	Thorium
Manganese	

Group 3: Estimated UNDISCOVERED (hypothetical and speculative) RESOURCES in quantities adequate to fulfill projected needs and in quantities significantly greater than IDENTIFIED SUBECONOMIC RESOURCES. Research efforts for these commodities should concentrate on geologic theory and exploration methods aimed at discovering new resources.

Iron	Platinum
Copper[a]	Tungsten
Zinc[a]	Beryllium[a]
Gold	Cobalt[a]
Lead[a]	Cadmium[a]
Sulfur	Bismuth[a]
Silver[a]	Selenium
Fluorine[a]	Niobium[a]

Group 4: IDENTIFIED SUBECONOMIC and UNDISCOVERED RESOURCES together in quantities probably not adequate to fulfill projected needs beyond the end of the century; research on possible new exploration targets, new types of deposits, and substitutes is necessary to relieve ultimate dependence on imports.

Tin	Antimony[a]
Asbestos	Mercury[a]
Chromium	Tantalum[a]

[a]Those commodities that may be in much greater demand than is now predicted because of known or potentially new applications in the production of energy.

Source: U.S. Geological Survey, 1975.

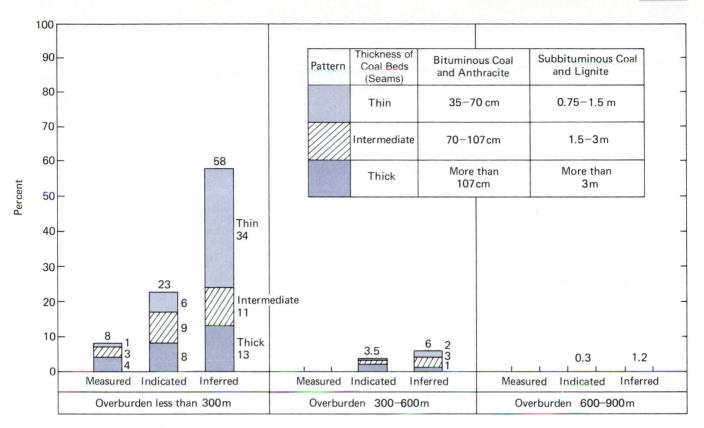

Figure 14.2
Major resource categories of identified U.S. coal resources. (After D. A. Brobst and W. P. Pratt, eds., U.S. Geological Survey Professional Paper 820, 1973.)

materials, minerals for the chemical industry, and minerals for agriculture. Metallic minerals can be classified according to their abundance. The abundant metals include iron, aluminum, chromium, manganese, titanium, and magnesium. Scarce metals include copper, lead, zinc, tin, gold, silver, platinum, uranium, mercury, and molybdenum.

Some mineral resources, such as salt (sodium chloride), are necessary for life. Primitive peoples traveled long distances to obtain salt when it was not locally available. Other mineral resources are desired or considered necessary to maintain a certain level of technology.

The basic issue with mineral resources is not actual exhaustion or extinction, but the cost of maintaining an adequate stock within an economy through mining and recycling. At some point, the costs of mining exceed the worth of the material. When the availability of a particular mineral becomes a limitation, four solutions are possible: find more sources; recycle what has already been obtained; find a substitute; or do without. Which choice or combination of choices is made depends on social, economic, and environmental factors.

The availability of a mineral resource in a certain form, in a certain concentration, and in a certain total amount at that concentration is determined by the earth's history and is a geological issue. What a resource is and

when it becomes limited are ultimately social questions. Before metals were discovered, they could not be considered resources. Before smelting was invented, the only metal ores were those in which the metals appeared in their pure form. Originally, gold was obtained as a pure or *native* metal. Now gold mines extend deep beneath the surface, and the recovery process involves reducing tons of rock to ounces of gold.

In reality then, mineral resources are limited, which raises important questions. How long will a particular resource last? How much short- or long-term environmental deterioration are we willing to concede to ensure that resources are developed in a particular area? How can we make the best use of available resources? These questions have no easy answers. We are now struggling with ways to better estimate the quality and quantity of resources.

We can use a particular mineral resource in several ways: rapid consumption, consumption with conservation, or consumption and conservation with recycling. Which option is selected depends in part on economic, political, and social criteria. Figure 14.3 shows the hypothetical depletion curves corresponding to these three options. Historically, with the exception of precious metals, rapid consumption has dominated most resource utilization. However, as more resources become in short

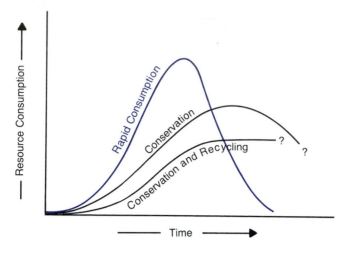

Figure 14.3

Diagram of three hypothetical depletion curves for the use of mineral resources.

supply, increased conservation and recycling are expected. Certainly the trend toward recycling is well established for such metals as copper, lead, and aluminum.

We usually think of mineral resources as the metals used in structural materials, but in fact (with the exception of iron) the predominant mineral resources are not of this type. Consider the annual world consumption of a few selected elements. Sodium and iron are used at a rate of approximately 0.1 to 1 billion tons per year. Nitrogen, sulfur, potassium, and calcium are used at a rate of approximately 10 to 100 million tons per year. These four elements are used primarily as soil conditioners or fertilizers. Zinc, copper, aluminum, and lead have annual world consumption rates of about 3 to 10 million tons, whereas gold and silver have annual consumption rates of 10,000 tons or less. Of the metallic minerals, iron makes up 95 percent of all the metals consumed, and nickel, chromium, cobalt, and manganese are used mainly in alloys of iron (as in stainless steel). Therefore, we can conclude that the nonmetallics, with the exception of iron, are consumed at much greater rates than elements used for their metallic properties.

As the world population and the desire for a higher standard of living increase, the demand for mineral resources expands at a faster and faster rate. From a global viewpoint, our limited mineral resources and reserves threaten our affluence. Approximately 10,000 kilograms of new mineral material (excluding energy resources) are required each year for each person in the United States (Figure 14.4). Predicted world increases in the use of iron, copper, and lead compared with population increases suggests that the rate of production of these metals would have to increase by several times if the world per capita consumption rate were to rise to the U.S. level. Such an increase is very unlikely; affluent countries will have to find substitutes for some minerals or use a smaller proportion of the world annual production.

Domestic supplies of many mineral resources in the United States and other affluent nations are insufficient for current use and must be supplemented by imports from other nations. Table 14.3 shows the deficiency of U.S. reserves for selected nonfuel minerals projected to the year 2000. The table also shows major foreign sources for the needed minerals. Of particular concern to industrial countries is the possibility that the supply of a much desired or needed mineral may become interrupted by political, economic, or military instability of the supplying nation. For example, the recent civil war in Zaire caused sufficient concern that the spot market in cobalt increased by 800 percent (4). Cobalt is a metal used to strengthen steel and thus is in great demand in many industrial countries. Today the United States, along with many other countries, is dependent on a steady supply of imports to meet the mineral demand of industries. Of course the fact that a mineral is imported into a country does not mean that it does not exist in quantities that could be mined within the country. Rather, it suggests that there are economic, political, or environmental reasons that make it easier, more practical, or more desirable to import the material.

Figure 14.4

Amount of new mineral materials required annually by each U.S. citizen. (After U.S. Bureau of Mines, *Mining and Mineral Policy,* 1975.)

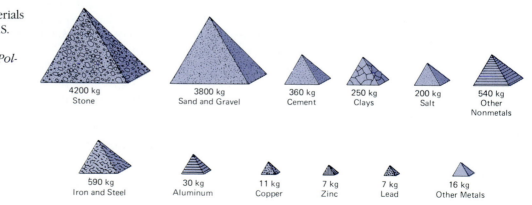

Table 14.3
Deficiency of U.S. reserves of selected nonfuel minerals. Foreign sources subject to potential interruption by political, economic, or military disruption are shown in lighter print.

	Commodity	Adequacy of U.S. Reserves for Cumulative U.S. Demand 1982–2000 (0 10 20 30 40 50 60 70 80 90 100%)	Major Foreign Source
Essentially No Reserves	Manganese		Gabon, Brazil
	Cobalt		Zaire
	Tantalum		Malaysia, Thailand, **Canada**
	Columbium		Brazil, **Canada**
	Platinum group		South Africa, USSR
	Chromium		USSR, South Africa
	Nickel		**Canada**, New Caledonia
	Aluminum		Jamaica, Australia
	Tin		Malaysia, Bolivia
	Antimony		South Africa, Bolivia
	Fluorine		**Mexico**, South Africa
	Asbestos		**Canada**, South Africa
	Vanadium		South Africa, Chile
Reserve Deficiency	Mercury		
	Silver		**Canada, Mexico**
	Tungsten		**Canada**, Bolivia
	Sulfur		**Canada, Mexico**
	Zinc		**Canada, Mexico**
	Gold		**Canada**, USSR
	Potash		**Canada**, Israel

▼ GEOLOGY OF MINERAL RESOURCES

The geology of mineral resources is intimately related to the entire geologic cycle, and nearly all aspects and processes of the cycle are involved to a lesser or greater extent in producing local concentrations or occurrences of useful materials. The term **ore** is sometimes used for those useful metallic minerals that can be mined at a profit, and locations where ore is found are anomalously high concentrations of these minerals. The necessary concentration factor of a particular mineral before the mineral can be classified as an ore varies with technology, economics, and politics, as well as with its natural concentration in the earth's crust. For example, the natural concentration of aluminum in the crust is about 8 percent, and needs to be concentrated only to about 35 percent (concentration factor of about 4), whereas the natural concentration of mercury in the crust is only a tiny fraction of 1 percent and must have a concentration factor of about 10,000 to be mined economically. Table 14.4 lists additional metallic elements and the current concentration factors necessary before mining becomes profitable. The concentration factors necessary for mining may vary as demand changes.

In a broadbrush approach to the geology of mineral resources, tectonic plate boundaries are related to the origin of such ore deposits as iron, gold, copper, and mercury (Figure 14.5). The basic ideas relate to processes operating at the diverging and converging plate boundaries.

Table 14.4
Approximate concentration factors of selected metals necessary before mining is economically feasible.

Metal	Natural Concentration (Percent)	Percent in Ore	Approximate Concentration Factor
Gold	0.0000004	0.001	2,500
Mercury	0.00001	0.1	10,000
Lead	0.0015	4	2,500
Copper	0.005	0.4 to 0.8	80 to 160
Iron	5	20 to 69	4 to 14
Aluminum	8	35	4

Source: Data from U.S. Geological Survey Professional Paper 820, 1973.

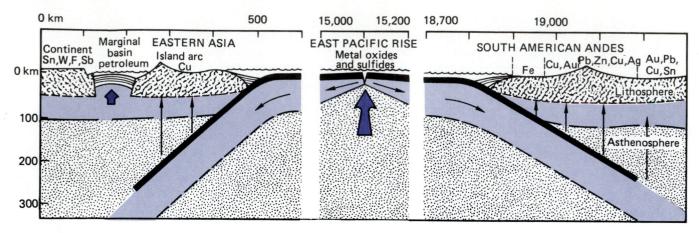

Figure 14.5

Diagram of the relationship between the East Pacific Rise (divergent plate boundary), Pacific margins (convergent plate boundaries), and metallic ore deposits. (After NOAA, *California Geology,* vol. 30, no. 5, 1977.)

The origin of metallic ore deposits at *divergent* plate boundaries is related to the migration (movement) of ocean water. Basically, cold dense ocean water moves down through numerous fractures in the basaltic rocks at oceanic ridges and is heated by contact with or heat from nearby molten rock (magma). The warm water is lighter and more chemically active and rises up (convects) through the fractured rocks, leaching out metals. The metals are carried in solution and deposited (precipitated) as metallic sulfides (5). Quite a few hot water vents with sulfide deposits have been discovered along oceanic ridges, and undoubtedly numerous others will be located.

The origin of metallic ore deposits at *convergent* plate boundaries is hypothesized to be the result of relationships between seawater-saturated oceanic rocks and partial melting of rocks associated with the descent of oceanic lithosphere at a zone of convergence (subduction zone). The idea is that, as the seawater-saturated rocks are subjected to elevated temperatures and pressures in the subduction zone, they partially melt. The combination of heat, pressure, and the resulting partial melting facilitates the mobilization of metals which are concentrated and ascend as components of the magma. The metal-rich fluids are eventually released (or escape) from the magma, and the metals are deposited in a host rock (5).

Perhaps the best example of this relation is the global occurrence of known mercury deposits (Figure 14.6). All the belts of productive deposits of mercury are associated with volcanic systems and are located near convergent plate boundaries (subduction zones). It has been suggested that the mercury, originally found in oceanic sediments of the crust, is distilled out of the downward-plunging plate and emplaced at a higher level above the subduction zone (2). The significant point, however, is that because mercury deposits are intimately associated

with volcanic and tectonic processes, convergent plate junctions characterized by volcanism and tectonic activities are likely places to find mercury. A similar argument can be made for other ore deposits, but there is danger in oversimplification, since many deposits are not directly associated with plate boundaries (Figure 14.7).

The geology of economically useful deposits of mineral and rock materials is as diversified and complex as the processes responsible for their formation or accumulation in the natural environment. Most deposits, however, can be related to various parts of the rock cycle within the influence of the tectonic, geochemical, and hydrologic cycles. The genesis of mineral resources with commercial value can be subdivided into several categories:

▼ Igneous processes, including crystal settling and late magmatic and hydrothermal activities
▼ Metamorphic processes associated with contact or regional metamorphism
▼ Sedimentary processes, including accumulation in oceanic, lake, stream, wind, and glacial environments
▼ Biological processes
▼ Weathering processes such as soil formations and *in situ* (in-place) concentrations of insoluble minerals in weathered rock debris

Table 14.5 lists examples of ore deposits from each of these categories.

Igneous Processes

Most ore deposits caused by igneous processes result from an enrichment process that concentrates an economically desirable ore of metals such as copper, nickel, or gold. In some cases, however, an entire igneous rock mass contains disseminated crystals that can be recovered economically. Perhaps the best-known example is

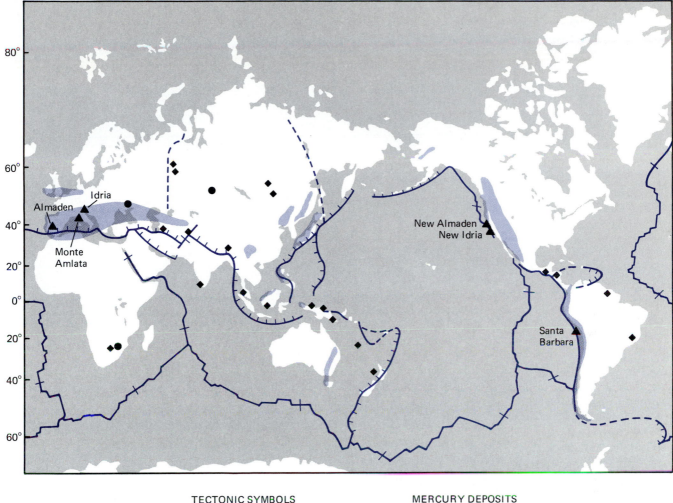

Figure 14.6
Relation between mercury deposits and recently active subduction zones. (From D. A. Brobst and W. P. Pratt, eds., U.S. Geological Survey Professional Paper 820, 1973.)

the occurrence of diamond crystals, found in a coarse-grained igneous rock called **kimberlite**, which characteristically occurs as a pipe-shaped body of rock that decreases in diameter with depth. Almost the entire kimberlite pipe is the ore deposit, and the diamond crystals are disseminated throughout the rock (Figure 14.8).

Ore deposits can also result from igneous processes that segregate crystals formed earlier from those formed later. For example, as magma cools, heavy minerals that crystallize early may slowly sink or settle toward the lower part of the magma chamber, where they form concentrated layers. Deposits of chromite (ore of chromium) have formed by this process (Figure 14.9).

Late magmatic processes occur after most of the magma has crystallized, and rare and heavy metalliferous materials in water- and gas-rich solutions remain. This late-stage metallic solution may be squeezed into fractures or settle into interstices (empty spaces) between earlier-formed crystals. Other late-stage solutions form coarse-grained igneous rock known as **pegmatite**, which is rich in feldspar, mica, and quartz, as well as certain rare minerals (6). Pegmatites have been extensively mined for feldspar, mica, spodumene (lithium mineral), and clay that forms from weathered feldspar.

Hydrothermal (hot-water) **ore deposits** are the most common type. They originate from late-stage magmatic processes and give rise to a variety of deposits, including

Figure 14.7
Occurrence of economic deposits of some selected metals in the United States. (Data from mineral resource maps of the U.S. Geological Survey.)

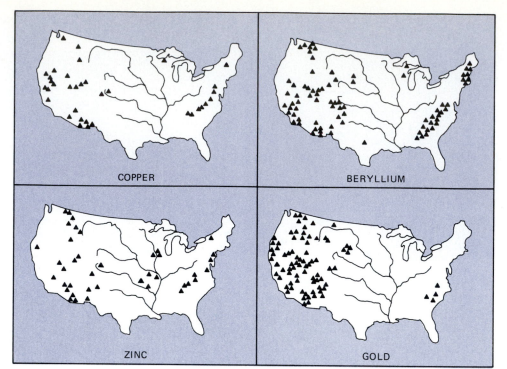

COPPER

BERYLLIUM

ZINC

GOLD

Table 14.5
Examples of different types of mineral resources.

Type	Example
Igneous	
Disseminated	Diamonds—South Africa
Crystal settling	Chromite—Stillwater, Montana
Late magmatic	Magnetite—Adirondack Mountains, New York
Pegmatite	Beryl and lithium—Black Hills, South Dakota
Hydrothermal	Copper—Butte, Montana
Metamorphic	
Contact metamorphism	Lead and silver—Leadville, Colorado
Regional metamorphism	Asbestos—Quebec, Canada
Sedimentary	
Evaporite (lake or ocean)	Potassium—Carlsbad, New Mexico
Placer (stream)	Gold—Sierra Nevada foothills, California
Glacial	Sand and gravel—northern Indiana
Deep-ocean	Manganese oxide nodules—central and southern Pacific Ocean
Biological	Phosphorus—Florida
Weathering	
Residual soil	Bauxite—Arkansas
Secondary enrichment	Copper—Utah

Source: Modified from Robert J. Foster, *General Geology*, 4th ed. Columbus, Ohio: Charles E. Merrill, 1983.

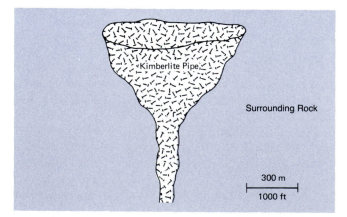

Kimberlite Pipe

Surrounding Rock

300 m

1000 ft

Figure 14.8
Idealized diagram showing a typical South African diamond pipe. Diamonds are scattered throughout the cylindrical body of igneous rock known as *kimberlite*. (After Foster, *General Geology,* 4th ed. [Columbus, Ohio: Charles E. Merrill, 1983].)

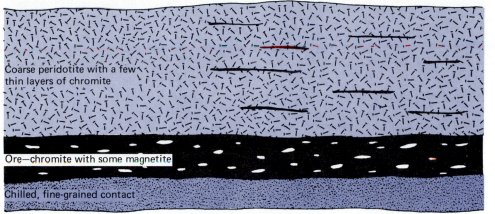

Figure 14.9
How chromite layers might form. The chromite crystallizes early, and the heavy crystals sink to the bottom and accumulate in layers. [From Foster, *General Geology*, 4th ed. (Columbus, Ohio: Charles E. Merrill, 1983).]

gold, silver, copper, mercury, lead, zinc, and other metals, as well as many nonmetallic minerals. The mineral material is either produced directly from the igneous parent rock, as shown in Figure 14.10, or altered by metamorphic processes as magmatic solutions intrude into the surrounding rock as veins or small dikes. Many hydrothermal deposits cannot be traced to a parent igneous rock mass, however, and their origin remains unknown. It is speculated that circulating groundwater, heated and enriched with minerals after contact with deeply buried magma, might be responsible for some of these deposits (6, 7).

Essentially, hydrothermal solutions that form ore deposits are mineralizing fluids that migrate through a host rock. Two types of deposits can be recognized: cavity-filling and replacement. *Cavity-filling deposits* are formed when hydrothermal solutions migrate along openings in rocks (such as fracture systems, pore spaces, or bedding planes) and precipitate ore minerals. *Replacement deposits,* on the other hand, form as hydrothermal solutions react with the host rock, forming a zone in which ore minerals precipitate from the mineralizing fluids and replace part of the host rock. Although replacement deposits are believed to dominate at higher temperatures and pressures than cavity-filling deposits, actually both may be found in close association as one grades into the other; that is, the filling of an open fracture by precipitation from hydrothermal solutions may occur simultaneously with replacement of the rock that lines the fracture (7).

Hydrothermal replacement processes are significant because, excluding some iron and nonmetallic deposits, they have produced some of the world's largest and most important mineral deposits. Some of these deposits result from a massive, nearly complete replacement of host rock with ore minerals that terminate abruptly; others form thin replacement zones along fissures; and still others form disseminated replacement deposits that may involve huge amounts of relatively low-grade ore (7).

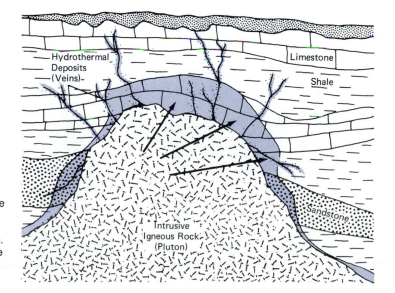

Contact metamorphic zone where mineral deposits may be present. Notice the zone is wider in the limestone rock than in the sandstone or shale. This results because limestone is chemically more active under contact metamorphism.

Figure 14.10
How hydrothermal and contact metamorphic ore deposits might form.

The actual sequence of geologic events leading to the development of a hydrothermal ore deposit is usually complex. Consider, for example, the tremendous, disseminated copper deposits of northern Chile. The actual mineralization is thought to be related to igneous activity, faulting, and folding that occurred 60 to 70 million years ago. The ore deposit is an elongated, tabular mass along a highly sheared (fractured) zone separating two types of granitic rock. The concentration of copper results from a number of factors:

▼ A source igneous rock supplied the copper
▼ The fissure zone tapped the copper supply and facilitated movement of the mineralizing fluids
▼ The host rock was altered and fractured, preparing it for deposition and replacement processes that produced the ore
▼ The copper was leached and redeposited again by meteoric water, which further concentrated the ore (8)

Metamorphic Processes

Ore deposits are often found along the contact between igneous rocks and the surrounding rocks they intrude. This area is characterized by **contact metamorphism**, caused by the heat, pressure, and chemically active fluids of the cooling magma interacting with the surrounding rock, called *country* rock. The width of the contact metamorphic zone varies with the type of country rock. The zone of limestone is usually thickest because limestone is more reactive; the release of carbon dioxide (CO_2) increases the mobility of reactants. The zone is generally thinnest for shale because the fine-grained texture retards the movement of hot, chemically active solutions; and the zone is intermediate for sandstone, as shown in Figure 14.10. Some of the mineral deposits that form in contact areas originate from the magmatic fluids and some from reactions of these fluids with the country rock.

Metamorphism can also result from regional increase of temperature and pressure associated with deep burial of rocks or tectonic activity. Regardless of its cause, **regional metamorphism** can change the mineralogy and texture of the preexisting rocks, producing ore deposits of asbestos, talc, graphite, and other valuable nonmetallic deposits (7).

Metamorphism has been suggested as a possible origin of some hydrothermal fluids. It is a particularly likely cause in high-temperature, high-pressure zones where fluids might be produced and forced out into the surrounding rocks to form replacement or cavity-filling deposits. For example, the native copper found along the top of ancient basalt flows in the Michigan copper district was apparently produced by metamorphism and alteration of the basalt, which released the copper and other materials that produced the deposits (8).

Our discussion of igneous and metamorphic processes has focused primarily on ore deposits. In addition, however, igneous and metamorphic processes are responsible for producing a good deal of stone used in the construction industry. Granite, basalt, marble (metamorphosed limestone), slate (metamorphosed shale), and quartzite (metamorphosed sandstone), along with other rocks, are quarried to produce crushed rock and dimension stone in the United States. Stone is used in many aspects of construction work; but many people are surprised to learn that, in total value, the stone industry is the largest nonfuel, nonmetallic mineral industry in the United States (9).

Sedimentary Processes

Sedimentary processes are often significant in concentrating economically valuable materials in sufficient amounts for extraction. As sediments are transported, wind and running water help segregate the sediment by size, shape, and density. Thus, the best sand or sand and gravel deposits for construction purposes are those in which the finer materials have been removed by water or wind. Sand dunes, beach deposits, and deposits in stream channels are good examples.

The sand and gravel industry amounts to over $1 billion per year, and by volume mined it is the largest nonfuel mineral industry in the United States. Currently, most sand and gravel is obtained from river channels and water-worked glacial deposits. The United States now produces more sand and gravel than it needs, but demand is likely to meet supply by the year 2000 (10).

Stream processes transport and sort all types of materials according to size and density. Therefore, if the bedrock in a river basin contains heavy metals such as gold, streams draining the basin may concentrate heavy metals to form **placer deposits** in areas where there is little water turbulence or velocity, such as in open crevices or fractures at the bottoms of pools, on the inside curves of bends, or on riffles (Figure 14.11). Placer mining of gold—which was known as a "poor man's method" because a miner needed only a shovel, a pan, and a strong back to work the streamside claim—helped to stimulate settlement of California, Alaska, and other areas of the United States. Furthermore, the gold in California attracted miners who acquired the expertise necessary to locate and develop other resources in the western conterminous United States and Alaska.

Placer deposits of gold and diamonds have also been concentrated by coastal processes, primarily wave action. Beach sands and near-shore deposits are mined in Africa and other places.

Rivers and streams that empty into the oceans and lakes carry tremendous quantities of dissolved material derived from the weathering of rocks. From time to time, geologically speaking, a shallow marine basin may be

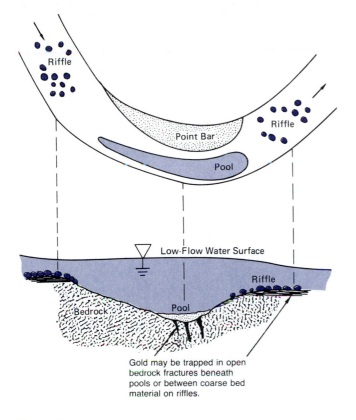

Figure 14.11
Diagram of a stream channel and bottom profile showing areas where placer deposits of gold are likely to occur.

Gold may be trapped in open bedrock fractures beneath pools or between coarse bed material on riffles.

isolated by tectonic activity (uplift) that restricts circulation and facilitates evaporation. In other cases, climatic variations during the ice ages produced large inland lakes with no outlets, which eventually dried up. In either case, as evaporation progresses, the dissolved materials precipitate, forming a wide variety of compounds, minerals, and rocks that have important commercial value. Most of these **evaporite** deposits can be grouped into one of three types: *marine evaporites* (solids)—potassium and sodium salts, gypsum, and anhydrite; *nonmarine evaporites* (solids)—sodium and calcium carbonate, sulfate, borate, nitrate, and limited iodine and strontium compounds; and *brines* (liquids derived from wells, thermal springs, inland salt lakes, and sea waters)—bromine, iodine, calcium chloride, and magnesium (10). Heavy metals (such as copper, lead, and zinc) associated with brines and sediments in the Red Sea, Salton Sea, and other areas are important resources that may be exploited in the future. Evaporite materials are widely used in industry and agriculture (11).

Extensive marine evaporite deposits exist in the United States (Figure 14.12). The major deposits are halite (common salt, NaCl), gypsum ($CaSO_4 \cdot 2H_2O$), anhydrite ($CaSO_4$), and interbedded limestone ($CaCO_3$). Limestone, gypsum, and anhydrite are present in nearly

all marine evaporite basins, and halite and potassium minerals are found in a few (11).

Marine evaporites can form stratified deposits that may extend for hundreds of kilometers with a thickness of several thousand meters. The evaporites represent the product of evaporation of seawater in isolated shallow basins with restricted circulation. Within many evaporite basins, the different deposits are arranged in broad zones that reflect changes in salinity and other factors controlling the precipitation of evaporites; that is, different materials may be precipitated at the same time in different parts of the evaporite basin. Halite, for example, is precipitated in areas where the brine is more saline, and gypsum where it is less saline. Economic deposits of potassium evaporite minerals are relatively rare but may form from highly concentrated brines.

Nonmarine evaporite deposits form by evaporation of lakes in a closed basin. Tectonic activity such as faulting can produce an isolated basin with internal drainage and no outlet. However, the tectonic activity must continue to uplift barriers across possible outlets or lower the basin floor faster than sediment can raise it, to maintain a favorable environment for evaporite mineral precipitation. Even under these conditions, economic deposits of evaporites will not form unless sufficient dissolved salts have washed into the basin by surface runoff from surrounding highlands. Finally, even if all favorable environmental criteria are present, including an isolated basin with sufficient runoff and dissolved salts, valuable nonmarine evaporites, such as sodium carbonate or borate, will not form unless the geology of the highlands surrounding the basin is also favorable and yields runoff with sufficient quantities of the desired material in solution (11).

Some evaporite beds are compressed by overlying rocks, mobilized, and then pierce or intrude the overlying rocks. Intrusions of salt, called **salt domes**, are quite common in the Gulf Coast of the United States and are also found in northwestern Germany, Iran, and other areas. Salt domes in the Gulf Coast are economically important because

▼ They are a good source for nearly pure salt
▼ Some have extensive deposits of elemental sulfur (Figure 14.13)
▼ Some have oil reserves on their flanks

They are also environmentally important as possible permanent disposal sites for radioactive waste, although because salt domes tend to be mobile, their suitability as disposal sites for hazardous wastes must be seriously questioned.

Evaporites from brine resources of the United States are substantial (Table 14.6), assuring that no shortage is likely for a considerable period of time. But many evaporites will continue to have a "place value" because transportation of these mineral commodities increases

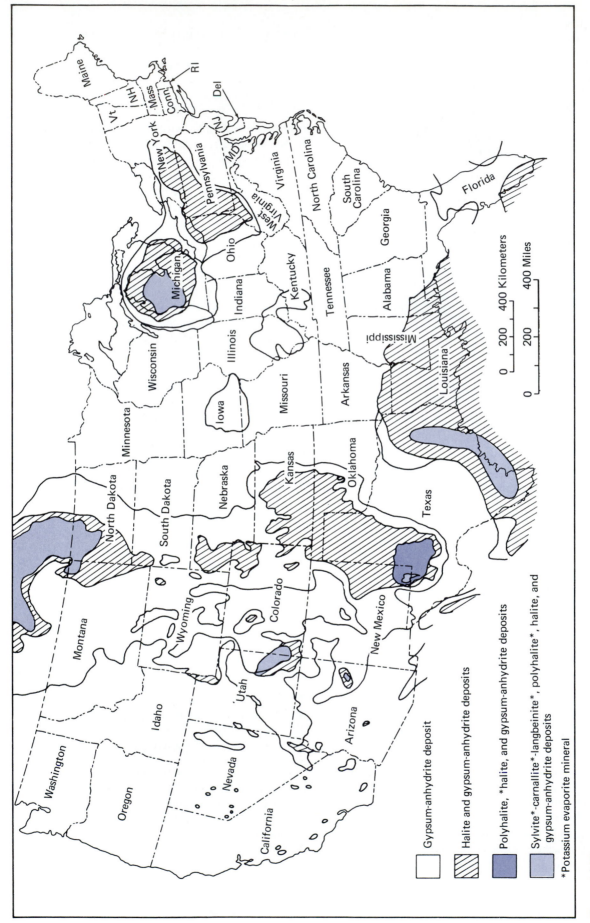

Figure 14.12
Marine evaporite deposits of the United States. (After D. A. Brobst and W. P. Pratt, eds., U.S. Geological Survey Professional Paper 820, 1973.)

Gypsum-anhydrite deposit

Halite and gypsum-anhydrite deposits

Polyhalite*, *halite, and gypsum-anhydrite deposits

Sylvite*-carnallite*-langbeinite*, polyhalite*, halite, and gypsum-anhydrite deposits

*Potassium evaporite mineral

400 Kilometers

400 Miles

200

200

0

0

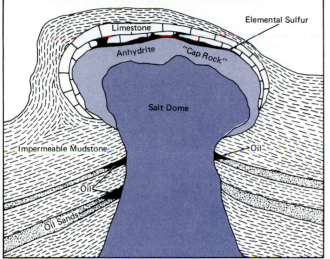

Figure 14.13
A cross section through a typical salt dome of the type found in the Gulf Coast of the United States.

their price, and therefore continued discoveries of high-grade deposits closer to where they will be consumed remains an important goal (11).

Biological Processes

Organisms are able to form many kinds of minerals, such as the calcium minerals in shells and phosphate in bones. Some of these minerals cannot be formed inorganically in the biosphere. Thirty-one different biologically produced minerals have been identified. Minerals of biological origin contribute significantly to sedimentary deposits (12).

An interesting example of mineral deposits produced by biological processes are phosphates associated with sedimentary marine deposition. Phosphorus-rich sedimentary rocks are fairly common in some of the western states as well as in Tennessee and Florida. The common phosphorus-bearing mineral is apatite which is a calcium phosphate associated with fish bones and teeth.

Table 14.6
Evaporite and brine resources of the United States expressed in years of supply at current rates of domestic consumption.

Commodity	Identified Resources[a] (Reserves[b] and Subeconomic Deposits)	Undiscovered Resources (Hypothetical[c] and Speculative[d] Resources)
Potassium compound	100 years	Virtually inexhaustible
Salt	1,000 + years	Unlimited
Gypsum and anhydrite	500 + years	Virtually inexhaustible
Sodium carbonate	6,000 years	5,000 years
Sodium sulfate	700 years	2,000 years
Borates	300 years	1,000 years
Nitrates	Unlimited (air)	Unlimited (air)
Strontium	500 years	2,000 years
Bromine	Unlimited (seawater)	Unlimited (seawater)
Iodine	100 years	500 years
Calcium chloride	100 + years	1,000 + years
Magnesium	Unlimited (seawater)	Unlimited (seawater)

[a]Identified resources: Specific, identified mineral deposits that may or may not be evaluated as to extent and grade, and whose contained minerals may or may not be profitably recovered with existing technology and economic conditions.

[b]Reserves: Identified deposits from which minerals can be extracted profitably with existing technology and under present economic conditions.

[c]Hypothetical resources: Undiscovered mineral deposits, whether of recoverable or subeconomic grade, that are geologically predictable as existing in known districts.

[d]Speculative resources: Undiscovered mineral deposits, whether of recoverable or subeconomic grade, that may exist in unknown districts or in unrecognized or unconventional form.

Source: G. I. Smith et al., U.S. Geological Survey Professional Paper 820, 1973.

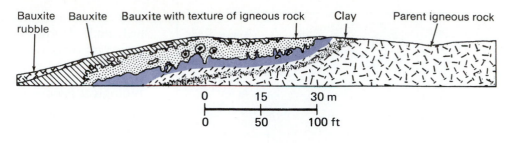

Figure 14.14
Cross section of the Pruden baux-ite mine, Arkansas. The bauxite was formed by intensive weather-ing of the aluminum-rich igneous rocks. (After G. Mackenzie, Jr., et al., U.S. Geological Survey Profes-sional Paper 299, 1958.)

The phosphate is extracted from seawater by fish and other marine organisms, and the mineral deposit results from sedimentary accumulations of the phosphate-rich debris. The richest phosphate mine in the world is known as "Bone Valley" located approximately 40 kilo-meters east of Tampa, Florida. The deposit is marine sedimentary rocks composed in part of fossils of marine animals that lived approximately 10 to 15 million years ago, when Bone Valley was the bottom of a shallow sea. That deposit now supplies approximately one-third of the world's phosphate production.

Another important source of phosphorus is guano (bird feces), which accumulates where there are large colonies of nesting sea birds and a climate dry enough for the guano to dry to a rocklike mass. Therefore, the formation of one of the major sources of phosphorus depends upon unique biological and geographical con-ditions.

Weathering Processes

Weathering is responsible for concentrating some mate-rials to the point that they can be extracted at a profit. For example, intensive weathering of residual soils (laterite) derived from aluminum-rich igneous rocks may concen-trate relatively insoluble hydrated oxides of aluminum and iron, while the more soluble elements, such as silica, calcium, and sodium, are selectively removed by soil and biological processes. If sufficiently concentrated, residual aluminum oxide forms an ore of aluminum known as **bauxite** (Figure 14.14). Important nickel and cobalt deposits are also found in laterite soils developed from ferromagnesian-rich igneous rocks.

Insoluble ore deposits such as native gold are generally residual and, unless removed by erosion, accumulate in weathered rock and soil. Accumulation is favored where the parent rock that contains insoluble ore minerals is relatively soluble, such as limestone (Figure 14.15). Care must be taken in evaluating a residual weathered rock or soil deposit because the near-surface concentration may be a much higher grade than ore in the parent, unweathered rocks (6).

Weathering is also involved in secondary enrichment processes to produce sulfide ore deposits from low-grade primary ore. Near the surface, primary ore con-taining such minerals as iron, copper, and silver sulfides is in contact with slightly acid soil water in an oxygen-rich environment. As the sulfides are oxidized, they are dissolved, forming solutions rich in sulfuric acid and silver and copper sulfate, which migrate downward, producing a leached zone devoid of ore minerals (Figure 14.16). Below the leached zone and above the ground-water table, oxidation continues, and sulfate solutions continue their downward migration. Below the water table, if oxygen is no longer available, the solutions are deposited as sulfides, enriching the metal content of the primary ore by as much as ten times. In this way, low-grade primary ore is rendered more valuable, and high-grade primary ore is made even more attractive (7, 8).

The presence of a residual iron oxide cap at the surface indicates the possibility of an enriched ore below, but is not always conclusive. Of particular importance to formation of a zone of secondary enrichment is the presence in the primary ore of iron sulfide (for example, pyrite). Without it, secondary enrichment seldom takes place, because iron sulfide in the presence of oxygen and water forms sulfuric acid, which is a necessary solvent. Another factor favoring development of a secondary-enrichment ore deposit is the primary ore being suffi-ciently permeable to allow water and solutions to migrate

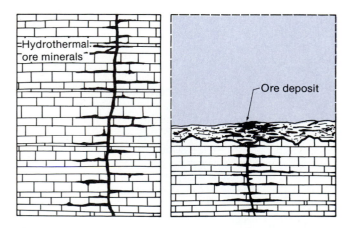

Figure 14.15
How an ore deposit of insoluble minerals might form by weathering and formation of a residual soil. As the limestone that contained the deposit weathered, the ore minerals be-came concentrated in the residual soil. [From Foster, *General Geology*, 4th ed. (Columbus, Ohio: Charles E. Merrill, 1983).]

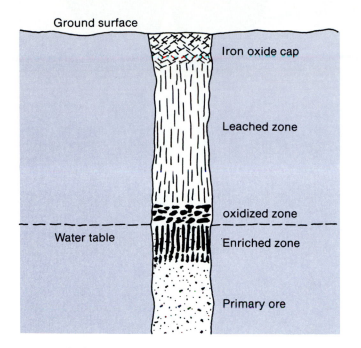

Ground surface

Iron oxide cap

Leached zone

oxidized zone

Water table

Enriched zone

Primary ore

Figure 14.16
Typical zones that form during secondary enrichment processes. Sulfide ore minerals in the primary ore vein are oxidized and altered, and then are leached from the oxidized zone and redeposited in the enriched zone. The iron oxide cap is generally a reddish color and may be helpful in locating ore deposits that have been enriched. [After Foster, *General Geology,* 4th ed. (Columbus, Ohio: Charles E. Merrill, 1983).]

freely downward. Given a primary ore that meets these criteria, the reddish iron oxide cap probably does indicate that secondary enrichment has taken place (7).

Several disseminated copper deposits have become economically successful because of secondary enrichment, which concentrates dispersed metals. For example, secondary enrichment of a disseminated copper deposit at Miami, Arizona, increased the grade of the ore from less than 1 percent copper in the primary ore to as much as 5 percent in some localized zones of enrichment (8).

Minerals from the Sea

Mineral resources in seawater or on the bottom of the ocean are vast and, in some cases, such as magnesium, are nearly unlimited. In the United States, magnesium was first extracted from seawater in 1940. By 1972, one company in Texas produced 80 percent of our domestic magnesium, using seawater as its raw material source. Companies in Alabama, California, Florida, Mississippi, and New Jersey are also extracting magnesium from seawater.

The deep-ocean floor may be the site of the next big mineral rush. Identified deposits include massive sulfide deposits associated with hydrothermal vents, manganese oxide nodules, and cobalt-enriched manganese crusts.

Massive sulfide deposits, containing zinc, copper, iron, and trace amounts of silver, are produced at divergent plate boundaries (oceanic ridges) by the forces of plate tectonics. Pressure created by several thousand meters of water at ridges forces cold seawater deep into numerous rock fractures, where it is heated by upwelling magma and emerges as hot springs, as illustrated in Figure 14.17. The circulating water leaches the rocks, removing metals that are deposited when hot mineral-rich water is ejected at temperatures up to 350°C into the cold sea. Sulfide minerals precipitate near vents, known as "black smokers" because of the color of the ejected mineral-rich water, forming massive towerlike formations rich in metals. The hot vents are of particular biologic significance because they support a unique assemblage of animals, including giant clams, tube worms, and white crabs. Communities of these animals base their existence on sulfide compounds extruded from black smokers, existing through a process called *chemosynthesis* as opposed to photosynthesis, which supports all other known ecosystems on earth.

The extent of sulfide mineral deposits along oceanic ridges is poorly known, and although leases to some possible deposits are being considered, it seems unlikely that such deposits will be extracted at a profit in the near future. Certainly potential environmental degradation, such as water quality and sediment pollution, will have to be carefully evaluated prior to any proposed mining activity in the future.

Study of the formation of massive sulfide deposits at oceanic ridges is helping geologists understand some of the mineral deposits on land. For example, massive sulfide deposits being mined in Cyprus are believed to have formed at an oceanic ridge and to have been later uplifted to the surface.

Manganese oxide nodules (Figure 14.18), which contain manganese (24 percent) and iron (14 percent) with secondary copper (1 percent), nickel (1 percent), and cobalt (0.25 percent), cover vast areas of the deep-ocean floor. The nodules are found in the Atlantic Ocean off Florida, but the richest and most extensive accumulations occur in large areas of the northeastern, central, and southern Pacific, where the nodules cover 20 to 50 percent of the ocean floor (13).

Manganese oxide nodules are usually discrete, but are welded together locally to form a continuous pavement. Although occasionally found buried in sediment, nodules are usually surficial deposits on the seabed. The average size of the manganese nodules varies from a few millimeters to a few tens of centimeters in diameter. Composed primarily of concentric layers of manganese in iron oxides mixed with a variety of other materials, each nodule formed around a nucleus of a broken nodule, fragment of volcanic rock, or sometimes a fossil. The estimated rate of growth is 1 to 5 millimeters per million years. The nodules are most abun-

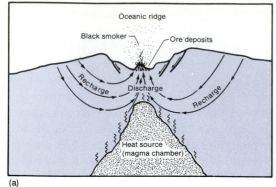

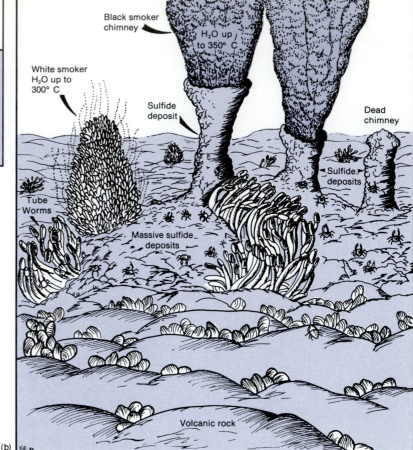

Figure 14.17
(a) Oceanic ridge hydrothermal environment; (b) detail of black smokers where massive sulfide deposits form.

dant in those parts of the ocean where sediment accumulation is at a minimum, generally at depths of 2500–6000 meters (14).

The origin of the nodules is not well understood; presumably they might form in several ways. The most probable theory is that they form from material weathered from the continents and transported by rivers to the oceans, where ocean currents carry the material to the deposition site in the deep-ocean basins. The minerals from which the nodules form may also derive from submarine volcanism, or may be released during physical and biochemical processes and reactions that occur near the water-sediment interface during and after deposition of the sediments (14).

Actual mining of nodules involves lifting the nodules off the bottom and up to the mining ship. French and Japanese researchers are experimenting with a continuous-line bucket dredge, in which a continuous rope with buckets attached at regular intervals is strung between two ships. The buckets drag along the bottom as the two ships move, and the nodules are dumped into one of the ships as the rope loop is reeled from one ship to the other. Other methods of recovery being examined are hydraulic lifting and use of airlift in conjunction with hydraulic dredging (15). Because the whole manganese oxide nodule industry is in its infancy, considerable change in methods and technology is likely. Nevertheless, based on current data, mining of the nodules appears technologically feasible and potentially profitable. However, prospective interest groups must cooper-

Figure 14.18
Manganese oxide nodules on the floor of the Atlantic Ocean. (Photo courtesy of R. M. Pratt.)

ate in management of the resources and must carefully evaluate the environmental impact of the mining on the ecology of the ocean bottom so that the seabed is not degraded.

Cobalt-enriched manganese crusts are present in the mid and southwest Pacific, on flanks of seamounts, volcanic ridges, and islands. Cobalt content varies with water depth; the maximum concentration of about 2.5 percent is found at water depths of 1000 to 2500 meters. Thickness of the crust averages about 2 centimeters. The processes of formation are not well understood. The nature and extent of the crusts, which also contain nickel, platinum, copper, and molybdenum, are being studied by U.S. Geological Survey scientists (16).

▼ ENVIRONMENTAL IMPACT OF MINERAL DEVELOPMENT

The impact of mineral exploitation on the environment depends upon such factors as mining procedures, local hydrologic conditions, climate, rock types, size of operation, topography, and many more interrelated factors. Furthermore, the impact varies with the stage of development of the resource. For example, the exploration and testing stage involves considerably less impact than the mining and processing stages.

Exploration activities for mineral deposits vary from collecting and analyzing remote-sensing data gathered from airplanes or satellites to field work involving surface mapping, drilling, and gathering of geophysical data. Generally, exploration has a minimal impact on the environment provided care is taken in sensitive areas, such as some arid lands, marshlands, and areas underlain by permafrost. Some arid lands are covered by a thin layer of pebbles over fine silt several centimeters thick.

The layer of pebbles, called *desert pavement,* protects the finer material from wind erosion. When the pavement is disturbed by road building or other activity, the fine silts may be eroded, impairing physical, chemical, and biological properties of the soil in the immediate environment and scarring the land for many years. In other areas, such as marshlands and the northern tundra, wet organic rich soils render the land sensitive to even light traffic.

Impacts on Other Resources

Mining and processing of mineral resources are likely to have a considerable adverse impact on land, water, air, and biologic resources and also can initiate social impacts because of increased demand for housing and services in mining areas. These impacts are part of the price we pay for the benefits of mineral consumption. It is unrealistic to expect that we can mine our resources without affecting some aspect of the local environment. We must, however, hold the adverse impact to a minimum. Minimizing environmental degradation caused by mining can be very difficult, though, since the demand for minerals continues to increase while the deposits of highly concentrated minerals decrease. Therefore, to provide even more material, we will need increasingly larger operations to mine everpoorer grades of ore. At the same time, mining ore deposits that are transitional into the surrounding rocks, allowing recovery of ever-lower grades of ore, is not always possible because some ore deposits terminate abruptly along geologic boundaries, as shown in Figure 14.19 (17).

One of the major practical issues is whether surface or subsurface mines should be developed in an area. Surface mining is cheaper but has more direct environ-

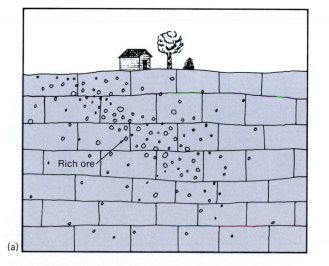 

Figure 14.19
Ore deposits: (a) the grade of the ore gradually lessens with distance from the greatest concentration; (b) the high-grade ore is concentrated in a definite zone with no outward gradation to lower the concentration of the ore.

mental effects. The trend in recent years has been away from subsurface mining and toward large, open-pit (surface) mines such as the Bingham Canyon copper mine in Utah (Figure 14.20) and Liberty Pit near Ruth, Nevada (Figure 14.21). The Bingham Canyon mine is one of the world's largest artificial excavations, covering nearly 8 square kilometers to a maximum depth of nearly 800 meters.

Surface mines and quarries today cover less than 0.5 percent of the total area of the United States. Figure 14.22 shows how much land is used for a number of selected activities—notice that nearly half of the land used for mining involves bituminous coal—and Figure 14.23 shows which states use the greatest amount of land for mining bituminous coal, copper, phosphate rock, and iron ore.

Even though the impact of a single operation is a local phenomenon, numerous local occurrences will eventually constitute a larger problem. Environmental degradation tends to extend beyond the excavation and surface plant areas of both surface and subsurface mines.

Large mining operations disturb the land by directly removing material in some areas, thus changing topography, and by dumping waste in others. At best these actions produce severe aesthetic degradation. Dust at mines may affect air resources, even though care is often taken to reduce dust production by sprinkling water on roads and other sites that generate dust. Water resources are particularly vulnerable to degradation even if drainage is controlled and sediment pollution reduced. Surface drainage is often altered at mine sites, and runoff from precipitation (rain or snow) may infiltrate waste material, leaching out trace elements and minerals. Trace elements (cadmium, cobalt, copper, lead, molybdenum, and others), when leached from mining wastes and concentrated in water, soil, or plants, may be toxic or may cause diseases in people and other animals who drink the water, eat the plants, or use the soil. Specially constructed ponds to collect such runoff can help but cannot be expected to eliminate all problems.

The white streaks in Figure 14.24 are mineral deposits apparently leached from tailings from a zinc

Figure 14.20
The Bingham Canyon, Utah, copper mine, one of the largest artificial excavations in the world. (Photo courtesy of Kennecott Copper Corporation.)

Figure 14.21
The Liberty Pit copper mine near Ruth, Nevada. (Photo courtesy of Kennecott Copper Corporation.)

mine in Colorado. Similar-looking deposits may cover rocks in rivers for many kilometers downstream from some mining areas. Thus, a potential problem associated with mineral resource development is the possible release of harmful trace elements to the environment.

Sometimes leaching is used as a mining technique. For example, some gold deposits contain such finely disseminated gold that extraction by conventional methods is not profitable. For some of these deposits a process known as *heap leaching* with a dilute cyanide solution is used. The cyanide solution is applied by sprinklers to a heap of crushed gold ore. As the solution seeps through the ore it dissolves the gold. Gold-bearing solutions are collected in a plastic-lined pond and treated to recover the gold. Because cyanide is extremely toxic, the mining process must be carefully controlled and monitored. The process has the potential (should an accident occur) to create a serious groundwater pollu-

tion problem. Research is also ongoing to develop in situ cyanide leaching to eliminate the need for removing ore from the ground. Control and monitoring of the leaching solution will probably be a difficult problem (18).

Groundwater may also be polluted by mining operations when waste comes into contact with slow-moving subsurface waters. Surface water infiltration or groundwater movement causes leaching of sulfide minerals that may pollute groundwater and eventually seep into streams to pollute surface water. Groundwater problems are particularly troublesome because reclamation of polluted groundwater is very difficult and expensive.

Even abandoned mines can cause serious problems. For example, subsurface mining for lead and zinc in the Tri-State Area (Kansas, Missouri, and Oklahoma), which started in the late nineteenth century and ceased in some areas in the 1960s, is causing serious water pollution problems in the 1980s and 1990s. The mines, extending to depths of 100 meters below the water table, were kept dry by pumping when the mines were in production. However, since mining stopped, some have flooded and started to overflow into nearby creeks. The water is very acidic because sulfide minerals in the mine react with oxygen and groundwater to form sulfuric acid, a problem known as "acid mine drainage." The problem was so severe in the Tar Creek area of Oklahoma that the Environmental Protection Agency in 1982 designated it as the nation's foremost hazardous waste site. Acid mine drainage is also a very widespread problem in many of the eastern coal fields.

Physical changes in the land, soil, water, and air associated with mining directly and indirectly affect the biological environment. Direct impacts include deaths of plants or animals caused by mining activity or contact with toxic soil or water from mines. Indirect impacts

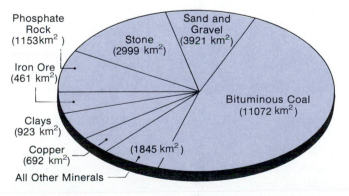

Figure 14.22
Land use in the U.S. by selected commodity. Total land utilized is 23,067 km². (After W. Johnson and J. Paone, 1982, U.S. Bureau of Mines Information Circular 8862.)

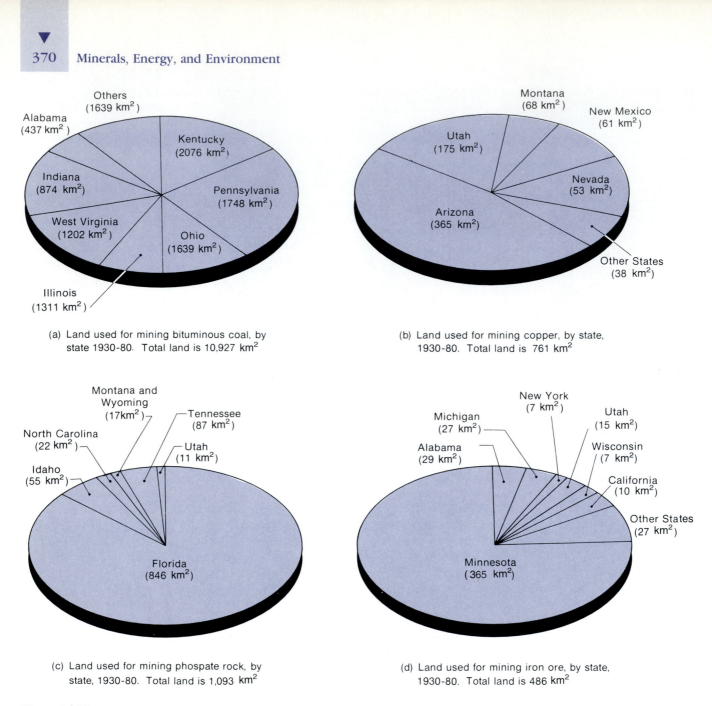

(a) Land used for mining bituminous coal, by state 1930-80. Total land is 10,927 km²

(b) Land used for mining copper, by state, 1930-80. Total land is 761 km²

(c) Land used for mining phospate rock, by state, 1930-80. Total land is 1,093 km²

(d) Land used for mining iron ore, by state, 1930-80. Total land is 486 km²

Figure 14.23
Amount of land (by state) mined for selected materials. (Data from W. Johnson and J. Paone, 1982, U.S. Bureau of Mines Information Circular 8862.)

include changes in nutrient cycling, total biomass, species diversity, and ecosystem stability due to alterations in groundwater or surface water availability or quality. Periodic or accidental discharge of low-grade pollutants through failure of barriers, ponds, or water diversions or through breach of barriers during floods, earthquakes, or volcanic eruptions also damages local ecological systems.

Social Impacts

Social impacts associated with large-scale mining result from a rapid influx of workers into areas unprepared for growth. Stress is placed on local services: water supplies,

sewage and solid waste disposal systems, schools, and rental housing. Land use shifts from open range, forest, and agriculture to urban patterns. More people also increase the stress on nearby recreation and wilderness areas, some of which may be in a fragile ecological balance. Construction activity and urbanization affect local streams through sediment pollution, reduced water quality, and increased runoff. Air quality is reduced as a result of more vehicles, dust from construction, and generation of power.

Adverse social impacts may occur when mines are closed or automation displaces miners, because towns surrounding large mines come to depend on the income shows, by area of activity, the amount of land the U.S. mining industry used and reclaimed from 1930 to 1980. These data suggest that approximately 48 percent of the land utilized by the mining industry has been reclaimed. Reclamation of land used for mining is common practice today and the methods of mine reclamation will be discussed in Chapter 15 when we consider the impact of coal mining on the environment.

Considering that the demand for mineral resources will increase, the only logical approach to the eventual environmental degradation is to minimize both on-site and off-site problems by controlling sediment, water, and air pollution through good engineering and conservation practices. Although these actions will raise the cost of mineral commodities and hence the price of all items produced from these materials, they will yield other returns of equal or higher value to future generations. We must realize, however, that even the most careful measures to control environmental disruption associated with mining will occasionally fail.

▼ RECYCLING OF RESOURCES

A diagram of the cycle of mineral resources (Figure 14.25) reveals that many components of the cycle are connected to waste disposal. In fact, the major environmental impacts of mineral resource utilization are related to waste products. Wastes produce pollution that may be toxic to humans, are dangerous to natural ecosystems and the biosphere, and are aesthetically undesirable. They may attack and degrade other resources such as air, water, soil, and living things. Wastes also deplete nonrenewable mineral resources with no offsetting benefits for human society. Recycling of resources is one way to reduce these wastes.

Materials (especially metals) that end up in land fills and other waste-management facilities are sometimes designated as *urban ore* because of the useful materials the waste may contain (19). The concept of "urban ore" was realized when it was discovered that ash from the incineration of sewage sludge in Palo Alto, California, contains large concentrations of gold (30 parts per million), silver (660 parts per million), copper (8000 parts per million), and phosphorus (6.6 percent). Each metric ton of the ash contains approximately 1 troy ounce of gold and 20 ounces of silver. The gold is concentrated above natural abundance by a factor of 7500 times, making the "deposit" double the average grade that is mined today. Silver in the ash has a concentration factor of 9400, similar to that of rich ore deposits in Idaho. Copper is concentrated in the ash by a factor of 145, similar to that of a common ore grade. Commercial phosphorus deposits vary from 2 to 16 percent, so the ash with 6.6 percent phosphorus has the potential of a high-value resource. The ash in the Palo Alto dump represents a silver and gold deposit with a value of about

Figure 14.24
A zinc mine in Colorado. The white streaks are mineral deposits apparently leached from tailings.

of employed miners. In the old American West, mine closures produced the well-known "ghost towns." Today, the price of coal and other minerals directly affects the lives of many small towns, especially in the Appalachian Mountain region of the United States where closures of coal mines are taking their toll. These mine closings result partly from lower prices for coal and partly from rising mining costs. One reason mining costs are rising is that environmental regulation of the mining industry has increased. Of course, regulations have also helped make mining safer and have facilitated land reclamation. Some miners, however, believe the regulations are not flexible enough, and there is some truth to their arguments. For example, some areas might be reclaimed for use as farmland following mining, if the original hills have been leveled. Regulations, however, may require that the land be restored to its original hilly state, even though hills make inferior farmland.

Following mining activities, land reclamation is necessary if the mining has had detrimental effects and if the land is to be used for other purposes. Table 14.7

Table 14.7
Land used and reclaimed by the mining industry in the U.S. (1930–1980) by area of activity.

Area of activity[1]	Utilized, km²	Percent of total land utilized	Reclaimed, km²	Percent reclaimed
Surface area mined (area of excavation only)	15,920	69%	8,735	55.0%
Area used for disposal of overburden waste from surface mining[2]	3,683	16%	1,910	52.0%
Surface area subsided or disturbed as a result of underground workings[3]	425	2%	24	5.6%
Surface area used for disposal of underground mine waste	770	3%	103	13.4%
Surface area used for disposal of mill or processing waste	2,242	10%	210	9.4%
Total[4]	23,067	100%	10,927	47.4%

[1]Excludes oil and gas operations.
[2]Includes surface coal operations for 1930–71 only.
[3]Includes data for 1930–71 only.
[4]Data may not add to totals shown because of independent rounding.
Source: Johnson, W., and Paone, J., 1982. U.S. Bureau of Mines Information Circular 8862.

$10 million, and gold and silver worth approximately $2 million are being concentrated and delivered each year (20).

The most likely sources of the metals in the Palo Alto sewage are the large electronics industry and the photographic industry located in the area. Gold in significant amounts has been found in the sewage of only one other city, and silver is usually present in much smaller concentrations than at Palo Alto. Thus Palo Alto's unique urban ore offers an unusual opportunity to study and develop methods to recycle valuable materials concentrated in urban waste (20). The city has now employed a private company to extract the gold and silver.

On the other hand, sewage sludge that contains high concentrations of heavy metals such as cadmium is a toxic material and precludes the application of the sludge for uses such as land reclamation. More efficient pretreatment of industrial wastewater and strict regulations are necessary to avoid production of toxic sewage sludge from urban areas.

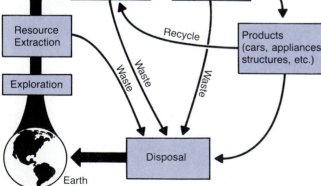

Figure 14.25
Simplified flow chart of the resource cycle.

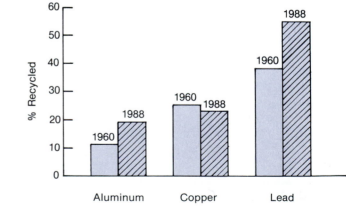

Figure 14.26
Approximate amount of recycled metals as a percent of U.S. consumption in 1960 and 1988. U.S. consumption for 1988 was: aluminum, 5.3 million metric tons; copper, 2.3 million metric tons; and lead, 1.2 million metric tons. (Data from U.S. Bureau of Mines.)

The practice of recycling metals is not new. Metals such as iron, aluminum, copper, and lead have been recycled for many years. Of the millions of automobiles discarded annually, nearly all are dismantled by auto wreckers and scrap processors for metals to be recycled (19). Recycling metals from discarded automobiles is a sound conservation practice, considering that 90 percent by weight of the average discarded automobile is metal.

Figure 14.26 shows the approximate amounts of aluminum, copper, and lead recycled as a percentage of consumption for 1960 and 1988. During that period the amount of recycled copper generally varied from about 20 percent to 25 percent, while the amount of recycled aluminum and lead has increased significantly. Today over 50 percent of the lead consumed is recycled and the corresponding value for aluminum is 20 percent.

Recycling may be one way to delay or partially alleviate a possible resource crisis caused by the convergence of a rapidly rising population and a finite resource base. However, as discussed in Chapter 12, the problem of integrated waste management is complex and before recycling can become more widespread, improved technology and more economic incentives are needed. Nevertheless, the trends are clearly in place and the volume of resources recycled will continue to grow.

▼ ▼ ▼ SUMMARY AND CONCLUSIONS

Mineral resources are generally extracted from naturally occurring, anomalously high concentrations of earth materials. These natural deposits allowed early peoples to exploit minerals while slowly developing technological skills.

Availability of mineral resources is one measure of the wealth of a society. In fact, modern technological civilization as we know it would not be possible without exploitation of mineral resources. However, it is important to recognize that mineral deposits are not infinite and that we cannot maintain exponential population growth on a finite resource base. The United States and many other affluent nations have insufficient domestic supplies of many mineral resources for current use and must supplement them by imports from other nations. In the future, as other nations industrialize and develop, such imports may be more difficult to obtain. Affluent countries may have to find substitutes for some minerals or use a smaller portion of the world's annual production.

An important concept in analyzing resources and reserves is that resources are not reserves. Unless discovered and captured, resources cannot be used to solve present shortages.

The geology of mineral resources is complex and intimately related to various aspects of the geologic cycle. In a broadbrush approach, many, but certainly not all, metallic mineral deposits may correlate with dynamic earth processes that occur at junctions of lithospheric plates. On a more practical scale, mineral resources for construction and industrial uses are concentrated in the geologic environment by: first, igneous or magmatic processes such as crystal settling or hydrothermal activity; second, metamorphic processes such as contact and regional metamorphism; third, sedimentary processes including oceanic and lake processes, running water, wind, and moving ice; fourth, biological processes such as biochemical precipitation; and fifth, weathering processes including soil-forming activity and in situ concentrations of insoluble minerals.

The environmental impact of mineral exploitation depends upon many factors, including mining procedures, local hydrologic conditions, climate, rock types, size of operation, topography, and many more interrelated factors. In addition, the impact varies with the stage of development of the resource. In general, the mining and processing of mineral resources greatly affect the land, water, air, and biological resources and initiate certain social impacts due to increasing demand for housing and services in mining areas. Because the demand for mineral resources is going to increase, we must strive to minimize both on-site and off-site problems by controlling sediment, water, and air pollution through good engineering and conservation practices.

Recycling of mineral resources appears to be one way to delay or partly alleviate a possible crisis caused by the convergence of a rapidly rising population and a limited resource base. Recycling the wide variety of materials found in urban waste is not an easy task, however, and innovative refinements in recycling methods will be necessary to ensure that the recycling trend continues.

▼ ▼ ▼ REFERENCES

1. McKELVEY, V. E. 1973. Mineral resource estimates and public policy. In United States mineral resources, eds. D. A. Brobst and W. P. Pratt, pp. 9–19. U. S. Geological Survey Professional Paper 820.
2. BROBST, D. A.; PRATT, W. P; and McKELVEY, V. E. 1973. Summary of United States mineral resources. U. S. Geological Survey Circular 682.
3. U.S. GEOLOGICAL SURVEY. 1975. Mineral resource perspectives 1975. U.S. Geological Survey Professional Paper 940.
4. ANONYMOUS. 1984. Yearbook, Fiscal Year 1983. U. S. Geological Survey.

5. NOAA. 1977. Earth's crustal plate boundaries: Energy and mineral resources. *California Geology* (May 1977): 108–9.

6. FOSTER, R. J. 1983. *General geology,* 4th ed. Columbus, Ohio: Charles E. Merrill.

7. BATEMAN, A. M. 1950. *Economic ore deposits.* 2nd ed. New York: Wiley.

8. PARK, C. F., Jr., and MacDIARMID, R. A. 1970. *Ore deposits.* 2nd ed. San Francisco: W. H. Freeman.

9. LAWRENCE, R. A. 1973. Construction stone. In *United States mineral resources,* ed. D. A. Brobst and W. P. Pratt, pp. 157–62. U.S. Geological Survey Professional Paper 820.

10. YEEND, W. 1973. Sand and gravel. In *United States mineral resources,* eds. D. A. Brobst and W. P. Pratt, pp. 561–65. U. S. Geological Survey Professional Paper 820.

11. SMITH, G. I.; JONES, C. L.; CUL-BERTSON, W. C.; ERICKSON, G. E.; and DYNI, J. R. 1973. Evaporites and brines. In *United States mineral resources,* eds. D. A. Brobst and W. P. Pratt, pp. 197–216. U. S. Geological Survey Professional Paper 820.

12. LOWENSTAM, H. A. 1981. Minerals formed by organisms. *Science* 211: 1126–30.

13. CORNWALL, H. R. 1973. Nickel. In *United States mineral resources,* eds. D. A. Brobst and W. P. Pratt, pp. 437–42. U. S. Geological Survey Professional Paper 820.

14. VAN, N.; DORR, J.; CRITTENDEN, M. D.; and WORL, R. G. 1973. Manganese. In *United States mineral resources,* eds. D. A. Brobst and W. P. Pratt, pp. 385–99. U. S. Geological Survey Professional Paper 820.

15. SECRETARY OF THE INTERIOR. 1975. *Mining and mineral policy, 1975.*

16. McGREGOR, B. A., and LOCKWOOD, M. (no date) *Mapping and research in the exclusive economic zone.* U. S. Geological Survey and NOAA.

17. FAGAN, J. J. 1974. *The earth environment.* Englewood Cliffs, New Jersey: Prentice-Hall.

18. SILVA, M. A. 1988. Cyanide heap leaching in California. *California Geology* 41(7): 147–56.

19. DAVIS, F. F. 1972. Urban ore. *California Geology,* May 1972: 99–112.

20. GULBRANDSEN, R. A.; RAIT, N.; DRIES, D. J.; BAEDECKER, P. A.; and CHILDRESS, A. 1978. *Gold, silver, and other resources in the ash of incinerated sewage sludge at Palo Alto, California—A preliminary report.* U. S. Geological Survey Circular 784.

Our discussion of mineral resources established that resources are not infinite, and that a finite resource base cannot support an exponential increase in population. The same applies to energy derived from mineral resources.

United States citizens encountered the effects of energy shortages for the first time in the 1970s, including increases in the prices of energy and products produced from petroleum. Nevertheless, to many people energy still seems unlimited, especially given the oil glut of the 1980s.

The United States continues to consume a disproportionate share of the total energy produced in the world. With only 5 percent of the world's population, the U.S. consumes about 25 percent of the total energy consumed in the world.

Nearly 90 percent of the energy consumed in the United States today is produced from coal, natural gas, and petroleum, sometimes called the fossil fuels because of their organic origin (Figure 15.1a), with small amounts of hydropower and, more recently, nuclear power. We still have huge reserves of coal, but there are restrictions: major new sources of natural gas and petroleum are becoming scarce; few new large hydropower plants can be expected; and planning and construction of new nuclear power plants have become uncertain for a variety of reasons. On the brighter side, alternative energy sources such as solar power for homes, farms, and offices are becoming economically more feasible and thus more common. It is likely, however, that people in industrialized countries will have to realize that the quality of life is not directly related to ever-expanding energy needs. Examination of Figure 15.1a provides some interesting information. From 1950 through 1974 there was a sharp increase in energy consumption. Following the shortages of the mid 1970s the rate of increase of energy consumed dramatically declined. From 1950 through 1975 the energy consumption increased from 30 to approximately 70 exajoules. Since 1975 energy consumption has increased only about 10 exajoules. This suggests that energy conservation policies such as requiring new automobiles to be more fuel-efficient and buildings to be better insulated have been at least partially successful.

Table 15.1 shows generalized energy flow for the United States in 1989. One fact stands clear: We import approximately the same amount of oil that we produce. As a result we are very vulnerable to changing world conditions that affect the available supply of crude oil.

Projections of supply and demand for energy are at best difficult, because technical, economic, political, and social assumptions that underlie the projections are constantly changing. Thus recent predictions state that energy consumption in the U.S. in the year 2010 may exceed 100 exajoules (10^{18} joules) or be as low as 63 exajoules (energy consumption in 1989 was about 81 exajoules); one exajoule is approximately equal to one

CHAPTER FIFTEEN

▼
▼
▼

Energy and Environment

quad (10^{15} Btu). The higher value assumes no change in energy policies, whereas the lower value assumes aggressive energy conservation policies.

Evaluation of all potential energy sources and conservation practices is necessary to ensure sufficient energy to maintain our industrial society and a quality environment. This difficult task is expected to become even more so; more innovations in energy production are likely to be forthcoming, and each innovation requires a new evaluation of the total available data (1). Of particular importance will be energy uses with applications below 100° C because a large portion of the total energy consumption (for uses below 300° C) in the United States is for space heating and water heating (Figure 15.1b). With these ideas in mind, we will cautiously explore some selected geologic and environmental aspects of such well-known energy resources as coal, petroleum, and nuclear sources, and other possibly important sources including oil shale, tar sands, and geothermal resources. We will also discuss briefly the expected increase in demand for water as a result of energy production and several other alternative energy sources such as hydropower (river and tidal) and solar power.

▼ COAL

Geology of Coal

Like other fossil fuels, coal is made up of organic materials that have escaped oxidation in the carbon cycle. Coal is essentially the altered residue of plants that flourished in ancient freshwater or brackish-water swamps, typically found in estuaries, coastal lagoons, and low-lying coastal plains or deltas (2).

Figure 15.1
(a) Energy consumption for the United States from 1949 to 1989. (b) Spectra of energy use below 300°C in the United States. (Data for [a] from Energy Information Administration, 1990. Part [b] from Los Alamos Scientific Laboratory [L.A.S.L. 78–24], 1978.)

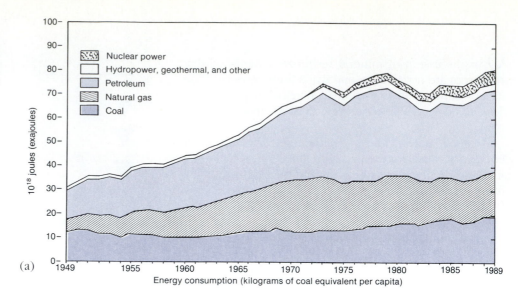

(a)

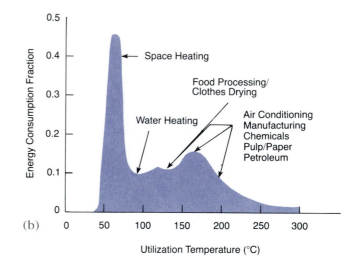

(b)

Table 15.1
Generalized energy flow for the United States in 1989.

Energy Source	Energy Production[a]	+ Imports	− Exports	± Adjustments	= Energy Consumed	Consumed by Sector
Coal	21.2	N	2.7			Residential/ Commercial 29.6
Natural Gas	19.7	1.4	N			
Oil	16.2	16.9	1.8			Industrial 29.5
Nuclear	5.7	N	N			
Hydropower	2.7	N	N			
Other	0.2	0.4	0.2			Transportation 22.1
Total	65.7	+ 18.7	− 4.7	+ 1.5[b]	= 81.2	81.2

N = none

[a]Exajoules (10^{18} joules)

[b]Balancing to account for a variety of items including unaccounted for supply, blending components, and changes in stock.

Source: *Annual Energy Review, 1989*, Energy Information Administration, 1990.

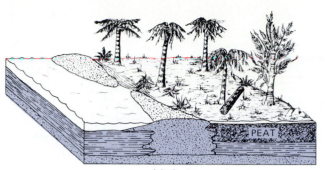

(a) *Coal swamp forms.*

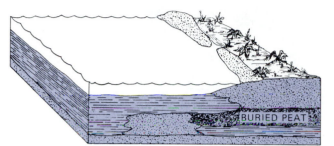

(b) *Rise in sea level buries swamp in sediment.*

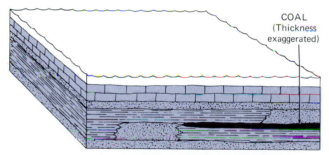

COAL
(Thickness
exaggerated)

(c) *Compression of peat forms coal.*

Figure 15.2
The processes by which buried plant debris (peat) is transformed into coal. Considerable lengths of geologic time must elapse before the transformation is complete.

Coal-forming processes (Figure 15.2) first require the development of a swamp rich in plants, which are partially decomposed in an oxygen-deficient environment and accumulate slowly to form a thick layer of peat. These swamps and accumulations of peat may then be inundated by a prolonged slow rise of sea level (a relative rise, as the land may be sinking) and covered by sediments such as sand, silt, clay, and carbonate-rich material. As more and more sediment is deposited, water and organic gases (volatiles) are squeezed out, and the percentage of carbon increases in the compressed peat. As this process continues, the peat is eventually transformed to coal. Because there are often several layers of coal in the same area, scientists believe the sea level may have alternately risen and fallen, allowing development and then drowning of coal swamps.

Classification and Distribution of Coal

Coal is commonly classified according to rank and sulfur content. The rank is generally based on the percentage of carbon, which increases from lignite to bituminous to anthracite, as shown in Figure 15.3. This figure also shows that heat content is maximum in bituminous coal, which has relatively few volatiles (oxygen, hydrogen, and nitrogen) and low moisture content. Heat content is minimum in lignite, which has a high moisture content. The distribution of the common coals (bituminous, subbituminous, and lignite) in the conterminous United States is shown in Figure 15.4.

The sulfur content of coal may be generally classified as low (zero to 1 percent), medium (1.1 to 3 percent), or high (greater than 3 percent). Most coal in the United States is of the low-sulfur variety (Table 15.2), and by far the most common low-sulfur coal is a relatively low-grade, subbituminous variety found west of the Mississippi River. The location of coal reserves has environmental significance because, with all other factors equal, the use of low-sulfur coal as a fuel for power plants causes less air pollution. Therefore, to avoid air pollution, thermal power plants on the highly populated East Coast will have to continue to treat some of the local coal to lower its sulfur content before burning or capture the sulfur after burning by a process such as scrubbing (see

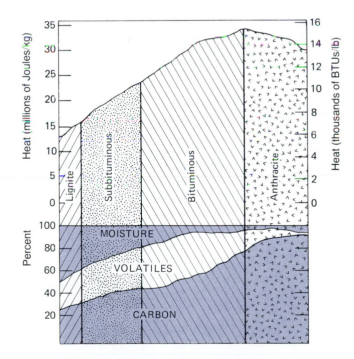

Figure 15.3
Generalized classification of different types of coal based upon their relative content (in percentage) of moisture, volatiles, and carbon. The heat values of the different types of coal are also shown. (After D. A. Brobst and W. P. Pratt, eds., U.S. Geological Survey Professional Paper 820, 1973.)

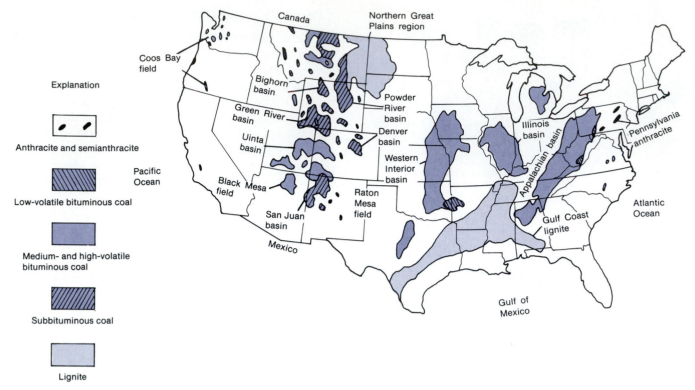

Figure 15.4

Coal areas of the conterminous United States. (From S. Garbini and S. P. Schweinfurth, U.S. Geological Survey Circular 979, 1986.

Chapter 16). These treatments increase the cost, but may be more economical than shipping low-sulfur coal long distances.

Impact of Coal Mining

Much of the coal mining in the United States is still done underground, but **strip mining** (open-pit), which started in the late nineteenth century, has steadily increased, whereas production from underground mines has stabilized. Strip mining is in many cases technologically and

Table 15.2

Distribution of United States coal resources according to their rank and sulfur content.

Rank	Sulfur Content (Percent)		
	Low 0–1	Medium 1.1–3.0	High 3+
Anthracite	97.1	2.9	—
Bituminous coal	29.8	26.8	43.4
Subbituminous coal	99.6	.4	—
Lignite	90.7	9.3	—
All ranks	65.0	15.0	20.0

Source: U.S. Bureau of Mines Circular 8312, 1966.

economically more advantageous than underground mining. The increased demand for coal will lead to more and larger strip mines to extract the estimated 40 billion metric tons of coal reserves that are now accessible to surface mining techniques. In addition, approximately another 90 billion metric tons of coal within 50 meters of the surface is potentially available for stripping if need demands.

The impact of large strip mines varies by region depending on topography, climate, and, most importantly, reclamation practices. In humid areas with abundant rainfall, mine drainage of acid water is a serious problem (Figure 15.5). Surface water infiltrates the spoil banks (material left after the coal or other minerals are removed) where it reacts with sulfide minerals such as pyrite (FeS_2) to produce sulfuric acid. The sulfuric acid then runs into and pollutes streams and groundwater resources. Although acid water also drains from underground mines and road cuts and areas where coal and pyrite are abundant, the problem is magnified when large areas of disturbed material remain exposed to surface waters. Acid drainage can be minimized through proper use of water diversion practices that collect surface runoff and groundwater before they enter the mined area and divert them around the potentially polluting materials. This practice reduces erosion, pollution, and water-treatment cost (3).

Figure 15.5
The stream in the foreground is flowing through waste piles of a Missouri coal mine. The water reacts with sulfide minerals and forms sulfuric acid. This is a serious problem in coal mining areas. (Photo by J. D. Vineyard, courtesy of Missouri Geological Survey.)

In arid and semiarid regions, water problems associated with mining are not as pronounced as in wetter regions, but the land may be more sensitive to mining activities such as exploration and road building. In some arid areas, the land is so sensitive that even tire tracks across the land survive for years. Soils are often thin, water is scarce, and reclamation work is difficult.

Common methods of strip mining include **area mining**, which is practiced on relatively flat areas (Figure 15.6), and **contour mining**, which is used in hilly terrain (Figure 15.7). The width of the cut in contour mining depends on the ease of excavation and topography. As the width increases, more and more overburden must be removed, which increases the cost. Topography thus limits the total width of the cut. The objective is to confine the cut to a given elevation (contour) and work around the hill while maintaining that elevation. Both methods of strip mining severely disturb the landscape by removing indigenous vegetation, overburden, and

desired minerals. Unless the mined land is reclaimed, unsightly conical piles or ridges of waste remain as a source material to pollute groundwater and surface-water. Area mining is used only on relatively flat ground where the potential velocity and erosion of runoff are low. Therefore, the potential to pollute streams by siltation is less than for contour mining. Potential groundwater pollution is greater for area mining, however, because the increased area allows more precipitation to infiltrate and slowly migrate through the spoil piles (3).

All methods of strip mining have the potential to pollute or destroy scenic, water, biologic, or other land resources, but good reclamation practices can minimize the damage (Figure 15.8). One potentially good method is to segregate the overburden to eventually replace the topsoil that is removed. This method has been used rather widely in the coal fields of the eastern United States, and experience has shown that it is a successful

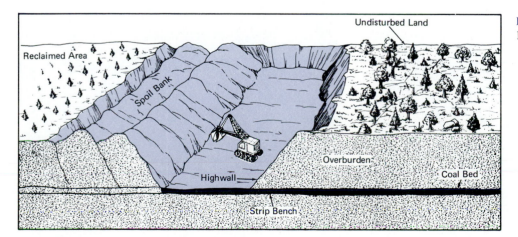

Figure 15.6
Diagram of area strip mining.

Undisturbed Land

Reclaimed Area

Spoil Bank

Highwall

Overburden

Coal Bed

Strip Bench

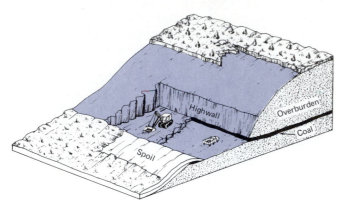

Figure 15.7
Diagram of contour strip mining.

way to control water pollution when combined with regrading and revegetation (3).

Low-sulfur coal deposits in the western United States, particularly in Montana, Wyoming, Colorado, Utah, New Mexico, and Arizona, will probably be mined by the area method of strip mining. Reclamation may be difficult even if the overburden is segregated: soils are thin, and revegetation is extremely difficult because water to establish new vegetation is scarce. Additional research into methods of reclamation for arid and semiarid regions is essential (3, 4).

Underground mining of coal and other resources has caused considerable environmental degradation from mine drainage of acid water and has produced serious hazards such as subsidence over mines or fires in mines (Figure 15.9). In the past, waste material (spoil) from underground mines has been piled on the surface, producing not only aesthetic degradation but also sedi-

ment and chemical pollution of water resources resulting from exposure to surface water and groundwater.

Federal guidelines now govern strip mining of coal in the United States. They require, basically, that mined land be restored to support its pre-mining use. Restoration includes disposing of wastes, contouring the land, and replanting vegetation. The hope is that after reclamation the mined land will appear and function as it did prior to extraction of the coal; however, as previously mentioned, this task will be difficult and probably not completely successful. The new regulations also prohibit mining on prime agricultural land and give farmers and ranchers the opportunity to restrict or veto mining on their land, even if they do not own the mineral rights.

Trapper Mine Near Craig, Colorado

Trapper Mine on the western slope of the Rocky Mountains in northern Colorado is a good example of a new generation of large coal strip mines. The main operation is designed to minimize environmental degradation during mining and to enable reclaiming the land for dry land farming and grazing of livestock and big game without artificial application of water.

The mine will produce 68 million metric tons of coal over a 35-year period, to be delivered to an 800 megawatt power plant adjacent to the mine. To meet this commitment, approximately 20 to 24 square kilometers of land will have to be strip-mined.

Four coal seams, varying from about 1 to 4 meters thick, each separated by various depths of overburden, will be mined. Depth of overburden to the coal varies from zero to about 50 meters. The steps in the actual mining are as follows:

Figure 15.8
A small open-pit mine before (left) and after (right) reclamation. (Photos by J. D. Vineyard, courtesy of Missouri Geological Survey.)

Figure 15.9
Escaping gases from holes drilled into a burning abandoned coal mine in western Pennsylvania. (Photo courtesy of U.S. Bureau of Mines.)

1. Vegetation and top soil are removed with dozers and scrapers, and the soil is stockpiled for reuse
2. Overburden along a cut up to 1.6 km long and 53 m wide is removed with a 23-cubic-meter dragline bucket (Figure 15.10)
3. Exposed coal beds are drilled and blasted to fracture the coal, which is removed with a backhoe and loaded into trucks (Figure 15.11)
4. The cut is filled, top soil is replaced, and the land is either planted in a crop or returned to rangeland (Figure 15.12)

At the Trapper Mine the land is reclaimed without artificially applying water. Precipitation (mostly snow) is about 35 cm per year, which is sufficient to reestablish vegetation provided there is adequate top soil to help hold the soil water. This factor emphasizes that reclamation is site specific; what works at one location or region may not apply to other areas.

Water and air quality are closely monitored at the Trapper Mine. Surface water is diverted around mine pits, and groundwater is intercepted while pits are open. Settling basins constructed downslope from pits trap suspended solids before discharging water into local streams. Air quality at the mine is degraded by dust produced from blasting, hauling, and grading of the coal. Dust is minimized by regularly watering or otherwise treating roads and other surfaces that may produce dust.

Reclamation of mined land at the Trapper Mine has been very successful during the first years of operation. Although the environmental protection techniques increase the cost of the coal by as much as 50 percent, the payoff will come in the long-range productivity of the land after termination of mining. Some might argue that the Trapper Mine is unique in that the fortuitous combination of geology, hydrology, and topography allow for successful reclamation; on the other hand, the

Figure 15.10
Removing overburden at the Trapper Mine, Colorado. The large dragline shown here has a 23-cubic-meter bucket.

Figure 15.11
Large backhoe at the Trapper Mine, Colorado, removing the coal which is then loaded in large trucks and delivered to a power plant just off the mining site.

mine shows that with careful site selection and planning, strip mining is not incompatible with other land uses.

Star Fire Mine in Eastern Kentucky

Coal has been one of the driving factors of the economy in eastern Kentucky for a long time. Eventually, however, the coal will be mined out and so there must be a shift to other uses of the land. The Star Fire Mine is in the midst of very interesting experiments concerning future land use and mine reclamation. Essentially, the mining company is involved with long-term land-use planning, so that land being mined now will be developed for other uses in the future. Valleys in this part of the world tend to be narrow with steep sides, with "flatland" for development relatively uncommon. Capitalizing on a variance of the 1977 Surface Mining Control and Reclamation Act, which in general requires restoration to approximate original contours following mining, Star Fire Mine lands are being heavily modified. The variance is allowed, provided the land that is reclaimed has an equal or better economic or public use following reclamation (5).

At Star Fire Mine the process of mining is known as mountain top removal/valley fill. The method involves filling the valleys with materials excavated from the surrounding mountains. As a result the landscape following mining is much different from what was originally there. In fact, following mining, the highest part of the land may be where the valleys were prior to mining. Mining at Star Fire is projected to continue until about the year 2010, and by that time more than 20 square kilometers of gently rolling land will have been produced. This will constitute the single largest parcel of potentially developable land in all of eastern Kentucky (5). Furthermore, the land does not have a flood hazard, which is unusual for flatland in eastern Kentucky.

It remains to be seen whether the experiment at Star Fire Mine will be successful. The mining company owns the mineral rights and the land outright, which provides some of the incentive for producing land with a potential for development. On the other hand, they are having to learn to work with huge quantities of fill. Valleys will be filled with more than 100 meters of material excavated from adjacent hills. The project has started with the development of a moderate-sized lake on top of about 80 meters of unconsolidated mine spoil. The lake is provid-

Figure 15.12
Reclaimed land at the Trapper Mine, Colorado. The site in the foreground has just had the soil replaced following mining whereas the vegetated sites have been entirely reclaimed.

ing an important aquatic habitat. Ongoing research is attempting to develop guidelines to determine if water lines, sewers, roads, and other structures necessary for development may be safely placed on the massive volumes of mine fill. The construction of the lake is viewed as a short-term goal compared to the long-term land-use planning that is being attempted. The importance of the Star Fire Mine Reclamation project is that it is challenging previously believed limitations on land development following mining (5).

Future Use of Coal

Limited resources of oil and natural gas are greatly increasing the demand for coal, and a significant change-over from burning oil and gas to burning coal in thermoelectric power plants and industrial heat-generating units is forthcoming. About 60 plants have been, or shortly will be, converted at a savings of about 250,000 barrels of oil per day. Total savings on the order of one million barrels per day could be realized if all thermo-electric power plants were to burn coal. There are sufficient coal reserves to meet the increased demand from such conversion and to supply coal for new plants for many hundreds of years (6).

The crunch on oil and gas supplies is still years away, but when it does come, it will put tremendous pressure on the coal industry to open more and larger mines in both the eastern and the western coal beds of the United States. This may have tremendous environmental impacts for several reasons. First, more and more land will be strip-mined and will thus require careful restoration. Second, unlike oil and gas, burned coal leaves ash (5 to 20 percent of the original amount of the coal) that must be collected and disposed of. Some ash can be used for landfill or other purposes, but about 85 percent is presently useless. Third, handling of tremendous quantities of coal through all stages—mining, processing, shipping, combustion, and final disposal of ash—will have potentially adverse environmental effects, such as aesthetic degradation, noise, dust pollution, and, most significant, release of trace elements that are likely to cause serious health problems into the water, soil, and air (6).

The transport of large amounts of coal, or energy derived from coal, from production areas with low energy demand to large population centers is a significant environmental issue. Coal can be converted on site to electricity, synthetic oil, or synthetic gas, all of which are relatively easy to transport, but with few exceptions these alternatives present problems. Transmission of electricity over long distances is expensive, and power plants in semiarid, coal-rich regions may have trouble finding sufficient water for cooling. Conversion of coal to synthetic oil or gas is expensive but the technology is available. However, conversion requires a tremendous amount of water and thus, as with the generation of electrical power, will place a significant demand on local water supplies in the coal regions of the western United States (1).

Methods of transporting large volumes of coal for long distances include freight trains and coal slurry pipelines. Trains have the advantage of a relatively low cost for new capital expenses and thus will continue to be used. Coal slurry pipelines, designed to use water to transport pulverized coal, have an economic advantage over trains if (1):

▼ Transport distance is long and a large volume of coal is shipped
▼ Inflation rates are high and interest rates low
▼ Mines are large and customers will purchase large volumes of coal over a long period of time
▼ Sufficient water is available at low cost

The economic advantages of the slurry pipeline are thus rather tenuous, especially in the western United States where large volumes of water will be difficult to obtain. For example, a pipeline that transports 30 million tons of coal requires about 20 million cubic meters of water per year, which is enough to meet the water needs for a city of about 85,000 people or to irrigate up to 40 square kilometers of farmland. Figure 15.13 is a simplified diagram of a slurry pipeline system.

Environmental problems associated with coal, while significant enough to cause concern, are not necessarily insurmountable, and careful planning could minimize them. At any rate, there may be few alternatives in the future to mining tremendous quantities of coal to feed thermoelectric power plants and to providing oil and gas by gasification and liquefaction processes.

▼ OIL AND GAS

Geology of Oil and Gas

Oil and natural gas (methane) are hydrocarbons. Like coal, they are fossil fuels in that they form from organic material that has escaped complete decomposition after burial. Next to water, oil is probably the most abundant fluid in the earth's crust, yet the processes that form it are only partly understood. Most earth scientists accept that oil and gas are derived from organic materials that are buried with marine or lake sediments. Favorable environments where organic debris might escape oxidation include *near-shore areas* characterized by rapid deposition that quickly buries organic material or *deeper-water areas* characterized by a deficiency in oxygen at the bottom which promotes an anaerobic decomposition. Beyond this, the locations where oil and gas form are

Figure 15.13
A coal slurry pipeline system.
(From Council on Environmental
Quality, 1979.)

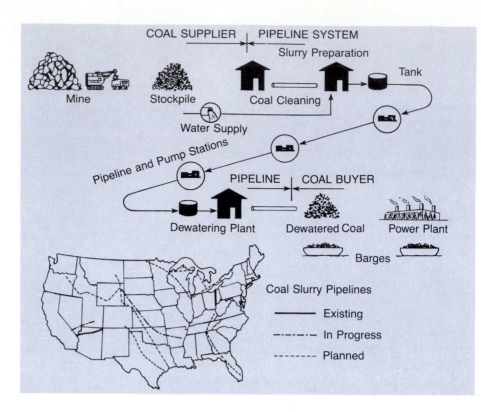

generally classified as subsiding, depositional basins in which older sediment is continuously buried by younger sediments, thus progressively subjecting the older, more deeply buried material to higher temperatures and pressures (7).

The major source material for oil and gas is a fine-grained, organic-rich sediment that is buried to a depth of at least 500 meters and subjected to increased heat and pressure that physically compress the source rock. The elevated temperature and pressure, along with other processes, start the chemical transformation of organic debris into hydrocarbons (oil and gas). As the pressure increases, the porosity of the source rock is reduced, and the higher temperatures thermally energize the hydrocarbons and induce them to begin an upward migration to a lower-pressure environment with increased porosity. The initial movement of the hydrocarbons upward through the source rock is termed *primary migration,* which merges into *secondary migration* as the oil and gas move more freely into and through coarse-grained, more permeable rock such as sandstone or fractured limestone. These porous, permeable rocks into which the oil and gas migrate are called *reservoir rocks.* If the path is clear to the surface, the oil and gas will migrate and escape there; this perhaps explains why most of the oil and gas is found in geologically young rocks. That is, hydrocarbons in older rocks have had a longer period of time in which to reach the surface and leak out (7). However, the lower amounts of hydrocarbon in very old rocks might also be explained by tectonic

processes that uplift rocks containing oil and gas, exposing them to erosion.

If in their upward migration, oil and gas are interrupted or stopped because they encounter a relatively impervious barrier, then they may accumulate. If the barrier *(cap rock)* has a favorable geometry (structure) such as a dome or an anticline, the oil and gas will be trapped in their upward movement at the crest of the dome or anticline below the cap rock (7). Figure 15.14 shows an anticlinal trap and two other possible traps caused by faulting or by an unconformity (buried erosion surface). These are not the only possible types of traps. Any rock that has a relatively high porosity and permeability and that is connected to a source rock containing hydrocarbons may become a reservoir, provided that the upward migration of the oil and gas is impeded by a cap rock so oriented that the hydrocarbons are entrapped at a central high point (7).

Petroleum Production

Production wells in an oil field recover petroleum through primary or enhanced recovery methods. *Primary recovery* uses natural reservoir pressure to move the oil to the well, but normally this pumping delivers only up to 25 percent of the total petroleum in the reservoir. To increase the recovery rate to 50 to 60 percent or more, *enhanced* (secondary or tertiary) *recovery* methods are necessary. The enhancement manipulates reservoir pressure and enables the reservoir to better

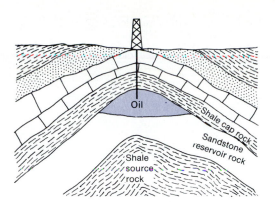

(a) Anticlinal trap

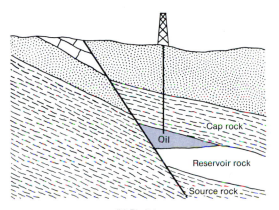

(b) Fault trap

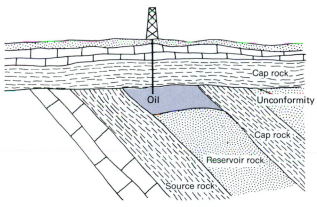

(c) Unconformity (stratigraphic) trap

Figure 15.14
Types of oil traps: (a) anticlinal, (b) fault, (c) unconformity.

transmit fluids by careful injection of natural gas, water, steam, and/or chemicals into the reservoir. Enhancement pushes petroleum to wells where it can be lifted to the surface by means of the familiar "horse head" bobbing pumps, submersible pumps, or other lift methods.

Petroleum production always brings to the surface a variable amount of salty water (brine) along with the oil. After separating the oil and water, the latter must be disposed of, because it is toxic to the surface environment. Disposal can be accomplished by injection as part of enhanced (secondary) recovery, evaporation in lined open pits, or deep-well disposal outside the field.

Distribution and Amount of Oil and Gas

The distribution of oil and gas in time (geologic) and space is rather complex, but in general, three principles apply. First, commercial oil and gas are produced almost exclusively from sedimentary rocks deposited during the last 500 million years of the earth's history (7). Second, although there are many oil fields in the world, approximately 85 percent of the total production plus reserves occurs in less than 5 percent of the producing fields; 65 percent occurs in about 1 percent of the fields; and, as remarkable as it seems, 15 percent of the world's known oil reserves is in two accumulations in the Middle East (7). Third, the geographic distribution of the world's giant oil and gas fields (Figure 15.15) shows that most are located near tectonic belts (plate junctions) that are known to have been active in the last 60 to 70 million years.

It is difficult to assess the petroleum and gas reserves of the United States, much less those of the entire world. Given this caveat, Figure 15.16 shows a recent estimate of the world's reserves of crude oil and natural gas. Recent estimates of proven oil and gas reserves of the United States suggest that, at present production rates, the oil and gas will last only a few decades. Using estimates of inferred and undiscovered resources, one can extend these projections; however, even the best data are only estimates subject to large errors. (8). The significance of energy projections is the implication that it will be extremely difficult for the United States to become economically self-sufficient in oil and gas (7). The last time the United States was essentially self-sufficient in oil was about 1950, and since then we have been importing oil.

What is the future of oil and gas? Unless large, new accumulations of petroleum are discovered in the United States—there is always this possibility, and in fact, large fields were recently discovered near Santa Barbara, California—known reserves and projected recoverable resources will last only a few more years before a shortage is evident. When a significant shortage does occur, we will most likely turn to large-scale gasification and liquefaction of our tremendous coal reserves, extract oil and gas from oil shale, perhaps rely more on atomic energy, and certainly rely more on solar energy. These changes will strongly affect our petroleum-based society, but there appears to be no insurmountable problem if we implement meaningful short- and long-range plans to phase out oil and gas and phase in alternative energy sources. Unfortunately, phasing in alternative energy

Figure 15.15
Giant oil and gas fields of the world relative to generalized tectonic belts. Active tectonic areas are shown by the stippled pattern, and regions not subjected to tectonic activity for the last 500 million years are shaded. Giant oil fields are denoted by the solid dots; giant gas fields by the open circles; and general areas of discovery either of less than giant size or under field development are indicated by an X. The numbers indicate how many fields are in a particular location. (After U.S. Geological Survey Circular 694, as adapted from C. L. Drake, *American Association of Petroleum Geologists Bulletin,* vol. 56, no. 2, 1972, with permission.)

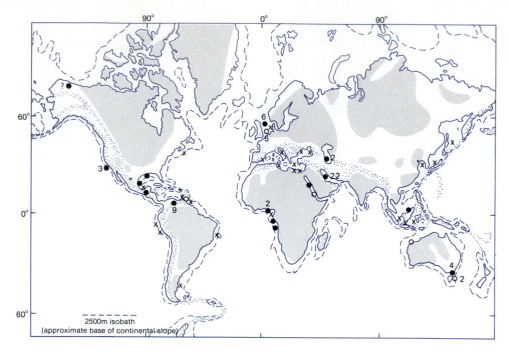

sources will require many years of research, exploration, and development, so it is crucial that we begin the task now. We are living in an interesting time from an energy standpoint, and it will be exciting to see new innovations, technology, and standards of living. Right now, no one knows the answer, or even all the questions.

Impact of Oil and Gas Exploration and Development

The environmental impact of exploration and development of oil and gas varies from negligible—for remote sensing techniques in exploration—to significant, un-

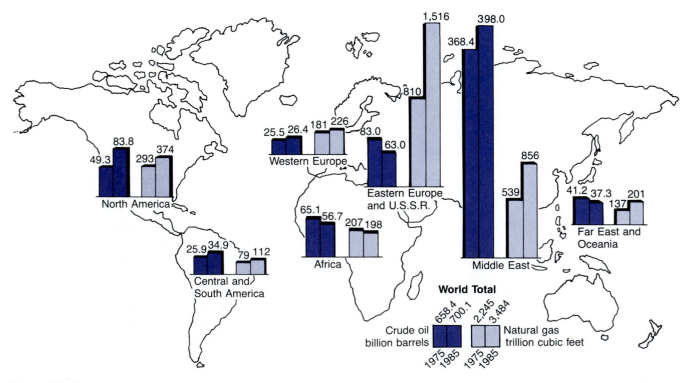

Figure 15.16
Estimated worldwide reserves of crude oil and natural gas. Bars are scaled in proportion to the Btu content of the reserves. One billion barrels of crude oil equals approximately 5.3 trillion cubic feet of wet natural gas. (After Energy Information Administration, 1986, *Annual Energy Review: 1985.*)

avoidable impact for projects such as the Trans-Alaska Pipeline. The impact of exploration for oil and gas can include building roads, exploratory drilling, and building a supply line (for camps, airfields, etc.) to remote areas. These activities, except in sensitive areas such as some semiarid-to-arid environments and some permafrost areas, generally cause few adverse effects to the landscape and resources compared to development and consumption activities. Development of oil and gas fields involves *drilling* wells on land or beneath the sea; *disposing* of waste water brought to the surface with the petroleum; *transporting* the oil by tankers, pipelines, or other methods to refineries; and *converting* the crude oil into useful products. All along the way, the possibility for environmental disruption from problems associated with disposal of waste water, accidental oil spills, shipwreck of tankers, air pollution at refineries, and other impacts is well documented. Tragic oil spills have affected the coastlines of Europe and America, spoiling beaches, estuaries, and harbors, killing marine life and birds, and causing economic problems for coastlines that depend on tourist trade. We need additional research and legislation to minimize these occurrences.

The most familiar serious impact associated with oil and gas use is air pollution, which is produced in urban areas when fossil fuels are burned to produce energy for electricity, heat, and automobiles. The adverse effects of smog on vegetation and human health are well documented and need not be restated here.

▼ OIL SHALES AND TAR SANDS

Geology of Oil Shale

Oil shale is a fine-grained sedimentary rock containing organic matter (kerogen). On heating (destructive distillation), oil shale yields significant amounts of hydrocarbons that are otherwise insoluble in ordinary petroleum solvents (9).

As with other fossil fuels, the origin of oil shale involves the deposition and only partial decomposition of organic debris. Favorable environments for oil shale are lakes, stagnant streams or lagoons in the vicinity of organic-rich swamps, and marine basins (10).

The best-known oil shales in the United States are those in the Green River Formation, which is about 50 million years old and underlies approximately 44,000 square kilometers of Colorado, Utah, and Wyoming (Figure 15.17). The Green River Formation consists of oil shale interbedded with variable amounts of sandstone, siltstone, claystone, and compacted volcanic ash (tuff). Variable amounts of halite (rock salt), trona (hydrous acid sodium carbonate), nahcolite (sodium bicarbonate), dawsonite (a basic aluminum sodium carbonate), and other materials or minerals are also found as nodules, lenses, thin beds, or disseminated crystals (9). Nahcolite is a potentially valuable source of sodium carbonate, and dawsonite is a possible source of aluminum. Recovering these materials with the oil could reduce the cost of mining.

The geology of the Green River Formation and the depositional environment in the large lakes obviously varied significantly as the nature and pattern of deposition varied. The Green River Formation varies in thickness from a few meters to more than 1,000 meters, and some beds are rich in organic materials while others are not. Generally, the accumulation of organic material is thickest in the central parts of the large, shallow lakes where microscopic algae and other micro-organisms flourished, died, and were buried with a mixture of sediment and other land-derived organic material. Toward the margins of the basins, less organic material accumulated; the oil shale there is of a lower grade (Figure 15.17) and is interbedded with land-derived, inorganic sediments (sand, silt, etc.) transported into the lake by streams, wind, and other surface processes. Few of these inorganic sediments reached the center of the lake, facilitating the deposition there of rich accumulations of organic material from which the oil shale formed (9).

The large, shallow, ancient lakes of Wyoming, Utah, and Colorado, in which the organic material accumulated, varied from fresh to alkaline water. The lake basins subsided slowly and irregularly, causing the center of deposition of the oil shale to shift slowly during the millions of years the deposition continued. For millions of years after the sediment was buried, little tectonic activity disturbed the sediments. More recently, uplift and local tilting of the rocks have exposed some of the oil shale to erosion (9).

Geology of Tar Sands

Tar sands are rocks that are impregnated with tar oil, asphalt, or other petroleum materials, from which recovery of petroleum products by usual methods such as oil wells is not commercially possible. The term *tar sand* is somewhat confusing because it includes several rock types, such as shale and limestone, as well as unconsolidated or consolidated sandstone. The one thing all these rocks have in common is that they contain a variety of semiliquid, semisolid, and solid petroleum products, of which some ooze from the rock outcrops and others are difficult to remove with boiling water (11).

Oil in tar sands is nearly the same as the heavier oil pumped from wells. The only real difference is that tar sand oil is much more viscous and therefore more difficult to recover. A possible conclusion concerning the geology of tar sands is that they form essentially the same way that the more fluid oil forms, but much more of the volatiles and accompanying liquids in the reservoir rocks have escaped, leaving the more viscous materials behind.

Figure 15.17
Distribution of oil shale in the Green River Formation of Colorado, Utah, and Wyoming. (After D. C. Duncan and V. E. Swanson, U.S. Geological Survey Circular 523, 1965.)

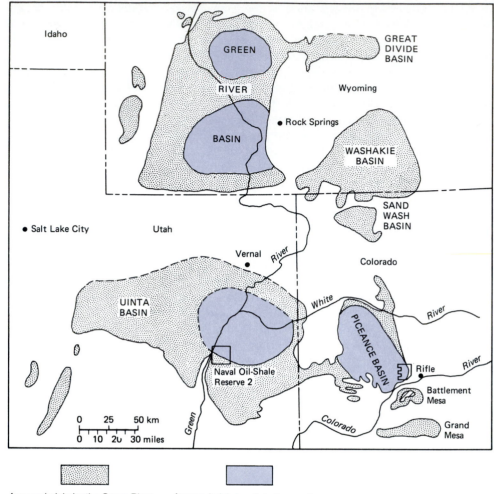

Area underlain by the Green River Formation in which the oil shale is unappraised or of low grade.

Area underlain by oil shale more than 3 m thick which yields 0.1 m³ or more oil per ton of shale.

Distribution of Oil Shale and Tar Sands

Many countries have deposits of oil shale, but there is a tremendous range in thickness of the rock, quality of the oil, and areal extent (12).

Total identified shale-oil resources of the world's land areas are estimated to contain about 3 trillion barrels of oil, but thorough evaluation of the grade and feasibility of economic recovery with today's technology and economic situation is incomplete. Shale-oil resources in the United States amount to about 2 trillion barrels of oil, or two-thirds of the total identified in the world; of this, 90 percent, or 1.8 trillion barrels, is located in the Green River Oil Shales.

Unfortunately, identified resources of oil shale are not recoverable with today's economics. This situation could change, however. We do know of large accumulations of tar sands; for example, the Athabasca Tar Sands of Alberta, Canada, cover an area of approximately 78,000 square kilometers and contain an estimated reserve of 300 billion barrels of oil that might be recovered (13). In addition, smaller tar-sand resources are known in Utah

(1.8 billion barrels), California (100 million barrels), Texas, Wyoming, and other areas (11).

Impact of Exploration and Development of Shale Oil and Tar Sands

Recovering petroleum from surface or near-surface oil shale and tar sands involves use of well-established exploration techniques. Numerous deposits are known— what we need are reliable techniques of developing the oil shale resources that will cause a minimum of environmental disruption. The really significant problem associated with mining oil shale is insufficient water supplies, which we will discuss in a later section on energy and water demand.

The Athabasca Tar Sands in Alberta are now yielding about 50,000 barrels of synthetic crude oil per day from one large open-pit mine operated by the Great Canadian Oil Sands Limited. The average thickness of the mined tar sands is about 42 meters; the amount of overburden removed varies considerably but is generally less than 50 meters. Approximately 126,000 metric tons of tar sand

and 117,000 metric tons of overburden are removed in each day's activity (13).

The methods of surface mining at the Canadian Tar Sands are atypical for four reasons. First, the indigenous vegetation is a water-saturated, organic matte (muskeg swamp) of decayed and decaying vegetation that can be removed easily only in the winter when it is frozen. Second, the muskeg must be drained before excavation, which takes about two years. Third, once the vegetation has been removed, the tar sands are extremely harsh on equipment and difficult to remove; furthermore, the overburden and other unwanted material with the tar and sand occupy considerably more volume than before they were removed, creating a disposal problem and a final land surface that may rise more than 21 meters above the original surface. Fourth, restoration of the land after mining is a serious problem because this fragile environment is so difficult to work with (13).

Fortunately, the actual process of recovering oil from tar sands is relatively easy and involves washing the viscous oil out of the sand with hot water. The oil then flows to the surface and is removed (11). Only about 10 percent of the total oil from the tar sands can be economically recovered by open-pit mining, however, so recovery of the oil in place (without removal by surface or subsurface mining) should become a necessary goal for obtaining additional oil (13). It is hoped that in-place recovery techniques will disturb the surface only minimally.

The environmental impact of developing oil shale resources will vary according to the recovery technique. Surface and subsurface mining as well as in-place (in situ) techniques have been considered.

Surface mining, either open pit or strip mine, is attractive because nearly 90 percent of the oil shale can be recovered, as opposed to less than 60 percent for underground mining. But waste disposal is a major problem with any mining, surface or subsurface, that requires that oil shale be processed at the surface for retorting (crushing and heating raw oil shale to about 540°C to obtain crude shale oil). This disposal problem results because the volume of waste will exceed the original volume of mined shale by 20 to 30 percent. Therefore, the mine from which the shale was removed will not be able to accommodate the waste, which will have to be piled up or otherwise disposed of (12, 14).

A process of oil shale recovery that is being seriously tested is known as *modified in situ,* or MIS, in which part of the oil shale (about 20 percent) is mined, and the remainder is highly fractured or rubbled to increase the permeability. A block of rubbled shale, the retort block, is then ignited, and the released oil and gas are recovered through wells (Figure 15.18).

The human environment will change significantly when oil shale is mined in Colorado, Wyoming, or Utah. Large scale mining will necessitate modifications for increased transportation, including roads, pipelines, and airports, as well as augmenting economic activity, construction of various industrial facilities, and rapid urbanization as the population increases—all of which affect the physical environment.

▼ FOSSIL FUEL AND ACID RAIN

Acid rain (or acid deposition, as it is sometimes called) encompasses both wet and dry acidic deposits that often occur near and downwind of areas where major emission of sulfur dioxide (SO_2) and nitrogen oxides (NO_x) occur as a result of burning fossil fuels. The term **acid rain** is a fairly recent one, even though the problem probably extends back at least as far as the beginning of the industrial revolution. In recent decades the problem of acid rain has gained more and more attention; today it is considered one of the major environmental problems facing industrialized society. A detailed discussion of acid rain is presented in Chapter 16 where global change and earth systems science are considered.

▼ NUCLEAR ENERGY: FISSION

The first controlled nuclear fission was demonstrated in 1942 and led the way to the development of the primary uses of uranium in explosives and as a heat source to provide steam for generation of electricity. It now appears that the energy from uranium (one kilogram of uranium oxide produces a heat equivalent of approximately 16 metric tons of coal) will continue as an important source of energy in the United States (15).

Fission is the splitting of uranium (U-235) by neutron bombardment (Figure 15.19). The reaction produces three more neutrons released from uranium, fission fragments, and heat. The released neutrons each strike other U-235 atoms, releasing more neutrons, fission products, and heat. The process continues in a chain reaction—as more and more uranium is split, it releases ever more neutrons.

Three types of uranium occur in nature: U-238, which accounts for approximately 99.3 percent of all natural uranium; U-235, which makes up about 0.7 percent; and U-234, which makes up about 0.005 percent. Uranium-235 is the only naturally occurring fissionable material and is therefore essential to the production of nuclear energy. However, uranium is processed to increase the amount of U-235 from 0.7 percent to about 3 percent before it is used in a reactor. The processed fuel is called *enriched uranium.* Uranium-238 is not naturally fissionable but is "fertile material" because upon bombardment by neutrons, it is converted to plutonium-239, which is fissionable (15).

Most reactors today consume more fissionable material than they produce and are known as *burner reactors*. The reactor itself (Figure 15.20) is part of the

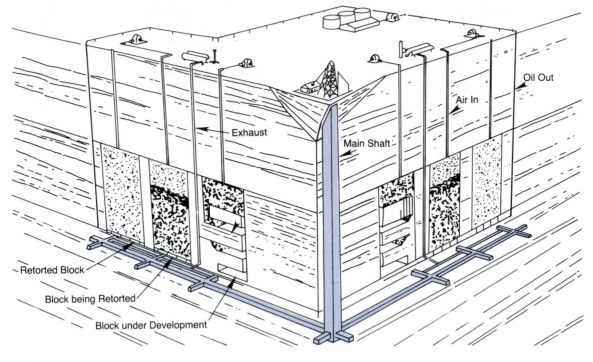

Figure 15.18
The modified in situ method of development of oil shale resources. (From U.S. Department of Energy.)

nuclear steam supply system which produces the steam to run the turbine generators that produce the electricity (16). The main components of the reactor shown in Figure 15.21 are the core, control rods, coolant, and reactor vessel. Fuel pins consisting of enriched uranium pellets placed into hollow tubes with a diameter less than about 1 cm are packed together (40,000 or more in a reactor) into fuel subassemblies in the core (Figure 15.22). A stable fission chain reaction in the core is maintained by controlling the number of neutrons that cause fission as well as the fuel concentration. A

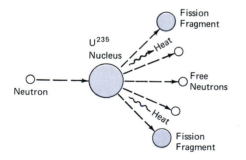

Figure 15.19
Fission of U-235. A neutron strikes the U-235 nucleus, producing fission fragments and free neutrons and releasing heat. The released neutrons may then each strike another U-235 atom, releasing more neutrons, fission fragments, and energy. As the process continues, a chain reaction develops.

minimum fuel concentration is necessary to keep the reaction critical (achieve a self-sustaining chain reaction).

Because the control rods contain materials that capture neutrons, they can thus be used to regulate the chain reaction. If the rods are pulled out, the chain reaction speeds up; if they are inserted into the core, the reaction slows down (17).

The function of the coolant is to remove the heat produced by the fission reactions. When the coolant is water, it not only removes heat but also acts as a moderator, slowing down the neutrons and facilitating efficient fission of uranium-235 (17).

The core of the reactor is contained in a heavy stainless steel reactor vessel. Then, for extra safety and security, the entire reactor is contained in a reinforced concrete building (17).

In addition to the reactor, other parts of the nuclear steam supply system are the *primary coolant loops* and pumps that circulate a coolant (usually water) through the reactor, extracting heat produced by fission, and *heat exchangers* or *steam generators* that use the fission-heated coolant to make steam. Figure 15.23 shows how these combine in a pressurized water reactor (PWR), which is a *light water reactor* (a type of burner reactor) so designated because the coolant is water. Water heated by the reactor core is circulated in the primary coolant loop (a closed system) through a steam generator (heat exchanger) turning the water in the secondary loop to steam which produces electricity (16).

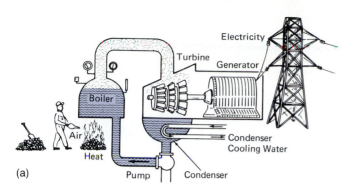

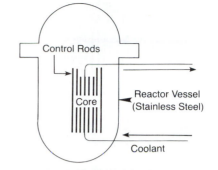

(a)

(b)

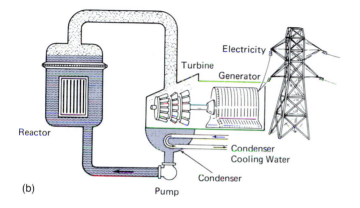

Figure 15.20
Comparison of (a) fossil fuel power plant and (b) nuclear power plant with a boiling water reactor. Notice that the nuclear reactor has exactly the same function as the boiler in the fossil fuel power plant. (Reprinted, by permission, from *Nuclear Power and the Environment,* American Nuclear Society, 1973.)

Figure 15.21
Diagram of the main components of a nuclear reactor. Below, inspecting the fuel element. (From U.S. Energy and Resource Development Administration, 1976. Photo courtesy of U.S. Department of Energy.)

A second type of light water reactor in use is the *boiling water reactor* (BWR), which is a direct cycle system because there is no heat exchanger. The primary coolant loop goes through the reactor core directly to a turbine, producing electricity (Figure 15.24). Examination of this figure reveals a major disadvantage of the BWR. The steam that turns the turbine comes directly from the reactor core rather than first going through a heat exchanger, as in the PWR system. Therefore, particular care must be taken to keep hazardous radiation from leaking into the turbine system that is outside the main containment structure. The primary coolant (water) will have significantly induced radioactivity, requiring that the turbines be heavily shielded, increasing construction and maintenance costs (16).

To conserve U-235, work is progressing on a reactor that actually produces more fissionable material (nuclear fuel) than it uses. These **breeder reactors,** using a fuel core of fissionable material (plutonium-239) surrounded by a blanket of fertile material (U-238), will produce or breed additional plutonium-239 from the U-238. The transformation from U-238 to P-239 occurs in the breeder reactor at the same time that the plutonium nuclei in the core are undergoing fission, providing heat that produces steam to drive turbines and generators that provide electricity. Development of the breeder reactor will greatly extend the limited supply of natural U-235. In fact, breeder reactors that use plutonium as a fuel can use uranium-238 tailings from enriched uranium or uranium recovered from spent light water reactor fuel. It has been estimated that the uranium now stockpiled as tailings and spent fuel, if used in breeder reactors, could supply the total electrical energy demands in the United States for up to 100 years (17).

In 1982, the demand for electric power decreased for the first time in many years. It will increase in the future, however, as will the demand for nuclear power, which produced about 19 percent of our electricity in 1989. By the year 2000, it is expected that heat from nuclear

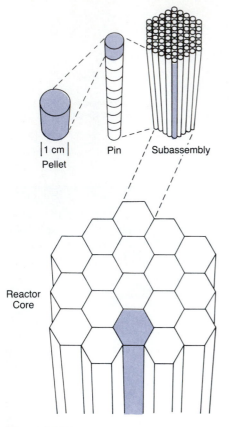

Figure 15.22
Fuel pellets of enriched uranium are placed in hollow tubes forming fuel pins that are bundled into fuel subassemblies, which are placed in the reactor as part of the core. (Modified from ERDA–76–107, 1976.)

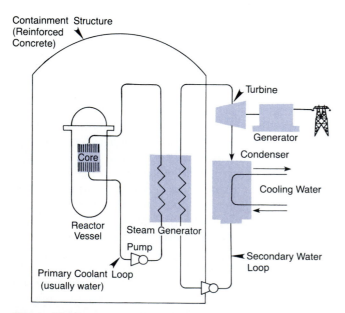

Figure 15.23
Pressurized water reactor (PWR) showing the three main parts of the nuclear steam supply system: reactor, primary coolant loop, and steam generator (heat exchanger). (Modified after Energy Research and Development Administration, 1976, ERDA–76–107.)

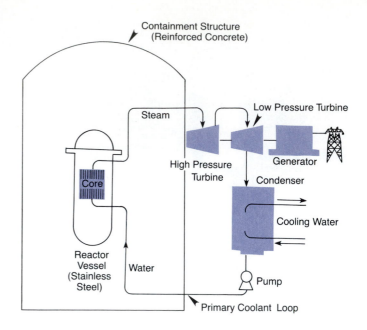

Figure 15.24
Boiling water reactor (BWR) showing the two main parts of the nuclear steam supply system: reactor and primary coolant loop. (Modified after Energy Research and Development Administration, 1976, ERDA–76–107.)

reactors may have the capacity to produce about 22 percent of the electrical power in the United States. About 110 reactors are now in operation (over 80 percent of which are in the eastern United States), but many more will be needed to realize the projected energy from uranium. Probably because of increased costs, environmental considerations, and other factors, the projected generating capacity is too high and will have to be revised. Thus, the full impact of what began in 1942 is still to be determined.

Geology and Distribution of Uranium

The natural concentration of uranium in the earth's crust is about 2 parts per million. Uranium originates in magma and is concentrated to about 4 parts per million in granitic rock, where it is found in a variety of minerals. Some uranium is also found with late-stage igneous rocks such as pegmatites. To be mined at a profit, uranium must have a concentration factor of 400 to 2,500 times the natural concentration (15).

Fortunately, uranium forms a large number of minerals, many of which can be found in high-grade deposits being mined today. Three types of deposits have produced most of the uranium in the last few years: sandstone impregnated with uranium minerals, veins of uranium-bearing materials localized in rock fractures, and placer deposits in river or delta deposits (now coarse-grained sedimentary rock) more than 2.2 billion years old (18).

Uranium in sandstone is often found in thin lenses interbedded with mudstone. It is hypothesized that the uranium in these deposits was derived by leaching from volcanic glass associated with the sedimentary rocks or from granitic rocks exposed along the margins of the sedimentary basins. Supposedly, the uranium was transported by groundwater and then precipitated into the pore spaces of the sandstone under reducing (oxygen-deficient) conditions (15).

Most of the uranium mined in the United States has been from sandstone deposits of two types: roll-type and tabular deposits. *Tabular* uranium deposits are discrete masses completely enclosed by altered (reduced) sandstone which is itself enveloped in oxidized sandstone, as shown in Figure 15.25a. *Roll-type* uranium deposits form at the interface between oxidizing and reducing conditions. Characteristically, the ore deposits are elongated bodies scattered like interconnected beads along kilometers of interface between altered (oxidized) sandstone and unaltered sandstone (Figure 15.25b) (15, 18).

Uranium deposits in veins generally fill fissures (rock fractures) in many types of rocks of varying geologic age. The veins usually vary from several centimeters to a few meters wide and several hundred meters long. Veins may coalesce to form vein systems that can extend for several thousand meters (15).

Uranium in ancient river or delta sediment was deposited before the occurrence of abundant free oxygen in the atmosphere. Therefore, it is believed, the uranium mineral was deposited as stream-rounded particles along with gold, pyrite, and other typical placer material (18). Although sandstone ores have been the main source of uranium in this country, vein deposits are a significant source in Australia, Canada, France, and Africa, and ancient placer deposits are being mined in Africa and Canada (18).

The amount of uranium from the known deposits is sufficient to last into the late 1990s. Beyond that, additional exploration for deposits and research of techniques to locate new reserves will be required (15). If the breeder reactor program is successful and shown to be environmentally safe, the supply of fissional material will be greatly extended.

Nuclear Energy (Fission) and the Environment

Nuclear energy and the possibly adverse effects associated with it have been subjects of vigorous debate. The

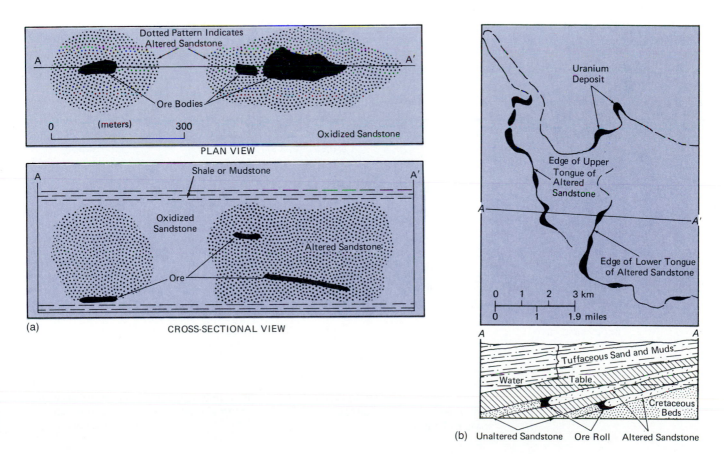

Figure 15.25
Geologic map and cross section of (a) tabular uranium deposits and (b) typical roll-type uranium deposits. (After U.S. Geological Survey, INF–74–14, 1974.)

debate is healthy because the number of reactors that produce heat to drive generators is expected to increase in the future, and we should examine the consequences carefully.

A first approach to evaluating the environmental effects of nuclear power might be to compare the effects of all common methods of generating electrical power by power source: coal, oil, gas, and uranium (Table 15.3). All methods affect the environment in various ways during normal operation. Disregarding the magnitude of a particular effect (as, for example, thermal pollution, which is greater for nuclear power because the generator capacity is generally larger and no heat is directly dissipated to the atmosphere), the real difference of nuclear power is radioactivity. The possible radiation hazard from nuclear energy production and waste disposal really concerns people.

Throughout the entire nuclear cycle, from mining and processing uranium to controlled fission, to reprocessing spent nuclear fuel, to final disposal of radioactive waste, various amounts of radiation enter and, to a lesser or greater extent, affect the environment.

The chance of a disastrous nuclear accident is estimated to be very low; nevertheless, the chance of an accident increases with every reactor put into operation.

Three Mile Island.   An important event in the history of radiation pollution occurred on March 28, 1979, at the Three Mile Island nuclear power plant near Harrisburg,

Pennsylvania. Malfunctions in the nuclear plant resulted in a release of radioisotopes into the environment as well as intense radiation release within one of the nuclear facilities. The actual release into the environment was at a low level per person exposed. Exposure from the plume emitted into the atmosphere has been estimated at 100 mrem, which is low in terms of the amount of radiation required to cause acute toxic effects. However, radiation levels were much higher near the site. On the third day after the accident, 1200 mrem/hr were measured at ground level near the site.

The Three Mile Island incident made clear that there are many problems with the way our society has dealt with nuclear power. Historically, nuclear power has been relatively safe, and the state of Pennsylvania was somewhat unprepared to deal with the accident. For example, there was no state bureau for radiation health, and the State Department of Health did not have a single book on radiation medicine (the medical library had been dismantled two years before for budgetary reasons). One of the major impacts of the incident was fear, yet there was no state office of mental health or any authority to allow anyone from the Department of Health to sit in on briefing sessions.

Because the long-term chronic effects of exposure to low levels of radiation are not well understood, the effects of the Three Mile Island exposure—although apparently small—are difficult to estimate. This case illustrates that our society needs to improve its ability to

Table 15.3

Comparison of environmental effects of coal, oil, gas, and uranium electrical power generation.

Energy Source	Effects on Land	Effects on Water	Effects on Air	Biological Effects	Supply
Coal	Disturbed land; large amounts of solid waste; leaching of soil and rocks by acid rain	Acid mine drainage; increased water temperature; acid rain	Sulfur oxides; nitrogen oxides; particulates; some radioactive gases	Respiratory problems from air pollutants; drainage from acid rain	Large reserves
Oil	Wastes in the form of brine; pipeline construction; leaching by acid rain	Oil spills; increased water temperature; acid rain	Nitrogen oxides; carbon monoxide; hydrocarbons	Respiratory problems from air pollutants; damage from acid rain	Limited domestic reserves
Gas	Pipeline construction	Increased water temperature	Some oxides of nitrogen	Few known effects	Limited domestic reserves
Uranium	Disposal of radioactive waste	Increased water temperature; some radioactive liquids	Some radioactive gases	None detectable in normal operation	Large reserves if breeders are developed

Source: Pennsylvania Department of Education. *The Environmental Impact of Electrical Power Generation: Nuclear and Fossil,* 1973.

handle the crises that could arise from sudden releases of pollutants from our modern technology. It also shows our lack of preparedness and apparent readiness to treat a nuclear power plant as an acceptable risk (19).

Chernobyl. Lack of preparedness to deal with a serious nuclear power plant accident was dramatically illustrated by events that began unfolding on the morning of Monday, April 28, 1986. Workers at a nuclear power plant in Sweden, frantically searching for the source of high levels of radiation near the plant, concluded that it was not their installation that was leaking radiation but that the radioactivity was coming from their Soviet neighbors by way of prevailing winds. Confronted, the Soviets on late Monday announced that there had been an accident at their nuclear power plant at Chernobyl. This was the first notice to the world that the worst accident in the history of nuclear power generation had occurred.

It is speculated that the system that supplies cooling waters for the reactor failed, causing the temperature in the reactor core to rise to over 3000°C, melting the uranium fuel. Explosions occurred that removed the top of the building over the reactor, and graphite surrounding the fuel rods used to moderate the nuclear reactions in the core ignited. The fires produced a cloud of radioactive particles that rose high into the atmosphere. Radiation killed about twenty people at and near the plant site in the days following the accident, and millions of other people were exposed to potentially harmful radiation as the cloud in the days after the start of the accident drifted first over parts of the Soviet Union, then north to Scandinavia, and then to eastern Europe.

During the next 25 to 30 years an increase in cancers will document the impact on humans from the accident. Beyond that, it is estimated that as much as 150 square kilometers of valuable farmland in the U.S.S.R. may be contaminated with radiation for decades (20).

One of the avoidable tragedies in the accident was the way the Soviet government handled the accident. Delays in warning people in their own country and outside resulted in unnecessary exposure to radiation. Although the Soviets have been accused of not giving attention to reactor safety and of using outdated equipment, people are now wondering if such an event could happen again. With about 400 reactors producing power in the world today, the answer has to be yes. In fact, the Chernobyl accident follows a history marked by about ten accidents that have released radioactive particles during the past 34 years. Therefore, while Chernobyl is the most serious nuclear accident to date, it certainly wasn't the first. Many countries such as Japan, Britain, and France that depend heavily on nuclear power plants for generating electric power are rethinking the risks of siting power plants near densely populated areas. While the probability of a serious accident is very small at a particular site, the consequences of the event may be great, perhaps resulting in an unacceptable risk. As a result of Chernobyl, risk analysis in nuclear power is now a real-life experience rather than a computer simulation.

As long as people build nuclear power plants and manage them, there will be the possibility of accidents. It's as much a problem of human nature as anything else. Considering the potential tremendous consequences of serious accidents like that at Chernobyl, we may have to recognize that we may not be ready as a people to handle nuclear power.

However, there is a move to design safer nuclear reactors. One of the ideas is to design smaller modular units that would be much less susceptible to serious accidents. Smaller modules at a particular site have another advantage: If one or more of them are down for maintenance the rest may still produce power.

Additional serious hazards are associated with transporting and disposing of nuclear material (see Chapter 12) as well as with supplying other nations with reactors. Terrorist activity and the possibility of irresponsible actions by governments add a risk that is present in no other form of energy production. Nuclear energy may indeed be the answer to our energy problems, and perhaps someday will provide unlimited cheap energy. But along with nuclear power comes the responsibility of ensuring that nuclear power is used for, not against, people, and that future generations will inherit a quality environment free from worry about hazardous nuclear waste.

▼ NUCLEAR ENERGY: FUSION

The long-range energy strategy for the United States is to develop and provide sufficient energy resources to meet energy demand in the twenty-first century and beyond. Energy sources capable of meeting long-term needs are those that are inexhaustible: solar, breeder reactor, and fusion reactor. Our discussion here will focus on fusion.

In contrast to fission, which involves splitting heavy atoms such as uranium, **fusion** involves combining light elements such as hydrogen to form a larger element such as helium. As fusion occurs, heat energy is released (Figure 15.26). Similar reactions are the source of energy in our sun and other stars. In a hypothetical fusion reactor, two isotopes of hydrogen (atoms with variable mass resulting from a different number of neutrons in the nucleus), deuterium (D) and tritium (T), are injected into the reactor chamber where necessary conditions (temperature, time, and density) for fusion are maintained. Products of the D-T fusion include helium, carrying 20 percent of the energy released, and neutrons, carrying 80 percent of the energy released (Figure 15.26) (21).

Three conditions are necessary for fusion: first, extremely high temperature (approximately 100 million

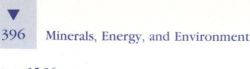

Figure 15.26
Deuterium-tritium (D-T) fusion reaction. (Modified from U.S. Dept. of Energy, 1980, DOE/ER–0059.)

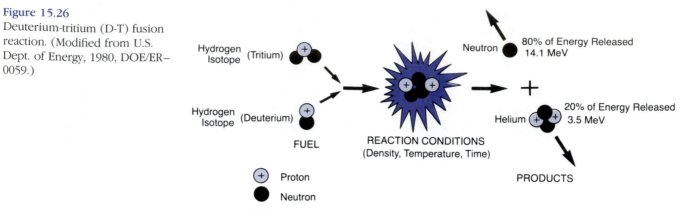

degrees Centigrade for D-T fusion); second, sufficiently high density of the fuel elements (at the necessary temperature for fusion nearly all atoms are stripped of their electrons forming a *plasma,* an electrically neutral material consisting of positively charged nuclei, ions, and negatively charged electrons); and third, confinement of the plasma for a sufficient time to ensure that the energy released by the fusion reactions exceeds the energy supplied to maintain the plasma (21, 22).

With development of fusion-reactor power plants, the potential available energy is nearly inexhaustible. One gram of D-T fuel (from a water and lithium fuel supply) has the energy equivalent of 45 barrels of oil. Deuterium can be extracted economically from ocean water, tritium can be produced in a reaction with lithium in a fusion reactor, and there are no problems obtaining lithium, as it is abundant and can be extracted economically.

Many problems remain to be solved before nuclear fusion is commercially available. The research is still in the first stage, which involves basic physics, testing possible fuels (mostly D-T), and magnetic confinement of plasma. Progress in fusion research has been steady in recent years, and there is optimism that useful power will eventually be produced from controlled fusion.

Energy from fusion, once released, has a variety of applications, including heating and cooling buildings and producing synthetic fuels, but producing electricity is probably the most important. It is expected (but not proven) that fusion power plants will be economically competitive with other sources of electric energy. From an environmental view, fusion certainly appears attractive: land use and transportation impacts are small compared to fossil fuel or fission energy sources. Compared to fission breeders, fusion produces no fission products and little radioactive waste, and has a much lower risk of potential hazard resulting from an accident (23). On the other hand, fusion power plants will probably use materials that are harmful to humans, such as lithium, which is toxic when inhaled or ingested. Other potential hazards are the strong magnetic fields and microwaves used in confining and heating plasma, and short-lived radiation emitted from the reactor vessel (21).

▼ GEOTHERMAL ENERGY

The useful conversion of natural heat from the earth's interior (**geothermal energy**) to heat buildings and generate electricity is an exciting application of geologic knowledge and engineering technology. The idea of harnessing the earth's internal heat is not new. As early as 1904, geothermal power was developed in Italy using dry steam, and natural internal heat is now used to generate electricity in the U.S.S.R., Japan, New Zealand, Iceland, Mexico, and California. The existing geothermal facilities use only a small portion of the total energy that might eventually be tapped from the earth's reservoir of internal heat.

Geology of Geothermal Energy

Natural heat production within the earth is only partly understood. We do know that some areas have a higher flow of heat from below than others, and that for the most part these locations are associated with the tectonic cycle. Oceanic ridge systems (divergent plate boundaries) and convergent plate boundaries, where mountains are being uplifted and volcanic island arcs are forming, are areas where this natural heat flow from the earth is anomalously high.

The distribution of geothermal temperature gradient (and thus, in a general way, heat flow) for the United States is shown in Figure 15.27. Moderate gradient is 30° to 45°C per kilometer, and this gradient is found over vast areas in the western United States. Therefore, the western region is, with limited exception, a good prospect for geothermal exploration. The data for the generalized classification of Figure 15.27 are not sufficient to accurately define the limits of the regions (24), so the map has limited value for locating specific sites where geothermal energy resources could be developed.

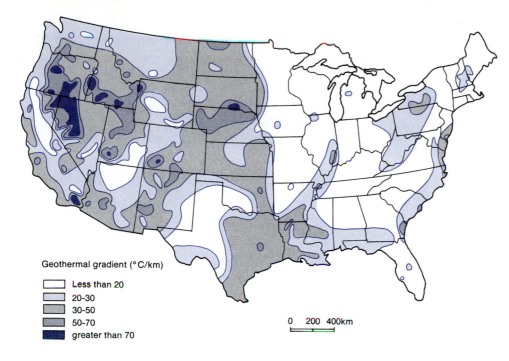

Figure 15.27
Distribution of geothermal temperature gradient for the United States. Almost all high gradients are located in the western part of the country. (After J. W. Tester, D. W. Brown, and R. M. Potter, *Hot Dry Rock Geothermal Energy.* Los Alamos National Laboratory, LA–11514–MS, 1989.)

Geothermal gradient (°C/km)

- [ ] Less than 20
- [ ] 20-30
- [ ] 30-50
- [ ] 50-70
- [ ] greater than 70

0 200 400km

The main region of high heat flow is concentrated in the western United States where tectonic and volcanic activity have been recent or are still active. The great width of the belt is somewhat of an anomaly, and for years geologists have vigorously debated the origin of the rock structure and tectonic activity of the region.

Within the region of relatively high heat flow, commercial development of geothermal power is most likely where a heat source such as convecting magma is relatively near the surface (3 to 10 kilometers) and in thermal contact with circulating groundwater. Thus, the likely sites for exploration are areas with naturally occurring hot springs and geysers that reflect near-surface hot spots; areas of recent volcanic activity, particularly those characterized by high-silica magma, because they are more likely to have stored heat near the surface where it might be accessible to drilling (24); and other localized hot spots with little or no near-surface expression that are discovered by direct and indirect (geophysical) subsurface exploration. The best location may not necessarily be that with the most prominent surface expression of geyser and hot-spring activity, however, because geothermal systems with a high rate of upflow, feeding geysers and hot springs, might not be as well insulated as a system with little or no leakage, as shown in Figure 15.28 (25). There is thus likely to be more stored heat at shallow depths in the well-insulated geothermal reservoir.

Based on geologic criteria, several geothermal systems may be defined: hydrothermal convection systems, hot igneous systems, and geopressured systems (26, 27). Each system has a different origin and different potential as an energy source.

Hydrothermal convection systems are characterized by a permeable layer in which a variable amount of hot water circulates. They are of two basic types: *vapor-dominated* systems and *hot-water* systems. Vapor-dominated hydrothermal convection systems are geothermal reservoirs in which both water and steam are present at depth. Near the surface, where pressure is less, the water flashes to superheated steam, which can be tapped and piped directly into turbines to produce electricity. These systems characteristically have a sufficiently slow recharge of groundwater so the hot rocks can convert the water to steam. That is, the heat supply is great enough to boil off more water than can be replaced by natural recharge (26).

Vapor-dominated systems are not very common. Only three have been identified in the United States: The Geysers, 145 kilometers north of San Francisco, California; Mt. Lassen National Park, California; and Yellowstone National Park, Wyoming. The parks are not available for energy development, but steam from The Geysers in California has been producing electric energy for years (Figure 15.29).

In the United States, hot-water systems are about twenty times more common than vapor-dominated systems (Figure 15.30). These systems, with surface temperatures greater than 150°C, have a zone of circulating hot water (no steam) that when tapped moves up with reduced pressure to yield a mixture of steam and water at the surface. The water must be removed from the steam before the steam can be used to drive the turbine (27). One problem with this system is disposal of the water. However, as ideally shown in Figure 15.31 the water could be injected back into the reservoir to be reheated.

Figure 15.28
(a) A hot-spring type of geothermal system with a high rate of
upflow characterized by vigorous
hot-spring or geyser activity at the
surface; (b) an insulated geothermal reservoir with little or no
leakage. (After U.S. Geological
Survey Publication GPO: 1969
O-339-536.)

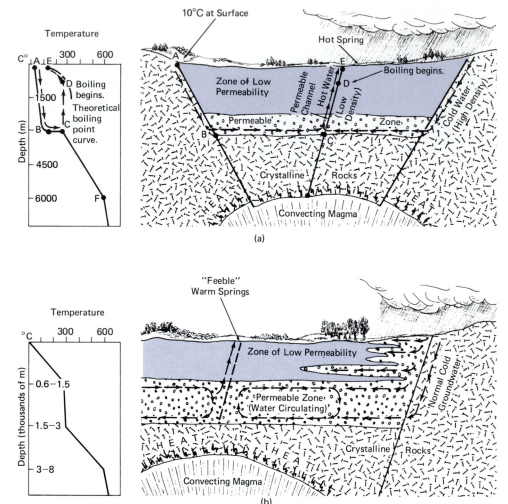

Hot igneous systems may involve the presence of molten magma at temperatures of 650° to about 1,200°C depending on the type of magma. Even if the igneous mass is not still molten, it may involve a large quantity of hot, dry rocks. These systems contain more stored heat per unit volume than any of the other geothermal systems; however, they lack the circulating hot water of the convection systems, and innovative methods to use the heat will have to be developed (26). Some of these geothermal reservoirs have hot, dry rocks accessible to drilling, in which case the rocks might first be drilled and then fractured with explosives or hydrofracturing tech-

Figure 15.29
The Geysers Power Plant north of San Francisco, California, with about 1000 megawatts installed capacity, is the world's largest geothermal electricity develop-ment. (Photo courtesy of Pacific Gas and Electric Company.)

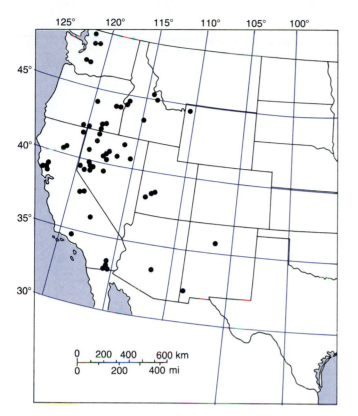

Figure 15.30
Distribution of hydrothermal convection systems in the conterminous United States where subsurface temperatures are thought to be in excess of 150 degrees Celsius. (After D. F. White and D. L. Williams, eds., U.S. Geological Survey Circular 726, 1975.)

niques. Then cold water might be injected into the rock at one location and pumped out at elevated temperatures at another location to recover the heat. The induced circulation of water would effectively mine the heat of the dry rock system (28). A pilot project by the Los Alamos Scientific Laboratory is currently under way in the Jemez Mountains of New Mexico. Approximately 3 to 10 megawatts (thermal energy) have been demonstrated using an innovative system that involves a nearly completely closed system of circulating water. Injected water is heated at a depth of about 3,000 meters to about 150°C. Pumped to the surface, the water goes through a heat exchanger, causing rapid expansion of freon which turns a turbine, generating electricity. The Jemez installation eventually may generate 35 megawatts while demonstrating the potential of hot, dry geothermal resources.

Testing of the hot, dry rock reservoir in New Mexico is ongoing and the Los Alamos National Laboratory plans a long-term test of at least one year. The objective of the test is to characterize power production and water loss as well as change in thermal characteristics of the reservoir over a relatively long period of time (29). The energy potential from geothermal sources are vast. Figure 15.32 compares the worldwide resource base for most of the nonrenewable energy sources. By comparison, present world consumption to date is approximately 330 exajoules. The hot, dry rock resource base alone is several hundred times that of the fossil fuels, but is much less than the potential of fusion, should that source ever come on line (29).

Recovery of heat directly from magma accessible to drilling (less than 10 kilometers deep) is an exciting prospect; however, the necessary technology is yet to be developed, and in fact, recovery may not be even remotely possible (30). Nevertheless, the possibility is

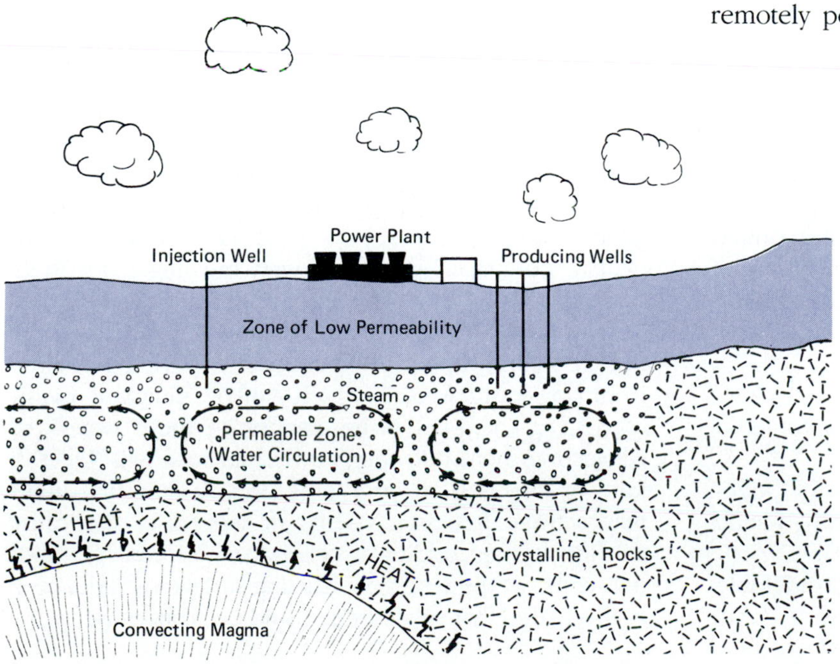

Figure 15.31
A hot-water geothermal system. At the power plant, the steam will be separated from the water and used to generate electrical power. The water will be injected back into the geothermal system by a disposal well. (Diagram courtesy of Pacific Gas and Electric Company.)

Figure 15.32
World energy resource base. Notice the large geothermal resource compared to fossil fuels. (Modeled after J. W. Tester, D. W. Brown, and R. M. Potter, *Hot Dry Rock Geothermal Energy*. Los Alamos National Laboratory, LA–11514–MS, 1989.)

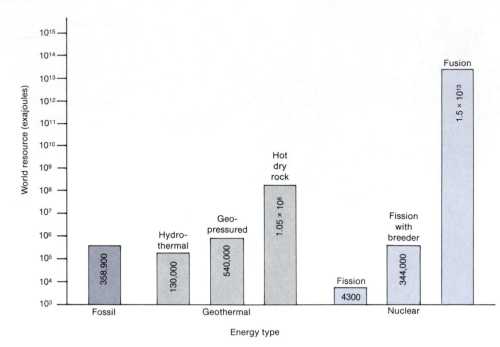

being studied and evaluated at sites in Hawaii and California. A generalized diagram of the concept to tap magma is shown in Figure 15.33. Some of the major problems to be solved include drilling and heat-extraction technology. Some important questions are how heat is transferred in magma, and how the introduction of a heat-exchange device would affect the proportion of crystals, liquid, vapor, and other physical and chemical properties of the magma in the vicinity of the device (31). Extraction of heat directly from magma, even if theoretically possible, is a long way off.

Geopressured systems exist where the normal heat flow from the earth is trapped by impermeable clay layers that act as an effective insulator. The favorable environment is characterized by rapid deposition of sediment and regional subsidence. Deeply buried water trapped in the sediments develops considerable fluid pressure and is heated (27).

Perhaps the best-known regions where geopressurized systems develop are in the Gulf Coast of the United States, where temperatures of 150° to 273°C at depths of 4 to 7 kilometers have been identified. These systems have the potential to produce large quantities of electricity for three reasons:

▼ They contain hot-water thermal energy that could be extracted
▼ They contain mechanical energy from the high-pressure water that could be used to turn hydraulic turbines to produce electricity
▼ As an added asset, the waters contain considerable amounts of dissolved methane gas, up to 1 cubic

meter per barrel of water, that could be extracted (27, 28)

Of the three, the energy potential from the heat and methane gas far exceeds that of the high-pressure water (32).

Environmental Impact of Geothermal Energy Development

Although the potentially adverse environmental impact of intensive geothermal energy development is perhaps not as extensive as that of other sources of energy, it is nevertheless considerable. Geothermal energy is developed at a particular site, and environmental problems include on-site noise, gas emissions, and industrial scars. Fortunately, development of geothermal energy does not require the extensive transportation of raw materials or refining that are typical of the fossil fuels. Furthermore, geothermal energy does not produce atmospheric particulate pollutants associated with burning fossil fuels; nor does it produce any radioactive waste. Geothermal development, except for the vapor-dominated systems, does, however, produce considerable thermal pollution from hot waste waters, which can be saline or highly corrosive. The plan is to dispose of these waters by reinjecting them into the geothermal reservoir, but that kind of disposal has problems because injecting fluids may activate fracture systems in the rocks and cause earthquakes. In addition, original withdrawal of fluids may compact the reservoir, causing surface subsidence. It is also feared that subsidence could occur because, as the

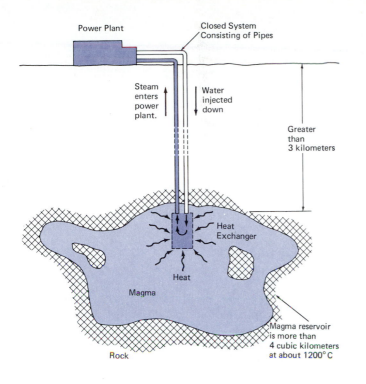

Figure 15.33
General diagram of a magma tap. (After J. L. Colp, Sandia
Laboratories Energy Report SAND–74–0253, 1974.)

heat in the system is extracted, the cooling rocks will contract (27).

Future of Geothermal Energy

Geothermal energy appears to have a bright future. Over a thirty-year period, the estimated yield from this vast resource, which is both identified and recoverable at this time (disregarding cost) far exceeds hundreds of modern nuclear power plants. In addition, it is expected that many more systems with presently recoverable energy are yet to be discovered (26). Furthermore, geothermal energy in the form of hot water (not steam) using normal heat flow may be a future source of energy to heat homes and other buildings along the Atlantic Coast, where alternative energy sources are scarce.

At present, geothermal energy supplies only a small fraction of one percent of the electrical energy produced in the United States. With the exception of the unusual vapor-dominated systems such as The Geysers in California, the production of electricity from geothermal reservoirs is still rather expensive, generally more so than from other sources of energy. For this reason commercial development of the energy will not proceed rapidly until the economics are equalized. Even so, the total energy output (electrical) from geothermal sources is not likely to exceed a few percent (10 percent at most) in the near future, even in states like California where it has

been produced and where expanding facilities are likely (28). Nevertheless, the growth in power produced from geothermal sources increased dramatically from 1960 to 1989 and then decreased a little (Figure 15.34). Continued growth can be expected.

▼ RENEWABLE ENERGY SOURCES

The energy transition from fossil fuels to renewable energy sources has begun. In the United States today, oil and gas still supply approximately three-fourths of our energy needs, but as oil and gas become scarce and/or more costly, we will slowly change to new energy sources. The three major sources of energy likely to be used are fossil fuels such as coal, oil shale, and tar sands; nuclear energy (fission today and perhaps fusion sometime in the future); and renewable energy, broadly defined to include direct solar energy, hydropower, biomass, and wind. Geothermal energy might also be added to this list, but in many instances is better thought of as a nonrenewable resource even though in specific situations the natural heat flow may be high enough that it can be replenished relatively quickly. An important point concerning the renewable energy sources is that in many parts of the world they may be the only energy sources indigenous to a given region.

The necessity for seriously considering renewable energy sources is emphasized by examining the environ-

Figure 15.34
Net generation of electricity from geothermal sources in the United States, 1960 through 1989. (Source: Energy Information Administration.)

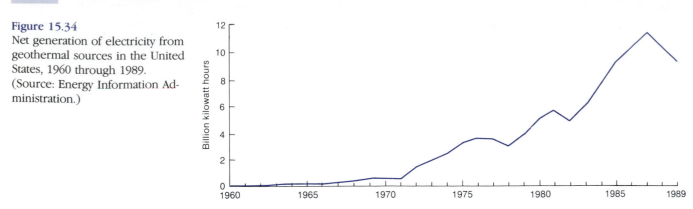

Direct Solar Energy

mental impact associated with other possible sources of energy. That is, the fossil fuels such as coal, although abundant, are often damaging to the environment throughout their entire fuel cycle, from mining to processing to eventual consumption. Fossil fuels also carry the threat of global climate modification through increased discharge of carbon dioxide, particulates, and other materials. Nuclear energy, while imposing no threat to climate modification, is associated with serious problems such as waste disposal, accidents, and weapons proliferation. Nuclear energy also releases waste heat into the environment through on-site cooling processes and when the electricity produced at the power plant is transported and used. Finally, we must also recognize that all fossil fuels and nuclear energy from fission are ultimately exhaustible, again emphasizing the need for eventual transition to the renewable energy sources (33).

Renewable energy sources are derived directly or indirectly from solar energy. Thus broadly defined, solar energy includes a number of different sources, as shown in Figure 15.35. The advantages of these sources are that they are inexhaustible and are generally associated with minimal environmental degradation. With the exception of burning biomass or urban waste, they do not pose the threat of adding carbon dioxide to the atmosphere and thus causing potential climate modification. One major disadvantage is that most forms of renewable energy, with the possible exceptions of hydropower, biomass, and ocean thermal conversion, are intermittent and spatially variable. Some of these sources, such as solar cells (photovoltaics) that produce electricity directly from solar energy, are presently much more expensive than fossil or nuclear energy sources.

Another important aspect of renewable resources is the lead time necessary to implement technology. Table 15.4 summarizes the technology, availability, and lead time necessary for several of the renewable resources. We emphasize that many of these are commercially available, and the construction lead time is often short relative to development of new sources or power plants using fossil or nuclear fuels.

The direct use of solar energy is not new. Nearly 2,500 years ago Greeks designed homes to capture sunlight, and ancient cliff-dwelling Indians in the southwestern United States used asymmetry of valleys and exposure to sunlight for natural winter heating and summer shading. Amory Lovins points out that throughout history there has been a steady evolution in solar architecture and technology, but progress has been periodically interrupted by the influx of apparently plentiful, cheap fuels such as new forests or large deposits of oil, natural gas, coal, and uranium. Earlier energy crises, beginning with shortages of wood during the Greek and Roman eras to coal strikes in the United States at the end of the nineteenth century, led to the rediscovery of earlier knowledge concerning practical applications of solar energy. These lessons should not be neglected in the current energy situation, since it appears that, with the possible exception of nuclear fusion, solar energy, broadly defined as shown in Figure 15.35, is the only long-term alternative at this time (34).

The total amount of solar energy that reaches the earth's surface is tremendous. On a global scale, two weeks' worth of solar energy is roughly equivalent to the energy stored in all known reserves of coal, oil, and natural gas on earth. In the United States, on the average, 13 percent of the sun's original energy entering the atmosphere arrives at the ground. The actual amount at a particular site is quite variable, however, depending upon the time of year and the cloud cover. The average value of 13 percent is equivalent to approximately 177 watts per square meter per hour (35).

Solar energy may be used directly through either passive or active solar systems. Passive solar systems often involve architectural design, without implementation of mechanical power, to enhance and take advantage of natural changes in solar energy that occur throughout the year. Many homes and other buildings in the southwestern United States and other parts of the country now use passive solar systems for at least part of their energy

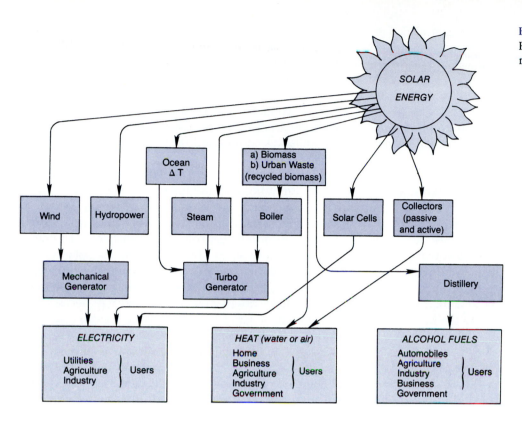

Figure 15.35
Routes of the various types of renewable solar energy.

Table 15.4
Commercial status and lead times of renewable energy technologies.

Technology	Commercially Available	Regulatory Plus Construction Lead Time (years)
Biomass		
Direct combustion, electric	Yes	2–4.5
Direct combustion, non-electric	Yes	2
Gasification	Yes	2
Anaerobic digestion	Yes	1–2
Municipal solid waste combustion	Yes	5–8
Wood stoves	Yes	Less than 1
*Wind turbines**		
Large	Yes	0.5–2
Small	Yes	0.5
Solar		
Passive	Yes	Less than 1
Water heating	Yes	Less than 1
Active space heating	Yes	Less than 1
Photovoltaics	Dispersed—Yes	0.5–2
	Central station—after 2000	5
Thermal electric	Yes	9
Industrial heat	Some applications	2

*Lead times given do not include resource verification, which requires a minimum of one year.

Source: Modified from California Energy Commission 1980. *Comparative Evaluation of Non-traditional Energy Resources.*

needs. Active solar systems require mechanical power, usually pumps and other apparatus to circulate air, water, or other fluids from solar collectors to a heat sink, where the heat is stored until used. Solar collectors are usually flat panels, a glass cover plate over a black background upon which water is circulated through tubes. Short-wave solar radiation enters the glass and is absorbed by the black background. As longer-wave radiation is emitted from the black material, the water in the circulating tubes is heated, typically to temperatures of 38 to 93°C (100–200°F) (Figure 15.36) (35). In the U.S., the number of solar systems that use collectors like these is rapidly growing. Changes in production of solar collectors from 1974 through 1988 are shown in Figure 15.37. Notice that the production of low-temperature collectors, mostly used for heating swimming pools, peaked in 1980, while those used for domestic water and space heating peaked in 1984 and then declined sharply. The decline was due in part to the fact that the federal energy tax credit for those collectors expired in 1985 (36).

Another potentially important aspect of direct solar energy involves solar cells or photovoltaics that convert sunlight directly into electricity. These cells are expensive, costing much more than electricity produced from traditional fossil fuel sources (37). On the other hand, the field of solar technology is changing quickly, and low-cost solar cells may become widely available in the future. During the period from 1985 to 1988 the number of photovoltaic modules shipped (in terms of potential power production) increased by about 67 percent (36). Recently a 6.5 megawatt solar plant utilizing photovoltaic cells was constructed near San Luis Obispo, California (38).

Questions often asked about direct solar energy are: where can it be used and how much energy will it save? These questions are not easy to answer. From an availability standpoint, Figure 15.38 shows in a general way the estimated year-round availability of solar energy in the United States. But this view of solar energy availability is analogous to trying to paint a picture with a paint roller. Locations for solar energy are site-specific and detailed observation in the field is necessary to evaluate the solar energy potential in a given area. As to how much energy might be saved by converting to direct solar power, this again is quite variable. Optimists hope that by the year 2000 direct solar energy may supply up to 20 percent of the total energy demand. Considering how fast the solar energy industry is growing and the short lead time necessary for conversion to solar energy, these

Figure 15.36
Details of a flat plate solar collector. (After Farallones Institute, 1979, *The Integral Urban House: Self-Reliant Living in the City,* San Francisco: Sierra Club Books.)

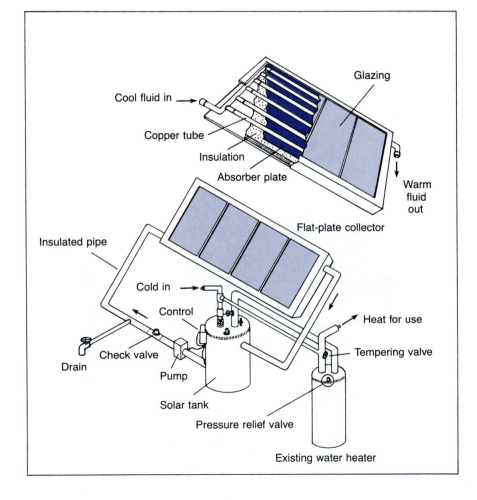

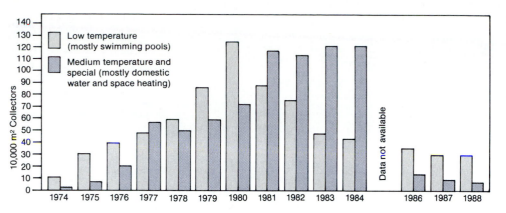

Figure 15.37
Production of solar collectors
from 1974 to 1988. (Dates from
Energy Information Administration, *Annual Energy Review,* 1985
and 1989.)

projections may prove accurate. On the other hand, conversions may be much slower if other sources of energy at cheaper prices become available in the next few years. Since this possibility seems unlikely, the future of solar energy looks secure.

During the 1980s solar energy power plants of several varieties were developed. One of the early plants was a "power tower" that worked by collecting solar energy as heat and delivering the energy in the form of steam to turbines to produce electric power (Figure 15.39). An experimental 10-megawatt power tower is located near Barstow, California. The tower is approximately 100 meters high surrounded by approximately 2,000 mirror modules, each with reflective area of about 40 square meters. The mirrors are adjusted continually as the earth rotates to reflect as much sunlight into the tower as possible. Other experimental power towers are being considered in Japan and other areas and continued research into the technology is worthwhile and should continue.

The most successful solar power experiment to date is also located in the Mojave Desert of California. The

project consists of eight "solar farms," each consisting of a power plant surrounded by hundreds of solar collectors (Figure 15.40). The system is called LUZ Solar Electric Generating Systems and as of 1990 was generating 275 megawatts of electricity for the southern California area. This amount of power is more than 90 percent of the world's solar-generated electricity that is connected to a utility system (38). The system works by utilizing solar collectors (mirrors) to heat a synthetic oil that flows through heat exchangers to drive steam turbine generators. Basic components of the system are shown in Figure 15.41. The company building the power plants anticipates generating as much as 600 megawatts by 1994, which is enough electricity to meet the needs of about 800,000 people. The LUZ system has a natural gas burner backup system that ensures uninterrupted power generation during periods of peak demand or on cloudy days. A major advantage of the modular nature of construction is that it allows for individual solar farms to be constructed in a period as short as about one year. The cost of the energy produced is close to the average cost of production of electricity from other sources and is less

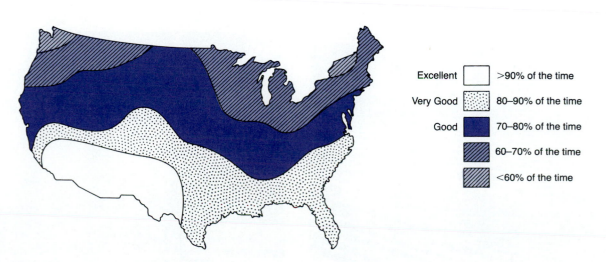

Figure 15.38
Estimated year-round usability of solar energy for the conterminous United States. (Modified after National Wildlife Federation, 1978.)

Figure 15.39
Idealized diagram showing how a solar power tower works. (Modified after drawing by Southern California Edison Company.)

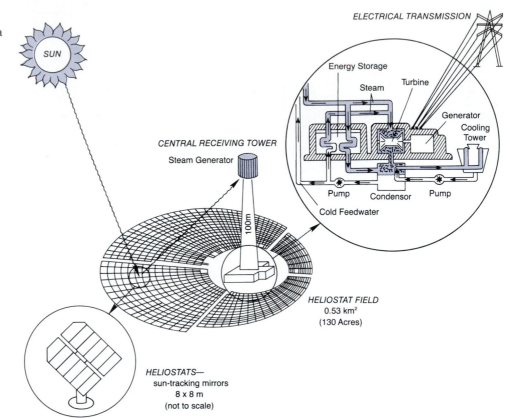

expensive than electricity produced from new nuclear power plants (38).

In summary, the LUZ system is a combination of solar technology and conventional power generation. The system looks very promising and further improvements in efficiency are probable. At present the LUZ system uses natural gas to generate approximately 25 percent of the power it produces. If the price of fossil fuels increases in the future, the profits from the combined system will increase substantially (38).

A last example of direct use of solar energy involves using part of the natural oceanic environment as a gigantic solar collector. The surface temperature of ocean water in the tropics is often about 28°C (82°F). At the bottom of the ocean, at a depth as shallow as about 600 meters, however, the temperature of the water may be 2–6°C (36–43°F). Low-efficiency heat engines can be designed to exploit this temperature differential either by using the seawater directly or by using an appropriate heat exchange system in a closed cycle in which a fluid

Figure 15.40
Photographs showing a "solar farm" in the Mojave Desert of California. (Courtesy of LUZ.)

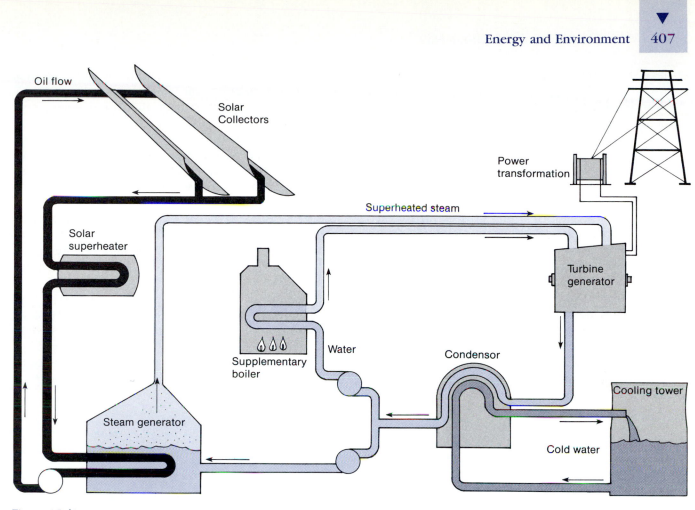

Figure 15.41
Idealized diagram showing how the LUZ solar electric generating system works. (Courtesy of LUZ.)

such as ammonia or propane is vaporized by the warm water. The expanding vapor is then used to propel a turbine and generate electricity. After generation of electricity, the vapor is cooled and condensed by means of the cold water. Whether or not large-scale ocean thermal plants are ever built will depend primarily upon whether locations are discovered close to potential markets and whether the plants are economically feasible (35). Answers to these questions are uncertain so the future of ocean thermal plants remains speculative.

Experiments with ocean thermal energy conversion on the big island of Hawaii have proved very interesting. The laboratory at Keahole Point on the Kona coast is ideally located because the sea floor slopes off very steeply to deep, cold waters. Although only a small amount of energy has actually been produced, the big payoff may be aquaculture. The cold, nutrient-rich water is being experimented with by several companies to grow seafoods such as abalone, salmon, oysters, and kelp. Several of these may be grown together in large polyculture tanks or ponds. The laboratory is also using the cold water for air conditioning, resulting in a significant saving on their electric bill.

Water Power

Water power is really a form of stored solar energy. Since at least the time of the Roman Empire, the use of water as a source of power has been a successful venture. Waterwheels that harness water power and convert it to mechanical energy were turning in Western Europe in the seventeenth century, and during the eighteenth and nineteenth centuries, large waterwheels provided the energy to power grain mills, sawmills, and other machinery in the United States. Today, hydroelectric power plants provide about 15 percent of the total electricity produced in the United States. Although the total amount of electrical power produced by running water will increase somewhat in the coming years, the percentage may be reduced as other energy sources such as nuclear, solar, and geothermal increase faster.

Most of the acceptable good sites for large dams to produce hydropower are probably already being utilized. On the other hand, small-scale hydropower systems may be much more common in the future. These are systems designed for individual homes, farms, or small industries. They will typically have power outputs of less than 100 kilowatts and are termed micro-hydropower systems (39). Micro-hydropower is one of the world's oldest and most common energy sources. Numerous sites in many areas have potential for producing small-scale electrical power. This is particularly true in mountainous areas, where potential energy from stream water is most available. Micro-hydropower development is by its nature very site-specific, depending upon local regulations, economic situation, and hydrologic limitations. Of particular importance is the fact that hydropower can be used to generate either electrical power or mechanical power to run machinery. Such plants may help cut the high cost of importing energy and help small operations become more independent of local utility providers (39).

An interesting aspect of hydropower is "pump storage" (Figure 15.42), the objective of which is to make better use of total electrical energy produced through energy management. The basic idea is that, during times when demand for power is low, electricity from oil, coal, or nuclear plants may be used to pump water up to a storage site or reservoir (high pool). Then, when demand for electricity is high, the stored water flows back down to the low pool through generators to supplement the power supply. It is important to keep in mind that pump storage systems are not as efficient as conventional hydroelectric plants. The bottom line is that about three units of energy from an oil, gas, or nuclear power plant are needed to produce two delivered units of energy from a pump storage facility. The advantage lies in the timing of energy production and use: the two units can be drawn during peak demand and the three units are used to pump the water to the high pool when the demand is low.

Tidal power. Another form of water power might be derived from ocean tides in a few places where there is favorable topography, such as the Bay of Fundy region of the northeastern United States and Canada. The tides in the Bay of Fundy have a maximum rise of about 15

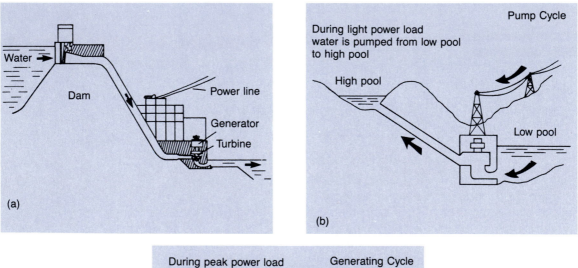

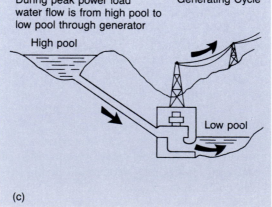

Figure 15.42
(a) Basic components of a hydroelectric power station. (b, c) How a pump storage system works. (Modified from Council on Environmental Quality, 1975, *Energy Alternatives: A Comparative Analysis*. Prepared by the Science and Public Policy Program, University of Oklahoma, Norman.)

meters. A minimum rise of about 8 meters is necessary to even consider developing tidal power.

The principle of tidal power is to build dams across the entrance to bays, creating a basin on the landward side so as to create a difference in water level between the ocean and the basin. Then, as the water in the basin fills or empties, it can be used to turn hydraulic turbines that will produce electricity (40).

Use of water power. Water power is clean power. It requires no burning of fuel, does not pollute the atmosphere, produces no radioactive or other waste, and is efficient. There is an environmental price to pay, however. Water falling over high dams may pick up nitrogen gas, which enters the blood of fish, expands, and kills them. Nitrogen has killed many migrating game fish in the Pacific Northwest. Furthermore, dams trap sediment that would otherwise reach the sea and replenish the sand on beaches. In addition, for a variety of reasons, many people do not want to turn all the wild rivers into a series of lakes.

For these reasons, and because many good sites for dams are already utilized, the growth of large-scale water power in the future appears limited. On the other hand, there seems to be an increase in interest for micro-hydropower to supply electricity or mechanical energy. Environmental impact of numerous micro-hydropower installations may be considerable in an area. The sites will change the natural stream flow, affecting the stream biota and productivity. Small dams and reservoirs also tend to fill more quickly with sediment, making their potential useful time period much shorter. Because micro-hydropower development can adversely affect stream environment, careful consideration must be given to its development over a wide region. A few such sites may cause little environmental degradation, but if the number becomes excessive, the impact over a wider region may be appreciable. This is a consideration that must be given to many forms of technology that involve

small sites. The impact of one single site may be nearly negligible in a broad region, but as the number of sites increases, the total impact may become very significant.

Wind Power

Wind power, like solar, has evolved over a long period of time, from early Chinese and Persian civilizations to the present. Wind has propelled ships and driven windmills for grinding grain or pumping water. More recently, wind has been used to generate electricity. The potential energy that might eventually be derived from the wind is tremendous; yet there will be problems because winds tend to be highly variable in time, place, and intensity (35).

Wind prospecting has become an important endeavor. On a national scale, regions with the greatest potential for development of wind energy are the Pacific Northwest coastal area, the coastal region of the northeastern United States, and a belt extending from northern Texas northward through the Rocky Mountain states and the Dakotas. There are also many other good sites, such as in mountain areas in North Carolina and the northern Coachella Valley in southern California.

At a particular site, the direction, velocity and duration of the wind may be variable, depending on local topography and regional to local magnitude of temperature differences in the atmosphere (41). For example, wind velocity often increases over hilltops or mountains, or may be funneled through a broad mountain pass (Figure 15.43). The increase in wind velocity over a mountain is caused by a vertical convergence of the wind, whereas through a pass, it is caused partly by horizontal convergence as well. Because the shape of a mountain or pass is often related to the local or regional geology, prospecting for wind energy is a geologic as well as geographic and meteorologic problem.

Significant improvements in the size of windmills and the amount of power they produce occurred from

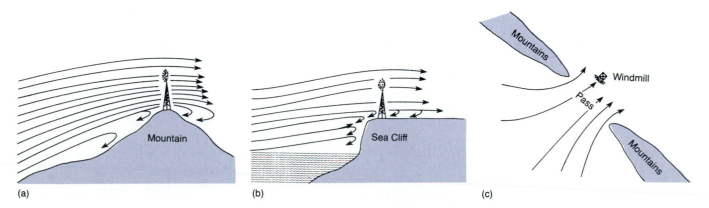

(a)   (b)   (c)

Figure 15.43
How wind may be converged (and velocity increased) vertically (a & b) or horizontally (c). Tall windmills are necessary on hilltops or on the top of a sea cliff to avoid rear surface turbulence.

the late 1800s through approximately 1950, when many countries in Europe and the United States became interested in larger-scale generators driven by the wind. In the United States, thousands of small wind-driven generators have been used on farms. Most of the small windmills generated about one kilowatt of power, which is much too small to be considered for central power generation needs. Interest in wind power declined during the several decades prior to the 1970s because of the abundance of cheap fossil fuels, but interest has revived in building larger windmills that might each supply a few megawatts of electrical power.

Small-scale producers. Small-scale power production from windmills in the United States was made more feasible by passage of the U.S. Public Utility Regulatory Policy Act (PURPA) in 1978. This act requires utilities to interconnect with small-scale independent producers of power and pay fair market price for the electricity produced. PURPA initiated an entrepreneurism by providing a market to small-scale power producers. This is nowhere better exemplified than in California, where in the last 10 years approximately 17,000 windmills, with a generating capacity of about 1,400 megawatts, have been installed. Individual windmills produce from 60 to 75 kilowatts of power and are arranged in wind farms consisting of clusters of windmills located in mountain passes. Electricity produced at these sites is connected to the general utility lines. These wind farms produced sufficient electricity to supply approximately 100,000 homes, making a significant contribution in the modern utility grid. Tax incentives have helped the wind-power industry become established, and in California it is expected that wind power may be the state's second least expensive source of power by the year 2012, second only to hydropower (42).

Use of wind power. Wind power will not solve all our energy problems, but it is yet another alternative energy source that is renewable and can be used in particular sites to reduce our dependency on foreign oil. Wind energy has a few detrimental qualities: first, demonstration projects suggest that vibrations from windmills may be a source of problems; second, windmills may cause interference with radio and television broadcast; and finally, windmills, particularly if there are many of them, may degrade scenic resources of an area. Still, everything considered, wind energy has a relatively low environmental impact and its continued use should be carefully evaluated. A number of large demonstration units are now being tested.

Energy from Biomass

Biomass fuel is a new name for the oldest human fuel. **Biomass** is organic matter that can be burned directly as a fuel or converted to a more convenient form and then burned. For example, we can burn wood in a stove or convert it to charcoal and then burn it. Biomass has provided a major source of energy for human beings throughout most of the history of civilization. When North America was first settled, there was more wood fuel than could be used. The forests often were cleared for agriculture by girdling trees (cutting through the bark all the way around the base of a tree) to kill them and then burning the forests.

Until the end of the nineteenth century, wood was the major fuel source in the United States. During the early mid-twentieth century, when coal, oil, and gas were plentiful, burning wood became old-fashioned and quaint; it was something done for pleasure in an open fireplace that conducted more heat up the chimney than it provided for space heating. Now, with other fuels reaching a limit in abundance and production, there is renewed interest in the use of natural organic materials for fuel.

Firewood is the best known and most widely used biomass fuel, but there are many others. In India and other countries, cattle dung is burned for cooking. Peat provides heating and cooking fuel in northern countries like Scotland, where peat is abundant.

On a global scale, more than 1 billion people in the world today still use wood as their primary source of energy for heat and cooking. Energy from biomass may take several routes: direct burning of biomass to produce either electricity or to heat water and air, and distillation of biomass to produce alcohol for fuel. Today in the United States, various biomass sources supply over 3 percent of the entire energy consumption (approximately 2.8 exajoules per year), primarily from the use of wood for industry and home heating (36, 43).

Sources of biomass fuel. There are three primary sources of biomass fuels in North America: forest products, otherwise unused agricultural products, and urban waste. Biomass can be recycled, and today there are numerous facilities in the United States that process urban waste for use in generating electricity or producing fuel. At processing plants such as those in Baltimore County, Maryland; Chicago, Illinois; Milwaukee, Wisconsin; Tacoma, Washington; and Akron, Ohio, municipal waste is burned and the heat energy is used to make steam for a variety of purposes ranging from space heating to industrial generation of electricity. The United States, due to lack of public acceptance, has been slower than other countries to use urban waste as an energy source. In Western Europe a number of countries now use between one-third to one-half of their municipal waste for energy production (33).

▼ **ENERGY AND WATER DEMAND**

The recent push to accelerate energy development rapidly has raised an important question: Is there enough water to support the projected development? This complex question has recently been addressed by a U.S. Geological Survey evaluation, and our discussion here summarizes some of the conclusions from that report (44).

The major withdrawals of water in the United States are for irrigation and for cooling water in electric power generation plants (fossil fuel, nuclear, or geothermal). Most of the water for power generation is withdrawn to cool and condense the steam on the outlet side of the turbine. The water is withdrawn from a river or lake and circulated once through the condenser, then it carries the waste heat to a point of discharge back in the river or lake. Where thermal pollution is unacceptable or there is no large lake or river from which to obtain abundant cooling water, cooling towers, evaporation ponds, and water-sprayed evaporation ponds can be used for the cooling process (44).

Water is needed in all stages of energy development, from mining and land reclamation, to on-site processing, refining, and converting fuels to other energy forms. Figure 15.44 compares water consumption in refining and conversion processes for present commercial energy sources and those likely to be used in the future.

The amount of water required for mining and reclamation varies for different fuels and different climatic conditions. The water demand for mining coal or uranium is generally low, but in arid regions, obtaining the water necessary to establish vegetation as part of the reclamation program may be a serious problem. On the other hand, reclaiming a surface mine in a humid climate would probably not have a problem of lack of water. Another example can be drawn from the oil industry. Little water is used in drilling for oil, but during the

recovery phase, a tremendous amount of water may be pumped down into the well to float the oil as part of a secondary recovery technique (44).

Mining oil shale, and particularly disposal of spent shale and surface reclamation in Colorado, Utah, and Wyoming, is also associated with high water demand. In fact, water requirements will probably limit production of shale oil to about one million barrels per day. A much larger production would require purchasing and transferring water that is now used for agricultural purposes (44).

The conclusion concerning the demand for water and production of energy in the United States is that in the entire nation, there is enough water to support accelerated energy development. In arid regions, however, limited water affects site location and design of energy recovery and conversion facilities (44).

▼ CONSERVATION AND COGENERATION

We established earlier in this chapter that we will have to get used to living with a good deal of uncertainty concerning the availability, cost, and environmental effects of energy use. Furthermore, we can expect that serious shocks will continue to occur, disrupting the flow of energy to various parts of the world.

Supply and demand for energy are difficult to predict because technical, economic, political, and social assumptions underlying projections are constantly changing. Large annual variations in energy consumption must also be considered: energy consumption peaks during the winter months (heating), with a secondary peak in the summer (air conditioning). Future changes in population or intensive conservation measures may change this pattern, as might better design of buildings and more reliance on solar energy. One recent prediction holds that energy consumption in the United States in the year 2010 may be as high as 137 exajoules or as low as 63 exajoules; such a wide disparity suggests the great potential for conservation in the emerging energy scenario. (Energy consumption in 1989 was about 81 exajoules.) The high value assumes no change in energy policies, whereas the low value assumes very aggressive energy conservation policies.

There has been a strong movement to change patterns of energy consumption through such measures as conservation, increased efficiency, and cogeneration. *Conservation* of energy refers to a moderation of energy use or simply getting by with less demand for energy. In a pragmatic sense this has to do, then, with adjusting our energy needs and uses to minimize the expenditure of energy necessary to accomplish a given task. *Efficiency* is related to designing equipment to yield more power from a given amount of energy (45). Finally, *cogeneration* refers to a number of processes, the objective of which is to capture waste heat rather than simply release it into the atmosphere or water as thermal pollution.

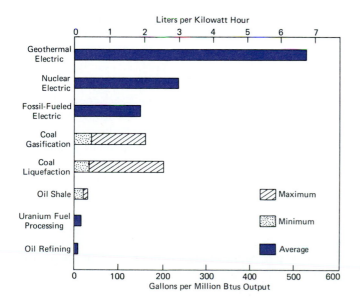

Figure 15.44
Consumption of water for various energy-related refining and conversion processes. (After G. H. Davis and L. A. Wood, U.S. Geological Survey Circular 703, 1974.)

The three concepts of energy conservation, increased efficiency, and cogeneration are very much interrelated. For example, when electricity is produced at large coal-burning power stations, large amounts of heat may be emitted into the atmosphere. Production of electricity typically burns three units of fuel to produce one unit of electricity. Cogeneration, which involves recycling of that waste heat, can increase the efficiency of a typical power plant from 33 percent to as high as 75 percent. Effectively, this reduces losses from 67 percent to 20 percent (45). Cogeneration of electricity also involves the production of electricity as a byproduct from industrial processes that normally produce and use steam as part of their regular operations. Optimistic energy forecasters estimate that we may eventually be able to meet approximately one-half of the electrical power needs of industry through cogeneration (46). In fact, it has been estimated that, by the year 2000, over 10 percent of the power capacity of the United States could be provided through cogeneration. Fuel savings through this would amount to as much as two million barrels of oil per day! Such a savings is really talking about conservation as well, showing how these two concepts are interrelated.

▼ ENERGY POLICY

Hard Path versus Soft Path

Energy policy today is at or near a crossroads. One road leads to development of so-called hard technologies, which involve finding ever greater amounts of fossil fuels and building larger power plants. Following the "hard path" means continuing as we have been for a number of years. In this respect, the hard path is more comfortable. That is, it requires no new thinking or realignment of political, economic, or social conditions. It also involves little anticipation of the inevitable depletion of the fossil fuel resources on which the hard path is built.

Those in favor of the hard path argue that, with the exception of the major oil-producing nations, environmental degradation has resulted because people in less developed countries have had to utilize so much of their local resources, such as wood, for energy rather than, say, for land conservation and erosion control. It is argued that the way to solve the environmental problems in less developed countries is to provide them with cheap energy that utilizes more heavy industrialization and technology, not less. Proponents of this argument point out that the United States and other countries with a sizeable energy resource in coal or petroleum should exploit that resource to make certain that the environmental degradation that is now plaguing other countries of the world will not strike here. In other words, the argument is to let the energy industry develop the available resources—that this freedom will ensure a steady supply of energy and less total environmental damage than if the government regulates the energy

industry. Proponents would argue that the present increase in the burning of firewood across the United States is an early indicator of the effects of strong governmental controls on energy supplies. Furthermore, they would argue that this is just the beginning and that the eventual depletion of forest resources will have an appreciably detrimental effect on the environment as it has done in so many less developed countries.

The other road is designated as the "soft path" (47). One of the champions of this choice has been Amory Lovins, who states that the soft path involves energy alternatives that are renewable, flexible, and environmentally more benign than those of the hard path. As defined by Lovins, these alternatives have several characteristics:

1. They rely heavily on renewable energy resources such as solar, wind, and biomass.
2. They are diverse, tailored for maximum effectiveness under specific circumstances.
3. They are flexible; that is, they are relatively low technologies that are accessible and understandable to many people.
4. They are matched in both geographical distribution and scale to prominent end-use needs (the actual use of energy).
5. There is good agreement or matching between energy quality and end-use.

The last point is of particular importance because, as Lovins points out, people are not particularly interested in having a certain amount of oil, gas, or electricity delivered to their homes; rather, they are interested in having comfortable homes, adequate lighting, food on the table, and energy for transportation (47). These latter uses are called end-uses, and only about 5 percent of the end-uses really require high-grade energy like electricity. Nevertheless, a lot of electricity is used to heat homes and water. For some purposes, continual use of electricity is very important; but for others, Lovins shows that there is an imbalance in using nuclear reactions at extremely high temperatures or in burning fossil fuels at high temperatures or in burning fossil fuels at high temperatures simply to meet needs where the temperature increase necessary may be only a few tens of degrees. Such large discrepancies are thought wasteful.

The bridge, or transition, from the hard to the soft path will presumably take place through the utilization of present energy sources such as coal and natural gas. Although there is sufficient coal to last hundreds of years, those in favor of the soft path would prefer to use this source as a transition rather than an ultimate long-term source.

Energy in the Future

Assuming that in the future we will be able to get by on a lower rate of energy consumption, it is interesting to

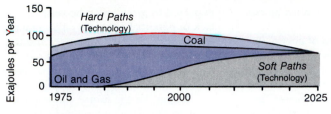

Figure 15.45
Low-energy scenario for the future, as envisioned by A. B. Lovins. (Modified from SOFT ENERGY PATHS by Amory Lovins. Copyright © 1977 by Friends of the Earth. Reprinted by permission of Harper Collins Publishers Inc.)

speculate on low-energy scenarios. One, published by Lovins (Figure 15.45), involves a moderate decrease in energy consumption following a maximum consumption of about 100 exajoules in the year 2000. The eventual reduction accompanies the shift from dependence on fossil fuels to the soft technologies (renewable energy resources).

Lovins emphasizes that the reduction need not be associated with a lower quality of life, but rather with increased conservation of energy and a more energy-efficient distribution of urban populations, agriculture, and industry. Alternatives that will help achieve lower energy consumption include: new settlement patterns that encourage "urban villages" with populations of about 100,000; constrained and minimized private transportation, and maximum accessibility of services; agricultural practices that emphasize locally grown and consumed food consisting largely of grains, beans, potatoes, and vegetables; and new industrial guidelines designed to conserve energy and minimize production and consumer waste to help reduce the total energy demand (46). To an extent, these alternatives are already taking form; that is, in some instances, highway construction has a lower priority than development of mass transit systems, agricultural lands near urban centers are preserved, and industry is more receptive to recycling and decreasing production of consumer wastes (such as unnecessary packaging).

There is debate as to whether our energy planning should seek reduced energy consumption through following the soft path or to continue to rely heavily on the fossil fuels, primarily coal. Both scenarios, however, do speculate that nuclear energy will play a reduced role in

the total energy picture by the year 2020. Certainly in the United States and in other parts of the world there is considerable public criticism of nuclear energy. This has resulted in the following (48):

1. Sweden has a referendum to terminate nuclear power stations by the year 2010.
2. In Germany and Switzerland, plans for additional nuclear power plants are at a standstill.
3. In the Soviet Union there is serious debate concerning the future of nuclear power.
4. In the United States there have been no new orders for nuclear plants since 1978.

A major problem concerning energy planning for tomorrow is that we are now conscious that burning of fossil fuels and even wood may be degrading our global environment. The energy path we take must be one capable of supplying the energy we require for human activities without endangering the planet that we live on (49). The energy option that can yield the greatest return in the near future is conservation. We have the technology now to save sufficient energy to eliminate or greatly reduce the need to import oil. Such a plan would reduce U.S. dependency on foreign sources of energy and improve our economy significantly.

As long ago as 400,000 years, early humans used fire for warmth and cooking. Throughout most of human history, few people have cared much about the source of energy, but rather have been concerned with the services it provides. We want to cook our food, heat our homes, drive our automobiles, and take hot showers. However, in recent years we have become conscious of the fact that our use of some sources of energy may be causing global changes potentially detrimental to our environment. These include, among others, possible global warming, which will be discussed in Chapter 16. Therefore, the path we follow in energy management should be one that minimizes adverse regional and global change. (49).

From an energy planning viewpoint, the next thirty years will be crucial to the United States and the rest of the industrialized world. The energy decisions we make in the near future will greatly affect both our standard of living and quality of life. Optimistically, we have the necessary information and technology to ensure a bright, warm, lighted, and moving future—but time is short and we need action now to build for the future an energy picture we can be proud of (46).

▼ ▼ ▼ SUMMARY AND CONCLUSIONS

The ever-increasing population and appetite of the American people and the world for energy are staggering. It is time to seriously question the need and desirability of an increasing demand for electrical and other sources of energy in industrialized societies. Quality of life is not necessarily directly related to greater consumption of energy.

The origin of fossil fuels (coal, oil, and gas) is intimately related to the geologic cycle. These fuels are, essentially, stored solar energy in the form of organic material that has escaped total destruction by oxidation. The environmental disruption associated with exploration and development of these resources must be weighed against the benefits gained from the energy; but this development is not an either-or proposition, and good conservation practices combined with pollution control and reclamation can help minimize the environmental disruption associated with fossil fuels.

Nuclear fission will remain an important source of energy. The growth of fission reactors in the United States will not be as rapid as projected, however, because of concern for environmental hazards and increasing costs to produce large nuclear power plants. To avert an eventual nuclear fuel shortage, work should continue on the breeder reactor.

Nuclear fusion is a potential energy source for the future, and may eventually supply a tremendous amount of energy from a readily available and nearly inexhaustible fuel supply.

Use of geothermal energy will become much more widespread in the western United States where natural heat flow from the earth is relatively high. Although the electric energy produced from the internal heat of the earth will probably never exceed 10 percent of the total electric power generated, it nevertheless will be significant. Geothermal energy also has an environmental price, however, and the possibility of causing surface subsidence through withdrawal of fluids and heat, as well as possibly causing earthquakes from injection of hot wastewater back into the ground, must be considered.

Renewable sources of energy depend upon solar energy and can take a variety of forms, including passive and active solar, wind and hydropower, and energy from biological processes (including recycled biomass from urban waste). These energy sources have varying attributes, but generally cause little environmental disruption and are indigenous to the land. These energy sources will not be depleted, so are dependable in the long term. There is an energy transition going on in the world today, and sooner or later we must make a complete transition from oil to other sources. Certainly we can mine more coal, but this will be expensive and could have unacceptable environmental consequences. A mixture of energy options, including coal and nuclear energy, along with renewable energy resources, seems a probable alternative. Of particular importance in the next few years will be the continued growth in direct solar energy as well as continued growth in recycling of urban waste to produce energy.

Hydropower will undoubtedly continue to be an important source of electricity in the future, but it is not expected to grow much because of lack of potential sites and environmental considerations. Wind power and use of solar cells along with various biomass alternatives are more speculative, so it is difficult to predict their growth.

Energy for the future will contine to be troubled by uncertainties. However, what appears to be certain is that we will continue to look more seriously at conservation, energy efficiency, and cogeneration. The most likely targets for energy efficiency and conservation as well as cogeneration are in the area of space heating for homes and offices, water heating, industrial processes that provide heat (mostly steam) for various manufacturing processes, and automobiles. These areas collectively account for approximately 60 percent of the total energy used in the United States today.

We may be at a crossroads today concerning energy policy. The choice is between the "hard path," characterized by centralized, high-technology energy sources, and the "soft path," characterized by decentralized, flexible, renewable energy sources. Perhaps the best path will be a mixture of the old and the new, ensuring a rational, smooth shift from depending so heavily on fossil fuels.

▼ ▼ ▼ REFERENCES

1. U.S. COUNCIL ON ENVIRONMENTAL QUALITY. 1978. *Environmental quality.* Ninth Annual Report of the Council on Environmental Quality.
2. AVERITT, P. 1973. Coal. In *United States mineral resources,* ed. D. A. Brobst and W. P. Pratt, pp. 133–42. U.S. Geological Survey Professional Paper 820.
3. U.S. ENVIRONMENTAL PROTECTION AGENCY. 1973. *Processes, procedures and methods to control pollution from mining activities.* EPA-430/9-73-001.
4. U.S. BUREAU OF MINES. 1971. *Strippable reserves of bituminous coal and lignite in the United States.* U.S. Bureau of Mines Information Circular 8531.
5. NIEMAN, T. J., and MESHAKO, D. 1990. The Star Fire Mine reclamation experience. *Journal of Soil and Water Conservation,* Sept.-Oct., 529–32.
6. COMMITTEE ON ENVIRONMENT AND PUBLIC PLANNING. 1974. Environmental impact of conversion from gas or oil to coal for fuel. *The Geologist,* Newsletter of the Geological Society of America. Supplement to vol. 9, no. 4.
7. McCULLOH, T. H. 1973. Oil and gas. In *United States mineral resources,* ed. D. A. Brobst and W. P. Pratt, pp. 477-96. U.S. Geological Survey Professional Paper 820.
8. FEDERAL ENERGY ADMINISTRATION. 1975. FEA reports on proven reserves. *Geotimes,* August 1975, pp. 22–23.
9. CULBERTSON, W. C., and PITMAN, J. K. 1973. Oil shale. In *United States mineral resources,* ed. D. A. Brobst and W. P. Pratt, pp. 497–503. U.S. Geological Survey Professional Paper 820.
10. DUNCAN, D. C., and SWANSON, V. E. 1965. *Organic-rich shale of the United States and world land areas.* U.S. Geological Survey Circular 523.
11. OFFICE OF OIL AND GAS, U.S. DEPARTMENT OF THE INTERIOR. 1968. *United States petroleum through 1980.* Washington, D.C.: U.S. Government Printing Office.
12. COMMITTEE ON ENVIRONMENTAL AND PUBLIC PLANNING. 1974. Development of oil shale in the Green River Formation. *The Geologist,* Newsletter of the Geological Society of America. Supplement to vol. 9, no. 4.
13. ALLEN, A. R. 1975. Coping with oil sands. In *Perspectives on energy,* ed. L. C. Ruedisili and M. W. Firebaugh. New York: Oxford University Press, pp. 386–96.
14. OFFICE OF TECHNOLOGY ASSESSMENT. 1980. *An assessment of oil shale technologies.* OTA-M-118. 517P.

15. FINCH, W. I. et al. 1973. Nuclear fuels. In *United States mineral resources,* ed. D. A. Brobst and W. P. Pratt, pp. 455–76. U.S. Geological Survey Professional Paper 820.

16. DUDERSTADT, J. J. 1977. Nuclear power generation. In *Perspectives on energy,* ed. L. C. Ruedisili and M. W. Firebaugh. New York: Oxford Univ. Press, pp. 249–273.

17. ENERGY RESEARCH AND DEVELOPMENT ADMINISTRATION. 1978. *Advanced nuclear reactors: An introduction.* ERDA-76-107. 24P.

18. U.S. GEOLOGICAL SURVEY. 1973. *Nuclear energy resources: A geologic perspective.* U.S.G.S. INF-73-14.

19. MacLEOD, G. K. 1981. Some public health lessons from Three Mile Island: A case study in chaos. *Ambio* 10: 18–23.

20. GREENWALD, J. 1986. *Time* 27(19): 39–52.

21. U.S. DEPARTMENT OF ENERGY. 1980. *Magnetic fusion energy.* DOE/ER-0059.

22. U.S. DEPARTMENT OF ENERGY. 1979. *Environmental development plan for magnetic fusion.* DOE/EDP-0052.

23. U.S. DEPARTMENT OF ENERGY. 1978. *The United States magnetic fusion energy program.* DOE/ET-0072.

24. SMITH, R. L., and SHAW, H. R. 1975. Igneous-related geothermal systems. In *Assessment of geothermal resources of the United States—1975,* ed. D. F. White and D. L. Williams, pp. 58–83. U. S. Geological Survey Circular 726.

25. WHITE, D. F. 1969. *Natural steam for power.* U.S. Geological Survey Publication, GP0-19690-339-S36.

26. WHITE, D. F., and WILLIAMS, D. L. 1975. In *Assessment of geothermal resources of the United States—1975,* ed. D. F. White and D. L. Williams, pp. 1–4. U.S. Geological Survey Circular 726.

27. MUFFLER, L. J. P. 1973. Geothermal resources. In *United States mineral resources,* ed. D. A. Brobst and W. P. Pratt, pp. 251–61. U.S. Geological Survey Professional Paper 820.

28. WORTHINGTON, J. D. 1975. *Geothermal development.* Status Report—Energy Resources and Technology, A Report of The Ad Hoc Committee on Energy Resources and Technology, Atomic Industrial Form, Incorporated.

29. TESTER, J. W.; BROWN, D. W.; and POTTER, R. M. 1989. *Hot dry rock geothermal energy—A new energy agenda for the 21st century.* Los Alamos National Laboratory. LA–11514–MS, UC–251. 30.

30. COLP, J. L. 1974. *Magma tap—the ultimate geothermal energy program.* Sandia Laboratories Energy Report, SAND74-0253.

31. COLP, J. L. et al. 1975. *Magma workshop: Assessment and recommendations.* Sandia Laboratories Energy Report, SAND75-0306.

32. PAPADOPULUS, S.S. et al. 1975. Assessment of onshore geopressured-geothermal resources in the northern Gulf of Mexico basin. In *Assessment of geothermal resources of the United States,* ed. D. F. White and D. L. Williams, pp. 125–46. U.S. Geological Survey Circular 726.

33. COUNCIL ON ENVIRONMENTAL QUALITY. 1979. *Environmental quality,* Tenth Anual Report of the Council on Environmental Quality.

34. LOVINS, A. B. 1979. Foreword in K. Butti and J. Perlin, *A golden thread.* Palo Alto, Calif.: Cheshire Books.

35. EATON, W. W. 1976. Solar energy. In *Perspectives on energy,* 2d ed., ed. L. C. Ruedisili and M. W. Firebaugh. New York: Oxford University Press, pp. 418–50.

36. ENERGY INFORMATION ADMINISTRATION. 1990. *Annual Energy Review 1989.*

37. RALOFF, J. 1978. Catch the sun. *Science News* 113, no. 16.

38. JOHNSON, J. T. 1990. The hot path to solar electricity. *Popular Science,* May, 82–85.

39. ALWARD, R.; EISENBART, S.; and VOLKMAN, J. 1979. *Micro-hydro power: Reviewing an old concept.* National Center for Appropriate Technology. U.S. Department of Energy.

40. COMMITTEE ON RESOURCES AND MAN, NATIONAL ACADEMY OF SCIENCE. 1969. *Resources and man.* San Francisco: W. H. Freeman.

41. NOVA SCOTIA DEPARTMENT OF MINES AND ENERGY. 1981. *Wind power.*

42. WEINBERG, C. J., and WILLIAMS, R. H. 1990. Energy from the sun. *Scientific American* 263 (3): 147–55.

43. OFFICE OF TECHNOLOGY ASSESSMENT. 1980. *Energy from biological processes,* OTA-E-124.

44. DAVIS, G. H., and WOOD, L. A. 1974. *Water demands for expanding energy development.* U.S. Geological Survey Circular 703.

45. DARMSTADTER, J.; LANDSBERG, H. H.; MORTON, H.C.; with CODA, M. J. 1983. *Energy today and tomorrow.* Englewood Cliffs, N.J.: Prentice-Hall.

46. STEINHART, J. S.; HANSON, M. E.; GATES, R. W.; DEWINKEL, C. C.; BRIODY, K.; THORNSJO, M.; and KABALA, S. 1978. A low energy scenario for the United States: 1975–2050. In *Perspectives on energy,* ed. L. C. Ruedisili and M. W. Firebaugh, pp. 553–88. New York: Oxford University Press.

47. LOVINS, A. B. 1979. *Soft energy paths: Towards a durable peace.* New York: Harper & Row.

48. HAFELE, W. 1990. Energy from nuclear power. *Scientific American,* 263 (3): 137–44.

49. DAVIS, G. R. 1990. Energy for planet Earth. *Scientific American* 263 (3): 55–74.

Change is a natural characteristic of our environment, and wise management accepts this fact. Chapter 16 presents principles of global change and earth systems science that foster better land use and decision making.

One of the controversial environmental issues today is determining the most appropriate uses of our land and resources. The controversy is a human issue that involves responsibility to future generations as well as compensation for landowners today. Before we decide how to plan the use of the land and its resources, we must develop reliable methods to evaluate the land and develop a legal framework within which we can make sound environmental decisions. Chapter 17 discusses several aspects of landscape evaluation such as land-use planning, site selection, landscape aesthetics, and environmental impact. Chapter 18 introduces the relatively new and expanding field of environmental law. Although environmental law is obviously not geological, there is a real need for applied earth scientists to know something about our legal system and the laws and legal theory that affect the natural environment.

PART FIVE

▼
▼
▼

Global Change, Land Use, and Decision Making

Photo by Ken Graham/Allstock

CHAPTER SIXTEEN

▼
▼
▼

Global Change, Earth Systems Science, and Urban Air

Preston Cloud, a famous scientist interested in the history of life on earth, the impact of human activity on our environment, and the use of resources, recently stated that two central goals of the earth sciences are:

> To understand how the earth works and how it has evolved from a landscape of barren rock to the complex landscape dominated by the life we see today.
>
> To apply that understanding to better manage our environment (1).

Cloud's comments emphasize that the earth is a planet characterized by a complex evolutionary history.

Interactions among the atmosphere, oceans, solid earth, and biosphere have resulted in development of a complex and abundant diversity of landforms at all scales, such as continents, ocean basins, mountains, lakes, plains and slopes; and life forms that inhabit a broad spectrum of environmental habitats. Of particular importance are interactions among physical, biological, and chemical processes that allow for the abundance and variety of life on earth. Early in this book we put forward the principle that the earth is a dynamic evolving system. We now return to that theme.

The concept of global change is not new or unique. Geologists have for over 100 years studied the history of the earth, and the changes that have occurred during its 4.6 billion years of history. Society today is very concerned with real and potential changes in our atmosphere that are taking place over a period of several decades, such as increase in carbon dioxide, hypothetically related to global warming, and reduction in the ozone layer that will affect our health.

Study of the earth's history suggests that the atmosphere has been changing since the earth first formed. Figure 16.1 shows an idealized cross section of the earth and atmosphere as it may have been during early Precambrian time, over three billion years ago. Volcanic outgassing produced water (H_2O) and carbon dioxide (CO_2), and in the upper atmosphere photodissociation produced an ozone (O_3) layer. At that time, however, the earth's atmosphere contained very little free molecular oxygen (O_2). Development of our oxygen-rich atmosphere was to wait about one billion years until photosynthesis by plants was widespread (2). When looking at changes in the global system we are often concerned with those that affect people today. Changes reported as first-order variations occur on a time scale of ten years to one hundred years and thus can be observed within a life time. Other important changes, sometimes termed second-order variability, can be detected during the historic record, which in some areas is several thousand years. Third-order variability relates to time periods of tens of thousands of years that may, for example, correlate with the phases of an ice age. Longer periods of change, such as fourth- or fifth-order variability, are related to the time necessary for an entire ice age or a major glacial event, which is on the order of approximately two to three million years (3). It is emphasized, however, that it is the first-order and second-order changes that are of most significance to people today. These are the time ranges over which we must find a way to plan and work with potential change.

In order to better understand global change and earth systems science (the science of studying the earth as a system) we will discuss: 1. goals for understanding global change, and earth systems science; 2. earth's atmosphere and energy budget; 3. the tools by which we may study change; 4. increase in atmospheric carbon dioxide and the greenhouse effect; 5. stratospheric ozone depletion; 6. acid deposition; and 7. urban air pollution. Following these topics we will end the chapter with a discussion of air pollution and the urban microclimate. This section is included because, as the earth becomes more populated, our urban areas are growing rapidly; as a result, the urban air has and is changing rapidly.

▼ GLOBAL CHANGE AND EARTH SYSTEMS SCIENCE

Until recently it was generally thought that human activity caused only local, or at most, regional environmental change. It is now generally recognized that the effects of human activity on the earth are of such an extent that we are involved in an unplanned planetary experiment. The main goal of the emerging science known as earth systems science is to obtain a basic understanding of how

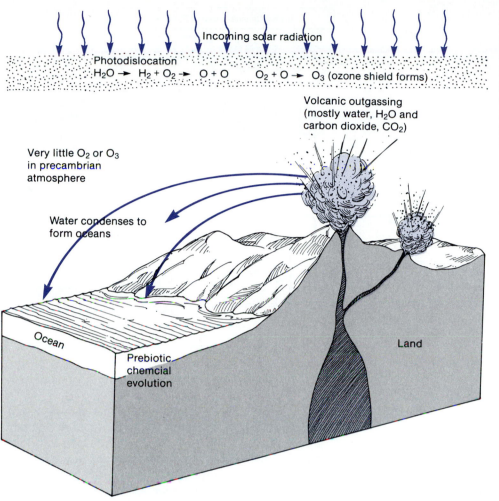

Incoming solar radiation

Photodislocation
$H_2O \rightarrow H_2 + O_2 \rightarrow O + O$ $O_2 + O \rightarrow O_3$ (ozone shield forms)

Volcanic outgassing
(mostly water, H_2O and
carbon dioxide, CO_2)

Very little O_2 or O_3
in precambrian
atmosphere

Water condenses to
form oceans

Ocean

Prebiotic
chemcial
evolution

Land

Figure 16.1
Idealized diagram showing selected global processes in the
early Precambrian (about 3 billion years ago). (Modified after
S. Kershaw, 1990, Evolution of the
earth's atmosphere and its geologic impact, *Geology Today,*
March-April, pp. 55–60. Blackwell
Scientific Publications Limited.)

the entire earth as a system works. Meeting this goal involves understanding on a global scale how the various components, such as the atmosphere, oceans, solid earth, and biosphere, are linked and interact to affect life on earth. Of particular importance is how the various components have evolved, presently function, and can be expected to function and evolve in the future, at a time scale significant to people (4). A major secondary goal is prediction of global changes that are likely to occur within the next few decades to a century. Of particular importance will be the distinction between human-induced changes, natural changes, and interactions be-tween the two (4).

Research priorities in studying global change are related to the various components, their linkages, and interactions. Figure 16.2 shows some of the major components that must be studied, including: solid earth, atmospheric chemistry, carbon cycle, hydrologic cycle, heat transport, and clouds and radiation. In many complex ways, these components interact with physical, chemical, and biological processes that may cause global change.

In order to reach the goals of earth systems science it is necessary to establish research priorities. These have been summarized (5) as:

▼ Establishment of worldwide measurement stations to better understand the physical, chemical, and biolog-ical processes significant for the evolution of our planet on a variety of time scales.
▼ Documentation of global changes, especially those changes that are of a time period of particular interest to people.
▼ Development of quantitative models that may be used to predict and anticipate future global change.
▼ Assistance in the gathering of essential information for decision making at the regional (national and international) and global levels.

▼ COMPOSITION OF THE ATMOSPHERE

Atmosphere

Our atmosphere can be thought of as a complex chemical factory with many little-understood reactions

Figure 16.2
Research areas and priorities for global change. (Source: NASA, 1990. *EOS: A Mission to Planet Earth.*)

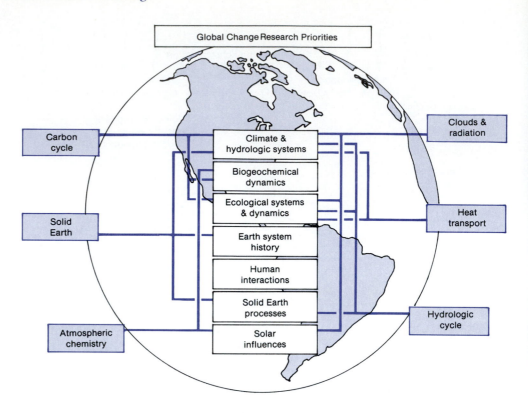

taking place within it. Many of the reactions that take place are strongly influenced by sunlight, and the compounds produced by life. The air we breathe is a mixture of nitrogen (78 percent), oxygen (21 percent), argon (0.9 percent), carbon dioxide (0.03 percent), and other trace elements (less than 0.07 percent), along with compounds such as methane, ozone, carbon monoxide, oxides of nitrogen and sulfur, hydrogen sulfide, hydrocarbons, and various particulates. The most variable part of the atmosphere's composition is water vapor (H_2O), which can range from approximately 0 to 4 percent by volume in the troposphere (the lower ten kilometers of the atmosphere) (7).

Energy Balance

Although the earth intercepts only a very tiny fraction of the total energy emitted by the sun, the sun is responsible for sustaining life on earth while driving many processes at or near the surface of the earth, including the hydrologic cycle, the waves in the ocean, and the circulation of air on a global scale. Figure 16.3 shows some of the important components of the earth's energy budget. Essentially, all of the energy that is available at the surface of the earth comes from the sun. Geothermal heat generated from the interior of the earth is only a very small fraction of one percent of the earth's energy budget. Nevertheless, it is this internal heat that drives the great global tectonic cycle and moves the tectonic plates of the lithosphere, thus generating earthquakes and creating volcanoes.

A modest acquaintance with the earth's energy budget and solar radiation is critical in understanding global processes and environmental problems associated with the atmosphere, such as the greenhouse effect and ozone depletion. At the planetary scale, the earth may be considered a part of a larger solar energy system. The earth receives energy from the sun, and this energy undergoes changes while affecting life, oceans, atmosphere, and land. The energy then is eventually emitted as heat back into the depths of space as terrestrial radiation.

Energy from the sun travels to the earth at the speed of light (approximately 300,000 kilometers per second) through the vacuum of space. Energy emitted from the sun is in the form of electromagnetic radiation. Different forms of electromagnetic radiation are distinguished by their wavelengths, and the collection of all possible wavelengths forms a continuous range known as the electromagnetic spectrum (Figure 16.4). The relatively long wavelengths (greater than 1 meter) include radio waves, and the shortest waves are X rays and gamma rays. Electromagnetic radiation to which our eyes respond is known as the visible electromagnetic radiation, and is only a very small fraction of the total spectrum (Figure 16.4). Other types of electromagnetic radiation with environmental significance include microwaves (with industrial uses), and ultraviolet radiation (important when considering ozone depletion).

The amount of energy per second radiated from a body such as the sun or earth varies with the fourth power of the temperature of that body. Therefore, if a

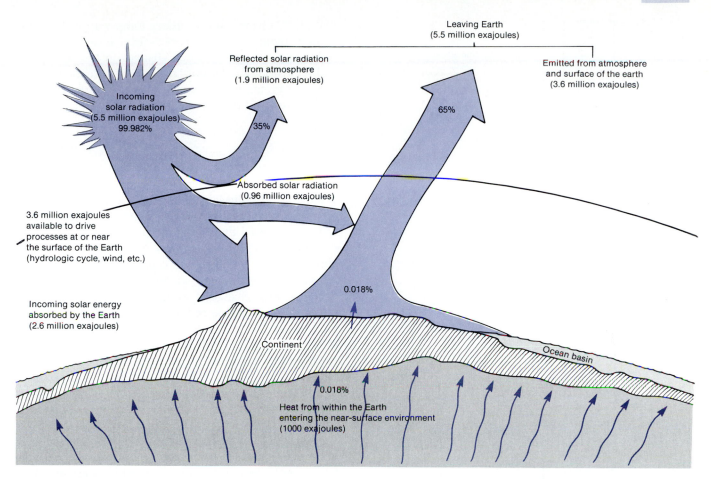

Figure 16.3
Annual energy flow to the earth from the sun. Also shown is the relatively small component of heat from the earth's interior to the near-surface environment. (Modified after Marsh and Dozier, *Landscapes: An Intro to Physical Geography.* Copyright © 1981. John Wiley & Sons, Inc. Reprinted by permission of John Wiley & Sons, Inc.)

body's temperature doubles, the energy radiated increases by sixteen times. This fact explains why the sun, with a surface temperature of 5800°C, radiates so much more energy than the earth at 15°C (Figure 16.5). Another important point shown in Figure 16.5 is that the sun's radiant energy is predominantly short-wave, compared to terrestrial radiation, which has a relatively long wavelength. The principle here is that the hotter an object, the more rapidly it radiates heat, and the shorter the wavelength of the predominant radiation. The surface of the earth, which includes the surfaces of plants, clouds, water, rocks, and animals, is so cool that heat is radiated predominantly in the infrared wavelengths, which are invisible to humans (8, 9).

When electromagnetic radiation encounters a material it may be transmitted, reflected, or absorbed (9). Which of these processes occurs can have important environmental ramifications. For example, the absorption of a portion of the ultraviolet radiation in the upper atmosphere by ozone (O_3) protects life at the surface of the earth. On the other hand, most visible radiation from the sun passes directly through the atmosphere without being absorbed, although some visible radiation is reflected by clouds back into space. It is not unusual for transmission, absorption, and reflection to all take place at the same time. For example, consider a skylight in a modern home. If you look at the skylight from outside, you may see a sparkle of light coming from it, indicating that some of the incoming solar radiation is reflected. If you touched the glass, it may feel warm, indicating some of the energy has been absorbed. If you feel warmth in the room below, you may conclude that some of the solar energy has been transmitted through the skylight (9).

The rate at which solar energy is either absorbed or reflected at the earth's surface depends, in part, upon the surface temperature and color of the material involved. If an object at the surface is cold and gives off less heat than it receives, it will warm up, but as it warms, it will also radiate heat more rapidly. As a result, for a constant input of energy, a physical object will eventually reach a temperature that will allow it to radiate heat energy at the same rate that it receives energy. This is why the earth receives approximately 5.5 million exajoules from the sun, but also loses the same amount of energy, although at different dominant wavelengths (Figure 16.5).

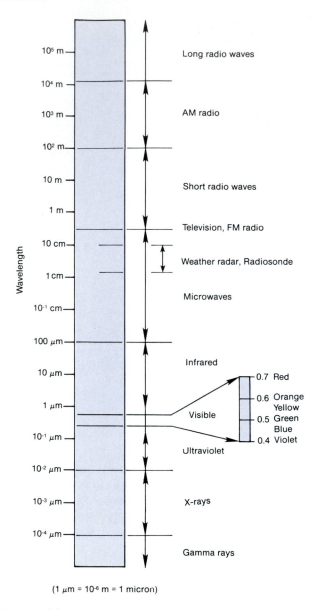

(1 μm = 10⁻⁶ m = 1 micron)

Figure 16.4
The electromagnetic spectrum.

White surfaces tend to reflect solar radiation (light) rather than absorbing and radiating heat. On the other hand, dark or black surfaces absorb and radiate heat much more readily. For example, ice reflects 80 to 95 percent of light, dry grassland 30 to 40 percent, and coniferous forest 5 to 15 percent. Ice that contains a good deal of rock and other dark material at the surface, however, will reflect much less light, and the dark objects will tend to absorb heat, which may partially melt the ice. This explains why darker glacial ice tends to be topographically lower than adjacent white ice.

It has been estimated that, given the present atmospheric conditions, a 1 percent change in the amount of sunlight reflected by the earth would cause an approxi-

mate 1.7°C change in the surface temperature. Thus, the surface temperature is very sensitive to change in reflectivity or *albedo*. Actual changes in the absorption and reflection of solar energy are due to such factors as cloud cover, amount of ice at the earth's surface, type of plant cover, and presence of water (7, 9).

▼ TOOLS FOR STUDYING GLOBAL CHANGE

The Geologic Record

Sediments deposited on floodplains or in lakes, bogs, or the ocean may be compared to pages of a history book. Organic material that is often deposited with sediment may be dated by a variety of methods to provide a chronology. In addition, the organic material itself can tell a story concerning the past climate, what lived in the area, and what changes have taken place.

Studies of marine sediments have helped delineate the history of ocean water temperature as well as biological and chemical changes occurring in the ocean basins over millions of years. Climatic information gleaned from ocean sediments has helped to establish the climatic change during the Quaternary (last 2 million years) in terms of the waxing and waning of continental ice sheets.

Study of river sediments has helped to delineate past flood events and to determine the probability of future high-magnitude floods. This is possible because, when the floods occur, sediment is deposited along with organic material, which may be dated by utilizing the carbon-14 (radiocarbon) method. The carbon-14 dating method utilizes the activity of radiogenic carbon-14, which undergoes radioactive decay at a known rate. All living things contain carbon-14, and when an organism dies it stops collecting carbon, which starts the atomic clock ticking. The method is good back to about 40,000 to 50,000 years before present. This technique has been used at Pallett Creek, in southern California, where the earthquake history for the past 2,000 years has been delineated through carbon-14 dating of organic material mixed with the flood deposits that have been disturbed by prehistoric earthquakes.

Changes in vegetation have also been studied by examining lake and floodplain deposits. A recent study (10) utilized analysis of pollen in floodplain deposits to establish the change in vegetation during the past 12,500 years in northeastern Iowa. That study used numerous carbon-14 dates to show that spruce forests present prior to 9,100 years ago, were followed by prairie vegetation from 5,400 to 3,500 years ago. The prairie vegetation was replaced about 3,500 years ago by oak savanna, which has been present until presettlement times. This type of information is very important in establishing climatic conditions in the past, and thus regional and perhaps even global change.

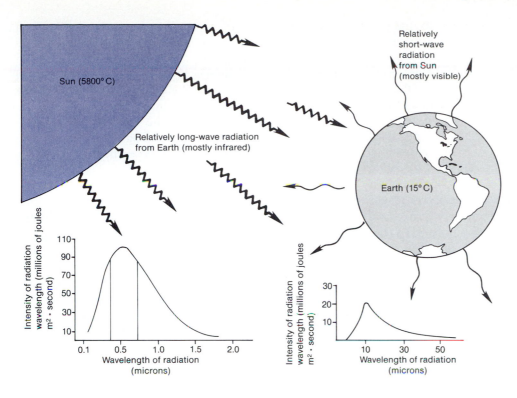

Figure 16.5
Idealized diagram comparing the emission of energy from the sun with that from the earth. Notice that the solar emissions have a relatively short wavelength, whereas those from the earth have a relatively long wavelength. (Modified after Marsh and Dozier, *Landscapes: An Intro to Physical Geography.* Copyright © 1981. John Wiley & Sons, Inc. Reprinted by permission of John Wiley & Sons, Inc.)

One of the more interesting uses of the geologic record has been the examination of glacial ice. Glacial ice contains trapped air bubbles that may be analyzed to provide information concerning atmospheric carbon dioxide (CO_2) concentrations when the ice formed. The trapped air bubbles in a long core of glacial ice are time capsules of the atmosphere from the past. This method has been used to analyze the carbon dioxide content of air as old as 160,000 years (11).

Another method of evaluating past earth history is *dendrochronology,* which is the study of tree rings. Many trees develop an annual growth ring, so counting and correlating the rings can establish a history of the area. The width of individual rings and other information can provide insights into the hydrologic and climatic conditions at the site. For example, careful analysis of the tree rings can reveal frequency of cycles of drought and wet years. This method has helped establish climatic changes during the past 12,000 years in many parts of the world.

Real-Time Monitoring

Monitoring is the regular collection of data for a specific purpose. For example, we may monitor the flow of water in rivers to evaluate water resources or flood hazard. In a similar way, samples of atmospheric gases can help establish trends or changes in the composition of the atmosphere, and measurements of temperature composition of the ocean may be used to look at changes there.

Methods of monitoring vary widely depending upon what is being measured. For example, deforestation may be monitored by evaluating remotely sensed data col-lected by satellite, or high-altitude aerial photographs taken by cameras mounted on airplanes. In particular, remote sensing utilizing satellite information has been useful in monitoring at the global level. Nevertheless, the best measurements are coupled with ground measurement to establish the validity of the airborne or satellite measurements. Gathering of real-time data is necessary for testing models and calibrating the extended prehistoric record derived from geological data.

Some of the most interesting evaluations concerning changes in ecosystems have resulted from long-term studies that gather specific data concerning interactions between physical and biological processes. For example, long-term studies of nutrient cycling in ecosystems of the eastern United States have helped establish principles of management for timber harvesting and other activities related to human use and interest in the land.

Mathematical Models

Mathematical models attempt to represent, through numerical means, real-world phenomena, and the linkages and interactions among processes operating. Models have been developed to predict flow of surface water and groundwater, erosion and deposition of sediment in river systems, ocean circulation, and atmospheric circulation. The models that have gained the most attention in terms of global change are the Global Circulation Models (GCM). The objective of the Global Circulation Models is to predict atmospheric changes (circulation) at the global scale (12). Variables used in the model include, among others: temperature, relative humidity, and wind conditions. Many of these variables are estimated for the past based on proxies, such as the tree-ring records discussed earlier. Data used in the calculations are arranged into large cells that are several degrees of latitude and longitude (typical cells measure on the order of the size of Oregon or Indiana and Ohio together). In addition, there are usually 6 to 20 levels of vertical data throughout the lower atmosphere. Equations that conserve mass energy and momentum are then used in the calculations to make predictions into the future. Global Circulation Models are complex and require supercomputers for their operation (12). Unfortunately, however, results from Global Circulation Models are relatively crude, and may not accurately define future conditions (13). The models are inaccurate in part because methods used to calculate variables such as temperature, relative humidity, and wind conditions over a very large region (the cell) may have unrealistic assumptions. Also, other variables such as cloud covers are difficult to estimate. These models must be viewed as a first approach to solving very complex problems. In spite of these limitations, the models provide information necessary for evaluating the earth as a system, and in pointing out which additional data are necessary to develop better models in the future. The models do predict in a relative sense which areas are likely to be wetter or drier if certain changes in the atmosphere occur. Thus, predictions are being taken seriously.

▼ INCREASING ATMOSPHERIC CO_2 AND GLOBAL WARMING

The temperature of the earth is determined, for the most part, by three factors: 1. the amount of sunlight the earth receives; 2. the amount of sunlight the earth reflects, and 3. atmospheric retention of heat (12). Sunlight that reaches the earth warms the atmosphere and surface, and then the earth's atmospheric system reradiates the heat as infrared radiation (12). Water vapor and several other gases (including carbon dioxide, methane, and chlorofluorocarbons) tend to trap heat and warm the earth because they absorb some of the heat energy radiating from the earth. This trapping or warming is somewhat analogous to a greenhouse which also traps heat. Most of the warmer air in the greenhouse is trapped because of reduced cooling by air circulation (wind) with a small amount due to trapped infrared radiation. The trapping of heat by the atmosphere is generally referred to as the *Greenhouse Effect*. It is important to understand that the effect is a natural phenomena that has been occurring for millions of years on earth as well as on other planets in our solar system. Were it not for the trapping of heat in the atmosphere, the earth would be approximately 33°C cooler than it is now, and all surface water would be frozen (14). The majority of natural "greenhouse" warming is due to water vapor in the atmosphere. However, potential global warming by anthropogenic processes is

related to carbon dioxide, methane, nitrous oxides, and chlorofluorocarbons. In recent years the atmospheric concentrations of these gases and others have been increasing due to human activities. These gases tend to absorb infrared radiation from the earth, and it has been hypothesized that the earth may be warming due to the anthropogenic increases in the "greenhouse gases." Figure 16.6 is a highly idealized diagram showing some important aspects of the greenhouse effect.

Changes in Greenhouse Gases

The major "greenhouse gases," in terms of their contribution to the anthropogenic portion of the greenhouse effect, are shown in Table 16.1. Notice that 60 percent of the anthropogenic greenhouse effect is attributed to carbon dioxide (CO_2). Measurements of carbon dioxide trapped in air bubbles of the Antarctic ice sheet suggest that during the past 160,000 years the atmospheric concentration of carbon dioxide has varied from a little less than 200 ppm to about 300 ppm (11). The highest levels are recorded during interglacial periods that occurred about 125,000 years ago and at the present. About 130 years ago, at the beginning of the industrial revolution, the atmospheric concentration of carbon dioxide was approximately 280 ppm, a level that was constant for at least the previous 700 years (11). Since about 1860, there has been an exponential growth of the concentration of carbon dioxide in the atmosphere. The change from approximately the year 1500 to 1990 is shown in Figure 16.7. Data prior to the mid-twentieth century are from measurements made from air bubbles trapped in glacial ice. The concentration of carbon dioxide in the atmosphere today is nearly 350 ppm, and is predicted to be at least 450 ppm (which is more than 1.5 times the preindustrial level) by the year 2050 (15). Figure 16.8 shows average annual concentrations of carbon dioxide in the atmosphere (in ppm) from 1958 to 1986, as well as global carbon dioxide emissions (by weight) for the same time period. It is obvious from these data that as carbon dioxide emissions have increased, so has the average concentration of carbon dioxide in the atmosphere. The rate of increase of carbon emissions from burning of fossil fuels is approximately 4.3 percent per year since the industrial revolution began. Given this rate, it would seem easy to suggest that the increase in carbon dioxide in the atmosphere since the industrial revolution is all due to human-induced processes such as burning of fossil fuels and deforestation. Unfortunately, proving this relationship has been difficult. The global carbon cycle is very complex, and the linkages and flow of carbon from the various natural sources and sinks are not well known. If all the carbon dioxide produced by human activities were to remain in the atmosphere, then the concentration should be even higher than it is today. Therefore, there must be sinks of carbon dioxide, in the oceans or on land, that are not well understood. In spite

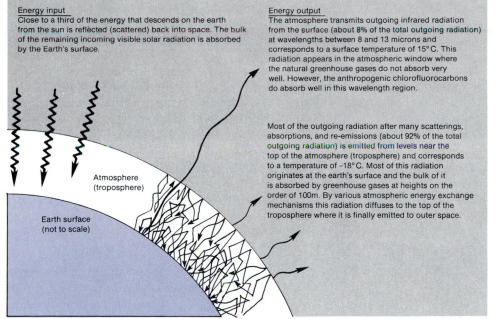

Energy input
Close to a third of the energy that descends on the earth from the sun is reflected (scattered) back into space. The bulk of the remaining incoming visible solar radiation is absorbed by the Earth's surface.

Energy output
The atmosphere transmits outgoing infrared radiation from the surface (about 8% of the total outgoing radiation) at wavelengths between 8 and 13 microns and corresponds to a surface temperature of 15°C. This radiation appears in the atmospheric window where the natural greenhouse gases do not absorb very well. However, the anthropogenic chlorofluorocarbons do absorb well in this wavelength region.

Most of the outgoing radiation after many scatterings, absorptions, and re-emissions (about 92% of the total outgoing radiation) is emitted from levels near the top of the atmosphere (troposphere) and corresponds to a temperature of −18°C. Most of this radiation originates at the earth's surface and the bulk of it is absorbed by greenhouse gases at heights on the order of 100m. By various atmospheric energy exchange mechanisms this radiation diffuses to the top of the troposphere where it is finally emitted to outer space.

Atmosphere (troposphere)

Earth surface (not to scale)

Figure 16.6
Idealized diagram showing greenhouse effect. Incoming visible solar radiation is absorbed by the earth's surface to be remitted in the infrared region of the electromagnetic spectrum. Most of this remitted infrared radiation is absorbed by the atmosphere, maintaining the greenhouse effect. (Developed by M. S. Manalis and E. A. Keller, 1990).

of these shortcomings, it is clear that carbon dioxide concentration in the atmosphere has significantly increased since the industrial revolution, and this increase may contribute to a global warming via the greenhouse effect.

Changes in other greenhouse gases are shown in Figure 16.9. Notice that chlorofluorocarbons and methane have also significantly increased since 1974. According to Table 16.1 these other gases account for approximately 40 percent of the anthropogenic greenhouse effect (16).

Methane is thought to contribute approximately 15 percent of the anthropogenic greenhouse effect. There is a controversy, however, concerning the sources and sinks of methane in the atmosphere and much more work needs to be done in this area. In spite of the uncertainties, it is fairly well understood that methane gas from anthropogenic sources is related to burning of biomass, production of coal and natural gas, and agricultural practices such as growing rice and raising cattle and sheep. The methane related to rice production is associated with anaerobic activity in flooded lands. Methane from cattle and sheep results from digestive processes and expulsion of gas by the animals. Over the past two decades methane concentrations in the atmosphere have been increasing rapidly, and to date that rate of increase is about 1 percent per year (3).

Approximately 12 percent of the anthropogenic greenhouse effect may be related to chlorofluorocarbons in the atmosphere (16). Furthermore, atmospheric concentrations of chlorofluorocarbons are increasing more rapidly than either carbon dioxide or methane. Because

Table 16.1
Rate of increase and relative contribution of several gases to the anthropogenic greenhouse effect.

	Rate of increase (% per year)	Relative contribution (%)
CO_2	0.5	60
CH_4	1	15
N_2O	0.2	5
O_3*	0.5	8
CFC-11	4	4
CFC-12	4	8

*In the troposphere.

Source: Data from H. Rodhe, 1990. A comparison of the contribution of various gases to the greenhouse effect. *Science* 248, 1218, Table 2. Copyright 1990 by the AAAS.

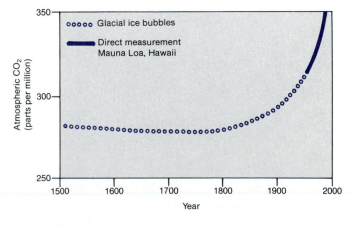

Figure 16.7
Average concentration of atmospheric carbon dioxide 1500–1990. (Data from W. M. Post et al., 1990, *American Scientist* 78[4]:210–26.)

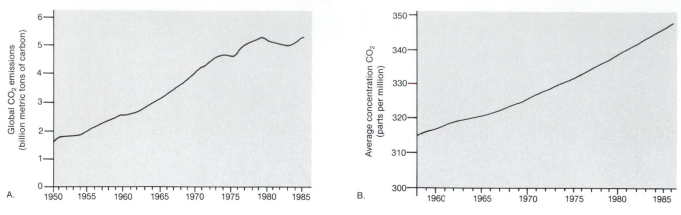

Figure 16.8

(a) Global emissions of carbon dioxide (CO_2) 1950–1986. (b) Average concentration of atmospheric CO_2 1958–1986. (After Council on Environmental Quality, 1990, *Environmental Trends 1989.*)

chlorofluorocarbons are highly stable compounds, their residence time in the atmosphere is long. The rate of increase of chlorofluorocarbons is about 5 percent per year, and even if production of these chemicals were drastically reduced, the elevated concentrations would be with us for many years, particularly in the stratosphere, where they may persist for more than a century (3).

Nitrous oxide (N_2O) is also increasing in the atmosphere, and may be contributing approximately 5 percent to the anthropogenic greenhouse effect (12). Most of the anthropogenic nitrous oxide results from the application of fertilizers in agricultural activities; another contributor is the burning of fossil fuels. Reduction in burning of fossil fuels and reduced use of fertilizers could reduce the emission of nitrogen dioxide into the atmosphere. However, nitrous oxide has a long residence time, and so if rates of emission are stabilized, elevated concentrations of the gas could persist for at least several decades (3).

Global Change in Temperature

During the past million years there have been major climatic changes involving swings in the world's mean annual temperature of several degrees celsius. The Pleistocene ice ages began approximately 2 million years ago, and since then there have been numerous changes in the earth's mean annual temperature (7). On a scale of 1 million years, times of high temperature reflect interglacial or ice-free periods, whereas times of low temperature times reflect major glacial events. As shown in Figure 16.10, on a scale of 150,000 years interglacial and glacial events become more prominent. On a scale of 1,500 years, several warming and cooling trends that have affected people are shown. For example, a major warming trend from about A.D. 800 to 1200 allowed the Vikings to colonize Iceland, Greenland, and northern North America. When glaciers made a minor advance during a cold period, known as the Little Ice Age, around A.D. 1400, the Viking settlements in North America and parts of Greenland were abandoned. Since approximately 1750,

an apparent warming trend lasted until approximately the 1940s when temperatures cooled slightly (7). At an even shorter time period, more changes are apparent, and the 1940s event is clearer (3). What is evident from the record from 1905 to 1985 is that global mean annual temperature has probably increased by approximately 0.5°C. Unfortunately, due to the complexities of the carbon cycle and the greenhouse effect, future warming trends remain uncertain (17). Nevertheless, projections of global warming are shown in Figure 16.11. The tremendous range in potential increase in temperature reflects the inadequacy of the various computer models to predict future climatic change. Nevertheless, all of the models predict that warming will occur, and probably accelerate, in the coming decades (17).

Potential Effects of Global Warming

If the greenhouse gases double in the future, it is estimated that the average global temperature will rise about 1.2°C (15), with significantly greater warming at the polar regions. Specific effects of this temperature rise are difficult to predict, but two of the main uncertainties being considered are change in global climate pattern, and sea level rise due to thermal expansion of warmer sea water and partial melting of glacial ice.

Global rise in temperature might significantly change rainfall patterns, soil moisture relationships, and other factors important in climate, and more specifically agriculture. It has been predicted that some more northern areas such as Canada and the U.S.S.R. may become more productive, whereas lands to the south will become more arid. It is emphasized that such predictions are very difficult in light of the uncertainties surrounding global warming. Actually, to some extent, the uncertainty is what makes people nervous. Global agricultural activities that are stable or expanding are crucial to people throughout the world who depend upon the food grown in the major grain belts. Hydrologic changes associated with climatic change resulting from global warming might seriously affect food supplies at the global level.

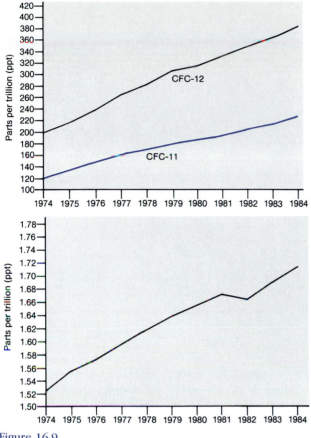

Figure 16.9

Average global concentrations (1974–1984) of atmospheric chlorofluorocarbons (a) and methane (b). (From Council on Environmental Quality, 1990, *Environmental Trends 1989.*)

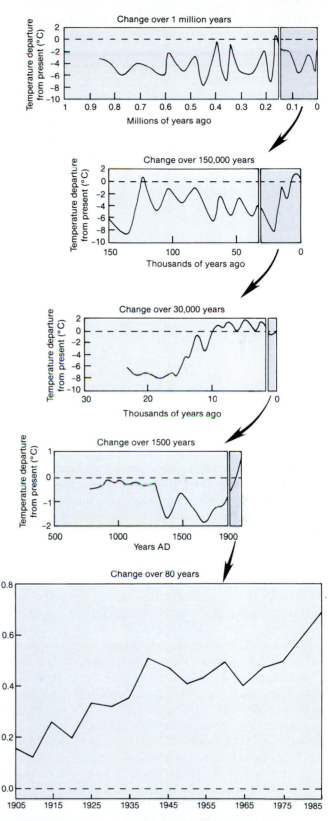

Figure 16.10

Change in temperature over different periods of time during the last million years. (From Committee for the Global Atmospheric Research Program, National Academy of Sciences, 1975; Marsh and Dozier, *Landscape,* © 1981, Figure 9.3; and Council on Environmental Quality, 1990, *Environmental Trends 1989.* Reprinted with permission of the National Academy of Sciences and Addison-Wesley, Reading, MA.)

Global warming might also change the frequency and intensity of violent storms, and this change may be more important than what areas become wetter, drier, hotter, or cooler. Warming oceans could feed more energy into high-magnitude storms such as hurricanes. More hurricanes or larger hurricanes would increase the hazard of living in low-lying coastal areas, many of which are experiencing rapid growth of human population.

Although there is an emerging general consensus that global warming may be occurring, studies are just beginning to try to verify if effects of warming can be recognized. The U.S. Geological Survey has initiated a pilot study of the Delaware River basin. A major objective is to evaluate what data requirements are necessary for verifying potential impacts of climate change (12). The Delaware River is a good study site because the basin crosses four distinct physiographic provinces (Figure 16.12), and may be investigated in terms of hydrologic response to existing management strategies. Questions related to storage of water for New York City, and ability to maintain stream flow requirements and control migration of salt water in the Delaware estuary associated with change in sea level, will all be monitored. The Delaware River basin does not, however, encompass nearly the full range of hydrologic variability, and so other basins across

Figure 16.11
Historical trends and range of predicted warming from mathematical models. All models predict substantial warming. (Adapted from P. D. Jones and T. M. L. Wigley, 1990 (modified) by John Deecken from "Global Warming Trends." *Scientific American* 263 [2]:84–91.) Copyright © 1990 by Scientific American, Inc. All rights reserved.

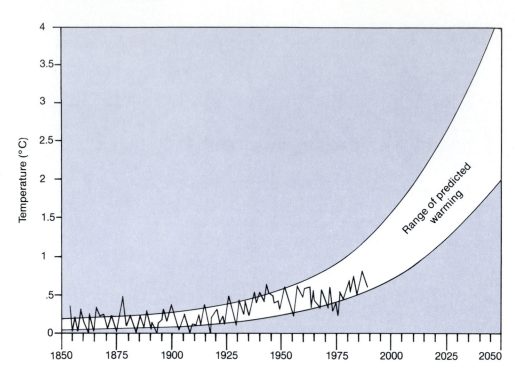

the country will also be studied as potential indicators of climatic change (12).

Rise of sea level is a potentially serious problem related to global warming. Estimates of the rise expected in the next century vary widely from approximately 40 to 200 centimeters, and precise estimates are not possible at this time. However, a 40-centimeter rise in sea level would have significant environmental impacts. Such a rise could cause increased coastal erosion on open beaches of up to 80 meters, rendering buildings and other structures more vulnerable to waves generated from high-magnitude storms. In areas characterized by coastal estuaries, a 40-centimeter sea level rise would cause a landward migration of existing estuaries, again putting pressure on human-built structures in the coastal zone (15). A rise of approximately 1 meter would have very serious consequences; significant alterations would be needed to protect investments in the coastal zone. Communities would have to choose between making very substantial investments in controlling coastal erosion, or allowing beaches and estuaries to migrate landward over wide areas (15). Coastal erosion is a serious problem today in many parts of the world, and substantial increase in sea level could double the current rate in some areas (15). It seems unavoidable that a substantial rise in sea level must lead to big investment for protecting cities in the coastal zone. Building of dikes, sea walls, and other erosion control structures will become common practice, and coastal erosion will continue to threaten urban property. In some areas where coastal development has been greatly restricted, erosion may proceed with little consequence to people. Because coastal erosion is so difficult to deal with, it seems prudent to allow erosion to continue wherever feasible, and fight it only where absolutely necessary.

Strategies To Reduce Potential Impacts of Global Warming

If the hypothesis is true that global warming is due, in part, to increase in the "greenhouse gases," then reduction of these gases in the atmosphere is a primary management strategy. Much of the carbon dioxide emissions from anthropogenic sources are related to burning of fossil fuels; therefore, energy planning that relies more heavily on alternative energy sources, such as wind power, solar, or geothermal, will reduce emissions of carbon into the atmosphere. A change to greater use of nuclear energy would also reduce the carbon loading in the atmosphere.

Another source of carbon dioxide in the atmosphere is related to deforestation. Burning of forest lands to convert them to agricultural lands puts a lot of carbon into the atmosphere. Management plans to protect the world's forests would be another strategy to help reduce the potential threat of global warming.

In summary, better management of the production of "greenhouse gases" would ultimately lead to a lower probability of global warming. There are large uncertainties as to whether global warming will, in fact, occur, and what its magnitude and effects will be. Nevertheless, a conservative plan would involve attempting to reduce emissions of "greenhouse gases" as much as economically and politically feasible in the short range, and preparing additional management strategies to further

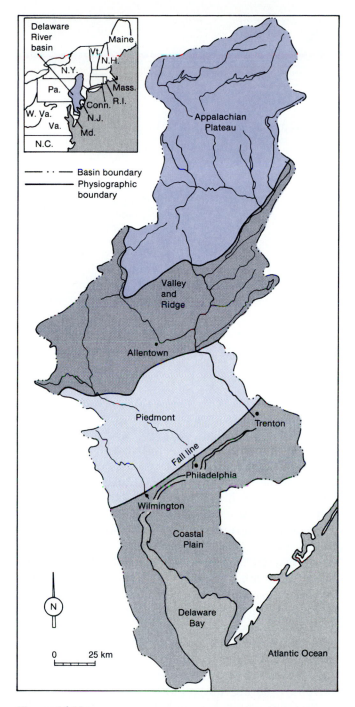

Figure 16.12
Delaware River basin with physiographic provinces shown: Appalachian Plateau (uplands), Valley and Ridge (alternating linear valleys and ridges), Piedmont (rolling hills), and Coastal Plain (low relief). (From M. E. Moss and H. F. Lins, U.S. Geological Survey Circular 1030.)

reduce emissions of the gases should the need arise. There is general consensus that the global climate will warm further, and because such warming might have very significant effects on our environment, we should take immediate steps to reduce the likelihood of future problems.

▼ GLOBAL OZONE DEPLETION

Ozone

Ozone (O_3) is a triatomic form of oxygen, in which three atoms of oxygen (O) are bonded. The oxygen we breathe is diatomic oxygen (O_2) consisting of two oxygen atoms bonded together. Ozone is relatively unstable, and releases an oxygen atom fairly easily. Ozone is a strong oxidant, and reacts with many different materials. Some of the reactions are useful; for example, ozone gas bubbling through water can purify it. In the lower atmosphere, ozone is considered a pollutant that may injure living organisms, including people.

Figure 16.13 shows the structure of the atmosphere. Ozone in the troposphere (lower 10 kilometers of the atmosphere, where human activity, weather, and urban air pollution occur) is produced in conjunction with sun-induced photochemical reactions related to emission of nitrogen dioxide and hydrocarbons.

Ozone (O_3) in the stratosphere is produced by photodissociation of the oxygen molecule (O_2) that contains two oxygen atoms into two single oxygen atoms (O). The single oxygen atoms may then combine with an oxygen molecule (O_2) to produce ozone (O_3). Natural processes in the stratosphere result in ozone production and destruction. The highest concentrations of stratospheric ozone occur from approximately 15 to 40 kilometers above the surface of the earth, forming the so-called ozone layer (Figure 16.13). This layer is sometimes referred to as earth's life or ozone shield, because it protects the troposphere from receiving harmful amounts of ultraviolet solar radiation. The total amount of ozone in the stratosphere is quite small. If all of it were brought down to the surface of the earth, it would form a thin layer of approximately 4 millimeters in thickness. Nevertheless, it is this small amount of ozone that is responsible for maintaining the life shield.

Removal of Stratospheric Ozone

The ozone layer in the stratosphere is naturally fragile. Natural processes are constantly forming, maintaining, and removing ozone. Absorption of ultraviolet radiation in the ozone layer is a natural mechanism of ozone removal. It was first suggested in 1974 that human-made chemicals might be responsible for destroying the ozone layer (18). The hypothesis presented in 1974 was that release of chlorine by photodissociation of chlorofluorocarbon molecules (CFCs) is responsible for triggering a chain reaction that significantly depletes stratospheric ozone. The importance of the hypothesis was dramatically heightened in 1985 with the discovery of the so-called Antarctic ozone hole (Figure 16.14).

Chlorofluorocarbons are contained in many commercial products, including refrigerants, cleaning solvents, and aerosol propellants. CFCs are extremely stable

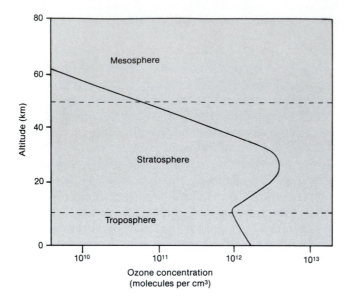

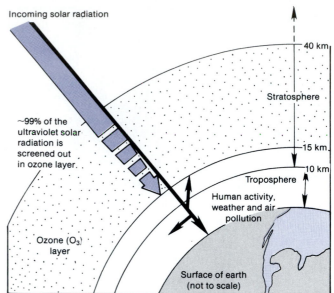

Figure 16.13

Structure of the atmosphere and ozone concentration. (Ozone concentrations from R. T. Watson, "*Atmospheric Ozone,*" in *Effects of Change in Stratospheric Ozone and Global Climate,* Vol. 1, *Overview,* ed. J. G. Titus, U.S. Environmental Protection Agency, p. 70.)

compounds (insoluble and nonreactive). The two most commonly used CFCs are trichlorofluoromethane (CFC-11) and dichlorodifluoromethane (CFC-12); they have been released into the atmosphere for many years. Figure 16.9 shows the global concentration of these two CFCs from 1974 to 1984. In recent years, the total emissions of these materials are in excess of 1 million tons per year.

When the stable CFCs are emitted into the atmosphere, they may remain there for a very long time, often exceeding 100 years. During that period, they randomly diffuse (gradually rise) to higher altitudes, and at the midstratospheric level may absorb ultraviolet radiation and release atomic chlorine. The stratospheric chlorine may then descend back through the ozone layer, and eventually be deposited as hydrochloric acid in rainfall. Prior to the formation of the acid the chlorine may participate in a chain reaction (18).

$$Cl + O_3 \rightarrow ClO + O_2$$
$$ClO + O \rightarrow Cl + O_2$$

The first equation involves the combination of chlorine (Cl) with ozone that leads to the formation of chlorine oxide (ClO), plus molecular oxygen (O_2). This reaction is the one that destroys ozone. However, in the second reaction, the chlorine oxide combines with an oxygen atom to produce chlorine, plus molecular oxygen again. The process becomes a chain reaction because the chlorine emerges again, and may combine with another ozone molecule to cause further ozone destruction. In

fact, it has been estimated that during the 1- to 2-year average residence time of one chlorine atom in the stratosphere, it may destroy approximately 100,000 molecules of ozone (19). Thus, the annual emission of approximately 1 million tons of CFCs may translate into a loss of ozone that may be about 100,000 times as great. It is also important to recognize that chlorine is not the only catalyst capable of entering into reactions that destroy ozone. Other materials such as nitrous oxide (N_2O) may enter into similar reactions. However, because CFCs are increasing in concentration in the atmosphere, and are being released in great amounts, the focus has been on chlorine in the stratosphere.

The first reports of significant ozone depletion were from land-based data in Antarctica. Approximately 30 years of data suggest that there was a significant springtime decrease in ozone that started about the 1970s and intensified during the mid-1980s (20). Satellite data confirm the ozone loss and the development of the Antarctic ozone hole. In 1987 the ozone decline above Antarctica was nearly 50 percent although in 1988 the decline was closer to 15 percent (21). The lesser decline in 1988 was thought to be a result of greater atmospheric mixing and a warmer stratosphere over Antarctica that produced fewer ice clouds to participate in ozone depletion reactions (21). Antarctic ozone depletion increased again in 1989 and 1990 to its historic maximum. Ozone holes also form over the Arctic, and this has important implications to the people in the northern hemisphere.

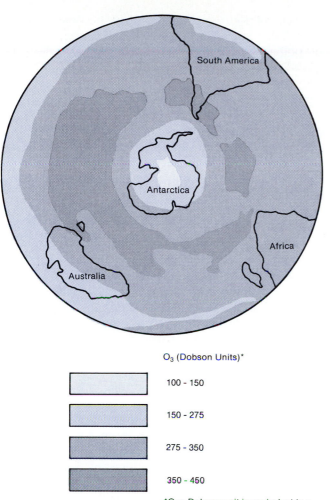

O₃ (Dobson Units)*

	100 - 150
	150 - 275
	275 - 350
	350 - 450

*One Dobson unit is equivalent to a concentration of 1 part per billion O₃

Figure 16.14
Antarctic "ozone hole" in early October 1990. (Data from NASA, Goddard Space Flight Center.)

Environmental Effects of Ozone Depletion

Depletion of stratospheric ozone allows greater amounts of ultraviolet radiation to reach the surface of the earth. Potential effects on humans include increases in all types of skin cancers, and greater severity of sunburns. There is concern also that ultraviolet radiation will affect productivity of food plants such as soybeans. In fact, it has been speculated that ozone depletion might adversely affect, through plant damage, the entire agriculture-based global food supply.

Increased ultraviolet radiation may also affect marine biologic processes. The radiation penetrates several meters of ocean water, and could potentially cause damage to planktonic life in the ocean. Plankton form the base of the food chain in the marine environment, and so any disruption here would cause potential problems throughout the entire food chain and eventually would

affect people who consume marine resources. Also, plankton are believed to be a major sink for carbon dioxide; hence, their disruption could increase atmospheric concentrations of carbon dioxide.

Management Issues Related to Ozone Depletion

Ozone depletion is going to be with us for many decades. Even if all uses of CFCs were stopped today, millions of tons of these chemicals are already in the lower atmosphere, moving upward towards the stratosphere. This movement will cause an increase in their concentration in coming decades, and reduction in the concentrations of stratospheric ozone. As a result, such reductions will have only a minor effect on the ozone problem before the end of this century (19). Nevertheless, it is very prudent to move forward with management plans that greatly restrict the use of CFCs and other chemicals that cause ozone depletion. Concern for ozone depletion has led the United Nation's Environmental Program to develop a strategy to protect stratospheric ozone. The objective of what is known as the Montreal Protocol is to facilitate 50 percent reduction in the emission of CFCs by the year 1999. In order to accomplish such a large reduction, new chemicals to replace the CFCs will have to be developed, and soon! Research to do this is ongoing and encouraging.

▼ ACID RAIN

Acid rain refers to both wet and dry acid deposition. Although *acid deposition* is a more correct term, *acid rain* is more commonly used. Many people are surprised to learn that all rainfall is slightly acidic: water reacts with atmospheric carbon dioxide to produce carbonic acid. Thus pure rainfall has a pH (a numerical value to describe the strength of an acid) of about 5.6. Acid rain is defined as precipitation in which the pH is below 5.6 (Figure 16.15). However, in some cases natural rainfall with pH between 4 and 5 has been observed in tropical rain forests, possibly related to acid precursors emitted by the trees (22). Because the pH scale is logarithmic, a pH value of 3 is 10 times more acidic than a pH value of 4 and 100 times more acidic than a pH value of 5. Automobile battery acid has a pH value of 1. It is alarming to learn that in Wheeling, West Virginia, rainfall has been measured with a pH value of 1.5, nearly as acidic as stomach acid and far more acidic than lemon juice or vinegar.

Perhaps more important than isolated cases of very acid rain is the apparent growth of the problem. Until quite recently it was believed that acid rain was primarily a European problem. It is now recognized that acid rain affects all industrial countries. In the United States it was believed that acid rain affected only a relatively small area

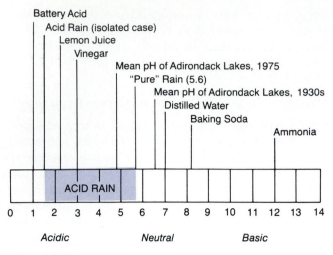

Figure 16.15
pH scale. (Modified after U.S. Environmental Protection Agency, 1980.)

in the northeastern United States. Now it is believed to affect nearly all of eastern North America, and West Coast urban centers such as Seattle, San Francisco, and Los Angeles are now beginning to record acid rainfall and acid fog events. The problem is also of great concern in Canada.

Causes of Acid Rain

Since 1900 the amounts of sulfur dioxide and nitrogen oxides released into the environment have increased significantly. Today about 20 million tons per year each of nitrogen oxide and sulfur dioxide are released into the atmosphere. Following emission of sulfur dioxide and nitrogen oxide into the atmosphere, they are transformed into sulfate (SO_4) or nitrate (NO_3) particles, which may combine with water vapor to form sulfuric and nitric acids. These acids may travel long distances with prevailing winds to be deposited as acid precipitation (Figure 16.16). Such precipitation may be in the form of rainfall, snow, or fog. Sulfate and nitrate particles may also be deposited directly on the surface of the land as dry deposition. These particles may later be activated by moisture to become sulfuric and nitric acids.

Sulfur dioxide is emitted primarily from stationary sources such as power plants that burn fossil fuels, whereas nitrogen oxides are emitted from both stationary and transport-related sources such as automobiles. Approximately 80 percent of the sulfur dioxide and 65 percent of the nitrogen oxides came from within states east of the Mississippi River. During the last 45 years sulfur dioxide emissions have increased approximately 50 percent while nitrogen oxides have increased by as much as 300 percent, probably reflecting the increased utilization of automobiles in urban areas. Throughout

that same period, emission stacks have become taller and taller. Taller stacks have reduced local concentrations of air pollutants, but have increased regional effects by spreading the pollution more widely. In this case, dumping waste into someone else's back yard has created more rather than fewer problems. For example, Swedish scientists have traced acid precipitation problems in Scandinavian lakes to airborne pollutants from Germany, France, and Great Britain; similarly, problems associated with acid precipitation in Canada may be traced to the emission of sulfur dioxide and other pollutants in the Ohio Valley.

The length of time that pollutants remain in the atmosphere depends upon weather patterns and the pollutants' chemical interactions with the atmosphere. Typically, the pollutants may return to the earth very quickly or may remain in the atmosphere for a week or longer. Prevailing winds in the eastern United States, particularly in the Ohio Valley where most of the sulfur dioxide is produced, tend to push the pollutants to the northwest and northeast. This accounts for the observation that acid precipitation problems are most prevalent in the northeastern United States and parts of Canada.

Analysis of the distances that sulfur compounds may be transported before deposition suggests that approximately one-third of the total amount deposited over the eastern United States originates from sources greater than 500 kilometers away. Another one-third comes from sources between 200 and 500 kilometers away, and the remainder comes from sources less than 200 kilometers away (23).

Effects of Acid Rain

Geology and climatic patterns as well as types of vegetation and soil composition all affect the potential impacts of acid rain. Figure 16.17, showing areas of the United States sensitive to acid rain, is based on some of these factors. Particularly sensitive areas are those in which the bedrock cannot buffer the acid input; such areas include terrain dominated by granitic rocks, as well as those in which the soils have little buffering action. Areas least likely to suffer damages are those in which the bedrock contains an abundance of limestone or other carbonate material or in which the soils contain a horizon rich in calcium carbonate. Soils in sensitive areas may be damaged from a fertility standpoint, either because nutrients are leached out by the acid or because the acid releases into the soil elements that are toxic to plants.

It has long been suspected that acid precipitation, whether in the form of snow, rain, fog, or dry deposition, adversely affects trees. Studies in Germany have led scientists to blame acid rain and other air pollution for the death of thousands of hectares of evergreen trees in Bavaria. Similar studies in the Appalachian Mountains of

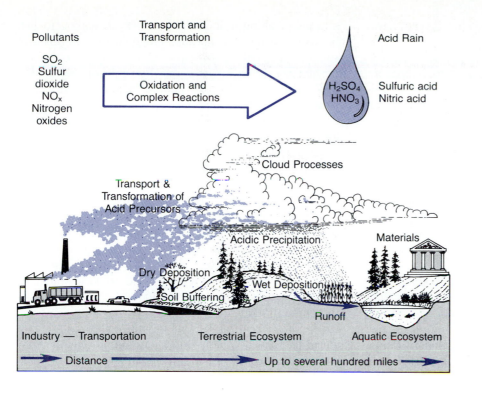

Figure 16.16
Paths and processes associated with acid rain. (Modified after D. L. Albritton, as presented in J. M. Miller.)

Vermont suggest that in some locations half of the red spruce trees have died in recent years. The damage is attributed in part to acid rain and fog with pH levels of 4.1 and 3.1, respectively. Acid solutions enter the soil and free toxic metals such as aluminum and cadmium from the minerals found there. These toxic metals may then move into the roots of plants and trees and weaken them.

Effect on lake ecosystems. In recent years fish that were once abundant and were used for food and recreation have disappeared from lakes in Sweden. Records of fifteen years or more from Scandinavian lakes show an increase in acidity accompanied by a decrease in

fish. The death of the fish has been traced to acid rain—the result of industrial processes far away in other countries, particularly in Germany and Great Britain.

Acid rain affects a lake ecosystem by dissolving chemical elements necessary for life and keeping them in solution so that they leave the lake with the water outflow. Elements that once cycled within the lake are thus lost from the system. Without these nutrients, the lake algae do not grow, and the small animals that feed on the algae have little to eat. The fish that typically prey on the small invertebrate animals also lack food. The acid water has other adverse effects on living organisms and their reproduction. For example, crayfish produce fewer

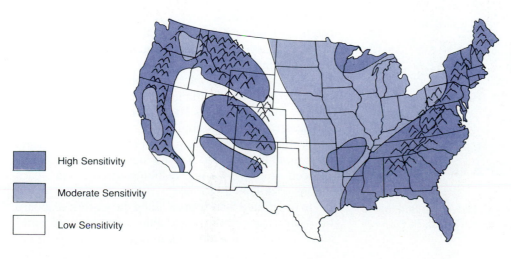

Figure 16.17
Areas in the United States sensitive to acid rain. (From U.S. Environmental Protection Agency, 1980.)

eggs in acid water, and those eggs produced often have malformed larvae.

To better study the effects of acidification on lakes, scientists in Canada added sulfuric acid to a lake in northwest Ontario over a period of years and observed the effects. When the experiment started, the pH of the lake was 6.8. The following year, owing to addition of the acid, the pH had dropped to 6.1. The initial drop in pH was not harmful to the lake, but as more and more acid was added the pH dropped first to 5.8 then to 5.6, 5.4, and finally, five years after the project started, to 5.1. The problems started when the pH was lowered to 5.8; some species disappeared and others experienced reproductive failure. At a pH of 5.8, the death rate among lake trout embryos increased. When the pH was lowered to 5.4, lake trout reproduction failed (24).

Those experiments proved valuable in pointing out what might be expected in thousands of other lakes that are now becoming acidified. The precise processes involved in the toxicity and damage to the lakes are poorly understood. However, it is known that acid rain leaches metals such as aluminum, lead, mercury, and calcium from the soils and rocks in a drainage basin and discharges them into rivers and lakes. Elevated concen-

trations of aluminum are particularly damaging to fish because the metal can cause clogging of the gills and suffocate the fish. The heavy metals may pose human health hazards because they may become concentrated in fish and be passed on to people, mammals, and birds when the fish are eaten. Of course, drinking water taken from acidic lakes and water may also have high concentrations of toxic metals.

Not all lakes are as vulnerable to acidification as the Ontario lakes. The acid is neutralized in waters with a high calcium or magnesium content—lakes on limestone or other rocks rich in calcium or magnesium carbonates can readily release the calcium and magnesium. This buffers the lakes against the addition of acids. Lakes with high concentrations of such elements are called hard-water lakes, whereas lakes on sand or igneous rocks such as granite tend to lack sufficient buffering to neutralize the acid and thus are more susceptible to acidification. In practice a simple and fast index of a lake's hardness or buffering capacity is its electrical conductivity. Pure water is a poor conductor of electricity; water high in dissolved elements is a good conductor (25).

Thousands of kilometers of rivers and thousands of lakes in the United States and Canada are currently in various stages of acidification. In Nova Scotia, for example, at least a dozen rivers have acid contents sufficiently high that they no longer support healthy populations of Atlantic salmon. In the northeastern United States about 200 lakes in the Adirondacks are no longer able to support fish, and thousands more are slowly losing the battle with acid rain.

One solution to lake acidification is rehabilitation by the periodic addition of lime. This has been done in New York state as well as in Sweden and Ontario, Canada. This solution is not satisfactory over a long period, however, because it is expensive and requires a continuing effort. The only practical long-term solution to the acid rain problem is to ensure that the production of acid-forming components in the atmosphere is minimized.

Effect on human society. Acid rain affects not only forests and lakes; it is capable also of damaging many building materials, including steel, paint, plastics, cement, masonry, galvanized steel, and several types of rock, especially limestone, sandstone, and marble (Figure 16.18). Classical buildings on the Acropolis in Athens, Greece, and other cities show considerable decay that has accelerated in this century as a result of air pollution. The problem has grown to such an extent that statues and other monuments need to have new protective coating replaced quite frequently, resulting in costs that reach billions of dollars a year (24).

In the United States, cities along the eastern seaboard are more susceptible to acid rain today because emissions of sulfur dioxide and nitrogen oxide are more

Figure 16.18
Detail of damage attributed to acid rain, Parliament Building, Ottawa, Ontario, Canada. (Photo courtesy of Canadian Embassy.)

abundant there. However, the problem is moving westward—acid precipitation has been recorded in California. Even more alarming is the discovery that acid fog events in Los Angeles may have a pH as low as 3, which is over ten times as acidic as the average acid rain in the eastern United States. In contrast to acid rain that may form relatively high in the atmosphere and travel long distances, acid fog forms when water vapor near the ground mixes with pollutants and turns into an acid. The acid evidently condenses around very fine particles of smog and, if the air is sufficiently humid, a fog may form. When the fog eventually burns off, nearly pure drops of sulfuric acid may be left behind. These acid fogs may be considerable health hazards because the tiny particles containing the acid may be inhaled deeply into people's lungs.

It is interesting that the geologic aspect of acid rain problems may help predict future effects and potential solutions. Since 1875 the Veteran's Administration in the United States has provided over 2.5 million tombstones to national cemeteries. These tombstones have come from only three rock quarries and have a standard size and shape. These stones, now located in various parts of the country, have been evaluated to assess damages caused by acid rainfall because they provide a variety of dates and locations to work from. Research will provide valuable data on air pollution and the meteorological patterns that contribute to acid rain (26).

▼ PARTICULATES IN THE ATMOSPHERE

Types and sizes of selected particulates are shown in Figure 16.19. Elevations that heavy particles may reach in the atmosphere depend on the specific transport processes involved. How particulate material in the atmosphere will affect global changes in the atmosphere's mean annual temperature is being debated. Certainly, particulates are being added to the atmosphere through human activity. Agriculture, forestry, power production, and other industrial activities all involve burning, which contributes particulates to the atmosphere. The fires deliberately set during the 1991 Persian Gulf War provide a dramatic example of atmospheric pollution produced by human activity. The Iraqi military set several hundred Kuwaiti oil wells ablaze, burning millions of barrels of oil each day for months. The fires produced an enormous amount of particulates in addition to carbon dioxide, carbon monoxide, and nitrogen oxides. These pollutants have created environmental problems on a regional scale but will have minor effect on the global environment. For example, the fires contribute less than 2 percent of the annual global production of carbon dioxide.

Particulates have two important effects. First, they act as condensation nuclei and therefore cause an increase in precipitation or fog. Second, they affect the amount of sunlight reaching the earth. As the total amount of

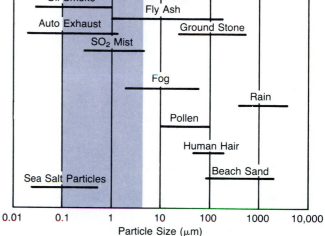

Figure 16.19
Types and sizes of selected particulates. (Modified from Giddings, 1973, *Chemistry, Man, and Environmental Change: An Integrated Approach,* Canfield Press; and Hidy and Brock, 1971, *Topics in Current Aerosol Research,* Pergamon Press.)

particulates in the atmosphere increases, a larger percentage of incoming solar radiation may be reflected away from the earth, causing mean annual temperature to decrease. On the other hand, some particles may absorb incoming solar radiation, causing an increase in the atmospheric and land-surface temperature. This second effect is observed in urban areas, where particulates are concentrated. Furthermore, if particles filter out of the atmosphere and are deposited on snow, then a greater portion of the solar radiation will be absorbed, making more radiation available to heat the atmosphere.

Slight global cooling and spectacular sunsets for up to a year or so have followed volcanic eruptions that blast volcanic ash and other material into the atmosphere. The 1883 eruption (the largest in historic time) of the volcano Krakatoa, a small island in the East Indies, blew about 2.6 cubic kilometers of volcanic ash and other material as high as 27 kilometers into the atmosphere. The hole left in the ocean floor was 300 meters deep and the blast could be heard 5,000 kilometers away! Dust in the stratosphere circled the earth for months before it settled out, causing a slight lowering in the mean temperature of the lower atmosphere. Recent research suggests that the amount of ash reaching the stratosphere is less important than the composition in determining climatic effects. For example, the eruptions of Mt. St. Helens in 1980 and El Chicón in southern Mexico in 1982 both ejected volcanic debris into the stratosphere. Mt. St. Helens caused little or no climatic disturbance, whereas El Chicón probably caused a measurable, significant effect, even though both events were of similar magnitude and much smaller than Krakatoa. The stratospheric cloud from El Chicón was sulfur rich and formed a cloud of aerosol sulfuric acid droplets about 100 times as dense as the stratospheric cloud from the Mt. St. Helens eruption, which had little sulfur in it. While fine volcanic ash settles out in a few weeks, an aerosol of sulfuric acid droplets may circle the earth for several years. It is hypothesized that the El Chicón eruption caused a drop in mean annual temperature of about 0.3–0.5°C during a three-year period following the eruption. The eruption also may have contributed to the formation of the strong 1982–83 El Niño by disrupting (weakening) atmospheric circulation that drives oceanic circulation (27). That is, the sulfurous cloud warmed the stratosphere by absorbing incoming solar radiation. Less radiation reaching earth cooled the lower atmosphere. These processes reduced normal atmospheric circulation, making it easier for oceanic currents to move west near the equator off South America rather than east as they normally do, helping produce the El Niño event.

The 1991 eruptions of Mt. Pinatubo in the Philippines were larger than the El Chicón events and also released tremendous quantities of sulfur dioxide that oxidize to sulfuric acid. The potential climatic events are being measured and modeled.

Yet it is not the particulate matter from volcanic eruptions that causes the most concern. The increase in atmospheric dust from human activities may offset the warming effects that may be occurring from the burning of fossil fuels and the addition of carbon dioxide to the atmosphere. Whether the cooling of the atmosphere from particulates will have a greater effect than the warming trends caused by carbon dioxide is unknown.

▼ COUPLING OF GLOBAL CHANGE PROCESSES

The major global change processes discussed in this chapter have interesting linkages. For example, ozone depletion is related to the anthropogenic emissions of chlorofluorocarbons; thus, when these molecules are released in the lower atmosphere they have a strong potential greenhouse effect. This occurs because CFCs absorb infrared radiation in an area of the spectrum where carbon dioxide does not absorb. One study (19) concluded that chlorofluorocarbons, on a per-molecule basis, are 10,000 times more efficient in absorbing infrared radiation than carbon dioxide. Nevertheless, carbon dioxide contributes much more to the total anthropogenic greenhouse effect than do CFCs because so much more carbon dioxide is being emitted.

The important point to be considered here is the coupling of the greenhouse and ozone problems via release of materials such as CFCs. Other couplings are related to processes such as burning of fossil fuels, which

releases precursors to acid rain as well as "greenhouse gasses." Thus, we see the principle of environmental unity in action. Many aspects of one problem are related to other problems. Likewise, solutions to the potential global warming problem through release of carbon dioxide will likely also affect other global processes such as formation and transport of acid precursors in the atmosphere.

▼ URBAN AIR: INTRODUCTION

Ever since life began on earth, the atmosphere has been an important resource for chemical elements and a medium for depositing wastes. The earliest photosynthetic plants dumped oxygen—the element that was their waste—into the atmosphere. The long-term increase in atmospheric oxygen, in turn, made possible the development and survival of higher life forms.

As the fastest moving fluid medium in the environment, the atmosphere has always been one of the most convenient places to dispose of unwanted materials. Ever since people first used fire, the atmosphere has all too often been a sink for waste disposal.

London Smog Crisis: 1952

In London, during the first week of December 1952, the air became stagnant and the cloud cover did not allow much of the incoming solar radiation to penetrate. The humidity climbed to 80 percent, and the temperature dropped rapidly until the noontime temperature was about −1°C. A very thick fog developed, and the cold and dampness increased the demand for home heating. Because the primary fuel used in homes was coal, emissions of ash, sulfur oxides, and soot increased rapidly. The stagnant air became filled with pollutants, not only from home heating fuels but from automobile exhaust. At the height of the crisis, visibility was greatly reduced and automobiles had to use their headlights at midday. Between December 4 and 10, an estimated 4000 people died from the pollution. Figure 16.20 shows the increase in sulfur dioxide and smoke and the accompanying deaths during this period. The siege of smog finally ended when the weather changed and the air pollution was dispersed. The environment, not human activities, finally solved this problem. Since the beginning of the industrial revolution and before, people had survived in London and other major cities in spite of the weather and of pollution. What had finally gone wrong?

During the London smog crisis, the stagnant weather conditions together with the number of homes burning coal and of cars burning gasoline exceeded the atmosphere's ability to remove or transform the pollutants; even the usually rapid natural mechanisms for removing sulfur dioxide were saturated. As a result, sulfur dioxide remained in the air and the fog became acid, adversely

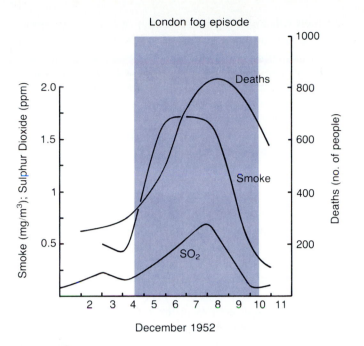

Figure 16.20

The relationship between the number of deaths and the London fog of 1952. (Modified from S. J. Williamson, 1973, *Fundamentals of Air Pollution,* Addison-Wesley.)

affecting people and other organisms, particularly vegetation. The health effects on people were especially destructive because small acid droplets became fixed on larger particulates, facilitating their being drawn deep into the lungs.

The 1952 London smog crisis was a landmark event. Finally, human activities had exceeded the natural abilities of the atmosphere to serve as a sink for the removal of wastes. The crisis was due in part to a positive, or reinforcing, feedback situation. Burning fossil fuels added particulates to the air, increasing the formation of fog and decreasing visibility and light transmission; the dense, smoggy layer increased the dampness and cold and accelerated the use of home heating fuels. The worse the weather and the pollution, the more people acted so as to further worsen the weather and the pollution.

Before 1952, London was well known for its fogs; what was relatively little known was the role of coal burning in intensifying fog conditions. Since 1952, London fogs have been greatly reduced because coal has been replaced to a large extent ' y much cleaner gas as the primary home heating fuel. A foggy day in London is now not so common and no longer seems quite so romantic.

▼ POLLUTION OF THE ATMOSPHERE

Chemical pollutants can be thought of as compounds that are in the wrong place or in the wrong concentrations at the wrong time. As long as a chemical is transported away

or degraded rapidly relative to its rate of production, there is no pollution problem. Pollutants that enter the atmosphere through natural or artificial emissions may be degraded not only within the atmosphere but also by natural processes in the hydrologic and geochemical cycles; on the other hand, pollutants that leave the atmosphere may become pollutants of water and of geological cycles.

People have long recognized the existence of atmospheric pollutants, both natural pollutants and those induced by humans. Acid rain was first described in the seventeenth century, and by the eighteenth century it was known that smog and acid rain damaged plants in London. Beginning with the industrial revolution in the eighteenth century, air pollution became more noticeable; by the middle of the nineteenth century, particularly following the American Civil War, concern with air pollution increased. The word "smog" was probably introduced by a physician at a public health conference in 1905 to denote poor air quality resulting from a mixture of smoke and fog.

Two major pollution events, one in the Meuse Valley in Belgium in 1930 and the other in Donora, Pennsylvania, in 1948, raised the level of scientific research on air pollution. The Meuse Valley event lasted approximately one week and caused 60 deaths and numerous illnesses. The Donora event caused 20 deaths and 14,000 illnesses. By the time of the Donora event, people recognized that meteorological conditions were an integral part of the production of dangerous smog events. This view was reinforced by the 1952 London smog crisis, after which regulations to control air quality began to be formulated.

General Effects of Air Pollution

Air pollution affects many aspects of our environment: visually aesthetic resources, vegetation, animals, soils, water quality, natural and artificial structures, and human health. Air pollutants affect visual resources by discoloring the atmosphere, reducing visual range and atmospheric clarity. We can't see as far in polluted air, and what we do see has less color contrast. Once limited to cities, these effects now extend even to the wide open spaces of the United States. For example, emissions from the Four Corners fossil fuel-burning power plant, near the joint border of New Mexico, Arizona, Colorado, and Utah, are altering visibility where one previously could see 80 kilometers from a mountain top on a clear day (28).

Effects of air pollution on vegetation include damage to leaf tissue, needles, or fruit; reduction in growth rates or suppression of growth; increased susceptibility to a variety of diseases, pests, and adverse weather; and the disruption of reproductive processes. Damage to entire terrestrial or aquatic ecosystems in turn can affect the vegetation (28).

Effects of air pollutants on vertebrate animals include impairment of the respiratory system; damage to eyes, teeth, and bones; increased susceptibility to disease, pests, or other stress-related environmental hazards; decrease in availability of food sources such as vegetation impacted by air pollutants; and reduction in ability to reproduce (28).

When pollutants from the air are deposited, soils and water may become toxic. Soils may also be leached of nutrients by pollutants that form acids.

Effects of air pollutants on human health include toxic poisoning, eye irritation, and irritation of the respiratory system. In urban areas people suffering from respiratory diseases are more likely to be affected by air pollutants, while healthy people tend to acclimate to pollutants relatively quickly. Nevertheless, urban air can cause serious health problems. Many pollutants have synergistic effects; for example, sulfate and nitrate may attach to particles in the air, facilitating their inhalation deep into lung tissue.

Effects of air pollution on buildings include discoloration, erosion, and decomposition of construction materials, as discussed in the section on acid rain.

Sources of Air Pollution

Many pollutants in our atmosphere have natural as well as human-related origins. Natural emissions of air pollutants include release of gases such as sulfur dioxide through volcanic eruptions; release of hydrogen sulfide from geyser and hot spring activities and by biological decay from bogs and marshes; increased concentration of ozone in the lower atmosphere as a result of unstable meteorological conditions, including violent thunderstorms; and emission of a variety of particles from wildfires and windstorms (28).

Major natural and human-produced air pollutants and sources are shown in Table 16.2. These data suggest that, with the exception of sulfur and nitrogen oxides, natural emissions of air pollutants exceed human-produced input. Nevertheless, the human component is most abundant in urban areas and leads to the most severe air pollution events for human health.

The two major kinds of air pollution sources are stationary and mobile. *Stationary sources,* those with a relatively fixed location, include point sources, fugitive sources, and area sources. *Point sources* are those stationary sources that emit air pollutants from one or more discrete controllable sites, such as smokestacks of industrial power plants. *Fugitive sources* generate air pollutants from open areas exposed to wind processes. Examples include dirt roads, construction sites, farmlands, storage piles, and surface mines. *Area sources* are locations that emit air pollutants from several sources within a well-defined area, as, for example, small urban communities or areas of intense industrialization within an urban complex. *Mobile sources* move from place to place while yielding emissions and include automobiles, aircraft, ships, and trains (28).

Table 16.2
Major natural and human-produced components of air pollutants.

Air Pollutant	Emissions (% of total)		Human-produced Component Major Sources	%
	Natural	Human-produced		
Particulates	89	11	Industrial processes	51
			Combustion of fuels (stationary sources)	26
Sulfur oxides (SO_x)	55	45	Combustion of fuels: (stationary sources, mostly coal)	78
			Industrial processes	18
Carbon monoxide (CO)	91	9	Transportation (mostly automobiles)	75
			Agricultural burning	9
Nitrogen dioxide (NO_2)		Nearly all	Transportation (mostly automobiles)	52
			Combustion of fuels (stationary sources, mostly natural gas and coal)	44
Ozone (O_3)	A secondary pollutant derived from reactions with sunlight, NO_2 and oxygen (O_2)		Concentration that is present depends on reaction in lower atmosphere involving hydrocarbons and thus automobile exhaust	
Hydrocarbons (HC)	84	16	Transportation (mostly automobiles)	56
			Industrial processes	16
			Evaporation of organic solvents	9
			Agricultural burning	8

▼ AIR POLLUTANTS*

The two main groups of air pollutants are primary and secondary. *Primary pollutants* are emitted directly into the air and include particulates, sulfur oxides, carbon monoxide, nitrogen oxides, and hydrocarbons. *Secondary pollutants* are produced when primary pollutants react with normal atmospheric compounds. As an example, over urban areas ozone forms through reactions among primary pollutants, sunlight, and natural atmospheric gases. Thus ozone becomes a serious pollution problem on bright, sunny days in areas with much primary pollution. While particularly well documented for southern California cities like Los Angeles, this occurs worldwide under appropriate conditions.

The primary pollutants that account for nearly all air pollution problems are particulates, hydrocarbons, carbon monoxide, nitrogen oxides, and sulfur oxides. Each year well over a billion metric tons of these materials enter the atmosphere from human-related processes. About half of this is carbon monoxide, and the other four each account for a few percent. At first glance this quantity of pollutants appears very large. However, if uniformly distributed in the atmosphere, this would amount to only a few parts per million by weight. Unfortunately, pollutants are not uniformly distributed but tend to be released, produced, and concentrated locally or regionally. For example, over large cities weather and climatic conditions combine with urbanization-industrialization to produce local air pollution problems.

The major air pollutants occur either in a gaseous form or as particulate matter (PM). The *gaseous pollutants* include sulfur dioxide (SO_2), nitrogen oxides (NO_x), carbon monoxide (CO), ozone (O_3), hydrocarbons (HC), hydrogen sulfide (H_2S), and hydrogen fluoride (HF). *Particulate-matter pollutants* are particles of solid or liquid substances and may be either organic or inorganic.

Sulfur Dioxide

Sulfur dioxide (SO_2) is a colorless and odorless gas under normal conditions at the earth's surface. One significant aspect of SO_2 is that once emitted into the atmosphere, it may be converted through complex reactions to fine particulate sulfate (SO_4). The major source for the anthropogenic component of sulfur dioxide is burning of fossil fuels, mostly coal in power plants. Industrial processes, ranging from refining of petroleum to production of paper, cement, and aluminum, are another major source.

*This section is summarized from *Air Resources Management Manual*, National Park Service, 1984.

Destructive effects associated with sulfur dioxide include corrosion of paint and metals and injury or death to plants and animals, especially to crops such as alfalfa, cotton, and barley. Sulfur dioxide can cause severe damage to human and other animal lungs, particularly in the sulfate form.

Nitrogen Oxides (NO_x)

Nitrogen oxides are emitted in several forms, the most important of which is nitrogen dioxide (NO_2), a visible yellow-brown to reddish-brown gas. Nitrogen dioxide may be converted by complex reactions in the atmosphere to fine particulate nitrate (NO_3). Additionally, nitrogen dioxide is one of the main pollutants contributing to the development of photochemical smog. Nearly all nitrogen dioxide is emitted from anthropogenic sources; the two major contributors are automobiles and power plants that burn fossil fuels such as coal and oil.

Environmental effects of nitrogen oxides are variable but include irritation of eyes, nose, and throat; increased susceptibility of animals and humans to infections; suppression of plant growth and damage to leaf tissue; and impaired visibility when the oxides are converted to their nitrate form in the atmosphere. On the other hand, when nitrate is deposited on the soil, it can promote plant growth.

Carbon Monoxide

Carbon monoxide (CO) is a colorless, odorless gas that at very low concentrations is extremely toxic to humans and other animals. This toxicity results from a striking physiological effect—namely, that carbon monoxide and hemoglobin attract each other. Hemoglobin in our blood will take up carbon monoxide nearly 250 times more rapidly than it will oxygen. Therefore, if any carbon monoxide is in the vicinity, a person will take it in very readily. Many people have been accidentally asphyxiated by carbon monoxide produced from incomplete combustion of fuels in campers, tents, and houses. Actual effects may range from dizziness and headaches to death. Carbon monoxide is particularly hazardous to people with known heart disease, anemia, or respiratory disease. Finally, the effects of carbon monoxide tend to be worse in higher altitudes, where oxygen levels are naturally lower.

Approximately 90 percent of the carbon monoxide in the atmosphere comes from natural sources and the other 10 percent comes mainly from fires, automobiles, and other sources of incomplete burning of organic compounds. Local concentrations of carbon monoxide can build up and cause serious health effects.

Ozone

Photochemical oxidants result from atmospheric interactions of pollutants (such as nitrogen dioxide) and sunlight. The most common photochemical oxidant is ozone (O_3), a colorless, unstable gas with a slightly sweet odor.

The major sources of oxidants, and particularly ozone, are automobiles, burning of fossil fuels, and industrial processes that produce nitrogen dioxide. Effects of oxidants, primarily ozone, are well known and include, among others, damage to rubber, paint, and textiles. Biological effects include damage to plants and animals.

The effects of ozone on plants can be subtle. At very low concentrations, ozone can reduce growth rates while not producing any visible injury. At higher concentrations, ozone kills leaf tissue, eventually killing entire leaves and, if the pollutant levels remain high, killing whole plants. The death of white pine trees planted along highways in New England is believed due to ozone from automobiles. Ozone also affects animals, including people, causing a variety of damage, especially to eyes and the respiratory system.

Hydrocarbons

Hydrocarbons are compounds composed of hydrogen and carbon. Thousands of such compounds exist, including natural gas or methane (CH_4), butane (C_4H_{10}), and propane (C_3H_8). Analysis of urban air has identified many different hydrocarbons, some of which react with sunlight to produce photochemical smog. Potential adverse effects of hydrocarbons are numerous because many are toxic to plants and animals or may be converted to harmful compounds through complex chemical changes that occur in the atmosphere. Approximately 85 percent of hydrocarbons (which are primary pollutants) entering the atmosphere are emitted from natural sources. The most important anthropogenic source is the automobile (see Table 16.2).

Hydrogen Sulfide

Hydrogen sulfide (H_2S) is a highly toxic corrosive gas easily identified by its rotten egg odor. Hydrogen sulfide is produced from natural sources, such as geysers, swamps, and bogs, and from human sources, such as industrial plants that produce petroleum or that smelt metals. Potential effects of hydrogen sulfide include functional damage to plants and health problems ranging from toxicity to death for humans and other animals.

Hydrogen Fluoride

Hydrogen fluoride (HF) is a gaseous pollutant released primarily by industrial activities such as production of aluminum, coal gasification, and burning of coal in power plants. Hydrogen fluoride is very toxic, and even a small concentration (as low as 1 ppb) may cause problems for plants and animals. Hydrogen fluoride is particularly dangerous to grazing animals because some forage plants can become very toxic.

Other Hazardous Gases

It's a rare month when the newspapers don't carry a story of a truck or train accident that releases toxic chemicals in a gaseous form into the atmosphere. As a result people are often evacuated until the leak is stopped or the gas dispersed to a nontoxic level. Chlorine gases or a variety of other materials used in chemical and agricultural processes may be involved.

Another source of air pollution is sewage treatment plants. Urban areas deliver a tremendous variety of organic chemicals, including paint thinner, industrial solvents, chloroform, and methyl chloride to treatment plants by way of sewers. These materials are not removed in the treatment plants; in fact, the treatment processes facilitate the evaporation of the chemicals into the atmosphere, where they may be inhaled by people. Many of the chemicals are toxic or are suspected carcinogens. It is a cruel twist of fate that the treatment plants designed to control water pollution are becoming sources of air pollution. While some pollutants can be moved from one location to another and even change form, as from liquid to gas, we really can't get rid of them as easily as we once thought.

Some chemicals are so toxic that extreme care must be taken to ensure they don't enter the environment. This was tragically demonstrated on December 3, 1984, when toxic gas (stored in liquid form) from a pesticide plant leaked, vaporized, and formed a toxic cloud that settled over a 64-square-kilometer area of Bhopal, India. The gas leak lasted less than one hour—yet over 2,000 people were killed and more than 15,000 were injured by the gas, which causes severe irritation (burns on contact) to eyes, nose, throat, and lungs. Breathing the gas in large quantities, in concentrations of only a few parts per million, causes violent coughing, swelling of the lungs, bleeding, and death. Less exposure can cause a variety of problems, including loss of sight.

The colorless gas, called *methyl isocyanate,* is an ingredient of the common pesticide known in the United States as Sevin and of at least two other insecticides used in India. Another plant located in West Virginia also makes the chemical; small leaks have evidently occurred there as they did prior to the catastrophic accident in Bhopal. The accident clearly suggests that hazardous chemicals that can cause catastrophic injuries and death should not be stored close to large population centers. Furthermore, more reliable accident-prevention equipment and personnel trained to control leaks or other problems at chemical plants are needed as well.

Particulate Matter

Particulate matter encompasses the small particles of solid or liquid substances that are released into the atmosphere by many activities. Modern farming adds considerable amounts of particulate matter to the atmosphere, as do desertification and volcanic eruptions. Nearly all industrial processes, as well as the burning of fossil fuels, release particulates into the atmosphere. Much particulate matter is easily visible as smoke, soot, or dust; other particulate matter is not easily visible. Included with the particulates are such materials as airborne asbestos particles and small particles of heavy metals such as arsenic, copper, lead, and zinc, which are usually emitted from industrial facilities such as smelters. Of particular importance are the very fine particle pollutants less than 2.5μm in diameter (2.5 millionths of a meter—very small indeed).

Among the most significant of the fine particulate pollutants are sulfates and nitrates. These are primarily secondary pollutants produced in the atmosphere through chemical reactions between normal atmospheric constituents and sulfur dioxide and nitrogen oxides. These reactions are particularly important in the formation of sulfuric and nitric acids in the atmosphere and thus lead to acid rain (28). When measured, particulate matter is often referred to as total suspended particulates (TSP).

Particulates affect human health, ecosystems, and the biosphere. Particulates that enter the lungs may lodge there and have chronic effects on respiration; asbestos is particularly dangerous in this way. Dust raised by road building and plowing and deposited on the surfaces of green plants may interfere with their absorption of carbon dioxide and oxygen and their release of water; heavy dust may affect the breathing of animals. Particulates associated with large construction projects may therefore kill organisms and damage large areas, changing species composition, altering food chains, and generally affecting ecosystems. In addition, modern industrial processes have greatly increased the total suspended particulates in the atmosphere. Particulates block sunlight and thus cause changes in climate. Such changes have lasting effects on the biosphere.

Asbestos. Asbestos particles have only recently been recognized as a significant hazard. In the past, asbestos was treated rather casually, and people working in asbestos plants were not protected from dust. Asbestos was used in building insulation and in brake pads for automobiles. As a result, asbestos fibers are found throughout industrialized countries, especially those of Europe and North America, and especially in urban environments. In one case, asbestos products were sold in burlap bags that eventually were reused in plant nurseries and other secondary businesses, thus further spreading the pollutant. Some asbestos particles are believed to be carcinogenic, or to carry with them carcinogenic materials, and so must be carefully controlled.

Lead. Lead is an important constituent of automobile batteries and other industrial products. When lead is added to gasoline, automobile engines burn more evenly. The lead in gasoline is emitted into the environment in the exhaust. In this way, lead has been spread widely around the world and has reached high levels in soils and waters along roadways.

Once released, lead can be transported through the air as particulates to be taken up by plants through the soil or deposited directly on plant leaves. Thus it enters terrestrial food chains.

When lead is carried by streams and rivers, deposited in quiet waters, or transported to the ocean or lakes, it is taken up by aquatic organisms and thus enters aquatic food chains.

Cadmium. Cadmium may enter the environment in part in the ash from burned coal. Cadmium, a trace element in the coal, exists in a very low concentration of 0.05 ppm. The ash is spread widely from stacks and chimneys and falls on plants, where the cadmium is incorporated into plant tissue and concentrated three to seven times. As the cadmium moves up the trophic levels through the food chains, each higher trophic level concentrates it approximately three times more. Herbivores have approximately three times the concentration of green plants, and carnivores approximately three times the concentration of herbivores.

▼ URBAN AREAS AND AIR POLLUTION

Wherever many sources are producing air pollutants over a wide area—automobile emissions in Los Angeles or smoke from wood-burning stoves in Vermont—there is potential for the development of smog. Whether air pollution develops depends on the topography and weather conditions because these factors determine the rate at which pollutants are transported away from their sources and converted to harmless compounds in the air. When the rate of production exceeds the rate of chemical transformations and of transport, dangerous conditions may develop.

Influence of Meteorology and Topography

Meteorological conditions can determine whether air pollution is only a nuisance or is a major health problem. The primary adverse effects of air pollution are damage to green plants and aggravation of chronic illnesses in people. Most of these effects are due to relatively low-level concentrations of toxins over a long period of time. Pollution periods in the Los Angeles basin or other

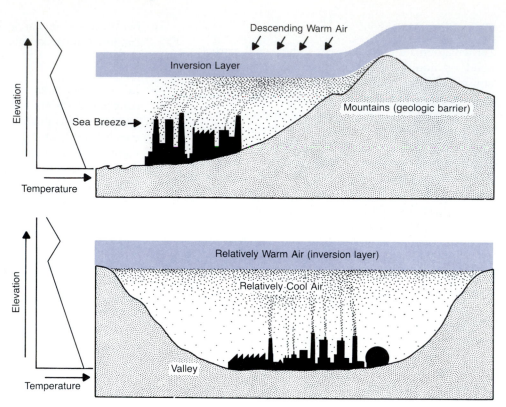

Figure 16.21
Two causes for the development of a temperature inversion, which may aggravate air pollution problems.

areas generally do not cause large numbers of deaths. However, serious pollution events can develop over a period of days and lead to an increase in deaths and illnesses.

In the lower atmosphere, restricted circulation associated with inversion layers may lead to pollution events. An *atmospheric inversion* occurs when warmer air is found above cooler air and is particularly a problem when there is a stagnated air mass. Figure 16.21 shows two types of developing inversions that may worsen air pollution problems. In the upper diagram, which is somewhat analogous to the situation in the Los Angeles area, descending warm air forms a semipermanent inversion layer. Because the mountains act as a barrier to the pollution, polluted air moving in response to the sea breeze and other processes tends to move up canyons, where it is trapped. The air pollution that develops occurs primarily during the summer and fall.

The lower part of Figure 16.21 shows a valley with relatively cool air overlain by warm air, a situation that can occur in several ways. When cloud cover associated with a stagnant air mass develops over an urban area, the incoming solar radiation is blocked by the clouds, which absorb some of the energy and thus heat up. On or near the ground, the air cools. If the humidity is high, the dewpoint is reached as the air cools and a thick fog may form. Because the air is cool, people living in the city burn more fuel to heat their homes and factories, thus delivering more pollutants into the atmosphere. As long as the stagnant conditions exist, the pollutants will build up.

Evaluating meteorologic conditions can be extremely helpful in predicting which areas will have potential smog problems. Figure 16.22 shows the number of days in a 5-year period for which conditions were favorable for reduced dispersion of air pollution for at least a 48-hour period. This illustration clearly shows that most of the problems have been located in the western United States.

Cities situated in a topographic "bowl" surrounded by mountains are more susceptible to smog problems than cities in open plains. Cities where certain kinds of weather conditions, such as temperature inversions, occur are also particularly susceptible. Both the surrounding mountains and the temperature inversions prevent the transportation of pollutants by the winds and weather systems.

Potential for Urban Air Pollution

The potential for air pollution in urban areas is determined by the following factors: the rate of emission of pollutants per unit area; the distance downwind that a mass of air may move through an urban area; the average speed of wind; and, finally, the height to which potential pollutants may be thoroughly mixed in the lower atmosphere (Figure 16.23) (29). The concentration of pollutants in the air is directly proportional to the first two factors. That is, as either the emission rate or downwind travel distance increases, so will the concentration of pollutants in the air. On the other hand, city air pollution decreases with increases in the wind velocity

Figure 16.24
Lingering air, an abundance of particulates, and the flow of air over heavily built-up areas create an urban dust dome and a heat island.

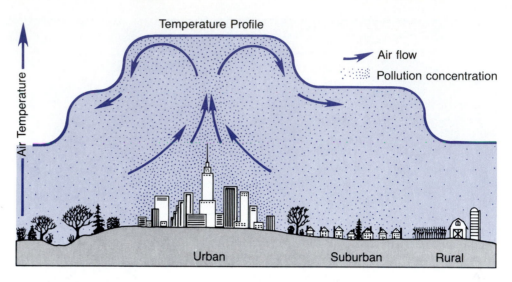

The Urban Microclimate

The very presence of a city affects the local climate, and as the city changes, so does its climate. For example, in the middle of the eighteenth century, Manhattan Island was "generally reckoned very healthy" (30), perhaps because of its nearness to the ocean and its relatively unobstructed ocean breezes. Today, the air pollution and the effects of tall buildings on air flow lead the average visitor to Manhattan to a quite different conclusion.

Although air quality in urban areas is in part a function of the amount of pollutants present or produced, it is affected also by the city's ability to ventilate and thus flush out pollutants. The amount of ventilation depends on several aspects of the urban microclimate.

Cities are warmer than surrounding areas. The observed increase in temperature in urban areas is approximately 1–2°C in the winter and 0.5–1.0°C in the summer for mid-latitude areas. The temperature increase results from increased production of heat energy; the heat emitted from the burning of fossil fuels and other industrial, commercial, and residential sources; and the decreased rate of heat loss—the dust in the urban air traps and reflects back into the city long-wave (infrared) radiation emitted from city surfaces. In the winter, space heating from the city is primarily responsible for heating the local air environment. For example, in Manhattan, the input of heat from industrial, commercial, and residential space heating in the winter has been measured to be about 2.5 times the solar energy that reaches the surface of the city; the annual average, however, is closer to 33 percent of the solar input. In large urban areas characterized by warmer winters, the heat input from artificial sources is much less (30). Concrete, asphalt, and roofs also tend to act as solar collectors and quickly emit heat, helping to increase the sensible heat in cities (7).

Cities are in general less windy than nonurban areas. Air over cities tends to move more slowly than in surrounding areas because buildings and other struc-

tures obstruct the flow of air. Thus, wind velocities are frequently reduced by 20–30 percent, and calm days are 20 percent more abundant in urban areas than in nearby rural areas (31).

Particulates in the atmosphere over a city are often 10 times or more as high as in surrounding areas. Although the particulates tend to reduce incoming solar radiation by up to 30 percent and thus cool the city, the effect of particulates is small relative to the effect of processes that produce heat in the city (31).

The combination of lingering air and abundance of particulates and other pollutants in the air produces the well-known urban dust dome and heat island effect (Figure 16.24). Also shown in Figure 16.24 is the general circulation pattern of air moving from the rural or suburban areas toward the inner city, where it flows up and then laterally out near the top of the dust dome. This circulation of air often occurs when a strong heat island develops over the city. For example, when a dust dome and heat island develop during a calm period in New York City, there is an upward flow of air over the heavily developed Manhattan Island, accompanied by a downward flow over the nearby Hudson and East rivers, which are green belts and thus have cooler air temperatures (7). Figure 16.24 also shows the air-temperature profile, which delineates the heat island. The dust dome effect explains why air pollution often tends to be most intense at city centers.

Particulates in the dust dome provide condensation nuclei, and thus urban areas experience 5–10 percent more precipitation and considerably more cloud cover and fog than surrounding areas. The formation of fog is particularly troublesome in the winter and may impede air traffic. If the pollution dome moves downwind, increased precipitation may occur outside the urban area. For example, in the mid-1960s, effluent from the southern Chicago-northern Indiana industrial complex apparently caused a 30 percent increase in precipitation at La

Porte, Indiana, 48 kilometers downwind to the south. La Porte also has almost 2.5 times as many hailstorms, 38 percent more thunderstorms, and less sunshine than the countryside not directly downwind (32). The La Porte case is extreme, but it is not unique. Atmospheric particulate matter from urban sources has undoubtedly altered local weather at numerous locations.

In summary, cities are cloudier, warmer, rainier, and less humid than their surroundings. Cities in middle latitudes receive about 15 percent less sunshine and 5 percent less ultraviolet light during the summer and 30 percent less ultraviolet light during the winter than nonurban areas. They are 10 percent rainier and 10 percent cloudier and have a 25 percent lower average wind speed, 30 percent more summer fog, and 100 percent more winter fog than nonurban areas. Average relative humidity is 6 percent less in cities, partly because they have large impervious surfaces and little surface water to exchange by evaporation with the atmosphere. The average maximum temperature difference between a city and its surroundings is about 3°C (33).

Smog

The two major types of smog are *photochemical smog,* which is sometimes called L.A.-type smog or brown air; and *sulfurous smog,* which is sometimes referred to as London-type smog or gray air. Solar radiation is particularly important in the formation of photochemical smog (Figure 16.25). The reactions that occur in the development of photochemical smog are complex and involve both nitrogen oxides (NO_x) and organic compounds (hydrocarbons).

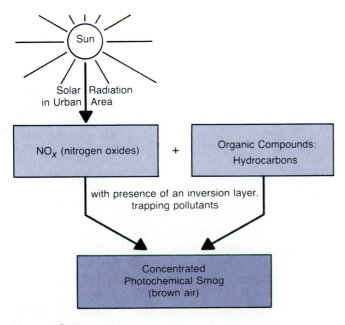

Figure 16.25
How photochemical smog may be produced.

The development of photochemical smog is directly related to automobile use. In southern California, for example, when commuter traffic begins to build up early in the morning, the concentrations of nitrogen oxide (NO) and hydrocarbons begin to increase. At the same time, the amount of nitrogen dioxide (NO_2) may decrease owing to the sunlight's action on NO_2 to produce NO plus atomic oxygen. The atomic oxygen is then free to combine with molecular oxygen to form ozone, so after sunrise the concentration of ozone also increases. Shortly thereafter, oxidized hydrocarbons react with NO to increase the concentration of NO_2 by midmorning. This reaction causes the NO concentration to decrease and allows ozone to build up, producing a midday peak in ozone and minimum in NO. As the smog matures, visibility may be greatly reduced (Figure 16.26) owing to light scattering by aerosols.

Sulfurous smog is produced primarily by burning coal or oil at large power plants. Sulfur oxides and particulates combine under certain meteorological conditions to produce a concentrated sulfurous smog (Figure 16.27).

Future Trends for Urban Areas

Air pollution levels in cities in developed countries have steadily improved, due to increased regulation of pollution. For example, sulfur oxide concentrations in New York and Chicago decreased markedly from the late 1960s to the early 1970s, and from 1980 to 1985 air pollution in the Los Angeles area was reduced by about 18 percent. A reduction in the number of pollution episodes in Los Angeles is due in part to regulatory efforts to control primary pollutants from automobiles. However, Los Angeles still has numerous air pollution episodes, in spite of increased pollution controls. Although air pollution has improved, the outlook may not be bright. If oil and gas become scarce and industrialized societies return to using coal, air quality in cities is projected to decline (34). In the United States, this could occur before the end of the century.

Exposure to air pollutants decreases with increased income; the city's poor often live where there is more pollution. For example, in St. Louis, Missouri, those with low income were subject in recent years to suspended particulates of 91.3 μg/ml of air; those with higher incomes lived where the air had a concentration of 64.9. Those who can afford it move away from pollution!

Cities in less developed countries with burgeoning populations are particularly susceptible to air pollution now and in the future; they don't have the financial base necessary to fight air pollution. They tend to be more concerned with basic survival and finding ways to house and feed their growing urban populations. Mexico City has a present population of 23 million people—projected to expand to 26 million by the end of the century,

(a)

(b)

Figure 16.26
Smog. Downtown Los Angeles photographed (a) on a relatively clear day and (b) on a smoggy day. (AP/Wide World Photos.)

making it the largest urban area in the world. Industry and power plants in Mexico City emit hundreds of thousands of tons of particulates and sulfur dioxide into the atmosphere. The city is at an elevation of about 7400 feet (2255 m) in a natural basin surrounded by mountains—a perfect situation for a severe air pollution problem. From Mexico City the mountains can rarely be seen now. Physicians report a steady increase in respiratory diseases, while headaches, irritated eyes, and sore throats are common.

▼ INDOOR AIR POLLUTION

In recent years buildings have been constructed more tightly to save energy. As a result, air is filtered through rather extensive systems in many buildings. Unless filters are maintained properly, indoor air can become polluted with a variety of substances, including smoke, chemicals, and disease-carrying organisms. For example, some investigators believe the virus responsible for Legionnaires' disease, a respiratory infection, multiplies and is transported through buildings by way of the air filters and the ventilation systems. In some buildings asbestos fibers are slowly released from insulation and other fixtures. People exposed to a particular type of asbestos fibers may develop a rare form of lung cancer. Both carbon monoxide and nitrogen dioxide may be released in homes from unvented or poorly vented gas stoves, furnaces, or water heaters. As final examples, consider formaldehyde, which is present in some insulation materials and wood products used in home construction, and the radioactive gas radon, which is present in some well water and building materials, such as concrete block and bricks if they are made from materials with a high radon concentration. Formaldehyde is known to cause irritation to ears, nose, and throat, and radon is suspected of causing lung and other cancers.

Modern urban structures are built of many substances, some of which release minute amounts of chemicals and other material into the nearby air. Buildings that lack a good system to recirculate the air with clean air are likely to have indoor pollution problems.

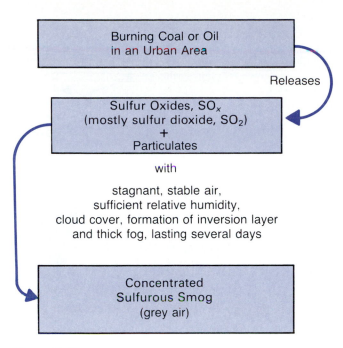

Figure 16.27
How concentrated sulfurous smog and smoke might develop.

Recommendations on how to improve indoor air quality will undoubtedly be forthcoming. Interestingly, people who lived centuries ago also suffered from indoor air pollution. In 1972 the body of a fourth-century Eskimo woman was discovered on St. Lawrence Island in the Bering Sea. The woman evidently was killed during an earthquake or landslide and her body frozen soon after death. Detailed autopsies showed that the woman suffered from black lung disease, which occasionally afflicts coal miners today. Anthropologists and medical personnel concluded that the woman breathed very polluted air for a number of years. They speculate that the air she breathed included hazardous fumes from lamps that burned seal and whale blubber, causing the black lung disease (35).

▼ CONTROL OF AIR POLLUTION

Reducing air pollution requires a variety of strategies tailored to specific sources and type of pollutants. For both stationary and mobile sources, the only reasonable strategies (excepting indoor pollution) have been to collect, capture, or retain pollutants before they enter the atmosphere. Other strategies may only make the problem worse, as happened at smelters in Sudbury, Ontario, Canada. The ores contain a high percentage of sulfur, and the smelter stacks emit large amounts of sulfur dioxide as well as particulates containing nickel, copper, and other toxic metals. Attempts to minimize the pollution problem close to the smelters by increasing the height of the smokestacks backfired by spreading the pollution over a larger area.

Pollution problems vary in different regions of the world, or even within the United States. For example, in the Los Angeles basin, nitrogen oxides and hydrocarbons are particularly troublesome because they combine in the presence of sunlight to form photochemical smog. Furthermore, in Los Angeles most of the nitrogen oxides and hydrocarbons are emitted from automobiles—a nonpoint source (32). In other urban areas, such as in Ohio and the Great Lakes region in general, air-quality problems result primarily from emissions of sulfur dioxide and particulates from industry and coal-burning power plants, which are point sources.

Because the problems vary so greatly from country to country and from region to region, it is often difficult to obtain both the monies and the political consensus for effective control. Nevertheless, effective control strategies do exist for many air pollution problems.

Control of Particulates

Particulates emitted from fugitive, point, or area stationary sources are much easier to control than are the very small particulates of primary or secondary origin released from mobile sources such as automobiles. As we learn more about these very small particles new control methods will have to be devised.

Control of coarse particulates from power plants and industrial sites (point or area sources) utilizes a variety of settling chambers or collectors, some of which are generalized in Figure 16.28. These methods settle out particles where they may be collected for disposal in landfills.

Particulates from fugitive sources (such as a waste pile) must be controlled on-site before they are removed (eroded) and carried by wind into the atmosphere. Control may involve protecting open areas, dust control, or reducing the effect of wind. For example, waste piles may be covered by plastic or other material and soil piles may be vegetated to inhibit wind erosion; water or a combination of water and chemicals may be spread to hold dust down; and structures or vegetation may be placed to lessen wind velocity near the ground, thus retarding wind erosion of particles.

Control of Automobile Pollution

Such pollutants as carbon monoxide, nitrogen oxides, and hydrocarbons in urban areas are best controlled by regulating automobile pollution, the source of most of the anthropogenic portion of these pollutants. Furthermore, control of these materials will also regulate the ozone in the lower atmosphere, where it forms from reactions with nitrogen oxides and hydrocarbons in the presence of sunlight.

Nitrogen oxide from automobile exhausts is controlled by recirculating exhaust gas, which dilutes the

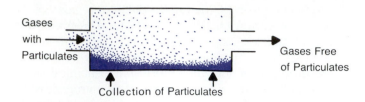

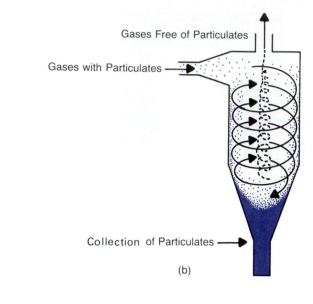

Figure 16.28
Devices being used to control emissions of particulates before they enter the atmosphere: (a) a simple settling chamber that collects particulates by gravity settling; (b) centrifugal collector, in which particulates are forced to the outside of the chamber by centrifugal force and then fall to the collection site.

air-to-fuel mixture being burned. The dilution reduces the temperature of combustion, decreases the oxygen concentration in the burning mixture (that is, it makes a richer fuel), and produces fewer nitrogen oxides. Unfortunately, this method increases hydrocarbon emissions. Nevertheless, exhaust recirculation to reduce nitrogen oxide emissions has been common practice in the United States since 1975 (36).

The two most common devices used to remove carbon monoxide and hydrocarbon emissions from automobiles are the catalytic converter and the thermal exhaust reactor (36). Both convert carbon monoxide to carbon dioxide, and hydrocarbons to carbon dioxide and water. In the thermal exhaust, or high-temperature, reactor, the addition of outside air (oxygen) helps achieve a more complete combustion of exhaust fumes. The catalytic converter works at a lower temperature and involves a metal catalyst material over which the exhaust gases are circulated with air. Oxidation then occurs, removing carbon monoxide and hydrocarbons from the exhaust. Catalytic converters have one major problem—the catalyst bed may be rendered ineffective by a number of substances, including lead additives in gasoline. As a result, the shift to nonleaded gasoline in recent years has been tremendous.

It has been argued that the automobile emission regulation plan in the United States has not been very effective in reducing pollutants. The pollutants may be reduced while a car is relatively new, but many people simply do not maintain their emission control devices over the life of the automobile. Also, some people disconnect smog control devices. It has been suggested

that effluent fees replace automobile controls as the primary method of regulating air pollution (37). Vehicles would be tested each year for emission control, and fees would be assessed on the basis of the test results. The fees encourage the purchase of automobiles that pollute less, and the annual inspections would ensure that pollution control devices are properly maintained. Although enforced pollution inspections are controversial, they are common in a number of areas and are expected to increase as air pollution abatement becomes imperative.

Control of Sulfur Dioxide

Sulfur dioxide emissions can be reduced by abatement measures performed before, during, or after combustion. The technology to clean up coal so that it will burn cleanly is already available, although the cost of removing the sulfur does make the fuel more expensive. However, if nothing is done, the consequences of burning sulfur-rich coal in the next decade or so will be very expensive indeed to future generations.

Changing from high-sulfur coal to low-sulfur coal seems an obvious solution to reducing sulfur dioxide emissions. Unfortunately, most low-sulfur coal is located in the western part of the United States, whereas most burning of coal occurs in the eastern part. Therefore, using low-sulfur coal is a solution only where it is economically feasible. Another possibility is cleaning relatively high-sulfur coal by washing it. By washing finely ground coal with water, the iron sulfide (the mineral pyrite) settles out because it is relatively more dense than coal. Although washing removes some of the sulfur, it is expensive. Another option is coal gasification, which converts relatively high-sulfur coal to a gas in order to remove the sulfur. The gas obtained from coal is quite clean and can be transported relatively easily, augmenting supplies of natural gas. The synthetic gas produced from coal is now fairly expensive compared to gas from other sources, but may become more competitive in the future (30).

During combustion of coal, a process known as fluidized-bed combustion can eliminate sulfur oxides. The process involves mixing finely ground limestone with coal and burning it in suspension. Sulfur may also be removed by a process known as limestone injection in multistage burners; this involves injecting ground limestone into a special burner. In both processes, the sulfur oxides combine with the calcium in the limestone to form calcium sulfides and sulfates, which may be collected.

Sulfur oxide emissions from stationary sources such as power plants can also be reduced by removing the oxides from the gases in the stack before they reach the atmosphere. Perhaps the most highly developed technology for the cleaning of gases in tall stacks is scrubbing (Figure 16.29). In this method the gases are treated with a slurry of lime (calcium oxide, CaO) or limestone (calcium carbonate, $CaCO_3$). The sulfur oxides react with the calcium to form insoluble calcium sulfides and sulfates, which are collected. The residue containing the calcium sulfides and sulfates, however, is a sludge that must be disposed of at a land disposal site. Because the

sludge can cause serious water pollution if it interacts with the hydrologic cycle, it must be treated carefully (30). Furthermore, scrubbers are expensive and add significantly (10–15%) to the total cost of electricity produced by burning coal.

Finally, an innovative approach has been taken at a large coal burning power plant near Mannheim, Germany. Smoke from combustion is treated with liquid ammonia (NH_3), which reacts with the sulfur to produce ammonium sulfate. The sulfur-contaminated smoke is cooled by outgoing clean smoke to a temperature that favors the reaction; then outgoing clean smoke is heated by incoming dirty smoke to force it out the vent. Waste heat from the cooling towers also heats nearby buildings, and the plant sells the ammonium sulfate to farmers as fertilizer in a solid granular form. The plant, finished in 1984, was built in response to tough pollution control regulations enacted to help abate acid precipitation, thought to contribute to pollution that is killing forests in Germany.

In Japan, pollution abatement has been successful. Japan had what was considered the most severe sulfur pollution problem in the world, and the health of the Japanese people was being directly affected. It was not uncommon to see residents of Yokohama and other areas wearing masks when out on the streets. The Japanese government in 1967 issued control standards, and between 1970 and 1975 the sulfur dioxide level was reduced by 50 percent, while the level of energy consumption more than doubled. New goals were set in 1973 and emission levels have steadily decreased. The

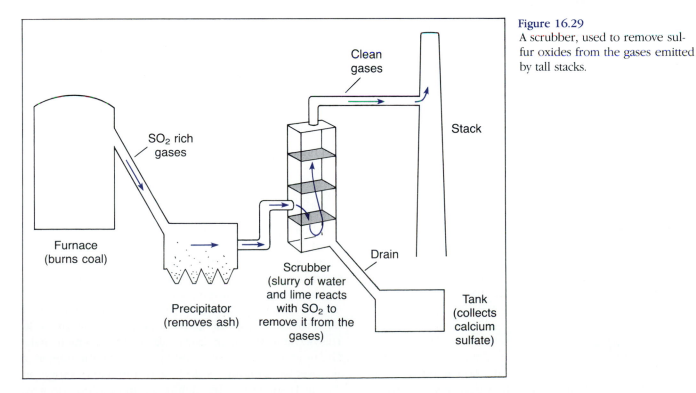

Figure 16.29
A scrubber, used to remove sulfur oxides from the gases emitted by tall stacks.

Table 16.3

Pollutant Standards Index[a]

PSI Index Value	Air Quality Level	Cautionary Statements	Health Effect Label	TSP (24-hour) μg/m³
500	Significant harm			1000
400	Emergency	All persons should remain indoors, keeping windows and doors closed. All persons should minimize physical exertion and avoid traffic.		875
300	Warning	Elderly and persons with diseases should stay indoors and avoid physical exertion. General population should avoid outdoor activity.	Hazardous (PSI > 300)	625
200	Alert	Elderly and persons with existing heart and lung disease should stay indoors and reduce physical activity.	Very unhealthful (PSI = 200 to 300)	375
100	NAAOS[b]	Persons with existing heart or respiratory ailments should reduce physical exertion and outdoor activity.	Unhealthful (PSI = 100 to 200)	260
50	50% of NAAQS[b]		Moderate (PSI = 50 to 100)	75[c]
0			Good (PSI = 0 to 50)	0

[a]One measure of air quality is the Pollutant Standards Index. It is a highly summarized health-related index based on five of the criteria pollutants: carbon monoxide, ozone, sulfur dioxide, total suspended particulates, and nitrogen dioxide. The PSI for one day will rise above 100 in a Standard Metropolitan Statistical Area when one of the five pollutants at one station reaches a level judged to have adverse short-term effects on human health.

[b]NAAQS = National Ambient Air Quality Standard.

[c]There are no index values reported at concentrations below those specified by Alert criteria.

[d]Annual primary NAAQS.

Source: Council on Environmental Quality.

Japanese have also begun to control nitrogen oxides as well. Power plants have met requirements for the most part by installing scrubbers known as flue gas desulfurization systems, which can remove over 95 percent of the sulfur from smokestacks. More than 1000 of these are in use in Japan today (38).

Air Quality Standards

Air quality standards are tied to emission standards that attempt to control air pollution. Countries that have developed air quality standards include France, Japan, Israel, Italy, Canada, the Soviet Union, Germany, Yugoslavia, Norway, and the United States. However, the standards for different countries vary considerably concerning acceptable levels of pollution over a particular time period. For example, the standard for sulfur oxides varies from 0.05/24 hr in the U.S.S.R. to 0.15/24 hr in Yugoslavia to 0.2/24 hr in Norway to 0.38/24 hr in Italy. The units are thousandths of a gram per cubic meter of air over a 24-hour period, and the large range in standards suggests a big difference in opinion concern-

SO² (24-hour) µg/m³	CO (8-hour) µg/m³	O₃ (1-hour) µg/m³	NO₂ (1-hour) µg/m³	General Health Effects
Pollutant level				
2620	57.5	1200	3750	
				Premature death of ill and elderly. Healthy people will experience adverse symptoms that affect their normal activity.
2100	46.0	1000	3000	
				Premature onset of some diseases in addition to significant aggravation of symptoms and decreased exercise tolerance in healthy persons.
1600	34.0	800	2260	
				Significant aggravation of symptoms and decreased exercise tolerance in persons with heart or lung disease, with widespread symptoms in the healthy population.
800	17.0	400	1130	
				Mild aggravation of symptoms in susceptible persons, with irritation symptoms in the healthy population.
365	10.0	240	(c)	
80ᵈ	5.0	120	(c)	
0	0	0	(c)	

ing acceptable levels. One problem in establishing air quality standards is the lack of agreement concerning what concentrations of pollutants cause environmental problems (37).

U.S. standards.

National Ambient Air Quality Standards (NAAQS) for the United States reflect the U.S. Clean Air Amendments of 1977, which define two levels or types of air quality standards. *Primary standards* are levels set to protect the health of people but which may not protect against damaging effects to structures, paint, and plants. *Secondary standards,* although designed to help prevent other environmental degradation, are the same, or nearly so, as the primary levels (37).

The 1977 amendments also set air quality standards for particular areas or land uses to prevent significant deterioration of air quality beyond baseline measurements (air quality before significant pollution is present). Three classes were established: national parks and wilderness areas (Class I); areas where moderate deterioration is allowed (Class II); and industrial areas (Class III). In practice all areas are considered Class II unless designated Class I or III (37).

To enforce U.S. air quality standards, states are required to submit to the Environmental Protection Agency (EPA) State Implementation Plans that indicate how and when they expect to conform to the standards. Revised plans may be submitted for areas where nonattainment of standards occurs. The EPA designates Air Quality Maintenance Areas for locations that may have problems maintaining air quality standards if potential new sources of air pollutants are not strictly regulated (37).

In the United States, air quality in urban areas is often reported as good, moderate, unhealthy, very unhealthy, or hazardous (Table 16.3). These levels are derived from monitoring the concentration of five major pollutants: total suspended particulates, sulfur dioxide, carbon monoxide, ozone, and nitrogen dioxide. During a pollution episode in Los Angeles, hourly ozone levels are reported and a first-stage smog episode begins if the primary National Ambient Air Quality Standard (NAAQS) (0.12 ppm) is exceeded. This corresponds to unhealthy air with a Pollutant Standard Index (PSI) between 100 and 300. A second-stage smog episode is declared if the PSI exceeds 300, a point   which the air quality is hazardous to all people. As the air quality decreases during a pollution episode, people are requested to remain indoors, minimize physical exertion, and avoid driving automobiles. Industry also may be requested to reduce emissions to a minimum during the episode.

Data from major metropolitan areas in recent years suggest a decline in the total number of unhealthful and very unhealthful days. Although air pollution has not been eliminated, the data indicate that the nation's air quality is improving. However, most urban areas such as New York and Los Angeles still have unhealthful air much of the time.

Cost of Controls

In the United States today over $25 billion per year is spent on air pollution controls; of this, about 40 percent is for facilities to reduce emissions from stationary sources. The nonfarm private-business sector alone spends about $5 billion/yr on capital outlays to support pollution control (36)—a great deal of money!

The cost and benefits of air pollution control are controversial subjects. Some argue that the present system of setting air quality standards is inefficient and unfair. They maintain that regulations are tougher for new sources than for existing ones, and even if benefits of pollution control exceed total costs, the cost of air pollution control varies widely from one industry to another. For example, consider the incremental control costs (cost to remove an additional unit of pollution beyond what is presently required) for utilities burning fossil fuels and for an aluminum plant. For a fossil fuel-burning utility, the cost is a few hundred dollars per additional ton of particulates removed compared to as much as several thousand dollars per ton for the aluminum plant (39). Some economists propose increasing the standards for utilities while relaxing or at least not increasing them for aluminum plants. This would lead to more cost-efficient pollution control while maintaining good air quality. However, the geographic distribution of various facilities will obviously determine the amount of prospective trade-offs (39).

Another economic consideration is that, as the degree of control of a pollutant increases, eventually a point is reached where the cost of incremental control (reducing additional pollution) is greater than the additional benefits. Because of this and other economic factors, it has been argued that fees and taxes for emitting pollutants might be preferable to attempting to evaluate uncertain costs and benefits, and that it makes more economic sense to enforce fees rather than standards. This debate is likely to go on for some time before any change is made in the present system.

Some variables that must be considered are shown in Figure 16.30. With increasing air pollution controls the capital cost to control air pollution increases, and as the controls for air pollution increase the loss from pollution damages decreases. The total cost is thus the sum of these two items, and the minimum is well defined in terms of a particular average pollution level. This graph indicates

Figure 16.30
Some of the relationships between economic cost and increasing air pollution controls. (Modified after S. J. Williamson, 1973, *Fundamentals of Air Pollution,* Addison-Wesley.)

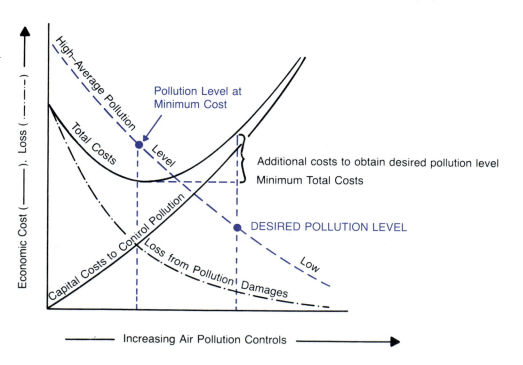

that if the desired pollution level is lower than that at which the minimum total cost occurs, then additional costs will be necessary. This type of diagram, while valuable for considering some of the major variables, does not consider adequately all of the loss from pollution damages. For example, long-term exposure to air pollution may aggravate or lead to chronic diseases in human beings, with a very high cost. How do we determine what portion of the cost is due to air pollution? Despite these drawbacks, it seems worthwhile

to reduce the air pollution level below some particular standard. Thus, in the United States, the ambient air quality standards have been developed as a minimum acceptable pollution level. However, alternatives such as charging fees or taxes for emissions should also be considered. If such charges are determined carefully and emissions are carefully monitored, the charges would encourage installation of control measures to avoid or lower fees or taxes. The end result would, we hope, be the same for both—better air quality.

▼ ▼ ▼ SUMMARY AND CONCLUSIONS

The concept of global change is not a new one. Geologists have for over 100 years studied the history of the earth, and the changes that have taken place during the several billion years of earth history. Today people are very concerned about changes taking place in our atmosphere, including: (1) increase in carbon dioxide, hypothetically related to global warming; (2) the ozone layer and depletion of ozone, which may affect our health; (3) increased emissions of sulfur and nitrogen oxides, which are precursors of acid rain; and (4) increasing urban air pollution.

The main goal of the emerging integrated study known as earth systems science is to obtain a basic understanding of how our planet works, and how the various components, such as the atmosphere, oceans, and solid earth, interact. An important secondary goal is to predict global changes that are likely to occur within a time frame of several decades, and thus are of particular importance to people.

Methods of studying global change include: (1) examination of the geologic record from lake sediments, glacial ice, tree rings, and other earth material; (2) gathering of real-time data from monitoring stations; and (3) development of mathematical models to predict change.

The trapping of heat by the atmosphere is generally referred to as the greenhouse effect. Water vapor and several other gases (including carbon dioxide, methane, and chlorofluorocarbons) tend to trap heat and warm the earth because they absorb some of the heat energy radiating from the earth. Since about 1860 there has been an exponential growth of the concentration of carbon dioxide in the atmo-

sphere, which is responsible for approximately 60 percent of the anthropogenic greenhouse effect.

It is estimated by some model studies that the average global temperature may rise about 1.2°C in the next several decades due to increases in the greenhouse gases and trapping of heat. Specific effects are difficult to predict, but include a rise in global sea level and potential significant changes in rainfall patterns, soil moisture relationship, and other factors important to agriculture.

It was first hypothesized in 1974 that release of chlorine by photodissociation of chlorofluorocarbon molecules was responsible for causing a chain reaction that could significantly deplete stratospheric ozone. The importance of this hypothesis was dramatically shown in 1985 with the discovery of the Antarctic ozone hole. The recent recognition of the Arctic ozone hole is an even greater concern for the world's population.

Depletion of stratospheric ozone allows greater amounts of ultraviolet radiation to reach the surface of the earth. The potential effect of this includes increase in all types of skin cancers and damage to plants and marine life. It is expected that international cooperation will lead to much tighter restrictions on the production and emission of chlorofluorocarbons (CFCs). However, because CFCs are so stable in the environment, such reductions will have only a minor effect on the ozone problem before the end of the century.

Acid rain, which refers to both wet and dry acid deposition, is a serious environmental problem in many parts of the world today. Emissions of sulfur and nitrogen oxides into the at-

mosphere undergo complex chemical processes, and transport to eventually become sulfuric or nitric acid. Effects of acid rain include: (1) damage to lake ecosystems, (2) damage to buildings and monuments, and (3) damage to forest ecosystems.

Some global changes, such as ozone depletion and the greenhouse effect, are coupled. For example, the chlorofluorocarbons released in the lower atmosphere have a strong greenhouse effect. It is these same CFC molecules that diffuse to the stratosphere, where chlorine is released by photodissociation, damaging the ozone layer.

Every year approximately 250–300 million metric tons of primary pollutants enter the atmosphere above the United States from processes related to human activity. Considering the enormous volume of the atmosphere, this is a relatively small amount of material. If it were distributed uniformly, there would be little problem with air pollution. Unfortunately, the pollutants are not generally evenly distributed, but rather are concentrated in urban areas, or in other areas where the air naturally lingers.

The two major types of pollution sources are stationary and mobile. Stationary sources have a relatively fixed position, and include point sources and area sources.

Two main groups of air pollutants are primary and secondary. Primary pollutants are those emitted directly into the air: particulates, sulfur oxides, carbon monoxide, nitrogen oxides, and hydrocarbons. Secondary pollutants are those produced through reactions among primary pollutants and other atmospheric compounds. A good example of a secondary pollutant is ozone, which forms over urban areas through photochemical reactions among primary pollutants and natural atmospheric gases.

Two major types of smog are photochemical and sulfurous. Each type causes particular environmental problems that vary with geographic region, time of year, and local urban conditions.

Meteorological conditions greatly affect whether polluted air is a problem in a particular urban area. In particular, restricted lower-atmosphere circulation associated with temperature inversion layers may lead to pollution events.

A city creates an environment different from that of the surrounding area. Cities change the local climate and, in general, are cloudier, warmer,

rainier, and less humid. A growing problem in urban areas is indoor air pollution.

Methods to control air pollution are tailored to specific sources and types of pollutants. These methods vary from settling chambers for particulates to catalytic converters to remove carbon monoxide and hydrocarbons from automobile exhaust to scrubbers or combustion processors that use lime to remove sulfur before it enters the atmosphere.

Air quality in urban areas is usually reported in terms of whether the quality is good, moderate, unhealthy, very unhealthy, or hazardous. These levels are defined in terms of the Pollutant Standards Index (PSI) and National Ambient Air Quality Standards

(NAAQS). The nation's air quality seems to have improved in recent years; however, in numerous areas urban air quality is still unhealthful a good deal of the year.

The relationships between emission control and environmental cost are complex. The minimum total cost is a compromise between capital costs to control pollutants and losses or damages resulting from such pollution. If additional controls are used to lower the pollution to a more acceptable level, then additional costs are incurred, which can increase quite rapidly. Some believe that our present system of setting air quality standards is inefficient and unfair and that we should change to a system that charges fees or taxes for emissions.

▼ ▼ ▼ REFERENCES

1. CLOUD, P. 1990. Personal written communication.
2. KERSHAW, S. 1990. Evolution of the earth's atmosphere and its geological impact. *Geology Today,* March-April, 55–60.
3. COUNCIL ON ENVIRONMENTAL QUALITY. 1990. *Environmental Trends 1989.*
4. NASA. 1990. *EOS: A mission to planet earth.* Washington, D.C.
5. EARTH SYSTEM SCIENCES COMMITTEE. 1988. *Earth system science: A preview.* Boulder, CO: University Corporation for Atmospheric Research.
6. HARDIN, GARRETT. 1990. University of California lecture (personal oral communication).
7. MARSH, W. M., and DOZIER, J. 1981. *Landscape.* Reading, Mass.: Addison-Wesley.
8. GATES, D. M. 1980. *Biophysical ecology.* New York: Springer Verlag.
9. ERLICH, P. R.; ERLICH, A. H.; and HOLDREN, J. P. 1970. *Ecoscience.* San Francisco: W. H. Freeman.
10. CHUMBLEY, C. A.; BAKER, R. G.; and BETTIS, E. A., III. 1990. Midwestern Holocene paleoenvironments revealed by floodplain deposits in northeastern Iowa. *Science* 249:272–74.
11. POST, W. M.; PENG, T.; EMANUEL, W. R.; KING, A. W.; DALE, V. H.; and DE ANGELIS, D. L. 1990. The global

carbon cycle. *American Scientist* 78(4): 310–26.
12. MOSS, M. E., and LINS, H. F. 1989. *Water resources in the twenty-first century.* U.S. Geological Survey Circular 1030.
13. NASA GODDARD SPACE FLIGHT CENTER, principal contributors Rind, D., and Lebedeff, S. 1984. *Potential climatic impacts of increasing atmospheric CO_2 with emphasis on water availability and hydrology in the United States.* U.S. Environmental Protection Agency.
14. TITUS, J. G., and SEIDEL, S. R. 1986. In *Effects of changes in the stratospheric ozone and global climate:* Vol. 1, edited by J. G. Titus. U.S. Environmental Protection Agency. 3–19.
15. TITUS, J. G.; LEATHERMAN, S. P.; EVERTS, C. H.; MOFFATT AND NICHOL ENGINEERS; KRIEBEL, D. L.; and DEAN, R. G. 1985. *Potential impacts of sea level rise on the beach at Ocean City, Maryland.* U.S. Environmental Protection Agency.
16. RODHE, H. 1990. A comparison of the contribution of various gases to the greenhouse effect. *Science* 248(6): 1217–19.
17. JONES, P. D., and WIGLEY, T. M. L. 1990. Global warming trends. *Scientific American* 263(2): 84–91.
18. MOLINA, M. J., and ROWLAND, F. S.

1974. Stratospheric sink for chlorofluoromethanes: Chlorine atom–catalyzed destruction of ozone. *Nature* 249:810–12.
19. ROWLAND, F. S. 1989. Chlorofluorocarbons and the depletion of stratospheric ozone. *American Scientist* 77:36–45.
20. FARMAN, J. C.; GARDINER, B. G.; and SHANKLIN, J. D. 1985. Large losses of total ozone in Antarctica reveal seasonal ClO_x/NO_x interaction. *Nature* 315:207–10.
21. KERR, R. A. 1988. A shallower ozone hole, as expected. *Science* 28 (October):515.
22. MELACK, J. M. 1989. Personal oral communication.
23. OFFICE OF TECHNOLOGY ASSESSMENT. 1984. Balancing the risks. *Weatherwise* 37: 241–9.
24. CANADIAN DEPARTMENT OF ENVIRONMENT. 1984. *The acid rain story.*
25. LIPPMANN, M., and SCHLESINGER, R. B. 1979. *Chemical contamination in the human environment.* New York: Oxford University Press.
26. U.S. ENVIRONMENTAL PROTECTION AGENCY. 1980. *Acid rain.*
27. RAMPINO, M. R., and SELF, S. 1984. The atmospheric effects of El Chicón. *Scientific American* 250: 48–57.
28. NATIONAL PARK SERVICE. 1984. Air resources management manual.

29. PITTOCK, A. B.; FRAKES, L. A.; JENS-SEN, D.; PETERSON, J. A.; and ZILL-MAN, J. W., eds. 1978. *Climatic change and variability: A southern perspective.* (Based on a conference at Monash University, Australia, 7–12 December, 1975.) New York: Cambridge University Press.

30. ANTHES, R. A.; CAHIR, J. J.; FRASER, A. B.; and PANOFSKY, H. A. 1981. *The atmosphere.* 3rd ed. Columbus, Ohio: Charles E. Merrill.

31. DETWYLER, T. R., and MARCUS, M. G., eds. 1972. *Urbanization and the environment.* North Scituate, Mass.: Duxbury Press.

32. GATES, D. M. 1972. *Man and his environment: Climate.* New York: Harper & Row.

33. LYNN, D. A. 1976. *Air pollution—Threat and response.* Reading, Mass.: Addison-Wesley.

34. COUNCIL ON ENVIRONMENTAL QUALITY AND THE DEPARTMENT OF STATE. 1980. *The global 2000 report to the President: Entering the twenty-first century.*

35. ZIMMERMAN, M. R. 1985. Pathology in Alaskan mummies. *American Scientist* 73: 20–5.

36. STOKER, H. S., and SEAGER, S. L. 1976. *Environmental chemistry: Air and water pollution.* 2nd ed. Glenview, Ill.: Scott, Foresman.

37. STERN, A. C.; BOUBEL, R. T.; TURNER, D. B.; and FOX, D. L. 1984. *Fundamentals of air pollution.* 2nd ed. Orlando: Academic Press.

38. ANONYMOUS. 1981. How many more lakes have to die? *Canada Today* 12 (2): 1–11.

39. CRANDALL, R. W. 1983. *Controlling industrial pollution.* Washington, DC. The Brookings Institution.

CHAPTER SEVENTEEN

Landscape Evaluation

Evaluating the landscape for such purposes as land-use planning, site selection, construction, and environmental impact is common practice. A primary role of geologists in landscape evaluation is to provide geologic information and analysis before planning, design, and construction. As a member of the landscape evaluation team, geologists commonly work with geographers, civil engineers, architects, planners, lawyers, and public administrators. Specific information geologists supply as part of the landscape evaluation process varies from area to area, and from project to project, but often centers on factors such as:

▼ Physical and chemical properties of earth materials, including soil and rock type, engineering properties of soil and rock, potential for building materials, and acceptability for waste disposal.

▼ Extent of natural hazards present, including seismic risk, slope instability, volcanic activity, flooding, erosion potential, and others.

▼ Depth to water table and groundwater flow characteristics.

▼ Depth to bedrock.

▼ LAND USE

Land use in the conterminous United States is dominated by agriculture and forestry, with only a small portion of the land (about 3 percent) used for urban purposes. Conversion of rural land to nonagricultural uses is currently about 9,000 square kilometers per year. About one-half of the conversion is for wilderness areas, parks, recreation areas, and wildlife refuges. The other half is for urban development, transportation networks and facilities. Although on a national scale, the conversion of rural lands to intensive urban uses may appear as a relatively small amount, it often occurs in rapidly growing urban areas where it may be viewed as destroying agricultural land, and intensifying existing urban environmental problems. Urbanization in more remote areas with high scenic and recreation value is often viewed as potentially damaging to important ecosystems.

An important contribution of earth scientists to landscape evaluation is emphasizing that not all land is the same, and that particular physical and chemical characteristics of the land may be more important to society than geographic location. Earth scientists recognize that there is a limit to our supply of land, and that we must strive to plan, so that suitable land is available for specific uses for this generation and those that follow (1).

The need for land near urban areas in some instances has led to application of the concept of sequential land use rather than permanent, exclusive use. The concept of sequential use of the land is consistent with the fundamental principle we discussed earlier: that effects of land-use are cumulative, and therefore, we have a responsibility to future generations. The basic idea is that after a particular activity (perhaps mining, or a landfill operation) has been completed, the land is reclaimed for another purpose.

There are several examples of sequential land use. Sanitary landfill sites have been planned so that when the site is completed, the land will be used for recreational purposes, such as a golf course. The city of Denver used abandoned sand and gravel pits for sanitary landfill sites that today are a parking lot and the Denver Coliseum (Figure 17.1) (2). Enormous underground limestone mines in the Missouri cities of Kansas City, Springfield, and Neosho have been profitably converted to warehousing and cold storage sites, offices, and manufacturing plants (2). Other possibilities also exist, such as use of abandoned surface mines for parking below shopping centers, chemical storage, and petroleum storage (Figure 17.2) (2).

▼ ENVIRONMENTAL GEOLOGY MAPPING

The primary goal of environmental geology mapping is to help the planner (3), so geologic information and engineering properties of earth materials must be presented in a way that is easy to interpret. That is, environmental geology maps must contain important information necessary for planning and eliminate extraneous data (3).

From a pragmatic view, the best environmental geology maps are those that combine geologic and hydrologic information in terms of engineering properties. The landscape may then be mapped in terms of a specific land use, as is currently done for soils. The idea

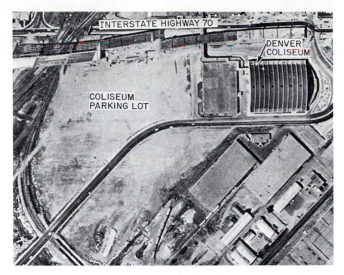

Figure 17.1

Examples of sequential land use in Denver, Colorado. Gravel pits were used as a sanitary landfill site as long ago as 1948. Today the Denver Coliseum and parking lots cover the landfill site. (Photos courtesy of Missouri Geological Survey.)

is to produce a series of maps, one for each possible land use. Thus, an *environmental geology map* is essentially a combination of geologic and hydrologic data expressed in less technical terms to facilitate general understanding by a large audience.

Interpretive environmental geology maps showing suitability for a particular land use may include a color code: green for "go" (favorable conditions); yellow for "caution"; and red for "stop" (unfavorable conditions or problem area). Ideally, interpretive maps should be prepared for geologic hazards such as land subject to flooding, landslide susceptibility, potential for ground rupture and seismic shaking from earthquakes, and soil conditions such as potential for liquefaction, potential to expand and contract, and potential for production of radon gas. Under ideal conditions a master hazard map

(a)

(b)

(c)

Figure 17.2

Examples of sequential land use of underground mines in Missouri, for (a) freezer storage and (b and c) warehousing. (Photos by J. D. Vineyard, courtesy of Missouri Geological Survey.)

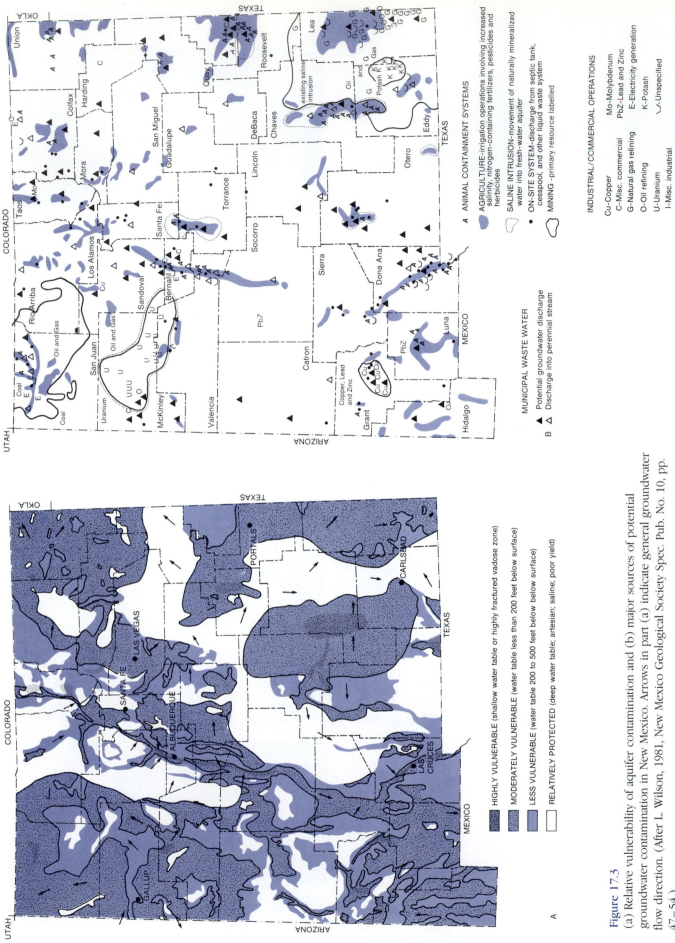

Figure 17.3

(a) Relative vulnerability of aquifer contamination and (b) major sources of potential groundwater contamination in New Mexico. Arrows in part (a) indicate general groundwater flow direction. (After L. Wilson, 1981, New Mexico Geological Society Spec. Pub. No. 10, pp. 47–54.)

HIGHLY VULNERABLE (shallow water table or highly fractured vadose zone)

MODERATELY VULNERABLE (water table less than 200 feet below surface)

LESS VULNERABLE (water table 200 to 500 feet below surface)

RELATIVELY PROTECTED (deep water table; artesian; saline; poor yield)

A

▲ ANIMAL CONTAINMENT SYSTEMS

A AGRICULTURE-irrigation operations involving increased salinity, nitrogen-containing fertilizers, pesticides and herbicides

⬚ SALINE INTRUSION-movement of naturally mineralized water into fresh-water aquifer

• ON-SITE SYSTEM-discharge from septic tank, cesspool, and other liquid waste system

◯ MINING-primary resource labelled

MUNICIPAL WASTE WATER

▲ Potential groundwater discharge

B △ Discharge into perennial stream

INDUSTRIAL/COMMERCIAL OPERATIONS

Cu-Copper

C-Misc. commercial

G-Natural gas refining

O-Oil refining

U-Uranium

I-Misc. industrial

Mo-Molybdenum

PbZ-Lead and Zinc

E-Electricity generation

K-Potash

◡-Unspecified

could be prepared. A prospective buyer or developer could consult the map and easily determine whether the proposed property has potential problems.

Interpretive maps can also be prepared for citing of facilities such as solid waste disposal. Such a map was prepared for De Kalb County, Illinois, by the Illinois Geological Survey. The evaluation considered topography, type and thickness of unconsolidated material, present and potential sources of surface water and groundwater, and type of bedrock. Sites with a minimum of problems are colored green, while less favorable areas are designated yellow. Acceptable waste disposal sites in areas colored yellow on the map can be found but require careful preliminary investigation. Areas shown by red on the maps are the least favorable for disposal of solid waste. However, even within the red areas there may be sites that would make satisfactory landfills following detailed evaluation and engineering (4). It is emphasized that interpretive maps are seldom drawn to appropriate scale for detailed evaluation. They serve as general guidelines to assist individuals and other parties interested in purchase of land.

Regional maps that delineate an area's relative vulnerability to environmental disruption are also valuable in planning future development and land use. Figure 17.3, for example, shows major existing sources of potential groundwater contamination and the relative vulnerability of aquifers to pollution from surface discharge in New Mexico. Relative vulnerability to groundwater pollution was predicted by considering aquifer type, depth to the water table, permeability, and presence or absence of natural barriers to migration of pollutants (5). The maps thus present geologic and hydrologic data as well as sources of potential groundwater pollution in a format useful to regional planners.

Environmental geology maps such as those in Illinois and New Mexico are valuable to the planner, but a larger-scale, more detailed map that shows accurate geologic boundaries is preferable, although not always practical—map scale is based on the accuracy and amount of available geologic data, and natural geologic boundaries are often gradual rather than sharp (4).

Another interesting approach to environmental mapping is the concept of an **environmental resource unit** (ERU). This method uses a multidisciplinary team approach to define and analyze the total natural environment—geologic, hydrologic, and biologic. A portion of the environment with a similar set of physical and biological characteristics is called an environmental resource unit. Ideally, every ERU is a natural division characterized by specific patterns or assemblages of structural components (rocks, soils, vegetation, etc.) and natural processes (erosion, runoff, soil genesis, productivity of wildlife, etc.) (3).

A study site of 10.4 square kilometers near Morrison, Colorado, illustrates the concept of defining and working with environmental resource units. Figure 17.4 shows the topography and the area, and Figure 17.5 shows the four ERUs. The first is the *mountain forest* ERU—ridge and ravine topography supporting a pine and Douglas fir forest. Running-water and mass-wasting processes are important in modifying the land, and the dominant ecological control is the moisture regime that varies with elevation and exposure. The second is the *floodplain forest* ERU—recent alluvial (stream-deposited) material deposited by channel and overbank stream flow. This ERU supports a cottonwood-willow forest that needs abundant water. The third is the *hogback wood* and *grassland* ERU—a series of tilted sedimentary rock units with steep scarp slopes, gentle dip slopes (Figure 17.4), and gentle debris slopes or level bottoms. Mass-wasting processes are controlled by slope and exposure. The scarp slopes and bottomlands support grass, and the dip slopes support junipers. The fourth is the *Pleistocene*

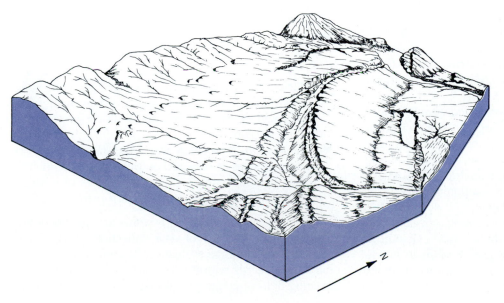

Figure 17.4
Physiographic diagram of the Morrison Test Site. (From Turner and Coffman, "Geology for Planning: A Review of Environmental Geology," *Quarterly of the Colorado School of Mines,* vol. 68, no. 3, 1973.)

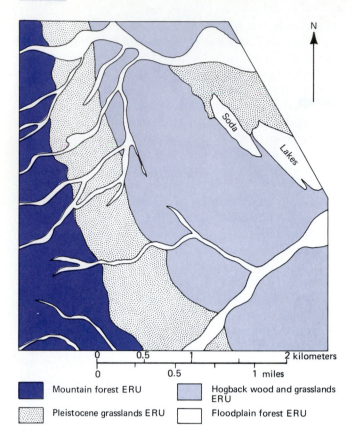

Soda

Lakes

N

	0	0.5	1		2 kilometers

| | 0 | 0.5 | | 1 miles |

■ Mountain forest ERU

▨ Hogback wood and grasslands ERU

▨ Pleistocene grasslands ERU

□ Floodplain forest ERU

Figure 17.5

Environmental resource units for the Morrison Test Site. (After Turner and Coffman, "Geology for Planning: A Review of Environmental Geology," *Quarterly of the Colorado School of Mines,* vol. 68, no. 3, 1973.)

grasslands ERU—a complex of older alluvial and mass-wasting material deposited by processes that are no longer operating. Natural vegetation is grass, but in the study area, much of the ERU is used for agricultural purposes (3).

Environmental resource units are used to establish patterns of structure and process and are valuable in establishing land capabilities and suitabilities. The ERU concept is most valuable in defining areas in a systems approach that involves the entire environment. These units may then be analyzed separately or collectively for a specific land use.

The ultimate value of ERUs is realized when multidisciplinary teams, including geologists, soil scientists, biologists, landscape architects, engineers, and planners, use remote sensing along with field survey techniques to identify different units and then analyze the data for planning purposes by means of computer-drawn maps for rapid data analysis and display.

The Franconia area in Fairfax County, Virginia, less than 15 kilometers southwest of the District of Columbia, offers a good example of the integration of environmental resource units with earth science maps to plan land use that is sensitive to the natural setting (Figure 17.6).

The synthesis resulting in the final land-use capability map is an example of computer composite mapping using input of topographic, geologic, and hydrologic data. This map has had a significant impact on decisions to rezone existing land or allow new development in the Franconia area.

▼ GEOGRAPHIC INFORMATION SYSTEM

The computer composite map that produced the land-use capability map for the Franconia area in Virginia is an early example of the application of an information system to environmental work. In recent years the concept has evolved to what is known as the *Geographic Information System* (GIS). The GIS is a form of new technology that is capable of storing, retrieving, transforming, and displaying spatial environmental data (6). The spatial data that is used in the GIS may be points, lines, or areas. The following are examples of the GIS spatial data: water wells and intersections of roads are points; rivers, roads, and faults are lines; and rock type or soil type define areas. It is important to recognize that the Geographic Information System is more than simply a way to categorize and store data. The system has the ability to manipulate data and thus create a new product. That product may be a land-use capability map, as shown in Figure 17.6, or it may be a sophisticated analysis of environmental variables.

The Geographic Information System has four essential elements (7):

▼ *Data acquisition:* collecting aerial photographs, maps, satellite imagery, census data, and so forth.

▼ *Data processing and management:* recording data from the maps and other sources in a systematic way. This process involves the creation of a computer data base where all the data are stored and accurately identified and located. Important here is developing consistent ways that data may be entered, deleted, updated, and retrieved.

▼ *Data manipulation and analysis:* this is the part of the GIS that can develop new information. For example, as shown earlier for the Franconia area (Figure 17.6), topographic, geologic, and hydrologic data may be combined and overlain with other information, such as data on landforms, to evaluate slope stability.

▼ *Generation of products:* this is the stage in GIS work that produces the desired output. This might include land capability maps, statistical analysis of environmental variables, or natural hazard maps.

Geographic Information Systems are fast becoming the dominant method of data collection, storage, and retrieval for landscape evaluation. Therefore, it behooves the serious earth scientist to become familiar with this exciting field of geographical research and application.

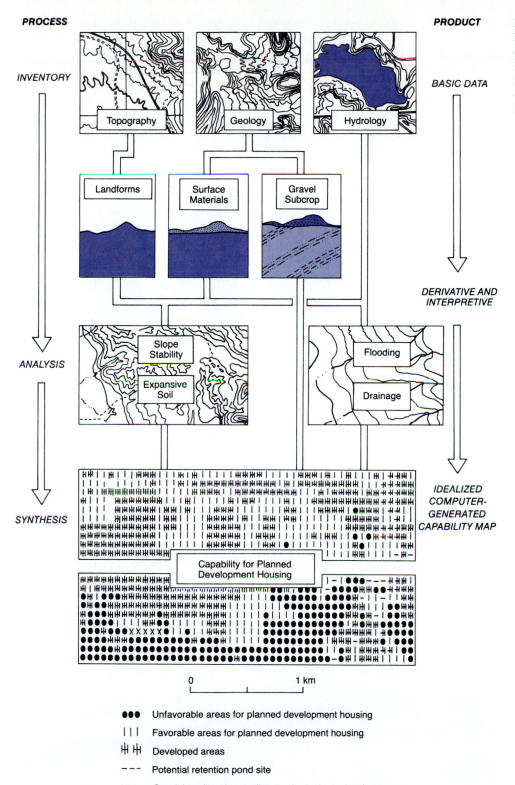

INVENTORY

Topography

Geology

Hydrology

Landforms

Surface
Materials

Gravel
Subcrop

ANALYSIS

Slope
Stability

Expansive
Soil

Flooding

Drainage

SYNTHESIS

Capability for Planned
Development Housing

0   1 km

●●●   Unfavorable areas for planned development housing

| | | Favorable areas for planned development housing

╫ ╫╫   Developed areas

– – –   Potential retention pond site

xxx   Special engineering studies required prior to development

PRODUCT

BASIC DATA

DERIVATIVE AND
INTERPRETIVE

IDEALIZED
COMPUTER-
GENERATED
CAPABILITY MAP

Figure 17.6
Flow chart showing the use of
earth science information for the
Franconia area, Virginia. (Modi-
fied from A. J. Frolich, and A. D.
Garnaas, U. S. Geological Survey
Professional Paper 950, 1978.)

▼ SITE SELECTION AND EVALUATION

Site selection is the process of evaluating a physical
environment that will support human activities. It is a task
shared by professionals in engineering, landscape architecture, planning, earth science, social science, and
economics, and thus involves a multidimensional approach to landscape evaluation.

In recent years, a philosophy of site evaluation
known as **physiographic determinism** has emerged (8).
The thrust of this philosophy is "design with nature" and
involves application of ecological principles to planning.

That is, rather than laying down an arbitrary design or plan for an area, it may be more advantageous to find a plan that nature has already provided (8). With this method, planners take advantage of a site's physical conditions in such a way as to maximize the amenities of the landscape while minimizing social and economic expenditures whenever possible. For example, if the soils of an area vary considerably in their ability to corrode, then the best corridor for an underground pipeline might not be the shortest route. Rather, the route might be designed so that the greatest possible length lies in the soils with the least corrosive potential. Although such a route is slightly longer and requires a higher initial cost, it may, by taking advantage of nature, produce tremendous savings in maintenance and replacement costs (9).

Although the philosophy of working in harmony with nature is obviously advantageous, it is often overlooked in the siting process. People still purchase land for various activities without considering whether the land use they have in mind is compatible with the site they have chosen. There are well-known examples of poor siting that have resulted in increased expense, limited production, or even abandonment of partially completed construction. Construction of a West Coast nuclear power facility was terminated when fractures in the rock (active faults) and possible serious foundation problems arose (Figure 17.7). The productivity of a large chicken farm in the southeast was greatly curtailed because the property was purchased before it was determined whether there was sufficient groundwater to meet the projected needs. A housing developer in northern Indiana purchased land and built country homes in one of the few isolated areas where bedrock (shale) is at the surface. Thus, septic-tank systems had to be abandoned and a surface sewage treatment facility had to be built. The rock also made it much more expensive to excavate for basements and foundations.

The goal of site selection for a particular land use is to ensure that site development is compatible with both the possibilities and limitations of the natural environment. The earth scientist's role may be significant in providing crucial geologic information to help accomplish this goal. The type of information geologists provided includes: soil and rock types, rock structure (especially fractures), drainage characteristics, groundwater characteristics, landform information, and estimates of possible hazardous earth events and processes, such as floods, landslides, earthquakes, and volcanic activity. The engineering geologist also takes samples, makes tests, and predicts the engineering properties of the earth materials.

Process of Site Selection

The process of site selection begins with a thorough understanding of what the general purposes and specific requirements of the designed activity are and for whom the site is being selected. The next step is an analysis of possible alternative sites in terms of physical limitations and cost. The final step is to recommend which site or sites are most advantageous for the desired land use (10).

The geologic aspect requires these steps:

▼ Collection of existing data, such as geologic reports, topographic maps, aerial photographs, soil maps, water surveys, engineering reports, and climate records

▼ Reconnaissance of likely sites to note potential problems and possibilities

▼ Collection of necessary new data, including soil, rock, and water analyses

▼ Determination of the magnitude and importance of geologic limitations and possibilities of the site or sites

The *magnitude* of geologic limitations and possibilities in site selection and evaluation refers to the degree,

Figure 17.7

Nuclear power development was abandoned due to an earthquake hazard at this site at Bodega Bay, California, after expenditure of millions of dollars and years of time in site preparation. This waste of time and money could have been avoided had data on the fault been available earlier. (From U.S. Geological Survey Circular 701, 1974.)

extensiveness, or scale, while the *importance* is a measure of the weight a particular limitation or possibility should be assigned. For example, if a site evaluation for the foundation of a large dam or nuclear facility reveals that the bedrock has several small fractures, the *magnitude* of the limitation is small. However, the *importance* may be great if the faults are active and less if they prove to be inactive. The presence of an active fault is obviously significant because it could cause the site to be abandoned, whereas inactive fractures can be treated by specific engineering procedures.

▼ METHODS OF SITE SELECTION AND EVALUATION

Although specific methods of determining a site and its suitability obviously vary with the intended purpose for the particular land use, a few methods are common for evaluating the desirability of a particular site for relatively large-scale land uses, such as construction of reservoirs, highways, and canals, and development of parks and other recreational areas. Two common methods are *cost-benefits analysis* and *physiographic determinism*, determining the best site by maximizing physical and social benefits while minimizing social costs. The first approach is strictly economical, while the second balances economics with less tangible aspects, such as aesthetics.

Cost-Benefits Analysis

Cost-benefits analysis is a way to assess the long-range desirability of the particular project, such as construction of a highway or development of a recreational site. The idea is to develop the benefits-to-cost ratio; that is, the total benefits in dollars over a period of time are compared to the total cost of the project, and the most desirable projects are those for which the benefits-to-cost ratio is greater than one (11).

Cost-benefits analysis assumes that all relevant costs and benefits can be determined. There are three difficult aspects: which costs and which benefits are to be evaluated; how they are to be evaluated; and how the intangible costs and benefits, such as aesthetic degradation or improvement, are to be evaluated (11).

Cost-benefits analysis is common for three general types of projects: *water resource* projects, such as irrigation systems, flood control, and hydroelectric power systems, as well as multipurpose schemes, such as reservoirs used for flood control, power generation, and recreation; *transportation* projects, such as highways, roads, railways, and river and canal work to facilitate navigation; and *land-use* projects, such as urban renewal and recreation (11).

Examining some of the costs and benefits for a hypothetical flood-control project will further illustrate the method. The initial tangible costs of a flood-control project are relatively straightforward, but evaluation of repair and maintenance costs and intangible costs over the life of the project is much more difficult. The main benefit from flood control is averting loss of property. (*Loss* here refers primarily to different types of assets, such as property, furnishings, and crops.) The general principle is to evaluate the annual flood loss and damage based on the most probable flood levels for several recurrence intervals, say, the 10-year and 25-year floods. This estimate of annual loss and damage is then taken as the maximum yearly amount that people would be willing to pay for flood control. Other benefits to consider include reducing the number of deaths by drowning and eliminating recurring costs of evacuating flood victims, emergency sandbag work, and incidence of sanitary failure (11).

Physiographic Determinism

An interesting graphic method of site selection based on physiographic determinism, or designing with nature, is structured such that the appropriate site literally selects itself (12). The method uses physical, social, and aesthetic data, and the objective is to maximize social benefit while minimizing social cost.

The actual methodology in selecting a site or a route for a highway, regional development plan, or other land use involves selecting a group of factors that reflect physical and social values. For each factor, three or more grades of the value are mapped on transparencies in tones from clear to dark grey. The darker tones indicate areas of conflict, where physical conditions are not appropriate or social values conflict with the desired land use. When the transparencies (one for each factor) are superimposed on one another, the darkest areas represent those places with the greatest physiographic obstructions and social costs, and the lighter areas represent places where the physical limitations (construction costs and social costs) are least. By this method, the best sites appear as the lightest areas on the set of superimposed transparencies. No matter how dark the entire area may look, there will always be relatively lighter areas (12). This method of overlays was an early form of information system, and can now be handled in a Geographic Information System as discussed earlier.

The method of determining lowest social cost in site selection is innovative in addressing social values as well as traditional physical-economic values, but it is not above criticism. It is extremely subjective, in that the person who is evaluating chooses the factors and thus essentially builds the formula used in the evaluation. For example, in evaluating the siting of a highway route, the evaluator may decide that property values are important and that the darker tone should be assigned to high property values. This decision essentially prescribes that the highway should be routed away from people, and most of all from wealthy people, a point some planners would vigorously argue against (8). Another criticism is that all factors, whether physical or social values, are weighted the same; that is, a geologic factor such as suitability of soil for foundation material or susceptibility to erosion is weighted equally with social values, such as scenic value or recreational property value. Regardless of these shortcomings, the method of physiographic determinism in evaluating both physical-economic factors and social values is a step in the right direction (8).

▼ SITE EVALUATION FOR ENGINEERING PURPOSES

The geologist has a definite role in the planning stages of site development for engineering projects, such as construction of dams, highways, airports, tunnels, and large buildings.

Dams

Construction of either masonry (concrete) dams or earth dams (Figure 17.8) requires careful and complete geologic mapping and testing early in the planning process. Testing should include evaluation of the valley and the dam site to determine the present and expected stability of the slopes; identification of possible problems, such as fracture zones in rocks or adverse soil conditions where the dam structure would be in contact with rock or soil (foundation); prediction of the rate of sediment accumulation; and assessment of the availability of building material for the construction of the dam. Continuous geologic mapping during construction may be important

because the geologic history of a river valley often involves several periods of alternating erosion and deposition, producing irregular or unexpected deposits that can cause problems in construction, operation, and maintenance of the dam and reservoir. We emphasize, however, that major irregularities or problems that might affect the dam or reservoir should be recognized by direct and indirect geologic investigation before the construction phase of the project.

Foundations for dams vary according to what rock types are encountered. Generally, igneous rocks, such as granite, and some pyroclastic rocks are satisfactory foundations for a dam site. Leakage, if any, will occur along fractures, and fractures can be grouted, that is, filled with a mixture of cement, sediment, and water. Metamorphic rocks may also be good foundation material, and the best sites are those in which the foliation of the rocks is parallel to the axis of the dam. Sedimentary rocks such as limestone and particularly compaction shale can be troublesome. Limestone may have large underground cavities, and compaction shales are noted for problems stemming from their tendency to deform and settle when loaded (13,14). Particularly important is the stability of sedimentary rocks following wetting and drying or freezing and thawing. Failure of the St. Francis Dam in California was caused in part by the deterioration upon wetting of the sedimentary rocks in the foundation.

Highways

Site evaluation for highways requires considerable geologic input, including mapping the geology along the proposed route and isolating problems of slope stability, topography, flooding, or weak earth materials that would cause foundation problems. It is also helpful to locate

Figure 17.8
Garrison Dam and Lake Sakakawea on the Missouri River in North Dakota. The earth fill dam is 61 meters high and 3,390 meters long. The width at the base is 1,020 meters. The construction required 50.54 million cubic meters of earth fill and 1.14 million cubic meters of concrete for the spillway, power plant, etc. The power plant generates 400,000 kilowatts. (Photo courtesy of U.S. Army Corps of Engineers, Omaha District.)

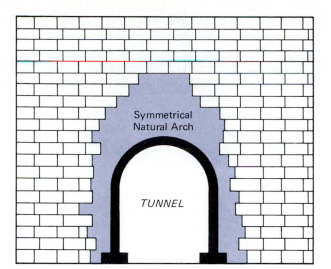 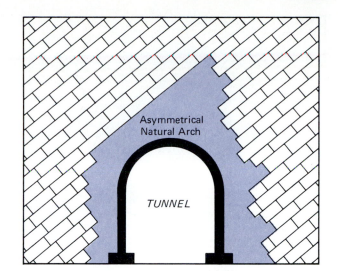

Figure 17.9
Overbreak that develops in (a) horizontally layered rocks and (b) inclined rocks.

and estimate the amount of possible construction materials in close proximity to the proposed route (14).

Airports

Important geologic aspects of site selection for airports are threefold:

▼ Topography as it relates to necessary grading, drainage, and surfacing
▼ Soils—the best are coarse-grained with high bearing capacities and good drainage
▼ Availability of construction materials

Good surface drainage is important, so areas subject to flooding should be avoided (13).

Tunnels

Construction of tunnels requires detailed knowledge of geology, and evaluation of geologic information is basic to selection of the route and determination of proper construction methods. Geology is a major factor that determines the economic feasibility of the project; therefore, a geologic profile along the center line of the proposed tunnel route is essential before tunneling. This profile is developed from mapping the surface geology and using subsurface data whenever possible (13,14). As with dam investigations, it is important to continue the geologic mapping as construction proceeds to ensure that unexpected hazards do not produce problems.

Two distinct types of tunnel conditions are defined by whether the tunnel is in solid rock or in cohesionless, or plastic, earth (soft ground) that may tend to flow and fill the tunnel (14). Both conditions may be present in one tunnel.

Potential problems in rock tunnels are determined primarily by the nature and abundance of fracturing or other partings in the rock. Fracturing greatly affects how much overbreak (excess rock that falls into the tunnel) is present. In general, the closer the fractures and other partings, the greater the overbreak (13). Tunnels in granite or horizontally layered rocks generally develop a symmetrical, naturally arch-shaped overbreak, whereas inclined rocks produce an irregular (asymmetric) roof (Figure 17.9). When possible, it is advisable to select tunnel routes through rock that has a minimum of fractures.

Rock structure, particularly folds and faults, is also significant in locating and orienting tunnels. When tunnels are built through synclines, water problems are likely, because the natural inclination of the rocks facilitates drainage into the tunnel (Figure 17.10). In addition, rock fractures in synclines tend to form inverted keystones, and thus the roof of the tunnel may require substantial support. On the other hand, tunnels through anticlines are less likely to have water problems, and the fractures tend to form normal keystones that are more nearly self-supporting (13).

Faults in tunnels usually cause water problems because water often migrates along open fractures associated with fault systems. Therefore, whenever possible, tunnels should be driven at right angles to faults to minimize the length of tunnel that will be in contact with the fault (13). Tunneling in soft ground is generally done at shallower depths, thus facilitating the collection of more detailed subsurface information before construction than is possible in deep rock tunnels (13,14).

Most problems in earth tunneling are compounded when groundwater saturates the material through which the tunnel is being driven. When it is saturated, sandy material may slowly flow and fill a tunnel or suddenly

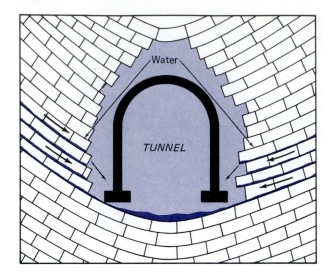

Figure 17.10
Possible groundwater hazard associated with tunnels driven through synclines.

develop a hazardous condition known as a "blow or boil," in which the material quickly fills the tunnel. Thorough evaluation of soil and hydrogeology is necessary when tunneling in soft ground (13,14).

Large Buildings

The geologist's role in evaluating a site for a large building is primarily to provide geologic information that can be used in designing the building's foundation. The site investigation includes drilling for soil or rock samples (Figure 17.11) to be evaluated for engineering properties to help engineers recognize unanticipated

Figure 17.11
Drilling for soil samples as part of site investigation. (Photo courtesy of Williams and Works.)

geologic problems that might arise during construction of the foundation. The fact that field investigation is in most cases very simple does not mean that detailed investigation can be routinely shortened (13,14).

Geologic investigation of foundation sites for nuclear facilities involves detailed geologic mapping of rock structure. Even small fracture systems must be carefully evaluated to ensure that there are no active or potentially active faults. In addition, the physical characteristics of old fractures, including their length, width, and degree of rock alteration (weathering) to clay and other material along the fractured planes, must be measured and evaluated for possible engineering treatment before construction. Figure 17.12 shows the McGuire Nuclear Power Plant in North Carolina during construction; the square boards bolted to the rock, along with dewatering points, stabilize the slope during excavation.

Hazardous Waste Disposal

Determining sites where hazardous waste may be safely disposed of requires careful geologic and hydrologic consideration. Because absolute containment is unlikely, a pragmatic objective is to design, construct, and regulate landfills so as to minimize potential problems associated with release of toxic materials into the environment. Siting hazardous waste disposal facilities to meet the objective of minimal environmental disruption begins with developing appropriate geologic and hydrologic criteria. Specific criteria vary regionally, according to environmental conditions such as climate and geology, and locally, according to variation in soils, hydrology, geology, and geochemistry. General site criteria for New Mexico, for example, include these:

▼ Major aquifers must not be present adjacent to or below the site, and the water table must be at least 50 meters below the site
▼ Rocks below the site should be clay-rich, with low permeability and high chemical adsorption properties
▼ Landscape surrounding the site should be stable, with little evidence of wind or water erosion, and at least 10,000 years old
▼ Climate at the site should be such that evaporation greatly exceeds precipitation
▼ Site should be in an area of low seismicity and low potential for landsliding, flooding, or volcanic activity
▼ Future land or other resource development should not be affected by disposal activities (15).

▼ LANDSCAPE AESTHETICS: SCENIC RESOURCES

Scenery in the United States has been recognized as a natural resource since 1864, when the first state park, Yosemite Valley in California, was established. The early

(a)   (b)

Figure 17.12
(a) Construction of the McGuire Nuclear Power Plant near Charlotte, North Carolina. (b) Stability of the slope was maintained during excavation by use of rock bolts and dewatering points, indicated by the arrows.

recognition and subsequent action primarily related scenic resources to outdoor recreation, focusing on preservation and management of discrete parcels of unique scenic landscapes. Public awareness of and concern for scenic values over a broader range of resource issues is relatively recent. Society now fully recognizes scenery as a valuable resource; recognition that the "everyday" nonurban landscape also has scenic value that enhances a region's resource base is also emerging, based on the assumption that there are varying scenic values just as there are varying economic values relating to other, more tangible resources (16).

Quantitative evaluation of tangible natural resources, such as water, forests, or minerals, is standard procedure before economic development or management of a particular land area. Water resources for power or other uses can be evaluated by flow in the rivers and storage in lakes; forest resources can be evaluated by the number, type, and size of the trees, and their subsequent yield in lumber; and mineral resources can be evaluated by estimating the number of tons of economically valuable mineral material (ore) at a particular location. We can state the quality and quantity of each tangible resource in comparison to some known low quality or quantity. Ideally, we would like to make similar statements about the more intangible resources, such as scenery; that is, we would like to compare scenery to specific standards (17). Unfortunately, this is a difficult task for which few standards are available.

A region's or area's **scenic resources** can be defined as the visual portion of an aesthetic experience; that is, the scenic base consists of those aspects of the landscape that produce visual amenities. Because areas with recog-

nized visual amenities are generally considered unique, they are of particular public concern and more likely to be protected.

The general philosophical framework for evaluating scenic resources includes certain concepts. First, scenic resources are visual amenities that can be evaluated in terms of an aesthetic judgment. Second, scenic resources, like soil and other resources, vary in quality from place to place. Third, topographic relief, presence of water, and diversity of form and color are three of the significant positive characteristics of scenic quality, whereas artificial change is often a negative characteristic. Fourth, unique landscape is more significant than common landscape. Fifth, although determination of what is aesthetically pleasing varies from person to person, the judgment of what is really displeasing is more universal.

Virtually all schemes to evaluate scenic resources involve measuring or observing specifically selected factors. Observation is thus admittedly subjective, but not necessarily a bad practice, as long as relevant factors based on sound judgment are chosen. A completely objective method of analyzing scenic resources, based on our present knowledge of how people perceive the landscape, is probably not possible and certainly not practical.

One approach to evaluating scenic resources is based on the premise that unique landscapes are more significant to society than common landscapes (18, 19). Explanation of how a unique landscape is mathematically determined is beyond the scope of this discussion; nevertheless, we can explain the major aspects. The method is most appropriate for corridors such as river valleys, coasts, and highway routes, and involves measur-

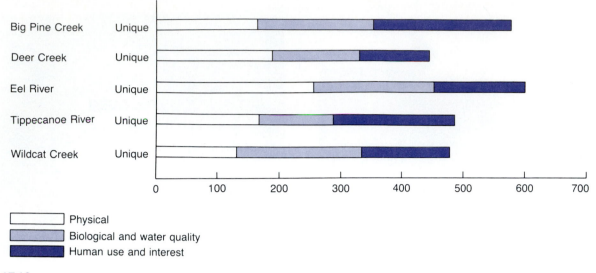

Figure 17.13
Bar graph of uniqueness indices for five Indiana streams. (From Melhorn, Keller, and McBane, Purdue University Water Resources Research Center, Technical Report No. 37, 1975.)

ing or observing variables that describe the physical, biological, and human use of the landscape under evaluation. The data are then analyzed to determine uniqueness indices showing how different, in a relative sense, one landscape is from others (18,19). Up to this point, the only subjective part of the method is selecting the variables to be used. Figure 17.13 shows uniqueness indices for five stream valleys in Indiana. This type of graph is valuable because it also shows what part of the total uniqueness results from the physical, biological, or human-use variables.

If only "relative uniqueness" were determined, the method would have little application. For example, the most unique landscape might be the only polluted and altered river valley, or it might be the most scenic. The second step of the evaluation therefore, requires determining why a particular landscape is unique (19). With scenic resources, we are most interested in determining whether the most unique landscapes are also the most scenic. To do so, we use our premise that the judgment of what is pleasing may vary, but what is really ugly to one person is usually ugly to many people. In the case of streams, the Wild and Scenic River Act, Public Law 90-542 of 1968, defines scenic rivers as "those rivers or sections of rivers that are free from impondments, with shorelines or watersheds still largely primitive and shorelines largely undeveloped, but accessible in places by roads." This definition implies that the stream is also clear-running, unpolluted, and unlittered. Using this information, we can compare the data for determining uniqueness with the idealized scenic river and lower a river valley's original uniqueness in proportion to how much of the original uniqueness is antithetical to our "ideal" scenic river (19). By this philosophy, even a "very

unique" stream, if it is heavily developed or polluted, will rank low in terms of a scenic river index. Figure 17.14 shows the scenic river indices for the same five streams in Indiana; the upper bar for each stream valley is the uniqueness index, and the lower is the scenic river index. This graph succinctly summarizes the uniqueness and scenic river indices.

Quantifying scenic resources is valuable because the results can be visually displayed and different landscapes hierarchically ranked. Quantification is basically another tool to help make decisions about land-use planning or analyze environmental impact. It also tends to separate facts from emotion while providing a way to balance intangibles with the more readily evaluated tangible aspects of landscape evaluation.

▼ ENVIRONMENTAL IMPACT

Philosophy and Methodology

The probable effects of human use of the land are generally referred to as environmental impact. This term became popular in 1969 when the National Environmental Policy Act (NEPA) required that all major federal actions that could possibly affect the quality of the human environment be preceded by an evaluation of the project and its impact on the environment. To carry out both the letter and the spirit of NEPA, the Council on Environmental Quality prepared guidelines to help in preparing **environmental impact statements** (EIS). The major components of the statement, revised in 1979 (20), are:

▼   A summary of the EIS.
▼   A statement concerning the purpose and need for the project.

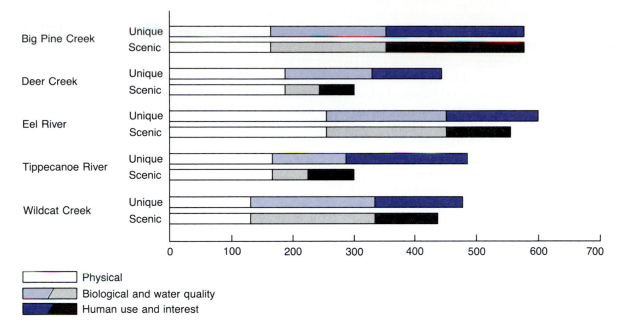

Figure 17.14

Bar graph of scenic and uniqueness indices for the same five Indiana streams shown in Figure 17.13. (Adapted from Melhorn, Keller, and McBane, Purdue University Water Resources Research Center, Technical Report No. 37, 1975.)

▼ A rigorous comparison of the reasonable alternatives.

▼ A succinct description of the environment of the area to be affected by the proposed project.

▼ A discussion of the environmental consequences of the proposed project and of the alternatives. This must include direct and indirect effects; energy requirements and conservation potential; possible depletion of resources; impact on urban quality and cultural and/or historical resources; possible conflicts with state or local land-use plans, policies, and controls; and mitigation measures.

In addition, a list of the names and qualifications of the persons primarily responsible for preparing the environmental impact statement, a list of agencies to which the statement has been sent, and an index are also required.

The environmental impact statement process was criticized during the first ten years under NEPA because it initiated a tremendous volume of paperwork by requiring detailed reports that tended to obscure important issues. In response, the revised regulations (1979) introduced two other important changes: *scoping* and *record of decision*.

Scoping is the process of identifying important environmental issues that require detailed evaluation early in the planning of a proposed project. As part of this process, federal, state, and local agencies, as well as citizen groups and individuals, are asked to participate by identifying issues and alternatives that should be addressed as part of the environmental analysis.

The **record of decision** is a concise statement by the agency that is planning the proposed project as to the alternatives considered and specifically, which alternatives are environmentally preferable. The record of decision also discusses both the rationale for making a particular decision and how environmental degradation may be avoided or minimized for the alternative chosen. The agency then has the responsibility to monitor the project to see that its decision is carried out. This is a significant point: for example, if, in an environmental impact statement for a proposal to construct an interstate highway, the agency commits itself to specific design and locations for the right-of-way to minimize environmental degradation, then the contracts authorizing the work must be conditional on incorporation of those designs and locations. Some projects are monitored by special consultants to ensure that procedures to minimize environmental degradation, outlined in the record of decision and the environmental impact statement, are carried out.

Another concept of importance to environmental impact analysis is that of **mitigation**. By definition, mitigation involves identification of actions that will avoid, lessen, or compensate for anticipated adverse environmental impacts of a particular project. For example, if a project involves filling in wetlands, a possible mitigation might be enhancement or creation of wetlands at another site. Mitigation is becoming a common feature of many environmental impact reports and statements. Unfortunately, it may be being overused. For example, sometimes no mitigations are possible for a particular environmental disruption. Furthermore, we often don't know enough about restoration and creation of habitats and environments such as wetlands to do a proper job. Requiring mitigation procedures may be useful in many instances, but must not be considered an across-the-board acceptable way to circumvent problems of adverse environmental impacts associated with a particular project.

There is no accepted methodology for assessing the environmental impact of a particular project or action. Furthermore, because of the wide variety of topics, such as hunting of migratory birds or construction of a large reservoir, no one method of impact assessment is appropriate for all situations. What is important is that those responsible for preparing the statement strive to minimize personal bias and maximize objectivity. The analysis must be scientifically, technically, and legally defensible and prepared according to highly objective scientific inquiry standards (21).

Environmental impact analysis for major projects or actions generally benefits from the combined effort of a task force or a team of investigators, each of whom is assigned specific resource topics or disciplines. It is important to remember that the specific function of the task force is to evaluate, not to decide the issue. The work of a task force is to provide information to enable those with authority to make just decisions. At this time in the development of our environmental awareness, knowledge of the nature and extent of information required to make decisions is still in the formative stage. As more and more work is done and more decisions are made, however, our concept of the critical elements of environmental information will be better understood, and eventually some will be acknowledged as requirements in the same way as cost and profit data are now accepted practice in economic analysis (21).

Following the National Environmental Policy Act (NEPA) of 1969, it was apparent that there was a need for state and local governments to better control activities not covered by the federal legislation. To date, approximately one-half of the states have enacted some sort of environmental review process. For example, states including Hawaii and California have enacted State Environmental Policy Acts (SEPAs), patterned, in general, after the federal legislation (NEPA) (22).

The California Environmental Quality Act (CEQA) of 1970 requires the completion of an environmental impact report (EIR) for a wide variety of projects that may affect the environment. In 1972, requirements for environmental review were extended to include all private, as well as public projects in California. The California law provides more protection for the environment than does the federal legislation. This can even be noted to some extent in the terms used to describe the acts. The federal legislation considers "environmental policy," whereas the California state law is concerned directly with "environmental quality." The contents of the environmental impact report required by CEQA is similar to that required by the federal legislation, except that CEQA requires sections in the report to deal with growth-inducing impacts and mitigation measures that are proposed to minimize the significant environmental effects that the project might cause. CEQA goes further than NEPA in environmental protection, because it may prohibit agencies from approving projects that will cause significant adverse environmental impacts for which alternatives are not feasible and mitigation is not possible (23). On the other hand, NEPA requires that agencies must consider the potential environmental impacts associated with the project, but does not necessarily require mitigation measures. As a result, it is possible that an agency might simply ignore conclusions drawn from an environmental impact statement, ignore potential alternatives and mitigation, and pursue actions that might cause environmental degradation (23). The potential for an agency to ignore the advice of an environmental impact statement, however, is diminished by review of draft EISs by other public agencies and private environmental institutions, such as the Sierra Club and Audubon Society. In general, the NEPA process has:

▼   Focused attention on potential environmental degradation.
▼   Provided a framework for evaluation of environmental consequences of a project.
▼   Greatly increased efforts to protect the environment in the United States.

The California Environmental Quality Act is similar to federal legislation in that it is possible to make what is known as a *negative declaration,* or a *mitigated negative declaration.* Negative declarations are filed when an agency determines that a particular project does not have a significant adverse impact on the environment following the environmental inventory and assessment. The negative declaration requires a statement that includes a description of the project and detailed information that supports the contention that the project will not have a significant effect on the environment. Negative declarations do not have to consider a wide variety of alternatives to the project, but should be a complete and comprehensive statement concerning potential environmental problems (23).

A mitigated negative declaration may be filed when the initial study of a project suggests that significant environmental problems will probably occur, but that the project could be modified in such a way as to reduce those effects to being nearly insignificant (23). An important step in the process of mitigated negative declaration is the assurance that the appropriate agency will implement the necessary mitigation measures. This involves the initiation of a monitoring program to ensure compliance to the agreed-upon mitigation measures (23).

Negative declarations and mitigated negative declarations are an important addition to the environmental review process. They provide a mechanism whereby projects that do not cause environmental disruption may avoid preparation of a lengthy environmental impact report. However, they do not avoid the important steps of public review and comment prior to adoption of the statement and approval of the project.

The spirit of NEPA and SEPAs arose in response to the recognized need to consider the environmental consequences of an action before its implementation. The objective is to recognize potential conflicts and problem areas as early as possible, so as to minimize regrettable and expensive environmental deterioration. The closing of a soda-ash operation in Saltville, Virginia, exemplifies environmental degradation that can be avoided by careful environmental impact analysis before initiation of projects. Three additional examples, the Trans-Alaska Pipeline, Cape Hatteras National Seashore, and the San Joaquin Valley, demonstrate the uses of environmental impact analysis.

Saltville, Virginia

Saltville is a small, historic Appalachian mountain town on the north fork of the Holston River. Before 1972, the community was strictly a company town—the Olin Chemical Works of the Olin Corporation built and maintained almost everything in the town. The plant, which produced soda ash (sodium carbonate), prospered for about 75 years before it finally closed down because it could not economically meet new water-quality standards of the Virginia Water Control Board. The plant disposed of the chloride wastes into the north fork of the Holston River (Figure 17.15), and reportedly up to 5,000 parts per million of chlorine was measured in

the river water. The salts from which the sodium was derived came from solution mining, a procedure in which subsurface salts are dissolved and pumped up from deep solution wells. The wells in Saltville were up to 900 meters deep, and the combination of available sodium and carbonate rock facilitated production of the sodium carbonate. The plant started production when the dilute and disperse philosophy of waste disposal was accepted, and, tragically for the town, ended production when this philosophy was no longer acceptable. Seventy-five years ago, people seldom worried about long-range environmental effects. Perhaps requirements for environmental-impact analysis will help alleviate future problems that might not be considered problems today.

The Trans-Alaska Pipeline

The controversy surrounding the Trans-Alaska Pipeline, which went into the construction phase in 1975 and was completed in 1977, provides a good example of environmental impact analysis. Experience gained from this project has helped set precedents and establish procedures for subsequent impact analyses.

In 1968, vast subsurface reservoirs of oil and gas were discovered near Prudhoe Bay in Alaska. Since the Arctic Ocean is frozen much of the year, it is impractical to ship the oil by tankers, and thus a 1,270-kilometer pipeline from Prudhoe Bay to Port Valdez, where tankers can dock and load oil the entire year, was suggested (21, 24).

The general route of the Trans-Alaska Pipeline is shown in Figure 17.16. More than 80 percent of the total length is across federal lands. Because of the sensitive nature of the Arctic environment and the certainty of an unknown amount of irreversible environmental degradation, a comprehensive impact analysis was required to

Figure 17.15
Abandoned chemical works plant in Saltville, Virginia, on the banks of the Holston River.

evaluate both the negative and positive aspects of the project. A summary of the evaluation published by the U.S. Geological Survey emphasizes the critical aspects of the natural and social-economic impacts of the project (21).

The corridor for the Trans-Alaska Pipeline (Figure 17.16) traverses rough topography, large rivers, areas with extensive permafrost, and areas with a high earthquake hazard. The pipeline crosses three major mountain ranges: the Brooks Range (which is the Alaskan continuation of the Rocky Mountains), the Alaska Range, and the tectonically very active Chugach Mountains. The pipeline also crosses the Denali fault on the north side of the Alaska Range, an active fault zone that has experienced recent displacement of the ground.

The corridor also traverses a number of large rivers, and it was feared that scour or lateral erosion at meander bends might damage the pipeline. A study was conducted to evaluate the river crossing (25), to predict areas that might experience excessive bank erosion or scour during the expected 30-year life of the pipeline.

The permafrost is also a potential hazard because, once frozen ground melts, it is extremely unstable. This aspect of the project has been thoroughly engineered, and it is hoped that problems with permafrost will be minimized. Nevertheless, a large hot-oil pipeline crossing hundreds of kilometers of permafrost is cause for concern.

Analysis of possible physical, biological, and social-economic impacts associated with the pipeline establishes three areas of concern (21): the construction, operation, and maintenance of the pipeline system, including access roads and highways; development of the oil fields; and operation of the marine tanker system at Port Valdez. Furthermore, some of the effects would

result in predictable, unavoidable disruptions, while other effects are more speculative and present a threat, such as the impact of an oil spill caused by a break in the pipeline. These threatened effects are not easy to evaluate and predict (21).

Evaluation of links between the effects that might have an impact on the environment was also done. For example, there are obvious possible links between oil spills or unavoidable release of oil in the tanker operation into the marine environment and the effect on marine resources. This is difficult to evaluate, however, because of the variable or unknown aspects of the hazard, the dynamic hydrologic environment, and seasonal and other changes in fish and marine resources. In general, the task force evaluating the impacts concluded that the impact linkage data were hard to obtain and evaluate and so were considered the largest problem area in quantitatively determining possible impacts (21).

Based on the environmental analysis, both unavoidable and threatened impacts of the Trans-Alaska Pipeline were determined (21). The main *unavoidable effects* were threefold:

1. Disturbances of terrain, fish and wildlife habitats, and human use of the land during construction, operation, and maintenance of the entire project, including the pipeline itself and access roads, highways, and other support facilities
2. Discharge of effluents and oil into Port Valdez from the tanker transport system
3. Increased human pressures on services, utilities, and many other areas, including cultural changes of the native population

Major *threatening effects* are accidental loss of oil from the oil field, pipeline, or tanker system. Accidental

Figure 17.16

Approximate corridor for the Trans-Alaska Pipeline. (After Brew, U.S. Geological Survey Circular 695, 1974.)

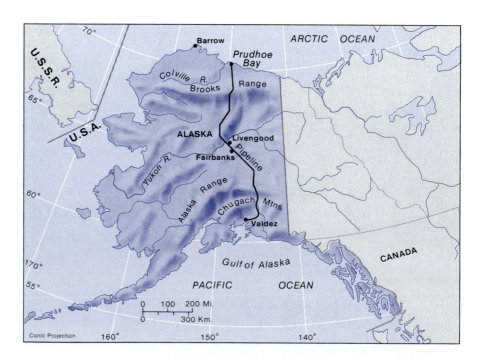

loss of oil from the pipeline could be caused by slope failure (landslides), differential settlement in permafrost areas, streambed or bank scour at river crossings, ground rupture during earthquakes, or destructive sea waves damaging the pipeline, causing a leak or rupture. Loss from tankers might be caused by shipwreck or by accident during transfer operations at Port Valdez. The potential loss of oil from the pipeline and tanker systems involves many variables, making predictions of oil loss difficult—but some loss is inevitable. Estimates place maximum oil loss at 1.6 to 6 barrels per day at Valdez for tanker operations, and 384 barrels per day from tanker accidents. The latter, of course, could be either a series of small spills or several large spills at unknown times, locations, and intervals (21). The prediction of 384 barrels per day from tanker accidents has so far proven to be conservative. The March 1989 rupture of the super-tanker *Exxon Valdez* released 250,000 barrels of crude oil into the Prince William Sound. Predicted release of oil from tanker accidents yields a ten-year amount of 1.4 million barrels, which is approximately six times that of the *Exxon Valdez*. However, other spills can be expected as long as human beings are operating supertankers.

Before the final impact statement was completed early in 1972, until construction began in 1975, the pipeline generated, and continues to generate, controversy. This controversy results because of the conflicts in balancing the need for resource development and the known or predicted environmental degradation. Although alternative routes and transport systems, including trans-Canada routes and railroads, other Alaska routes, and marine routes, were extensively evaluated, the earlier proposed route to Port Valdez was ultimately approved, perhaps because this route led to the most rapid resource development while maintaining national security. We emphasize, however, that no one route is superior in all respects to the others (21).

Based on many evaluations and analyses (21), it was concluded that, to avoid the marine environment and earthquake zones, and to place both an oil and a gas pipeline in one corridor, the trans-Alaska–Canada route to Edmonton, Canada, would cause the least environmental impact. This is past history, since the route is to Port Valdez, but it shows the alternatives that were considered and the obligation of scientists to state their opinion based on sound scientific information, even though that opinion may be either unpopular or likely to be overridden in the final balancing of alternatives; and it illustrates the significant power of diverse political maneuvering at all levels in deciding among alternatives.

Cape Hatteras National Seashore

The Outer Banks of North Carolina has for generations been inhabited by people living and working in a marine-dominated environment. Until recently, the way of life had depended on raising livestock, fishing, hunting, boat building, and other marine pursuits (26).

The landscape of the Outer Banks characteristically can change in a very short time in response to major storms such as hurricanes and northeasters that periodically strike the islands. Of the two types of storms, the more frequent northeasters probably cause the most erosion. On the other hand, infrequent hurricanes can cause major changes, including extensive overwash and formation of new inlets.

Historically, the people of the Outer Banks have philosophically lived with and adjusted to a changing landscape. In recent times, however, this philosophy has changed because of economic pressure to develop coastal property. A new philosophy of coastal protection has arisen in an attempt to stabilize the coastal environment. Stabilization, or a constant-appearing landscape, is a prerequisite to commercial development.

Congress approved in 1937 and amended in 1940 an act establishing the Cape Hatteras National Seashore as the first national seashore. The park consists of 115 square kilometers along a 120-kilometer portion of the Outer Banks (Figure 17.17) and includes portions of three islands of the more than 240 kilometers of the

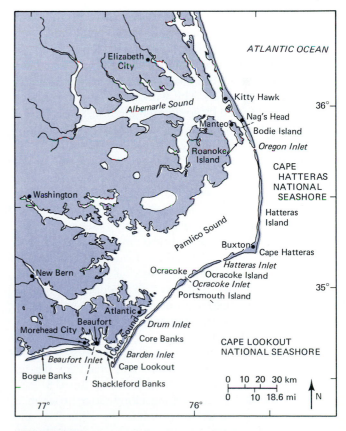

Figure 17.17
Map showing the Outer Banks of North Carolina and the Cape Hatteras National Seashore. (After Godfrey and Godfrey, in D. R. Coates, ed., *Publications in Geomorphology* [New York: State University of New York, 1971].)

Barrier Island system. The Barrier Islands essentially bound and protect the largely undeveloped coastal plains lowland of North Carolina (26).

Eight unincorporated villages are bounded by the Cape Hatteras National Seashore and are spaced along nearly the entire length of the seashore. Legislation authorizing the park provided for the continued existence of these villages, including the beach in front of each of them facing the Atlantic Ocean. This legislation has been interpreted by many as an obligation of the federal government to maintain and stabilize beach access and frontage of these villages. Unfortunately, stabilization has been difficult, and in some villages, less than 60 meters of beach remains because of recent coastal erosion. This philosophy of protection is contrary to that during the early development of the islands, which was mainly on the sound (inland) side of the islands. With construction of the first dune systems in the 1930s and opening of the road link in the 1950s, however, the basic configuration of the villages began to change as communities began to spread toward the ocean, away from the more protected sites along the sound. This migration has persisted and has grown at an ever-increasing pace (26). For example, in 1953, there were only 75 oceanside subdivision lots, compared to more than 4,000 today.

At one time it was assumed that much of the Outer Banks was heavily forested and that logging and overgrazing had destroyed these forests. Evidence for this is controversial, and in fact extensive forests may have been developed on only a few islands with an orientation more protected from storms. The wooded areas were located in an area protected by a natural dune line, so the logical conclusion was to construct an artificial barrier dune system to help the area return to a natural state by inhibiting beach erosion. Of course, the dune line would also protect the roads and communities. As a result, an artificial dune system now extends along the entire Cape Hatteras National Seashore at an average dune height of 5 meters and an average distance of 100 meters from the ocean (26). This erosion-control measure, along with programs to artificially keep sand on the beach, could cost about $1 million per year if continued. This expensive alternative, along with an essentially geological argument as to whether artificial dune lines will jeopardize the future of the Barrier Islands, has led to a controversy on how to best manage the park.

The geologic problem and controversy of how dynamic earth processes create and maintain barrier islands are at the philosophical heart of the U.S. Park Service's dilemma in developing a long-range program to maintain the natural environment for the use and enjoyment of present and future generations. Although there are several ideas and known ways that coastal processes can develop barrier islands, debate on the processes needed to maintain them is heated.

The present Barrier Islands may have developed about 5,000 years ago in response to complex wind and water processes associated with a slowly rising sea level. One hypothesis is that, since many of the capes (protruding barrier islands) apparently coincide with mouths of rivers, development of these capes has been facilitated by sediment deposited as deltas during times of relatively low sea levels during the Pleistocene era (the last two million years). As Pleistocene sea levels retreated because water was held in the continental glaciers, deltas developed seaward of the present coastline. Then, when the ice melted a few thousand years ago and sea levels began to rise, these deltas became the nucleus of the Barrier Islands and capes. Submergence was accompanied by coastal erosion until it produced the present island system, which is still eroding at a rate of 1.2 to 1.8 meters per year in response to continuing rise in sea level (26, 27).

Although there is debate about the true role of a rising sea level in forming the barrier islands, most investigators do agree that the rise going on today is closely associated with tremendous changes due to erosion of the coastline and that this is a real and immediate hazard to future coastal development, which demands a static shoreline.

Debate on the nature and extent of geologic processes that maintain the "naturally changing" Barrier Islands centers on whether periodic overwash of the frontal dune system by storm waves is important in maintaining the islands. If the overwash is important, then building a dune line is contrary to natural processes, and the final result over a period of years may be deterioration of the islands as a natural system. This would be contrary to the Park Service's stated objectives to preserve the islands in a natural state.

The argument that all barrier islands are maintained in natural landward migration by periodic overwash of frontal dunes, or inlet breaching that allows landward transportation of sediment, is unconvincing. Most of the energy expended by storm waves is concentrated in the surf zone rather than splashing over low dunes. In fact, one recent study concluded that the present rise in sea level is causing erosion on both the seaward and landward sides of many U.S. Atlantic Coast barrier islands. Rather than migrating toward the land, these barrier islands are becoming narrower in response to rising sea level (28). On the other hand, overwash of dunes on some barrier islands is a real process that does tend to facilitate a more "natural appearing" coastline. However, it has been argued that frequent overwash is atypical and occurs only on islands that are already in an unnatural state from overgrazing and subsequent lowering of dunes resulting from loss of protective vegetation. Regardless of the cultural influence and historic role of overwash, it is apparent that overwash is a natural process and as such is probably significant in the geologic history of barrier-island migration. This does not mean, however, that it is necessarily always bad to attempt to selectively control rapid coastal erosion by maintaining a protective frontal dune system. Proper placement of dunes and subsequent sound conservation practice in maintaining them remain a feasible alternative to short-range selected coastal erosion problems, particularly near settlements and critical communication lines.

Faced with the problem of selecting a management policy for the Cape Hatteras National Seashore, the Park Service in 1974 presented five possible alternatives (Table 17.1) ranging from essentially no control of natural coastline processes (Alternative 1) to attempting complete protection (Alternative 5). Each alternative was analyzed to determine the entire spectrum of possible impacts. Few people want complete protection or complete lack of control of the seashore (26); the position of the Park Service is to attempt to strike a compromise in which the Barrier Islands for the most part may be preserved in a natural state, while the need to maintain a transportation link with the mainland is recognized. Thus, the communities will have to live with and adjust to the dynamic high-risk environment they choose to live in, much as the people of the Outer Banks historically have. In contrast to prevailing trends in coastal development that assume a more stable environment is desirable, this acknowledges that natural processes play a significant role in preserving the natural environment (26).

A general management plan for the Cape Hatteras National Seashore was recently completed after several years of careful environmental impact work and public review (29). The plan proposes several actions:

▼ Allowing natural seashore processes and dynamics to occur except in instances when life, health, or significant cultural resources on the major transportation link are jeopardized
▼ Controlling use of offroad vehicles
▼ Expanding and improving access to recreation sites on the beach and sound
▼ Controlling the spread of exotic vegetation species
▼ Ensuring that significant natural and cultural resources are preserved and maintained
▼ Cooperating with state and local governments in mutually beneficial planning endeavors (28)

For planning purposes, the national seashore was divided into four environmental resource units (ERUs): ocean/beach, vegetated sand flats, interior dunes/maritime forest, and marsh/sound. Table 17.2 summarizes planning objectives for each ERU, and Figure 17.18 illustrates the principle of management zoning for Hatteras Island.

The major impact of the proposed management program will be threefold. First, the Cape Hatteras National Seashore will be preserved in a natural state. Second, residents of the villages will have to live with and

adjust to the effects of natural events to a greater extent. Third, economic development of the villages will change, and even though the road will be maintained, it will be subject to more periodic damage.

San Joaquin Valley

The San Joaquin Valley in central California is one of the richest farm regions in the world. The farming on the west side of the valley has been made possible through extensive irrigation from deep wells and water transported by canals to the agricultural area from the Sierra Nevada Mountains.

On the west side of the valley (Figure 17.19), soils have developed on alluvial fan deposits shed from the Coast Ranges. With an annual average precipitation of less than 25 cm., irrigation is necessary for agricultural productivity. A drainage problem exists on the west side because at shallow depths of 3 to 15 meters there are often clay barriers that restrict the downward migration of water and salts. When irrigation waters are applied, these clay barriers produce a local zone of saturation near the surface that is well above the regional groundwater level (a condition known as *perched groundwater*). As more and more irrigation water is applied, the saturated zone may rise to the crop-root zone and drown the plants (30). To avoid this hazard a subsurface drainage system may be installed to remove the excess water (Figure 17.20).

Irrigation water contains a variable amount of dissolved salts and as soil water evaporates and plants remove water through transpiration, the salts are left behind in the soil. Periodic leaching by applying large amounts of fresh water will remove the salts, but subsurface drains are necessary to remove the salty water (Figure 17.20). Tile drains (pipes with holes in them to let the water in) beneath each field connect to larger drains (ditches or canals) that eventually join a master drainage system, usually a canal.

The San Luis Drain (a 135-kilometer-long concrete-lined canal) was constructed between 1968 and 1975 to convey the agricultural drainage water northward toward the San Francisco Bay area. The drain to the bay was never completed because of limited funding and uncertainty over the potential adverse environmental effects of discharging drain water into the bay. As a result, Kesterson Reservoir (a series of 12 ponds with average depth of 1.3 meters and total surface area of about 486 hectares) is the terminal point for the drainage water (Figure 17.21).

During the first few years following construction of Kesterson Reservoir, the flow in the drain was primarily fresh water that was purchased to provide water for the Kesterson Wildlife Refuge, established primarily for waterfowl. The reservoir started receiving agricultural drainage in 1978, and by 1981 this was its main source of

Table 17.1

Possible alternative management plans for the Cape Hatteras National Seashore.

Alternative 1	
Management objectives:	To manage Cape Hatteras National Seashore as a primitive wilderness so as not to impede natural processes.
Proposed action:	Let nature take its course, and oppose any attempts to control natural forces or stabilize the shoreline.
Legislative changes:	Classify Cape Hatteras as a recreation area, and officially designate the Seashore a wilderness area by administrative decree.
Major impacts:	A natural environment would be restored; the existing highway could not be maintained; property within the villages would be lost; historic structures would be destroyed unless relocated; and the economic base and economic development of the village enclaves would decline.
Costs of alternative:	No direct federal costs except possible relocation of Cape Hatteras Lighthouse and other historic structures. State costs of maintaining the road would be excessive.

Alternative 2	
Management objectives:	To manage Cape Hatteras National Seashore in accordance with the policies established for natural areas of the National Park Service, and to preserve the area in a natural state.
Proposed action:	Develop an ongoing resource management program, revegetate major overwash areas following destructive storms, and investigate alternative transportation modes.
Legislative changes:	Classify Cape Hatteras as a recreation area, and officially designate the Seashore a natural area by administrative decree.
Major impacts:	The Seashore would be preserved in a natural state, and residents of the villages would have to live with and adjust to natural events. Some property would be lost, and economic development of the villages would be retarded. Historic structures would be relocated or abandoned. The road would be damaged or washed out in places but could be maintained over the next several years.
Costs of alternative:	Estimated at $125,000 annually, or $3,125,000 over a 25-year period.

Alternative 3	
Management objectives:	Continue present shore erosion-control practices.
Proposed action:	Management practices of the past—dune building, dune repair, and beach nourishment would continue.
Legislative changes:	No changes would be required.
Major impacts:	The Seashore environment would become increasingly artificial as efforts were concentrated on protecting private development. The village enclaves would continue to grow and the highway would have to be widened to serve these areas.
Costs of alternative:	Estimated at $1 million annually, or $25 million over a 25-year period.

Alternative 4	
Management objectives:	To acquire threatened private property within the village enclaves to preserve a public beach fronting the villages and avoid having to protect such property.
Proposed action:	Identify threatened private property, determine fair market value, and purchase the property.
Legislative changes:	Authorize legislation for Cape Hatteras to allow for the acquisition of additional land.
Major impacts:	The economic base of the village enclaves would be reduced, and as the shoreline continues to recede, future acquisition would be required at much greater cost. This alternative would establish the precedent of compensating unwise development.
Costs of alternative:	Very high—over $25 million for Kinnakeet Township alone (1973 prices). The land involved is the costliest on the Seashore. Loss of tax revenue and revenue derived from visitor expenditures is not included.

Alternative 5	
Management objectives:	To protect the development taking place within the village enclaves and protect the highway running through the Seashore.
Proposed action:	Stabilize the shoreline fronting threatened private property and protect exposed sections of highway by relocating or elevating the road or by stabilizing shoreline structures or building dunes.
Legislative changes:	Amend the authorizing legislation for Cape Hatteras, deleting the section pertaining to managing the area as a primitive wilderness.
Major impacts:	The natural setting of the Seashore would be destroyed and the resource base would be seriously degraded. The village enclaves would develop into urban complexes and dominate the landscape. This alternative would acknowledge a Federal responsibility to subsidize unwise economic development.
Costs of alternative:	Overall initial cost of construction would be $40 to $56 million with annual maintenance of $3.2 to $6.4 million.

Source: National Park Service.

ERU	Characteristics	Planning Objectives
Ocean/Beach	Shifting sands, frequent overwash, limited vegetation on dunes	Allow natural processes to continue unhampered; allow for wide range of unstructured recreational activities by visitors; no construction allowed
Vegetated sand flats	Located between dune line and edge of salt-water marsh	Continue use as transportation corridor; allow development necessary to support visitor activities and resource protection
Interior dunes/ Maritime forest	Found in relatively few locations; variable topography, remoteness, dense vegetation	Maintain in natural state; allow passive recreation; design any construction to minimize impact on natural processes and systems
Marsh/Sound	Includes the sound, sound shore, and associated marshes	Maintain in natural state; provide limited access to the sound; allow limited development to support passive recreational activities

Source: National Park Service, 1984.

Table 17.2
Planning objectives for Cape Hatteras National Seashore Environmental Resource Units (ERU)

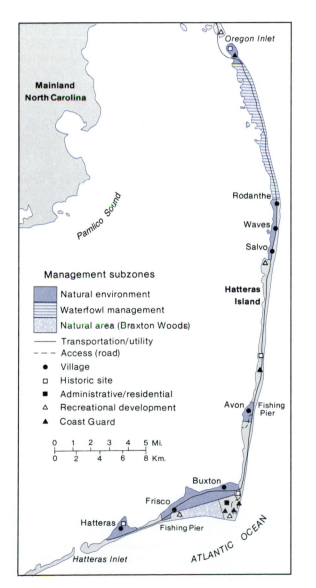

Figure 17.18
Proposed management subzones for part of the Cape Hatteras National Seashore. (After National Park Service, 1984).

water. The canal and reservoir seemed like a good system to get rid of the wastewater from farming activities, but unfortunately, as the water infiltrated the soil, it picked up not only salts and agricultural chemicals but also the heavy metal selenium. The selenium is derived from weathering of sedimentary rock in the Coast Ranges on the west side of the valley. The sediments, including the selenium, are carried by streams to the San Joaquin Valley and deposited there.

In 1982 a United States Fish and Wildlife study revealed unusually high selenium levels in fish from Kesterson Reservoir. Dead and deformed chicks from water birds and other wildfowl were reported from 1983 to 1985. Analysis of water samples revealed selenium in the ponds ranged from 60 to 390 parts per billion (ppb). Selenium concentrations as high as 4000 ppb were measured in the agricultural drain waters. These concentrations of selenium are many times the drinking water standard of 10 ppb set by the Environmental Protection Agency. As a result of the studies it was inferred that selenium toxicity was the probable cause of the deaths and deformities of the wildfowl. The selenium was clearly shown to come from the San Luis Drain, and as the water spread out through the Kesterson Wildlife Refuge, the heavy metal was incorporated in increasingly higher concentrations from the water to the sediment, to the vegetation, and eventually to the wildlife (a process known as *biomagnification*).

In 1985 Kesterson wastewater was classified as hazardous and a threat to public health. At that time the state Water Resources Control Board ordered the United States Bureau of Reclamation, which was running the facility, to alleviate the hazardous condition at the reservoir. They were given 5 months to develop a cleanup plan and 3 years to comply with it (30). The immediate strategy was three-fold: first, initiate a monitoring program; second, take steps to ensure that wildfowl would no longer use the wetlands (this was

Figure 17.19

The San Joaquin Valley and San Luis Drain that terminates at Kesterson Reservoir. (Modified from Tanji, Lauchli and Meyer, 1986).

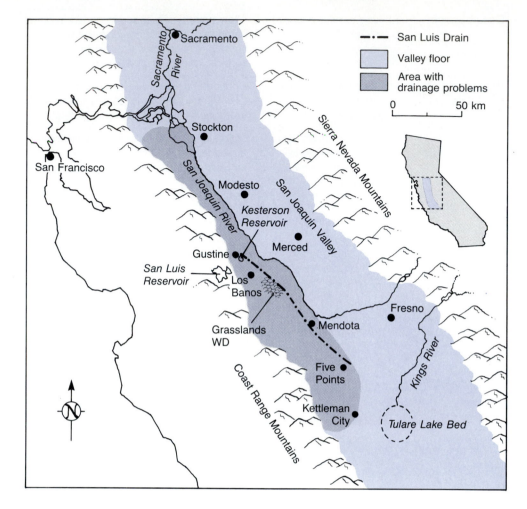

Figure 17.20

Diagram indicating the need for subsurface drains on the west side of the San Joaquin Valley. (Modified after U.S. Department of the Interior, Bureau of Reclamation, 1984, *Drainage and Salt Disposal*, Information Bulletin 1, San Luis Unit, Central Valley Project, California.)

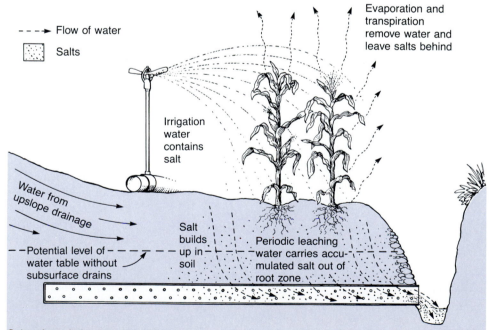

Subsurface drains remove saline water, lower water table. Average depth 2.5m. Without drains, salt continues to accumulate.

Drains empty into collector system that transports drainage water to discharge point (ditch or canal)

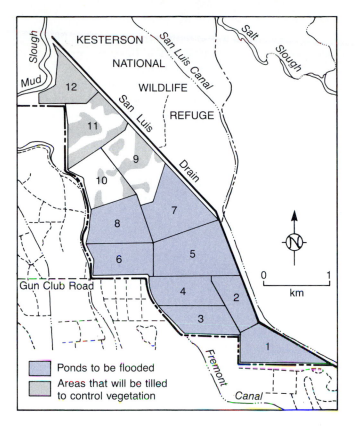

Figure 17.21
The Kesterson Reservoir ponds and the proposed Flexible Response Plan. (After U.S. Department of Interior, 1987, *Kesterson Program,* Fact Sheet No. 4.)

done by scaring off the birds); and third, reduce public exposure to the hazard.

The Kesterson case history is interesting because it provides insight into the environmental planning and review process. Primary environmental concerns and impacts were identified by the scoping process, which identified the following issues:

▼ Disposition of Kesterson Reservoir and the San Luis Drain following cleanup
▼ Environmental, social, and economic impacts of closing the Kesterson Reservoir to agricultural drain water discharge
▼ Potential public health impacts of selenium and other contaminants in the drain water
▼ Consideration of alternative methods to clean up the water, soils, sediment, and vegetation
▼ Potential for migration of contaminated groundwater away from the Kesterson area

The process of environmental analysis at Kesterson will continue for a number of years. Alternatives are being considered for the future of the reservoir and the drain, as well as for methods of cleanup. In addition, both primary and secondary as well as third order impacts are being analyzed in the search for a long-range solution to the San Joaquin Valley drainage problem. Because selenium occurs naturally in the Coast Ranges, other areas in the valley have similar problems that are being discovered and evaluated.

In October of 1986 the U.S. Department of Interior, Bureau of Reclamation, issued a final environmental impact statement and record of decision that addressed the impacts of alternative methods to clean up Kesterson Reservoir and the San Luis Ditch. That statement suggested a phased approach including three elements: the Flexible Response Plan, the Immobilization Plan, and the On-Site Disposal Plan. A no-action alternative was also considered but was not acceptable. The Flexible Response Plan has both a wet and a dry phase (Figure 17.21). The ponds with the lowest contamination from selenium (ponds 9–12, Figure 17.21) will be drained and vegetation plowed under. The more contaminated ponds (1–8, Figure 17.21) will be flooded with fresh clean groundwater (less than 2 ppb selenium) in the hope that the selenium contamination can be sealed within the fine grained sediments at the bottom of the pond. The Flexible Response Plan is the least expensive of those proposed, costing several million dollars (31).

If the monitoring suggests that the cleanup from the Flexible Response Plan is not working, then the second phase, or Immobilization Plan, will be initiated if its feasibility has been demonstrated through experiments performed in Pond 4 during the first phase. The Immobilization Plan includes increasing water depths to the flooded ponds and harvesting vegetation.

If neither of these procedures is successful, the third phase, the On-Site Disposal Plan, would be initiated. All sediment with a selenium concentration exceeding 4 ppb and all above-ground vegetation would be removed and placed on-site in a lined and capped landfill (31). Costs for the On-Site Disposal Plan would greatly exceed those of the other plans.

Government agencies and individuals disagree as to what action should be taken to clean up Kesterson. In late March of 1987 the state Water Resources Control Board rejected the phased plan and ordered the Bureau of Reclamation to excavate all contaminated vegetation and soil and dispose of it on-site in a monitored, lined, and capped landfill. This is the Bureau's Phase Three and will produce a toxic waste disposal site. The Board also required the Bureau to mitigate the loss of Kesterson Wildlife Refuge by providing alternative wetlands for wildlife. The disposal plan is estimated to cost as much as $50 million. Although the disposal plan may help solve the Kesterson problem, if the selenium problem is widespread in the western San Joaquin Valley (as it probably is) then a more general long-term solution is needed. People in the valley can't afford to construct and maintain a series of toxic waste dumps for contaminated soil and vegetation. Research to evaluate the selenium problem is continuing, so the initial plans may change with time as more information is obtained.

Some of the farmers who use the San Luis Drain are taking independent action, including constructing water treatment plants that utilize bacterial action to remove the selenium. If the initial small treatment plants are successful then larger, more costly ones may be constructed. The selenium problem must be solved because without agricultural drainage the approximate 17,000 hectares (42,000 acres) of farmland that depend upon the San Luis Drain would eventually be lost from production. Another alternative being considered is deep-well disposal of selenium-contaminated water.

The environmental work and research at Kesterson clearly illustrates the importance and impact of environmental analysis at several levels of state and federal government in evaluating alternatives and mitigation measures to minimize environmental degradation and hazardous conditions. These are integral parts of all environmental planning, whether it is for site selection, land use planning, or environmental impact analysis.

▼ LAND-USE PLANNING

Land-use planning is an important environmental issue. Good land-use planning is essential for sound economic development. Good land-use planning is also essential if conflicts between land uses are to be avoided, and communities wish to maintain a high quality of life. It has been pointed out that when a business manages its capital and resources in an efficient manner, we call that good business. When a city or county efficiently manages its land and resources, we call that good planning (32). The basic philosophy from an earth science perspective of good land-use planning is to avoid hazards, conserve natural resources, and generally protect the environment through the use of sound ecological principles. This involves understanding earth systems as outlined in Chapter 2, and planning for an ever-changing environment.

The land-use planning process shown in Figure 17.22 includes several steps (33):

▼ Identify and define issues, problems, goals, and objectives.
▼ Collect, analyze, and interpret data (including inventory of environmental resources and hazards).
▼ Develop and test alternatives.
▼ Formulate land-use plans.
▼ Review and adopt plans.
▼ Implement plans.
▼ Revise and amend plans.

The role of the earth scientist in the planning process is most significant in the data collection and analysis stage. Depending on the specific task or plan, the earth

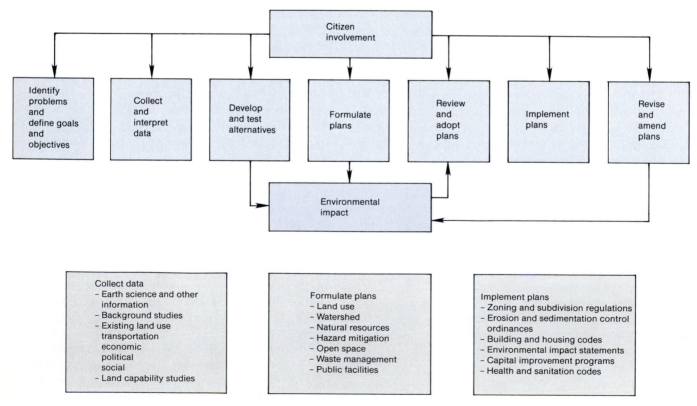

Figure 17.22
The land-use planning process. (Modified after U.S. Geological Survey Circular 721, 1976.)

Process	Description of Hazard
Flooding	Overtopping of river and stream banks by water produced by sudden cloudbursts, prolonged rains, tropical storms, or seasonal thaws; breakage or overtopping of dams; ponding or backing up of water because of inadequate drainage
Erosion and sedimentation	Removal of soil and rock materials by surface water and depositing of these materials on floodplains and deltas
Landsliding	Perceptible downslope movement of earth masses
Faulting	Relative displacement of adjacent rock masses along a major fracture in the earth's crust
Ground motion	Shaking of the ground caused by an earthquake
Subsidence	Sinking of the ground surface caused by compression or collapse of earth materials; common in areas with poorly compacted, organic, or collapsible soils and commonly caused by withdrawal of groundwater, oil, or gas; or collapse over underground openings, such as mine workings or natural caverns
Expansive soils	Soils that swell when they absorb water and shrink when they dry out
High water table	Upper level of underground water close to ground surface causing submergence of underground structures, such as septic tank systems, foundations, utility lines, and storage tanks
Seacliff retreat	Recession of seacliffs by erosion and landsliding
Beach destruction	Loss of beaches owing to erosion and/or loss of sand supply
Migration of sand dunes	Wind-induced inland movement of sand accelerated by disturbance of vegetative cover
Salt-water intrusion	Subsurface migration of seawater inland into areas from which fresh water has been withdrawn, contaminating fresh water supplies
Liquefaction	Temporary change of certain soils to a fluid state, commonly from earthquake-induced ground motion causing the ground to flow or lose its strength

Source: U.S. Geological Survey Circular 721, 1976.

scientist will use available earth science information, collect necessary new data, prepare pertinent technical information such as interpretive maps and texts, and assist in the preparation of land-capability maps that, ideally, should match the natural capability of a land unit to specific potential uses (34).

As an example, consider the role of the earth scientist in planning for reduction of natural hazards. Many natural hazards are also natural processes, and in a hazards reduction program the first task is to identify problem areas. Table 17.3 summarizes many of the natural processes that continue to be potentially hazardous to human use of the land. For a particular site or area, individual potentially hazardous processes must be evaluated carefully in light of intended land use. Siting of critical facilities, such as power plants, dams, or highway overpasses, should obviously be prohibited in high-risk areas, as should hospitals, schools, high-rise buildings, and disaster relief facilities.

The applied earth scientist is also qualified to advise about developing hazard-mitigation methods. There are several ways that hazardous risks from natural processes can be reduced (34). First, land-use can be regulated to restrict occupancy of potentially hazardous areas, such as floodplains and unstable slopes. Second, attempts can be made to partially control the potentially hazardous processes; for example, by constructing retaining walls to stabilize slopes or on-site storm water holding ponds in urban areas to minimize flooding. Third, rather than trying to control the process directly, efforts can be initiated to reduce its impact; for example, special construction methods for foundations and buildings may reduce the expansive soils or earthquake hazard. Fourth, hazardous processes can be monitored and studied in detail for help in developing early warning systems to decide when a hazardous area should be evacuated. It is particularly important to develop procedures for releasing predictions of hazardous events such as earthquakes.

Finally, in helping to select a particular set of procedures or plans to minimize natural hazards, the earth scientist is an active member of a responsible decision-making team that must consider the physical, environmental, social, political, and economic aspects of hazard mitigation.

Comprehensive Planning

The *comprehensive plan,* by definition, is an official document adopted by local government that states general and long-range policies on how the community will deal with future development (32). What is unique to the comprehensive plan is the development of an environmental inventory (including geologic resources and hazards) as a basis for planning and zoning by local governments. The older term *general plan* did not usually involve development of an inventory of the natural environment as a basis for planning. However, although potentially misleading, other names for the comprehensive plan might include the older names *master plan* or *general plan.* The process may be termed comprehensive planning and the product, most appropriately, *comprehensive plan.*

Only a few states in the United States have developed statewide planning programs that require development and approval of a comprehensive plan at the state level. Oregon has one of the more rigorous state-wide planning programs. In Oregon, state law requires cities and counties to develop comprehensive plans that implement a set of state-wide goals. For example, one goal is concerned with citizen involvement in the planning; another protects against natural hazards; and still another addresses quality of air, water, and land resources. In all, comprehensive planning in Oregon addresses 19 state-wide goals developed by the Oregon Land Conservation and Development Commission (LCDC). Furthermore, the Department of Land Conservation and Development (DLCD) reviews and approves local plans. State law in Oregon requires that all land be planned and zoned on the basis of inventories of environmental resources and hazards, but it is the responsibility of the cities and counties to prepare their plans according to state goals and guidelines and to issue permits for development (32). State-mandated planning in Oregon has been labeled as innovative but controversial concerning individual property rights, but it seems to be working, and other states such as Florida have adopted similar programs.

Other states, including California, have developed a system of recommending but not requiring comprehensive planning. Comprehensive plans for a city or county in California must address several elements identified by state law as being important (33). These required elements include land use, housing, conservation, open space, and safety (natural hazards). Cities and counties in

California may add elements that they feel are relevant to their specific comprehensive plan. California also has a State Environmental Quality Act, and environmental review is an integral part of the planning process. In fact, the State of California has recommended that local governments should consider combining the comprehensive plan and the environmental impact report on the plan to save both time and cost (33). This is a major difference from comprehensive planning in Oregon, where a commission approves the plans developed by cities and counties.

The three elements that must, by state law, be included in the comprehensive general plans of California that are of most interest to earth scientists are land use, conservation, and safety. The land-use element is the heart of the plan. It serves as a statement of land-use policy that:

▼ Provides a guide for public investment in land, while protecting environmental resources and alerting planners and the public to environmental hazards.
▼ Provides a useful framework for budgeting and planning construction of facilities such as schools, roads, and sewer and water systems.
▼ Helps coordinate regulatory policies and decisions.

The land-use element contains maps that designate intensity and uses of the land for purposes such as housing, business, industry, open space, recreation, public facilities, waste disposal sites, and other uses (33). What is shown on the land-use maps reflects the goals, objectives, and policies of the other elements in the comprehensive plan. Thus, the land classification maps of the land-use element are a very visible part of the comprehensive plan and often are the most-used part (33).

The conservation element of the comprehensive general plan is intended to set goals and policies concerning development and use of natural resources such as water, soils, wildlife, forests, and mineral deposits (33). Because this element addresses so many different resources, it is crucial that the element be carefully coordinated with a number of local, state, and federal agencies that deal with resources and their use.

The safety element is of particular interest to the earth scientist. The purpose is to establish goals, policies and programs to protect people from natural hazards such as earthquakes, landslides, floods, and wildfires (33). Included with this element are discussions of the various hazards, and maps that help delineate the hazards and areas where particular development should be avoided. This would include maps that show active faults, landslides, and floodplains. The safety element also includes establishment of evacuation routes, emergency procedures, and other issues related to disaster preparedness.

The safety element may be expanded when necessary to include issues such as disposal of hazardous waste, transport of hazardous materials, and failure of utility services (33).

▼ PRINCIPLES OF LAND MANAGEMENT

Land management and land-use planning are related to one another in that management often follows development under the guidance of land-use plans. That is, once land has been designated for a particular use, as for example urban development or recreation, then management of that land follows. In evaluating management we should first consider what actually needs to be managed. At this level we can identify several aspects of environmental impact including:

▼ Impact of natural processes (for example flooding, landslides, erosion).
▼ Impact of human use and interest on physical, chemical, and biological processes and on natural resources such as soil, water, wildlife, and vegetation.

Another important point concerning management is to define goals. For example, if we are interested in managing and urbanizing a watershed the goals might include:

▼ Protection of human lives and property.
▼ Protection of water quality and supplies.
▼ Protection of wildlife resources.
▼ Ecosystem protection.
▼ Increasing public access and recreation potential to natural portions of the drainage basin.

The overlying management approach for many areas is to try to understand the physical, biological, and hydrologic processes of the ecosystem. An important principle is that we must learn to manage complex systems that change frequently as a result of human-induced and natural disturbance. For example, if we are considering an urbanizing drainage basin, as discussed above, then some of the management approaches might be:

▼ Delineating floodplains, landslide-prone areas, and areas susceptible to excessive erosion.
▼ Delineating water processes and sediment processes in the drainage basin.
▼ Identifying sources of potential soil and water pollution.
▼ Identifying useful functions of the drainage basins.
▼ Identifying sensitive environmental areas.
▼ Delineating upstream processes.
▼ Understanding the role of extreme events.

Integration of land management goals with appropriate management approaches must include an understanding of the system being managed, linkages between physical, chemical, and biological processes, and potential impacts due to human use and interests in the area. For our urbanizing watershed the management might include elements such as:

▼ Floodplain avoidance.
▼ Control of sediment and storm-water runoff.
▼ Development of pollution abatement measures.
▼ Protection of natural habitats.
▼ Development of guidelines to allow for restoration following wildfires, high-magnitude storms, and other disturbances.

▼ EMERGENCY PLANNING

In recent years, two types of planning have emerged: projects in which engineering design and environmental impact is an integral part; and emergency projects following catastrophic events such as hurricanes or volcanic eruptions that cause widespread damage. Emergency planning is always in response to pressing needs and an influx of emergency money. Too often, the authorized work is either overzealous and beyond what is necessary, or is not carefully thought out. As a result, emergency projects can cause further environmental disruptions instead of help. Acknowledgment of these problems does not imply that all emergency work is poor or unnecessary; nevertheless, when millions of dollars of emergency money arrive, care must be taken to ensure that it is used for the best possible purposes. We will discuss two examples: Hurricane Camille in 1969 and the eruption of Mt. St. Helens in 1980.

Following severe storms and floods that struck Virginia in the aftermath of Hurricane Camille, emergency federal aid was used to channelize streams in hopes of alleviating future floods. Although much necessary work was performed, in some instances local bulldozer operators with little or no instruction or knowledge of streams were contracted to clear and straighten stream channels. Results of this unplanned and unsupervised channel work have been disastrous to many kilometers of streams. Since catastrophic storms can seriously damage roads, farms, and homes, emergency channel work is often needed; however, such emergency work should be confined to stream channels that require immediate attention. Emergency funds should not be considered a license for wholesale modification of any stream in the damaged area at the request of property owners or others.

The eruption and catastrophic landslide-debris avalanche of Mt. St. Helens delivered 2.5 cubic kilometers of debris into the Toutle River. Fearing continued downstream sediment pollution, two emergency catch dams were constructed to filter the water through rockfill and gravel barriers, thus maintaining the flow of water in the river while trapping the sediment in the catch dams. Unfortunately, the dams were constructed before reliable estimates could be made of the volume of sediment likely to be delivered to the dams. On the North Fork of the Toutle River, the catch dam had the capacity to hold 0.0065 cubic kilometers of debris, and the Corps of Engineers estimated that the sediment load would be 0.01 to 0.02 cubic kilometers per year, thus requiring 2 to 3 dredgings per year for the dam to remain in service. The U.S. Geological Survey, on the other hand, estimated, after a small August flood, that the sediment yield would

be closer to 0.3 to 0.38 cubic kilometers per year, therefore, the catch dams were only about 1 percent of the size needed to survive expected winter storms and floods. This estimate proved true during a storm in late October when the dams failed. Thus, while the idea of catch dams was worthwhile, they were constructed too quickly and were too small.

Emergency planning is important, and even if there is not enough time to prepare an environmental impact statement, all projects should be considered carefully to determine if they are really necessary and will not cause future problems.

▼ ▼ ▼ SUMMARY AND CONCLUSIONS

A primary role of the earth scientist in landscape evaluation is to provide information and analysis before planning, design, and construction. An important contribution of earth scientists and landscape evaluation is emphasizing that all land is not the same, and that particular physical and chemical characteristics of the land can greatly affect appropriate land use.

Environmental geology mapping often involves the preparation of interpretive maps that show suitability for particular land use, or the nature and extent of geologic hazards, such as flooding, landslide, and seismic risk.

The Geographic Information System is a new emerging technology capable of storing, retrieving, transforming, and displaying environmental data. An important aspect of the Geographic Information System is that it has the ability to manipulate data and create new products, such as a land capability map or hazard map. Geographic Information Systems are being widely used in a variety of fields related to land-use planning.

Site selection and evaluation is the process of evaluating the physical environment to determine its capability of supporting human activity and conversely the possible effects of human activity on the environment. A philosophy of site evaluation based on physiographic determinism, or "design with nature," has emerged as a philosophical framework to partially balance traditional economic aspects of site evaluation. From a geologic view, this philosophy requires determining the magnitude and importance of geologic

limitations of a particular site for a particular use.

Site evaluation for engineering purposes, such as construction of dams, highways, airports, tunnels, and large buildings, requires careful geologic study before planning and designing the project. The role of the geologist is to work with engineers and indicate possible adverse or advantageous geologic conditions that might affect the project.

Evaluation of scenic resources and other environmental intangibles is becoming important in landscape evaluation. The significance lies in balancing intangibles with the more obvious economic aspects of the environment.

Probable effects of human use of the land are generally referred to as environmental impact. The National Environmental Policy Act (NEPA) of 1969 requires the preparation of an environmental impact statement (EIS) for all major federal actions that could significantly affect the quality of human environment. The major components of the environmental impact statement are:

▼ Declaration of the purpose and need for the project.
▼ Comparison of reasonable alternatives.
▼ Succinct description of the environment that is to be affected.
▼ Discussion of the environmental consequences of the proposed project and of potential alternatives.

Two other important processes in environmental impact work are: scoping,

which is the process of early identification of important environmental issues that require detailed evaluation in the planning of a proposed project; and mitigation, which is the identification of actions that will avoid, lessen, or compensate for anticipated adverse environmental impact of a particular project.

Following implementation of the National Environmental Policy Act of 1969, it became apparent that there was a need for state and local governments to control activities not covered by the federal legislation. In response to this need many states have enacted legislation that requires environmental review. Some states have enacted State Environmental Policy Acts (SEPAs) patterned after the federal legislation.

Land-use planning is an important environmental issue. The basic philosophy of good land-use planning, from an earth science perspective, is to plan to avoid hazards, conserve natural resources, and generally protect the environment through the use of sound ecological principles. The land-use planning process includes several steps: identification of issues, problems, goals, and objectives; collection and analysis of data; development of alternatives; formulation of land-use plans; review and adoption of plans; implementation of plans; and procedures to revise and amend plans.

The process of comprehensive planning involves development of an official planning document that states general and long-range policies on how the community will deal with future development. In some states, such as

Hawaii, Oregon, and California, the state government has enacted laws that explicitly state what must be covered in local comprehensive plans. The state of Oregon has developed and required a number of planning goals, and the state of California has outlined specific elements that are recommended to be included in comprehensive plans. Review of the plans by a state agency or other environmental review, such as development of an environmental impact report, are important parts of the overall planning process.

Land management and land-use planning are related to one another, because management often follows development of land-use plans. The overlying management approach in many instances is to attempt to understand the physical, biological, and hydrologic processes of the ecosystem. An important principle is that we must learn to manage complex systems that change frequently as a result of natural disturbance and human processes.

Emergency planning follows catastrophic events, such as hurricanes or volcanic eruptions, that cause widespread damage. Emergency planning is an important endeavor, and projects must be considered carefully to determine whether they are necessary and to ensure that they will not cause future problems.

▼ ▼ ▼ REFERENCES

1. BARTELLI, L. J.; KLINGEBIEL, A. A.; BAIRD, J. V.; and HEDDLESON, M. R., eds. 1966. *Soil surveys and land use planning.* Soil Science Society of America and American Society of Agronomy.

2. HAYES, W. C., and VINEYARD, J. D. 1969. *Environmental geology in town and country.* Missouri Geological Survey and Water Resources, Educational Series No. 2.

3. TURNER, A. K., and COFFMAN, D. M. 1973. Geology for planning: a review of environmental geology. *Quarterly of the Colorado School of Mines* 68.

4. GROSS, D. L. 1970. *Geology for planning in DeKalb County, Illinois.* Environmental Geology Notes No. 33, Illinois State Geological Survey.

5. WILSON, L. 1981. Potential for ground-water pollution in New Mexico. In *Environmental geology and hydrology in New Mexico,* S. G. Wells and W. Lambert. New Mexico Geological Society Special Publication No. 10, pp. 47–54.

6. PARKER, H. D. 1987. What is a geographic information system? In *GIS '87: Second annual international conference, exhibits and workshops on Geographic Information Systems.* American Society for Photogrammetry and Remote Sensing. pp. 72–79.

7. STAR, J., and ESTES, J. 1990. *Geographic Information Systems.* Englewood Cliffs, N.J.: Prentice-Hall.

8. WHYTE, W. H. 1968. *The last landscape.* Garden City, New York: Doubleday.

9. FLAWN, P. T. 1970. *Environmental geology.* New York: Harper & Row.

10. LYNCH, K. 1962. *Site planning.* Cambridge, Massachusetts: M.I.T. Press.

11. PREST, A. R., and TURVEY, R. 1965. Cost-benefit analysis: a survey. *The Economic Journal* 75: 683–735.

12. McHARG, I. L. 1971. *Design with nature.* Garden City, New York: Doubleday.

13. SCHULTZ, J. R., and CLEAVES, A. B. 1955. *Geology in engineering.* New York: John Wiley & Sons.

14. KRYNINE, D. P., and JUDD, W. R. 1957. *Principles of engineering geology and geotechniques.* New York: McGraw-Hill.

15. LONGMIRE, P. A.; GALLAHER, B. M.; and HAWLEY, J. W. 1981. Geological, geochemical, and hydrological criteria for disposal of hazardous wastes in New Mexico. In *Environmental geology and hydrology in New Mexico,* ed. S. G. Wells and W. Lambert. New Mexico Geological Society Special Publication No. 10, p. 93–102.

16. ZUBE, E. H. 1973. Scenery as a natural resource. *Landscape Architecture* 63: 126–32.

17. LINTON, D. L. 1968. The assessment of scenery as a natural resource. *Scottish Geographical Magazine* 84: 219–38.

18. LEOPOLD, L. B. 1969. *Quantitative comparison of some aesthetic factors among rivers.* U.S. Geological Survey Circular 620.

19. MELHORN, W. N.; KELLER, E. A.; and McBANE, R. A. 1975. *Landscape aesthetics numerically defined (land system): Application to fluvial environments.* Purdue University Water Resources Research Center Technical Report No. 37.

20. COUNCIL ON ENVIRONMENTAL QUALITY. 1979. *Environmental quality,* Annual Report.

21. BREW, D. A. 1974. *Environmental impact analysis: The example of the proposed Trans-Alaska Pipeline.* U.S. Geological Survey Circular 695.

22. CALLIES, D. L. 1984. *Regulating paradise.* Honolulu: University of Hawaii Press.

23. REMY, M. H.; THOMAS, T. A.; and MOOSE, J. G. 1991. *Guide to the California Environmental Quality Act.* 5th ed. Point Arena, Calif.: Solano Press Books.

24. NATIONAL RESEARCH COUNCIL. 1972. *The earth and human affairs.* San Francisco: Canfield Press.

25. BRICE, J. C. 1971. *Measurements of lateral erosion at proposed river crossing sites of the Alaskan Pipeline.* U.S. Geological Survey, Water Resources Division, Alaska District.

26. NATIONAL PARK SERVICE. 1974. Cape Hatteras Shoreline Erosion Policy Statement, Department of Interior.

27. HOYT, J. H., and HENRY, V. J., JR. 1971. Origin of capes and shoals along the southeastern coast of the United States. *Geological Society of America Bulletin* 82: 59–66.

28. LEATHERMAN, S. P. 1983. Barrier dynamics and landward migration with Holocene sea-level rise. *Nature* 301 (5899): 415–18.

29. NATIONAL PARK SERVICE. 1984. *General management plan, development concept plan, and amended environmental assessment, Cape Hatteras National Seashore.*

30. TANJI, K.; LAUCHLI, A.; and MEYER, J. 1986. Selenium in the San Joachin

Valley. *Environment* 28 (6): 6–11, 34–39.

31. U.S. DEPARTMENT OF THE INTERIOR, BUREAU OF RECLAMATION. 1987. *Kesterson Program.* Fact Sheet No. 4.

32. ROHSE, M. 1987. *Land-use planning in Oregon.* Corvalis, Oregon: Oregon State University Press.

33. CURTIN, D. J., JR. 1991. *California land-use and planning law.* 11th ed. Point Arena, Calif.: Solano Press Books.

34. WILLIAM SPANGLE AND ASSOCIATES; F. BEACH LEIGHTON AND ASSOCIATES; and BAXTER, McDONALD AND COMPANY. 1976. *Earth-science information in land-use planning—guidelines for earth scientists and planners.* U.S. Geological Survey Circular 721.

A land, or environmental, ethic is becoming an important part of our government institutions and is being incorporated into our legal and economic principles. No longer is the best use of land that which always returns the greatest profit. The new land ethic establishes that the human race is part of a land community that includes trees, rocks, animals, and scenery, and that we are morally bound to assure the community's continued existence. Thus, this ethic affirms our belief that this earth is our only suitable habitat (Figure 18.1) and recognizes the rights of people to breathe clean air, drink unspoiled water, and generally exist in a quality environment.

According to the National Environmental Policy Act of 1969, the intention of national environmental policy is to encourage productive and enjoyable harmony between people and their environment, to promote efforts to eliminate or at least minimize degradation of the environment, and to continue investigating relations between ecological systems and important natural resources. This policy suggests that each generation has a moral responsibility to provide the next generation with healthful, productive, and aesthetically pleasing surroundings. It implies a doctrine of trust, asserting that all public land, and, to a lesser extent, some private land, is essentially held in trust for the public.

The Trust Doctrine as a theory relates closely to constitutional issues involving people's right to a quality environment. Although nowhere in the Constitution does it state this right, some students of law believe that such a right can be inferred from the Ninth Amendment to the Constitution, which recognizes that the list of specific rights in the Bill of Rights does not deny the existence of other unlisted rights—that is, there are many rights not specifically listed but always held by the people (1).

The fundamental rights entitled to protection under the Ninth Amendment are considered so basic and important to our society that it is inconceivable that they are not protected from unwarranted interference (2). It can therefore be argued that this amendment encompasses the right of people to have clean air, clean water, and other resources necessary to ensure a quality environment.

It is important to recognize a certain amount of friction between the Trust Doctrine, which establishes that particularly significant land resources are held in public trust, and the Fifth Amendment, which establishes that land may not be taken from an individual without due process of law and just compensation. Problems result in interpreting law and determining how to measure just compensation. An important issue is defining at what point the private use of land infringes on the rights of other people and future generations.

CHAPTER EIGHTEEN

Environmental Law

▼ MONO LAKE AND THE PUBLIC TRUST DOCTRINE

Mono Lake, located in the Mono basin at the foot of the Sierra Nevada east of Yosemite National Park in California (Figure 18.2), is the focus of recent controversy, which centers around the very existence of the lake. From the lake's watershed, approximately 100,000 acre feet of water per year is diverted south to the city of Los Angeles. Mono Lake is large, measuring approximately 21 by 13 kilometers, with an average depth of about 17 meters. These dimensions make it the largest lake by volume contained entirely within the state of California.

During the last million years, a number of important geologic events associated with active uplift of the Sierra Nevada, volcanic activity, and glaciation have affected the lake and consequently there is now no natural outlet from the lake. The lake is fed by a number of streams from the Sierra Nevada and some groundwater flow as well. Because there is no natural outlet, the lake is salty, having a salinity approximately three times that of sea water.

Mark Twain visited Mono Lake in the 1860s and had little good to say about it except that the alkaline waters made laundry work easy. In *Roughing It,* Twain wrote, "Half a dozen little mountain brooks flow into Mono Lake but not a stream of any kind flows out of it. What it does with its surplus water is a dark and bloody mystery" (3). What happens to the water, of course, is that it evaporates. In fact, approximately 22 cm/yr evaporate from the surface of Mono Lake. Under natural conditions this loss would be matched from streams that feed the lake system (3).

Figure 18.1
This earth is indeed our only suitable habitat, and there are few alternatives to maintaining a quality environment. (Photo courtesy of NASA.)

Mono Lake and the basin it is in have a long and interesting history going back at least to 1853, when Yosemite Indians were pursued by the military to the shores of the lake. About that time, gold was discovered in the area, initiating a small gold rush that lasted until approximately 1889. Then in 1913 the city of Los Angeles considered importing water into the growing urban area, and by 1930 funds had been approved for the construction of dams, reservoirs, and a tunnel to divert water from the eastern Sierra and Mono Lake area. In 1941 diversion of water from the Mono basin began in earnest and by 1981 the lake level had dropped approximately 15 meters. This decreased the volume of the lake by approximately one-half and resulted in increasing the salinity by 100 percent.

Brine shrimp grow in great abundance in the lake and provide the major food source for migrating birds. If the salinity were to become too high, the brine shrimp would die and the birds would have no food during a crucial stage in their migration.

The lowering of the lake level also exposed nearly 9000 hectares of highly alkaline lake bed. During windy periods alkali dust may rise into the atmosphere several thousand feet and be transported both around and out of the basin, causing air pollution (3).

More significantly, lowering of the lake formed a land bridge to several volcanic islands in the lake that are major breeding grounds for California gulls. In 1979, after the land bridge had formed, coyotes entered the nesting area and routed all 34,000 nesting birds (3). Extremely wet years in 1983 and 1984 caused the lake level to rise a bit, but it was still much lower than the

1941 level. Figure 18.2 shows the 1980 situation with inflow and a diversion of waters.

People interested in the preservation of Mono Lake and its ecosystem would like to see the lake level stabilized approximately 3 meters above that necessary to support the healthy ecosystem. They advocate a wet year/dry year plan that would limit diversion to the dry years when the city of Los Angeles really needs the water. They further advocate a statewide program to conserve urban and agricultural water.

No one disagrees with the advocacy of water conservation. The city of Los Angeles, however, which receives approximately 17 percent of its water supply from the Mono basin, would like to see diversions continue at a rate greater than that advocated by those who would like to see the lake preserved. The people in favor of continued diversion point out that the project produces a good deal of energy (approximately 300 million kilowatt hours per year, which saves approximately half a million barrels of oil per year). They would like to see the diversions continued and the lake level eventually stabilized at about 15 meters below the 1981 level. One of their arguments is that the city of Los Angeles has invested more than $100 million in the area since the 1930s and it really needs the water.

The Mono Lake story is an important one in environmental law because in 1983 the California Supreme Court reaffirmed the public interest in protecting natural resources through what is known as the *Public Trust Doctrine*. The 1983 decision states that it is the duty of the state to protect the people's common heritage, including streams, lakes, marshlands and tidelands. In

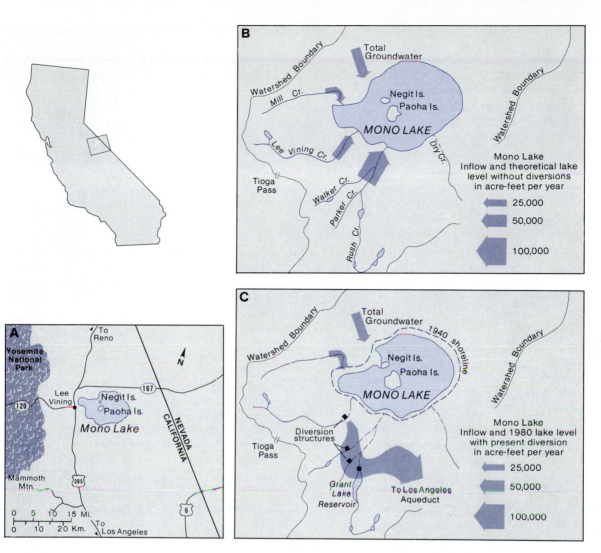

Figure 18.2
(a) Location of Mono Lake. (b) Situation without water diversion, and (c) with water diversion. (From Mono Lake Committee, 1985.)

essence the court decided that public trust obligates the State of California to protect lakes such as Mono as much as possible, even if this means reexamining past water allocations (3).

The case of Mono Lake has not been finally resolved, and eventually the courts may have to weigh the public trust values at Mono Lake against the water needs of the people of Los Angeles. Ideally, the resolution will reflect a balance between concern for environmental issues and the water needs of urban people.

▼ ENVIRONMENTAL LAW

Environmental law is becoming an important part of our jurisprudence. The many works devoted to environmental law attest to the fact that the subject is becoming

increasingly popular among lawyers and environmental law societies, and environmental courses have been established at leading law schools.

We often take for granted the air, water, and other resources necessary for our survival. In America, we still tend to suffer from the myth of superabundance and to think of resources as inexhaustible. In large cities with accompanying large populations of people, cars, and industries, however, resources are deteriorating. As a result, regulations and laws concerning pollution of air and water are necessary. This situation is not new; by 1306 large amounts of smoke from burning coal polluted the air in London to such an extent that a royal proclamation was issued to curtail use of coal. Violations of the law were punishable by death (4).

Air pollution was proclaimed a public nuisance as early as 1611 when an English court ruled, in essence, that property owners have the right to breathe and smell air that is free of pollutants. The case involved a plaintiff who asked for injunction and damages because his neighbor, the defendant, raised hogs, whose odor the plaintiff considered a nuisance. The defendant was found to be creating a nuisance even though he pleaded that raising hogs was necessary for his subsistence and that neighbors should not have such delicate noses that they could not bear the smell of hogs (5). The English Nuisance Law of 1536 involved a type of common law still used in the United States, the main principle of which is that if other people suffer equally from a particular pollution, an individual cannot bring suit against the polluter. In the case of the hog farmer, it is obvious that the effect of the nuisance varied with proximity to the hog pens (4).

The Process of Law

It is beyond the scope of our discussion to consider in detail the process of law and how it relates to the environment. Suffice it to say that law is a technique for the ordered accomplishment of economic, social, and political purposes, and the most desirable legal technique generally is one that most quickly allows ends to be reached. Law serves primarily the major interest that dominates the culture, however, and in our sophisticated culture, the major concerns are wealth and power. These concerns stress society's ability to use the resource base to produce goods and services, and the legal system provides the vehicle to ensure that productivity (6).

Some environmental lawyers today believe that the process of law as it has generally been practiced is not working satisfactorily when environmental issues are concerned. In general, when two views conflict, adversarial confrontation occurs. Emotional levels are often high and it is difficult for disagreeing parties to see positions other than their own. An emerging view in environmental law is one which stresses problem solving. This may take the form of mediation through negotiation. For example, in the 1970s the Environmental Protection Agency often announced new environmental regulations that were thrust upon various sectors of society without warning. Not surprisingly, many of these regulations ended up in lawsuits and lengthy litigation. Then in the 1980s the U.S. Environmental Protection Agency began a practice of consulting interested parties prior to regulation. Consultation, negotiation, and mediation may well prove much more successful than earlier strategies that produced unproductive adversarial reactions.

One of the major problems with mediation and negotiation is how to get all the important players in a given case to sit down and talk about the issues in a meaningful way. Whichever party seems to have the upper hand may try to make the negotiations particularly difficult for opposing views. Increasingly, however, both parties in environmental issues are recognizing that it is advantageous to work together towards a solution that is satisfactory for everyone. This is basically a collaborative process that seeks solutions that favor the environment while allowing activities and projects to go forward. It is important to recognize that the process of collaboration is different from and broader than compromise, which often requires giving something up to get something else. Collaboration is more comprehensive in that it includes, in fact necessitates, various parties working together to create opportunities for mutual gain. It creates a climate of joint problem solving. What is required for negotiation and mediation to work is for both parties to clearly and honestly state their positions and then work to see where common ground might be drawn. Relationships built upon mutual trust are then developed. It should come as no surprise that almost all issues are negotiable and alternatives can often be worked out that avoid or at least minimize costly litigation and lengthy delays.

▼ CASE HISTORIES

A discussion of particular case histories is useful for understanding some legal aspects of environmental problems. Our selection of cases does not imply a judgment of any particular activity but rather indicates the considerable variability and possibilities in environmental law. The cases we will discuss include areas of air pollution, aesthetic pollution, and land use.

Ducktown, Tennessee

The Ducktown story starts in 1843 with what was thought to be a gold rush in Tennessee that turned out to be a copper rush. By 1855 approximately 30 companies in the Ducktown area transported copper ore by mule to a place called Copper Basin. Trees were cut from the surrounding area to fuel giant open ovens used to separate the copper from zinc, iron, and sulfur. These huge pits were up to about 200 meters wide and 30 meters deep. Eventually an area of over 100 square kilometers was deforested to fuel these giant smelters. The open fires produced tremendous clouds of noxious sulfur dioxide gas. At times the smoke was so thick that mules had to wear bells to keep from colliding with each other and people. People reportedly got lost at midday! Deposition of the sulfur, combined with a wet climate, resulted in massive acid precipitation and acid dust deposition, killing the vegetation surrounding the smelters. This plus the deforestation led to massive soil erosion (Figure 18.3). Adding insult to injury, grazing cattle consumed plants that managed to survive (7).

Not surprisingly, people in the surrounding area were not pleased by the mining operations and pollution

Figure 18.3
Fumes from the smelter at Ducktown, Tennessee, killed nearby vegetation and initiated a rapid erosion cycle. (Photo by A. Keith, courtesy of U.S. Geological Survey.)

that prevented them from using their farms and homes as they did before the construction and operation of the smelting and refining operations. In what was an early environmental case in 1904 *(Madison* v. *Ducktown Sulphur, Copper, and Iron Company)*, the court chose to balance the equities in favor of the mining industry, and declined to grant injunctive relief (an order to stop operations). The court reasoned that since it was not possible for the industry to reduce the ores in a different way or move to a remote location, forcing them to stop air pollution would compel the companies to stop operating their plants. Such an order would cause 10,000 people to lose their jobs, destroy the tax base of the county, and make plant properties practically worthless (5). Therefore, although damages were awarded the plant was allowed to continue operation.

Legal action again rose to the surface in 1907 when the state of Georgia sued the Tennessee Copper Company in an effort to stop the pollution from drifting across the nearby border and damaging Georgia's land. In 1915 the copper smelter was ordered to limit sulfur emissions to 20 tons per day during the summer months. This was a very significant reduction and might have caused the industry to flounder. During the case the famous Supreme Court Justice Oliver Wendell Holmes argued that the state of Georgia was well within its rights to decide whether degradation of the land and air pollution were acceptable. Limiting the emissions of sulfur dioxide led to development of a new technology that converted the sulfur in the smelting operations to a useful product (sulfuric acid). Ironically, the sulfuric acid became the main product that Tennessee Copper sold (7, 8).

The legacy from the mining era with open-pit smelting left an area of approximately 140 square kilometers that was essentially an eastern desert. Many people in the Ducktown area are proud of their mining heritage and have even tried to turn the area into a tourist attraction. Various companies have owned the mines

over the years, and since 1982 it has been the Tennessee Chemical Company. Ducktown was a typical mining town in that the work in the mines was a family tradition; sons worked in mines where their fathers and grandfathers had worked. Until the 1960s the company store provided everything that the people needed from birth to death (7).

Import of low-cost, high-grade copper ore from countries such as Chile, Peru, and Zambia put pressure on the mines and they finally closed in July of 1987. The economic future of the area was in trouble. Revegetation of the area devastated by the earlier smelting and refining activities started in the 1930s and today approximately two-thirds of the area has some vegetation. Some people, however, want to leave part of the area as a desert. Because the landscape is unique in the eastern United States there is still hope of developing tourism as an industry (7).

The legal issues and court battles over copper mining, smelting, and refining in the Ducktown area have important historical meaning for the practice of environmental law. The cases involved states' rights, and helped establish important aspects of the **Balancing Doctrine**, which asserts that the public benefit from and the importance of a particular action should be balanced against potential injury to certain individuals. The Supreme Court case in 1907 was one of the nation's first environmental-rights decisions, and the state of Georgia came out the winner (7, 8).

The important lessons from the Ducktown experience will, it is hoped, help alleviate or lessen similar activities in other parts of the world. Certainly, Ducktown is not unique. Refineries and smelters in the Sudbury, Ontario, area have damaged an area much larger than the Copper Basin. Furthermore, there are similar activities operating today in the Third World. Destruction of vegetation in countries such as Brazil and Mexico is occurring where environmental regulations are far behind industrial activities (7).

Aesthetic Resources

Water, land, and air pollution are relatively easy to measure and evaluate compared to the intangible aesthetic pollution of visual, audio, and other senses that affect one's well-being. We do recognize some of the possible harmful effects of aesthetic pollution, however, and the courts are forming attitudes on these matters. The precedent that aesthetics is a valid subject of legislative concern was set in New York courts in cases that involved aesthetic degradation, particularly loss of value of property when the view of a lake or mountain was spoiled. It was recognized that reduction in property value from loss of a view that otherwise increased the value of the property should not be borne by the owner whose land was taken for public purposes without permission (4).

Two examples from New York concerning road construction emphasize that the law of aesthetics is becoming more important and that intangible factors that were formerly considered outside the public interest are now as important as other more easily measured factors. In *Clarance* v. *State of New York,* the court awarded the plaintiff $10,000 because a road degraded his view of Seneca Lake and denied him easy access to the lake (4). In a similar case, *Dennison* v. *State of New York,* a property owner was awarded damages because a new highway caused loss of privacy and seclusion, loss of view of forests and mountains, traffic noise, lights, and odors. The court concluded that all these factors tended to cause damage to the plaintiff's property (4).

Storm King Mountain Case

The Storm King Mountain dispute is a classic example of possible conflict between a utility company and conservationists. In 1962, the Consolidated Edison Company of New York announced plans for a hydroelectric project approximately 64 kilometers north of New York City in the Hudson River Highlands, an area considered by many to have unique aesthetic value because nowhere else in the eastern United States is there a major river eroded through the Appalachian Mountains at sea level, giving the effect of a fjord (9). Early plans for the project called for a powerhouse to be constructed aboveground, requiring a deep cut into Storm King Mountain. The project was redesigned to site the powerhouse entirely underground, eliminating the cut on the mountain (Figure 18.4). Regardless, conservationists continued to oppose the project, and the issues broadened to include possible damage to fishery industries. The argument was that the high rate of water intake from the river, 31,200 cubic meters per minute, would draw many fish larva into the plant where they would be destroyed by turbulence and abrasion. The most valuable sport fish is the striped bass, and one study showed that 25 to 75 percent of the annual bass hatch might be destroyed if the plant were operating, The fish return from the ocean to tidal water to spawn, and since the Hudson River is the only estuary north of Chesapeake Bay where the striped bass spawn, concern for the safety of the fisheries is justified. The problem is even more severe because the proposed plant is located near the lower end of the 13-kilometer reach in which the fish spawn (9).

The Storm King Mountain controversy is interesting because it emphasizes the difficulty of making decisions about multidimensional issues. On the one hand, a utility company is trying to survive in New York City where unbelievably high peak-power demands are accompanied by high labor and maintenance costs. On the other hand, conservationists are fighting to preserve a beautiful landscape and fishery resources. Both have legitimate arguments, but in light of their special interests, it is difficult to resolve the conflicts. Present law and procedures are sufficient to resolve the issues, but trade-offs are necessary. Ultimately, an economic and environmental price must be paid for any decision, a price that reflects our desired life-style and standard of living.

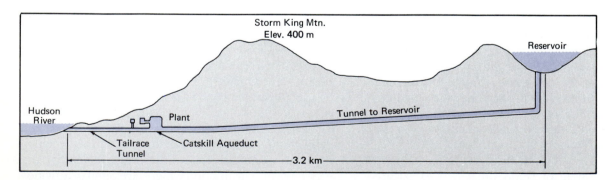

Figure 18.4

Diagram showing how the entire Storm King Mountain hydroelectric project might be placed underground. (After L. J. Carter, *Science,* vol. 184, June 1974, pp. 1353–58. Copyright 1974, American Association for the Advancement of Science.)

The first lawsuit in the Storm King dispute was filed in 1965, and following sixteen years of intense courtroom battles, the dispute was settled in 1981. The total paper trail exceeded 20,000 pages and in the end the various parties used an outside mediator to help them settle their differences. This famous case has been cited as a major victory for environmentalists, but the real question is, could the outcome have been decided much earlier? Had the various parties been able to sit down and talk about the issues openly, then the process of negotiation and mediation might have been completed much earlier and at much less cost to the individual parties and society in general (10). This case emphasizes our earlier discussion that environmental law issues may best be approached from a problem-solving viewpoint rather than by various sides taking adversarial positions.

Florissant Fossil Beds Case

A case reported by Yannacone, Cohen, and Davison in Colorado emphasizes the significance of the Trust Doctrine and the Ninth Amendment as they pertain to land management (2). The conflict surrounded the use of 7.3 square kilometers of land near Colorado Springs. The land is part of the Florissant Fossil Beds where insect bodies, seeds, leaves, and plants were deposited in an ancient lake bed about 30 million years ago. Today, they are remarkably preserved in thin layers of volcanic shale. Unfortunately, the fossils are delicate, and unless protected, tend to disintegrate when exposed. Many people consider the fossils unique and irreplaceable. At the time of the controversy, a bill had been introduced into Congress to establish a Florissant Fossil Beds National Monument. The bill had passed the Senate, but the House of Representatives had not yet acted on it.

While the House of Representatives was deliberating the bill, a land development company that had contracted to purchase and develop recreational homesites on 7.3 square kilometers of the ancient lake bed announced that it was going to bulldoze a road through a portion of the proposed national monument site to gain access to the property it wished to develop. A citizens' group formed to fight the development until the House acted on the bill. The group tried to obtain a temporary restraining order, which was first denied because no law prevented the owner of a property from using that land in any way he wished provided that existing laws were upheld. The conservationists then argued before an appeals court that the fossils were subject to protection under the Trust Doctrine and the Ninth Amendment. Their argument was that protection of an irreplaceable, unique fossil resource was an unwritten right retained by the people under the Ninth Amendment, and that furthermore, since the property had tremendous public interest, it was also protected by the Trust Doctrine. An analogy used by the plaintiffs was that if a property owner were to find the

Constitution of the United States buried on the land and wanted to use it to mop the floor, certainly that person would be restrained. After several more hearings on the case, the court issued a restraining order to halt development; shortly thereafter, the bill to establish a national monument was passed by Congress and signed by the president (2).

The court order prohibiting destruction of the fossil beds may have deprived a landowner of making the most profitable use of the property, but it does not prohibit all uses consistent with protecting the fossils. For instance, the property owners are free to develop the land for tourism or scientific research. While this might not result in the largest possible return on the property owner's investment, it probably would return a reasonable profit.*

▼ ENVIRONMENTAL LEGISLATION

The ultimate goal of those who are concerned with how we treat and use our natural environment and resources is to ensure that ecologically sound, responsible, socially acceptable, economically possible, and politically feasible legislation is passed (2). To attain this, we need professional people to assist legislators at all levels of government in drafting the needed laws and regulations.

Environmental legislation has already had a tremendous impact on the industrial community. New standards in regulations limiting the discharge of possible pollutants into the environment have placed restrictions on industrial activity. Furthermore, any new activity that directly or indirectly involves the federal government must be preceded by an evaluation of the environmental impact of the proposed activity. Beyond this, federal legislation has set an example that many states are following in passing environmental protection legislation.

Examining all federal legislation that has environmental implications is beyond the scope of our discussion; however, discussing some of the major acts is valuable in understanding the basics of such legislation. For this purpose, we will discuss the Refuse Act of 1899, the Fish and Wildlife Coordination Act of 1958, the National Environmental Policy Act of 1969, the Water Quality Improvement Act of 1970, the Resource Conservation and Recovery Act of 1976, the Surface Mining Control and Reclamation Act of 1977, the Clean Water Act of 1977, the Comprehensive Environmental Response, Compensation and Liability Act of 1980, the Hazardous and Solid Waste Amendments of 1984, the Water Quality Act of 1987, and the Clean Air Act Amendments of 1990.

---

*Yannacone et al., *Environmental Rights and Remedies* (San Francisco: Bancroft-Whitney, 1972), pp. 39–46.

The Refuse Act of 1899

This act states that it is unlawful to throw, discharge, or deposit any type of refuse from any source except that flowing from streets and sewers into any navigable water. Furthermore, the act implies that it is unlawful to discharge refuse into tributaries of navigable water. For all practical purposes, this means that it is against the law to pollute any stream in the United States. However, the Secretary of the Army can allow the discharge of refuse into a stream if a permit is first applied for.

The Fish and Wildlife Coordination Act of 1958

This act establishes a national policy that recognizes the important contribution of wildlife resources and specifically provides that conservation of wildlife shall be balanced with other factors in water resources development planning (2).

Wildlife resources in the act are broadly defined to include birds, fish, mammals, and all other types of animals, as well as aquatic and land vegetation upon which the wildlife depends. The act also provides for coordination of wildlife aspects of resource development and requires that projects to develop power, control flooding, or facilitate navigation must first consult with the U.S. Fish and Wildlife Service, as well as with the head of the state agency exercising administrative control of wildlife resources in the particular state where the project is planned. This applies, with the exception of impoundments of less than 4 hectares, to the waters of any stream or other body of water to be impounded, ditched, diverted, or otherwise modified for any purpose by any department or agency of the United States (2).

The object of the Fish and Wildlife Coordination Act is to prevent damage or loss of wildlife resources from water resources development and, at the same time, provide for development and improvement of wildlife and necessary habitat. Reports by the Fish and Wildlife Service, along with those of state agencies, provide details of expectable damage to wildlife resources from a particular project and include recommendations as to how to reduce the projected damages and develop and improve the fish and wildlife resources. According to the act, agencies that receive the reports must consider these recommendations fully and include in the project plans ways and means to achieve wildlife conservation (2).

The National Environmental Policy Act of 1969

The philosophical purposes of the Environmental Policy Act are: to declare a national policy that will encourage harmony with our physical environment; to promote efforts that prevent or eliminate environmental degradation, thereby stimulating human health and welfare; and to improve our understanding of relations between ecological systems and important natural resources. To promote interest, research, and authority to achieve these purposes, the act establishes the Council on Environmental Quality. The Council is in the Executive Office of the President and is responsible for preparation of a yearly Environmental Quality Report to the Nation. It also provides advice and assistance to the president on environmental policies.

The most significant aspect of the act is that it requires an environmental impact statement before major federal actions are taken that could significantly affect the quality of the human environment. This requirement extends to such activities as construction of nuclear facilities, airports, federally assisted highways, electric power plants, and bridges; release of pollutants into navigable waters and their tributaries; and resource development on federal lands, including mining leases, drilling permits, and other uses. Since enactment of the National Environmental Policy, many thousands of environmental impact statements have been prepared.

The Water Quality Improvement Act of 1970

This act, a comprehensive water-pollution control law, essentially gives more power to the Federal Water and Pollution Control Act of 1956. The purpose of the 1956 act was to enhance the quality and value of our water resources and to establish a national policy to prevent, control, and abate pollution of the country's water resources. The 1970 act provides for control of oil pollution by vessels and offshore and onshore oil wells, control of hazardous pollutants other than oil, control of sewage from vessels, research and development methods to control and eliminate pollution of the Great Lakes, research grants to universities and scholarships to students to train people in water quality control, and projects to demonstrate methods for eliminating and controlling mine drainage of acid water from both active and abandoned mines (2).

The main regulatory function of the Water Quality Improvement Act of 1970 requires that all facilities or activities that involve discharge into navigable water and that require a federal license must obtain a certificate of reasonable assurance that the proposed activity will not violate the state's water quality standards as approved by the Environmental Protection Agency.

Resource Conservation and Recovery Act of 1976

To control hazardous wastes and protect human health, this act provides for "cradle-to-grave" control of hazardous wastes. At the heart of the act is the identification of hazardous wastes and their life cycles. Regulations will then require stringent record keeping and reporting to ensure that wastes do not present a public nuisance or health problem.

Surface Mining Control and Reclamation Act of 1977

The purpose of the act is to control the environmental effects of strip mining. The act prohibits mining practices that have in the past led to environmental degradation. Reclamation of the land after mining is required. Although the act is making great improvements in the way strip mining is conducted, it has been criticized because it does not sufficiently allow for site-specific conditions in the regulations that control mining and reclamation.

Clean Water Act of 1977

The purpose of this act is to clean up the nation's water. In particular, most municipal sewage treatment plants are to achieve secondary treatment or use the best practicable waste treatment technology. Billions of dollars in federal grants have been awarded to meet these goals. The act clearly encourages innovative and alternative techniques in water treatment and waste disposal, such as land application of sewage sludge, aquifer recharge of treated wastewater, and energy recovery. Results of water treatment nationwide have been encouraging. For example, in the 1950s and 1960s, the Detroit River was considered dead. As a result of water treatment, fishermen now catch walleye pike, muskellunge, smallmouth bass, salmon, and trout. Although the Detroit River is still not a really clean river, improvements continue as the discharge of pollutants is reduced.

Comprehensive Environmental Response, Compensation, and Liability Act of 1980

This act established a revolving fund (popularly called the "Superfund") to clean up several hundred of the worst abandoned hazardous waste disposal sites. Implementation of the fund by the Environmental Protection Agency has been criticized, and the funds available for cleanup are not sufficient. Nevertheless, some sites have been treated and the legislation is a step in the right direction.

Hazardous and Solid Waste Amendments of 1984

This legislation amended and strengthened the Resource Conservation and Recovery Act of 1976. In particular, it provided for implementation of a comprehensive regulatory program for underground storage tanks that contained substances defined as hazardous under the 1976 legislation. The 1984 legislation is particularly important because it regulates underground gasoline storage tanks. As a result, the legislation has initiated a tremendous amount of environmental work to evaluate underground tanks and develop new technology to minimize the potential for hazardous material such as gasoline to pollute groundwater resources.

Water Quality Act of 1987

This Act established national policy for programs to control nonpoint sources of water pollution. Nonpoint pollution sources result from processes such as land runoff, atmospheric deposition, and precipitation, and are not nearly so easy to regulate as point sources that emanate from a particular site such as a pipe. This legislation has been important in the development of management plans by states to control nonpoint water pollution sources.

Clean Air Act Amendments of 1990

This important legislation imposes much tighter controls on air quality. In particular, the legislation places tighter controls on emission of sulfur dioxide produced from coal-burning power plants that are partly responsible for the acid rain problem; places tighter limitations on automobile emissions that produce urban smog; and limits emissions of chemical compounds that cause ozone depletion. With respect to ozone depletion, the legislation sets a target date of the year 2000 to terminate production of all chlorofluorocarbons. Because the provisions of the 1990 Act are very broad, impacts from the pollution abatement measures will be significant and widely felt throughout many sectors of American industry and society.

State and Local Environmental Legislation

Federal environmental legislation has set many good examples for state and local governments to follow. A large number of states have enacted legislation analogous to federal laws, and this trend will continue. Areas of particular public concern at the state and local levels are environmental impact, floodway regulation, sediment control, land use, and water and air quality. Many effective programs at the local level result because acceptance and enforcement, when combined with education and communication, are probably most effective where environmental problems are experienced first-hand.

For example, in 1973 the state of North Carolina passed a sediment-pollution control act that established that it is vital to the public interest and necessary for public health and welfare to control erosion and sedimentation. The North Carolina legislation recognizes the need for local participation and encourages local government to draft ordinances consistent with the state act. Macklenburg County, part of the fourth-largest urbanizing area in the southeastern United States, drafted a sediment-control ordinance that calls for submission of erosion-sediment control plans before land-disturbing activities are initiated. These plans must be approved by the County Engineer and reviewed by the Soil and Water Conservation District Office for comments and recommendations. An example of such a plan is shown in Figure 18.5.

Communication and education are keys to success with ordinances such as that in North Carolina to control sediment pollution. Therefore, a significant number of activities, including urban field trips and workshops on sediment control, are part of an ongoing program. The basic idea is not to punish developers into sediment control, but to indicate the benefits of sound conservation practices.

▼ WATER LAW

Water is so necessary to all aspects of human use and interest that water resources may be the most legislated and discussed commodity in the arena of environmental law. Struggles for sufficient water by populous areas such as New York City and southern California are examples. Intrastate and legislative contracts were necessary to obtain water for New York City from the Delaware River waters in the Catskill Mountains. In California, a fight between California and Arizona over Colorado River water has continued for 50 years (4).

There will always be problems in allocating water resources, and the problem is most severe in areas such as the southwestern United States where water is scarce. California provides a good example of the type of conflict that arises when a large population with accompanying industrial and agricultural activity is concentrated in an area with a natural deficiency of water. Approximately two-thirds of the state's water supply comes from the northern third of California, but the greatest need for water is in the southern two-thirds of the state where the vast majority of people live and where most of the industry and agricultural activity takes place (11).

The deficiency of water in southern California, combined with an almost insatiable demand for water, resulted in construction of the California aqueduct to

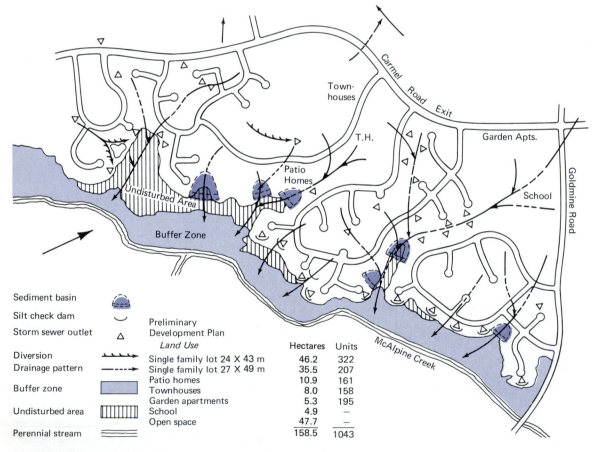

Land Use	Hectares	Units
Single family lot 24 X 43 m	46.2	322
Single family lot 27 X 49 m	35.5	207
Patio homes	10.9	161
Townhouses	8.0	158
Garden apartments	5.3	195
School	4.9	—
Open space	47.7	—
	158.5	1043

Figure 18.5
Erosion- and sediment-control plan for a housing development including single-family homes, patio homes, townhouses, garden apartments, a school, and open space. (Courtesy of Braxton Williams, Soil Conservation Service.)

move water from the northern and southeastern parts of the state to the Los Angeles area. The legal grounds that allowed California citizens to vote for and pass a state bond of nearly $2 billion for the aqueduct that now transports water from the northern part of the state to the southern (Figure 18.6) was derived from the Constitution of the State of California: "The general welfare requires that water resources of the state be put to beneficial use to the fullest extent of which they are capable . . . that the conservation of such water is to be exercised with a view to the reasonable and beneficial use thereof in the interest of the people and for the public welfare" (4). From a philosophical viewpoint, we might raise the question, "Is continued development in southern California warranted?" Perhaps people should be located where the water is, rather than moving huge quantities of water hundreds of kilometers over rough terrain and active faults, at tremendous cost to an area already overcrowded and suffering from pollution and other environmental problems. Value judgments in these matters have

Figure 18.6
California aqueduct in the San Joaquin Valley, California. (Photo courtesy of State of California, Department of Water Resources.)

little validity, and the fact that the aqueduct was constructed indicates the price people are willing to pay to support their standard and style of living. We should point out, however, that many of the people in the Feather River and Owens River areas from which the water is diverted and transported to Los Angeles are not so pleased as those who receive the water.

A rather elaborate framework of law surrounds the use of surface water, two major aspects of which are the Riparian Doctrine and the Appropriation Doctrine.

Riparian rights to water are restricted to owners of the land adjoining a stream of standing water. The word *riparian* comes from the Latin word meaning *bank*, and traditional Riparian Doctrine is a common-law concept. It holds, essentially, that each landowner has the right to make reasonable use of water on his or her land, provided that the water is returned to its natural stream channel before it leaves the property. The property owner also has the right to receive the full flow of the stream undiminished in quantity and quality, but is not entitled to make withdrawals of water that infringe upon the rights of other riparian owners (12).

The Riparian Doctrine was the prevailing water law in most states before 1850, and is still used in all the states east of the Mississippi and in the first tier of states immediately west of the river (Figure 18.7). The right to use water is considered real property, but the water itself does not belong to the property owner. Riparian water rights are considered natural rights and property that enter into the value of land. They may be transferred, sold, or granted to other people (13).

The **Appropriation Doctrine** in water law holds that prior usage is a significant factor. That is, the first to use the water for beneficial purpose is prior in right. This right is perfected by use and is lost if beneficial use ceases (13).

Appropriation water law is common in the western part of the United States, and, generally, states with the poorest water supply manage their water most closely. Arizona is a good example: with an average precipitation of less than 38 centimeters per year, Arizona must, of necessity, manage its water very closely. The state constitution says that riparian water rights are not authorized, and the state's comprehensive water code declares that all water is subject to appropriation. Preferred uses are domestic, municipal, and irrigation. Colorado also has a limited water supply and has declared that all streams are considered public property subject to appropriation (13).

Comparison of the two doctrines suggests that management of water resources is considerably more effective when the principles of appropriation are applied. Because riparian law requires a judicial decision, it is therefore subject to possible variations and interpretations in different courts. As a result, property owners are never sure of their position. The riparian system also

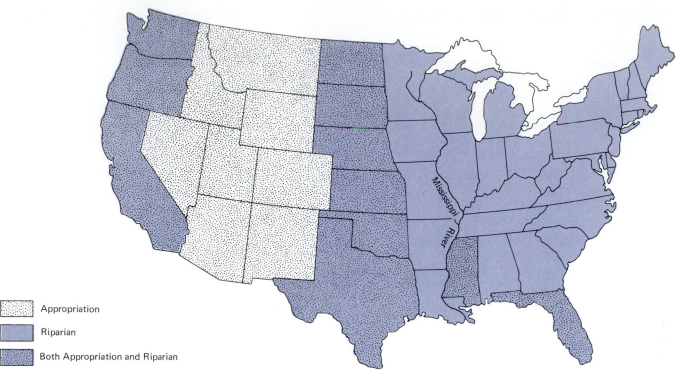

Figure 18.7
Surface water laws for the conterminous United States. (Data from New Mexico Bureau of Mines, Circular 95, 1968.)

tends to encourage nonuse of water and is thus counter-productive in times of shortage. On the other hand, states with an appropriation system have the power to make and enforce regulations based on sound hydrologic principles, which is more likely to lead to effective management of water resources (13).

▼ LAW OF THE SEABED

For more than twenty years, controversy has surrounded mining of the seabed. Industrial nations with international companies interested in mining the seabed are at odds with developing countries who argue that mineral wealth on the bottom of the sea is a common heritage to all people on earth, consistent with a United Nations resolution on the issue of seabed resources. Five international consortia of companies have spent hundreds of millions of dollars on research and exploration in preparation for mining nodules containing manganese, cobalt, nickel, and copper (see Chapter 14), and naturally want to be sure that their investment is returned. Nevertheless, developing nations, wanting a relatively large share of the mineral wealth, have generally taken a slow-growth-development position at Law of the Sea Conferences, a United Nations' sponsored group of 150 nations.

Lack of action in ratifying a treaty (international law) concerning mining of the seabed has led to threats by

industrial nations to begin mining without an agreement. Some developing nations have responded with threats to cut off oil, metals, and other resources to industrial nations that mine the seabed without approval of the nations in the United Nations group. In spite of this, the United States Congress has passed legislation authorizing American companies to mine the seabed. Although the controversy will go on, it is expected that plans to mine will continue and that developing countries will eventually modify their demands in light of the tremendous amount of money industrial nations are spending on mining technology and exploration.

▼ LAND-USE PLANNING AND LAW

Few environmental topics are as controversial as land-use planning legislation. The controversy results from several factors. First, unlike air and water pollution that can be measured, evaluated, and possibly corrected, it is very difficult to determine the "highest and best use" of the natural environment as opposed to the "most profitable use." Second, landowners fear that land-use planning will take away their right to decide what to do with their property; that is, the idea of individual ownership of property could be converted to a social property owner-ship in which the individual would have a caretaker role.

People in the United States greatly value private ownership of land, so a law that requires rural landowners to use their lands in very restricted ways is not always popular. Some people argue that private ownership of property is a sacred right, and only in those countries where private ownership is permitted are there also other personal rights. Everything considered, it is extremely unlikely that private ownership in the United States will be abolished. On the other hand, it is increasingly apparent that private ownership of land does not mean the owner has the right to deliberately or inadvertently degrade the environment. To be effective, land-use planning should leave the property owner alternatives of land use from which to choose freely.

It seems inevitable that some form of federal land-use planning act will eventually become law, with the purpose of assuring that all the land in the nation will be used in such a way as to facilitate harmonious existence with our natural environment, so that environmental, economic, social, and other requirements of present or future generations are not degraded. The argument in favor of a federal land-use planning act stresses that such planning is urgently needed to minimize degradation of land resources of statewide or national importance. Those in favor of such an act point out that land use is the most important aspect of environmental quality control that remains to be stated as national policy (14).

Many states have implemented some form of land-use legislation. The most common programs include comprehensive planning, coastal zone management, wetlands management, siting of power plants, surface mining regulation, and floodplain management. However, only a relatively small number of states have enacted mandatory local planning.

In the 1970s both North Carolina and California enacted coastal zone management. The North Carolina legislation was passed in recognition of the need to preserve for the people an opportunity to enjoy the aesthetic, cultural, and recreational quality of natural coastlines. The first two objectives of the act provided a management system that allows for preservation of estuaries, barrier islands, sand dunes, and beaches, such that natural productivity and biological, economic, and aesthetic values are safeguarded. Another purpose was to ensure that development in the coastal area did not exceed the capabilities of available land and water resources. The California legislation was voted in by the citizens of the state and it requires permits for development to occur in the coastal zone. Of particular importance and controversy has been the requirement that developments must provide for public access to the coast.

The state of Maine is a real leader in land-use regulation. In the early 1970s state legislators enacted a law, the objective of which is to maintain the highest and best use of the natural environment. The law was designed to regulate development and place the burden of proof on the developer that the project will not cause adverse environmental degradation. This philosophy has been used by other states in development of environmental legislation that requires that environmental impact be analyzed prior to development of a project. Land-use regulation, as exemplified by the Maine legislation, is significant because it strongly supports the right of states to limit land use so as to preserve the environment for future generations (15); it was from such ideas that mandatory comprehensive planning in states such as Oregon and Florida developed.

▼ ▼ ▼ SUMMARY AND CONCLUSIONS

A land ethic is emerging in our national environmental policy. Productive and enjoyable harmony with our physical environment is now encouraged, and legal theories based upon the Trust Doctrine or the Ninth Amendment are being used to argue in court for the right of this generation and future generations to an unspoiled land.

The term *environmental law* has only recently gained common usage, but the subject is becoming increasingly popular among lawyers and promises to be more significant in the future. Topics of particular concern to individuals and society are degradation of air, water, scenery, and other natural resources.

The process of law is undergoing some important changes. There is a move toward problem solving and mediation rather than confrontation and adversarial position when dealing with environmental matters. When negotiation replaces inflexibility, progress in solving problems is enhanced.

Environmental legislation is having a tremendous impact on the industrial community. In addition to regulations controlling emissions of possible pollutants, perhaps the most significant and far-reaching piece of environmental legislation is the National Environmental Policy Act of 1969. The act requires that before any environment-affecting activity that is directly or indirectly involved with federal government can begin, a statement evaluating the environmental impact must be completed. Other important legislation includes the Refuse Act of 1899, the Fish and Wildlife Coordination Act of 1958, the Water Quality Improvement Act of 1970, the Resource Conservation and Recovery Act of 1976, the Clean Water Act of 1977, the Comprehensive Environmental Response, Compensation, and Liability Act of 1980, the Hazardous and Solid Waste Amendments of 1984, the Water Quality Act of 1987, and the Clean Air Act Amendments of 1990. Many states are following the federal government's leadership and are developing environmental and land-management legislation.

Areas of particular concern at the state level are environmental impact, floodway regulation, sediment control, land use, and air and water quality.

Water law remains a significant issue, particularly in regions with deficiencies of water. Eastern states generally have what are known as *riparian rights,* whereby owners of land that adjoins water have the right to reasonable use. In the western states, water law is generally governed by the *Appropriation Doctrine*, which holds that prior beneficial usage is the key to water rights. In some states, such as Arizona and Colorado, where water is especially valuable, all water is appropriated on the basis of prior and preferred uses. In comparing the two systems, one can conclude that water appropriation is superior because it leads to better management of water resources.

Perhaps the most controversial and potentially significant of all environmental legislation is land-use planning. Conflicts arise because it is often difficult to determine the highest and best use of land. Also, landowners fear that land-use planning will cause property owners to lose the right to control their property, and there is considerable concern over the idea that federal and state governments should initiate local planning.

The argument in favor of land-use planning legislation stresses that planning is urgently needed to control degradation of the land, and that land use is the remaining important aspect of environmental quality control that has not yet been stated as national policy.

The significant effect of land-management legislation is that it shifts to the developer the burden of proof that an activity will not degrade the land. We will have to live with the concept that the most profitable use of land is not always necessarily the best use. This concept stems directly from the Trust Doctrine and an emerging land ethic.

▼ ▼ ▼ REFERENCES

1. LANDAN, N. J., and RHEINGOLD, P. D. 1971. *The environmental law handbook*. New York: Ballantine Books.
2. YANNACONE, V. J., JR., COHEN, B. S., and DAVISON, S. G. 1972. *Environmental rights and remedies* 1. San Francisco: Bancroft-Whitney.
3. MONO LAKE COMMITTEE. 1985. *Mono Lake: Endangered oasis*. Lee Vining, California.
4. COATES, D. R. 1971. Legal and environmental case studies in applied geomorphology. In *Environmental Geomorphology*, ed. D. R. Coates, pp. 223–42. Binghamton, New York: State University of New York.
5. JUERGENSMEYER, J. C. 1970. Control of air pollution through the assertion of private rights. *Environmental Law*: 17–46. Greenvale, New York: Research and Documentation Corporation.
6. MURPHY, E. F. 1971. *Man and his environment: Law*. New York: Harper & Row.
7. BARNHARDT, W. 1987. The death of Ducktown. *Discover,* Oct., 35–43.
8. RODGERS, W. H., JR. 1977. *Handbook on Environmental Law*. St. Paul, Minn.: West.
9. CARTER, L. J. 1974. Con Edison: Endless Storm King dispute adds to its troubles. *Science* 184:1353–58.
10. BACOW, L. S., and WHEELER, M. 1984. *Environmental Dispute Resolution*. New York: Plenum Press.
11. CARGO, D. N., and MALLORY, B. F. 1974. *Man and his geologic environment*. Menlo Park, California: Addison-Wesley.
12. Private Remedies for Water Pollution. 1970. *Environmental Law*: 47–69. Greenvale, New York: Research and Documentation Corporation.
13. Legal Approach to Water Rights. 1972. In *Water quality in a stressed environment,* ed. W. A. Pettyjohn, pp. 255–76. Minneapolis: Burgess.
14. HEALY, M. R. 1974. National land use proposal: land use legislation of landmark environmental significance. *Environmental Affairs* 3:355–95.
15. WAINRIGHT, J. K., JR. 1974. Spring Valley: Public purpose and land use regulation in a "taking" context. *Environmental Affairs* 3:327–54.

A **soil horizon**   Uppermost soil horizon, sometimes referred to as the *zone of leaching*.

Absorption   The process of taking up, incorporating, or assimilating a material.

Acid rain   Rain made artificially acid by pollutants, particularly oxides of sulfur and nitrogen. (Natural rainwater is slightly acid owing to the effect of carbon dioxide dissolved in the water.)

Adsorption   Process in which molecules of gas or molecules in solution attach to the surface of solid materials with which they come in contact.

Aerobic   Characterized by the presence of free oxygen.

Aesthetics   Originally a branch of philosophy, defined today by artists and art critics.

Age element   Elements that characteristically accumulate in tissue with age.

Aggregate   Any hard material such as crushed rock, sand, gravel, or other material that is added to cement to make concrete.

Albedo   A measure of reflectivity or the amount (by decimal or percent) of electromagnetic radiation reflected by a material or surface.

Alkaline soil   Soil found in arid regions that contains a large amount of soluble mineral salts (primarily sodium) that in the dry season may appear on the surface as a crust or powder.

Alluvium   Unconsolidated sediments, including sand, gravel, and silt, deposited by streams.

Anaerobic   Characterized by the absence of free oxygen.

Angle of repose   The maximum angle that loose material will sustain.

Anhydrite   Evaporite mineral ($CaSO_4$) calcium sulfate.

Anthracite   A type of coal characterized by a high percentage of carbon and low percentage of volatiles, providing a high heat value. Anthracite often forms as a result of metamorphism of bituminous coal.

Anticline   Type of fold characterized by an upfold or arch. The oldest rocks are found in the center of the fold.

Appropriation doctrine, water law   Holds that prior usage of water is a significant factor. The first to use the water for beneficial purposes is prior in right.

Aquifer   Earth material containing sufficient groundwater that the water can be pumped out. Highly fractured rocks and unconsolidated sands and gravels make good aquifers.

Aquitard   Earth material that retards the flow of groundwater.

Area (strip) mining   Type of strip mining practiced on relatively flat areas.

Argillic *B* soil horizon   Designated as B_t, a soil horizon enriched in clay minerals that have been translocated downward by soil-forming processes.

Artesian   Refers to a groundwater system in which the groundwater is isolated from the surface by a confining layer and the water is under pressure. Groundwater that is under sufficient pressure will flow freely at the surface from a spring or well.

Asbestos   Fibrous mineral material used as insulation. It is suspected of being either a true carcinogen or a carrier of carcinogenic trace elements.

Ash, volcanic   Unconsolidated volcanic debris, less than 44 mm in diameter, physically blown out of a volcano during an eruption.

Ash fall   Volcanic ash eruption that blows up into the atmosphere and then rains down on the landscape.

Ash flow   Mixture of volcanic ash, hot gases, and fragments of rock and glass that flows rapidly down the flank of a volcano. May be an extremely hazardous event.

Atmosphere   Layer of gases surrounding the earth.

Avalanche   A type of landslide involving a large mass of snow, ice, and rock debris that slides, flows, or falls rapidly down a mountainside.

Azonal soil   Recent surface materials such as floodplain deposits that do not have distinctive soil layering.

B **soil horizon**   Intermediate soil horizon; sometimes known as the *zone of accumulation*.

B_k **soil horizon**   Soil horizon characterized by accumulation of calcium carbonate that may coat individual soil particles and fill some pore spaces but does not dominate the morphology of the horizon.

Balancing doctrine   Asserts that public benefits and importance of a particular action should be balanced against potential injury to certain individuals. The method of balancing is changing, and courts are considering possible long-range injury caused by certain activities to large numbers of citizens other than the immediate complainants.

Barrier island   Island separated from the mainland by a salt marsh. It generally consists of a multiple system of beach ridges and is separated from other barrier islands by inlets that allow exchange of seawater with lagoon water.

Basalt   A fine-grained extrusive igneous rock; one of the most common igneous rock types.

Basaltic   Engineering geology term for all fine-grained igneous rocks.

Bauxite   A rock composed almost entirely of hydrous aluminum oxides. It is a common ore of aluminum.

Bed material   Sediment transported and deposited along the bed of a stream channel.

Bedding plane   The plane that delineates the layers of sedimentary rocks.

Bentonite A type of clay that is extremely unstable; upon wetting, it expands to many times its original volume.

Biochemical oxygen demand (BOD) A measure of the amount of oxygen necessary to decompose organic materials in a unit volume of water. As the amount of organic waste in water increases, more oxygen is used, resulting in a higher BOD.

Biomass Organic matter. As a fuel, biomass can be burned directly (as wood) or converted to a more convenient form (such as charcoal or alcohol) and then burned.

Biosphere The zone adjacent to the surface of the earth that includes all living organisms.

Biotite A common ferromagnesian mineral, a member of the mica family.

Bituminous coal A common type of coal characterized by relatively high carbon content and low volatiles; sometimes called *soft coal.*

Blowout Failure of an oil, gas, or disposal well resulting from adverse pressures that can physically blow part of the well casing upward. May be associated with leaks of oil, gas, or, in the case of disposal wells, harmful chemicals.

Braided river A river channel characterized by an abundance of islands that continually divide and subdivide the flow of the river.

Breakwater A structure (as a wall), which may be attached to a beach or located offshore, designed to protect a beach or harbor from the force of the waves.

Breccia A rock or zone within a rock composed of angular fragments. Sedimentary, volcanic, and tectonic breccias are recognized.

Breeder reactor A type of nuclear reactor that actually produces more fissionable (fuel) material than it uses.

Brine Water that has a high concentration of salt.

British thermal unit (Btu) A unit of heat defined as the heat required to raise the temperature of one pound of water one degree Fahrenheit.

Brittle Material that ruptures before any plastic deformation.

Bulk element Any of the common elements that make up the bulk of living material.

C soil horizon Lowest soil horizon, sometimes known as the *zone of partially altered parent material.*

C_{ca} soil horizon A soil horizon rich in calcium carbonate.

Cadmium A metallic element with an atomic number of 48 and an atomic weight of 112.4. As a trace element, it has been associated with serious health problems.

Calcite Calcium carbonate ($CaCO_3$); common carbonate mineral that is the major constituent of the rock limestone. It weathers readily by solutional processes, and large cavities and open weathered fractures are common in rocks that contain the mineral calcite.

Caldera Giant volcanic crater produced by very rare but extremely violent volcanic eruption or by collapse of the summit area of a shield volcano following eruption.

Caliche A white-to-gray irregular accumulation of calcium carbonate in soils of arid regions.

Calorie The quantity of heat required to raise the temperature of one gram of water from 14.5° to 15.5° C.

Cambic soil horizon Soil horizon diagnostic of incipient soil profile development, characterized by slightly redder color than the other horizons.

Capillary action The rise of water along narrow passages, facilitated and caused by surface tension.

Capillary fringe The zone or layer above the water table in which water is drawn up by capillary action.

Capillary water Water that is held in the soil through capillarity.

Carbonate A compound or mineral containing the radical (CO_3^-). The common carbonate is calcite.

Carcinogen Any material known to produce cancer in humans or other animals.

Catastrophe An event or situation causing sufficient damage to people, property, or society in general that recovery and/or rehabilitation is long and involved. Natural processes most likely to produce a catastrophe include floods, hurricanes, tornadoes, tsunamis, volcanoes, and large fires.

Cesium-137 A fission product produced from nuclear reactors with a half-life of 33 years.

Channelization An engineering technique to straighten, widen, deepen, or otherwise modify a natural stream channel.

Circum-Pacific belt One of the three major zones where earthquakes occur. This belt is essentially the Pacific plate. It is also known as the *ring of fire,* as many active volcanoes are found on the edge of the Pacific plate.

Clay May refer to a mineral family or to a very fine-grained sediment. It is associated with many environmental problems, such as shrinking and swelling of soils and sediment pollution.

Clay skins Oriented plates of clay minerals surrounding soil grains and filling pore spaces between grains.

Coal A sedimentary rock formed from plant material that has been buried, compressed, and changed.

Colluvium Mixture of weathered rock, soil, and other, usually angular, material on a slope.

Columnar jointing System of fractures (joints) that break rock into polygons of typically five or six sides; the polygons form columns. This type of fracturing is common in basalt and most likely is caused by shrinking during cooling of the lava.

Common excavation Excavation that can be accomplished with an earth mover, backhoe, or dragline.

Composite volcano Steep-sided volcanic cone produced by alternating layers of pyroclastic debris and lava flows.

Compressibility (soil) Measure of a soil's tendency to decrease in volume.

Conchoidal fracture A shell-like or fan-shaped fracture characteristic of the mineral quartz and natural glass.

Cone of depression A cone-shaped depression in the water table caused by withdrawal of water at rates greater than those at which the water can be replenished by natural groundwater flow.

Confined aquifer An aquifer that is overlain by a confining layer (aquitard).

Conglomerate A detrital sedimentary rock composed of rounded fragments, 10 percent of which are larger than 2 mm in diameter.

Connate water Water that is no longer in circulation or in contact with the present water cycle; generally, saline water trapped during deposition of sediments.

Contact metamorphism Type of metamorphism produced when country rocks are in close contact with a cooling body of magma below the surface of the earth.

Continental drift Movement of continents in response to sea-floor spreading. The most recent episode of continental drift supposedly began about 200 million years ago with the breakup of the supercontinent Pangaea.

Continental shelf Relatively shallow ocean area between the shoreline and the continental slope that extends to approximately a 600-foot water depth surrounding a continent.

Contour (strip) mining Type of strip mining used in hilly terrain.

Convection Transfer of heat involving movement of particles; for example, the boiling of water in which hot water rises to the surface and displaces cooler water that moves toward the bottom.

Convergent plate boundary Boundary between two lithospheric plates in which one plate descends below the other (**subduction**).

Corrosion A slow chemical weathering or chemical decomposition that proceeds inward from the surface. Objects such as pipes experience corrosion when buried in soil.

Cost-benefits analysis A type of site selection in which benefits and costs of a particular project are compared. The most desirable projects are those for which the benefits-to-cost ratio is greater than one.

Creep A type of downslope movement characterized by slow flowing, sliding, or slipping of soil and other earth materials.

Crystal settling A sinking of previously formed crystals to the bottom of a magma chamber.

Crystalline A material with a definite internal structure such that the atoms are in an orderly, repeating arrangement.

Crystallization Processes of crystal formation.

Debris flow Rapid downslope movement of earth material often involving saturated, unconsolidated material that has become unstable because of torrential rainfall.

Deep-well disposal Method of waste disposal that involves pumping waste into subsurface disposal sites such as fractured or otherwise porous rocks.

Dendrochronology Study of tree rings.

Desertification Conversion of land from a more productive state to one more nearly resembling a desert.

Detrital Mineral and rock fragments derived from preexisting rocks.

Diamond Very hard mineral composed of the element carbon.

Dilatancy of rocks Inelastic increase in volume of a rock that begins after stress on the rock has reached one-half the rock's breaking strength.

Discharge The quantity of water flowing past a particular point on a stream, usually measured in cubic feet per second (cfs) or cubic meters per second (cms).

Disseminated mineral deposit Mineral deposit in which ore is scattered throughout the rocks; examples are diamonds in kimberlite and many copper deposits.

Divergent plate boundary Boundary between lithospheric plates characterized by production of new lithosphere; found along oceanic ridges.

Dose dependency Refers to the effects of a certain trace element on a particular organism being dependent upon the dose or concentration of the element.

Drainage basin Area that contributes surface water to a particular stream network.

Drainage net System of stream channels that coalesce to form a stream system.

Dredge spoils Solid material, such as sand, silt, clay, or rock deposited from industrial and municipal discharges, that is removed from the bottom of a water body to improve navigation.

Driving forces Those forces that tend to make earth material slide.

Ductile Material that ruptures following elastic and plastic deformation.

_E_ soil horizon A light-colored horizon underlying the _A_ horizon that is leached of iron-bearing compounds.

Earthquake Natural shaking or vibrating of the earth in response to the breaking of rocks along faults. The earthquake zones of the earth generally correlate with lithospheric plate boundaries.

Earth system science The study of the earth as a system.

Ease of excavation (soil) Measure of how easily a soil may be removed by human operators and equipment.

Ecology Branch of biology that treats relationships between organisms and their environments.

Economic geology Application of geology to locating and evaluating mineral materials.

Effluent Any material that flows outward from something; examples include wastewater from hydroelectric plants and water discharged into streams from waste-disposal sites.

Effluent stream Stream in which flow is maintained during the dry season by groundwater seepage into the channel.

Elastic deformation Type of deformation in which the material returns to its original shape after the stress is removed.

Engineering geology Application of geologic information to engineering problems.

Environment That which surrounds an individual or a community; both physical and cultural surroundings. **Environment** also sometimes denotes a certain set of circumstances surrounding a particular occurrence, for example, environments of deposition.

Environmental geology Application of geologic information to environmental problems.

Environmental geology map A map that combines geologic and hydrologic data expressed in nontechnical terms to facilitate general understanding by a large audience.

Environmental impact statement A written statement that assesses and explores the possible impacts of a particular project that may affect the human environment. The statement is required by the National Environmental Policy Act of 1969.

Environmental law A field of law concerning the conservation and use of natural resources and the control of pollution.

Environmental resource unit (ERU) A portion of the environment with a similar set of physical and biological characteristics, a supposedly natural division characterized by specific patterns or assemblages of structural components (such as rocks, soils, vegetation) and natural processes (such as erosion, runoff, soil processes).

Ephemeral Temporary or very short-lived. Characteristic of beaches, lakes, and some stream channels that change rapidly (geologically).

Epicenter   The point on the surface of the earth directly above the hypocenter (area of first motion) of an earthquake.

Erodability (soil)   Measure of how easily a soil may erode.

Evaporite   Sediments deposited from water as a result of extensive evaporation of seawater or lake water; dissolved materials left behind following evaporation.

Exponential growth   A type of compound growth in which a total amount or number increases at a certain percentage per year, and each year's rate of growth is added to the total from the previous year; characteristically stated in terms of a particular doubling time, that is, the time in years it will take the original number to double. Commonly used in reference to population growth.

Extrusive igneous rocks   Igneous rock that forms when magma reaches the surface of the earth; a volcanic rock.

Fault   A fracture or fracture system that has experienced movement along opposite sides of the fracture.

Fault gouge   A clay zone formed by pulverized rock during an earthquake, which may create a groundwater barrier.

Feldspar   The most abundant family of minerals in the crust of the earth; silicates of calcium, sodium, and potassium.

Ferromagnesian mineral   Minerals containing iron and magnesium, characteristically dark in color.

Fertile material   Material such as uranium-238, which is not naturally fissionable but upon bombardment by neutrons is converted to plutonium-239, which is fissionable.

Fission   The splitting of an atom into smaller fragments with the release of energy.

Floodplain   Flat topography adjacent to a stream in a river valley, produced by the combination of overbank flow and lateral migration of meander bends.

Floodway district   That portion of a channel and floodplain of a stream designated to provide passage of the 100-year regulatory flood without increasing elevation of the flood by more than one foot.

Floodway fringe district   Land located between the floodway district and the maximum elevation subject to flooding by the 100-year regulatory flood.

Fluorine   Important trace element, essential for nutrition.

Fluvial   Concerning or pertaining to rivers.

Fly ash   Very fine particles (ash) resulting from the burning of fuels such as coal.

Fold   Bend that develops in stratified rocks because of tectonic forces.

Foliation   Property of metamorphic rock characterized by parallel alignment of the platy or elongated mineral grains; environmentally important because it can affect the strength and hydrologic properties of rock.

Formation   Any rock unit that can be mapped.

Fossil fuels   Fuels such as coal, oil, and gas formed by the alteration and decomposition of plants and animals from a previous geologic time.

Fracture zone   A fracture system that may or may not be active and may or may not have an alteration zone along the fracture planes. Fracture zones are environmentally important because they greatly affect the strength of rocks.

Fumarole   A natural vent from which fumes or vapors are emitted, such as the geysers and hot springs characteristic of volcanic areas.

Fusion, nuclear   Combining of light elements to form heavy elements with the release of energy.

Gaging station   Location at a stream channel where discharge of water is measured.

Gasification   Method of producing gas from coal.

Geochemical cycle   Migratory paths of elements during geologic changes and processes.

Geologic cycle   A group of interrelated cycles known as the hydrologic, rock, tectonic, and geochemical cycles.

Geomorphology   The study of landforms and surface processes.

Geopressured system   Type of geothermal energy system resulting from trapping the normal heat flow from the earth by impermeable layers such as shale rock.

Geothermal energy   The useful conversion of natural heat from the interior of the earth.

Glacial surge   A sudden or quick advance of a glacier.

Glacier   A landbound mass of moving ice.

Gneiss   A coarse-grained, foliated metamorphic rock in which there is banding of light and dark minerals.

Gravel   Unconsolidated, generally rounded fragments of rocks and minerals greater than 2 mm in diameter.

Gravitational water   Water that occurs in pore spaces of a soil and is free to drain from the soil mass under the influence of gravity.

Greenhouse effect   Trapping of heat in the atmosphere by water vapor, carbon dioxide, methane, and CFCs.

Groin   A structure designed to protect shorelines and trap sediment in the zone of littoral drift, generally constructed perpendicular to the shoreline.

Groundwater   Water found beneath the surface of the earth within the zone of saturation.

Grout   A mixture of cement and sediment that is sufficiently fluid to be pumped into open fissures or cracks in rocks, thereby increasing the strength of a foundation for an engineering structure.

Gypsum   An evaporite mineral, $CaSO_4 \cdot 2H_2O$.

Half-life   The amount of time necessary for one-half of the atoms of a particular radioactive element to decay.

Halite   A common mineral, NaCL (salt).

Hardpan soil horizon   Hard, compacted or cemented soil horizon, most often composed of clay but sometimes cemented with calcium carbonate, iron oxide, or silica. This horizon is nearly impermeable and often restricts the downward movement of soil water.

Hematite   An important ore of iron, a mineral (Fe_2O_3).

High-value resource   Materials such as diamonds, copper, gold, and aluminum. These materials are extracted wherever they are found and transported around the world to numerous markets.

Hot igneous system   Type of geothermal energy system in which heat is supplied by the presence of magma.

Hot spot   Assumed stationary heat source located below the lithosphere that feeds volcanic processes near the earth's surface.

Humus   Black organic material in soil.

Hydraulic conductivity   Measure of the ability of groundwater to move through a particular earth material.

Hydrocarbon   Organic compounds consisting of carbon and hydrogen.

Hydroconsolidation   Consolidation of earth materials upon wetting.

Hydrofracturing   Pumping of water under high pressure into subsurface rocks to fracture the rocks and thereby increase their permeability.

Hydrogeology   Discipline that studies relationships between groundwater, surface water, and geology.

Hydrograph   A graph of the discharge of a stream over time.

Hydrologic cycle   Circulation of water from the oceans to the atmosphere and back to the oceans by way of precipitation, evaporation, runoff from streams and rivers, and groundwater flow.

Hydrologic gradient   The driving force for both saturated and unsaturated flow of groundwater. Quantitatively, it is the slope or rate of change of the hydraulic head, which at the point of measurement is the algebraic sum of the elevation head and pressure head.

Hydrology   The study of surface and subsurface water.

Hydrosphere   The water environment in and on earth and in the atmosphere.

Hydrothermal convection system   Geothermal energy system characterized by the circulation of hot water. May be dominated by water vapor or hot water.

Hydrothermal ore deposit   A mineral deposit derived from hot water solutions of magmatic origin.

Hygroscopic water   Refers to water absorbed and retained on fine-grained soil particles. This water may be held tenaciously.

Hypocenter   The point in the earth where an earthquake originates; also known as the focus.

Igneous rocks   Rocks formed from solidification of magma; **extrusive** if they crystallize on the surface of the earth and **intrusive** if they crystallize beneath the surface.

Impermeable   Earth materials that greatly retard or prevent movement of fluids through them.

Infiltration   Movement of surface water into rocks or soil.

Influent stream   Stream that is everywhere above the groundwater table and flows in direct response to precipitation. Water from the channel moves down to the water table, forming a recharge mound.

Intrusive igneous rock   Igneous rock that forms when magma solidifies below the surface of the earth; a volcanic rock.

Iodine   A nonmetallic element needed in trace amounts by the human body. Iodine is necessary for the normal functioning of the thyroid gland; a shortage can hinder growth or cause goiter.

Island arc   A curved group of volcanic islands associated with a deep-oceanic trench and subduction zone (**convergent plate boundary**).

Isotopes   Forms of the same element having a variable atomic weight.

Itaiitai disease   Extremely painful disease that attacks bones, causing them to become very brittle so that they break easily. Associated with heavy metals, especially cadmium, in concentrations of a few parts per million in the soil or in food consumed by victims of the disease.

Jetty   Often constructed in pairs at the mouth of a river or inlet to a lagoon, estuary, or bay, a jetty is designed to stabilize a channel, control deposition of sediment, and deflect large waves.

Juvenile water   Water derived from the interior of the earth that has not previously existed as atmospheric or surface water.

***K* soil horizon**   A calcium-carbonate-rich horizon in which the carbonate often forms laminar layers parallel to the surface. Carbonate completely fills the pore spaces between soil particles.

Karst topography   A type of topography characterized by the presence of sinkholes, caverns, and diversion of surface water to subterranean routes.

Kimberlite pipe   An igneous intrusive body that may contain diamond crystals disseminated (scattered) throughout the rock type.

Land ethic   Ethic that affirms the right of all resources, including plants, animals, and earth materials, to continued existence and, at least in some locations, continued existence in a natural state.

Landslide   Specifically, rapid downslope movement of rock and/or soil; also a general term for all types of downslope movement.

Land-use planning   Complex process involving development of a land-use plan to include a statement of land-use issues, goals, and objectives; summary of data collection and analysis; land-classification map; and report describing and indicating appropriate development in areas of special environmental concern. An extremely controversial issue.

Laterite   Soil formed from intense chemical weathering in tropical or savanna regions.

Lava   Molten material produced from a volcanic eruption, or rock that forms from solidification of molten material.

Leachate   Obnoxious liquid material capable of carrying bacteria, produced when surface water or groundwater comes into contact with solid waste.

Leaching   Process of dissolving, washing, or draining earth materials by percolation of groundwater or other liquids.

Lignite   A type of low-grade coal.

Limestone   A sedimentary rock composed almost entirely of the mineral calcite.

Limonite   Rust, hydrated iron oxide.

Liquefaction   Transformation of water-saturated granular material from the solid state to a liquid state.

Lithosphere   Outer layer of the earth approximately 100 kilometers thick of which the plates that contain the ocean basins and continents are composed.

Littoral   Pertaining to the near-shore and beach environments.

Loess   Angular deposits of windblown silt.

Low-value resource   Resources such as sand and gravel that have primarily a place value, economically extracted because they are located close to where they are to be used.

Magma   A naturally occurring silica melt, much of which is in a liquid state.

Magma tap   Attempt to recover geothermal heat directly from magma. Feasibility of such heat extraction is unknown.

Magnetite   A mineral and important ore of iron, Fe_3O_4.

Manganese oxide nodules   Nodules of manganese, iron with secondary copper, nickel, and cobalt, that cover vast areas of the deep-ocean floor.

Marble   Metamorphosed limestone.

Marl   Unconsolidated clays, silts, sands, or mixtures of these materials that contain a variable content of calcareous material.

Meanders   Bends in a stream channel that migrate back and forth across the floodplain, depositing sediment on the inside of the bends, forming point bars, and eroding outsides of bends.

Metamorphic rock   A rock formed from preexisting rock by the effects of heat, pressure, and chemically active fluids beneath the earth's surface. In *foliated* metamorphic rocks, the mineral grains have a preferential parallel alignment or segregation of minerals; *nonfoliated* metamorphic rocks have neither.

Meteoric water   Water derived from the atmosphere.

Methane   A gas, CH_4, the major constituent of natural gas.

Mica   A common rock-forming silicate mineral.

Mineral   A naturally occurring, solid, crystalline substance with physical and chemical properties that vary within known limits.

Mitigation   Identification of actions that will avoid, lessen, or compensate for anticipated adverse impacts of a particular project. Mitigation is becoming an important process in environmental-impact work.

Mudflow   A mixture of unconsolidated materials and water that flows rapidly downslope or down a channel.

Myth of superabundance   The myth that land and water resources are inexhaustible and management of resources is therefore unnecessary.

National Environmental Policy Act of 1969 (NEPA)   Act declaring a national policy that harmony between man and his physical environment be encouraged. Established the Council on Environmental Quality. Established requirements that an environmental impact statement be completed prior to major federal actions that significantly affect the quality of the human environment.

Neutron   A subatomic particle having no electric charge, found in the nuclei of atoms. Neutrons are crucial in sustaining nuclear fission in a reactor.

Nonrenewable resource   A resource cycled so slowly by natural earth processes that, once used, will be essentially unavailable during any useful time frame.

Nuclear reactor   Device in which controlled nuclear fission is maintained; the major component of a nuclear power plant.

O soil horizon   Soil horizon that contains plant litter and other organic material. Found above the *A* soil horizon.

Oil shale   Organic-rich shale containing substantial quantities of oil that can be extracted by conventional methods of destructive distillation.

Ore   Earth material from which a useful commodity can be extracted profitably.

Osteoporosis   Disease characterized by reduction in bone mass.

Outcrop   A naturally occurring or human-caused exposure of rock at the surface of the earth.

Overburden   Earth materials (spoil) that overlie an ore deposit, particularly material overlying or extracted from a surface (strip) mine.

Oxidation   Chemical process of combining with oxygen.

Ozone   Triatomic oxygen (O_3)

P wave   One of the seismic waves produced by an earthquake; the fastest of the seismic waves, it can move through liquid and solid materials.

Pathogen   Any material that can cause disease; for example, microorganisms, including bacteria and fungi.

Pebble   A rock fragment between 4 and 64 mm in diameter.

Ped   An aggregate of soil particles. Peds are classified by their shapes as spheroidal, blocky, prismatic, etc.

Pedology   Study of soils.

Pegmatite   A coarse-grained igneous rock that may contain rare minerals rich in elements such as lithium, boron, fluorine, uranium, and others.

Percolation test   A standard test for determining rate at which water will infiltrate into the soil. Primarily used to determine feasibility of a septic-tank disposal system.

Permafrost   Permanently frozen ground.

Permeability   A measure of the ability of an earth material to transmit fluids such as water or oil. (See *hydraulic conductivity.*)

Petrology   Study of rocks and minerals.

Physiographic determinism   Site selection based on the philosophy of designing with nature.

Physiographic province   Region characterized by a particular assemblage of landforms, climate, and geomorphic history.

Placer deposit   Ore deposit found in material transported and deposited by such agents as running water, ice, or wind; for example, gold and diamonds found in stream deposits.

Plastic deformation   Deformation that involves a permanent change of shape without rupture.

Plate tectonics   A model of global tectonics that suggests that the outer layer of the earth known as the **lithosphere** is composed of several large plates that move relative to one another; continents and ocean basins are passive riders on these plates.

Plutonium-239   A radioactive element produced in a nuclear reactor; has a half-life of approximately 24,000 years.

Point bar   Accumulation of sand and other sediments on the inside of meander bends in stream channels.

Pollution   Any substance, biological or chemical, in which an identified excess is known to be detrimental to desirable living organisms.

Pool   Common bed form produced by scour in meandering and straight stream channels with relatively low channel slope; characterized at low flow by slow-moving, deep water. Generally, but not exclusively, found on the outside of meander bends.

Porosity   The percentage of void (empty space) in earth material such as soil or rock.

Potable water   Water that may be safely drunk.

Pyrite   Iron sulfide, a mineral, commonly known as fool's gold. Environmentally important because, in contact with oxygen-rich water, it produces a weak acid that can pollute water or dissolve other minerals.

Pyroclastic activity   Type of volcanic activity characterized by eruptive or explosive activity in which all types of volcanic debris, from ash to very large particles, are physically blown from a volcanic vent.

Quartz   Silicon oxide, a common rock-forming mineral.

Quartzite   Metamorphosed sandstone.

Quick clay   Type of clay which when disturbed, as by seismic shaking, may experience spontaneous liquefaction and lose all shear strength.

R **soil horizon**   Consolidated bedrock that underlies the soil.

Radioactive waste   Type of waste produced in the nuclear fuel cycle, generally classified as *high-level* or *low-level.*

Radioisotope   A form of a chemical element that spontaneously undergoes radioactive decay, changing from one isotope to another and emitting radiation in the process.

Radon   A colorless, radioactive, gaseous element.

Reclamation, mining   Restoring land used for mining to other useful purposes, such as agriculture or recreation, after mining operations are concluded.

Record of Decision   A concise statement by an agency planning a proposed project as to which alternatives were considered and, specifically, which alternatives are environmentally preferable. The record of decision is becoming an important part of environmental impact work.

Recycling   The reuse of resources reclaimed from waste.

Regional metamorphism   Wide-scale metamorphism of deeply buried rocks by regional stress accompanied by elevated temperatures and pressures.

Renewable resource   A resource such as timber, water, or air that is naturally recycled or recycled by human-induced processes within a useful time frame.

Reserves   Known and identified deposits of earth materials from which useful materials can be extracted profitably with existing technology under present economic and legal conditions.

Resisting forces   Forces that tend to oppose downslope movement of earth materials.

Resistivity   A measure of an earth material's ability to retard the flow of electricity; the opposite of conductivity.

Resources   Includes reserves plus other deposits of useful earth materials that may eventually become available.

Richter magnitude   A measure of the amount of energy released by an earthquake, determined by converting the largest amplitude of the shear wave to a logarithmic scale in which, for example, 2 indicates the smallest earthquake that can be felt and 8.5 indicates a devastating earthquake.

Riffle   A section of stream channel characterized at low flow by fast, shallow flow; generally contains relatively coarse bed-load particles.

Riparian rights, water law   Right of the landowner to make reasonable use of water on his land, provided the water is returned to the natural stream channel before it leaves his property; the property owner has the right to receive the full flow of the stream undiminished in quantity and quality.

Rippable excavation   Type of excavation that requires breaking up soil before it can be removed.

Riprap   Layer or assemblage of broken stones placed to protect an embankment against erosion by running water or breaking waves.

Riverine environment   Land area adjacent to and influenced by a river.

Rock   *Geologic:* An aggregate of a mineral or minerals. *Engineering:* Any earth material that must be blasted to be removed.

Rock cycle   Group of processes that produce igneous, metamorphic, and sedimentary rocks.

Rock salt   Rock composed of the mineral halite.

Rotational landslide   Type of landslide that develops in homogeneous material; movement is likely to be rotational along a potential slide plane.

S **wave**   Secondary wave, one of the waves produced by earthquakes.

Saline   Salty; characterized by high salinity.

Salinity   A measure of the total amount of dissolved solids in water.

Salt dome   A structure produced by upward movement of a mass of salt; frequently associated with oil and gas deposits on the flanks of a dome.

Sand   Grains of sediment with a size between 1/16 and 2 mm in diameter. Often, sediment composed of quartz particles of this size.

Sand dune   Ridge or hill of sand formed by wind action.

Sandstone   Detrital sedimentary rock composed of sand grains that have been cemented together.

Sanitary landfill   Method of solid-waste disposal that does not produce a public health problem or nuisance; confines and compresses waste and covers it at the end of each day with a layer of compacted, relatively impermeable material, such as clay.

Saturated flow   A type of subsurface or groundwater flow in which all the pore spaces are filled with water.

Scarp   Steep slope or cliff commonly associated with landslides or earthquakes.

Scenic resources   The visual portion of an aesthetic experience; scenery is now recognized as a natural resource with varying values.

Schist   Coarse-grained metamorphic rock characterized by foliated texture of the platy or elongated mineral grains.

Schistosomiasis   Snail fever, a debilitating and sometimes fatal tropical disease.

Scoping   Process of identifying important environmental issues that require detailed evaluation early in the planning of a proposed project. Scoping is an important part of environmental impact analysis.

Seawall   Engineering structure constructed at the water's edge to minimize coastal erosion by wave activity.

Secondary enrichment   Weathering process of sulfide ore deposits that may concentrate the desired minerals.

Sedimentary rock   A rock formed when sediments are transported, deposited, and then lithified by natural cement, compression, or other mechanism. *Detrital* sedimentary rock is formed from broken parts of previously existing rock; *chemical* sedimentary rock is formed by chemical or biochemical processes removing material carried in chemical solution.

Sedimentology   Study of environments of deposition of sediments.

Seismic   Refers to vibrations in the earth produced by earthquakes.

Seismograph   Instrument that records earthquakes.

Selenium   Important nonmetallic trace element with an atomic number of 34.

Sensitivity (soil)   Measure of loss of soil strength due to disturbances such as human excavation and remolding.

Septic tank   Tank that receives and temporarily holds solid and liquid waste. Anaerobic bacterial activity breaks down the waste, solid wastes are separated out, and liquid waste from the tank overflows into a drainage system.

Sequential land use   Development of land previously used as a site for the burial of waste. The specific reuse must be carefully selected.

Serpentine   A family of ferromagnesian minerals; environmentally important because they form very weak rocks.

Sewage sludge   Solid material that remains after municipal wastewater treatment.

Shale   Sedimentary rock composed of silt- and clay-sized particles; the most common sedimentary rock.

Shield volcano   A broad, convex volcano built up by successive lava flows; the largest of the volcanoes.

Shrink-swell potential (soil)   Measure of a soil's tendency to increase and decrease in volume as water content changes.

Silicate minerals   The most important group of rock-forming minerals.

Silt   Sediment between 1/16 and 1/256 mm in diameter.

Sinkhole   Surface depression formed by solution of limestone or collapse over a subterranean void such as a cave.

Sinuous channel   Type of stream channel (not braided).

Slate   A fine-grained, foliated metamorphic rock.

Slump   Type of landslide characterized by downward slip of a mass of rock, generally along a curved slide plane.

Soil   *Soil science:* Earth material so modified by biological, chemical, and physical processes that the material will support rooted plants. *Engineering:* Earth material that can be removed without blasting.

Soil chronosequence   A series of soils arranged in terms of relative soil profile development from youngest to oldest.

Soil fertility   Capacity of a soil to supply nutrients (such as nitrogen, phosphorus, and potassium) needed for plant growth when other factors are favorable.

Soil horizons   Layers in soil (A, B, C, etc.) that differ from one another in chemical, physical, and biological properties.

Soil profile   Set of soil horizons that characterize a particular soil as viewed in cross section from the surface downward through the entire thickness of the soil.

Soil survey   A survey consisting of a detailed soil map and descriptions of soils and land-use limitations; usually prepared by the Soil Conservation Service in cooperation with local government.

Solar energy   Energy that is collected directly from the sun. Broadly, includes also energy that is collected indirectly.

Solid waste   Material such as refuse, garbage, and trash.

Spoils, mining   Banks or piles that are accumulations of overburden removed during mining processes and discarded on the surface.

Storm surge   Wind-driven oceanic waves.

Strain   Change in shape or size of a material as a result of applied stress; the result of stress.

Strength (soil)   Ability of a soil to resist deformation. Results from cohesive and frictional forces in the soil.

Stress   Force per unit area; may be compressive, tensile, or shear.

Strip mining   A method of surface mining.

Subduction   Process in which one lithospheric plate descends beneath another.

Subsidence   Sinking, settling, or other lowering of parts of the crust of the earth.

Subsurface water   All the waters within the lithosphere.

Surface water   Waters above the surface of the lithosphere.

Surface wave   One type of wave produced by earthquakes; these waves generally cause most of the damage to structures on the surface of the earth.

Suspended load   Sediment in a stream or river carried off the bottom by the fluid.

Syncline   Fold in which younger rocks are found in the core of the fold; rocks in the limbs of the fold dip inward toward a common axis.

System   Any part of the universe that is isolated in thought or in fact for the purpose of studying or observing changes that occur under various imposed conditions.

Tar sand   Naturally occurring sand, sandstone, or limestone that contains an extremely viscous petroleum.

Tectonic   Referring to rock deformation.

Tectonic creep   Slow, more or less continuous movement along a fault.

Tectonic cycle   Group of processes that collectively produce external forms on the earth, such as ocean basins, continents, and mountains.

Tephra   Any material ejected and physically blown out of a volcano; mostly ash.

Texture, rock   The size, shape, and arrangement of mineral grains in rocks.

Tidal energy   Electricity generated by tidal power.

Till   Unstratified, heterogeneous material deposited directly by glacial ice.

Toxic   Harmful, deadly, or poisonous.

Transform fault   Type of fault associated with oceanic ridges; may form a plate boundary, such as the San Andreas Fault in California.

Translation (slab) landslide   Type of landslide in which movement takes place along a definite fracture plane, such as a weak clay layer or bedding plane.

Tropical cyclone   Severe storm generated from a tropical disturbance; called *typhoons* in most of the Pacific Ocean and *hurricanes* in the Western Hemisphere.

Tsunami   Seismic sea wave generated by submarine volcanic or earthquake activity; characteristically has very long wave length and moves rapidly in the open sea; incorrectly referred to as *tidal wave*.

Tuff   Volcanic ash that is compacted, cemented, or welded together.

Unconfined aquifer   Aquifer in which there is no impermeable layer restricting the upper surface of the zone of saturation.

Unconformity   A buried surface of erosion representing a time of nondeposition; a gap in the geologic record.

Unified soil classification system   Classification of soils, widely used in engineering practice, based on amount of coarse particles, fine particles, or organic material.

Uniformitarianism   Concept that the present is the key to the past; that is, we can read the geologic record by studying present processes.

Unsaturated flow   Type of groundwater flow that occurs when only a portion of the pores is filled with water.

Vadose zone   Zone or layer above the water table in which some water may be suspended or moving in a downward migration toward the water table or laterally toward a discharge point.

Volcanic breccia, agglomerate Large rock fragments mixed with ash and other volcanic materials cemented together.

Volcanic dome Type of volcano characterized by very viscous magma with high silica content; activity is generally explosive.

Water budget Analysis of sources, sinks, and storage sites for water in a particular area.

Water table Surface that divides the vadose zone from the zone of saturation; the surface below which all the pore space in rocks is saturated with water.

Watershed Land area that contributes water to a particular stream system. (See *drainage basin.*)

Weathering Changes that take place in rocks and minerals at or near the surface of the earth in response to physical, chemical, and biological changes; the physical, chemical, and biological breakdown of rocks and minerals.

Zinc An important trace element necessary in life processes.

Zone of saturation Zone or layer below the water table in which all the pore space of rock or soil is saturated.

Other Conversion Factors

1 ft³/sec = .0283 m³/sec = 7.48 gal/sec = 28.32 liters/sec

1 acre-foot = 43,560 ft³ = 1,233 m³ = 325,829 gal

1 m³/sec = 35.32 ft³/sec

1 ft³/sec for one day = 1.98 acre-feet

1 m/sec = 3.6 km/hr = 2.24 mi/hr

1 ft/sec = 0.682 mi/hr = 1.097 km/hr

1 billion gallons per day (bgd) = 3.785 million m³ per day

Strength of Common Rock Types

	Rock Type	Range of Compressive Strength (Thousands of psi)	Comments
Igneous	Granite	23 to 42.6	Finer-grained granites with few fractures are the strongest. Granite is generally suitable for most engineering purposes.
	Basalt	11.8 to 52.0	Brecciated zones, open tubes, or fractures reduce the strength.
Metamorphic	Marble	6.7 to 34.5	Solutional openings and fractures weaken the rock.
	Gneiss	22.2 to 36.4	Generally suitable for most engineering purposes.
	Quartzite	21.1 to 91.2	Very strong rock.
Sedimentary	Shale	Less than 1 to 33.5	May be a very weak rock for engineering purposes; careful evaluation is necessary.
	Limestone	5.3 to 37.6	May have clay partings, solution openings, or fractures that weaken the rock.
	Sandstone	4.8 to 34.1	Strength varies with degree and type of cementing material, mineralogy, and nature and extent of fractures.

Source: Data primarily from *Handbook of Tables for Applied Engineering Science,* ed. R. E. Bolz and G. L. Tuve (Cleveland, Ohio: CRC Press, 1973).